Advances in Sustainability Research from the University of Oradea

Advances in Sustainability Research from the University of Oradea

Editors

Constantin Bungău
Alina Badulescu
Dorina Camelia Ilieș
Cosmin Mihai Vesa
Delia Mirela Tit

MDPI

Basel • Beijing • Wuhan • Barcelona • Belgrade • Novi Sad • Cluj • Manchester

Editors
Constantin Bungău
Department of Engineering and Management, Faculty of Management and Technological Engineering
University of Oradea
Oradea
Romania

Alina Badulescu
Department of Economics and Business, Faculty of Economic Sciences
University of Oradea
Oradea
Romania

Dorina Camelia Ilieș
Department of Geography, Tourism and Territorial Planning, Faculty of Geography, Tourism and Sport,
University of Oradea
Oradea
Romania

Cosmin Mihai Vesa
Department of Preclinical Disciplines, Faculty of Medicine and Pharmacy,
University of Oradea
Oradea
Romania

Delia Mirela Tit
Department of Pharmacy, Faculty of Medicine and Pharmacy
University of Oradea
Oradea
Romania

Editorial Office
MDPI
St. Alban-Anlage 66
4052 Basel, Switzerland

This is a reprint of articles from the Special Issue published online in the open access journal *Sustainability* (ISSN 2071-1050) (available at: www.mdpi.com/journal/sustainability/special_issues/S0449P7K6M).

For citation purposes, cite each article independently as indicated on the article page online and as indicated below:

Lastname, A.A.; Lastname, B.B. Article Title. *Journal Name* **Year**, *Volume Number*, Page Range.

ISBN 978-3-7258-0838-0 (Hbk)
ISBN 978-3-7258-0837-3 (PDF)
doi.org/10.3390/books978-3-7258-0837-3

Contents

About the Editors

Constantin Bungău

Constantin Bungău is the Rector of the University of Oradea, as well as a Doctor Honoris Causa of two universities in Romania. With a teaching career of over 33 years, he dedicated his activity as a professor to both the development of the field of engineering and management, as well as university management. He has been a PhD supervisor since 2015. He has coordinated projects and grants with a total value of over EUR 7 million. He has published over 100 scientific papers, 55 of which are indexed by Web of Science. Moreover, he was a Reviewer of numerous journals and scientific conferences. His scientific reputation promotes him as a member of 14 editorial and scientific committees of specialized journals and conferences at both national and international levels. Mr. Bungău is a member of the Industrial Engineering and Management Commission of the National Council for Awarding University Degrees, Diplomas and Certificates. His expertise is recognized in various fields, such as production system management and engineering, strategic management, integrated production systems, lean manufacturing, machine tools, hydraulic drives, optimization methods and algorithms, numerical modeling and simulations, sustainability and green efficiency.

Alina Badulescu

Alina Badulescu is a Professor of Economics (since 2006) and PhD coordinator of Economics (since 2007) at the Faculty of Economics and Doctoral School of Economic Sciences, University of Oradea, Romania. She graduated at the Bucharest University of Economic Studies (1992) and holds a PhD in Economics from Babes-Bolyai University (1997) and a Habilitation Certificate in Business Administration from Bucharest University of Economic Studies (2017). During her academic career, which started in 1992, she has authored and co-authored more than 150 journal articles and books, and participated in over 50 international conferences. She is an Editorial Member, Guest Editor and reviewer for more than 40 journals, including 30 WoS indexed journals, and an international evaluator for research grants and competitions.

Dorina Camelia Ilieș

Dorina Camelia Ilies is a Professor at the Department of Geography, Tourism and Territorial Planning at the University of Oradea in Romania. She has over 30 years of experience in didactic and research activities. Her field of interest is the investigation and mapping of natural and cultural heritage and better valorization for tourism in times of climate change. Her research activity has culminated to over 100 scientific papers published in leading national and international scientific journals. She has coordinated several national and international research projects in the domain of geography and tourism. She is also a Collaborator of the Crișana Maramureș, a Geographical Atlas of Tourism Heritage (2014) initiative, that won the Romanian Academy Award "Simion Mehedinți" in 2016; Geography of tourism of Romania, in Volume: The Geography of Tourism of Central and Eastern Europe Countries (2016); and The Atlas of local horizon of Bihor County in 2020. She is currently the Coordinator of the Romanian International Association for the Promotion of Geoethics.

Cosmin Mihai Vesa

Cosmin Mihai Vesa became an associate professor at the University of Oradea, Faculty of Medicine and Pharmacy, in 2022. He teaches physiology of the renal, respiratory and cardiovascular systems. He has been a PhD supervisor since 2023. As a medical doctor specialized in diabetes, nutrition and metabolic diseases, his research interests include the pathophysiology of diabetes mellitus complications, insulin resistance, metabolic syndrome, obesity, health impact of the environment, public health and active compounds for treatment. He served as the Guest Editor for six MDPI Special Issues and is an active reviewer of several journals.

Delia Mirela Tit

Delia Mirela Tit is a former student of the University of Oradea, climbing the steps of her teaching career since 2005 and becoming a full professor in 2020. Her research skills were crowned by the diploma of Doctor of Medicine in 2014 and the presentation of her Habilitation Thesis in 2019, with her entire career taking place at the same university where she studied. Currently, she is a PhD supervisor in the field of Pharmacy at the Doctoral School of Biomedical Sciences, having more than 100 works indexed in Web of Science and 4 books published by publishing houses in Romania. She is also a reviewer of a number of journals. Her primary areas of interest are focused on topics relating to major pharmaceutical disciplines, public health and sustainability.

Preface

The demand for sustainability in all fields has steadily grown as a topic of major interest to policy makers and practitioners around the world. Moreover, the concept of sustainability has been extended to various non-economic fields, including educational institutions such as universities. The idea of the "triple bottom line" highlights the three dimensions (economic, social and environmental) that must be considered in all activities.

Moreover, it must be emphasized that sustainable development goes hand in hand with digitization, innovation and performance, creating the environment and appropriate conditions to ensure a more sustainable development, strong economy and prosperous, fair and peaceful society in order to achieve a safer, better and cleaner future. Approaching coherent strategies for sustainable development, relevant research directions and policies focused on the implementation of sustainability is thus a necessity.

Through its projects, involvement and development, as well as scientific activities, The University of Oradea highlights constant, deep and extensive consideration of this field.

Constantin Bungău, Alina Badulescu, Dorina Camelia Ilieș, Cosmin Mihai Vesa, and Delia Mirela Tit
Editors

sustainability

MDPI

Editorial

Advances in Sustainability Research from the University of Oradea

Constantin Bungău [1,*], Alina Badulescu [2], Dorina Camelia Ilieș [3], Cosmin Mihai Vesa [4,*] and Delia Mirela Tit [5]

1 Department of Engineering and Management, Faculty of Management and Technological Engineering, University of Oradea, 410087 Oradea, Romania
2 Department of Economics and Business, Faculty of Economic Sciences, University of Oradea, 410087 Oradea, Romania; abadulescu@uoradea.ro
3 Department of Geography, Tourism and Territorial Planning, Faculty of Geography, Tourism and Sport, University of Oradea, 410073 Oradea, Romania; dilies@uoradea.ro
4 Department of Preclinical Disciplines, Faculty of Medicine and Pharmacy, University of Oradea, 410073 Oradea, Romania
5 Department of Pharmacy, Faculty of Medicine and Pharmacy, University of Oradea, 410028 Oradea, Romania; dtit@uoradea.ro
* Correspondence: bungau@uoradea.ro (C.B.); cosmin.vesa@csud.uoradea.ro (C.M.V.)

Interest in the topic of sustainability has been constantly increasing across all domains, among scholars, policy makers, and practitioners all over the world. Moreover, the quest for greater sustainability has extended to cover various non-economic arenas, including educational institutions and universities. The concept of the "triple bottom line" highlights the three dimensions, i.e., economic, social, and environmental, that must be considered in all human activities. Sustainable development goes hand in hand with digitalization, innovation, and performance to create a suitable environment and conditions for a more sustainability-driven, prosperous, fair, and peaceful economy and society for the future. When approaching coherent strategies for sustainable development, relevant research directions and policies focusing on the implementation of sustainability have become a necessity. The main topics addressed in this Special Issue and the contributions within each field are showcased in the following papers.

Regarding the field of education, a paper from Tătar et al. (Contribution 1) examines the extent to which international students' gender, the development level of their country of origin, and gender equality affect their choice of field of study at the University of Oradea, Romania. Buhas R et al. (Contribution 2) demonstrated the importance of sports events in students' personal and professional development. Another research study explores sports event organization and the impact on dual-career student–athletes. A survey of 139 participants in a national championship revealed that academic studies enhance professional competencies, while participation in sports events contributes to students' socio-professional development. The conclusions emphasize the significance of sports events for dual-career students' socio-professional growth, calling for the development of sustainable strategies to support dual-career students (Contribution 2).

In the domain of agriculture, Budau et al. proposed a model for selecting an adequate agroforestry system to be introduced to plain sites in Bihor County (Contribution 3) Ghitea et al. analyzed the composition of grape seed oil sourced from different organic cultivated wine varieties, identifying their high nutrient value in polyunsaturated acids and high antioxidant capacity (Contribution 4). Another important contribution from Budau's team is the identification of the beneficial influence of Driver and Kuniyuki Woodyon culture media on *Robinia pseudoacacia* development (Contribution 5). Timar AV et al. performed quality analysis on different sources of wheat, demonstrating that the Romanian variety Crisana recorded good parameters, being comparable to the Hungarian variety Bekes from Hajdu Bihar County (Contribution 6). Venig and Stănica focused on the influences of irrigation and fertilization on qualitative indices in tow plum varieties (Contribution 7).

Citation: Bungău, C.; Badulescu, A.; Ilieș, D.C.; Vesa, C.M.; Tit, D.M. Advances in Sustainability Research from the University of Oradea. *Sustainability* **2024**, *16*, 2712. https://doi.org/10.3390/su16072712

Received: 14 March 2024
Accepted: 22 March 2024
Published: 26 March 2024

In the field of hydrology, mineral water sources from Tămășeu, Sîntimreu, Pădurea Neagră were evaluated from a microbiological/chemical point of view, and the results confirmed that they have beneficial hydraulic and therapeutic proprieties (Contribution 8).

Culture, heritage, and construction are richly represented in this Special Issue. El-hosiny et al. studied the "Night of the museums" event in Oradea, Romania, revealing that the physical environment and situational interactions played a decisive role in the contextual perception of the importance of the event, while the most important motivations were curiosity, gratuity, and the need to learn and spend one's free time in the most enjoyable way (Contribution 9). In the analysis performed by Deac et al., a correlation was established between the ethno-religious and ethno-cultural heritage elements in Crișana Region, Romania, at the territorial-administrative-unit level, emphasizing their mutual influences, which can lead to the preservation and promotion of ethno-confessional and ethno-cultural features (Contribution 10). Another study aimed to monitor the microclimate and air quality within visitable heritage buildings and accurately correlate recorded parameters with human health and with the degree of conservation of heritage objects, as well as identifying interconnections between indoor microclimate and outdoor climate changes (Contribution 11). A scientometric analysis performed by Bungau CC et al. demonstrates the increased interest among researchers worldwide in the field of building energy efficiency and the correlations between state developments and research outputs (Contribution 12). Bungau C.C. and his team also analyzed the impacts of complex interventions, in the form of the elements of a master plan for the ecological restoration of buildings on the Campus of the University of Oradea, Oradea, Romania, emphasizing numerous benefits regarding energy consumption (Contribution 13). In the field of old buildings, Tudorica et al. performed a sustainability study on a historical building in Arad County, Romania, with the study's main topic being the consolidation of the resistance structure of the building. The conclusions revealed the weakest points in the building, offering more sustainable solutions related to carbon dioxide emissions and embodied energy (Contribution 14). In the sustainable construction field, Tudorica and Bob demonstrate that reinforced soil retaining walls are far more sustainable than reinforced concrete in a paper evaluating a sustainable slope design while accounting for environmental, economic, and safety variables (Contribution 15). To improve our understanding of steel trusses in architecture, engineering, and construction, Savu et al. propose a methodology that represents significant advancement in 3D modeling optimization. They show that the use of BIM and 3D scanning significantly reduces non-physical waste in the fireproofing process of steel trusses, and they help determine a more precise budget (Contribution 16).

In the energy field, Secui et al. propose a novel algorithm to address the issue of economically dispatching emissions for power system optimization, incorporating wind power. According to the findings, adding wind units lowers costs by 10% and reduces emissions by 45% (Contribution 17).

In the field of waste management, Indrie et al. propose an algorithm for textile waste management, focused on visual design rather than compatibility checks and constraints, to be used for the creation of interior decorative parts (Contribution 18). Another study explored various modalities to reduce the environmental impact of medical waste from upper digestive endoscopies (Contribution 19). Costea et al. used the comet test and three cellular bioindicators to quantify the possible genotoxicity of the sludge produced at sewage treatment plants. The findings suggest that to attain a higher degree of relevance, it is necessary to combine several genotoxicity assays in a test battery (Contribution 20). In another paper, Gavrilă et al. (Contribution 21) present a sustainable approach to metal coin canceling methods using 3D modeling and finite element method analysis.

The research performed by Iconaru et al. draws attention to sustainability in the field of healthcare by applying an online questionnaire and establishing several correlations between physical activity level and health status in Romanian adults in the post-COVID-19 period (Contribution 22).

In the economic sphere, Trip et al. analyses the innovative potential of businesses operating in the tourism sector in the Baile Felix Spa Resort area near Oradea, Romania by surveying managers and business owners active in health and spa tourism services to investigate the differences between their perspectives. Considering the ongoing concern around sustainable innovation and modernization in both groups, significant achievements and strategies have been implemented by large companies (Contribution 23). Felea et al. emphasize the importance of Romania adopting the Green Deal Strategy and 0 emissions until 2050 (Contribution 24). Matei et al. have looked at the difficulties in sustainable human resource management (HRM), especially when it comes to financial restraints. They also offer a useful framework for evaluating the viability of initiatives for human resource development from a financial standpoint, as well as for promoting sustainable HRM practices (Contribution 25). In a further study, Badulescu et al. examine the connection between employment, spending, and economic growth in Romania's research and development sector, concentrating on the Northwest Region, a specific development region, and two of its member counties (Contribution 26).

Regarding public transportation, Oargă et al. address the elements involved in putting into practice a modular autonomous vehicle solution to extend travel routes, linking major urban hubs, such as train or airport terminals, to other major hubs, such as the city center, significant infrastructure, or peripheral neighborhoods (Contribution 27). For another study, using data collection methods, the energy efficiency of a fuel cell electric bus operating in an urban setting was assessed by Cărăușan et al. (Contribution 28).

Regarding the sustainable development of border regions, Chirodea and colleagues conducted a study on the mechanisms of microregional integration, which is seen as a workable and sustainable option for cross-border areas with resources that may be used to draw in investment and cultivate prosperity and wellbeing. For regional public bodies searching for workable and long-term solutions for the growth of cross-border microregions, the paper provides a model of study and work (Contribution 29).

In the textile industry, Tripa et al. gathered data on the mechanical, physical, and dyeability characteristics of wet-spun hemp yarn in both its natural and bleached forms. The outcomes show that bleach did not influence the yarn count, but it had a beneficial impact on the yarn twist as its value rose (Contribution 30).

Darabaneanu et. al studied the influence of environmental perception on place attachment in Romanian rural areas (Contribution 31). Another paper is related to the administrative aspects regarding the valorisation of geothermal waters for balneological purposes in Bihor County, Romania (Contribution 32).

The papers published in this Special Issue on "Advances in Sustainability Research from the University of Oradea" provide an overview of the context of developments in all types of research devoted to investigating the complex nexus among education, research, innovation, and sustainability.

Conflicts of Interest: The authors declare no conflicts of interest.

List of Contributions:

1. Tătar, C.F.; Tătar, M.I.; Pénzes, J.; White, G.W. How Gender, Culture, and Economy Influence Field of Study Preferences in Higher Education: Exploring Gender Gaps in STEM, AHSS, and Medicine among International Students. *Sustainability* **2023**, *15*, 15820. https://doi.org/10.3390/su152215820.
2. Buhaș, R.; Ilieș, A.; Săveanu, S.; Szabo-Alexi, P.; Szabo-Alexi, M.; Buhaș, S. Socio-Professional Implications of Sports Events: A Perspective from Dual-Career Students. *Sustainability* **2023**, *15*, 7813. https://doi.org/10.3390/su15107813.
3. Budău, R.; Apăfăian, A.; Caradaică, M.; Bratu, I.A.; Timofte, C.S.C.; Enescu, C.M. Expert-Based Assessment of the Potential of Agroforestry Systems in Plain Regions across Bihor County, Western Romania. *Sustainability* **2023**, *15*, 15724. https://doi.org/10.3390/su152215724.
4. Gitea, M.A.; Gitea, D.; Mirela Tit, D.; Bungau, S.G.; Bogdan, M.A.; Radu, A.-F.; Dulf, F.V.; Pasca, M.B. Organically Cultivated Vine Varieties—Distinctive Qualities of the Oils Obtained from Grape Seeds. *Sustainability* **2023**, *15*, 11037. https://doi.org/10.3390/su151411037.

5. Budău, R.; Bei, M.; Onet, C.; Agud, E.; Mintas, O.S.; Timofte, A.I.; Rosan, C.A.; Laslo, V.; Vicas, S.I. In Vitro Propagation of Several Valuable Selections of Robinia pseudoacacia L. as a Fast and Sustainable Source for Wood Production. *Sustainability* **2023**, *15*, 15243. https://doi.org/10.3390/su152115243.
6. Timar, A.V.; Teusdea, A.C.; Purcarea, C.; Vuscan, A.N.; Memete, A.R.; Vicas, S.I. Chemometric Analysis-Based Sustainable Use of Different Current Baking Wheat Lots from Romania and Hungary. *Sustainability* **2023**, *15*, 12756. https://doi.org/10.3390/su151712756.
7. Venig, A.; Stănică, F. Irrigation and Fertilization: A Comprehensive Analysis of Their Influences on Qualitative Indices in Two Plum Varieties. *Sustainability* **2024**, *16*, 2496. https://doi.org/10.3390/su16062496.
8. Linc, R.; Pantea, E.; Șerban, E.; Ciurba, A.-P.; Serban, G. Hydrochemical and Microbiological Investigations and the Therapeutic Potential of Some Mineral Waters from Bihor County, Romania. *Sustainability* **2023**, *15*, 15640. https://doi.org/10.3390/su152115640.
9. Elhosiny, S.M.; Hassan, T.H.; Josan, I.; Salem, A.E.; Abdelmoaty, M.A.; Herman, G.V.; Wendt, J.A.; Janzakov, B.; Mahmoud, H.M.E.; Abuelnasr, M.S. Oradea's Cultural Event Management: The Impact of the 'Night of the Museums' on Tourist Perception and Destination Brand Identity. *Sustainability* **2023**, *15*, 15330. https://doi.org/10.3390/su152115330.
10. Deac, L.A.; Herman, G.V.; Gozner, M.; Bulz, G.C.; Boc, E. Relationship between Population and Ethno-Cultural Heritage—Case Study: Crișana, Romania. *Sustainability* **2023**, *15*, 9055. https://doi.org/10.3390/su15119055.
11. Ilies, D.C.; Herman, G.V.; Safarov, B.; Ilies, A.; Blaga, L.; Caciora, T.; Peres, A.C.; Grama, V.; Bambang, S.W.; Brou, T.; et al. Indoor Air Quality Perception in Built Cultural Heritage in Times of Climate Change. *Sustainability* **2023**, *15*, 8284. https://doi.org/10.3390/su15108284.
12. Bungau, C.C.; Hanga Prada, F.I.; Bungau, T.; Bungau, C.; Bendea, G.; Prada, M.F. Web of Science Scientometrics on the Energy Efficiency of Buildings to Support Sustainable Construction Policies. *Sustainability* **2023**, *15*, 8772. https://doi.org/10.3390/su15118772.
13. Bungau, C.C.; Bungau, C.; Toadere, M.T.; Prada-Hanga, I.F.; Bungau, T.; Popescu, D.E.; Prada, M.F. Solutions for an Ecological and Healthy Retrofitting of Buildings on the Campus of the University of Oradea, Romania, Built Starting from 1911 to 1913. *Sustainability* **2023**, *15*, 6541. https://doi.org/10.3390/su15086541.
14. Tudorica, M.R.; Toadere, M.T.; Bob, C.I. The Sustainability Study Done for a Consolidation Work on a Historical Building. *Sustainability* **2023**, *15*, 15285. https://doi.org/10.3390/su152115285.
15. Tudorica, M.R.; Bob, C.I. Intervention Works Conducted to Ensure the Stability of a Slope: A Sustainability Study. *Sustainability* **2024**, *16*, 1544. https://doi.org/10.3390/su16041544.
16. Savu, C.; Pescaru, A.-H.; Zsak, I.-G.; Durgheu, A.-M.; Frent, A.-P.; Suba, N.-S.; Buda, A.S.; Nistor, S. Analysis on Using 3D Scanning and BIM to Reduce the Physical and Non-Physical Construction Waste for Sustainable Fireproofing of Steel Trusses. *Sustainability* **2024**, *16*, 1832. https://doi.org/10.3390/su16051832.
17. Secui, D.C.; Hora, C.; Bendea, C.; Secui, M.L.; Bendea, G.; Dan, F.C. Modified Social Group Optimization to Solve the Problem of Economic Emission Dispatch with the Incorporation of Wind Power. *Sustainability* **2024**, *16*, 397. https://doi.org/10.3390/su16010397.
18. Indrie, L.; Ilieva, J.; Zlatev, Z.; Tripa, S.; Sturza, A. Development of an Algorithm for Textile Waste Arrangement. *Sustainability* **2023**, *15*, 11399. https://doi.org/10.3390/su151411399.
19. Ilias, T.I.; Hocopan, C.S.; Brata, R.; Fratila, O. Current and Future Sustainability Traits of Digestive Endoscopy. *Sustainability* **2023**, *15*, 15872. https://doi.org/10.3390/su152215872.
20. Costea, M.A.; Rosan, C.A.; Laslo, V.; Agud, E.; Purcarea, C.; Vicas, S.I. The Comet Assay as a Sustainable Method for Evaluating the Genotoxicity Caused by the Soluble Fraction Derived from Sewage Sludge on Diverse Cell Types, Including Lymphocytes, Coelomocytes and *Allium cepa* L. Cells. *Sustainability* **2024**, *16*, 457. https://doi.org/10.3390/su16010457.
21. Gavrilă, C.C.; Lateş, M.T.; Grebenişan, G. Sustainable Approach to Metal Coin Canceling Methods, Using 3D Modeling and Finite Element Method Analysis. *Sustainability* **2024**, *16*, 2322. https://doi.org/10.3390/su16062322.
22. Iconaru, E.I.; Tarcau, E.; Ciucurel, M.M.; Draghici, L.; Ciucurel, C. The Relationship between Physical Activity Level and Sociodemographic Factors in Romanian Adults in the Post-COVID-19 Pandemic Period. *Sustainability* **2023**, *15*, 13488. https://doi.org/10.3390/su151813488.
23. Trip, D.-T.; Simut, R.; Badulescu, D. Do Size and Ownership Determine the Willingness for Sustainable Innovations in Spa and Health Tourism? A Case Study on Baile Felix Spa Resort, Romania. *Sustainability* **2023**, *15*, 14501. https://doi.org/10.3390/su151914501.

24. Felea, A.I.; Felea, I.; Hoble, C.R. Multicriteria Quantification of the Compatibility of the Targets from Romania's Relevant Strategies with the European Green Deal. *Sustainability* **2023**, *15*, 13386. https://doi.org/10.3390/su151813386.
25. Matei, M.-C.; Abrudan, L.-C.; Abrudan, M.-M. Financial Perspectives on Human Capital: Building Sustainable HR Strategies. *Sustainability* **2024**, *16*, 1441. https://doi.org/10.3390/su16041441.
26. Badulescu, D.; Gavrilut, D.; Simut, R.; Bodog, S.-A.; Zapodeanu, D.; Toca, C.-V.; Badulescu, A. The Relationship between Sustainable Economic Growth, R&D Expenditures and Employment: A Regional Perspective for the North-West Development Region of Romania. *Sustainability* **2024**, *16*, 760. https://doi.org/10.3390/su16020760.
27. Oargă, I.-T.; Varga, B.O.; Moldovanu, D.; Cărăușan, H.; Prunean, G. Modular Autonomous Vehicles' Application in Public Transport Networks: Conceptual Analysis on Airport Connection. *Sustainability* **2024**, *16*, 1512. https://doi.org/10.3390/su16041512.
28. Cărăușan, H.; Varga, B.O.; Moldovanu, D.; Prunean, G.; Oargă, I.-T. Energy Efficiency Analysis of a Fuel Cell Bus Model Using Real Scenarios Generated by Data Collection. *Sustainability* **2024**, *16*, 1863. https://doi.org/10.3390/su16051863.
29. Chirodea, F.; Soproni, L.; Marian, M. European Union Tools for the Sustainable Development of Border Regions. *Sustainability* **2024**, *16*, 388. https://doi.org/10.3390/su16010388.
30. Tripa, S.; Kadınkız, N.; Kanwal, A.; Nazeer, M.A.; Nazir, A.; Tripa, F.; Uzun, M. Analysing the Impact of the Bleaching Process on Wet Spun Hemp Yarn Properties. *Sustainability* **2023**, *15*, 16894. https://doi.org/10.3390/su152416894.
31. Darabaneanu, D.; Maci, D.; Oprea, I.M. Influence of Environmental Perception on Place Attachment in Romanian Rural Areas. *Sustainability* **2024**, *16*, 1106. https://doi.org/10.3390/su16031106.
32. Ciurba, A.-P.; Haidu, I.; Ianc, D. Administrative Aspects Regarding the Valorisation of Geothermal Waters for Balneological Purposes in Bihor County, Romania. *Sustainability* **2023**, *15*, 10320. https://doi.org/10.3390/su151310320.

sustainability

MDPI

Article

Web of Science Scientometrics on the Energy Efficiency of Buildings to Support Sustainable Construction Policies

Constantin C. Bungau [1], Francesca Ioana Hanga Prada [1,2,*], Tudor Bungau [3], Constantin Bungau [4,5,*], Gabriel Bendea [6] and Marcela Florina Prada [1]

1 Department of Architecture and Constructions, Faculty of Constructions, Cadaster and Architecture, University of Oradea, 410058 Oradea, Romania; constantin.bungau@csud.uoradea.ro (C.C.B.); mfprada29@gmail.com (M.F.P.)
2 Doctoral School, Technical University of Cluj-Napoca, 400114 Cluj-Napoca, Romania
3 Civil, Industrial and Agricultural Constructions Program of Study, Faculty of Constructions, Cadastre and Architecture, University of Oradea, 410058 Oradea, Romania; tudor.bungau@gmail.com
4 Doctoral School of Engineering Sciences, University of Oradea, 410087 Oradea, Romania
5 Department of Engineering and Management, University of Oradea, 410087 Oradea, Romania
6 Department of Energy Engineering, Faculty of Energy Engineering and Industrial Management, University of Oradea, 410058 Oradea, Romania; gbendea@uoradea.ro
* Correspondence: pradaifrancesca@gmail.com (F.I.H.P.); bungau@uoradea.ro (C.B.)

Abstract: A variety of strategies intended to support environmentally friendly and resource-efficient building processes comprise sustainable construction policies. The limited number of bibliometric analyses in the field may hinder the ability to assess the efficacy and impact of research efforts, impede the potential for collaboration, and even limit the dissemination of best practices. Therefore, the present study aims to analyze the impact of published data on the topic of energy efficiency of buildings using the Web of Science core collection database. We perform a bibliometric analysis and science mapping research that assesses significant parameters for the field. A total of 28,555 papers were analyzed using the VOSviewer program. The data was divided into two periods to determine the evolution of trends in this field. The most prolific countries in this field were China, the United States, and England. Following the analysis of the collaboration maps, it was determined that there is a strong collaborative relationship between these countries in the development of papers. The most prolific papers of the first period were published in *Energy Policy* and *Energy and Buildings*, which also ranked first in the second period, followed by *Energies*. It was observed that the most frequent terms used in literature searches in the field differ according to the periods analyzed. In the beginning, the most frequent term was "energy efficiency and performance", and between 2011 and 2023, the terms "applied energy" and "renewable and sustainable energy" increased considerably with technological development. The results of this research demonstrate the significant and expanding scientific interest in this area and serve as a valuable asset for researchers studying the energy efficiency of buildings.

Keywords: building envelope; building management system; energy efficiency; renewable energy sources; scientometrics; sustainable construction; Web of Science; zero energy building

Citation: Bungau, C.C.; Hanga Prada, F.I.; Bungau, T.; Bungau, C.; Bendea, G.; Prada, M.F. Web of Science Scientometrics on the Energy Efficiency of Buildings to Support Sustainable Construction Policies. *Sustainability* **2023**, *15*, 8772. https://doi.org/10.3390/su15118772

Academic Editor: Thanikanti Sudhakar Babu

Received: 28 April 2023
Revised: 24 May 2023
Accepted: 26 May 2023
Published: 29 May 2023

1. Introduction

In recent years, energy demand and energy consumption have grown rapidly as a result of the ever-increasing needs of humankind in relation to the economy, industries and agriculture. Depending on the region, global warming/cooling is occurring, and the seasons are extremely variable due to climate change. These changes are leading to significant lifestyle changes, including trends towards a healthy built environment [1,2]. It is undeniably true that increases in energy consumption are driven by the intrinsic structure of the economy, and these trends in integrating the economy into the European context lead to the need to adapt the national legislation of European countries to European energy efficiency legislation and policies [3].

Existing buildings are responsible for up to 40% of final energy consumption, but especially for a significant percentage of greenhouse gas emissions of CO_2 [4]. Therefore, constructions and their energy efficiency through innovative solutions [5] have become a point of interest for the European Union, focusing on current and future trends and concerns. For the period until 2050 [6], Europe will be focused on climate neutrality and the decarbonization of the construction sector. This change is of great interest, and it can be observed that a real revolution is taking place in this direction [7], in conjunction with the existing path towards digitization [8]. In this context, the approaches to planning new buildings are different compared with those for existing buildings, with the results converging towards the same objective, namely that of energy efficiency. When planning and developing sustainable buildings, numerous financial, cultural, and environmental factors are considered, as well as their influence on the adjacent ecosystem. Implementation at a practical level requires the active participatory collaboration of the most important stakeholders in the field, including governments, universities, professional organizations, and private industry [9].

In the case of new buildings, the general concern is achieving nearly zero energy buildings (NZEB), which are buildings with a high energy performance and extremely low overall energy consumption; the objective may even be to achieve zero energy buildings (ZEB) [10,11]. Moreover, buildings are frequently referred to as "consumers", but the tendency is for them to be "prosumers" (both producers and consumers of energy from renewable sources—sun and wind) and supply generated energy to the network (the surplus of the generated energy in addition to that consumed) [12]. This can be achieved in the case of SMART constructions that aim for suitable design of buildings, ensuring that energy losses are almost zero and that appropriate equipment is installed. Such constructions have become both a necessity and a requirement of SMART cities [13,14].

The suitable design of such buildings refers to both efficient insulation and the correct approach to the building envelope in relation to the shape, surface and stratification. The envelope of the building is its enclosing surface, which delimits the interior spaces from the exterior ones, and therefore, the heated from the unheated volumes. (It is the surface that delimits the building and through which the thermal transfer takes place.) The more efficient the building envelope, the lower the heat loss [15,16].

Providing the building with high-performance equipment is mandatory for both NZEB and "prosumer" constructions. This includes, on the one hand, equipment for the provision of utilities to produce heating/cooling of the environment and domestic hot water, ventilation equipment (with heat recovery), lighting, shading equipment, and appliances, etc. On the other hand, moving to a superior design, the building is provided with equipment that, based on the technology of their design and operation, produces energy. Renewable energy is based on non-fossil sources (i.e., wind, solar, aerothermal, geothermal, hydrothermal, ocean/hydropower, biomass, waste gas/biogas, wastewater treatment) [17–19]. In contrast, energy from non-renewable sources is obtained from resources that are depleted by exploitation (such as energy from fossil fuels) [20]. Energy can be either consumed in real time, stored in energy storage batteries (to be thus usable long after its production), or can be supplied to the network.

Another category of equipment is that used for monitoring, control, and controlled operation of all equipment in the building for environmental comfort, security and safety. Such equipment is known as a building management system or building monitoring system (BMS) [21] and ensures efficient operation of the building, specific to SMART buildings.

In the case of existing buildings, the general concern is their thermal and energy efficiency. This is addressed by rehabilitating building envelopes using suitably arranged, thermally performant materials and working with the existing envelope elements, either maintaining them or replacing the existing elements [22]. The energy efficiency of existing buildings involves the rehabilitation of the systems adding utilities or upgrading of these systems to provide the building with new equipment in a unified efficient design, possibly with energy production equipment (preferably using renewable sources). The next step of efficiency is the monitoring and control of the existing equipment and the comfort provided (indoor climate) with monitoring and control equipment, a BMS [21].

Depending on the particular objective and nature of the study, the critical parameters in the context of the energy efficiency of buildings can vary. Architecture, insulation, and building envelope materials have an important effect on energy efficiency. Factors such as thermal efficiency and air permeability play an essential role when assessing the building's overall energy performance [23]. The heating, ventilation, and air conditioning systems of buildings are significant energy consumers. System effectiveness, maintenance, sizing, and control practices are a few examples of factors that can have a significant impact on energy efficiency [24].

The efficacy of lighting systems has a direct effect on energy consumption. In building structures, energy efficiency is affected by illumination technologies, management systems, daylighting approaches, and illumination levels [25]. Utilizing sources of renewable energy, such as solar photovoltaic systems, can help improve energy efficiency. Important considerations include the capacity of renewable energy systems, their integration within the building's energy infrastructure, and their general efficiency [26].

Occupant behavior, occupancy trends, and user participation in energy-efficient strategies can all have a significant impact on energy consumption. Relevant variables include occupant behavior characteristics such as inhabitant density, utilization structures, and knowledge of energy-efficiency measures [27]. For a comprehensive knowledge of energy efficiency, it is essential to consider the entire life cycle of a structure from construction to operation to eventual demolition. For a comprehensive evaluation, parameters such as embodied energy, operational electricity consumption, and end-of-life aspects are essential [28].

The main factors describing the energy consumption of a building and the energy performance indicators are summarized in Figure 1.

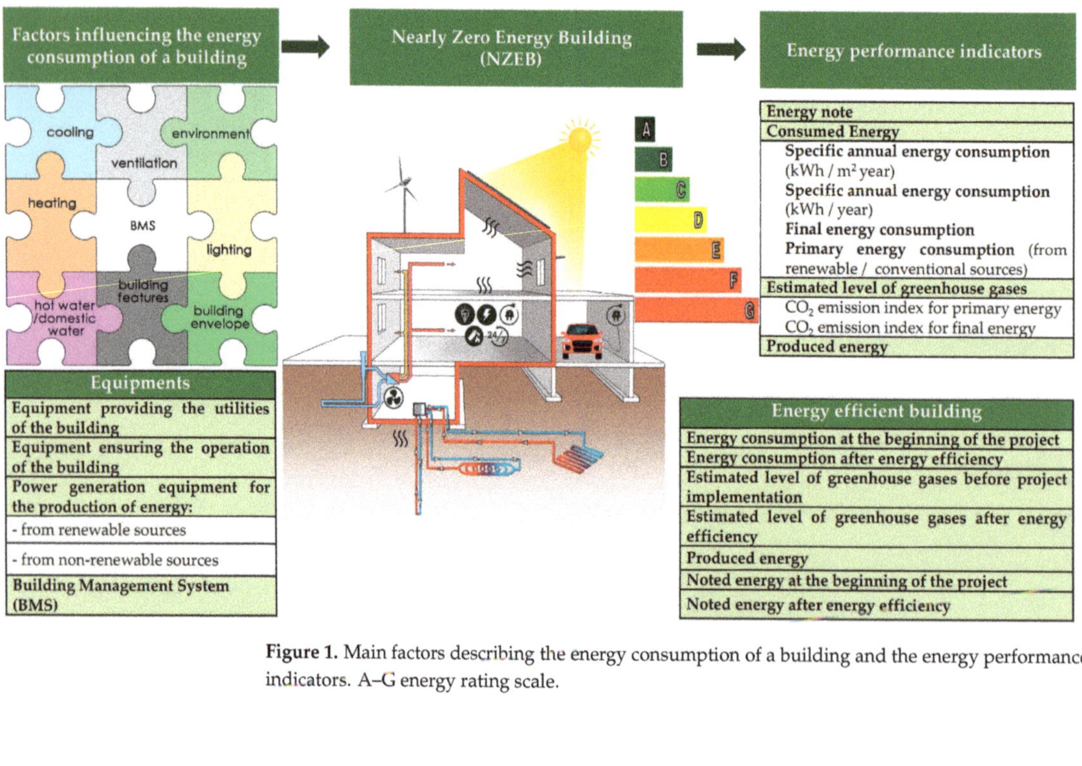

Figure 1. Main factors describing the energy consumption of a building and the energy performance indicators. A–G energy rating scale.

Analyzing the development potential of the study topic, as well as some of the current unmet needs in terms of bibliometric analysis and mapping studies focused on subfields, led to the establishment of the objectives of the present study:

- To identify bibliographic resources in the published scientific literature in the field of energy efficiency and analyze the impact of already published data on the topic, classified according to Web of Science (WoS) domains;
- To contribute in a new and unique way to revealing the importance of this field by centralizing the most prolific journals, collaborative networks, publications etc.;
- To highlight and rank the most prolific/important authors, journals, and countries by time intervals to better assess scientific trends;
- To open up new possibilities for collaboration between authors;
- To provide a conceptual overview of the research topics and practical approaches for future authors to target different journals, topics, publications, and collaborative networks;
- To highlight the most frequently cited articles so that the most significant data can be retrieved more rapidly and precisely;
- To demonstrate the rapidly rising interest in energy efficiency;
- To provide time-saving tools for researchers interested in the topic by performing a visualization and bibliometric analysis of the available research on renewable energy.

Future researchers can, therefore, approach the most prominent journals, as well as the most relevant authors in the domain of energy efficient buildings, by observing the current work. The novelty of this topical approach resides in the fact that our research topic is less focused on science mapping and analysis than such studies in other scientific areas (i.e., economics, chemical compounds, medicine). Instead, we provide a complex bibliometric analysis showing the increasing trend of studies in this field concomitant with technological development. Moreover, the present paper addresses in a distinct and targeted way, the field of energy efficiency in buildings because the field of energy efficiency in general is much broader. Therefore, the results of the present research will provide a clearer understanding of the evolution of energy efficiency within the framework of sustainable construction and considering the context of the most recent national and international energy regulations and policies.

2. Materials and Methods

2.1. Web of Science Search and Filter Algorithm

Due to the advantages offered by WoS in terms of filtering and export possibilities, as well as the validity of the articles indexed in this database, it was selected for the present bibliometric analysis. Figure 2 shows the search algorithm in WoS, with the Boolean OR operator maximizing the number of results provided by the database. The papers that were identified were written in a total of 25 different languages, with the majority of the papers being written in English. Other languages identified in higher proportions were Spanish and German, and the rest of the languages presented fewer than 300 papers in total. The present study includes original research articles and review articles written in English, thus limiting the number of papers to 28,555.

Figure 3 depicts the classification made by WoS in terms of fields. The remaining categories were assigned fewer than 6000 papers.

2.2. Algorithm Analysis and Explanatory Detailing of Graphical Elements

The papers required for this study were downloaded using the Export function available in the WoS interface. They were downloaded in tab-delimited file format, and the "Record count" was selected as "Full Record and Cited References".

Dividing the analysis into two periods enables a clearer understanding the evolution of the research field over time, which is in close correlation with the development of technology. The first period covers articles from 1978 to 2010, and the second period covers data from 2011 to 2023. For each period the following datasets were analyzed:

most productive charges in the field, most prolific journals, citation analysis of publication in the evaluated period, analysis of most productive organizations, and most prolific research areas. In addition to these analyses, the science mapping studies focused on co-authorship by country, the average year of publication of source and citation maps, and term maps and network maps of term co-occurrence. These were performed using VOSviewer version 1.6.19. [29].

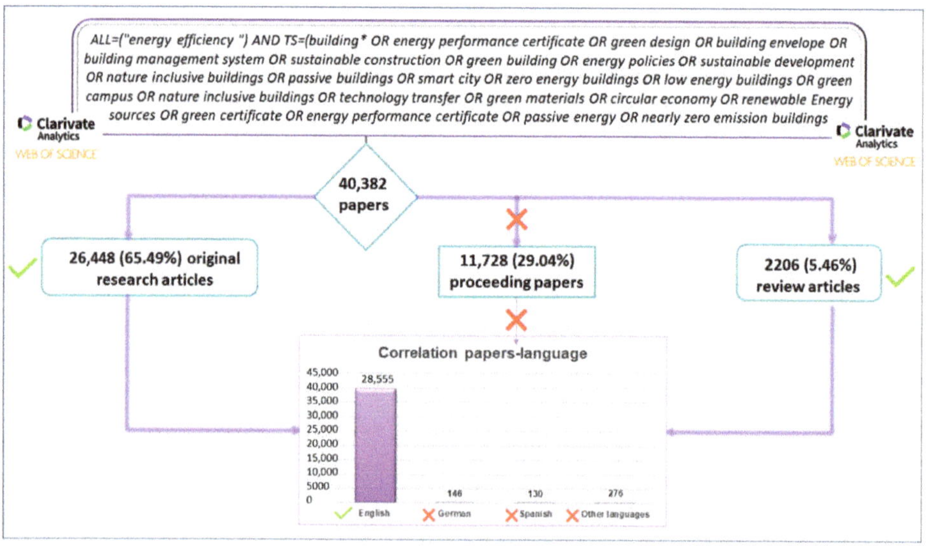

Figure 2. Flow diagram showing the search and sorting algorithm. building*, symbol used to find also the results containing the term buildings.

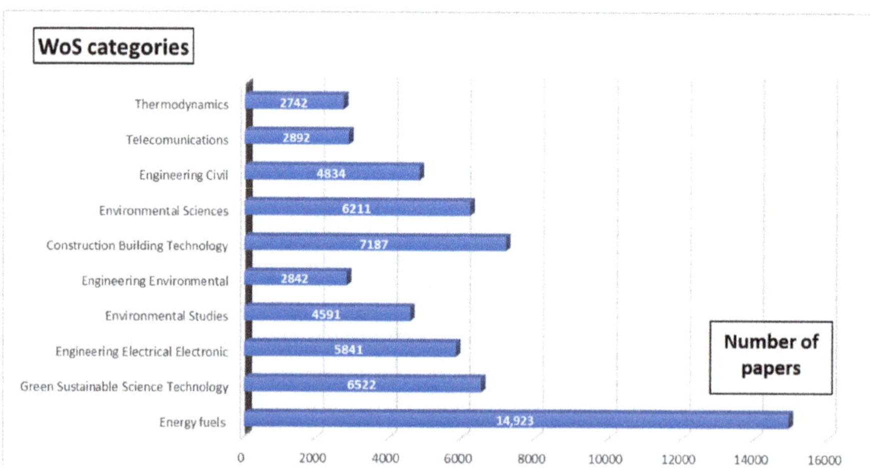

Figure 3. 3-D stacked bar chart of the top 10 WoS categories.

On the network map of the co-authorship by country, the color indicates the cluster in which the respective country is located. Countries that have stronger collaborative links are represented closer together on the map. The size of the node/bubble is directly proportional to the number of articles published, and the thickness of a line connecting two bubbles/nodes is directly proportional to the degree of collaboration between the countries.

For the bubble map of the average publication year, the color of the bubble indicates the average year of publication of the articles in the respective journal. The years are represented as year.decimal, where the decimal represents a fraction of the year. In the case of the citation network maps, the size of the bubble is directly proportional to the number of papers published by the respective journal, and the color of the bubble represents the cluster in which the respective journal is distributed. Typically, journals that are in the same cluster and are closer to each other in the graphical representation contain papers that often cite each other. The thickness of a line connecting two bubbles is directly proportional to the degree of citation between the two journals.

In the bubble map of high-frequency terms, the color of the bubble is influenced by the average citation/paper and the size of the bubble is directly proportional to the number of occurrences of that word. In the case of the network map of term co-occurrence, the size of the bubble is directly proportional to the number of occurrences of the respective word, and the color of the bubble indicates the cluster in which the word is embedded; usually, words that occur more often in the same article are grouped in a cluster. The thickness of a line linking two words is directly proportional to the degree of co-occurrence of these words in an article.

3. Results

3.1. *Period 1978–2010*

3.1.1. Evaluation of the Most Productive Countries in the Field

Researchers from a total of 90 countries have written papers that fit the search terms presented above. Moreover, 41 (45.55%) have had at least 10 papers published. The most prolific country was the United States, with 518 (22.02%) published papers and a total of 34,422 citations. Although Canada ranks 6th in terms of the number of papers published, it stands out with a high average citation/manuscript, indicating that papers published by authors from this country had a high impact. Table 1 details the top 10 countries in terms of the number of papers published.

Table 1. Top 10 prolific countries in the field of energy efficiency (period 1978–2010).

Country	Papers	Citations	Average Citation/ Manuscript	TLS
United States	518	34,422	66.45	147
China	223	11,457	51.38	66
England	213	16,142	75.78	83
Germany	103	7160	69.51	77
Sweden	93	6416	68.99	41
Canada	93	9259	99.56	32
The Netherlands	90	4506	50.07	47
Italy	69	4227	61.26	41
France	66	4669	70.74	51

TLS, total link strength value attributed by VOSviewer.

3.1.2. Assessment of the Most Prolific Journals in the Field

During this period, 557 journals were identified that published at least one paper in the evaluated field. Of these journals, 29 (5.20%) had at least 10 papers published. The most prolific journal was *Energy Policy* with 415 (17.64% of the total) papers published. This journal has an impact factor (IF) of 7576, or 7014 without self-citations, and on average, each paper had 60.59 citations, indicating that papers published in this journal have had a significant impact in this field. The second journal in terms of number of papers published was *Energy and Buildings*. This journal published 179 (7.61%) papers (a significant difference from *Energy Policy* with 415 papers published) and has an IF of 7201 or 6147 if self-citations are excluded. Table 2 details the most prolific journals in this period, and Figure 4 shows the journals with at least 10 papers published.

Table 2. Top 10 prolific journals in the field of energy efficiency (period 1978–2010).

Source	No.	Citations	Average Citation/ Paper	Impact Factor/2021	IF without Self-Citations	Publisher
Energy Policy	415	25,146	60.59	7.576	7.014	Elsevier
Energy and Buildings	179	14,259	79.66	7.201	6.147	Elsevier
Renewable Energy	122	1854	15.20	8.634	7.711	Pergamon-Elsevier
Energy	103	4619	44.84	8.857	7.271	Pergamon-Elsevier
Building and Environment	66	3432	52.00	7.093	5.741	Pergamon-Elsevier
Energy Conversion and Management	66	3395	51.44	11.533	9.932	Pergamon-Elsevier
Applied Energy	55	2822	51.31	11.446	10.305	Elsevier
Energy Efficiency	46	2090	45.43	3.134	2.890	Springer
Renewable and Sustainable Energy Reviews	42	6482	154.33	16.799	15.532	Pergamon-Elsevier
Building Research and Information	41	2195	60.59	4.799	4.483	Routledge Journals, Taylor and Francis Ltd.

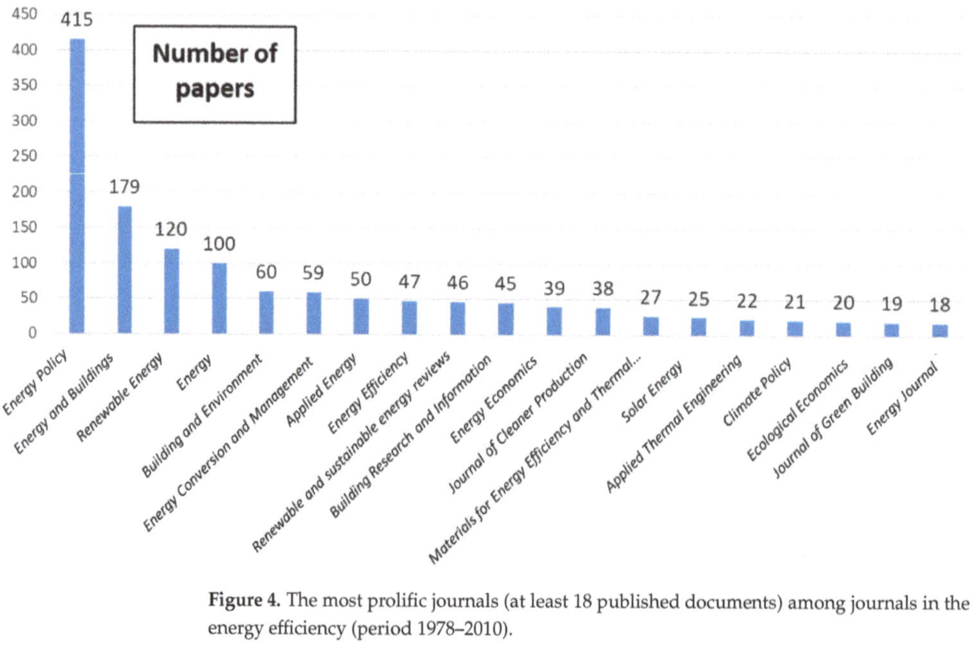

Figure 4. The most prolific journals (at least 18 published documents) among journals in the field of energy efficiency (period 1978–2010).

3.1.3. Citation Analysis of Publications in the Evaluated Period

A total of 2352 papers were published during the period under examination. The top 10 most cited papers of this period that fell within the search terms are shown in Table 3.

Table 3. The most influential manuscripts in the field of energy efficiency (period 1978–2010).

First Author (Year)	Title	Journal	Impact Factor	Citations	Ref.
Perez-Lombard (2008)	A review on buildings energy consumption information	*Energy and Buildings*	7.201	3701	[30]
Ibrahim, H. (2008)	Energy storage systems—Characteristics and comparisons	*Renewable and Sustainable Energy Reviews*	16.799	1325	[31]
Granqvist, Claes G. (2007)	Transparent conductors as solar energy materials: A panoramic review	*Solar Energy Materials and Solar Cells*	7.305	1231	[32]
Swan, LG (2009)	Modeling of end-use energy consumption in the residential sector: A review of modeling techniques	*Renewable and Sustainable Energy Reviews*	16.799	1133	[33]
Greening, LA (2000)	Energy efficiency and consumption—the rebound effect—a survey	*Energy Policy*	7.576	1125	[34]
Song, CS (2002)	Fuel processing for low-temperature and high-temperature fuel cells—Challenges, and opportunities for sustainable development in the 21st century	*Catalysis Today*	6.562	950	[35]
Jones, AP (1999)	Indoor air quality and health	*Atmospheric Environment*	5.755	935	[36]
Omer, AM (2008)	Energy, environment and sustainable development	*Renewable and Sustainable Energy Reviews*	16.799	922	[37]
Yu, Wenhua (2008)	Review and comparison of nanofluid thermal conductivity and heat transfer enhancements	*Heat Transfer Engineering*	2.431	858	[38]
Dietz, T	Household actions can provide a behavioral wedge to rapidly reduce US carbon emissions	*Proceedings of the National Academy of Sciences of The United States of America*	12.779	843	[39]

The first ranked article, "A review on buildings energy consumption information", was published by Perez-Lombard in 2008 in the journal *Energy and Buildings*, which has an IF of 7.201. This article received the highest number of citations in this period (3701). The second article in terms of number of citations (1325), "Energy storage systems—Characteristics and comparisons", was published by Ibrahim H. in 2008 in the journal *Renewable and Sustainable Energy Reviews*, which has an IF of 16,799.

3.1.4. Bibliometric Analysis of the Most Active Organizations in the Field

A total of 1623 affiliations were identified for papers published during this period. The most active organization was the United States Department of Energy with 202 published papers (8.59% of the total), followed by Lawrence Berkeley National Laboratory with 137 (5.82%) published papers, and University of California System with 132 (5.61%) published papers (Table 4).

Table 4. The most active organizations in the field of energy efficiency (period 1978–2010).

Organization	Papers	%
United States Department of Energy (DOE)	202	8.59
Lawrence Berkeley National Laboratory	137	5.82
University of California System	132	5.61
University of California Berkeley	105	4.46
Hong Kong Polytechnic University	37	1.57
Tsinghua University	31	1.32
Swiss Federal Institutes of Technology Domain	29	1.23
Utrecht University	27	1.15
Lund University	26	1.10
Oak Ridge National Laboratory	26	1.10
Chongqing University	22	0.93
City University of Hong Kong	21	0.89
Fraunhofer Gesellschaft	21	0.89
University of Cambridge	20	0.85

Most Prolific Research Areas

The papers were classified by WoS into 74 categories. Figure 5 shows the categories in which most of the papers were classified. The category with the most articles (1271) was "Energy Fuels", followed by the category "Environmental Sciences Ecology" with 769 articles.

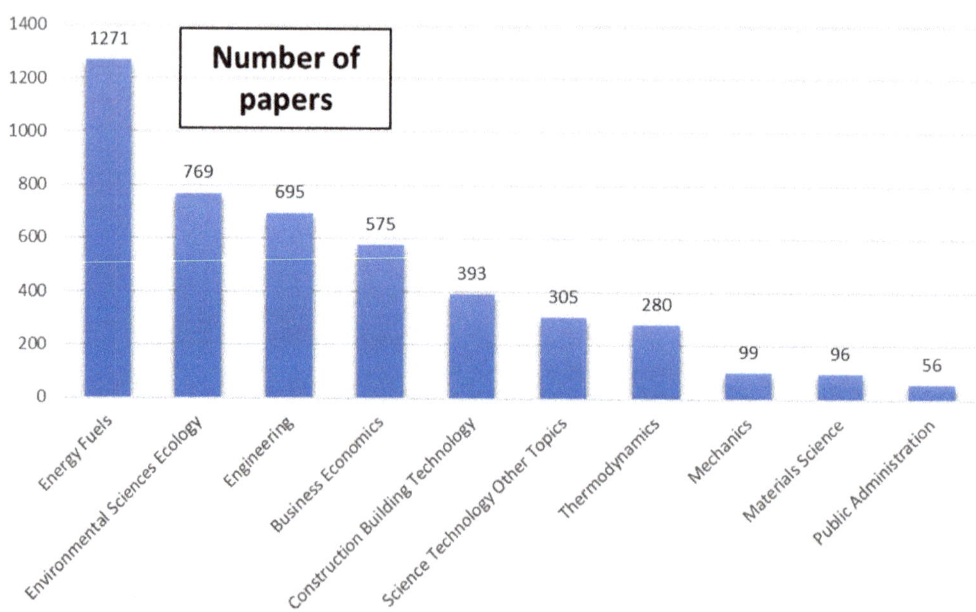

Figure 5. The most prolific research areas in the field of energy efficiency (period 1978–2010).

3.2. Science Mapping Analysis Tools

3.2.1. Country Co-Authorship Assessments

Figure 6 shows a map of co-authorship by country. In this figure, the countries that have at least 10 published papers (41 or 45.55%) are represented. These countries are grouped into five clusters: the red cluster contains 22 countries and is led by England (by number of articles); the green cluster includes 7 countries and is led by the United States; the blue cluster contains 5 countries and is led by Spain; the yellow cluster contains 4 countries and is led by Australia; and the purple cluster contains 2 countries and is led by Thailand. The countries that have collaborated most often are the United States and China, the United States and England, and the United States and Germany.

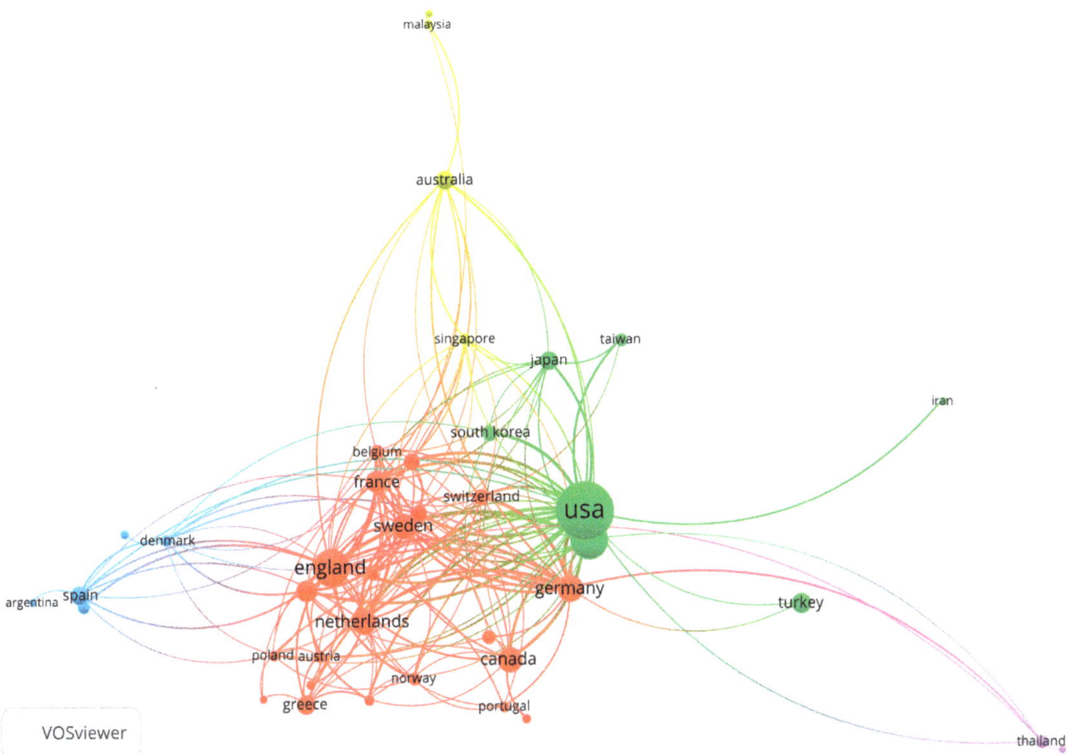

Figure 6. Network map of country co-authorship between 1978 and 2010 (VOSviewer).

3.2.2. Source Average Publication Year and the Citation Network Map

Figure 7 shows the average publication years of the 29 journals that had at least 10 papers published. *Energy Policy*, the most prolific journal of this period with 415 papers published, had an average publication year of 2004.79. The average publication year of *Energy and Buildings*, the second most prolific journal, was 2005.72. *Renewable Energy* had an average publication year of 1999.11 and *Energy*, 2002.50. The journals that published the most articles towards the end of the period analyzed were *Renewable and Sustainable Energy Reviews* (2007.95), *Journal of Green Building* (2008.71), and *Energy Efficiency* (2008.87).

The parameters used for filtering the included journals were also used in the analysis shown in Figure 8, which depicts the citation network map between journals. According to Figure 8, the journals are grouped into three clusters. The red cluster contains 15 journals and is led by *Energy and Buildings*; the green cluster contains 8 journals and is led by *Energy Policy*; and the blue cluster contains 4 journals and is led by *Climate Policy*. The journals most strongly connected by frequently cited papers are *Energy Policy* and *Energy*, *Energy Policy* and *Energy Economics*, *Energy Policy* and *Energy Efficiency*, *Energy and Buildings* and *Building and Environment*, and *Energy and Buildings* and *Energy Policy*.

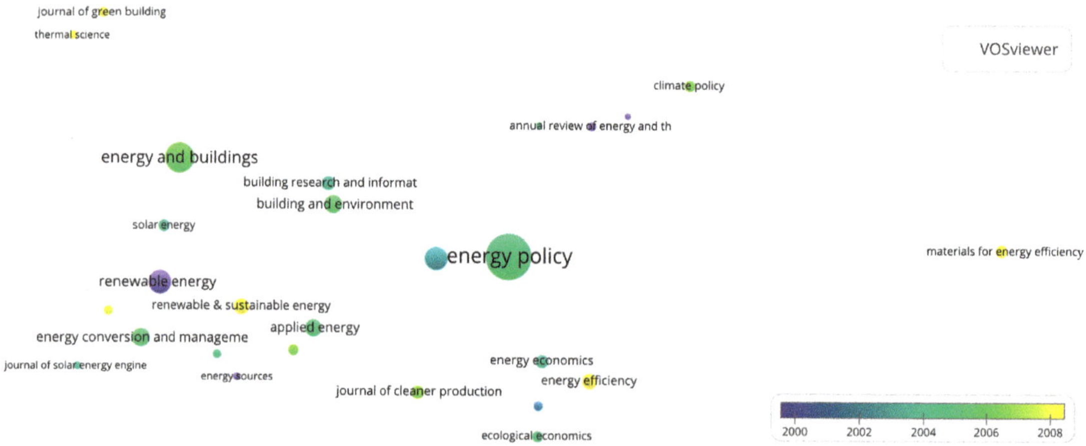

Figure 7. Bubble map of the average publication year 1978–2010.

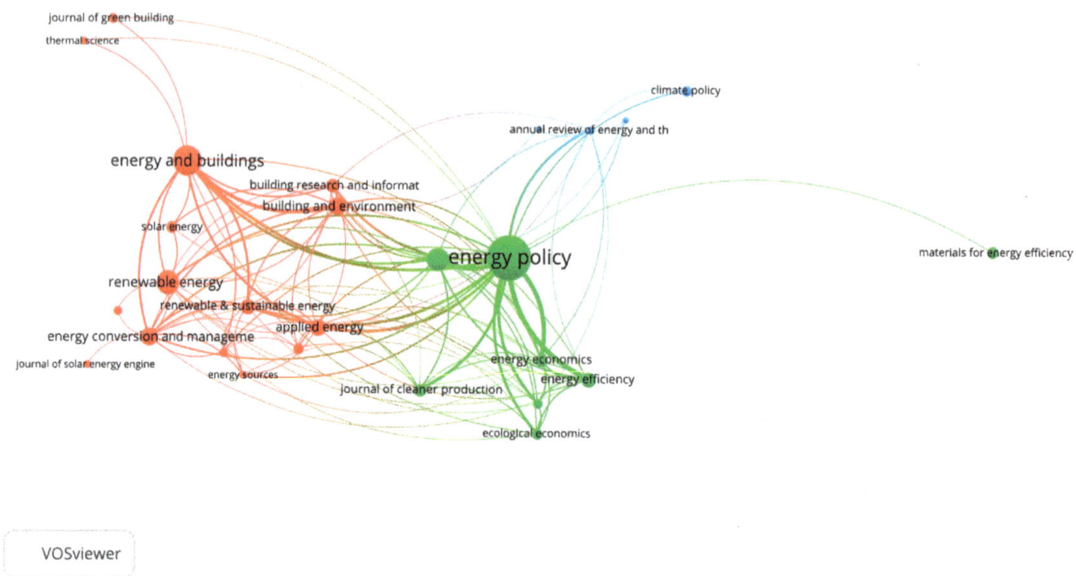

Figure 8. Source citation network map between 1978 and 2010 (VOSviewer).

3.2.3. Term Map and Network Map of Term Co-Occurrence

Figure 9 shows the bubble map of high-frequency terms for the period 1978–2010. In this figure, only words with a minimum frequency of 25 are shown. The terms that had a high average citation/paper are "conservation" (47 occurrences, average citations 100.49), "consumption" (74, 90.99), and "electricity" (35, 93.66). The words with the highest occurrence, such as "energy efficiency", have an occurrence of 732 and an average citation/paper 53.99. Other keywords including "sustainability" (53, 66.02), "renewable energy" (109, 59.58), "sustainable development" (83, 63.80), and "energy" (127, 72.22) also have a high prevalence.

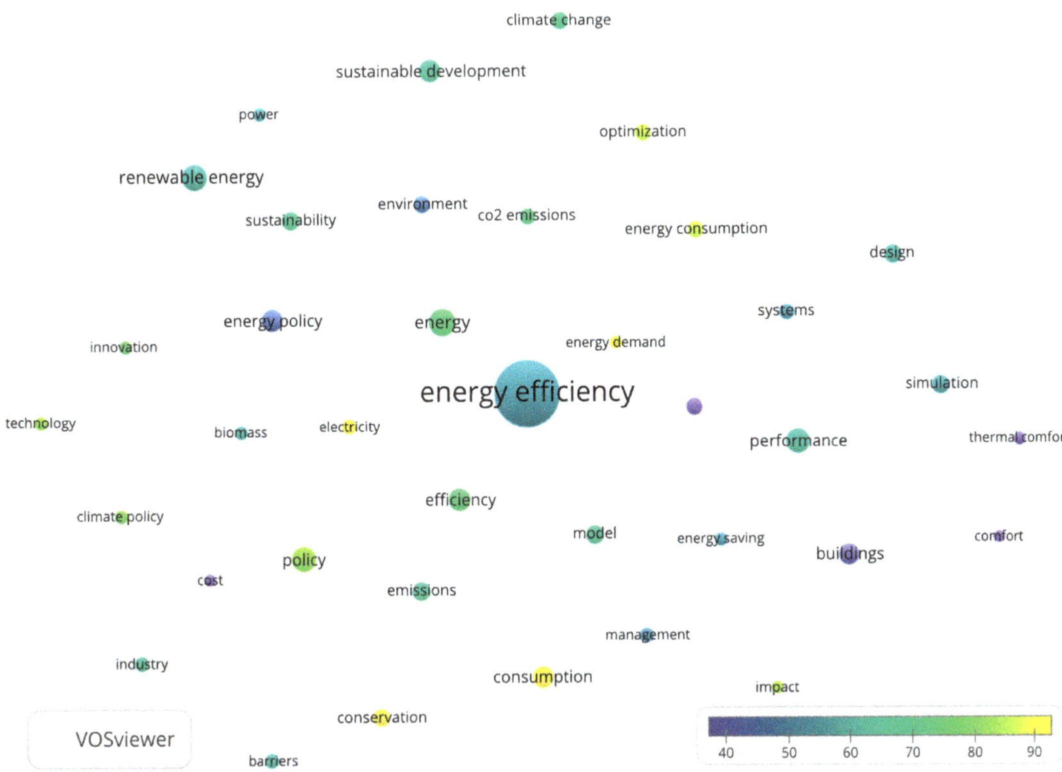

Figure 9. Bubble map of high-frequency terms (39 terms) (1978–2010).

The method of filtering the words included in Figure 10 was the same as in the previous assessment. This figure shows the network map of term co-occurrences. The terms are grouped into three clusters: the red cluster includes 17 terms, the green cluster includes 11 terms, and the blue cluster contains 11 terms. The following terms show an increased co-occurrence:

- "energy efficiency" with "performance", "energy policy", "renewable energy", "performance", "buildings" and "consumption";
- "policy" with "energy", "efficiency", "consumption", "conservation" and "renewable energy";
- "renewable energy" with "power", "sustainable development", and "energy policy".

3.3. *Period 2011–2023*

3.3.1. Evaluation of the Most Productive Countries in the Field

The total number of countries contributing to the development of papers increased to 150 from 90 in the previous period. Furthermore, 55 countries (36.66%) had more than 100 papers published. China was the country with the most published papers (6126) followed by the United States and England. In terms of the number of citations, the United States ranked first, followed by China and England, indicating that papers published by authors from the United States had a higher impact with an average citation/manuscript of 33.97. Table 5 shows in detail the top 10 countries in terms of the number of papers published.

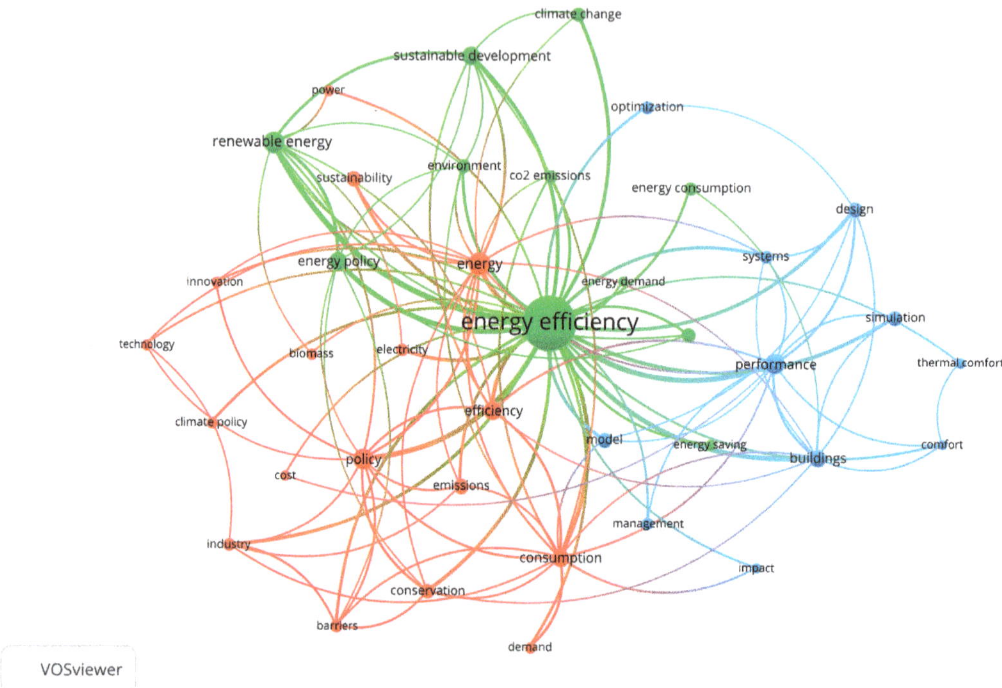

Figure 10. Co-occurrence network map of field related terms between 2001 and 2011.

Table 5. Top 10 prolific countries in the field of energy efficiency (period 2011–2023).

Country	Papers	Citations	Average Citation/Manuscript	TLS
China	6126	152,483	24.89	3343
United States	4792	162,762	33.97	3125
England	1956	60,583	30.97	2271
Italy	1867	50,175	26.87	1409
Spain	1521	32,196	21.17	999
Germany	1171	36,555	31.22	1461
Australia	1100	29,858	27.14	1183
India	995	16,805	16.89	711
South Korea	866	21,408	24.72	632
Canada	823	27,942	33.95	943

TLS, total link strength value attributed by VOSviewer.

3.3.2. Assessment of the Most Prolific Journals in the Field

A total of 2888 journals were identified as having published papers in this domain, which is a significant increase from the first period (557), indicating that interest in this area has increased considerably. The number of journals that had at least 50 papers published was 60 (2.07%). The most prolific journal was *Energy and Buildings* with 1877 papers published (7.17% of the total output in this period). This journal has an IF of 7.201 and an IF score without self-citations of 6.147, with an average of 30.77 citations/paper. The second journal in terms of number of articles published, *Energies* (1358, 5.19%), has an IF of

3.252 and an IF without self-citations of 2.466. Table 6 details the top 10 journals ranked by number of papers published, and Figure 11 shows journals with at least 100 papers published.

Table 6. Top 10 prolific journals in the field of energy efficiency (period 2011–2023).

Source	No.	Citations	Average Citation/Paper	IF/2021	IF without Self-Citations	Publisher
Energy and Buildings	1877	57,763	30.77	7.201	6.147	Elsevier
Energies	1358	12,176	8.97	3.252	2.466	MDPI
Sustainability	983	8365	8.51	3.889	3.008	MDPI
Journal of Cleaner Production	962	34,286	35.64	11.072	9.707	Elsevier Sci Ltd.
Energy Policy	950	33,101	34.84	7.576	7.014	Elsevier
Applied Energy	875	39,770	45.45	11.446	10.305	Elsevier
Energy	808	24,959	30.89	8.857	7.271	Pergamon-Elsevier
Renewable and Sustainable Energy Reviews	676	39,635	58.63	16.799	15.532	Pergamon-Elsevier
Energy Efficiency	606	7062	11.65	3.134	2.89	Springer
Building and Environment	484	14,828	30.64	7.093	5.741	Pergamon-Elsevier

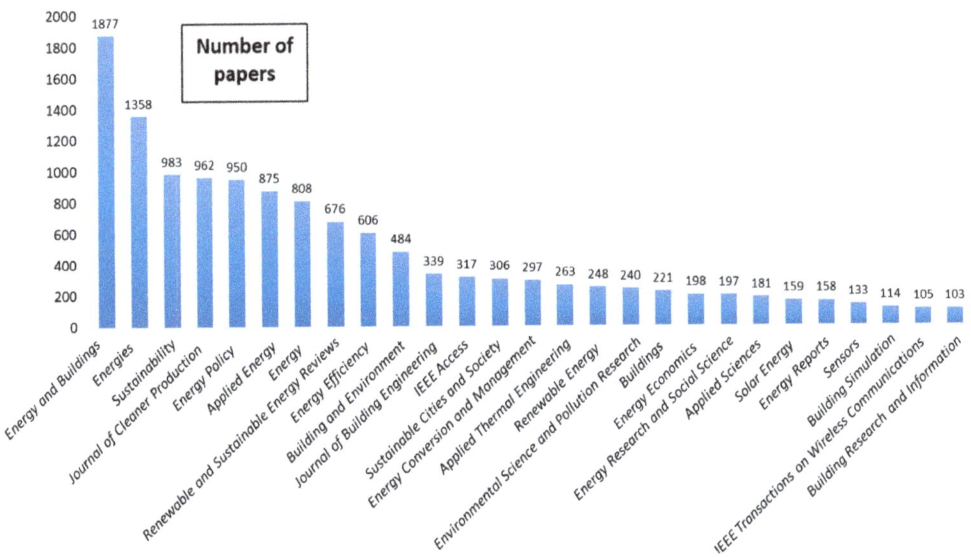

Figure 11. The most prolific journals (at least 100 published documents) in journals in the field of energy efficiency (period 2011–2023).

3.3.3. Citation Analysis of Publications in the Evaluated Period

During the period under investigation, a total of 26,203 papers were published. The top 10 most cited papers of this period that fell within the search terms are shown in Table 7.

Table 7. The most influential manuscripts in the field of energy efficiency in 2011–2023.

First Author (Year)	Title	Journal	IF	Citations	Ref.
Yang, ZG (2011)	Electrochemical Energy Storage for Green Grid	*Chemical Reviews*	72.087	3600	[40]
Luo, X (2015)	Overview of current development in electrical energy storage technologies and the application potential in power system operation	*Applied Energy*	11.446	2032	[41]
Palensky, P (2011)	Demand Side Management: Demand Response, Intelligent Energy Systems, and Smart Loads	*IEEE Transactions on Industrial Informatics*	11.648	1782	[42]
Bocken, NMP (2014)	A literature and practice review to develop sustainable business model archetypes	*Journal of Cleaner Production*	11.072	1465	[43]
Raman, AP (2014)	Passive radiative cooling below ambient air temperature under direct sunlight	*Nature*	69.504	1334	[44]
Gielen, D (2019)	The role of renewable energy in the global energy transformation	*Energy Strategy Reviews*	10.01	1224	[45]
Dincer, I (2015)	Review and evaluation of hydrogen production methods for better sustainability	*International Journal of Hydrogen Energy*	7.139	1173	[46]
Cabeza, LF (2011)	Materials used as PCM in thermal energy storage in buildings: A review	*Renewable and Sustainable Energy Reviews*	16.799	1084	[47]
Di Renzo, M (2014)	Spatial Modulation for Generalized MIMO: Challenges, Opportunities, and Implementation	*Proceedings of the IEEE*	14.91	1030	[48]
Wicklein, B (2015)	Thermally insulating and fire-retardant lightweight anisotropic foams based on nanocellulose and graphene oxide	*Nature Nanotechnology*	40.523	865	[49]

The first article, by number of citations, "Electrochemical Energy Storage for Green Grid", was published by Yang, Z.G., in *the Journal Chemical Reviews*, which has an IF of 72,087; this article has gathered a total of 3600 citations. In second place is Luo, X., with the article entitled "Overview of current development in electrical energy storage technologies and the application potential in power system operation", which was published in t *Applied Energy* with an IF of 11.446 and a total of 2032 citations. Due to the high number of citations gathered by these articles, they have had a significant impact in the field of sustainability.

3.3.4. Bibliometric Analysis of the Most Active Organizations in the Field

A total of 11,453 affiliations of authors publishing during the period evaluated were identified. The most active organization was the United States Department of Energy, with 1393 published documents (5.32% of the total). Furthermore, the Lawrence Berkeley National Laboratory with 606 (2.31%) published papers and the University of California System with 576 (2.20%) published papers were the next most prolific. The most active organizations in the evaluated domain are presented in Table 8.

The papers were classified by WoS into 107 categories. Figure 12 shows the categories in which most papers were classified. The category with the most articles (10,686) is "Energy Fuels" followed by "Engineering" with 10,642 articles. It is important to note that a paper can be classified into more than one WoS category.

Table 8. The most active organizations in the field of energy efficiency (period 2011–2023).

Organization	Papers	%
United States Department of Energy (DOE)	1393	5.32
Lawrence Berkeley National Laboratory	606	2.31
University of California System	576	2.20
Tsinghua University	450	1.72
Chinese Academy of Sciences	411	1.57
National Renewable Energy Laboratory USA	361	1.38
University of California Berkeley	360	1.37
University of London	290	1.11
Swiss Federal Institutes of Technology Domain	264	1.01
Udice French Research Universities	257	0.98
Polytechnic University of Milan	254	0.97
Hunan University	253	0.97
Centre National de la Recherche Scientifique (CNRS)	251	0.96
Hong Kong Polytechnic University	245	0.94
Chongqing University	222	0.85
University College London	221	0.84
Aalto University	206	0.79
National University of Singapore	206	0.79
Norwegian University of Science Technology (NTNU)	202	0.77
City University of Hong Kong	200	0.76
Tianjin University	200	0.76

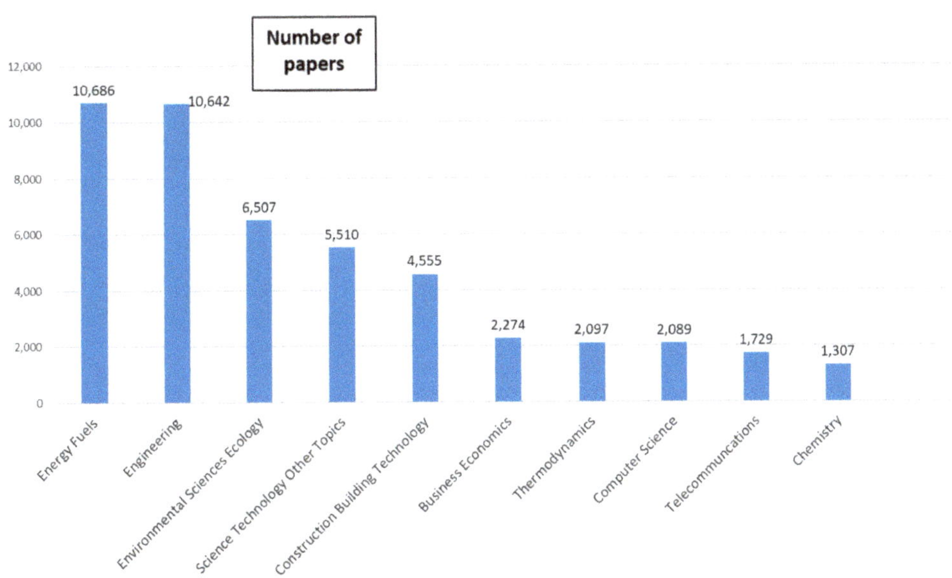

Figure 12. The most prolific research areas in the field of energy efficiency (period 2011–2023).

3.4. *Science Mapping Analysis Tools*

3.4.1. Country Co-Authorship Assessments

Figure 13 shows the map of the co-authorship by country. In this figure shows the countries that have at least 100 published papers (55 or 36.66%). The countries are grouped into four clusters: the red cluster contains 24 countries and is led by Italy (by number of articles); the green cluster includes 17 countries and is led by China; the blue cluster contains 11 countries and is led by England; the yellow cluster is led by Spain. The countries showing a strong collaborative relationship are represented by China, the United States, Australia, Canada, and England. The United States collaborated often with Germany, Canada, South Korea, Italy, and England.

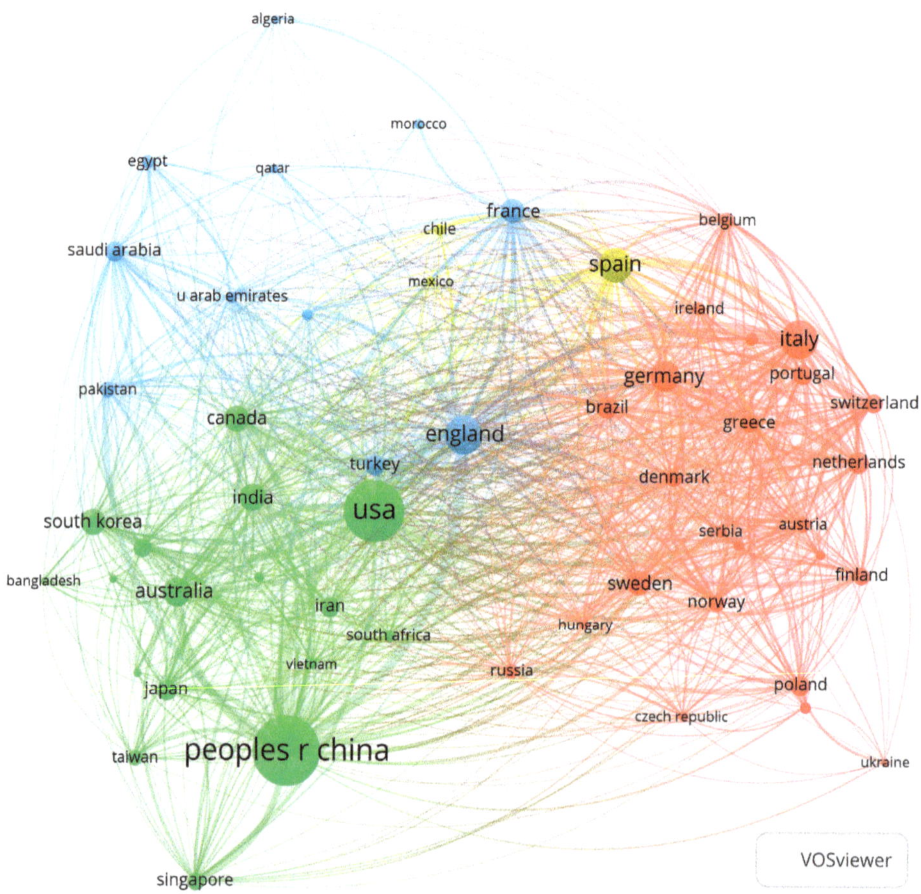

Figure 13. Network map of country co-authorship between 2011 and 2023 (VOSviewer).

3.4.2. Source Average Publication Year and the Citation Network Map

Figure 14 shows the average years of publication for the 60 (2.07%) journals that had at least 50 papers published. The average publication year of the most prolific journal in this period, *Energy and Buildings*, is 2017.20; the journal *Energies* ranked second in terms of the number of papers published in this area and has an average publication year of 2019.83; *Sustainability*, which is ranked third, has an average publication year of 2019.73. The journals that published the most articles at the end of this period are *Buildings* (2021.00) and *Energy Reports* (2021.30).

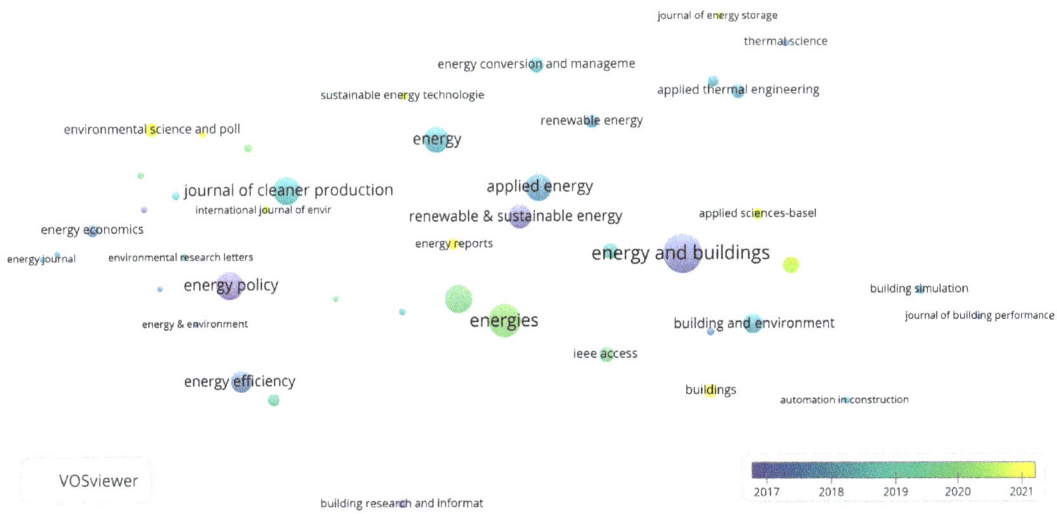

Figure 14. Bubble map of the average publication year for the period 2011–2023.

The filter parameters for the journals are the same as those in the previous figure. Figure 15 shows the citation network map between journals. According to Figure 15, the journals are grouped in three clusters differentiated by color. The red cluster contains 20 journals and is led by the *Journal of Cleaner Production*, the green cluster contains 13 journals and is led by the most prolific journal of this period, *Energy and Buildings*, and the blue cluster contains 11 journals and is led by *Applied Energy*. The following journals are strongly connected by frequently cited papers: *Energy and Buildings* and *Applied Energy*, *Renewable and Sustainable Energies* and *Building and Environment*. Another group of journals linked by frequent citations includes *Energy Policy*, *Journal of Cleaner Production*, *Energy Efficiency*, *Energy*, and *Energies*.

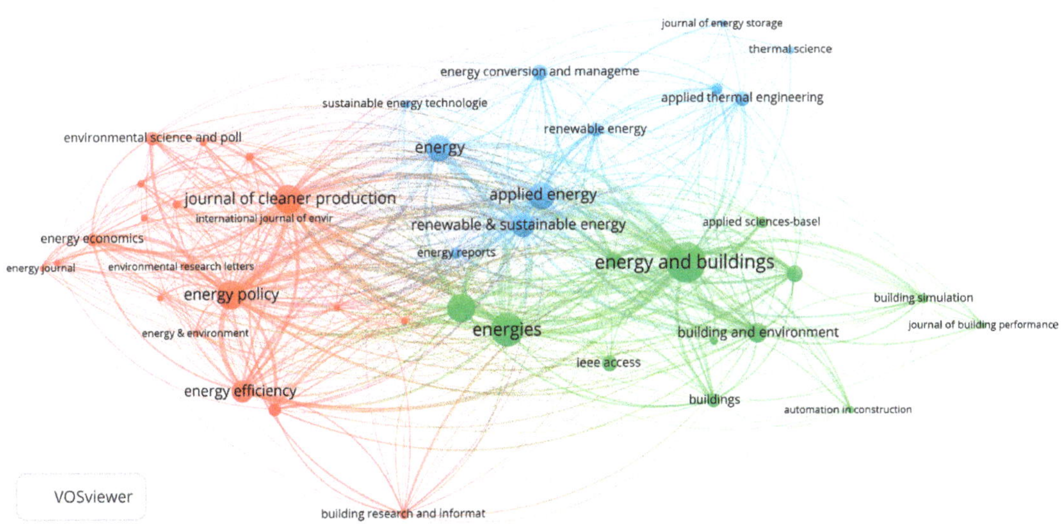

Figure 15. Source citation network map between 2011 and 2023 (VOSviewer).

3.4.3. Term Map and Network Map of Term Co-Occurrence

Figure 16 shows a bubble map of high-frequency terms for the period 2011–2023. In this figure, only the words with a minimum occurrence of 250 (81 terms) are shown. The terms with high average citations/paper were "office buildings" (398, 32.74), "life cycle assessment" (500, 32.67), "electricity consumption" (299, 31.95), "greenhouse gas emissions" (253, 36.66), "carbon emissions" (379, 30.04), "economic growth" (523, 29.70), and "electricity" (393, 30.49). Other keywords for the topic of interest were "renewable energy" (1390, 25.31), "residential buildings" (898, 27.60), "sustainability" (1123, 19.05), and "efficiency" (1709, 22.18).

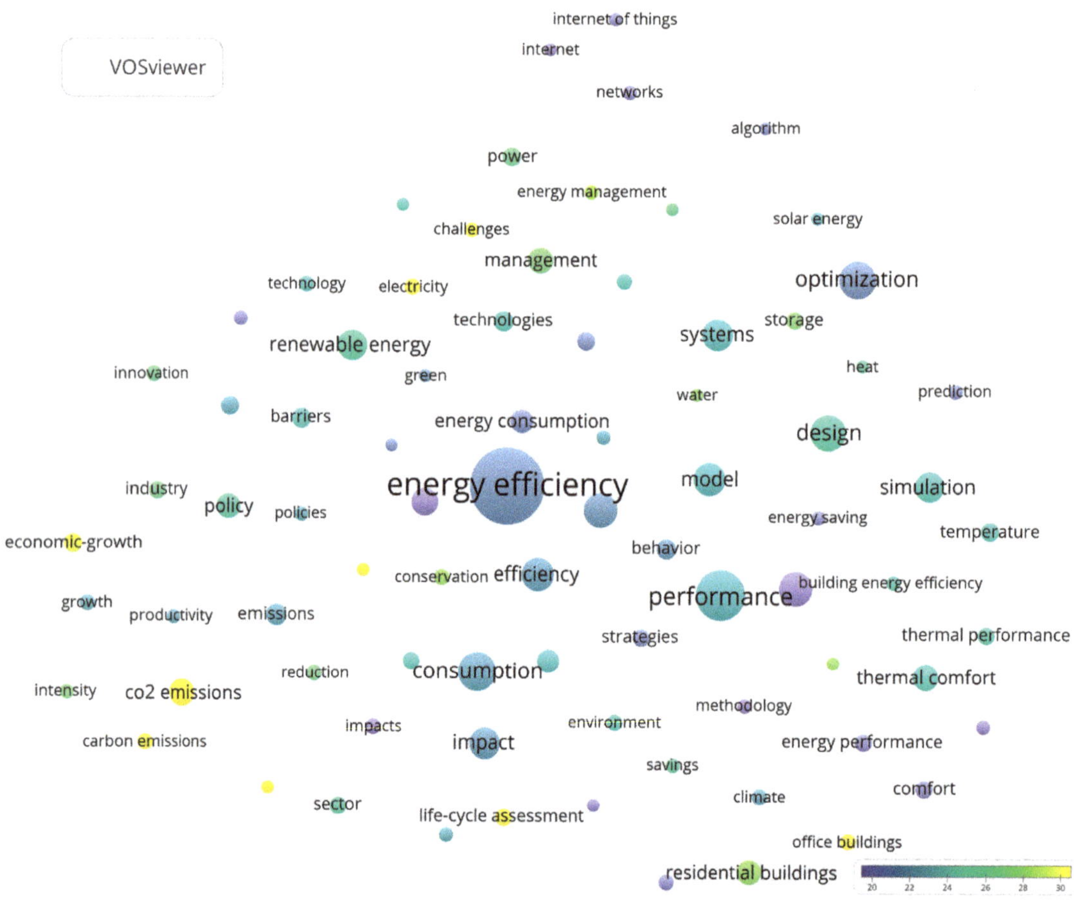

Figure 16. Bubble map of 39 high-frequency terms (1978–2010).

The method of filtering the words included in Figure 17 remains unchanged from the previous figure. In this figure, the network map of term co-occurrence is shown. The terms are grouped into three clusters: the red cluster includes 34 terms, the green cluster includes 25 terms, and the blue cluster contains 22 terms. The following terms show a re-estimated co-occurrence:

- "energy efficiency with "performance", "consumption", "buildings", "design", "optimization" and "impact";
- "performance" with "consumption", "simulation", "buildings", "design", "thermal comfort" and "optimization";

- "optimization" with "renewable energy", "thermal comfort", "residential buildings", and "buildings".

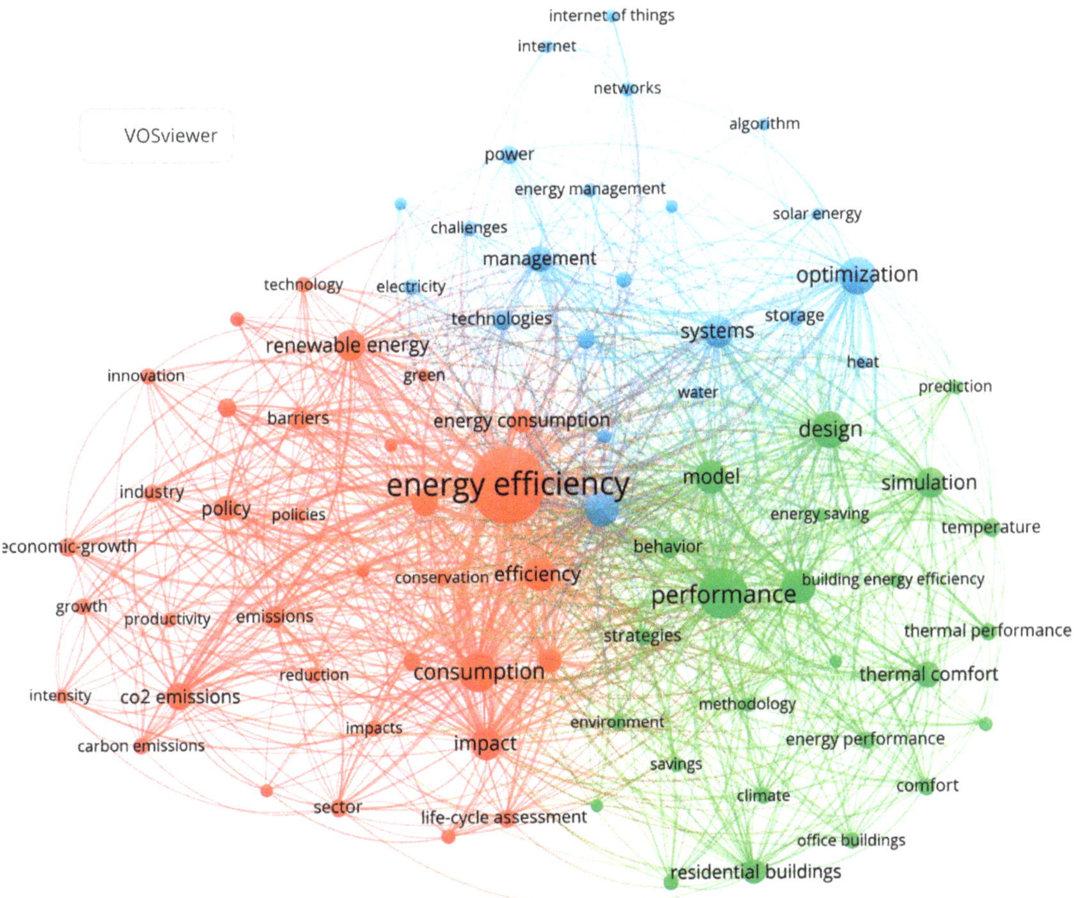

Figure 17. Co-occurrence network map of RA related terms between 2001 and 2011.

4. Discussion

Global scientific and economic interest has grown in recent years towards solving unmet energy consumption needs by applying sustainable green building concepts and practices or retrofitting old buildings for increasing energy efficiency [28].

In the framework of environmentally friendly growth and the efficiency of energy in buildings, policies and initiatives implemented by the government play an essential part in the process of supporting sustainable practices and setting energy-efficient transitions. The incorporation of pertinent energy policies into research not only provides useful insights into the regulatory system but also opens the way for additional investigation [50]. The assessment of the energy consumption of both existing buildings (rehabilitated/or not) and newly built buildings Is done through the energy certification of the building [51]. The certification entity must be independent of the building owner, investor, constructor, architect, and material supplier so that the entire building certification system becomes an essential tool for sustainable development [52].

In Romania, an Energy Performance Certificate (EPC) provides information on the energy consumption of a building from conventional or renewable sources to provide utilities, including heating, cooling, domestic hot water, ventilation and lighting, and the amount of CO_2 emissions is also assessed. This system assigns each building to an energy class (indicated with letters between A and G) and a grade between 1 and 100. In the EPC, the auditor provides recommendations for reducing energy consumption. The EPC system is regulated by Law no. 101/1, July 2020 [53] and Law no. 37/13, December 2005 [54].

Governments frequently enact building energy codes and standards, which typically specify the minimum energy efficiency requirements for newly constructed buildings and significant renovations. These building regulations and norms can be examined in order to determine the effect that they have on the amount of energy that is consumed by buildings as well as the spread of energy-efficient technology and practices [55].

Numerous governments and enterprises that work directly together offer financial incentives, grants, and subsidies in order to encourage the construction of energy-efficient buildings, retrofits, and the incorporation of renewable energy sources. Researchers are able to study the efficacy of these incentives in promoting investments in energy efficiency and the consequent effects on lowering energy usage [56].

Green building certification systems, such as LEED (Leadership in Energy and Environmental Design) and BREEAM (Building Research Establishment Environmental Assessment Method), provide voluntary frameworks for evaluating and recognizing the sustainability performance of buildings. Evaluating the degree to which these certification systems have been adopted and how much of an impact they have had may offer perspectives on the important function that they play in advancing sustainable construction practices [57].

Modern facilities and associated tools such as the Internet of Things (IoT) and smart infrastructure have transformed the design, construction, and operation of buildings. The combined application of cutting-edge technologies and architectural principles seeks to improve energy efficiency, comfort for residents, and sustainability as a whole [58].

With IoT-enabled intelligent facilities, buildings may incorporate multiple systems, including lighting, heating/cooling, security, and energy management, allowing for centralized control and efficient administration. In this regard, sensors may track occupancy levels, humidity, temperature, and illumination conditions, optimizing the use of energy while preserving comfort. Intelligent lighting systems may adjust luminance dynamically based on natural light, occupancy, and customer preferences, thereby reducing wasteful energy consumption [59]. Furthermore, contemporary building design prioritizes the use of eco-friendly materials and construction methods. Energy-efficient building envelopes with improved insulation, efficient glazing, and adequate shade serve to reduce heat gain and loss. Green roofs, rooftop solar panels, and precipitation collection systems assist in the production of energy and the preservation of water [59,60].

Energy management in newly formed smart cities is an important area of research and practice. It attempts to maximize the efficiency of energy use, lower the amount of carbon emissions produced, and improve the overall sustainability of metropolitan environments. Smart cities work toward the goal of effective management of energy resources by incorporating emerging forms of technology, data analytics, and artificial intelligence into their infrastructure [61].

In order to obtain insights into patterns of energy usage, smart cities gather and examine large volumes of data from a variety of sources, such as sensors, smart meters, and devices connected to the IoT. Identifying energy usage patterns, locating anomalies, and developing forecasting algorithms for energy demand may all be accomplished with the help of advanced data analytics techniques such as machine learning and intelligent technology. These methods, which are driven by data, make it possible to arrive at more accurate energy management plans and decision-making procedures. The demand response solutions that are implemented in smart cities enable residents to make adjustments to the amount of energy they use depending on real-time price or demand signals [62].

The energy storage systems that are integrated into smart cities are used to store the surplus of renewable energy that is generated. This stored energy is available for use during times of intense demand or when there is a shortage of renewable generation. Battery storage systems, hydroelectric power storage, and other cutting-edge technologies are being implemented to increase the proportion of sources of clean energy while simultaneously improving the dependability and adaptability of the energy grid [63].

Every single smart city varies in terms of its innovations, regulations, and strategies, but they all use cutting-edge technology and environmentally friendly techniques to enhance the quality of life for inhabitants while reducing adverse environmental effects. There are several examples of recently developed smart urban areas around the world which demonstrate innovative technologies and environmentally responsible practices: Delhi and Stockholm [64], Singapore [65], Masdar City [60], and Barcelona [66].

The present bibliometric analysis highlights a subject that is less addressed in the literature by investigating the growing trend towards optimizing energy efficiency and sustainable development with renewable energy (in the management of sustainable buildings). In this respect, the possibilities for collaboration, setting search patterns in the scientific literature, and identifying the current state of knowledge for detecting current unmet needs and future research directions become highly important.

Overall, China has published the most papers in this field (6349), followed by the United States with 5310 published papers and England with 2179 published papers. In total, these countries published 13,838 (48.46%) of the total number of papers, indicating that sustainability is a topic of primary interest in these countries. Figure 18 shows in detail the evolution over time of the number of papers published by these countries. The number of articles published by these countries has gradually increased, and since 2010, a much faster increase can be observed in correlation with technological development. The first year in which China surpasses the United States in the number of articles published (343 vs. 325) is 2016.

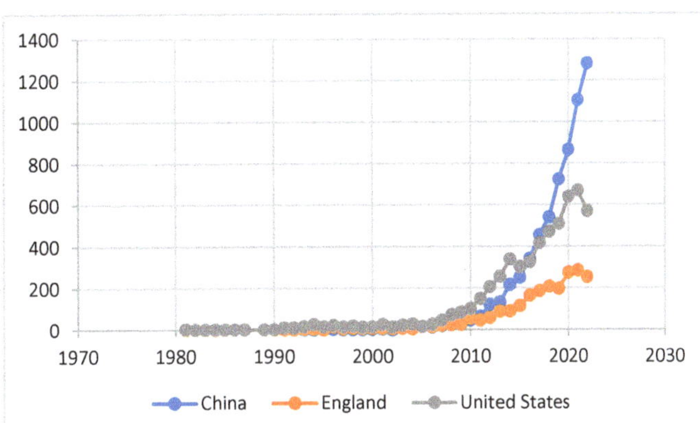

Figure 18. Evolution over time of the number of articles published by the most prolific countries.

Following the analysis shown in the tables highlighting the most prolific journals in the period under analysis, we can deduce that in the first period, the most prolific journals were *Energy Policy*, *Energy and Buildings* and *Renewable Energy* with 415, 179 and 122 papers published, respectively. In the second period, the most prolific journals were *Energy and Buildings* with 1877 papers, followed by *Energies* and then *Sustainability* with 1358 and 983 papers. The journal that stands out in both periods due to its high IF and high average citations per paper is *Renewable and Sustainable Energy Reviews*, indicating that papers published in this journal have a significant impact in this field.

Based on the analysis of the papers with the highest citation rate, it can be determined that they cover several topics in the field of energy and sustainability. The paper ranked first in the first period has the highest number of citations, but this may also be because it was published much earlier compared with the papers published in Table 2. The papers that received the most citations were published in journals such as *Chemical Reviews* and *Nature*, which highlights the impact of these journals in the field.

The analysis of the collaboration maps between countries indicates that there is an increasing trend. These maps allow us to analyze which countries collaborate frequently with each other to produce quality papers and can serve as a guide for authors who are interested in finding collaborators interested in the same topic. In both periods, there was a strong collaborative link among the top three countries in terms of the number of papers (the United States, China, and England).

In both periods, three clusters are formed around journals whose articles are frequently cited by articles in other journals. In the first period, the clusters are formed around the following journals: *Energy policy*, *Energy and Buildings* and *Annual Review of Energy and the Environment*. In the second period, clusters are formed around *Energy and Buildings*, *Journal of Cleaner Production* and *Applied Energy*.

The present paper applies bibliometric approaches to determine the most productive nations, prolific journals, highly cited publications, active organizations, collaborative networks, term co-occurrence maps, high-frequency terms, and research areas in the field. This yields significant knowledge about the research landscape and contributes to the field's fundamental development.

The outcomes of the bibliometric analysis are equally practical and academically valuable. Moreover, policymakers and industry professionals may use the data on the most prosperous states and active organizations to pinpoint potential collaborators for international partnerships and knowledge sharing. In addition, being aware of the most prolific journals in the field enables stakeholders to keep abreast of the most recent research and developments in the field. This aids in the formulation of sustainable building regulations and the promotion of energy-efficient building approaches.

From an academic point of view, the research findings provide an overview of the existing knowledge and current developments in the field of energy efficiency of buildings from 1978 (the year of the first publication that matched the search algorithm) to the present. Furthermore, the terms co-occurrence mapping and high-frequency terms provide scientists with a conceptual overview of the research topics and practical approaches, enabling them to identify popular research areas and direct their studies accordingly. The recognition of collaborative networks and research areas facilitates multidisciplinary cooperation and the transfer of knowledge between researchers working on multiple facets of sustainable construction policies.

The paper also makes significant contributions to the field's fundamental research. It offers an in-depth examination of the research landscape by conducting the first bibliometric analysis based on a complex sorting algorithm on the energy efficiency of buildings in the context of sustainable building policies. This evaluation offers a foundation for subsequent studies by providing an initial comprehension of the current state, key contributors, and current research trends in the field. Based on this analysis, researchers may proceed deeper into specific sub-topics, investigate integrative connections, and identify areas of investigation requiring additional study.

The results of this bibliometric evaluation can direct further study in multiple ways. Initially, researchers can concentrate on countries that have demonstrated significant productivity and collaboration potential, allowing for the formation of international research networks and the exchange of best practices. In addition, an analysis of the identified highly cited papers may offer insights into the most significant works in the field, highlighting areas where future research can build on or challenge existing theories and methodologies. In the rapidly evolving scientific landscape of the present day, scientists strive to keep abreast of the most recent developments in order to enhance their own work and contribute to the field. In addition, the term co-occurrence map and high-frequency terms emphasize emerging research areas and evolving trends in time (1978–present), which can assist researchers in identifying novel research questions and inter-disciplinary links in a way that is less time-consuming than every researcher individually searching through all the scientific literature to determine the topic of study, journals to publish in, and identify current trends, etc.

Although bibliometric analyses have certain advantages related to the method of analysis, they also have several limitations. Their most significant disadvantage related to the method of analysis itself, is that due to the large number of documents, the results cannot be manually checked, so that certain articles may be included that are not exactly on the subject being researched. Another disadvantage is that only articles written in English were included in this article; thus, quality and influential documents written in another language may be omitted.

5. Conclusions

Based on the outcomes of the bibliometric analysis, the present research paper has several significant implications. Initially, it indicates the most productive states, emphasizing the global distribution of building energy efficiency research efforts. Policymakers and funding agencies can use this knowledge to target certain nations for cooperation and the allocation of resources. Secondly, the analysis identifies the most prolific journals, which signifies the most relevant platforms for disseminating research results and facilitates the identification of the most significant publications in the field. In addition, the centralized identification of highly cited publications highlights significant contributions that have considerably influenced the energy efficiency of buildings research topic.

The present study on the energy efficiency of buildings involves bibliometric and visualization analysis in combination with specialized software applications that provide a quantitative measurement of articles and reports completed to date, bridging chronological gaps, and delivering a useful resource for academics interested in this area. Moreover, 28,555 papers were returned by the complex algorithm used in searching and sorting data over a long period of time and constituted the resource for bibliometric analysis. Most of the analyzed papers (91.76% or 26,203) were published in the second period analyzed (2011–2023), indicating that this topic has gained high interest in this latter period due to the advantages arising from technological advancement.

Scientists can make informed decisions regarding the topics they choose to research, the working partnerships they form, and the journals they publish in by considering the factors examined in this analysis and the displayed results. This aids in streamlining the research procedure, accelerating the dissemination of knowledge, and maximizing the impact of their work in the field of study.

Author Contributions: Conceptualization, C.C.B. and M.F.P.; Data curation, C.C.B. and T.B.; Formal analysis, F.I.H.P.; Investigation, C.C.B., T.B. and M.F.P.; Methodology, F.I.H.P. and G.B.; Project administration, C.B.; Resources, T.B.; Software, T.B. and G.B.; Supervision, C.B. and M.F.P.; Visualization, F.I.H.P. and M.F.P.; Writing—original draft, C.C.B., C.B. and G.B.; Writing—review and editing, F.I.H.P. and T.B. All authors have read and agreed to the published version of the manuscript.

Funding: This research was funded by the Romanian Ministry of Research, Innovation and Digitization through Program 1—Development of the National Research and Development System, Subprogram 1.2—Institutional Performance—Projects for funding the excellence in RDI, Contract No. 29 PFE/30.12.2021 with the University of Oradea.

Data Availability Statement: Information presented in this paper are supported by Web of Science data base and by the references mentioned in the text.

Conflicts of Interest: The authors declare no conflict of interest.

References

1. Butera, F.M. Climatic change and the built environment. *Adv. Build. Energy Res.* **2011**, *4*, 45–75. [CrossRef]
2. Bungău, C.C.; Prada, I.F.; Prada, M.; Bungău, C. Design and Operation of Constructions: A Healthy Living Environment-Parametric Studies and New Solutions. *Sustainability* **2019**, *11*, 6824. [CrossRef]
3. Prada, M.; Popescu, D.E.; Bungau, C.; Pancu, R. Parametric Studies on European 20-20-20 Energy Policy Targets in University Environment. *J. Environ. Prot. Ecol.* **2017**, *18*, 1146–1157.
4. Wang, W.; Zmeureanu, R.; Rivard, H. Applying multi-objective genetic algorithms in green building design optimization. *Build. Environ.* **2005**, *40*, 1512–1525. [CrossRef]
5. Prada, M.; Prada, I.F.; Cristea, M.; Popescu, D.E.; Bungău, C.; Aleya, L.; Bungău, C.C. New solutions to reduce greenhouse gas emissions through energy efficiency of buildings of special importance—Hospitals. *Sci. Total Environ.* **2020**, *718*, 137446. [CrossRef]
6. Energy Roadmap. 2050. Available online: https://ec.europa.eu/energy/sites/ener/files/documents/2012_energy_roadmap_2050_en_0.pdf (accessed on 13 February 2023).
7. Prada, M.; Popescu, D.; Bungau, C. Building Education, Source of Energy Saving in Romania. In Proceedings of the 15th National Technical-Scientific Conference on Modern Technologies for the 3rd Millennium, Oradea, Romania, 27–28 November 2015; pp. 157–162.
8. Herman, G.V.; Caciora, T.; Ilies, D.C.; Ilies, A.; Deac, A.; Sturza, A.; Sonko, S.M.; Suba, N.S.; Nistor, S. 3D Modeling of the Cultural Heritage: Between Opportunity and Necessity. *J. Appl. Eng. Sci.* **2020**, *10*, 27–30. [CrossRef]
9. Bungau, C.C.; Bungau, T.; Prada, M.F.; Prada, I.F.; Moleriu, R.D. Sustainable development through green buildings: Updated bibliometric analysis of the literature in the field. *Rom. J. Mater.* **2023**, *53*, 82–93.
10. Hernandez, P.; Kenny, P. From net energy to zero energy buildings: Defining life cycle zero energy buildings (LC-ZEB). *Energy Build.* **2010**, *42*, 815–821. [CrossRef]
11. Kapsalaki, M.; Leal, V.; Santamouris, M. A methodology for economic efficient design of Net Zero Energy Buildings. *Energy Build.* **2012**, *55*, 765–778. [CrossRef]
12. Caballero, V.; Vernet, D.; Zaballos, A. A Heuristic to Create Prosumer Community Groups in the Social Internet of Energy. *Sensors* **2020**, *20*, 3704. [CrossRef]
13. Bhutta, F.M. Application of smart energy technologies in building sector—Future prospects. In Proceedings of the International Conference on Energy Conservation and Efficiency (ICECE), Lahore, Pakistan, 22–23 November 2017; Institute of Electrical and Electronics Engineers Inc.: Piscataway, NJ, USA, 2017; pp. 7–10.
14. Popescu, D.E.; Bungau, C.; Prada, M.; Domuta, C.; Bungau, S.; Tit, D.M. Waste management strategy at a public university in smart city context. *J. Environ. Prot. Ecol.* **2016**, *17*, 1011–1020.
15. Iwaro, J.; Mwasha, A. The impact of sustainable building envelope design on building sustainability using Integrated Performance Model. *Int. J. Sustain. Built Environ.* **2013**, *2*, 153–171. [CrossRef]
16. Mwasha, A.; Williams, R.G.; Iwaro, J. Modeling the performance of residential building envelope: The role of sustainable energy performance indicators. *Energy Build.* **2011**, *43*, 2108–2117. [CrossRef]
17. Owusu, P.A.; Asumadu-Sarkodie, S. A review of renewable energy sources, sustainability issues and climate change mitigation. *Cogent Eng.* **2016**, *3*, 1167990. [CrossRef]
18. Panwar, N.L.; Kaushik, S.C.; Kothari, S. Solar greenhouse an option for renewable and sustainable farming. *Renew. Sustain. Energy Rev.* **2011**, *15*, 3934–3945. [CrossRef]
19. Tiwari, G.N.; Mishra, R.K. *Advanced Renewable Energy Sources*; Royal Society of Chemistry: London, UK, 2012.
20. Methodology for Calculating the Energy Performance of Buildings. Part I-Building Envelope. Code Mc 001/1-2006. Available online: https://www.mdlpa.ro/userfiles/reglementari/Domeniul_XXVII/27_11_MC_001_1_2_3_2006.pdf (accessed on 14 February 2023).
21. Minarčík, P.; Procházka, H.; Gulan, M. Advanced Supervision of Smart Buildings Using a Novel Open-Source Control Platform. *Sensors* **2020**, *21*, 160. [CrossRef]
22. Shahi, S.; Esnaashary Esfahani, M.; Bachmann, C.; Haas, C. A definition framework for building adaptation projects. *Sustain. Cities Soc.* **2020**, *63*, 102345. [CrossRef]
23. Asdrubali, F.; Baldassarri, C.; Fthenakis, V. Life cycle analysis in the construction sector: Guiding the optimization of conventional Italian buildings. *Energy Build.* **2013**, *64*, 73–89. [CrossRef]
24. Simpeh, E.K.; Pillay, J.P.G.; Ndihokubwayo, R.; Nalumu, D.J. Improving energy efficiency of HVAC systems in buildings: A review of best practices. *Int. J. Build. Pathol. Adapt.* **2022**, *40*, 165–182. [CrossRef]
25. Al-Ghaili, A.M.; Kasim, H.; Al-Hada, N.M.; Othman, M.; Saleh, M.A. A Review: Buildings Energy Savings—Lighting Systems Performance. *IEEE Access* **2020**, *8*, 76108–76119. [CrossRef]
26. Sailor, D.J.; Anand, J.; King, R.R. Photovoltaics in the built environment: A critical review. *Energy Build.* **2021**, *253*, 111479. [CrossRef]
27. Schweiker, M. Understanding Occupants' Behaviour for Energy Efficiency in Buildings. *Curr. Sustain. Energy Rep.* **2017**, *4*, 8–14. [CrossRef]

28. Bungau, C.C.; Bungau, C.; Toadere, M.T.; Prada-Hanga, I.F.; Bungau, T.; Popescu, D.E.; Prada, M.F. Solutions for an Ecological and Healthy Retrofitting of Buildings on the Campus of the University of Oradea, Romania, Built Starting from 1911 to 1913. *Sustainability* **2023**, *15*, 6541. [CrossRef]
29. Van Eck, N.J.; Waltman, L. Software survey: VOSviewer, a computer program for bibliometric mapping. *Scientometrics* **2010**, *84*, 523–538. [CrossRef] [PubMed]
30. Pérez-Lombard, L.; Ortiz, J.; Pout, C. A review on buildings energy consumption information. *Energy Build.* **2008**, *40*, 394–398. [CrossRef]
31. Ibrahim, H.; Ilinca, A.; Perron, J. Energy storage systems—Characteristics and comparisons. *Renew. Sustain. Energy Rev.* **2008**, *12*, 1221–1250. [CrossRef]
32. Granqvist, C.G. Transparent conductors as solar energy materials: A panoramic review. *Sol. Energy Mater. Sol. Cells* **2007**, *91*, 1529–1598. [CrossRef]
33. Swan, L.G.; Ugursal, V.I. Modeling of end-use energy consumption in the residential sector: A review of modeling techniques. *Renew. Sustain. Energy Rev.* **2009**, *13*, 1819–1835. [CrossRef]
34. Greening, L.A.; Greene, D.L.; Difiglio, C. Energy efficiency and consumption—The rebound effect—A survey. *Energy Policy* **2000**, *28*, 389–401. [CrossRef]
35. Song, C. Fuel processing for low-temperature and high-temperature fuel cells: Challenges, and opportunities for sustainable development in the 21st century. *Catal. Today* **2002**, *77*, 17–49. [CrossRef]
36. Jones, A.P. Indoor air quality and health. *Atmos. Environ.* **1999**, *33*, 4535–4564. [CrossRef]
37. Omer, A.M. Energy, environment and sustainable development. *Renew. Sustain. Energy Rev.* **2008**, *12*, 2265–2300. [CrossRef]
38. Yu, W.; France, D.M.; Routbort, J.L.; Choi, S.U.S. Review and comparison of nanofluid thermal conductivity and heat transfer enhancements. *Heat Transf. Eng.* **2008**, *29*, 432–460. [CrossRef]
39. Dietz, T.; Gardner, G.T.; Gilligan, J.; Stern, P.C.; Vandenbergh, M.P. Household actions can provide a behavioral wedge to rapidly reduce US carbon emissions. *Proc. Natl. Acad. Sci. USA* **2009**, *106*, 18452–18456. [CrossRef]
40. Yang, Z.; Zhang, J.; Kintner-Meyer, M.C.W.; Lu, X.; Choi, D.; Lemmon, J.P.; Liu, J. Electrochemical energy storage for green grid. *Chem. Rev.* **2011**, *111*, 3577–3613. [CrossRef]
41. Luo, X.; Wang, J.; Dooner, M.; Clarke, J. Overview of current development in electrical energy storage technologies and the application potential in power system operation. *Appl. Energy* **2015**, *137*, 511–536. [CrossRef]
42. Palensky, P.; Dietrich, D. Demand side management: Demand response, intelligent energy systems, and smart loads. *IEEE Trans. Ind. Inform.* **2011**, *7*, 381–388. [CrossRef]
43. Bocken, N.M.P.; Short, S.W.; Rana, P.; Evans, S. A literature and practice review to develop sustainable business model archetypes. *J. Clean. Prod.* **2014**, *65*, 42–56. [CrossRef]
44. Raman, A.P.; Anoma, M.A.; Zhu, L.; Rephaeli, E.; Fan, S. Passive radiative cooling below ambient air temperature under direct sunlight. *Nature* **2014**, *515*, 540–544. [CrossRef]
45. Gielen, D.; Boshell, F.; Saygin, D.; Bazilian, M.D.; Wagner, N.; Gorini, R. The role of renewable energy in the global energy transformation. *Energy Strategy Rev.* **2019**, *24*, 38–50. [CrossRef]
46. Dincer, I.; Acar, C. Review and evaluation of hydrogen production methods for better sustainability. *Int. J. Hydrog. Energy* **2015**, *40*, 11094–11111. [CrossRef]
47. Cabeza, L.F.; Castell, A.; Barreneche, C.; De Gracia, A.; Fernández, A.I. Materials used as PCM in thermal energy storage in buildings: A review. *Renew. Sustain. Energy Rev.* **2011**, *15*, 1675–1695. [CrossRef]
48. Di Renzo, M.; Haas, H.; Ghrayeb, A.; Sugiura, S.; Hanzo, L. Spatial modulation for generalized MIMO: Challenges, opportunities, and implementation. *Proc. IEEE* **2014**, *102*, 56–103. [CrossRef]
49. Wicklein, B.; Kocjan, A.; Salazar-Alvarez, G.; Carosio, F.; Camino, G.; Antonietti, M.; Bergström, L. Thermally insulating and fire-retardant lightweight anisotropic foams based on nanocellulose and graphene oxide. *Nat. Nanotechnol.* **2014**, *10*, 277–283. [CrossRef]
50. Akadiri, P.O.; Chinyio, E.A.; Olomolaiye, P.O. Design of A Sustainable Building: A Conceptual Framework for Implementing Sustainability in the Building Sector. *Buildings* **2012**, *2*, 126–152. [CrossRef]
51. Di Foggia, G. Energy efficiency measures in buildings for achieving sustainable development goals. *Heliyon* **2018**, *4*, e00953. [CrossRef]
52. Hanga-Farcas, I.F.; Bungau, C.C.; Scurt, A.A.; Cristea, M.; Prada, M. The building certification system—A tool of sustainable development of university campuses. *J. Appl. Eng. Sci.* **2023**, *13*, 105–112.
53. Law No. 101 of 1 July 2020 Amending and Supplementing Law No 372/2005 on Energy Performance of Buildings. Available online: https://legislatie.just.ro/Public/DetaliiDocumentAfis/227538 (accessed on 15 February 2023).
54. Law No. 372 of 13 December 2005 (*Republished*) on the Energy Performance of Buildings. Available online: https://legislatie.just.ro/Public/DetaliiDocument/66970 (accessed on 16 February 2023).
55. Evans, M.; Roshchanka, V.; Graham, P. An international survey of building energy codes and their implementation. *J. Clean. Prod.* **2017**, *158*, 382–389. [CrossRef]
56. Qadir, S.A.; Al-Motairi, H.; Tahir, F.; Al-Fagih, L. Incentives and strategies for financing the renewable energy transition: A review. *Energy Rep.* **2021**, *7*, 3590–3606. [CrossRef]

57. Grzegorzewska, M.; Kirschke, P. The Impact of Certification Systems for Architectural Solutions in Green Office Buildings in the Perspective of Occupant Well-Being. *Buildings* **2021**, *11*, 659. [CrossRef]
58. Belli, L.; Cilfone, A.; Davoli, L.; Ferrari, G.; Adorni, P.; Di Nocera, F.; Dall'Olio, A.; Pellegrini, C.; Mordacci, M.; Bertolotti, E. IoT-Enabled Smart Sustainable Cities: Challenges and Approaches. *Smart Cities* **2020**, *3*, 1039–1071. [CrossRef]
59. Bellini, P.; Nesi, P.; Pantaleo, G. IoT-Enabled Smart Cities: A Review of Concepts, Frameworks and Key Technologies. *Appl. Sci.* **2022**, *12*, 1607. [CrossRef]
60. Sankaran, V.; Chopra, A. Creating Global Sustainable Smart Cities (A Case Study of Masdar City). *J. Phys. Conf. Ser.* **2020**, *1706*, 012141. [CrossRef]
61. Calvillo, C.F.; Sánchez-Miralles, A.; Villar, J. Energy management and planning in smart cities. *Renew. Sustain. Energy Rev.* **2016**, *55*, 273–287. [CrossRef]
62. Almihat, M.G.M.; Kahn, M.T.E.; Aboalez, K.; Almaktoof, A.M. Energy and Sustainable Development in Smart Cities: An Overview. *Smart Cities* **2022**, *5*, 1389–1408. [CrossRef]
63. Kolokotsa, D.; Kampelis, N.; Mavrigiannaki, A.; Gentilozzi, M.; Paredes, F.; Montagnino, F.; Venezia, L. On the integration of the energy storage in smart grids: Technologies and applications. *Energy Storage* **2019**, *1*, e50. [CrossRef]
64. Thynell, M.; Mohan, D.; Tiwari, G. Sustainable transport and the modernisation of urban transport in Delhi and Stockholm. *Cities* **2010**, *27*, 421–429. [CrossRef]
65. Cavada, M.; Tight, M.R.; Rogers, C.D.F. A smart city case study of Singapore—Is Singapore truly smart? In *Smart City Emergence. Cases from Around the World*; Anthopoulos, L., Ed.; Elsevier: Amsterdam, The Netherlands, 2019; pp. 295–314. ISBN 978-0-12-816169-2.
66. Bakıcı, T.; Almirall, E.; Wareham, J. A Smart City Initiative: The Case of Barcelona. *J. Knowl. Econ.* **2013**, *4*, 135–148. [CrossRef]

sustainability

MDPI

Article

Irrigation and Fertilization: A Comprehensive Analysis of Their Influences on Qualitative Indices in Two Plum Varieties

Adelina Venig [1,*] and Florin Stănică [2]

1 Department of Environmental Engineering, Faculty of Environmental Protection, University of Oradea, 26 Magheru Street, 410048 Oradea, Romania
2 Department of Bioengineering of Horti-Viticultural Systems, Faculty of Horticulture, University of Agronomic Sciences and Veterinary Medicine of Bucharest, 59 Mărăști Blvd, District 1, 011464 Bucharest, Romania; florin.stanica@usamv.ro
* Correspondence: adelina.venig@uoradea.ro

Abstract: The chief aim of this study is to provide information regarding the value and effectiveness of localized irrigation applied to plum trees grown in nurseries; this study also emphasizes how irrigation impacts various qualitative indices in the context of different fertilization treatments. By increasing production in the nursery, the application of differentiated rules for fertilization and irrigation is expected to yield vigorous, healthy fruit tree planting material. As in the case of all cultivated plants, fruit trees in nurseries are primarily dependent on soil and climatic conditions. This research was carried out in a private fruit tree nursery in the northwestern part of Romania. The soil taxonomic unit identified on the research field was arable, weakly glaciated loamy clay on fluvial deposits. The two plum cultivars that were studied were Stanley and Cacanska Lepotica, both of which are valued for the high caliber of their fruit. This research was conducted using a 4 × 2 × 4 trifactorial experiment, with irrigation acting as the primary factor, cultivar as the secondary factor, and fertilization as the tertiary factor. During this research, the fertilization treatments proved to have the most significant impact (34.50%) on stem diameter compared with irrigation (20.67%) and cultivar (5.63%), given that the cultivar had no discernible influence on the increase in the diameters of the grafted trees.

Keywords: fruit tree nursery; planting material; fertilization; irrigation; cultivar; stem diameter

Citation: Venig, A.; Stănică, F. Irrigation and Fertilization: A Comprehensive Analysis of Their Influences on Qualitative Indices in Two Plum Varieties. *Sustainability* **2024**, *16*, 2496. https://doi.org/10.3390/su16062496

Academic Editors: Jan Hopmans and Georgios Koubouris

Received: 5 January 2024
Revised: 19 February 2024
Accepted: 25 February 2024
Published: 18 March 2024

1. Introduction

In upcoming years, it is expected that fruit cultivation will emerge as one of Romania's main horticultural sectors. Fruit production will need to supply a significant amount of product to meet demand in international markets, in addition to fulfilling domestic needs. Fruit farming may considerably boost national income if adequate ecological, material, and human resources are available [1]. Obtaining tree planting material with both high biological value and favorable economic factors requires careful control of the rootstocks used in the nursery [2]. High-vigor generative rootstocks have been previously used in Romania, particularly in traditional orchards. Recently, there has been increasing demand for rootstocks with vegetative propagation due to the intensification of fruit tree plantations. Romania is currently among the European countries with exemplary achievements in vegetative rootstocks, which are requested for testing in many European nurseries. The interest in using rootstocks has increased in line with the development of fruit growing as a commercial activity, which requires the production of many trees in specialized nurseries. The cultivation of fruit trees and shrubs requires planting material meant to restore orchard areas that have been cleared by using specific replacement rate; expanding the orchard heritage by establishing new plantations; restoring the density of existing plantations by filling the gaps; promoting the production of cultivars recommended for each area and culture area; and multiplying and promoting the culture of valuable genotypes, such as newly created or imported cultivars, newly selected clones, and general materials with high biological value [3]. The recent occurrence of drought in Romania is a worrying phenomenon. In horticulture, drought causes the senescence of the leaves at the base of plants, their fall, and thus reductions in leaf surface and total transpiration. Common symptoms of more severe water deficit in plants include the loss of cell turgor, the cessation of growth, the reduction in gas exchange, and the intensification of the respiration process to the point it exceeds the intensity of photosynthesis [4]. Because water is redistributed from plant tissues, plants can survive relatively short periods of time without soil water, depending on the species. However, the effects are not severe. The water from the roots of plants is transported to the aerial organs, and the water from the apoplast penetrates through endosmosis into the cells of leaves, where it restores the turgor pressure and ensures normal conditions for the development of metabolic processes [5]. The increase in the thickness of the cuticle, the decrease in the number of stomata, and the accumulation of hemicelluloses in the form of a gel that retains water in the cell walls are additional means whereby, in times of water stress, plants can prevent significant water loss [6].

Agriculture is not only a main driver of the economy but also a provider of food for all humanity. However, climate change can have a significant impact on this domain of activity [7]. In addition, low agricultural production can be caused by the excessive fragmentation of properties and the reduction in the degree of mechanization of tasks, irrigation, and chemical treatment [8]. Plants can adapt to specific environmental conditions, which allows them to carry out their vital processes under optimal conditions and ensure the perpetuation of their species. Plant metabolism is influenced by climatic changes, which can impact plant production and quality, as well as the capacity of plants to provide a population with plant-based food sources [9]. Plants respond differently to the actions of stress factors, depending on the characteristics of the species, cultivar, etc.; from this point of view, there are species or cultivars that are either resistant or sensitive to the action of a stress factor [10]. Extended periods of drought, reduced river flow, lower lake levels, and falling water tables all have a major detrimental impact on the economy in general and on the agricultural sector in particular. Precipitation provides the water needed for agricultural crops on more than 80% of the globe's surface, but much of this surface is exposed to drought stress. Of the total cultivated land on the Earth, only about 18% is irrigated. The action of stress factors can result in losses reaching up to 40–50% of production [11]. Current climate changes increase the demand for irrigating horticultural crops and intensify the competition between agricultural and non-agricultural water needs. For these reasons, it is imperative to identify new methods for increasing the efficiency of water use in plants [12]. The rational use of water in agriculture implies the prioritization of water consumption in critical situations, the adoption of technologies with reduced water consumption, the adoption of measures that require the application of reference models, the application of innovative solutions to reduce water loss, and the quality control of water with the view of reducing environmental pollution [13]. In addition, improving irrigation technology and promoting drip irrigation are effective ways to enhance the efficiency of using the water resources [14].

Similar to other areas of agricultural production, modern fruit growing can only be imagined in the context of a water regime that corresponds to the requirements of the cultivated species and the culture system used [15]. Through strong root systems that make it possible for plants to explore a large volume of soil and due to the increased absorption capacity of roots, many fruit tree species can achieve favorable results even in areas with a lower rainfall regime or in plantations located on sloping land and on dry sands, where water is more difficult to retain [16]. However, because fruit trees are plants that have increased specific water consumption for the development of growth and fruiting processes at the appropriate level, compensating for water deficit through irrigation in fruit plantations becomes a necessary if not indispensable measure [17]. An optimal water supply level of 65–75% of the soil's total holding capacity must be ensured without creating excess moisture (>80%) through irrigation.

Due to its geographical location at the confluence of the continental and Mediterranean climates, Romania generally offers both favorable climatic and soil conditions for many fruit tree nurseries. Initially, fruit tree nurseries were concentrated in areas with greater rainfall, so that rootstock capture depended to a greater extent on the rainfall regime, with human intervention being initially limited. The uneven distribution of precipitation throughout the year, which causes prolonged periods of drought in some areas (periods longer than 10 days during the vegetation period and 14 days during the rest period when rainfall is no greater than 5 mm), is a characteristic shortcoming of our country's climatic regime and is reflected quite significantly in fruit growing. Considering these factors, along with the tendency to develop significant fruit-growing centers in normally arid regions, on zonal soils, and on sands, irrigation ought to be a top priority in Romania's fruit-growing industry. However, the specifics of this concern vary based on the pedo-climatic zone, kind of rootstock, etc. [18].

Where annual precipitation is less than 500 mm, fruit trees need irrigation; in regions where annual precipitation is between 500 and 700 mm, additional irrigation is used [19]. As is the case with all cultivated plants, fruit trees in nurseries grow primarily in response to the soil and climate in which they are placed [20]. An amount of 75–85%, and occasionally even more, of the weight of the tree's various organs is composed of water [21]. In addition to the fact that water ensures the circulation of fertilizing elements from the soil to the plant, water is an essential element in synthesizing all the organic substances that make up the tissues of rootstocks and trees. Trees need water so that their growth processes might occur as intensely as possible [22].

The fertilization system in nurseries includes long-term activities aimed at ensuring improvement in the physical and chemical properties of the soil and increasing its fertility, ensuring that the demand for assimilable nutrients is met in accordance with the needs of the species, rootstocks, and cultivar/rootstock associations in relation to age and the vegetation phases of plants [23].

Practicing intensive horticulture requires significant investment; thus, production must be neither mediocre nor fluctuating. This requires more professional competence; permanent supervision; and control over vegetation factors, among which the nutritional ones play a predominant role. The system for applying fertilizer consists of a set of guidelines and procedures that deal with setting fertilizer doses; applying them annually and in a full rotation (or off-rotation for non-rotating species); and applying organic fertilizers and minerals in a way that is coordinated with the biological characteristics of each species, the properties of the soil, technological advancements, and economic factors, emphasizing the ways in which fertilizers are administered for each crop [24]. Since each harvest removes a certain amount of mineral and organic matter from the soil, the fertilizer application system in nurseries needs to address issues such as replenishing the soil's reserve of readily assimilated nutrients in a balanced ratio to meet the needs of seedlings and rootstocks [25], which otherwise have to compete with higher plants for mineral nutrition. In addition to the above, fertilizers provide soil microflora with organic matter [26]. The following features need to be considered when designing the nursery fertilizer application system: the soil's hydro-physical characteristics; the differences in nutrients according to rootstocks, species, cultivar, the stage of vegetation, harvest level, weather (temperature, rainfall, and light intensity); the unique technological characteristics of every species; the agrochemical properties and content of the applied fertilizers; the optimal selection of the technique for determining fertilizer dosage and managing the condition of nutrient delivery; establishing the appropriate financial and administrative frameworks for the use, storage, and supply of fertilizers [27–29]. To increase the synthesis of organic matter and achieve large, cost-effective production with superior quality indices, fertilizers are used to optimize nutritional conditions without polluting the environment or weakening plants' resistance to disease and pest attack [30].

Water is directed to the area where it can be most effectively consumed by the plant and easily dosed according to plant requirements, thereby achieving water economy. Losses due to infiltration and evaporation are also nearly eliminated in this process. At the same time, harmful salts are transported to the surface soil beneath the root zone by drip irrigation. A faster growth rate is observed as a direct effect on the plant, leading to earlier harvests than those obtained by other methods of irrigation [31]. Some benefits of localized irrigation include avoiding soil settlement by allowing machinery to move freely during maintenance tasks, minimizing the need for weed-control treatments, and using significantly less energy than with sprinkling (one bar at the end of the irrigation pipe). The drip irrigation method has an additional advantage over the conventional sprinkler irrigation method, in that it can lower the amount of disease treatments and modify the humidity ceiling [4].

Based on the research carried out so far, it can be concluded that watering only along the rows of trees does not harm the development of the root system. Under such conditions, roots are grouped in a narrower space without reducing too much of the total length, so that the absorption capacity is practically not hindered. The ability to use mineralized irrigation water is another benefit of the drip irrigation technique. It has been discovered that salts migrate from the supply point to the boundaries of the wetting zone, resulting in small amounts of salts being found in the root system development zone in the greatest quantities [32–34]. On the other hand, the concentration of salts in the moistened root zone is kept lower due to the maintenance of high humidity throughout the irrigation period. Thus, plants find better conditions for development in salinized soils. The danger of salinization when using mineralized water is also reduced since a smaller volume of water is used per hectare compared with other irrigation methods. The above-mentioned advantages justify the efforts to increase the irrigation capacity. However, it must be remembered that these are potential advantages that can materialize only if rational irrigation is applied, considering the plants' requirements and the soil's physical properties [35]. When comparing the amount of water used for irrigation with that obtained from precipitation and groundwater intake, it becomes evident that crops need to be irrigated. Conversely, a similar comparison can be made with the production obtained in slightly rainier years or between irrigated and non-irrigated crops. To determine whether irrigation is necessary, one must consider not only the amount of precipitation but also the evolution of temperature, relative air humidity, frequency, and wind intensity. This is because the damaging effects of drought are exacerbated during dry spells that are marked by high temperatures, low relative air humidity, and hot and dry winds. In these situations, there is atmospheric drought in addition to soil drought [36,37].

Irrigation is a technique meant to supplement precipitation-derived water supply when this is inadequate for crop needs, in the context of weather conditions characteristic of Romania. It is a way to address a natural factor that causes significant variations in harvest from one year to another. The goal of irrigation is to achieve production that is as stable as possible, nearing the plants' potential for productivity under the specific phytoclimatic conditions. Research conducted in our country revealed that there are years when harvests are severely reduced, even compromised, because of insufficient rainfall during certain periods.

This research addresses an important and current issue, namely, the production of quality fruit tree planting material associated with economically efficient activity at the level of nurseries, which requires considering both the pedo-climatic conditions in the area where the fruit tree planting material is produced and the elements of applied technology. This research aimed to obtain information about the necessity and efficiency of applying localized irrigation in plum culture in a nursery against the background of different fertilization treatments.

2. Materials and Methods

2.1. Location of the Experimental Field

This research was completed in a private nursery in Girișu de Criș, Bihor County, Romania.

Girișu de Criș is a commune situated in the lower area of the Crisurilor Plain. The low plain is the result of hydrographic networks that drift down from Bihor County's higher region as a result of accumulation and erosion.

In the region of Girișu de Criș, the plain is predominant, which determines a uniform distribution of meteorological values throughout the year. The rivers that drain the plain have shallow beds and are not accompanied by terraces.

The low area of the Crișurilor Plain has an average altitude of 110 m.

The annual precipitation was higher (636 mm) in 2020 and lower (498 mm) in 2021, given that during a period of three years, the annual precipitation was lower than the multi-year average. In 2020, the monthly precipitation showed a variation of 114.9 mm, from 12.7 mm in October to 127.6 m in July. For the year 2021, the level of precipitation registered a variation (110.6 mm) was close to that of the previous year, against the background of high precipitation of 119.3 mm in June and very low of only 8.7 mm in February. The amplitude of the variation in monthly precipitation in 2022 was higher (119.7 mm) than in the previous period, varying between 13.5 mm in January and 133.2 mm in July. The amount of precipitation in the April–September vegetation period recorded values between 337.5 mm in 2020 and 370.1 mm in 2022 (Figure 1).

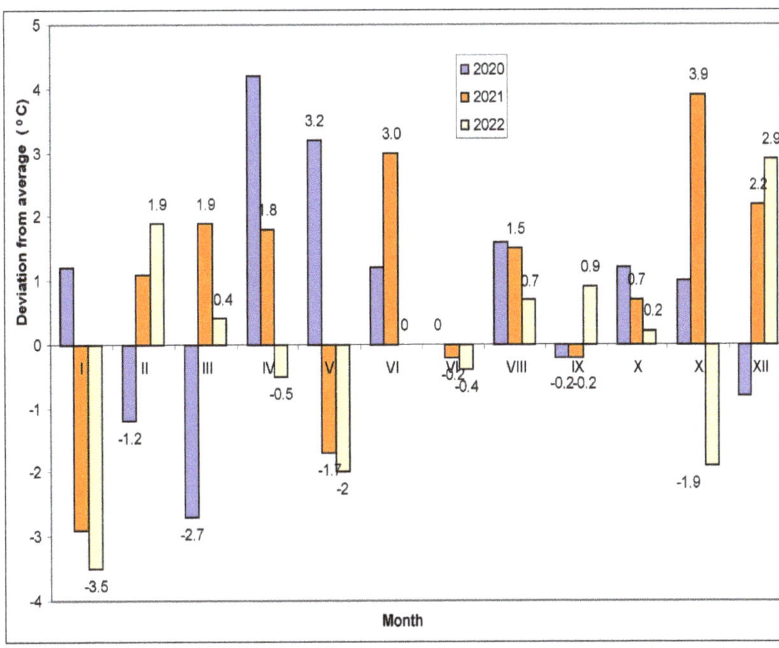

Figure 1. Variations in monthly average rainfall from 2020 to 2022 compared with multi-annual average monthly temperatures in Girişu de Criş.

To give a thorough account of the climatic conditions during the research period, a number of aridity indices were used (Table 1). According to the UNEP aridity index (IAU), Girişu de Criş's area falls into the dry–sub-humid climate class, with average values for the vegetation period over the three years ranging from 0.52 in 2020 to 0.59 in 2021. This index shows that the driest months in 2020 and 2021 were August and September, and July and August, respectively.

Table 1. Hydroclimatic characterization of conditions in Girişu de Criş during 2020–2022.

Year		Month						
	Index	IV	V	VI	VII	VIII	IX	Average
2020	IA_U	0.50	0.48	0.63	1.00	0.25	0.26	0.52
IA_{TH} = 46.16	IA_{DM}	17.88	23.27	32.15	47.85	12.99	9.54	23.95
2021	IA_U	0.74	1.00	0.94	0.16	0.08	0.39	0.55
IA_{TH} = 39.55	IA_{DM}	23.69	51.19	43.38	8.26	4.06	14.24	24.14
2022	IA_U	0.32	0.75	0.96	0.97	0.21	0.34	0.59
IA_{TH} = 33.41	IA_{DM}	8.80	29.39	40.60	50.58	10.48	12.87	25.45

IA_U—UNEP aridity index; IA_{DM}—Martonne aridity index; Bh—water balance; IA_{TH}—Thornthwaite aridity index.

According to Formula (1), the values of the Martonne index (IADM) for the three years (23.95–25.45) characterize the climate of Girişu de Criş as semi-humid. Thus, during the 2020 vegetation period, July (IADM = 47.85) was considered the wettest month, and September (IADM = 9.54) was the driest one. Under the conditions of 2021, May and June (IADM = 43.38–51.19) were the wettest months, and July and August (IADM = 4.06–8.26) were the driest ones. In 2022, the index had values between 8.8 in April and 50.58 in July, according to Formulas (2) and (3).

According to the Thornthwaite aridity index (IATH), Formula (4), the climate in Girişu de Criş in 2021 and 2022, with values between 20 and 40, was considered semi-arid, while the climate in 2020, with a value of 46.16, was considered arid.

Following the analyses, the soil taxonomic unit identified was arable, weakly glaciated loamy clay on fluvial deposits, having the profile Ap-A0-AC-Cg. The depth of groundwater was 2–3 m.

2.2. Experimental Design

This research was carried out based on a trifactorial experiment of the 4 × 2 × 4 type, organized in five repetitions, with plots comprising four trees planted in 0.7 × 0.25 m, with irrigation as the primary factor (a1—non-irrigated; a2—10 mm irrigation; a3—20 mm irrigation; a4—30 mm irrigation), the cultivar as the secondary factor (b1—Stanley; b2—Cacanska Lepotica), and fertilization as the tertiary factor (c1—$N_0P_0K_0$; c2—$N_8P_8K_8$; c3—N_{16} P_{16} K_{16}; c4—$N_{24}P_{24}K_{24}$). The PAST 4.10 program was used for statistical interpretation.

Regarding the technology applied, the first field of the nursery was prepared by deep plowing (35 cm) in August 2020, followed by the leveling work by discus and harrowing in October.

The planting of rootstocks in Field I was carried out in the fall of 2020. In 2021, two mechanical slings between rows and two manual slings per row were applied in Field I. Additionally, phytosanitary treatments using the insecticide Decis (0.02%) and fungicide Dithane (0.2%) were carried out. The complex fertilizer 16:16:16 (Azomureş) was applied concurrently with the mechanical nets in the following amounts (kg/ha): 50 kg for $N_8P_8K_8$; 100 kg for $N_{16}P_{16}K_{16}$; 150 kg for $N_{24}P_{24}K_{24}$. An irrigation control system was installed to measure the amount of water used (Figure 2).

To ensure that the moisture requirement of the soil was met, watering was performed one time in July and three times in August in accordance with amounts related to the three experimental variants, i.e., 100 m^3/ha, 200 m^3/ha, and 300 m^3/ha, respectively. The grafting took place in August 2021.

In the second field of the nursery, the cutting of the cone was carried out in the spring of 2022, followed by the dragging of two mechanical harrows between rows and four manual harrows per row. Also, the weeding out of wild plants was performed four times and the pinching of side growths two times (Figure 3).

Figure 2. Irrigation control system in the nursery for measuring the amount of water used.

Three treatments were applied with the fungicide Dithane (0.2%) and the insecticide Fastak (0.02%) to combat diseases and pests (Figure 4). In August, the soil water deficit was compensated for by applying irrigation three times in accordance with amounts related to the three experimental variants, namely, 100 m^3/ha, 200 m^3/ha, and 300 m^3/ha, respectively. The complex fertilizer 16:16:16 was used in the amounts (kg/ha) of 50 kg for $N_8P_8K_8$, 100 kg for $N_{16}P_{16}K_{16}$, and 150 kg for $N_{24}P_{24}K_{24}$ to obtain the NPK doses associated with the fertilization treatments.

Figure 3. Second field of the nursery.

Figure 4. Second field of the nursery.

2.3. Calculations

Throughout the study, descriptive statistical analysis was employed. The data were gathered, analyzed, and summarized by the authors. The following aridity indices were calculated in order to describe the weather during the study period:

- UNEP aridity index (1992):

$$IA_U = P/FTE \quad (1)$$

where P—precipitation (mm); ETP—potential evapotranspiration (mm).

According to this index, the climate varied from hyper-arid ($IA_U < 0.03$) to humid ($IA_U > 65$).

- Martonne aridity index:

$$\text{annually:} AIDM = P/(T + 10); \quad (2)$$

$$\text{monthly:} AIDM = 12p/(t + 10); \quad (3)$$

where T—average annual temperature (°C); t—mean monthly temperature (°C).

This index allows for the determination of the degree of aridity of a region for certain periods (a year or a month), with climatic variations from hyper-arid (IADM = 0–5) to humid (IADM > 30).

- Thornthwaite aridity index:

$$IATH = 100 \times \Sigma(P - ETP)/\Sigma ETP; \quad (4)$$

This index characterizes a climate as arid at values above 40 and as semi-temperate at values of 0–10.

To determine water consumption directly, the soil water balance was established according to the soil water reserve at the beginning and end of each month from April to September. Soil moisture was determined by the gravimetric method; for this, soil samples were collected from the field at the beginning and middle of each month. These were weighed before and after having been dried in an oven. By using the following formula, which considers the difference between the two weighing instances, soil moisture was calculated:

$$W = 100 \times (B - A)/A \quad (5)$$

where W—soil moisture (%); A—mass of sample with wet soil (g); B—mass of sample with dry soil.

The soil water reserve (R) was established with the following formula:

$$R = 100 \times DA \times H \times W \quad (6)$$

where DA—apparent density (t/m^3); H—depth of active soil layer (m); W—soil moisture (%).

The minimum ceiling—the lowest level of humidity that is readily accessible to plants—was determined to establish the exact moment at which irrigation was required to provide water to the soil. The minimum ceiling (%) for medium soils—which includes the soil from the experimental land—was determined by using the following formula:

$$Pmin = CO + \frac{1}{2}(CC - CO) \quad (7)$$

where CO—withering coefficient (%); CC—field capacity for water (%).

The minimum volumetric ceiling (m^3/ha) was calculated with the following formula:

$$\text{Pmin.vol.} = 100 \times \text{DA} \times \text{H} \times \text{Pmin} \tag{8}$$

As regards the version watered with the amount of 10 mm, the moment for applying irrigation was established in relation to the time when the minimum ceiling was reached. In this sense, watering was performed several times in some months, based on the three amounts discussed in this study.

Potential evapotranspiration was estimated using the Thornthwaite method, so as to determine water consumption indirectly. This method estimates potential evapotranspiration as a function of air temperature by using the following formula:

$$ETP = 160\left(\frac{10 \cdot t}{I}\right)^{a} \cdot K \tag{9}$$

where *ETP*—potential monthly evapotranspiration (m^3); *tn*—average monthly temperature for which ETP is calculated (°C); *I*—annual thermal index, determined as

$$I = \sum_{n=1}^{n=12} i = \sum_{n=1}^{n=12} \left(\frac{t_n}{5}\right)^{1.514} \tag{10}$$

a—empirical coefficient, which is determined as

$$a = 0.000000675I^2 - 0.000077I + 0.01279211\,I + 0.49239 \tag{11}$$

K—brightness coefficient depending on latitude of land (for the period April–September: 1.135, 1.3, 1.32, 1.133, 1.225, 1.045).

The effective precipitation (Pe) during the vegetation months was calculated according to the total precipitation (Pt) that fell in that month, using the USBR method:

$$\text{Pe} = \text{Pt}\,(125 - 0.2\,\text{Pt})/125, \text{ when Pt} < 250 \text{ mm} \tag{12}$$

$$\text{Pe} = \text{Pt}(125 + 0.1\,\text{Pt})/125, \text{ when Pt} > 250 \text{ mm} \tag{13}$$

For each combination of the three factors (irrigation × cultivar × fertilization), determinations were made in relation to the morphological characteristics below, which are connected to the stem diameter (cm).

The growth rate for different characteristics under the effect of different watering amounts and fertilization doses was determined based on an exponential function represented by the following equation:

$$y = \alpha \cdot e^{\beta x} \tag{14}$$

where α—initial value; x—time; y—growth rate.

The accuracy of the respective estimates was evaluated by means of the coefficient of determination (R^2).

3. Results

3.1. Analysis of Variance

The statistical analysis of the factors under investigation revealed significant effects on the stem diameter. The fertilization treatments had the highest impact (34.50%) on this characteristic, compared with irrigation (20.67%) and cultivar (5.63%). In addition, the combination of the three factors showed a significant influence on the stem diameter, but this was considerably lower than their separate effects. The interaction between cultivar and fertilization was highlighted by a major effect of 5.03%, followed by the irrigation x cultivar interaction. At the level of experience, the obtained results were influenced to a degree of 16.6% by other sources of variation not included in the experimental device (Table S1).

3.2. Main Effects of Watering, Cultivar, and Fertilization on Stem Diameter

Considering the unilateral effect of irrigation, it was observed that the stem diameter registered a variation of 2.46 cm, with values between 7.09 cm in the non-irrigated version and 9.55 cm in the situation where 30 mm watering was used (Table 2).

Table 2. Stem diameter under main effects of watering, cultivar, and fertilization.

Irrigation	Mean (cm)	Cultivar	Mean (cm)	Fertilization	Mean (cm)
0 mm	7.09 c	Stanley	8.09 b	$N_0P_0K_0$	6.51 d
10 mm	7.98 b	Cacanska Lepotica	8.45 a	$N_8P_8K_8$	7.90 c
20 mm	8.47 b			$N_{16}P_{16}K_{16}$	8.93 b
30 mm	9.55 a			$N_{24}P_{24}K_{24}$	9.74 a
LSD5% = 0.56		LSD5% = 0.36		LSD5% = 0.54	

Means with different letters are significant at $p < 0.05$.

As such, the three watering amounts showed major and strongly statistically assured influences, determining progressive increases in the diameter of the stem between 12.59 and 34.75%.

Changing the watering amount from 10 to 20 mm had a negligible impact linked to an increase in diameter of only 6.17%. Instead, an increase in this characteristic of 12.73% was found when irrigation was increased from 20 to 30 mm.

Regarding the individual effect of the cultivars, the stem diameter registered a variation of 0.36 cm and low variability, from 8.09 cm in the case of Stanley seedlings to 8.45 cm in the case of Cacanska Lepotica (Table 2).

Therefore, over the course of the entire experiment, it was confirmed that the cultivar had no discernible impact on the seedlings' diameter increase under the 2020 climate.

Considering the many fertilization techniques, the stem diameter showed a variation amplitude of 3.24 cm, with values between 6.51 cm under unfertilized agricultural conditions and 9.74 cm in the case where the 24 kg NPK dose was applied, with a variability of 8.49% among treatments.

The application of NPK fertilization variants led to the recording of significant increases in this characteristic of 21.51–49.79% compared with the non-fertilized agricultural conditions. The seedlings successfully used the additional fertilizer, which increased from 8 to 16 kg and from 16 to 24 kg, respectively, to produce notable gains of 9.18–12.92%.

3.3. Effect of Irrigation × Cultivar Interaction on Stem Diameter

In the case of the Cacanska Lepotica cultivar, it appears that the three watering amounts did not produce substantial changes compared with the non-irrigated variant, considering the influence of the interaction between cultivar and irrigation on the stem diameter of the seedlings. Only the application of amounts of 20 and 30 mm to Stanley cultivar seedlings led to the identification of substantial increases; the influence of the 10 mm amount was negligible and smaller than that in the Cacanska Lepotica cultivar. Considering the interaction between irrigation and the diameter of the plot for the seedlings of the Stanley cultivar, it was observed that only 20–30 mm irrigation allowed for a significant variation of 12.26–27.45%. By contrast, the influence of the 10 mm watering amount was negligible. Additionally, merely increasing the watering rate from 20 to 30 mm resulted in a notable 13.53% increase in diameter in the cultivar's seedlings (Table 3).

Under the effect of different watering amounts, the seedlings of the Cacanska Lepotica cultivar recorded a stem diameter from 6.98 cm in the case of the non-irrigated variant up to 9.93 cm in the 30 mm variant, with a variability between treatments of 20.53%.

Compared with the non-irrigated agricultural group, this cultivar's samplings successfully benefited from all three watering amounts, exhibiting notable increases ranging from

15.33 to 42.29%. When the irrigation standard was gradually adjusted by 10 mm, notable variations in this characteristic, ranging between 10.14 and 12.02%, were observed.

Table 3. Stem diameter under effect of cultivar × irrigation interaction.

	Watering Amounts					
Cultivar	**0 mm**	**10 mm**	**20 mm**	**30 mm**	$\bar{x} \pm s_{\bar{x}}$	**CV**
Stanley	z 7.20 a	yz 7.91 a	y 8.08 b	x 9.17 b	8.45 ± 0.20	21.66
Cacanska Lepotica	u 6.98 a	z 8.04 a	y 8.86 a	x 9.93 a	8.09 ± 0.18	20.53
$\bar{x} \pm s_{\bar{x}}$	7.09 ± 0.24	7.09 ± 0.21	8.47 ± 0.26	9.55 ± 0.25	8.27 ± 0.14	
CV	21.87	16.43	19.02	16.54	21.13	

Cultivar LSD5% = 0.73 cm; means with different letters (a and b) are significant at $p < 0.05$ in vertical comparisons of cultivars. Irrigation LSD5% = 0.74 cm; means with different letters (x, y, z, and u) are significant at $p < 0.05$ in horizontal comparisons of watering amounts.

When the stem diameters of the two cultivars were compared under varying watering rates, it was found that the Stanley cultivar exhibited better behavior under agricultural conditions where 20–30 mm of water was applied. This was demonstrated by a growth spurt of 8.23–9.69% (Table 3).

In the absence of irrigation, the stem diameter of the Cacanska Lepotica cultivar was reduced by approximately 3%, without this difference reaching the level of statistical significance. The two seedling types benefited from irrigation to comparable degrees, with the typical irrigation amount of 10 mm.

Regarding the Stanley cultivar, the impact of watering on the growth of seedlings could be estimated by means of an exponential regression, with a precision of 93.78%.

As a result, the value increased at an average rate of 0.07 cm/mm of watering with a variation of 7.2 cm for the non-irrigated kind, with values ranging from 0.02 to 0.11 cm/mm of watering. The relationship between the watering rate and the increase in the stem diameter in the seedlings of the Cacanska Lepotica cultivar was calculated with a precision of approximately 99.5% by means of an exponential function.

Thus, compared with the average value of 7.05 cm in the absence of irrigation, the average rate of increase in this characteristic was 0.1 cm/mm of watering, with different values among the different irrigation amounts (0.08–0.11 cm/mm of watering). It was observed that the Cacanska Lepotica cultivar benefited more from irrigation (Figure 5).

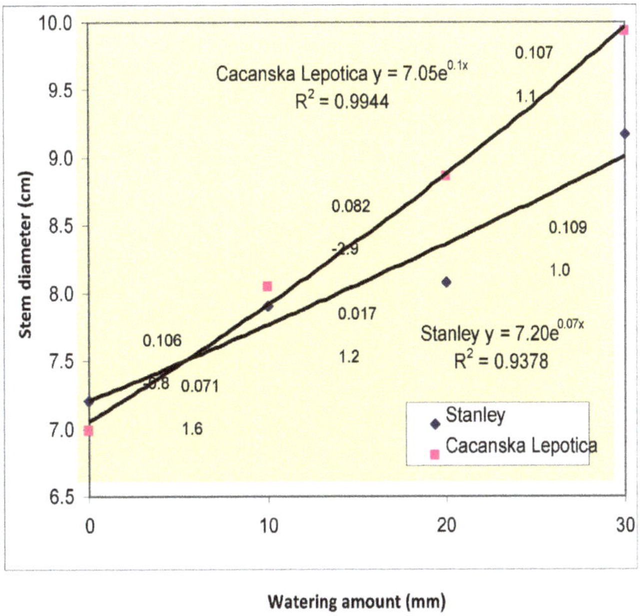

Figure 5. Variation in stem diameter for the two cultivars under effect of different watering norms.

3.4. Effect of Irrigation × Fertilization Interaction on Stem Diameter

When examining how fertilization affected the diameter increase attained at a given watering rate, the non-irrigated cultivar had the largest variation amplitude, while 30 mm irrigation displayed the lowest variation amplitude. There was small variation among NPK dosages (Table 4).

Table 4. Stem diameter under effect of irrigation x fertilization interaction.

	NPK Doses					
Irrigation	**$N_0P_0K_0$**	**$N_8P_8K_8$**	**$N_{16}P_{16}K_{16}$**	**$N_{24}P_{24}K_{24}$**	$\bar{x} \pm s_{\bar{x}}$	**CV**
0 mm	z 4.92 c	y 6.84 c	x 8.13 b	x 8.46 c	7.09 ± 0.24	21.87
10 mm	z 6.53 b	yz 7.45 bc	y 8.42 b	x 9.52 b	7.09 ± 0.21	16.43
20 mm	z 6.65 b	y 8.44 ab	y 8.62 b	x 10.18 ab	8.47 ± 0.26	19.02
30 mm	y 7.93 a	y 8.90 a	x 10.54 a	x 10.83 a	9.55 ± 0.25	16.54
$\bar{x} \pm s_{\bar{x}}$	6.51 ± 0.24	7.90 ± 0.17	8.93 ± 0.19	9.74 ± 0.20	8.27 ± 0.14	
CV	23.23	13.83	13.48	12.67	21.13	

Irrigation LSD5% = 1.06 cm; means with different letters (a, b, and c) are significant at $p < 0.05$ in vertical comparisons of watering amounts. Fertilization LSD5% = 1.07 cm; means with different letters (x, y, and z) are significant at $p < 0.05$ in horizontal comparisons of NPK doses.

Compared with the version without irrigation, the three watering amounts made it possible to achieve very noticeable increases in seedling growth under these conditions, with values ranging from 32.76 to 61.34%. Changing watering amounts from 10 to 20 mm did not generate significant increases, while increasing irrigation from 20 to 30 mm was connected with a notable increase in stem diameter of 19.25%.

In the absence of fertilization, the investigated watering amounts allowed for the acquisition of stem diameter values ranging from 4.92 cm under the non-irrigated agricultural conditions to 7.93 cm under 30 mm irrigation, with a variance of 21.87% between the watering amounts (Table 4).

This characteristic varied under 8 kg of NPK fertilization, ranging from 6.84 cm for the non-irrigated cultivar to 8.9 cm under 30 mm watering. When compared with the non-irrigated alternative, only the 20 and 30 mm watering amounts in this agricultural area showed statistically significant increases of 23.41–30.21%.

Furthermore, increasing the watering rate from 10 to 30 mm resulted in a significant variation of 19.50% in stem diameter, whereas changing the rate by 10 mm had no notable effect.

As a result of applying the treatment with 16 kg of NPK fertilizer, the stem diameter was between 8.13 and 10.54 cm. In this case, only the watering amount of 30 mm showed a significant positive effect of 29.58% greater growth and thickness of the seedlings compared with the non-irrigated version. Instead, employing 20 to 30 mm irrigation allowed for a significant increase of 22.29% in stem diameter.

Considering the fertilization process with 24 kg of NPK fertilizer, the application of irrigation with the three watering amounts determined a significant increase in this characteristic of 12.47–27.96%. In the case of this agricultural experiment, the differences between the watering amounts were small and insignificant; only the increase in watering from 10 to 30 mm showed high efficiency, evidenced by a 13.77% increase in diameter.

Considering the impact of fertilization on stem diameter in seedlings under different watering rates, it was observed that in the absence of irrigation, the seedlings recorded values ranging from 4.92 cm for the unfertilized variant up to 8.46 cm for the variant with 24 kg of NPK application, with a variability among treatments of 21.87%. Compared with the non-fertilized agricultural conditions, the applied treatments had significantly higher efficiency, evidenced by increases in this characteristic of 39.02–71.95%. Only the change in the dose of NPK from 8 to 16 kg was associated with significant effects of a 18.86% increase in thickness in the seedlings.

Under 10 mm irrigation, the applied fertilization options allowed for the acquisition of stem diameters ranging between 6.53 cm for the unfertilized agricultural specimens and 9.52 cm for specimens to which the dose of 24 kg of NPK fertilizer was applied, with an amplitude of variation of 2.99 cm and a variability of 16.43% among treatments.

Under these soil moisture conditions, only the application of 16–24 kg of NPK fertilizer generated significant increases in seedling diameter of 28.94–45.79% compared with the unfertilized version. Additionally, compared with doses of 8–16 kg, the treatment with 24 kg of NPK fertilizer had a significantly greater impact on stem diameter, resulting in increases ranging from 13.06 to 27.79%.

Under the effect of 20 mm irrigation, fertilization with NPK showed a significant effect on the increase in seedling diameter, associated with increases of 26.92–53.08%. The increase in NPK fertilization from 8 to 16 kg had a small and insignificant effect, but the application of 24 kg of NPK fertilizer generated significant increases of 18.1–20.62% in thickness in seedlings compared with the other two doses.

Under the circumstances where 30 mm watering was applied, the seedlings showed stem diameter values from 7.93 cm in the unfertilized variant up to 10.83 cm in the variant with 240 kg of NPK fertilizer, with a variability among treatments of 16.54%.

Compared with the unfertilized agricultural background, only the application of 16–24 kg of NPK fertilizer had significantly higher efficiency, evidenced by increases of 32.91–36.57%. The progressive modification of the dose of NPK fertilizer from 8 to 16 and 24 kg was associated with significant effects of 18.43–21.69% greater thickness in seedlings. The increase in NPK fertilization from 16 to 24 kg had a small and insignificant influence on stem diameter.

In the case of the non-irrigated version, the exponential regression indicates that the increase in stem diameter revealed a mean rate of 0.15 cm for each kg of NPK fertilizer applied, with values from 0.04 between the doses of 16 and 24 kg to 0.24 cm/kg of NPK between the unfertilized version and the 8 kg dose (Figure 6).

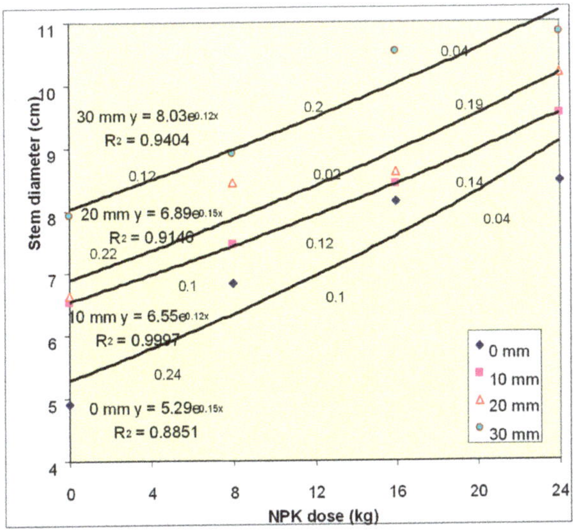

Figure 6. Variation in stem diameter under effect of different watering amounts and NPK doses.

With a stem diameter of roughly 5.3 cm, the corresponding estimates had an accuracy of 88.5% in the absence of fertilization.

For the watering rate of 10 mm, the impact of fertilization on the thickness of the seedlings was expressed by a regression with a precision of 99.97% and indicates an average increase in this characteristic at a relatively constant rate of 0.12 cm/kg of NPK.

Under 20 mm irrigation, a variation in stem diameter associated with an average rate of 0.15 cm/kg of NPK was found, with a coefficient of determination of 91.5% and an initial value of this characteristic of 6.89 cm for non-irrigated agricultural land.

With an estimation accuracy of 94 percent, the average rate of increase in diameter under 30 mm irrigation was 0.12 cm/kg of NPK fertilizer applied, with values among doses (0.04–0.2 cm/kg of NPK).

3.5. Effect of Fertilization × Cultivar Interaction on Stem Diameter

Concerning the impact of fertilization on and the variation in the diameter of the stem, only in the Stanley cultivar, the treatments with NPK showed significantly positive influences, with small variations among the treatments. In the case of seedlings of the Cacanska Lepotica cultivar, fertilization did not significantly influence the diameter of the stem (Table 5).

Table 5. Stem diameter under effect of cultivar × fertilization interaction.

Cultivar	NPK Doses					
	$N_0P_0K_0$	$N_8P_8K_8$	$N_{16}P_{16}K_{16}$	$N_{24}P_{24}K_{24}$	$\bar{x} \pm s_{\bar{x}}$	CV
Stanley	z 5.81 b	y 7.86 a	x 9.13 a	x 9.56 a	8.45 ± 0.20	21.66
Cacanska Lepotica	z 7.20 a	z 7.95 a	y 8.73 a	x 9.93 a	8.09 ± 0.18	20.53
$\bar{x} \pm s_{\bar{x}}$	6.51 ± 0.24	7.90 ± 0.17	8.93 ± 0.19	9.74 ± 0.20	8.27 ± 0.14	
CV	23.23	13.83	13.48	12.67	21.13	

Cultivar LSD5% = 0.74 cm; means with different letters (a and b) are significant at $p < 0.05$ in vertical comparisons of cultivars. Fertilization LSD5% = 0.76 cm; means with different letters (x, y, and z) are significant at $p < 0.05$ in horizontal comparisons of NPK doses.

Considering the impact of the variant on stem diameter under various fertilization conditions, there was a variation from 0.09 cm at the dose of 8 kg of NPK fertilization to 1.39 cm for the unfertilized variant.

Considering this variation, in the non-fertilized variant of the cultivar Cacanska Lepotica, the stem diameter grew at a substantially higher rate, accounting for 23.92% of the seedling diameter. The seedlings of the two cultivars benefited from comparable levels of fertilization, with the Stanley cultivar experiencing greater gains in diameter under agricultural conditions with 16–24 kg of NPK fertilization. However, the differences between the two cultivars were not statistically significant.

In terms of how fertilization affected the increase in stem diameter in each cultivar, in Stanley plants, the results ranged from 5.81 cm for the unfertilized version to 9.56 cm when 24 kg of NPK fertilizer was applied. The three treatments significantly increased this property by 35.28–64.54% compared with the unfertilized variant. Additionally, it was discovered that a notable increase in thickness in the seedlings of 16.16–21.63% was related to the gradual increase in dosage from 8 to 16 and 24 kg. The increase in NPK fertilization from 16 to 24 kg did not significantly influence the diameter of the stem of this cultivar.

According to the results of the exponential regression, the diameter in the Cacanska Lepotica cultivar varied at an average rate of 0.11 cm for every kg of NPK fertilizer, with values from 0.09 cm/kg of NPK fertilization for initial doses and up to 0.15 cm/kg of NPK fertilization for the 24 kg dose (Figure 7).

When stems were not fertilized, the estimates showed a precision of 99.5% based on a stem diameter of roughly 7.16 cm.

The Stanley cultivar showed a stronger response to fertilization in terms of increased seedling diameter, with an average growth rate of 0.16 cm/kg of NPK fertilizer, with variations from 0.05 cm/kg of NPK fertilization between the two larger treatments and 0.25 cm/kg of NPK fertilization between the 8 kg dose and the unfertilized version.

Based on an initial value of 6.21 cm for unfertilized soil, the logarithmic regression between the fertilization dosage and the increase in seedling thickness in the Stanley cultivar showed 89.1% predictability. The difference in fertilization application among the cultivars can also be seen from the slope of the regression lines, which is higher for the Stanley cultivar (Figure 7).

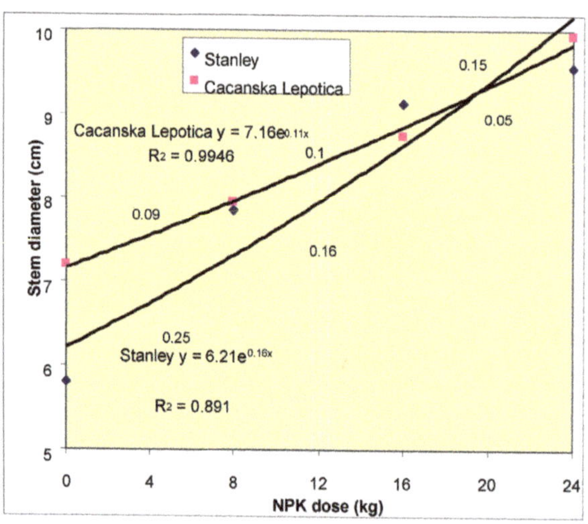

Figure 7. Variation in stem diameter under the effect of different fertilization doses.

3.6. *Effect of Irrigation × Fertilization × Cultivar Interaction on Stem Diameter*

The variation in Cacanska Lepotica cultivar seedling diameter was less correlated with the effects of fertilization, with a variation of 2.73 cm, ranging from 7.2 cm under unfertilized agricultural conditions to 9.93 cm at the dosage of 24 kg of NPK fertilizer. The NPK dose of 8 kg showed a small and insignificant influence of 0.75 cm on the increase in seedling diameter. Instead, the other two treatments generated significant increases of 21.25–37.92%. The addition of fertilization by progressively increasing the doses determined a significant increase of 9.81–13.75% in the development of seedlings of the Cacanska Lepotica cultivar. Considering the combined effect of the three factors, fertilization had a more pronounced effect on the diameter of seedlings of the Cacanska Lepotica cultivar, especially on non-irrigated farmland, compared with the seedlings of the Stanley cultivar (Table 6).

Fertilization resulted in a notable increase in the diameter of Stanley cultivar seedlings on the non-irrigated farmland of 2.64–4.13 cm, with negligible differences among the three doses administered.

When 10 mm irrigation was used, the diameter of seedlings fertilized with 24 kg of NPK increased significantly more than that of seedlings fertilized with 8 kg of NPK or the untreated cultivar. Fertilization had a smaller impact on the diameter of the seedlings when 20 mm irrigation was used. This was accompanied by considerable increments of 3.53–4.75 cm in comparison to the non-fertilized version, as well as negligible variations among the three doses. When 30 mm irrigation was applied, when comparing the treatments with 16–24 kg of NPK fertilizer to the other two fertilization methods, it was found that there was a noticeable increase in stem diameter. When 16–24 kg of NPK fertilizer was applied to Cacanska Lepotica cultivar non-irrigated seedlings, the diameter of the stem increased significantly by 2.3–3.6 cm compared with the untreated group. Furthermore, a far higher efficiency was found for the 24 kg dose compared with the 8 kg NPK dose.

In contrast to other variations, the seedlings of this cultivar demonstrated a notable increase in stem diameter under the influence of 10 mm irrigation, demonstrating an excellent utilization of the 16–24 kg NPK treatments. Only 24 kg of NPK fertilizer, with minor differences among the three dosages, enabled a considerable increase in the stem diameter by 2.3 cm under 20 mm irrigation. Under the conditions of 30 mm irrigation, the seedlings efficiently utilized 16–24 kg fertilization, registering significant increases in this characteristic of 1.6–2.6 cm compared with the non-fertilized version.

Table 6. Stem diameter under effect of irrigation × fertilization × cultivar interaction.

Watering Amount: 0 mm				
	NPK dose			
Cultivar	$N_0P_0K_0$	$N_8P_8K_8$	$N_{16}P_{16}K_{16}$	$N_{24}P_{24}K_{24}$
Stanley	y 4.63 a	x 7.27 a	x 8.76 a	x 8.12 a
Cacanska Lepotica	z 5.20 a	yz 6.40 a	xy 7.50 a	x 8.80 a
Watering Amount: 10 mm				
	NPK dose			
Cultivar	$N_0P_0K_0$	$N_8P_8K_8$	$N_{16}P_{16}K_{16}$	$N_{24}P_{24}K_{24}$
Stanley	z 6.25 a	yz 7.42 a	y 8.14 a	x 9.83 a
Cacanska Lepotica	z 6.80 a	yz 7.48 a	xy 8.70 a	x 9.20 a
Watering Amount: 20 mm				
	NPK dose			
Cultivar	$N_0P_0K_0$	$N_8P_8K_8$	$N_{16}P_{16}K_{16}$	$N_{24}P_{24}K_{24}$
Stanley	y 5.10 b	x 8.63 a	x 8.73 a	x 9.85 a
Cacanska Lepotica	y 8.20 a	y 8.24 a	y 8.50 a	x 10.50 a
Watering Amount: 30 mm				
	NPK dose			
Cultivar	$N_0P_0K_0$	$N_8P_8K_8$	$N_{16}P_{16}K_{16}$	$N_{24}P_{24}K_{24}$
Stanley	y 7.26 a	y 8.10 b	x 10.87 a	x 10.45 a
Cacanska Lepotica	y 8.60 a	xy 9.70 a	x 10.20 a	x 11.20 a

Cultivar LSD5% = 0.74 cm; means with different letters (a and b) are significant at $p < 0.05$ in vertical comparisons of cultivars. Fertilization LSD5% = 0.76 cm; means with different letters (x, y, and z) are significant at $p < 0.05$ in horizontal comparisons of NPK doses.

On non-fertilized agricultural land, the seedlings of the Cacanska Lepotica cultivar made more efficient use of 20 mm irrigation, registering a notable increase in the stem diameter compared with the Stanley cultivar. Also, in the case of 8 kg NPK fertilization and 30 mm irrigation, the diameter of the stem in the seedlings of the Cacanska Lepotica cultivar had a significantly higher value (1.6 cm higher) than the seedlings of the Stanley cultivar. There were no discernible differences between the two cultivars in this trait under the other fertilization and irrigation combinations.

4. Discussion

In the agrotechnical field of environmental protection, this research may be considered very up to date, especially in light of the apparent changes in global climate and the food and energy crises. The results of the undertaken research confirm the working hypothesis and complement the current knowledge in the field. The positive influence of the watering amounts and the doses of fertilizers on the number of plum seedlings obtained from the two analyzed cultivars has been highlighted. The current research has theoretical and practical implications in the domain of fruit growing, with regard to developing horticultural fields and increasing their productive level, ensuring the increase in economic efficiency and profit. It also provides information on obtaining fruit planting material through the application of various irrigation and fertilization doses to plum seedlings in nurseries.

In the climate context of recent years, Romania is facing the phenomenon of complex agricultural drought, which represents a climatic hazard phenomenon that induces the most serious consequences in agriculture.

Changes in climate have an impact on the rainfall regime, with the volume of annual precipitation and monthly distribution favoring the appearance of dry periods or, on the contrary, periods with excessive precipitation [38]. Global warming intensifies the processes of plant transpiration and water absorption, as well as that of water evaporation from the soil surface, and reduces the amount of water available for plants [39]. High temperatures, strong insolation, drought, and excessive irrigation also increase soil salinity, which has a negative impact on plants [40]. The simultaneous action of these stress factors induces numerous morphological, physiological, biochemical, and molecular changes in crop plants, unfavorably affecting their growth, development, and production [41].

A characteristic shortcoming of the climatic regime in Romania, which is reflected quite significantly in fruit growing, is the defective distribution of precipitation during the year, resulting in prolonged periods of drought in some areas. Considering these aspects, associated with the tendency to develop important fruit-growing centers in typically dry areas, on zonal soils, and on sands, irrigation must be a concern of prime importance for the fruit-growing sector in our country. However, the specifics of this concern vary based on the pedo-climatic zone, type of rootstock, etc.

Research on the application of fertilizer and irrigation in nurseries is limited, since most studies have focused on fruit tree orchards. The present study aims to investigate a current issue, namely, the production of quality fruit tree planting material associated with economically efficient activity at the level of nurseries, which requires considering both the pedo-climatic conditions in the area where the planting material is produced and the components of applied technology.

5. Conclusions

The results of this research show that the fertilization treatments had the highest impact on stem diameter compared with irrigation and the plant variety, considering that the cultivar did not significantly influence the increase in the diameter of grafted trees. When compared with the non-fertilized agricultural conditions, the stem diameter dramatically increased by 21.51–49.79% after various doses of NPK were applied. The combined application of irrigation and fertilization with NPK is recommended to obtain high-quality planting material production. It is also necessary to monitor the soil water reserve and apply irrigation in periods of water deficit when the trees need high levels of water. As future research directions, it is necessary to establish the influence of irrigation, fertilization, and cultivar on other morphological and physiological characteristics, as well as on the production of grafted trees and their economic efficiency.

Supplementary Materials: The following supporting information can be downloaded at: https://www.mdpi.com/article/10.3390/su16062496/s1, Table S1: Analysis of variance regarding the effect of cultivar, irrigation, and fertilization on stem diameter.

Author Contributions: Conceptualization, A.V. and F.S.; methodology, A.V. and F.S.; software, A.V.; validation A.V. and F.S.; formal analysis, A.V.; investigation, F.S.; resources, F.S.; data curation, A.V.; writing—original draft preparation, A.V.; writing—review and editing, A.V.; visualization, A.V.; supervision, F.S. All authors have read and agreed to the published version of the manuscript.

Funding: This work was supported by University of Oradea, within the Grants Competition "Scientific Re-search of Excellence Related to Priority Areas with Capitalization through Technology Transfer: INO—TRANSFER—UO—2nd Edition", project Nos. 233/2022 and 234/2022.

Institutional Review Board Statement: Not applicable.

Informed Consent Statement: Not applicable.

Data Availability Statement: Data are contained within the article or the references.

Conflicts of Interest: The authors declare no conflicts of interest.

References

1. Wallin, C. *Growing Trees for Profit*, 2nd ed.; Publishing House: London, UK, 2020; p. 36.
2. Abbate, C.; Scavo, A.; Gaetano, R.P.; Fontanazza, S.; Restuccia, A.; Mauromicale, G. Soil Bioplastic Mulches for Agroecosystem Sustainability: A Comprehensive Review. *Agriculture* **2023**, *13*, 197. [CrossRef]
3. Prabhu, N. *Hydrology and Management*; Unser Wissen Publishing House: Berlin, Germany, 2020; pp. 100–102.
4. Singh, V.P.; Su, Q. *Irrigation Engineering: Principles, Processes, Procedures, Design and Management*; Cambridge University Press: Cambridge, UK, 2022; pp. 61, 65–67.
5. Waller, P. *Irrigation and Drainage Engineering*; Springer Publishing House: New York, NY, USA, 2016; pp. 75–76.
6. Lăcătușu, R. *Agrochemistry*; Terra Noastra Publishing House: Iași, Romania, 2006; p. 144.
7. Vasilescu, S. *Recognition of the Main Pests of Fruit Orchards*; Ars Libri Publishing House: Buzoești, Romania, 2020; pp. 90–91.
8. Mayer, J. *My Little Fruit Tree*; Franckh Kosmos Publishing House: München, Germany, 2019; p. 49.
9. Papouschek, E. *Fruit Trees in Our Garden*; Servus Publishing House: Berlin, Germany, 2022; pp. 64–65.
10. Ruiz-Espin, A.; Garcia-Caparros, P.; Llanderal, A.; Colunje, J.; Francisco Moreira, J.; Lao, M.T. Physiological and Nutritional Responses to Ozone Application in Tomato Seedling Plants. *Agriculture* **2023**, *13*, 60. [CrossRef]
11. Rajemahadik, V.A.; Chavan, V.G. *Textbook of Irrigation and Water Management*; Brillion Publishing House: London, UK, 2019; p. 132.
12. Gurcharan, S. *Irrigation Engineering*; Standard Book House: New Delhi, India, 2018; p. 39.
13. Guy, S. *Fertilization and Irrigation*; Cropaia Publishing House: Hasharon, Israel, 2021; p. 73.
14. Wheeler, J. *Growing Fruit Trees*; University of Chicago Press: Chicago, IL, USA, 2023; p. 52.
15. Stănică, F. *Micropropagation of Horticultural Plants*; Invel Multimedia Publishing House: Bucharest, Romania, 2004; p. 96.
16. Cronin, D. *Irrigation and Soil Nutrition*; Syrawood Publishing House: New York, NY, USA, 2019; pp. 37–38.
17. Klock, P. *Grafting: Fruit Trees and Ornamental Plants*; BLV Publishing House: München, Germany, 2019; pp. 29–31.
18. Megh, R.G.; Chavan, V.K.; Tripathi, V.K. *Innovations in Micro Irrigation Technology*; Apple Academic Press: Palm Bay, FL, USA, 2016; p. 89.
19. Eisenhauer, D.; Derrel, M.; Hoffmann, G. *Irrigation Systems Management*; ASABE Publishing House: St. Joseph, MI, USA, 2021; pp. 95–96.
20. Vlaic, R.A. *Traceability of Biochemical Compounds in Plum Fruit*; Mega Publishing House: Cluj Napoca, Romania, 2019; p. 68.
21. Pinske, J. *Irrigation in Garden*; BLV Publishing House: Berlin, Germany, 2017; pp. 164–165.
22. Faulkner, N. *Principles of Irrigation*; Syrawood Publishing House: New York, NY, USA, 2022; pp. 127–128.
23. Santos, A.F.; Alvarenga, P.; Gando-Ferreira, L.-M.; Quina, M.J. Urban Wastewater as a Source of Reclaimed Water for Irrigation: Barriers and Future Possibilities. *Environments* **2023**, *10*, 17. [CrossRef]
24. Venig, A. *Technologies to Produce Tree Planting Material*; University of Oradea Press: Oradea, Romania, 2006; pp. 57–58.
25. Megh, R.G. *Micro Irrigation Scheduling and Practices*; Apple Academic Press: Palm Bay, FL, USA, 2021; p. 28.
26. Maurer, J. *Handbook for Obtaining Organic Fruit Cultivar*; Löwenzahn Publishing House: Innsbruck, Austria, 2016; p. 57.

27. Venig, A.; Venig, A.; Sabău, N.C. Research Regarding the Effect of Irrigation on the Leaf Surface of Plum Seedlings in the Second Field. *Analele Univ. Oradea Fasc. Protecţia Mediu.* **2022**, *XXXVIII*, 47–51.
28. Cimpoieș, G. *Fruit Trees Cultivars*; Arc Publishing House: Chisinău, Republic of Moldova, 2020; pp. 45–47.
29. Peña, D.; Martín, C.; Fernández-Rodríguez, D.; Terrón-Sánchez, J.; Vicente, L.A.; Albarrán, Á.; Rato-Nunes, J.M.; López-Piñeiro, A. Medium-Term Effects of Sprinkler Irrigation Combined with a Single Compost Application on Water and Rice Productivity and Food Safety. *Plants* **2023**, *12*, 456. [CrossRef] [PubMed]
30. Gipson, A.E. *Horticulture by Irrigation*; Wentworth Press: Surry Hills, Australia, 2018; p. 39.
31. Mitre, V. *General Horticulture*; AcademicPres: Cluj Napoca, Romania, 2020; p. 64.
32. Rubio-Asensio, J.S.; Abbatantuono, F.; Ramírez-Cuesta, J.M.; Hortelano, D.; Ruíz, J.L.; Parra, M.; Martínez-Meroño, R.M.; Intrigliolo, D.; Buesa, I. Effects of Cover Crops and Drip Fertigation Regime in a Young Almond Agroecosystem. *Agronomy* **2022**, *12*, 2606. [CrossRef]
33. Orin, M.; Manjula, M.; Waters, A. *Fruit Trees for Every Garden: An Organic Approach to Growing Apples, Pears, Peaches, Plums, Citrus and More*; Ten Speed Press: Berkeley, CA, USA, 2019; pp. 19–22.
34. Amory, P.F. *Water Supply Engineering, the Designing, Construction and Maintenance of Water Supply Systems*; Sagwan Press: New Haven, CT, USA, 2015; p. 119.
35. Carr, M.K.V. *Advances in Irrigation Agronomy: Fruit Crops*; Cambridge University Press: Cambridge, UK, 2014; pp. 78–81.
36. Jensen, M.; Allen, R. *Evaporation, Evapotranspiration and Irrigation Water Requirements*; American Society of Civil Engineering Press: Reston, VA, USA, 2016; pp. 70–71.
37. Otto, S.; Glenn, W.; Hatch, P. *The Backyard Orchardist: A Complete Guide to Growing Fruit Trees in the Home Garden*; Ottographics Press: Oakland Park, FL, USA, 2015; pp. 26–28.
38. Rickmann, M. *Irrigation in Agriculture*; Erling Press: Berlin, Germany, 2014; p. 119.
39. Polak, P. *Handbook of Water in the Garden*; Löwenzahn Publishing House: Vienna, Austria, 2018; p. 39.
40. Schmid, H. *Grafting Plants*; Ulmer Publishing House: Ulm, Germany, 2019; p. 12.
41. Patzwahl, W. *Water Management and Irrigation in Viticulture*; Ulmer Publishing House: Ulm, Germany, 2023; p. 118.

sustainability

Article

Sustainable Approach to Metal Coin Canceling Methods, Using 3D Modeling and Finite Element Method Analysis

Cornel Cătălin Gavrilă [1,*], Mihai Tiberiu Lateş [1] and Gavril Grebenişan [2]

[1] Department of Product Design, Mechatronics and Environment, Transilvania University of Braşov, 500036 Braşov, Romania; latesmt@unitbv.ro
[2] Department of Industrial Engineering, University of Oradea, 410087 Oradea, Romania; grebe@uoradea.ro
* Correspondence: cgavrila@unitbv.ro; Tel.: +40-721-948-185

Abstract: Over time, many minted coins were withdrawn from circulation, being replaced with new ones. The returned, obsolete metal coins were melted in order to ensure sustainable reuse of the alloy for other purposes. Between the withdrawal and melting, some of the metal coins were canceled by the destruction of their original shape and dimensions using adequate tools. The first part of this paper is focused on presenting some insights into the canceling method used on old Romanian nickel coins; also, some examples are presented. The introduction also includes a literature review in the field of coin manufacturing, covering subjects such as metal behavior under striking load and aspects of 3D modeling and FEM analysis as well as explaining some striking errors. The main purpose of this paper is to study the particularities of canceling methods applied to coins, which is conducted on relatively valuable collection metal pieces. In the second part of the paper, an adequate 3D model is computed for the canceling dies and the coin. Then, the assembled models are introduced, corresponding to each canceling case, consisting of the obverse and reverse canceling dies with coins inside them. For each model, the finite element analysis is realized and is achieved for different initial conditions. The final part of the paper presents the analysis results as well as the discussion and conclusions.

Keywords: metal coin; coin withdrawn; coin canceling method; metal sustainable recycling; 3D modeling; FEM analysis; collectible item

Citation: Gavrilă, C.C.; Lateş, M.T.; Grebenişan, G. Sustainable Approach to Metal Coin Canceling Methods, Using 3D Modeling and Finite Element Method Analysis. *Sustainability* **2024**, *16*, 2322. https://doi.org/10.3390/su16062322

Academic Editors: Alina Badulescu, Constantin Bungău, Delia Mirela Tit, Cosmin Mihai Vesa and Dorina Camelia Ilieș

Received: 28 December 2023
Revised: 2 February 2024
Accepted: 23 February 2024
Published: 11 March 2024

1. Introduction

1.1. Coin Minting and Withdrawal

Coin minting is a manufacturing process that consists of pressing a coin blank using a high load and with both hardened steel obverse and reverse negative dies, which, together with the collar, form a closed space to be filled by coin metal [1,2]. Usually, the relief figures on the negative dies are engraved; after the coin striking, the figures are embossed and engraved on the final metal coin [3]. Of course, the coin model also contains the circulating nominal value and the issuing authority [4].

Technical coin characteristics such as diameter, weight, alloy content and represented figures are provided by adequate legal provisions [5]. Also, the issuing terms are provided for the wider public. Following the circulation period, coin withdrawal is necessary, and it is also mandated by legal provisions, terms and conditions [5]. Valueless coins are retracted and, following the sustainable reuse of the metal alloy, are subjected to melting [6].

But not all issued coins are returned to the issuing authority. Some are retained by the public for different reasons: the metal pieces can be kept as a valuable memory from different times and places, as possible collector's items or without any reason [7]. Some other pieces were lost in different circumstances and found after many years [8].

Most of the obsolete metal coins, especially those with former small denominations, remain on the market after the withdrawal period as scrap metal pieces without any chance to be sustainably recycled. Different metal quantities, such as brass, nickel, bronze, aluminum and steel, can be found scattered as valueless metal discs in many places.

In the best case, over time, a valueless coin can acquire another value as a collector's item, depending on its age, state, rarity or metal; its price can subsequently fluctuate [4,9].

1.2. Coin Canceling

Countries around the world have been canceling the metal coins withdrawn from circulation. This is mainly carried out to prevent the metal coin from returning to circulation. Metal coin canceling (or metal coin defacing) consists of destruction if a coin via pressing between different die patterns [4]: lines, circles or other figures. A variety of these metal coin canceling figures are also known in the literature as waffle designs [10]. Some of the metal coins were canceled by perforating holes. Usually, the canceling is followed by melting in order to realize sustainable recycling of the coin's metal content, especially if the subject metal is precious. But, for different reasons, some of these canceled coins are placed on the collector's market. Due to the pieces' state of conservation, visual appeal and rarity, some are more valuable than others.

The literature indicates many canceled coins from different countries, such as the United Kingdom, Germany, the United States, Malaysia and the Philippines [10–12]. In Europe, the introduction of the euro currency generated a large hoard of canceled coins in the form of the former currency of the involved countries. For commercial purposes, some of them are presented in lots or sealed numismatic sets and are highly appreciated by collectors; for example, in Figure 1, there is a coin set containing the canceled coins of Belgium's former franc. After the euro currency introduction, some of the metal coins were defaced because of their inconsistent manufacture or eventual deterioration; these pieces are also recorded by specialty catalogs and internet sources [12,13].

Figure 1. Collectible numismatic set of canceled coins, obverse view.

In Romania, metal coin canceling was introduced by authorities during the interwar period and was applied to some withdrawn coins in order to recycle the precious metal but also to the discovered coin fakes to prevent their reintroduction into circulation before melting [5]. For example, in Figure 2, there are some canceled Romanian coins from the interwar period.

(a) (b) (c)

Figure 2. A few canceled Romanian metal coins: (**a**) a counterfeit 100 lei from 1932 which imitates the original made of silver, (**b**) perforated withdrawn silver 100 lei from 1932, (**c**) silver 250 lei from 1935.

The best-known Romanian coins that have been widely canceled are the nickel coins of 50 and 100 lei, minted between the years 1936–1938. These coins made of fine nickel, weighting 5.38 and 8.2 g, respectively, were circulated until 1941 when, due to the country's political changes and outbreak of war, they were withdrawn from circulation [4,5]. Their metal, fine nickel, was considered important to supply the war industry needs, and these obsolete coins had to be returned and replaced by other currency [14]. In order to prevent their reintroduction into circulation, their cancellation was mandated by law and had to be realized by those companies who indented to recycle the metal [14]. As a result, for each nominal value, more than 80% of the initial mintage was retracted [5], as shown in Table 1, where the valueless remaining pieces are also indicated. It has to be mentioned that the remaining undamaged pieces, less appreciated in the past than their undamaged counterparts, became valuable decades later on the collector's market [4,9]. But, to date, the exact number of melted pieces of the withdrawn coins which remained on market is still unknown.

Table 1. Issued, withdrawn and remaining pieces of 50 and 100 lei nickel coins.

Coin Nominal Value	Year	Issued Pieces	Withdrawn Pieces	Remaining Pieces [1]
50 lei	1937 1938	12,000,000 8,000,000	16,731,147	3,268,853
100 lei	1936 1938	16,750,000 3, 250,000	17,030,101	2,969,899

[1] Counted at the end of withdrawal period in early 1945.

These factors may lead collectors or interested customers to experience, on one hand, apprehension about acquiring an item of little value that they do not truly require [6]; on the other hand, they must also take into account the inclination to purchase the piece with the intention of capitalizing on it later at a higher price.

Today, the specialty literature [4,10,15] records a large number of canceled 50 and 100 lei nickel coins, as presented in Figure 3. Some of those canceling methods were previously used on other previous Romanian coins, as already presented in Figure 2.

The high demand for canceled coins to be melted and also the low maintenance costs of this operation led to a relatively deep print being obtained for the same type of cancellation; while on properly canceled pieces, the original coin pattern is hard to recognize, on superficially canceled ones, the coin pattern is almost unaffected by the cancellation method applied. Figure 4 presents well and poorly manufactured examples of the parallel lines canceling pattern.

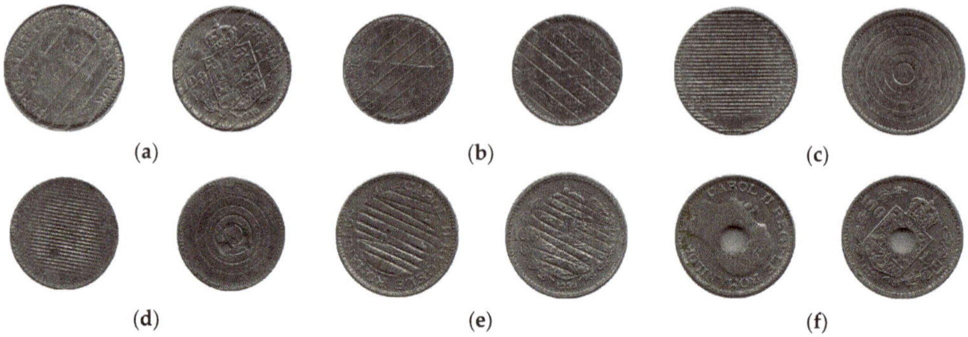

Figure 3. Various canceled Romanian nickel coins: (**a**,**b**) with parallel lines, (**c**,**d**) with circles and fine parallel lines, (**e**) with inscription and (**f**) hole perforations.

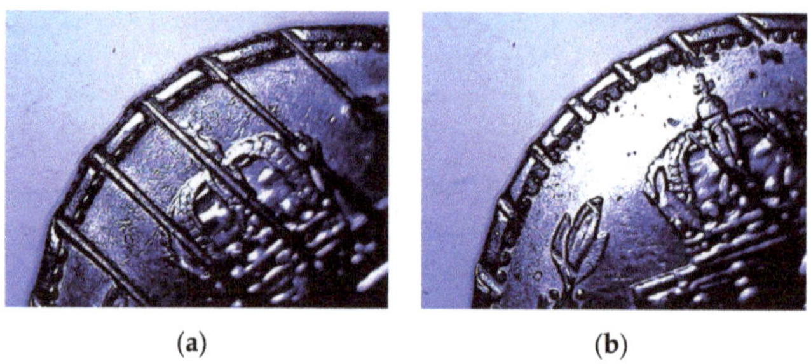

Figure 4. The parallel lines canceling pattern: (**a**) well and (**b**) poorly manufactured.

It has to be mentioned that the recorded canceling pattern fits both coin sizes: 24 mm for the 50 lei coin, and 27 mm for the 100 lei coin [4]. Also, the coin position between the canceling dies was random. It is also true that the coins were canceled in different places, using different tools and press machines [4,14]. The mentioned manufacturing conditions led to some questions about what caused some particularities of canceling patterns and which was the most productive, so the real motivation behind which canceling pattern was used may be revealed using modern study techniques.

In the literature, the coin-themed studies are focused mainly on striking procedures by considering theoretical or practical approaches. Insufficient attention has been given to studying the remaining and obsolete coins, particularly regarding specific methods for material recycling or potential cancellation procedures. A significant number of these pieces are still prevalent in the market as scrap metal, necessitating the implementation of sustainable recycling practices. Due to these aspects, this paper presents theoretical research regarding some aspects of the canceling procedure of the coins, as well as the canceling patterns.

2. Using of 3D Modeling and FEM Analysis in Coinage Field

The coin industry is guided through special regulations by most of the world's governments [1], so specific studies about the coining process are quite rare in the literature. Despite these inconveniences, the studied literature reveals important, comprehensive contributions in the coining field that should be mentioned as follows.

Brekelmans et al. [16] estimated in their work the conditions of coin material flow and also the adequate coining pressure in order to realize a conical relief feature in the central portion of a piece. The finite element method was applied in order to obtain accurate information regarding the corresponding stresses and strains. Also, the upper bound theorem was used, which offered the possibility to obtain more general results concerning the coining process; it was applied to an adequate quasistatic formulation with constitutive elastoplastic equations suitable for research into different parameters. Both methods were extensively described then evaluated, and a quantitative comparison with developed experiments was used for verification; as a result, excellent agreement was discovered. The work is considered by subsequent authors as the first application of the modern finite element method in the coin minting process.

Alexandrino et al. [2] consider that the previously studied research in the science and technology of coin minting can be structured in three main periods that cover the past five decades. In their research paper, the authors proposed a finite element method design procedure to correct the striking die relief in order to obtain an optimized pressure distribution and also the desired alignment for the resultant vertical force, measured at the end of the die stroke. This innovative design procedure for correcting the engraved die model, by tilting both the obverse and reverse die reliefs, shows that the numerical simulation is able to be used to optimize the die model shapes, in order to reduce the forces on coin minting and extend the die lifetime. The presented procedure was successfully applied at the state mint of Portugal to produce some collection coins, dedicated to the ancient age of iron and glass in Europe and also to the Portuguese ethnography. The coins were minted in 2017 [17].

In their research, Zhong et al. [1] studied the mechanism of flash-line defect appearance in the coining process. The flash line is considered by authors as one of the most important surface failures which can appear on manufactured coins. It is mostly unwanted on the silver commemorative coin. Usually, due to a lack of research into the causes of this defect, the failure requires a long time to be eliminated, through many die tryouts, leading to significant disruptions and increasing costs in the coin manufacturing process. Within the work, the mechanism of the flash line defect is revealed by the authors through metal flow analysis, using the modern finite element method. The authors concluded that the radial components of friction, which appeared between the die and the coin blank during striking, can be considered as the main reason for this defect. The authors also disclosed that the areas where defects easily manifest are those on the model field plane, which facilitates the compression and horizontal extrusion of metal flow. The die hardening effect and stress were also considered to be significant in the defect areas.

The mechanism of the flash-line defect which appeared in the coin manufacturing process was also studied by Xu et al. [18]; they estimated that the distribution of the flash-line surface defect is obtained by incrementally increasing the radial friction on the work model down to the elemental level. The large elastoplastic behavior of considered porous materials undergoing deformations is explored, and the constitutive minimization level updates are the result of the local variation problem. The material flow was studied on different die strokes during the complete coining, and it was observed that the change in the flow material direction in the coin outer rim portion may contribute to the formation of flash lines. A new method to investigate the rim geometry of the coin blank was proposed, in order to reduce the flash-line defects, which also resulted in strong alignment with the experimental findings.

In another work [3], Xu et al. developed commercial software, named CoinForm, which is able to analyze the material flow which appears during the coin manufacturing procedure. The presented software allows the prediction of the working force and the geometry optimization for the working dies.

Li et al. in their article [19] proposed an eight-node hexahedral element based on multi-point integration in a dynamic explicit framework, with a newly adopted adaptive subdivision method. Based on detailed numerical examples, the locking-free and also the hourglass-free properties of the introduced hexahedral element are validated; the accuracy of the striking simulation algorithm is also demonstrated. In the authors' opinion, this successful performance of the introduced element was validated by numerical examples without the locking phenomenon and hourglass problems taking place. Some practical coining simulations are also presented. In the first example, a simple round shape coin was designed, made from 99.9% Ag. In the second example, a complex key-shaped commemorative coin (also made from 99.9% Ag) was designed for 2010 World Expo in Shanghai. In both studied instances of practical coining, the simulation results closely matched the experimental findings, which led to the accuracy and also the stability of the proposed algorithm being determined with adaptive element subdivision.

The work of Keran et al. [20] is focused on providing an accurate estimation model of the applied force in the closed die coin striking manufacturing procedure. The authors consider that, in micro-forming procedures, such as coin striking, the microstructure of the metal and reduced size of the coined geometry may have an important influence on the coin metal deformation phenomenon. To determine the accuracy of the proposed force estimation model, the experimental and modeled data are statistically analyzed and also presented graphically by the authors.

The article of Peng et al. [21] presents adequate estimations related to stress distribution and material flow in the coin manufacturing procedure, applied in the case of a bimetallic commemorative coin, using professional software, named Deform 3D. The stress distribution and also the material flow during the coining procedures are studied and compared for a single metal and for bimetallic commemorative coins. The developed numerical examples reveal that there are three main typical stages during the entire procedure for both single and bimetallic coins. The stress concentrators appear on the striking die corners and, in addition, in the case of the bimetallic coin, on the material interface. Since the two metals with different hardness are used to produce the bimetallic coin core and also the bimetallic coin ring, the numerical results reveals that the large strains occur both at the coin's round edge and also at the interface between metals. These led to the conclusion that deep adhesion occurs on the metal's interface in the case when the soft material is used in the core. The authors suggest that adopting the hard material in the inner core and the soft one in the outer ring may cause the subsequent fall of the coin core. The findings of the article are also applicable to bimetal coins, which are eventually sustainably recycled to split the different metal cores and outer rings.

The specific literature on the studied coinage focuses on using modern methods such as 3D modeling and FEM analysis in a few main directions: creating theoretical support for the general real striking condition, resolving some particular issues appearing in striking of certain coins and studying and explaining the appearance of manufacturing errors in some coins. For each work, the obtained results are notable.

At the same time, the literature examined lacks depth in addressing coin cancellation techniques as well as the coin metal recycling issues.

The literature also lacks comprehensive studies using modern methods to explore intriguing and valuable coins as collectible items from past eras. This highlights the necessity to use the available modern techniques to study the particularities of the canceling methods used on certain former coins in order to simplify the value estimation process of the pieces as collector's items. In these cases, the collection items that have an adequate estimated catalogue value are able to increase customer satisfaction [22] and promote sustainable consumption practices in this field [7].

3. Computing the Virtual Model

As already presented, there are different types of imprinted patterns on canceled 50 and 100 lei coins: the same parallel lines or inscription applied on both coin faces and fine parallel lines applied on one face and concentric circles on the other face. The studied model contains both coin face canceling dies and also the coin. Due to the simplistic nature of the canceling die patterns, the studied coin exhibits numerous intricate details on its surface that cannot be accurately replicated in the virtual model [23,24]. Therefore, a simplified model of the 100 lei coin was developed, featuring only the main contour represented on both faces, as per the proper model. The coin angle between obverse–reverse figures, 180°, is measured: while the obverse figure is placed in normal position, the reverse figure is placed upside down. Taking account of this, the computing of the virtual model followed for all needed parts using the facilities offered by the module Part Design in CATIA V5 software [25,26]. Each canceling die model consists of a cylinder with an engraved negative contour pattern, as presented in Figure 5.

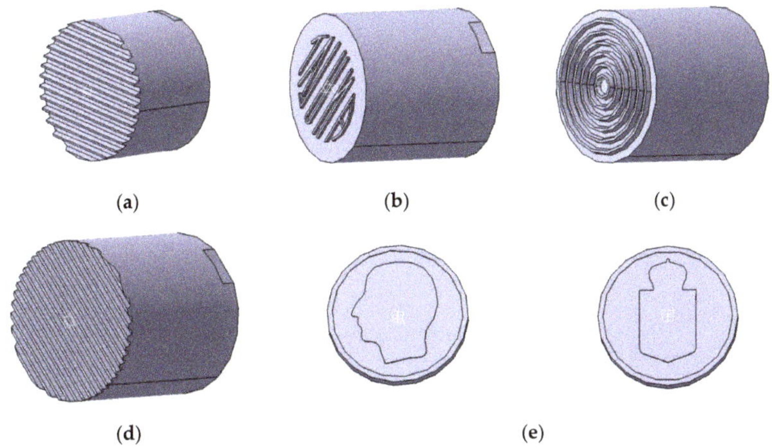

Figure 5. The virtual model parts: (**a**) canceling die with parallel lines, (**b**) canceling die with inscription, (**c**) canceling die with circles, (**d**) canceling die with fine parallel lines and (**e**) the coin from two perspectives.

To obtain each ensemble, the obtained individual parts must be combined: there are always two canceling dies having between them the introduced coins. Using the CATIA software's Assembly Design module [25,26], the device's assemblies were computed, as presented in Figure 6.

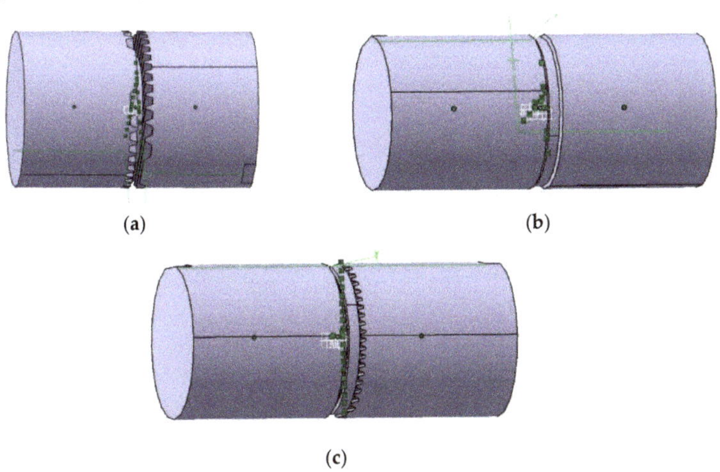

Figure 6. The ensemble models with the canceling die in contact with the coin: (**a**) both canceling dies with parallel lines, (**b**) both canceling dies with inscription and (**c**) one canceling die with circles and another with fine parallel lines.

Following the first contact surface between the pressed die pattern and coin samples, the ensemble constraints were adequately defined. The defined contact area between the canceling dies and coin samples encompasses, in the depicted figures, the entire highest common area. This situation occurs presumptively when there are no misalignments inside the pressing machine and the relief is flat [24,27]. For the first and second ensemble models from Figure 6, corresponding to both faces parallel lines and the inscription on both faces, respectively, the canceling die position in the model reproduces the real pattern position, when the obverse–reverse angle between the lines or inscription is 90^0. For the third ensemble model in Figure 6, corresponding to the fine parallel lines on one face and concentric circles on the other face, the canceling die position in the model is not relevant.

4. Finite Element Model, Analysis and Simulation

To perform the analysis, the ANSYS 15.0 software was used [28]. The evaluation aims to ascertain the behavior of the pressed canceling dies on the coin ensemble when subjected to a load. In the analysis, the previously computed virtual assembled models were used. In the following figure, Figure 7, the adequate finite element model view and geometry are presented. They were previously computed for all of three studied cases: the model with the parallel lines canceling pattern on both faces, the model with the inscription "ANULAT" on both faces and the model with fine parallel lines on one face and concentric circles on the other face as canceling patterns.

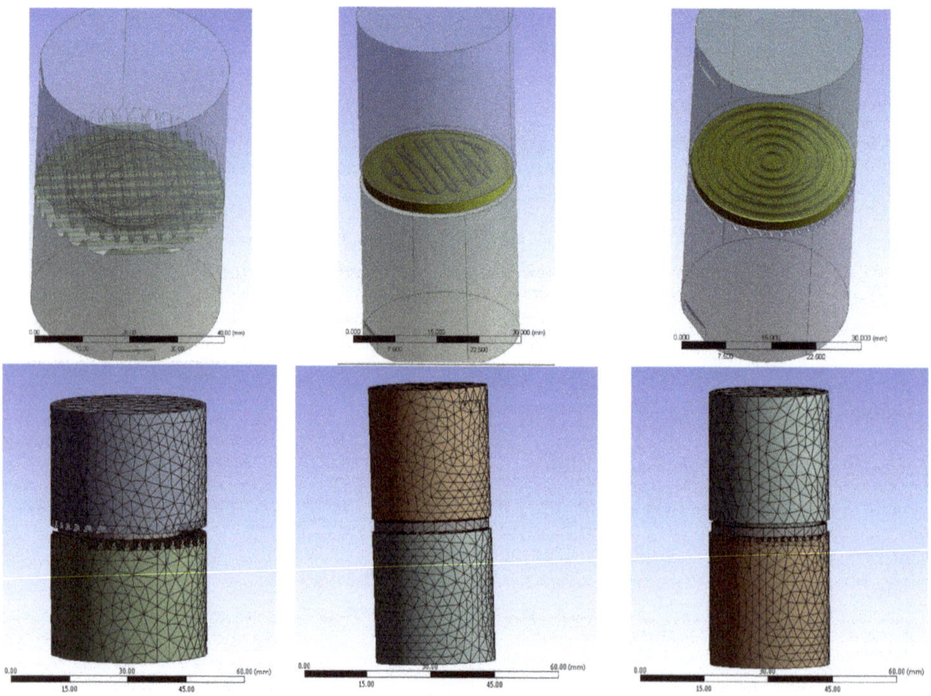

(a) (b) (c)

Figure 7. The obtained finite element model: (**a**) for both canceling dies with parallel lines, (**b**) for both canceling dies with inscription and (**c**) for one canceling die with circles and another with fine parallel lines.

The chosen material for canceling dies is hardened steel; it has the following mechanical properties: *density of* 7850 kg/m^3, *tensile yield strength* of 250 MPa, *tensile ultimate strength* of 460 MPa, a *Young's modulus* of 200,000 MPa and a *Poisson's ratio* of 0.3 [23,28,29]. The coin material is nickel, with the following mechanical properties: *density* of 8900 kg/m^3, *tensile yield strength* of 59 MPa, *tensile ultimate strength* of 317 MPa, a *shear modulus* of 76,000 MPa and a *Poisson's ratio* of 0.31 [23,28,29]. In the studied contact area, a smooth mesh with the minimum edge length equal to 0.001 mm was chosen. Taking into account that the canceling dies do not form a closed space around the pressed coin, the literature indicates that the chosen coin material's allowable stress should be decreased by 30–50%, related to the value corresponding to the closed space coin striking [30,31].

The size of the finite elements varies between 0.15 mm and 1.2 mm, with small values in the contact region between the assembled parts. Tetrahedral-type finite elements are used. The tetrahedral shape finite element is the best 3D-type finite element, which assures the smallest discretization error [28]. The values for the size of the finite elements were established in concordance with the dimensions of the engraving patterns. In the finite element model, a bonded contact type was employed to reflect practical conditions. This decision was made because the construction assembly does not permit horizontal motion of the coins or canceling dies.

The applied normal force is equal to 60 KN in order to obtain high contact pressures, over the canceled coin material's *allowable stress*, 1000 MPa [5,30–32].

5. Results and Discussion

For all three studied cases, the results are presented in Table 2 and in Figures 8–15 and consist of the maximum contact pressure values and also the maximum penetration values on the each canceling die material. Therefore, the values presented in Table 2 should be viewed as relative and are useful when evaluating the different studied cases.

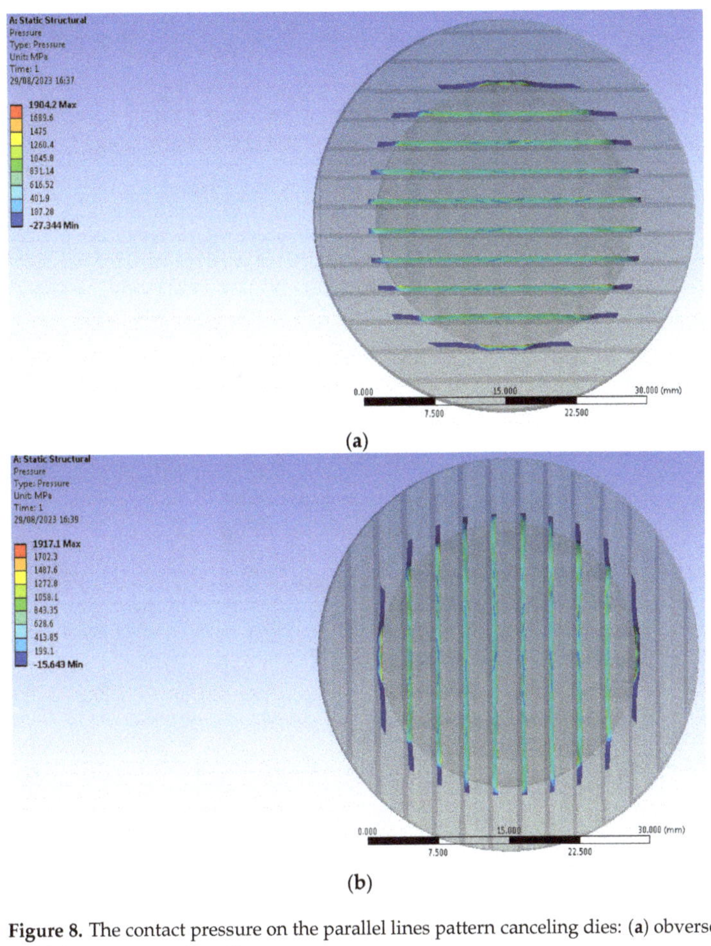

Figure 8. The contact pressure on the parallel lines pattern canceling dies: (**a**) obverse canceling die and (**b**) reverse canceling die.

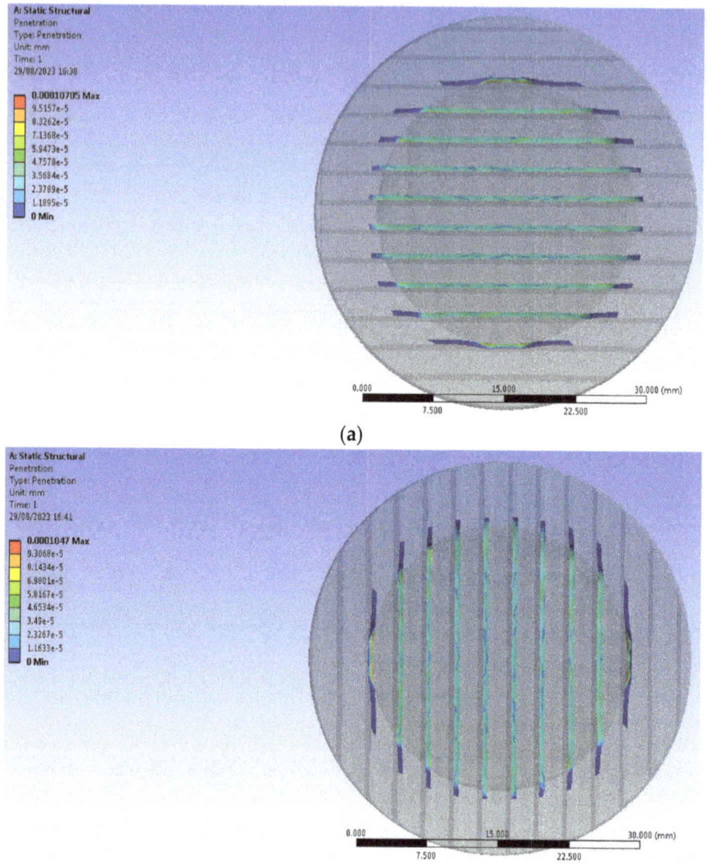

(a) (b)

Figure 9. The penetration on the parallel lines pattern canceling dies: (**a**) obverse canceling die and (**b**) reverse canceling die.

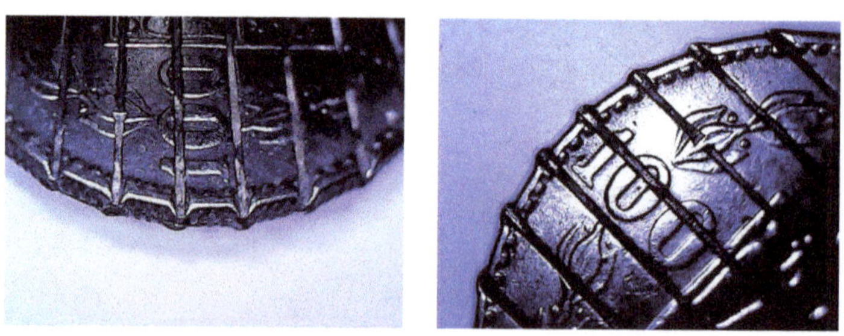

(**a**) (**b**)

Figure 10. The dislocated edges on lines pattern canceled coin: (**a**) perspective and (**b**) normal view.

(**a**)

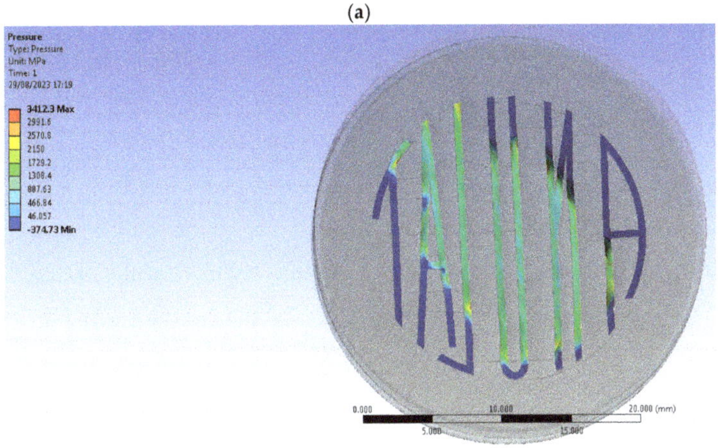

(**b**)

Figure 11. The contact pressure on "ANULAT" pattern canceling dies: (**a**) obverse canceling die and (**b**) reverse canceling die.

(**a**)

Figure 12. *Cont.*

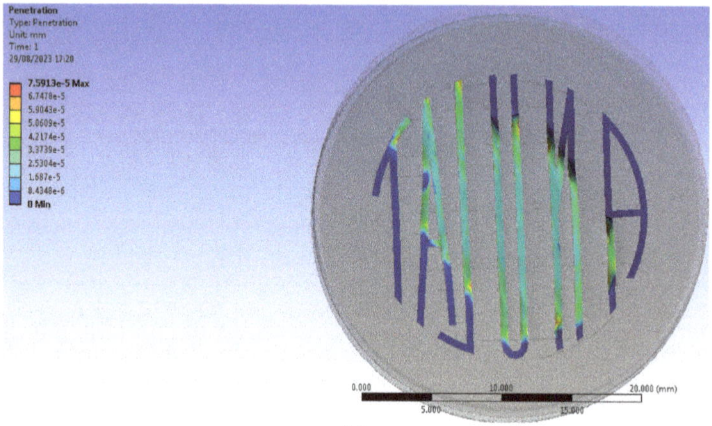

(**b**)

Figure 12. The penetration on "ANULAT" pattern canceling dies: (**a**) obverse canceling die and (**b**) reverse canceling die.

(**a**) (**b**)

Figure 13. The "ANULAT" pattern canceled coin: (**a**) obverse view and (**b**) reverse view.

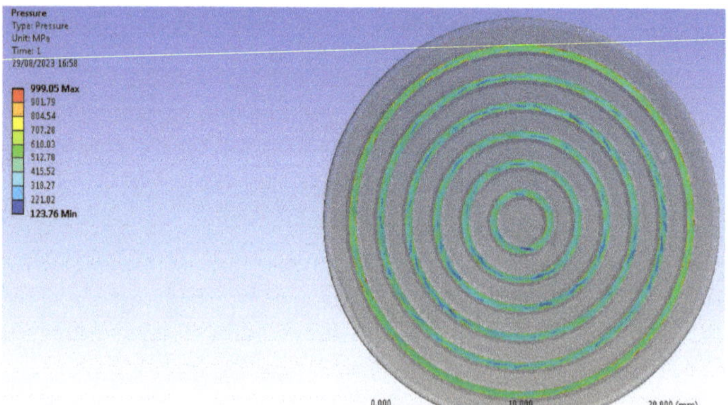

(**a**)

Figure 14. *Cont.*

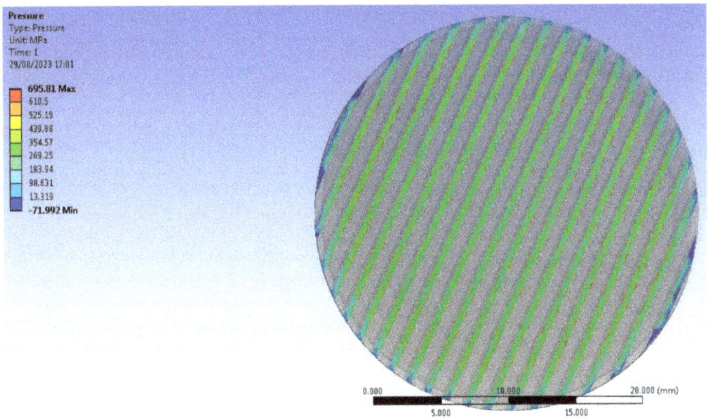

(**b**)

Figure 14. The contact pressure on the concentric circles and fine parallel lines pattern canceling dies: (**a**) obverse canceling die and (**b**) reverse canceling die.

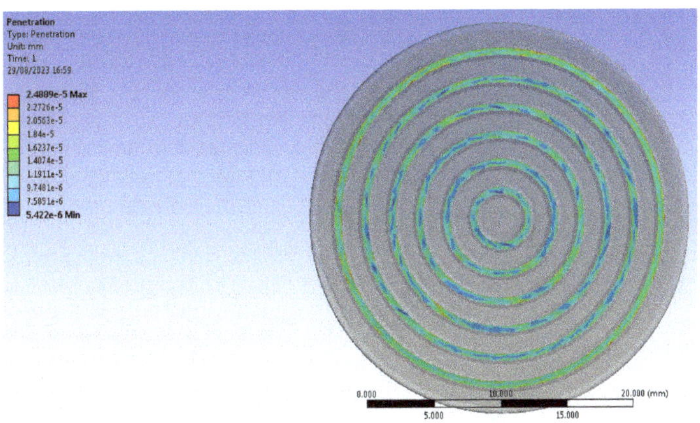

(**a**)

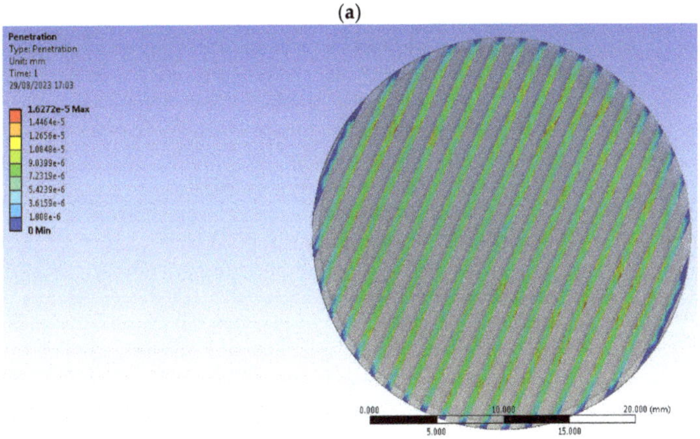

(**b**)

Figure 15. The penetration on the concentric circles and fine parallel lines pattern canceling dies: (**a**) obverse canceling die and (**b**) reverse canceling die.

Table 2. The contact pressure and the penetration on canceling dies, maximum values, for the loading case of 60 KN.

The Canceling Dies	Contact Pressure, MPa		Penetration in the Material, mm	
	Coin Obverse Canceling Die	Coin Reverse Canceling Die	Coin Obverse Canceling Die	Coin Reverse Canceling Die
On both faces, parallel lines	1904.2	1917.1	0.00010705	0.0001047
On both faces, the inscription "ANULAT"	3543	3412.3	0.000079472	0.000075913
Concentric circles on coin obverse combined with fine parallel lines on coin reverse	999.05	695.81	0.000024889	0.000016272

In the initial model under examination, which features a parallel lines canceling pattern on both faces, the maximum values of related contact pressure are similar to each other. Minor discrepancies arise due to variations in the shapes of the represented contours on the coin's obverse and reverse sides. For slightly increased contact areas, the maximum contact pressure values decrease slightly. From Figure 8 it can be observed that those maximum values are recorded close to the coin outer edge, where, due to the coin metal flow, the stress concentrators are increased, as described in [18,19,21]. The other values appear in the main contact area and exceed the coin material's allowable stress. The maximum penetration values are also recorded near the coin's outer edge and are close for both obverse and reverse canceling die patterns, as presented in Figure 9. Due to the open space pressing procedure, the pieces are deformed from the original round shape; on the piece edges, the dislocated metal along the canceling lines can be observed, as is detailed in Figure 10. On the well-canceled pieces, the resulting edges are quite prominent.

For the second model, with the inscription "ANULAT" on both faces, the maximum values of the related contact pressure are closely aligned; additionally, minor variations result from the differing shapes of the depicted contours on both the obverse and reverse sides of the coin.

Because the canceling dies are smaller than the coin's outer ring, it is evident from Figure 11 that the maximum values are recorded close to each face model edge contour, as is also described in [20,21]; because of the stiffness of the coin material in the direction of the applied load, the contours of the faces influence their contact pressure in the contact area [20].

The other values appear in the main contact area and exceed the coin material's allowable stress. But some of the contact pressure values also exceed the die material's allowable stress, so the dies are damaged. This led to the conclusion that this type of canceling was applied using manually operated small pressing machines. The maximum penetration values are also recorded near the coin figures edge contour and are close for both obverse and reverse canceling die patterns, as presented in Figure 12. Due to the open space pressing procedure, the coin metal is dislocated only in the surrounding area by the canceling die letters' contours, as described in [33]. For finished pieces, the coin's round shape and dimensions remained unchanged.

The third model has the canceling pattern of concentric circles applied on the coin obverse and fine parallel lines applied on the coin reverse. The varying patterns applied to the coin faces result in significant differences in the contact area; in this case, the coin face's model differences are less important. The decreased area under the circles pattern led to increased maximum contact pressure values; also, the increased area under the fine lines pattern led to decreased maximum contact pressure values, as presented in Figure 14.

Since the circles pattern is close to exceeding the coin material's allowable stress, the fine lines pattern remains significantly below it. The maximum penetration values, presented in Figure 15, have the same trend: for the circles pattern, the value is higher than for the fine lines pattern. It can be concluded that the load value, 60 KN, was not adequate for this canceling method. The required load must be obtained from a larger pressing machine.

Consequently, this model was again simulated with an increased load, 120 KN [5,30–32]. The results are presented in Table 3 and also in Figures 16 and 17. The increased load led to increased contact pressure and maximum penetration values, but, due to the different contact area, the differences between the obverse and reverse values are maintained. The contact pressure on the fine line pattern eventually surpasses the material's allowable stress level for the pressed coin. But, at the same time, the values are elevated, exceeding the allowable stress of the die material. So, the circular pattern die is subjected to the most damage.

Table 3. The contact pressure and the penetration on canceling dies, maximum values, for loading case 120 KN.

The Canceling Dies	Contact Pressure, MPa		Penetration in the Material, mm	
	Coin Obverse Canceling Die	Coin Reverse Canceling Die	Coin Obverse Canceling Die	Coin Reverse Canceling Die
Concentric circles on coin obverse combined with fine parallel lines on coin reverse	2035.5	1347.5	0.000050711	0.000031489

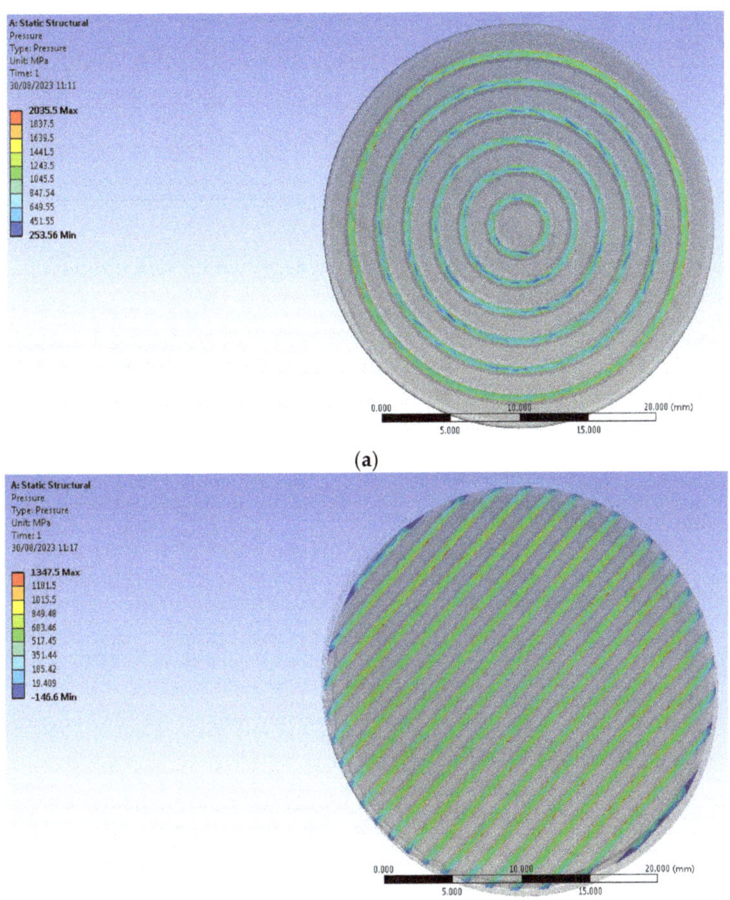

(**a**)

(**b**)

Figure 16. The contact pressure for the increased load on the concentric circles and fine parallel lines pattern canceling dies: (**a**) obverse canceling die and (**b**) reverse canceling die.

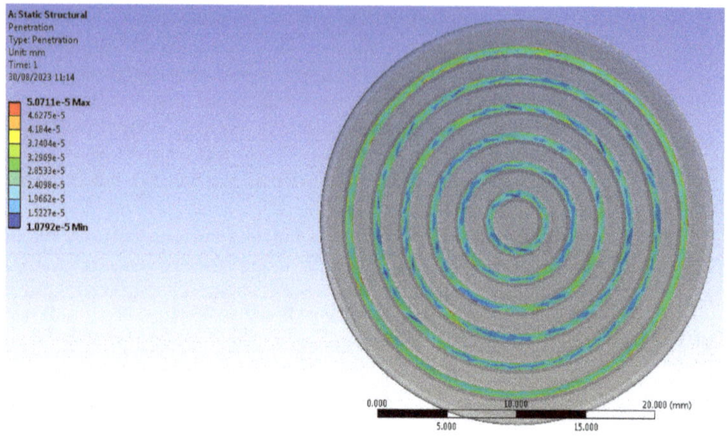

(**a**)

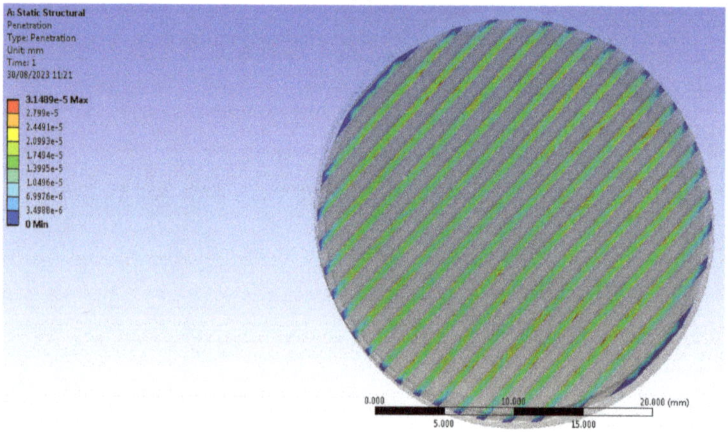

(**b**)

Figure 17. The penetration for the increased load on the concentric circles and fine parallel lines pattern canceling dies: (**a**) obverse canceling die and (**b**) reverse canceling die.

During the open space pressing procedure, the coin metal flow led to the piece's deformation on the outside, and the involved canceling dies behave like closed space striking dies. As a result, some of the coins canceled with these pattern types have flash-line defects impressed on both faces, as described in [1,18] and presented in Figures 18 and 19. In addition, the die cracks, as presented in [11,32], may occur on the circle's devaluated face, as detailed in Figure 18. It seems that this canceling die combination was not resistant to the large numbers of manufactured pieces; some sources indicate that the circle design's weak part was eventually replaced with the fine parallel lines pattern [12].

(**a**) (**b**)

Figure 18. The concentric circles and fine parallel lines pattern canceled coin: (**a**) the unaffected obverse fine lines die, (**b**) the circled reverse die, affected by the initial fine flash line on edge and a radial die crack.

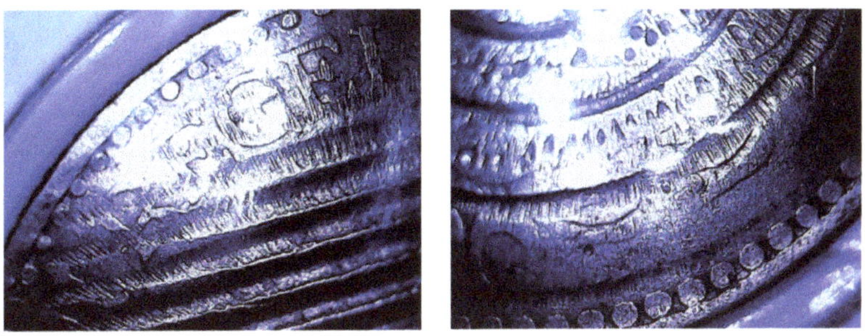

Figure 19. Another coin canceled with the concentric circles and fine parallel lines pattern, with advanced flash-line defects on both faces.

6. Conclusions

In this paper, the particularities of three different canceling methods applied on withdrawn nickel coins are studied. The research led to some conclusions, as follows:

- The withdrawal of more than 80% issued nickel coins and their canceling and metal recycling, even during wartime, can be considered today as a sustainable procedure for that time;
- The three studied models indicate that the coin canceling was carried out using different types of pressing machines. These generated different loads, not necessarily well adapted to the applied canceling method;
- The research reveals that the similar applied pattern used for both coin faces assured a longer die lifetime, and the different patterns used on the coin faces led to the destruction of overloaded dies;
- The improper mating and usage of canceling dies resulted in defects being impressed on the resulting pieces. These defects are closely related to those described by other authors for properly struck coins.
- It was found that it is easier to relate the impressed canceling pattern to the adequate value estimation for the resulting piece;
- By the proposed sustainable approach, this paper obtained results which are useful in further research on other different metal coin canceling and recycling techniques, as well as for establishing the market value of the pieces;
- The research based on the finite element analysis presented in this paper offers flexibility regarding the inputs (striking loads, materials of the coins and dies, discretization), but also there are some limitations to be improved regarding the geometrical modeling, given the size and the detailed complexity of the studied coins.

Author Contributions: Conceptualization, C.C.G.; methodology, C.C.G. and M.T.L.; software, C.C.G. and M.T.L.; validation, C.C.G., M.T.L. and G.G.; formal analysis, C.C.G. and M.T.L.; investigation, C.C.G.; resources, C.C.G., M.T.L. and G.G.; data curation, C.C.G. and M.T.L.; writing—original draft preparation, C.C.G.; writing—review and editing, C.C.G.; visualization, C.C.G. and M.T.L.; supervision, C.C.G., M.T.L. and G.G. All authors have read and agreed to the published version of the manuscript.

Funding: This research received no external funding.

Informed Consent Statement: Not applicable.

Data Availability Statement: Data are contained within the article.

Conflicts of Interest: The authors declare no conflicts of interest.

References

1. Zhong, W.; Liu, Y.; Hu, Y.; Li, S.; Lai, M. Research on the Mechanism of Flash Line Defect in Coining. *Int. J. Adv. Manuf. Technol.* **2012**, *63*, 939–953. [CrossRef]
2. Alexandrino, P.; Leitão, P.J.; Alves, L.M.; Martins, P.A.F. Finite Element Design Procedure for Correcting the Coining Die Profiles. *Manuf. Rev.* **2018**, *5*, 3. [CrossRef]
3. Xu, J.P.; Liu, Y.Q.; Li, S.Q.; Wu, S.C. Fast Analysis System for Embossing Process Simulation of Commemorative Coin–CoinForm. *Comput. Model. Eng. Sci.* **2008**, *38*, 201–215.
4. Buzdugan, G.; Luchian, O.; Oprescu, C.C. *Romanian Coins and Banknotes*; Editura Sport Turism: Bucureşti, Romania, 1977; pp. 256–259.
5. Finance Ministry. *National Mint, Ten Years of Activity*; Monitorul Oficial şi Imprimeriile Statului: Bucureşti, Romania, 1945; pp. 81–87.
6. Petean, I.; Paltinean, G.A.; Taut, A.C.; Avram, S.E.; Pripon, E.; Barbu Tudoran, L.; Borodi, G. Ag and Sn Implications in 3-Polker Coins Forgeries Evidenced by Non destructive Methods. *Materials* **2023**, *16*, 5809. [CrossRef] [PubMed]
7. Dabija, D.-C.; Csorba, L.M.; Isac, F.-L.; Rusu, S. Managing Sustainable Sharing Economy Platforms: A Stimulus–Organism–Response Based Structural Equation Modelling on an Emerging Market. *Sustainability* **2023**, *15*, 5583. [CrossRef]
8. Petean, I.; Paltinean, G.A.; Pripon, E.; Borodi, G.; Barbu Tudoran, L. Silver Depreciation in 3-Polker Coins Issued during 1619–1627 by Sigismund III Vasa King of Poland. *Materials* **2022**, *15*, 7514. [CrossRef] [PubMed]
9. Krause, C.; Mishler, C. *Standard Catalog of World Coins*; Krause Publications: Iola, KS, USA, 2001.
10. Snyder, B. World Mints Deface Coins with "Waffle Designs". *Mint Errors News* **2004**, *5*, 10.
11. de Olivier, F.; Michel, P. *Euro 4, Monnaies et Billets*; Edition Les Chevau-Légers: Paris, France, 2007; p. 205.
12. Coin World. Available online: https://www.coinworld.com/news/us-coins/canceled-coins-can-sometimes-return-to-the-co.html (accessed on 15 December 2023).
13. Error-Ref. Available online: https://www.error-ref.com/canceled-waffled/ (accessed on 20 September 2023).
14. Romanian Government. *The Official Monitor of Romania*; Monitorul Oficial şi Imprimeriile Statului: Bucureşti, Romania, 1941; No. 189; Part I; pp. 4708–4709.
15. Romanian Coins. Available online: https://romaniancoins.org/ (accessed on 20 September 2023).
16. Brekelmans, W.A.M.; Mulders, L.H.G.; Ramaekers, J.A.H. The Coining Process: Analytical Simulations Evaluated. *CIRP Ann. Manuf. Technol.* **1988**, *37*, 235–238. [CrossRef]
17. Catalogue Euro. *Coins and Banknotes*; Leuchtturm Albenverlag GmbH & Co. KG: Geesthacht, Germany, 2019; p. 529.
18. Xu, J.P.; Khan, K.A.; Sayed, T.E. A Novel Method to Alleviate Flash-Line Defects in Coining Process. *Precis. Eng.* **2012**, *37*, 389–398. [CrossRef]
19. Li, Q.; Zhong, W.; Liu, Y.; Zhang, Z. A New Locking-Free Hexahedral Element with Adaptive Subdivision for Explicit Coining Simulation. *Int. J. Mech. Sci.* **2017**, *128–129*, 105–115. [CrossRef]
20. Keran, Z.; Kondic, Z.; Piljec, P.; Runje, B. Accuracy of Model Force Prediction in Closed Die Coining Process. *J. Mech. Eng.* **2018**, *64*, 225–232.
21. Peng, Y.; Xu, J.; Wang, Y. Predictions of Stress Distribution and Material Flow in Coining Process for Bi-Material Commemorative Coin. *Mater. Res. Express* **2022**, *9*, 066505. [CrossRef]
22. Huang, T.-C.; Ho, C.-T. Are Customers Always Right? The Importance of Sincerity and Keenness in Creating Retail Sustainable Development. *Sustainability* **2023**, *15*, 5579. [CrossRef]
23. Arndt, C.; Crusenberry, C.; Heng, B.; Butler, R.; TerMaath, S. Reduced-Dimension Surrogate Modeling to Characterize the Damage Tolerance of Composite/Metal Structures. *Modelling* **2023**, *4*, 485–514. [CrossRef]
24. Gavrilă, C.C.; Lateş, M.T. 3D Modelling and FEM Analysis on Die Clash Mint Error. *IOP Conf. Ser. Mater. Sci. Eng.* **2020**, *898*, 012040. [CrossRef]
25. Ghionea, I.G. CATIA v5. In *Application in Mechanical Engineering*; Editura Bren: Bucureşti, Romania, 2009.
26. Ghionea, I.G. CATIA v5. In *Parametric Design and Programming Applications*; Editura Printech: Bucureşti, Romania, 2021.
27. Hilbert, H. *Stanzereitechnik*; Carl Hanser Verlag: Munchen, Germany, 1938; Volume I, pp. 14+196.

28. Lee, H.H. *Finite Element Simulations with ANSYS Workbench 14. Theory, Applications, Case Studies*; Schroff Development Corporation: Mission, KS, USA, 2012.
29. Geru, N.; Chircă, D.; Bane, M.; Ripoşan, I.; Marin, M.; Coşmeleaţă, G.; Biolaru, T. *Metallic Materials. Structure, Properties, Uses*; Editura Tehnică: Bucureşti, Romania, 1985.
30. Romanovski, V.P. *Stamping and Cold Molding*; Editura Tehnica: Bucureşti, Romania, 1970; pp. 427–428.
31. Iliescu, C.; Tureac, O. *Cold Pressing Technology*; Editura Universităţii: Braşov, Romania, 1987; pp. 262–267.
32. Gavrilă, C.C.; Lateş, M.T. 3D Modelling and FEM Analysis on Holed Metal Coin Striking Die Mint Error. *IOP Conf. Ser. Mater. Sci. Eng.* **2022**, *1256*, 012001. [CrossRef]
33. Gavrilă, C.C.; Lateş, M.T. 3D Modelling and FEM Analysis on Metal Coin Edge Punching Error. *IOP Conf. Ser. Mater. Sci. Eng.* **2021**, *1169*, 012006. [CrossRef]

sustainability

MDPI

Article

Energy Efficiency Analysis of a Fuel Cell Bus Model Using Real Scenarios Generated by Data Collection

Horațiu Cărăușan [1,2], Bogdan Ovidiu Varga [1,2], Dan Moldovanu [1,2], Gabriel Prunean [1,2,*] and Ioan-Tudor Oargă [1,2]

1 Department of Automotive Engineering and Transports, Technical University of Cluj-Napoca, B-dul Muncii 103-105, 400641 Cluj-Napoca, Romania; horatiu.carausan@auto.utcluj.ro (H.C.); dan.moldovanu@auto.utcluj.ro (D.M.); tudor.oarga@auto.utcluj.ro (I.-T.O.)

2 EMARC—Electric Mobility Applied Research Center, Muncii Bvd. 103–105, Room C205-a 2, 400641 Cluj-Napoca, Romania

* Correspondence: gabriel.prunean@auto.utcluj.ro

Abstract: Modernizing public transportation is crucial, given the ongoing call for sustainable mobility. Growing concerns about climate change and the increasingly stringent emissions standards have compelled public transport operators to embrace alternative propulsion vehicles on a broader scale. For the past years, the Battery Electric Buses (BEBs) have been the vehicle of choice for public transportation. However, an emerging contender in this sector is the Fuel Cell Electric Bus (FCEB). This paper aims to evaluate the way one such vehicle would perform in terms of energy efficiency while being exploited in an urban scenario generated from collected data.

Keywords: fuel cell buses; battery electric buses; energy efficiency; energy analysis; hydrogen; data collection; public transportation; urban scenario

Citation: Cărăușan, H.; Varga, B.O.; Moldovanu, D.; Prunean, G.; Oargă, I.-T. Energy Efficiency Analysis of a Fuel Cell Bus Model Using Real Scenarios Generated by Data Collection. *Sustainability* **2024**, *16*, 1863. https://doi.org/10.3390/su16051863

Academic Editors: Alina Badulescu, Constantin Bungău, Delia Mirela Tit, Cosmin Mihai Vesa and Dorina Camelia Ilieș

Received: 14 January 2024
Revised: 19 February 2024
Accepted: 22 February 2024
Published: 24 February 2024

1. Introduction

Due to the existing paradigm, the transportation industry is facing a critical juncture regarding one of the most pressing issues that humanity needs to tackle. Worries about climate change have been rapidly increasing in recent decades, leading to a major change in the focus of development across all industries. The alarming rate of carbon footprint expansion is a significant indicator of this ongoing trend. Regrettably, society's emphasis on rapid progress and heightened consumption has resulted in a dearth of energy efficiency. As a result, civilization is already on a perilous trajectory that will have substantial long-term environmental consequences for our development [1].

In spite of the fact that the transportation industry has been one of the primary contributors to this concern, one of the most effective solutions to this problem is the construction of extensive sustainable public transportation systems in conjunction with the implementation of traffic restrictions. A number of innovative and environmentally friendly transportation alternatives are being presented as viable alternatives to the use of personal vehicles in an effort to increase the use of public transit as a primary mode of transportation. This process is being carried out in order to facilitate the growth of the public transportation systems.

Sustainability in the transportation industry has undoubtedly become a major issue across Europe in light of the current environmental challenges. The European Union has taken a number of actions to solve this problem, such as imposing stricter emission limits and continually updating the fleets of public transportation vehicles [2]. The BEBs and the FCEBs are two of the most popular solutions for environmentally friendly public transportation in Europe. The way in which these solutions manage and store the energy necessary for propulsion is where they diverge the most from one another [3,4].

At the other end of the spectrum, hydrogen is used by the FCEBs, in addition to smaller battery packs, as an energy buffer. Fuel cells use hydrogen as a source of energy to produce the necessary electricity for propulsion. By managing energy delivery and storage using hydrogen, this dual-energy strategy lessens the burden on the battery, thus reducing the rate at which they degrade over time while reducing the total operational costs [3–5].

As a result of Europe's dedication to sustainability in the transportation industry, both the BEBs and the FCEBs have been adopted, with the latter providing a special benefit in reducing battery-related problems (e.g., in cold regions) and enhancing the long-term viability of sustainable bus transportation. These cutting-edge strategies are essential first steps in tackling the pressing environmental issues.

As hydrogen is one of the most effective energy carriers on the market right now, it plays an especially important role in this context. It can be obtained in several ways and used in conjunction with fuel cells to store the energy needed to run long-distance electric motors. Right now, the two most common processes for producing hydrogen are steam reformation of natural gas and electrolysis, with the latter being the more ecologically favorable option [6].

There is a wide variety of fuel cells available, each with its own advantages and disadvantages. The proton exchange membrane (PEM) fuel cell is the most widely used type of fuel cell, employing an exchange membrane as a solid electrolyte. The PEM fuel cells have the capability to produce a maximum power output of one hundred kilowatts with an efficiency ranging from 40% to 60%. These fuel cells operate within a specific temperature range of 80 to 100 degrees Celsius. Within the mobility sector, these devices are extensively utilized due to their small dimensions, lightweight construction, and rapid initialization period. Nevertheless, the PEM fuel cells are vulnerable to fluctuations in temperature, the level of salt in the water, moisture, and dehydration. Phosphoric acid (PAFC), alkaline (AFC), solid oxide (SOFC), and molten carbonate (MCFC) fuel cell technologies are less frequently utilized in mobility applications due to their extended startup times and higher production and operating expenses [7–9].

Considering the technologies used in the FCEBs, as well as the use of hydrogen in propulsion systems, the outlook offers a remarkable energy efficiency perspective as well as a very viable option to reduce emissions. Because of their alteration in the powertrain, the FCEBs are able to go farther than the traditional BEBs. While fuel cells boast zero emissions and impressive range, they have a weakness: they cannot react as quickly as batteries to sudden changes in power demand. To overcome this limitation, fuel cell vehicles often team up with batteries or supercapacitors. This synergy ensures a smooth and responsive driving experience while maintaining the benefits of fuel cells.

The Municipality of Cluj-Napoca is currently investing in an extensive hydrogen production infrastructure, the findings of this study is one of the starting points in the future of fuel cell bus fleet management. By having a perspective regarding the consumption and usage patterns of such vehicles, the hydrogen demand can be estimated, as well as the refueling strategy on the considered route. Furthermore, this study can serve as a basis for expansion, taking into consideration further data sets from different routes, and optimizing the use of such vehicles based on energy efficiency.

2. Materials and Methods

Fuel cell electric buses (FCEBs) have emerged as promising alternatives to regular diesel-powered buses in the pursuit of eco-friendly transportation options. This study examines the operational features of a Fuel Cell Electric Bus (FCEB) when used in a public transportation system, particularly in a highly populated urban area. The main goal is to measure the energy management patterns and overall efficiency of the vehicle by evaluating its hydrogen consumption, distance traveled, and vehicle velocity.

AVL Cruise M (R2021 R2), software developed by AVL List GmbH, was used to simulate and assess the performance of the FCEB accurately. AVL Cruise M is a versatile software tool that can simulate several elements of vehicle operation, such as engine dynamics, energy management, and emissions.

A simulation model of the Solaris Urbino 12 bus was created. Urbino 12 is a multifunctional public transit bus available in both electric and fuel cell variations. The primary objective of this study is to examine the FCEB version, with a specific emphasis on its range and hydrogen consumption.

Before commencing the simulation, it was essential to determine the vehicle's structural and functional characteristics. The specs were carefully and thoroughly defined and organized in Table 1, offering a comprehensive summary of the vehicle's essential characteristics.

Table 1. Main characteristics of the Solaris Urbino 12 Fuel Cell bus [10].

Model	
Urbino 12 Fuel Cell	
Kerb mass	11,032 kg
Maximum authorized mass	19,000 kg
Length	12,000 mm
Width	2550 mm
Frontal area	1.97 m^2
Friction coefficient	0.8
Battery power	100 kW
Motor	2 × ZF AVE 130
Motor power	2 × 150 kW
Fuel cell	Ballard HD 60
Range	350 km
Tank capacity	28–37.5 kg H_2

The AVL Cruise M simulation was performed using a thorough model and detailed input data. The simulation entailed duplicating authentic driving events, encompassing diverse conditions, patterns of acceleration and deceleration, and stop-and-go scenarios. Through the simulation of these situations, the FCEB's energy consumption and range were observed under different operating conditions.

The modeling technique involved a sequence of parametrization procedures. The FCEB exhibits a modular design, wherein each functional component is represented by a separate module. To ensure the acquisition of accurate findings, all elements were simulated according to the specifications provided by the manufacturer. The electric model consists of several key components, including a battery pack, a set of 2 ZF AVE 130 hub-mounted electric motors, a consumer module, and a control functions subsystem. In addition to the aforementioned components, the Urbino 12 Fuel Cell model is outfitted with a Ballard HD60 fuel cell.

The battery pack was designed and arranged based on the output power, voltage, and current specifications, with the aim of achieving a high level of precision. The tractive system comprises two electric motors that are designed based on the actual machinery employed by Solaris and produced by ZF. According to ZF [11], every motor exhibits a maximum power output of 250 kW, operates at a nominal voltage of 650 V, and can sustain a maximum current of 340 A. The control function subsystem encompasses algorithms employed for the regulation of individual motors, with the computational functions responsible for determining the range and performance characteristics of the vehicle.

The functions utilized for the implementation of the test cycle are likewise encompassed inside this subsystem. The fuel cell model incorporates the Ballard HD60 fuel cell together with a specialized mechanism responsible for regulating the energy transfer between the battery and the fuel cell [12]. The model can be observed in Figure 1.

Figure 1. Urbino 12 Fuell Cell model.

The powertrain in the model consists of the essential components mentioned earlier. The fuel cell is responsible for the production of energy through the utilization of hydrogen as fuel. Subsequently, the generated energy is transmitted to the battery pack, which serves the purpose of supplying power to the electric motors [13]. The model includes various types of links, including a mechanical link that represents the physical connections of the vehicle, such as those found in the powertrain architecture and braking system. The electrical circuit serves a dual purpose: it facilitates the transmission of electrical energy throughout the powertrain while also enabling the communication of information from the management system to all the components. In a real fuel cell vehicle, the system would naturally be split into two separate systems based on voltage level: a high voltage system, which handles voltages above 48 V, and a low voltage system consisting of electronic control units and driver input panels, responsible for data transfer.

The management system includes specific sections designed to govern the behavior of the fuel cell. A key feature of this management system is the capability to program a specific state of charge for the battery. This state triggers the activation of the fuel cell, allowing for the optimization of energy consumption between the two integrated energy buffers in the powertrain [14]. This management subsystem has the ability to control the overall energy consumption of auxiliary components, including the high-voltage heating system and the air conditioning system.

The running gear subsystem has been accurately modeled based on the dimensions of the actual buses used in the City of Cluj-Napoca, including tire and brake specifications. The data on the final drive ratio has been obtained from technical information provided by the manufacturer.

The outcomes derived from the simulations were produced from authentic data gathered from the operational electric buses within the municipality of Cluj-Napoca. The data obtained from the buses encompasses several factors, including the GPS coordinates of the bus, the velocity of the vehicle, the date and time of the data collection, and additional metrics such as brake lining usage and the total amount of energy charged and discharged.

Precise replication of a bus's real-life performance necessitates comprehensive information regarding its driving patterns, velocity, and position. The data can be acquired by different methodologies, including the connection of a CANedge 2 device to the OBD (On Board Diagnostics) port of the bus. The device (depicted in Figure 2), along with a manufacturer-provided adapter, captures available data. There are two methods to carry out the data gathering process: on-site collection using a memory card or utilizing an API (Application Programming Interface) to manage larger amounts of data. The collected data was thoroughly filtered and sorted in order to eliminate any sort of errors.

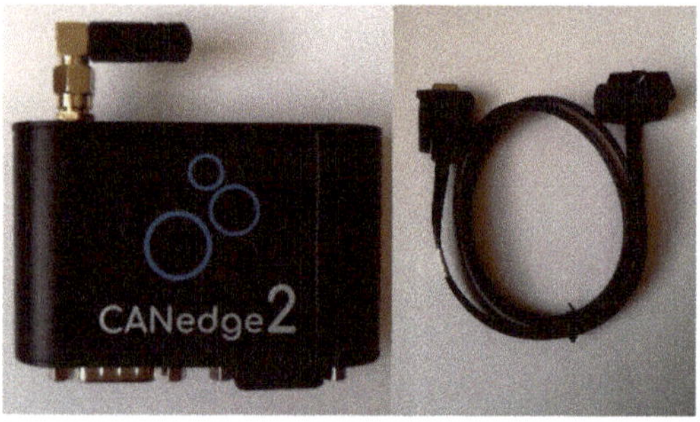

Figure 2. CANedge2 device used for bus data collection.

An important aspect to note is the challenges faced during the data acquisition process. The bus manufacturer has implemented a security key to protect the entire data flow of the bus in compliance with highly stringent data protocols. The CANedge2 device utilized for data acquisition exclusively provided unprocessed, encoded data directly from the controller area network of the bus. To acquire usable data, a lengthy process of data identification needed to be conducted. Via the fleet management platform used by The Cluj-Napoca Public Transport Company (CTP) (Cluj-Napoca, Romania), we were able to execute this intricate procedure. The platform facilitated the elucidation of the parameter identification numbers, allowing us to establish a correlation between the recorded data and their true significance. The collected data encompasses information on energy consumption, GPS coordinates (which can be correlated with the actual route), vehicle velocity (which was utilized to generate the depicted driving cycles in Figure 3), as well as brake lining deterioration, potential bus powertrain errors, and various yet-to-be-identified parameters.

This entire process was carried out over an extended period of time due to difficulties in establishing the connection between the recorded data and the data provided via the fleet management platform. Once we had discovered the actual correlations, a series of tests were performed in order to confirm that our correlation algorithm was correct. During this process, we have stumbled upon a plethora of errors generated by the data collection device as well as the bus itself. We have also tried to create our own API so that we could remotely access the collected data, which turned out to be quite difficult due to the unstable network connection in certain areas, with local storage of the data becoming the most accessible option.

The study comprised the collection of real-world driving data from Line 47 in the Municipality of Cluj-Napoca, Romania for an entire work week. This method guaranteed the incorporation of nuanced fluctuations in the bus's usage pattern, enabling a more authentic simulation.

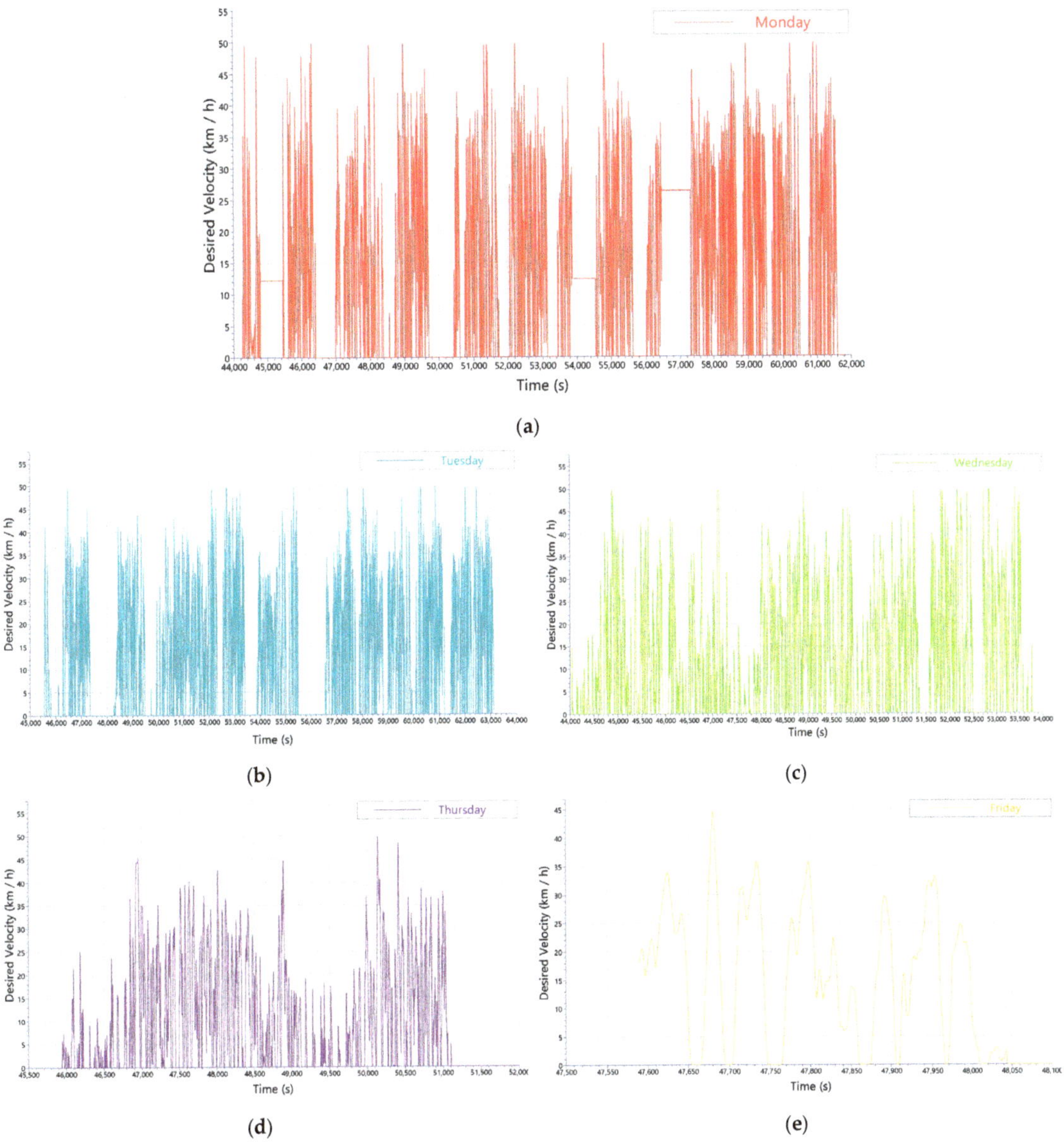

Figure 3. Data generated driving cycles: Monday (**a**); Tuesday (**b**); Wednesday (**c**); Thursday (**d**); Friday (**e**).

The collected data were thoroughly examined and sorted, and subsequently transferred into AVL Cruise M, a specialized data-driven simulation software tailored for vehicle operation modeling. During this procedure, a total of five driving cycles were created, with each cycle corresponding to a specific workday of the week. The cycles effectively depicted fluctuations in bus utilization, with Monday displaying the highest level of demand and subsequent days showing a declining pattern as other buses also travel the same route.

The next step is the implementation of the raw, collected data as a driving cycle in AVL Cruise M. In order to achieve this, a data extraction stage has been carried out to generate the driving cycles used in this paper. The block used for cycle implementation in AVL Cruise M is called the Driving Cycle and includes consecrated and industry-standard cycles such as WLTP and NEDC. However, the software allows the parametrization of such a cycle using two main metrics: time and vehicle velocity. Five simulation cases were generated, in which the main variable was the actual driving cycle.

One fact worthy of being mentioned is that the model is based on blocks that include specific mathematical equations in order to work together inside the solver architecture. In the presented model, the Fuel Cell has been adopted from a unit created by AVL, and we have intervened in the parametrization process of the block, according to data available online from the manufacturer [15].

The Fuel Cell Block contains a module for the Stack and a converter. The block responsible for the mathematical modeling of the dependencies and energy efficiency is the Stack block. The mathematical equations contained in the Stack block are briefly presented in the technical manual of the simulation software under section 6.17.19.5. With regard to energy efficiency and fuel consumption, the main mathematical equations used in the generation of the results are to be presented below [15–17].

The mass flow of the reacted hydrogen and oxygen can be calculated by

$$m_{reacted,\ H_2} = \frac{I_{stack} * n_{cells}}{2 * F} * M(H_2)$$

and

$$m_{reacted,\ O_2} = \frac{I_{stack} * n_{cells}}{4 * F} * M(O_2)$$

where $M(H_2)$ and $M(O_2)$ are the molar masses of hydrogen and oxygen. Using the stoichiometric ratio (*SR*), the mass flow of total amount of hydrogen and oxygen fed to the fuel cell is equal to

$$m_{flow,\ H_2} = m_{reacted,\ H_2} * SR_{H_2}$$

and

$$m_{flow,\ O_2} = m_{reacted,\ O_2} * SR_{O_2}$$

The mass flow of the demanded air fed to the cathode of the fuel cell is equal to

$$m_{flow,\ air} = \frac{I_{stack} * n_{cells} * M(Air)}{4 * F * O_2(Air)}$$

where F is the Faraday constant, I_{stack} is the stack current, n_{cells} is the number of cells [17].

The driving cycles generated, as shown in Figure 3, illustrate the unique patterns of utilization seen throughout the week. Over the course of a two-way journey between the two ends of the defined transport line, the cycles provide a description of the movement of the vehicle's velocity in relation to the passage of time. This aspect makes it possible to study the behavior of vehicles as they travel toward the city center, as well as the behavior of vehicles as they travel back to a less densely populated area of the city.

One thing that should be brought up is the fact that the cycles that are depicted might be different depending on the driver's profile, as well as the temperature outside and the patterns of ridership. Although the results that are presented in this paper have a solid foundation in reality, it is possible that the results that are obtained could be different depending on a number of factors that were discussed earlier.

The meticulous process of collecting data and simulation yields important insights into the operational dynamics of the bus, allowing for the assessment of many factors such as energy usage, route planning, and driver behavior. These simulations are crucial for optimizing bus operations and improving overall efficiency.

The cycles express a series of energy consumption peaks due to the periods of heavy traffic registered in the city of Cluj-Napoca. These peaks correspond to the early morning rush hour, while the population starts daily activities (07:30–08:30). The second peak of energy consumption is registered during the afternoon rush hour, when people return home from work (16:30–17:30). The final rush hour is registered in the evening (21:00–22:00). These peaks are easily observable in the graphs represented in Figure 3.

The continuous variations of vehicle speed correspond to the successive inputs made by the drivers throughout the day. These inputs are a proper indication of the usage patterns observed in the monitored vehicle. The resulting and generated drive cycles indicate a significant difference in contrast to the industry standard drive cycles used for vehicle testing and homologation. Therefore, a significant difference in vehicle velocity and driver input can be observed.

The cycle on Mondays depicted in Figure 3a includes many stops, which are a result of the strong demand from passengers and the concomitant requirement for prolonged breaks to charge the vehicles and switch drivers. Throughout the week, the frequency of stops decreases, suggesting a decrease in the number of passengers and an increase in the utilization of alternative buses on the same route. This evolution is depicted in Figure 3b–e. The usage time of the bus also decreases as the week progresses. This is due to the management strategy of the Cluj-Napoca Public Transport Company, which implies the use of secondary vehicles to take the load off the bus considered for data collection.

3. Results

The first analyzed metric is the total distance traveled in each of the modeled cycles. The main indication of the difference between the traveled distances is the usage rate of the bus. During the first day, the bus is subjected to a series of charge–discharge cycles. As the week progresses, the usage rate of the bus reduces, thus causing the traveled distance to decrease. The total driven distance was calculated by the simulation software as well as logged from the bus. The difference between the sources has been less than 2%. The total traveled distance is depicted in Figure 4. The data can be seen in Table 2.

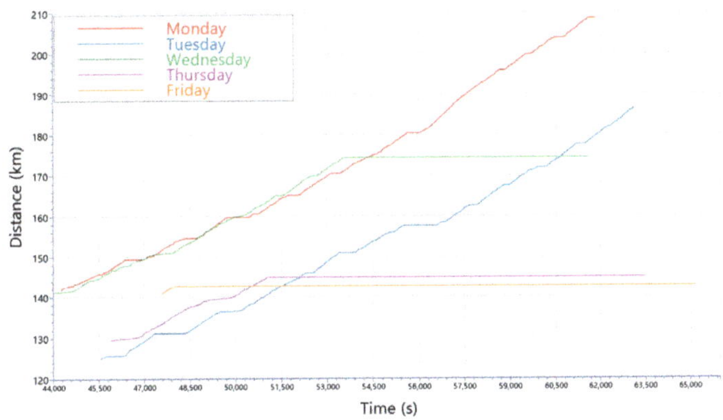

Figure 4. Total distance traveled.

Table 2. Total traveled distance.

	Simulation Case				
	Monday	**Tuesday**	**Wednesday**	**Thursday**	**Friday**
Distance traveled (km)	218.2	196	185.1	148.8	131.5
Average traveled distance (km)			175.9		
Total traveled distance (km)			879.6		

The process of logging the data and mapping the data results in a gradual reduction in the total distance traveled, which can be observed after the process is complete. The total distance traveled is 879.6 km, with an average distance traveled of 175.9 km that was actually traveled. On Monday and Tuesday, the progression of this parameter is relatively consistent throughout the day. On the other days, however, most of the vehicle exploitation occurs in the first part of the day, with the route being served by other electric buses that have exploitation cycles that are comparable to the one being employed by this particular vehicle.

The second metric that is analyzed is the velocity of the vehicle, which is captured during the entire process of generating the driving cycles. The variations that were observed are a result of the inputs made by the driver, in addition to being constrained by the traffic conditions in the surrounding area. In the process of generating the drive cycles, the prime hours were the most significant time periods that should be considered. There is information regarding the typical speed of vehicles that can be found in Table 3.

Table 3. Average vehicle velocity.

	Simulation Case				
	Monday	**Tuesday**	**Wednesday**	**Thursday**	**Friday**
Average vehicle velocity (km/h)	13.6	12.5	12.8	13.2	13.7

The hydrogen consumption of the bus is the most significant parameter that emerged as a result of the extensive simulations. This metric is demonstrated by the total mass of hydrogen consumed, in addition to the flow of hydrogen mass. It is stated by the manufacturer that the maximum range of the bus would be up to 350 km if it were to be used with its full storage capacity of 37.5 kg. According to the information that was gathered from the simulations, it is quite challenging to achieve this range when the conditions are in their natural state. There are, however, a few facts that are worthy of being mentioned. There is no consideration given to the number of passengers who ride the buses in the tests that are carried out by the manufacturer. Within the framework of the model that was developed, the bus was arranged in such a way that it reached its maximum carrying capacity in each of the cycles. The fact that each of the cycles takes into account every single decrease in velocity, whether it be due to the input of the driver or to the external traffic conditions, is another aspect that deserves to be mentioned. When these conditions are present, the powertrain is subjected to a significant amount of strain. Every cycle that was modeled represents a charging cycle for the electric bus, and the longer brakes that were applied in each scenario are representative of this. It would not be necessary to perform these kinds of charging cycles in the event that a fuel cell bus was used. The simulation model, on the other hand, records a consumption of hydrogen during these periods of inactivity, which influences the overall results. The evolution of the hydrogen mass that has been consumed can be seen in Figure 5a, and the flow of hydrogen mass can be seen in Figure 5b. The data are widely represented in Table 4.

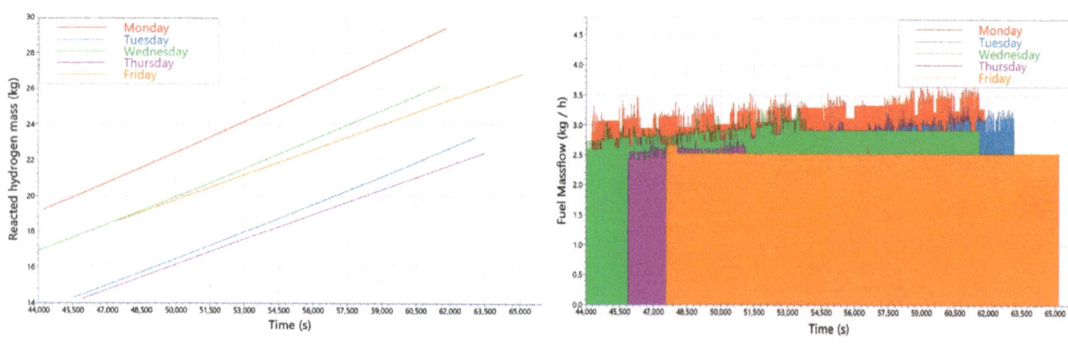

(a) (b)

Figure 5. Hydrogen consumption: (**a**) Consumed hydrogen mass; (**b**) Hydrogen mass flow.

Table 4. Hydrogen consumption.

	Simulation Case				
	Monday	Tuesday	Wednesday	Thursday	Friday
Consumed hydrogen mass (kg)	29.4	23.3	26.1	22.3	26.8
Average consumed hydrogen mass (kg)			25.5		
Average hydrogen mass flow (kg/h)	2.08	1.84	1.89	1.67	1.68
Global hydrogen mass flow average (kg/h)			1.83		

The results were generated in a scenario in which the bus is permanently operating with the auxiliary systems functioning as well as at maximum ridership capacity. Due to this fact, the energy consumption might, at times, be over some values that are considered to be typical. This effect can also be influenced by the fuel cell never being turned off.

4. Discussion

In view of the aspects mentioned in this paper, as well as the data provided following the comparative analysis of the five driving cycles, the following can be concluded:

- Since the model was constructed using data that are widely available and because it has not yet been validated using an actual vehicle, the results that were obtained are specific to the current model and do not represent the overall behavior of vehicles. This is due to the fact that the model was constructed using data that are widely available.
- The information obtained provides a useful perspective on the typical operation of a vehicle that is used for public transportation. A conclusion was drawn from the simulation of a fuel cell vehicle within five of these scenarios. In contrast to a BEB, an FCEB does not require charging cycles throughout the day and is able to deal readily with the demanding requirements of the environment.
- Following the simulations, a series of comprehensive results were discovered. The final results are observable in Table 5.

Table 5. Data overview.

	Simulation Case				
	Monday	Tuesday	Wednesday	Thursday	Friday
Distance traveled (km)	218.2	196	185.1	148.8	131.5
Average vehicle velocity (km/h)	13.6	12.5	12.8	13.2	13.7
Consumed hydrogen mass (kg)	29.4	23.3	26.1	22.3	26.8
Average hydrogen mass flow (kg/h)	2.08	1.84	1.89	1.67	1.68
Average traveled distance (km)			175.9		
Average consumed hydrogen mass (kg)			25.5		
Global hydrogen mass flow average (kg/h)			1.83		
Total traveled distance (km)			879.6		

- Taking into account the comparison between Monday and Tuesday (visible in Figure 3), it is easily observable that the driver inputs vary significantly. The driver on Tuesday adopts a more energy-efficient driving style, giving less aggressive inputs, correlated with fewer instances of achieving the top speed of the vehicle. On Tuesday, there are also more periods of time in which the bus travels at lower speeds, correlated with pauses between each endpoint of the route. The obtained driving cycles are also heavily influenced by the traffic patterns in which the bus has been integrated through the day, which in turn influences the inputs to the model and, thus, offers varying hydrogen consumption values. Another explanation for this phenomenon is the way in which the model manages the fuel cell mass flow relative to the battery SoC, e-motor load (lower in the case of Tuesday), as well as alternating between braking and acceleration.

- Given the average traveled distance of 175.9 km, and the average hydrogen mass consumption of 25.5 kg, the general estimate of range with a full tank of 37.5 kg, one FCEB would be able to cover around 258.6 km under the circumstances modeled throughout the simulations.
- Given the fact that the maximum declared range of one such vehicle is 350 km, obtained in controlled settings, an FCEB would be more than capable of offering a viable alternative to the BEBs in use today.

5. Conclusions and Further Research

Following the discussion, it can be concluded that the behavior of an FCB in a simulation environment could indicate an increase in overall energy efficiency. In contrast, the BEBs generally rely on battery packs with adequate capacity for both storing and delivering the electrical energy needed for propulsion. Although using batteries offers a series of benefits for the environment, their degradation over time cannot be ignored. Degradation is heavily influenced by a multitude of factors, such as temperature and usage patterns relative to the state of charge. This, in turn, equates to larger operating costs and more tedious maintenance operations [18–21]. The consumption results obtained in the case of FCB vary widely depending on a series of factors, such as battery state of charge, driver inputs, and vehicle speed, among others. One such variation can be observed between the cases highlighted on Monday and Tuesday.

It is possible to adjust the use of one such FCEB so that, with an optimal usage rate, it would successfully be able to cover two consecutive days of usage, and it would only need to be charged once every two days. This recommendation is based on the data that were obtained from the simulations. The hydrogen consumption numbers that were obtained are slightly higher than the ones that were declared by the manufacturer. This is due to the fact that the series of test cycles that were carried out were more demanding, and they were also carried out with the maximum carrying capacity applied. Given the current exploitation strategy that is in place for the BEBs, the incorporation of such vehicles into the fleet of public transportation would necessitate a modification to the charging strategy for such vehicles. This would result in a longer usage time for each individual bus, as well as a reduction in the number of refueling stops that can be carried out during non-critical time intervals.

The model used in this study proves itself useful in energy consumption estimation with respect to dimensioning a future integrated infrastructure for the production, storage, and distribution of hydrogen. In terms of further research, one main goal is to gather data from an actual FCB in real conditions and further optimize the model in order to obtain a valid solution. This, in turn, would allow for a simulation solution capable of giving an accurate dimension regarding the implementation of such vehicles.

Author Contributions: Conceptualization, H.C., D.M. and B.O.V.; methodology, H.C., D.M. and B.O.V.; software, H.C., G.P. and I.-T.O.; validation, H.C., D.M. and B.O.V.; formal analysis, H.C. and B.O.V.; resources, H.C. and B.O.V.; data curation, H.C.; writing—original draft preparation, G.P., I.-T.O. and H.C.; writing—review and editing, H.C., G.P. and I.-T.O.; supervision, B.O.V. and D.M.; project administration, B.O.V. and D.M.; funding acquisition, B.O.V. All authors have read and agreed to the published version of the manuscript.

Funding: This paper is supported by the European Union's Horizon 2020 research and innovation programme under grant agreement No. 101036871, project hOListic Green Airport (OLGA).

Institutional Review Board Statement: Not applicable.

Informed Consent Statement: Not applicable.

Data Availability Statement: Data are contained within the article.

Conflicts of Interest: The authors declare no conflicts of interest.

References

1. Nevskaya, M.A.; Raikhlin, S.M.; Vinogradova, V.V.; Belyaev, V.V.; Khaikin, M.M. A Study of Factors Affecting National Energy Efficiency. *Energies* **2023**, *16*, 5170. [CrossRef]
2. Pietrzak, K.; Pietrzak, O. Environmental Effects of Electromobility in a Sustainable Urban Public Transport. *Sustainability* **2020**, *12*, 1052. [CrossRef]
3. Abu-Eisheh, S.; Kuckshinrichs, W.; Dwaikat, A. Strategic Planning for Sustainable Transportation in Developing Countries: The Role of Vehicles. *Transp. Res. Procedia* **2020**, *48*, 3019–3036. [CrossRef]
4. Skeete, J.-P.; Wells, P.; Dong, X.; Heidrich, O.; Harper, G. Beyond the EVent horizon: Battery waste, recycling, and sustainability in the United Kingdom electric vehicle transition. *Energy Res. Soc. Sci.* **2020**, *69*, 101581. [CrossRef]
5. Fakhreddine, O.; Gharbia, Y.; Derakhshandeh, J.F.; Amer, A.M. Challenges and Solutions of Hydrogen Fuel Cells in Transportation Systems: A Review and Prospects. *World Electr. Veh. J.* **2023**, *14*, 156. [CrossRef]
6. Mo, T.; Li, Y.; Luo, Y. Advantages and Technological Progress of Hydrogen Fuel Cell Vehicles. *World Electr. Veh. J.* **2023**, *14*, 162. [CrossRef]
7. Mo, S.; Du, L.; Huang, Z.; Chen, J.; Zhou, Y.; Wu, P.; Meng, L.; Wang, N.; Xing, L.; Zhao, M.; et al. Recent Advances on PEM Fuel Cells: From Key Materials to Membrane Electrode Assembly. *Electrochem. Energy Rev.* **2023**, *6*, 28. [CrossRef]
8. Efficiency of Fuel Cell: Calculation Formula & Equation. Available online: https://www.linquip.com/blog/wp-content/uploads/2021/10/linquip.com-Efficiency-of-Fuel-Cell-Calculation-Formula-Equation.pdf (accessed on 15 September 2023).
9. U.S. Department of Energy. Comparison of Fuel Cell Technologies. Energy Efficiency & Renewable Energy. 2016. Available online: https://www.energy.gov/sites/default/files/2016/06/f32/fcto_fuel_cells_comparison_chart_apr2016.pdf (accessed on 15 September 2023).
10. Bus, S.; Coach, S.A. Solaris Urbino 12 Catalogue. 2021. Available online: https://www.solarisbus.com/pl/serwis-i-uslugi/czesci-zamienne-i-dokumentacja (accessed on 20 October 2023).
11. ZF Product Overview. Available online: https://www.zf.com/products/media/en/pim/tu___axle___transmission_systems_for_buses___coaches/chassis_7/TU_Product_Overview_202206_DE_EN_LowRes_Opt_pdf.pdf (accessed on 20 October 2023).
12. Iclodean, C.; Cordoș, N.; Todoruț, A. Analysis of the Electric Bus Autonomy Depending on the Atmospheric Conditions. *Energies* **2019**, *12*, 4535. [CrossRef]
13. Guo, L.; Li, Z.; Outbib, R.; Gao, F. Function approximation reinforcement learning of energy management with the fuzzy REINFORCE for fuel cell hybrid electric vehicles. *Energy AI* **2023**, *13*, 100246. [CrossRef]
14. Wang, Z.; Liu, Z.; Fan, L.; Du, Q.; Jiao, K. Application progress of small-scale proton exchange membrane fuel cell. *Energy Rev.* **2023**, *2*, 100017. [CrossRef]
15. Pei, J.; Wang, Z.; Li, X.; Wang, X.V.; Yang, B.; Zheng, J. Energy consumption prediction and optimization of industrial robots based on LSTM. *J. Manuf. Syst.* **2023**, *70*, 137–148. [CrossRef]
16. Li, X.; Lan, Y.; Jiang, P.; Cao, H.; Zhou, J. An Efficient Computation for Energy Optimization of Robot Trajectory. *IEEE Trans. Ind. Electron.* **2021**, *69*, 11436–11446. [CrossRef]
17. Available online: http://avl.com/CruiseMManual (accessed on 10 February 2023).
18. Zhang, R.; Li, X.; Sun, C.; Yang, S.; Tian, Y.; Tian, J. State of Charge and Temperature Joint Estimation Based on Ultrasonic Reflection Waves for Lithium-Ion Battery Applications. *Batteries* **2023**, *9*, 335. [CrossRef]
19. Xu, J.; Sun, C.; Ni, Y.; Lyu, C.; Wu, C.; Zhang, H.; Yang, Q.; Feng, F. Fast Identification of Micro-Health Parameters for Retired Batteries Based on a Simplified P2D Model by Using Padé Approximation. *Batteries* **2023**, *9*, 64. [CrossRef]
20. Fanoro, M.; Božanić, M.; Sinha, S. A Review of the Impact of Battery Degradation on Energy Management Systems with a Special Emphasis on Electric Vehicles. *Energies* **2022**, *15*, 5889. [CrossRef]
21. García-Miguel, P.L.C.; Alonso-Martínez, J.; Arnaltes Gómez, S.; García Plaza, M.; Asensio, A.P. A Review on the Degradation Implementation for the Operation of Battery Energy Storage Systems. *Batteries* **2022**, *8*, 110. [CrossRef]

sustainability

MDPI

Article

Analysis on Using 3D Scanning and BIM to Reduce the Physical and Non-Physical Construction Waste for Sustainable Fireproofing of Steel Trusses

Cristian Savu *, Adrian-Horatiu Pescaru, Ivett-Greta Zsak, Anca-Maria Durgheu, Alexandra-Paula Frent, Norbert-Szabolcs Suba, Aurelian Stelian Buda and Sorin Nistor

Faculty of Construction, Cadastre and Architecture, University of Oradea, 410058 Oradea, Romania; adrian.pescaru@didactic.uoradea.ro (A.-H.P.); zsak.greta@didactic.uoradea.ro (I.-G.Z.); durgheu.anca.maria@didactic.uoradea.ro (A.-M.D.); afrent@uoradea.ro (A.-P.F.); nsuba@uoradea.ro (N.-S.S.); aurelianbuda@uoradea.ro (A.S.B.); sonistor@uoradea.ro (S.N.)

* Correspondence: savu.cristian@didactic.uoradea.ro

Abstract: Embracing the archaic methods used in supervision and/or management in the field of architecture, engineering, and construction (AEC) creates the imperative need to adopt and/or develop sustainable methods to reduce construction physical waste, as well as the non-physical waste. Thus, the articles present a method that convenes three different specializations, namely geodesy, civil engineering, and architecture, that make use of the 3D terrestrial laser scanner (TLS), 3D reconstruction, and the 3D analysis to make an accurate bill of quantity (BOQ) to be able to assess the quality of the construction in compliance with the idea of sustainable development in the AEC industry. This article describes the three major parts: obtaining the point cloud, the 3D modeling, and extracting the relevant data. The novelty of the research is threefold: (i) a viable methodology for obtaining an accurate BOQ by reducing the non-physical waste to avoid cost and time overruns; (ii) the significance and impact of a (a) strictly automated and (b) automatic with traditional survey to obtain the 3D model of the point cloud on the resulting area that requires rehabilitation or to be covered with intumescent paint; and (iii) a detailed analysis of the specific elements that can blunder the final 3D model, such as the shadowing effect that can appear in very complex construction structures. Moreover, the proposed methodology represents a significant advancement in optimizing 3D modeling to improve the comprehension of steel trusses in the field of AEC sector.

Keywords: 3D terrestrial laser scanner; Building Information Modeling (BIM); 3D modeling; waste reduction; sustainable methodology

Citation: Savu, C.; Pescaru, A.-H.; Zsak, I.-G.; Durgheu, A.-M.; Frent, A.-P.; Suba, N.-S.; Buda, A.S.; Nistor, S. Analysis on Using 3D Scanning and BIM to Reduce the Physical and Non-Physical Construction Waste for Sustainable Fireproofing of Steel Trusses. *Sustainability* **2024**, *16*, 1832. https://doi.org/10.3390/su16051832

Academic Editor: Marijana Hadzima-Nyarko

Received: 29 December 2023
Revised: 12 February 2024
Accepted: 20 February 2024
Published: 23 February 2024

1. Introduction

The AEC industry is one of the most important industries in the global economy, but it is very inefficient because productivity in this sector has been on a downward trend for more than 30 years [1], and also because about 57% of costs in this industry are generated by different types of waste [2]. Waste represents losses caused by the consumption of material or non-material resources that generate direct or indirect costs without adding value to the final product [3]. Construction waste is classified into physical waste that includes material waste and non-physical waste that is generated by time and cost overruns [4]. Reducing this waste is very important to reduce the environmental impact of the construction industry. Studies conducted by different authors demonstrate that, in general, 3D laser scanning facilitates fast and accurate data acquisition, and that these data can be used to produce a 3D model in BIM software to automatically generate accurate areas or bills of quantities (BOQ) [5].

1.1. Steel Fireproofing

Steel trusses are structures composed of multiple small-section profiles that have the advantage of being able to cover very large spans in an economical manner and with low material consumption. The configuration of these trusses generates several challenges in terms of fire resistance, surveying, modeling in BIM programs, and then calculating the coverage areas of the steel profiles for fireproofing. Regarding the fire performance of steel elements, it should be noted that they are non-combustible, but without protection, they can resist fire for only 15 min. The main reason is that, at temperatures above 550 °C, steel loses more than 50% of its strength, and in case of fires and standard tests, these temperatures are reached in about 5 min [6]. Fire protection of steel structures is achieved through active systems involving the use of fire-fighting systems (extinguishers and sprinklers) and passive systems involving the covering of steel elements with different materials that increase fire resistance (intumescent coating, cement-based coatings, and fire-resistant board) [7]. Intumescent coatings are often chosen as a fire protection solution because they are applied similarly to paint and are therefore quick to apply, thin, lightweight, do not take up space, and do not affect the appearance of the metal structure [8]. The fire resistance of these paints is determined by the thickness of the coating, and this is related to the cross-section of the steel elements and the design temperature. Depending on this thickness, the intumescent coating must be applied in one or more layers. In the case of multilayer applications, the correct calculation of the coverage areas becomes very important, as it leads to a multiplication of the amount of waste or losses generated as a result of purchasing an excessive or insufficient amount of paint.

Statistics presented by Eurostat in 2020 highlight that the construction industry is the main waste generator in Europe, being responsible for 37.5% of the waste generated. These statistics also highlight that, from 2004 to 2020, the amount of waste produced by the construction industry increased by 12.5% [9]. At the same time, the studies carried out in the UK revealed that 13% of the materials that arrive at construction sites are thrown away, without being used [10]. Also, in the construction sector, cost overruns are a common phenomenon generated by several factors. Most often, cost overruns are caused by time overruns, frequent project changes, and inaccurate time and cost estimates [11]. Part of these estimation errors are caused by the fact that the areas underlying the estimates are calculated incorrectly.

Areas and BOQs are traditionally extracted manually from 2D drawings, but this process is very slow, requires higher effort, and presents high risks in terms of results' accuracy because this process involves the interpretation of information extracted from 2D drawings that often do not capture all the data necessary to understand complex structures like steel trusses [12]. For projects that involve interventions on existing buildings, these 2D drawings are often made using traditional survey techniques, which, in the case of complex structures, including the steel trusses studied in this paper, involve another process of abstraction and approximation of the structure. Therefore, all of these phases that take place before establishing the BOQ assume a series of data interpretations that can cause the accumulation of errors, the obtainment of wrong quantities, and, ultimately, physical or non-physical waste.

To reduce or even eliminate these deficiencies, the traditional survey process can be replaced by terrestrial laser scanning surveys to generate point clouds that can be processed in Building Information Modeling (BIM) software to create more accurate 3D models, 2D drawings, and quantities.

1.2. Terrestrial Laser Scanning (TLS)

This advanced surveying technique utilizes laser beams to capture highly detailed and precise three-dimensional information about the surveyed environment. The statistics associated with 3D terrestrial laser scanning underscore its growing significance in the construction industry and quantity surveying practices [13]. One key statistic highlighting the efficiency of 3D laser scanning is its rapid data acquisition speed. Traditional surveying methods using pen, paper, and measuring tapes, and also methods involving the use of total station class surveying equipment, often require significant time and resources to collect data, particularly in large and complex construction sites. In contrast, 3D laser scanning can capture thousands of points per second, allowing for swift and comprehensive data acquisition. This efficiency not only accelerates the surveying process but also minimizes disruptions to ongoing construction activities [14–16].

Another notable statistic pertains to the level of detail achievable through 3D laser scanning. This technology can produce highly accurate point clouds, representing millions of individual points in a surveyed space. This wealth of data enables quantity surveyors to create precise and detailed models of existing structures or construction sites, facilitating more accurate quantity takeoffs and cost estimations. The ability to capture intricate details with minimal human intervention enhances the reliability of the surveying process.

Three-dimensional terrestrial laser scanning has emerged as a revolutionary technology in the field of quantity surveying, offering unprecedented capabilities for accurate and efficient data collection. With the help of a 3D laser scanner, a 3D point cloud is generated by using a time-of-flight technique for computing the 3D coordinates of consistent areas, which are then modeled to obtain a 3D as-built replica of the real world. Due to the capability to obtain a massive volume of data in a relatively short time, this approach is suitable for large-scale and infrastructure projects [17,18], such as the measurement and modeling of bridges and tunnels (without disrupting ongoing traffic and endangering the lives of the surveyors), the modeling of complex industrial sites, and the digital conservation of buildings with valuable architectural details.

In terms of safety, 3D terrestrial laser scanning also presents compelling statistics. The non-contact nature of laser scanning reduces the need for surveyors to physically access hazardous or hard-to-reach areas. This enhances overall project safety by minimizing the risks associated with on-site data collection, particularly in complex construction environments [19]. Regarding the surveying of steel trusses, it should be noted that there are several challenges due to their geometrical complexity and the fact that these structures are often positioned at great heights, making it unsafe, difficult, or even impossible to understand the configuration and details of these structures to be surveyed manually using conventional techniques. The non-contact nature of laser scanning reduces the need for surveyors to physically access hazardous or hard-to-reach areas. This enhances overall project safety by minimizing the risks associated with on-site data collection, particularly in complex construction environments [7]. However, laser scanning also presents some challenges in terms of data acquisition due to the complex configuration of the steel truss, which results in shadowed areas where the laser beam cannot reach [20]. This problem can be overcome by using multiple terrestrial stations or positioning them according to this limitation by using UAVs (Unmanned Aerial Vehicles) equipped with scanning instruments to complement the data acquired by terrestrial scanning [21], or by collecting data using conventional techniques (manual surveying and photographs) when possible.

1.3. Building Information Modeling (BIM)

Three-dimensional models contain robust data about construction projects that are essential to the BIM (Building Information Modeling) process and are far superior to traditional construction approaches. BIM workflows provide the ability to classify and organize project-related data, such as construction typology, building geometry, and material properties, that can be used to make informed decisions [22–25].

BIM is a complex process that involves, among other things, a software component that facilitates the production of 3D models, their analysis, and communication between all the stakeholders involved in the project [26,27]. Thus, improving the level of automation in the AEC domain progressively becomes a standard in many countries [28].

Studies comparing traditional quantity surveying techniques with those based on the automatic extraction of quantities from the BIM model demonstrate that the use of this methodology has multiple advantages, such as streamlining calculation processes by increasing accuracy and productivity [29]. At the same time, some studies highlight that, by using BIM software instead of the traditional method, the time required to make quantity takeoffs and estimates can be reduced by 80%, and the results obtained can have deviations of up to 3% [30].

It is important to note that these advantages depend, to the greatest extent, on the level of detail (LOD) of the BIM model [19,31,32]. The fact that detailing the BIM model leads to increased design times and costs means that the BIM model is rarely detailed enough to be used in extracting BOQs [33,34]. The study carried out by Olson and Taylor highlights the fact that BIM models that reach general contractors contain only 50% of the information needed to extract quantities because they do not contain finishes; temporary structures; mechanical, electrical, and plumbing installations; and landscaping [30,33,35]. A similar situation occurred in the project presented in this article. The initial project that was carried out for the modernization of the University of Oradea auditorium building did not have the BOQ required for contracting the steel trusses' fireproofing, and to get these quantities, it was necessary to carry out the research presented in this article.

Taking all of these aspects into account, the research presented in this paper has a few distinct goals that represent the main research motivation and significance of the paper: (a) analyzing the implications of using 3D laser scanning and BIM software on the calculation of steel truss-fireproofing coverage areas; (b) a sustainable methodology of an efficient and accurate combination that integrates the TLS technology and traditional survey to rigorously calculate the areas to avoid the production of non-physical waste; and (c) a comprehensive analysis of a few key elements that are prone to the shadowing effect that appears in highly complex construction structures that can blunder the 3D model. Furthermore, the proposed methodology for quantity surveying using the 3D TLS and traditional survey techniques for assessing the non-physical waste represents a novelty in advanced research in sustainability in the AEC sector.

Although the research shows no material waste, the innovation within the article presents a more unpleasant type of waste—the so-called non-physical waste which is generated because of cost and time overruns. The cost overruns imply that the actual cost of fireproofing is higher than the approved budget for this work, and that an additional budget needs to go through all administrative procedures from the beginning to be able to complete the work. Thus, it results in a chain effect, resulting in time overruns caused by the fact that, for public investments, approvals for budget supplements involve a time-consuming bureaucratic process, and the construction team that performs the fireproofing works has to stop and relocate both materials and human assets until the necessary budget is approved.

2. Materials and Methods

2.1. The Case Study: Aula Magna Hall of the University of Oradea

The Aula Magna Hall, situated within the campus of the University of Oradea, stands as a distinguished facility within this esteemed institution of higher learning. Renowned as the most expansive hall within the university, it is frequently used for a plethora of significant events, encompassing ceremonies, conferences, seminars, and a variety of other scholarly activities. Architecturally, it is designed to accommodate a capacity of 300 individuals. The Aula Magna is a component of Building F (Figure 1a,b), which was constructed in the period spanning from 1993 to 1998. This particular building underwent rehabilitation as a part of the "SMART Campus—University of Oradea" project. Within this framework, the present research emerged, especially since one of the rehabilitation's objectives encompassed fire safety.

(**a**) (**b**)

Figure 1. (**a**) Building F—inner courtyard façade. (**b**) Building F—University Street façade.

The campus's contemporary character is defined by several key elements:

- The university buildings have undergone an organic evolution, incorporating the architectural nuances of the Secessionist style. This is exemplified by 12 heritage buildings of significant value, constructed between 1911 and 1913, under the guidance of architect Jozsef Vago. The project, while retaining the essence of their spatial organization, originally commenced as a Gendarmerie School [36].
- The implementation of initiatives like "SMART Campus—University of Oradea" demonstrates a dedication to modernization, technological integration, and enhanced accessibility. These efforts underscore a strategic pivot towards innovation and technological advancement within the university's framework.
- The EU GREEN project, spearheaded by the University of Oradea, places a strong emphasis on sustainability and educational approaches to sustainable development. The university's active participation in this project signifies a robust commitment to fostering sustainable development principles. Through this project, there is a concerted effort to elevate the level of awareness and engagement among faculty and students concerning sustainable development concepts.

The interdisciplinary nature of these initiatives has fostered a project that serves as a testament to the interpretation of field data, aiming to convert this information into the most precise quantifiable measures for the hall's rehabilitation project. Additionally, this rehabilitation effort has been augmented by incorporating adjustments to align with the latest fire safety standards.

The metal roof truss over the Aula Magna Hall is a spatial structure composed of transversally arranged truss beams, supported by the reinforced concrete structure, with a span of 20.45 m. The spatial structure is stiffened longitudinally through roof panels fixed over the top chord, at the nodes of the truss beams, and by cross bracings arranged in the plane of the roof. At the lower part, the stiffening of the trusses is achieved through longitudinal beams. Additionally, the bottom horizontal chord of the truss beam is equipped with a fastening system for the suspended ceiling, which uses tension rods, over which mineral wool thermal insulation is laid.

The members of the truss beam are composed of sections made from two equal-flange angles slightly spaced apart, joined together by gusset plates, and fixed through welding. Laminated steel profiles with standardized sections were used. Subsequently, the metal structure was protected against corrosion by at least two layers of paint, which increased the dimensions of the metal profiles identified in the survey. To verify the accuracy of the manual survey and the 3D scanning, it was necessary to establish the initial size of the sections. The initial section of the metal elements was determined based on the standard dimensions of the steel profiles, as given in technical catalogues for hot-rolled steel profiles.

2.2. Traditional Surveying Methods and Terrestrial Laser Scanner Surveying

In the context of our project, the execution of the traditional survey necessitated a specific set of tools, comprising a clipboard, A4 paper sheets, a graphite pencil, a standard tape measure, and a caliper. The methodological approach to the surveying process involved several critical stages.

This methodical process was pivotal in ensuring the collection of precise and comprehensive data necessary for the accurate 3D modeling of the elements in the subsequent phases of the project.

The initial phase involved manually surveying and identifying the profiles through the use of standard catalogues. The traditional surveying process of a steel frame truss structure was made manually, using simple tools like a measuring tape for assessing the dimensions of the steel profiles and a caliper to measure their thickness. Several sketches of sections of the roof structure had to be made on paper on-site in order to mark the specific dimensions of the different measured members. Firstly, the profiles were marked with the actual measured dimensions. Secondly, after completing the manual survey, the steel sections identified were compared with the standard steel sections from technical catalogues for hot-rolled steel profiles.

The steel profile catalogues give the exterior surface of a standard profile based on its section dimensions. The steel structure was protected against rust with at least two coats of paint. In addition, in the corners of the flanges and especially on the lower joints, layers of cemented dust increased the dimensions of the steel sections in comparison with the initial standard section. Thus, in order to manually determine the actual fireproofing surface, the steel sections had to be approximated to a standardized section. Regarding this project, it was not possible to carry out a full survey using traditional techniques because of the lack of safe access to most of the roof elements, but also because the process would not have been time-efficient due to the complexity of the structure. Therefore, the survey process relied mostly on terrestrial laser scanning (TLS) technology.

Terrestrial laser scanning (TLS) is a geospatial surveying technique that utilizes laser technology to capture three-dimensional spatial information of objects and environments from a stationary position on the Earth's surface. TLS is recognized for its accuracy, efficiency, and versatility in capturing complex geometries and has found application in fields such as architecture, engineering, geology, forestry, cultural heritage preservation, and quantity surveying [37]. The resulting point cloud data can be further processed and analyzed to derive valuable insights, making TLS a fundamental tool in modern geospatial research and spatial data acquisition [38]. The TLS is highly efficient if the resulting point cloud has an accuracy considerably higher than the changes from the real-word surface; however, the necessary changes in the real-word surface that have to be captured by the TLS have to be properly defined by the architect or the engineer. If the subtle changes that have to be measured by the TLS are in the order of a few millimeters, then the required TLS equipment has to be properly chosen because the point position is somewhere between ± 2 and ± 50 mm [39]. Although, if we are modeling the point cloud, we can improve the accuracy up to 20 times compared to single point accuracy, according to [40].

In general, there are four fundamental intricate steps to obtain a 3D geometric model from the resulting 3D point cloud [17]:

1. Data collection—using the 3D scanner to obtain the necessary point cloud of the specified construction site from different stations,
2. Data post-processing—referencing and geo-referencing of the assembly of all the stations from which we made the scanning and applying the necessary computational models for noise reduction and adjustment,

3. Geometric modeling of the refined point cloud to generate the 3D model as a mesh or as an object,
4. Generating the digital documentation.

For obtaining the 3D point cloud, we used the Trimble X7 laser scanner, which is a professional-grade terrestrial laser scanner designed for high-precision 3D scanning and data acquisition applications in fields such as construction, surveying, and building documentation. Trimble X7's key features are as follows:

➢ High-speed scanning: The Trimble X7 is known for its fast and efficient data capture capabilities. It can rapidly collect dense point cloud data, allowing for the quick and comprehensive 3D scanning of structures and environments. The TLS is capable of working at speeds up to 500 kHz (thus capturing up to half a million points per second).
➢ Integrated imaging: The scanner typically comes equipped with integrated imaging capabilities, such as high-resolution cameras. This allows users to capture colored panoramic images alongside the 3D point cloud data, providing additional visual context.
➢ Automated operation: The X7 is designed to streamline the scanning process with automation features. Automated workflows and onboard software—we used the Trimble Perspective software 1.1.3 to assist in simplifying data capture, making it more accessible for users with varying levels of expertise.
➢ User-friendly interface: The device is often designed with a user-friendly interface to enhance the overall user experience. This includes a touch screen or other intuitive controls for easy operation in the field.
➢ Lightweight and portable: While still being a professional-grade scanner, the Trimble X7 is typically designed to be relatively compact and lightweight compared to some other laser scanning solutions, with the scanner weighing just 5.8 kg and measuring 178 mm (W) × 353 mm (H) × 170 mm (D) (both values without tripod), according to the data provided in Table 1. This enhances its portability and ease of transportation to different job sites.
➢ Accuracy and range: The scanner is engineered to provide high accuracy in point cloud data. It offers a range suitable for various applications, from close-range detailed scans to capturing data from a distance, with the range accuracy (measured distance between the scanner and object) being 2 mm and the absolute point accuracy of the 3D model being dependent on the scanner–object distance (Table 1, 3D point accuracy section).

The Trimble X7 is commonly used for applications such as building documentation, construction site monitoring, quality control, clash detection, and creating accurate as-built models.

Each individual scan is defined by a station and, thus, is identified by a number, a specified color, and a marked position; hence, the point cloud registration can be performed automatically in the field. This can be performed also with the help of the built-in Inertial Measurement Unit (IMU), which has the ability to orient the scanner when we move it from one station to another so that the initial cloud alignment can be obtained. The aforementioned point cloud registration or auto-registration is achieved typically with a high degree of success without any user input or intervention into the Trimble specific software.

The main surveying equipment that was used to obtain the 3D point cloud was the Trimble X7 scanner. The relevant technical specifications are listed in Table 1.

Table 1. The technical specification of Trimble X7 laser scanner according to [41].

Scan Parameters	Trimble X7 Specifications
Range principle	High speed, digital time-of-flight distance measurement data
Range noise	<2.5 mm @ 30 m
Range	0.6–80 m

Table 1. *Cont.*

Scan Parameters	Trimble X7 Specifications
Field of view (degree)	360° × 282°
Scan speed	Up to 500 kHz
Range accuracy	2 mm
Angular accuracy	21″
3D point accuracy	2.4 mm @ 10 m, 3.5 mm @ 20 m, 6.0 mm @ 40 m
Scanning EDM laser class	Laser class 1, eye safe in accordance with IEC EN 60825-1 [42]
Laser wavelength	1550 nm, invisible
Weight	5.8 kg
Dimensions	178 mm (W) × 353 mm (H) × 170 mm (D)

2.3. Three-Dimensional Modeling BIM Software

The process of 3D scanning modeling was performed using the Revit 2021 software, which encompassed the following stages:

1. Data importation into Revit: After the processing of the raw data obtained from the scanning process, wherein errors and noise were eliminated, the 3D data were imported into Autodesk Revit. Revit is Building Information Modeling (BIM) software that is extensively used for the digital representation and management of building data. In Figure 2a, the 3D point cloud is shown as it appears after being imported into the program. This contains all the information about the building, such as the structural parts, partition walls, ventilation equipment, furniture, textile materials, and position of the windows. Additionally, the presence of undesirable substances, including dust, grime, and remnants of construction materials, was noted. In this specific case, as is observable in Figure 2b, these accumulations obstructed the precise identification of horizontal elements located directly on the floor.

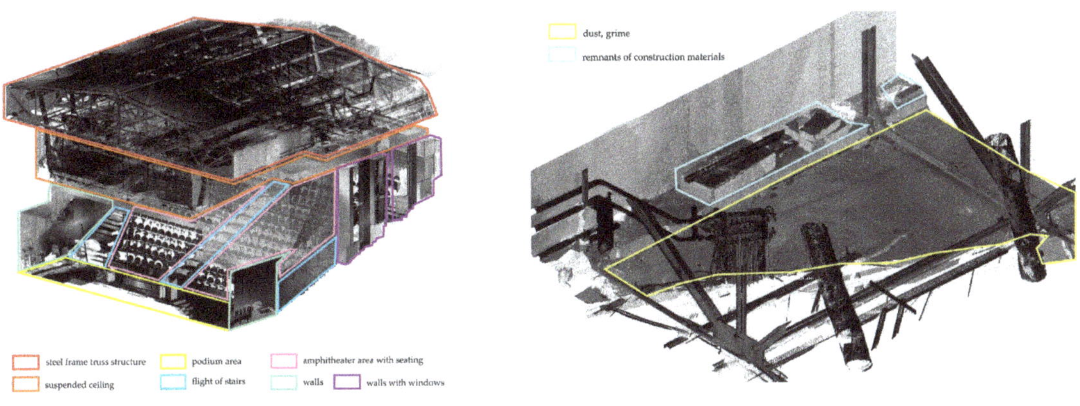

(**a**) (**b**)

Figure 2. *Cont.*

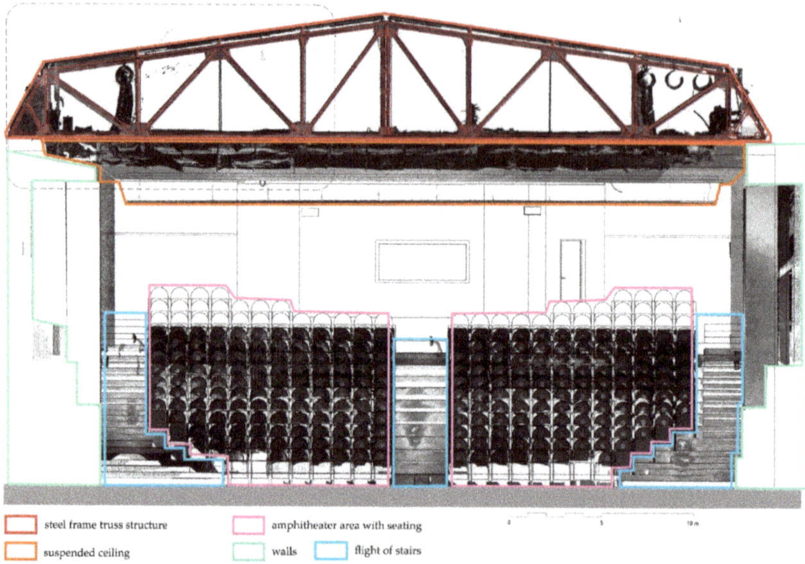

(c)

Figure 2. (**a**) 3D point cloud. (**b**) Identification of obstructed areas. (**c**) Section through the 3D scan depicts a two-dimensional representation of structural elements in red, and the representation of contextual elements (walls, windows, steps, and furniture) is made with black lines.

2. Two- and three-dimensional modeling in Revit: The initial phase of modeling involved scrutinizing the scanned data to discern elements constituting the roof framing. Initial efforts included segmenting the scanned data to ascertain the profiles present within the framing structure. In Figure 2c, the identification of structural components is presented, followed by their representation in a two-dimensional format. The scan is shown in grayscale, with the red lines representing the identified contours of the structure. These profiles were initially drawn in a two-dimensional format, recognizing the existence of some unclear areas in the scan. Additionally, other blurred regions, as shown in Figure 3a,b, were later identified as ventilation ducts in the 3D modeling phase. In Figure 3c, the components constituting the truss beam are presented. These elements are as follows: bottom and top chord, webs, and gusset plates. Along with areas that could not be scanned, these are marked in red in Figure 4. The comprehensive 3D model was developed using the "Component/Model in-Place" tool in Revit, employing techniques such as Extrusion, Blend, and Sweep for this purpose [43].

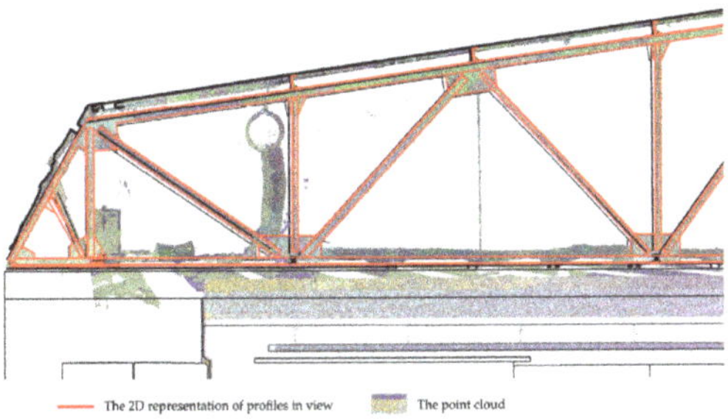

(**a**)

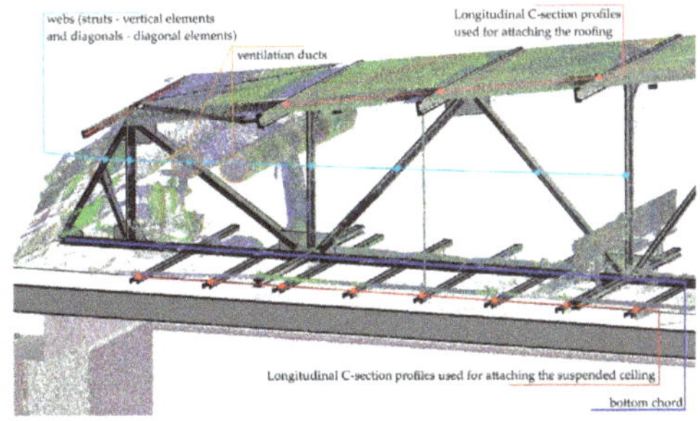

(**b**)

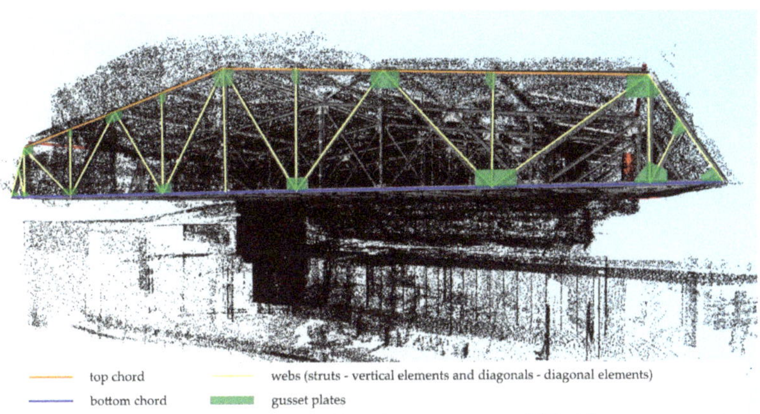

(**c**)

Figure 3. Detailed area: (**a**) representation in section—2D model. In this section the point cloud is colored green, yellow and purple, and the outline of the profiles is represented with red lines; (**b**) representation in axonometric view with 3D model; and (**c**) representation of a metallic profile in cross-section from a 3D point cloud.

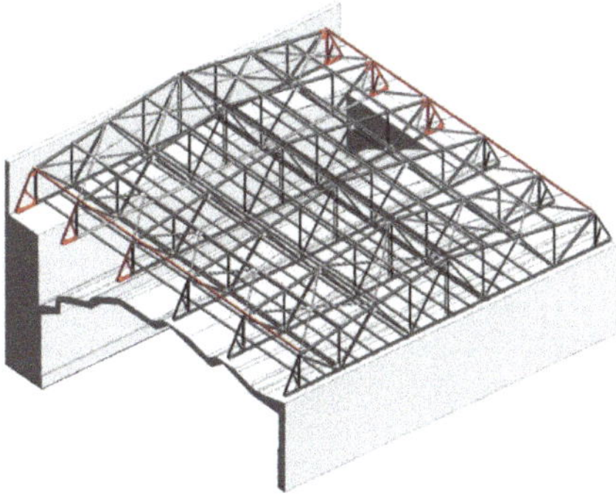

(**a**)

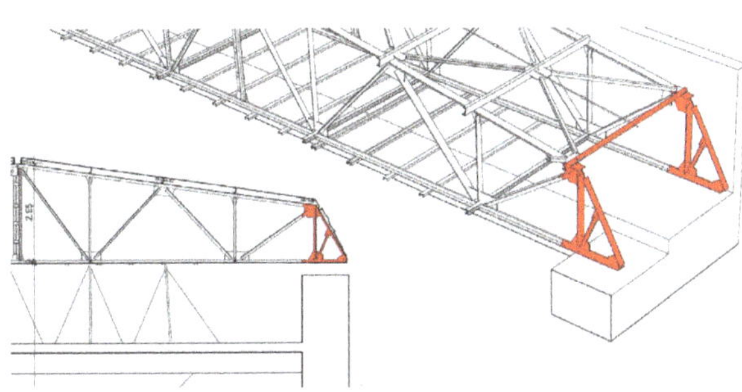

(**b**)

Figure 4. (**a**) The 3D model created in Revit® and the marking of elements inferred from the repetitiveness of the structure. (**b**) Virtual reconstruction of unscanned elements (shown in red), based on existing geometry.

3. Analysis: All structural elements were created using the "Structural Framing" feature, and the connecting elements were created with components categorized as "Structural Connections". Elements identified based on the point cloud were assigned a material designated as "cloud", while those created outside the point cloud were labeled "red". Using these settings enabled the execution of a differentiated area calculation, distinguishing between elements identified from the scan and those created to complete the structure.
4. Collaboration and sharing: In order to share the model with other colleagues, it was exported in the IFC (Industry Foundation Classes) and DWG (Drawing) formats. The 3D model created in Revit served as a basis for verifying the existing structure's strength according to current standards. For this purpose, the openings and overall dimensions were required, rather than detailing all profiles. The detailing was to be added subsequently.

2.4. Methodology

The methodology used for this research is similar to that used in real practice. At the same time, a multidisciplinary team composed of surveyors, architects, and civil engineers contributed to this project, with each member of the team having a role and tasks adapted to his/her specialization.

Surveyors oversaw carrying out and processing the 3D scan for the architects. The architects were in charge of the 3D modeling, extracting the fireproof coated surface area from the BIM software and providing 2D drawings for the civil engineers. The civil engineers were in charge of the manual surveying of the roof structure, including the identification of steel profiles from standard profile catalogues and the manual calculation of the fireproof coated surface area, using the 2D drawings.

The whole process was divided into 3 main steps: surveying, modeling, and area calculation for quantity surveying. The entire process followed in this study and the data flow are illustrated graphically in Figure 5.

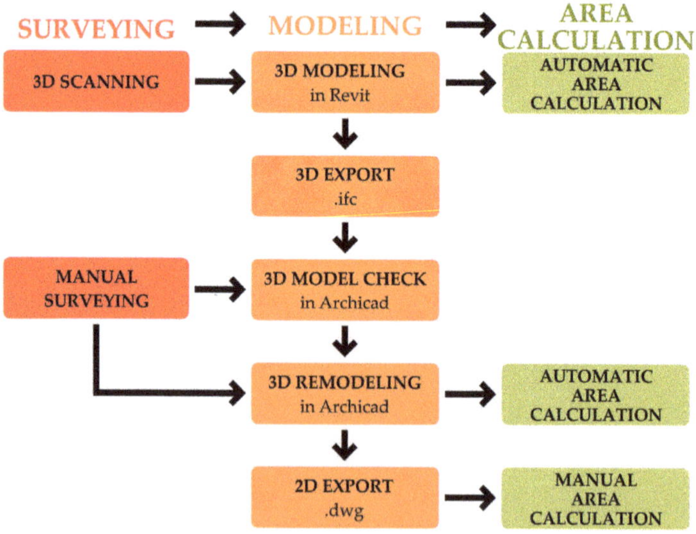

Figure 5. Graphic illustration of the process followed to determine the areas needed to calculate the lists of quantities.

The first step consisted of surveying the existing steel truss structure of the roof, using traditional techniques and tools (measuring tape, caliper, paper, and pen), and also using 3D scanning technology. It was necessary to survey the structure partially manually, using traditional techniques, to complete the data collected by TLS because the configuration of the roof structure did not allow for the positioning of sufficient scanning stations to eliminate the shadowing effect, nor did it allow for the introduction of a UAV equipped with 3D scanning equipment. So, there is a risk that the data collected using only 3D scanning are insufficient to model the composite elements of steel trusses correctly. Thus, the traditionally collected data could also be used to check and correct the initial 3D model which was made exclusively from the point cloud generated by the 3D scan.

The second step consisted of the 3D modeling of the steel trusses, using exclusive data collected from the point cloud collected by TLS. This model was exported in .ifc format to be checked and later completed with data collected through a manual survey. After correcting the initial 3D model, the 2D drawings needed to manually calculate the coverage areas of the elements that required fireproofing were extracted and exported in .dwg format.

In the third step, the coverage areas of the profiles to be fireproofed were calculated. In order to check, analyze and evaluate how the structure survey technique and the area calculation technique influence the results, three areas were generated: one was automatically generated using the 3D model obtained with the help of 3D scanning, one was automatically generated from the 3D model obtained using manual surveying, and one was manually calculated using traditional quantity surveying techniques based on 2D drawings and standard profile catalogues.

The automatic generation of the areas was performed in ArchiCAD 26 by creating a custom Surface Schedule that was configured to automatically generate a list of coverage areas of the steel elements that had to be fireproofed. In this list, the elements were automatically grouped by element type, area subtotals were generated by element type to allow checking and identifying errors, and finally, the total fireproofing area was calculated.

The manual calculation consisted in the first phase of extracting from the 2D drawings the lengths of the elements that had to be fireproofed and calculating the total lengths for each element type. Subsequently, the coverage area (AL) of the profiles used in the project was extracted from the standard profile catalogues, and this area was multiplied by the previously calculated lengths. All of these calculations were performed in Google Sheets to be shared with all team members for verification and evaluation.

After the evaluation of these data, it becomes possible to identify the most efficient method of data acquisition, 3D modeling, and coverage area calculation of steel trusses, while ensuring construction waste reduction.

3. Results

The terrestrial laser scanning process consisted of mounting the Trimble X7 scanner in a total of 33 different scanning positions (stations), which covered the necessary data acquisition for both the roof section and the Aula Magna. Twelve of these stations were used to scan the roof area, yielding 340.341.222 points. The total scanning time from all 33 stations was just under an hour (58 min). This remarkable speed was achieved also due to the fact that this type of equipment does not need scanning targets (specific spherical or other types of targets) for the point cloud registration (assembling the whole 3D model resulting from different stations). The whole registration process is based on identifying common points in scans from different stations, with the maximum error for registration being 1.5 mm. Due to this, the uncertainty regarding the correct position of the scanned points is kept to a near minimum, with the reported average confidence level being around 97.6% and the rest being accounted as noise.

Following the modeling process, it can be observed in Figure 6 that 91.45% of the surface area of the structural framing was identifiable from the point cloud. The remaining 8.55% was deduced from the completely identified trusses.

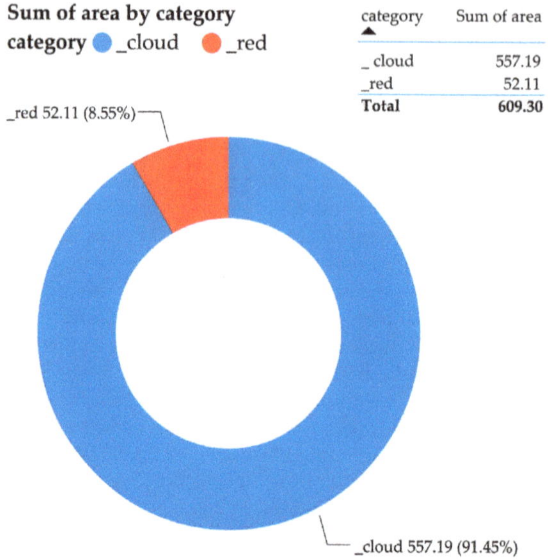

Figure 6. Chart illustrating the modeled surfaces based on the point cloud (blue) and those completed for areas missing from the 3D scan (red).

The 3D model made in Revit was exported in *.ifc format and imported into ArchiCAD to be checked by the team members who carried out the survey using traditional techniques. This model was also used to automatically calculate the areas that needed to be fireproofed.

After checking the 3D model obtained using only the information gathered by the 3D scanner, it was easy to notice that most of the steel elements were identified and modeled incorrectly because they had different sizes and/or sections compared to the real ones. However, an automatic calculation of the covering areas of the entire structure was carried out using the functions of ArchiCAD because we wanted to analyze and compare these values with those obtained using other calculation methods to identify the most efficient method to obtain the correct results, reduce waste, and also perfect a way of integrating the 3D scanning techniques into the design process.

Based on the calculations automatically performed in ArchiCAD on the exported model from Revit, the resulting surface area was found to be identical to that in Revit. The determined surface area requiring intumescent paint coverage was 609.30 square meters in both programs.

The first difference we observed is that most of the elements that were composed of two elements were modeled as a single element. In reality, all the elements of the steel trusses have a cross-section composed of two L-type profiles with equal angles spaced apart. These elements were modeled as a T-section profile. For example, the bottom chords of the trusses are composed of two L-type profiles with equal angles that have a 90 × 90 × 10 mm section and are spaced apart by 12 mm (Figure 7a), but they were modeled as a T-section profile that has 194 × 93 mm (Figure 7c). The real cross-section has an AL coverage area of 0.702 sqm/m, while the 3D modeled cross-section has an AL coverage area of only 0.574 sqm/m. So, in this case, the coverage area of the real profile is 22.3% higher than the profile modeled using 3D scanning. This means that, for the fireproofing of the real profile, a 22.3% larger amount of paint is needed than the one resulting from the model made after 3D scanning data. At the same time, it was observed that there is a difference of 0.052 m^2/m (9.49%) between the coverage area of two L-type profiles with equal sides of 90 mm (Figure 7b) and the coverage area of the profile modeled in the 3D scan (Figure 7c).

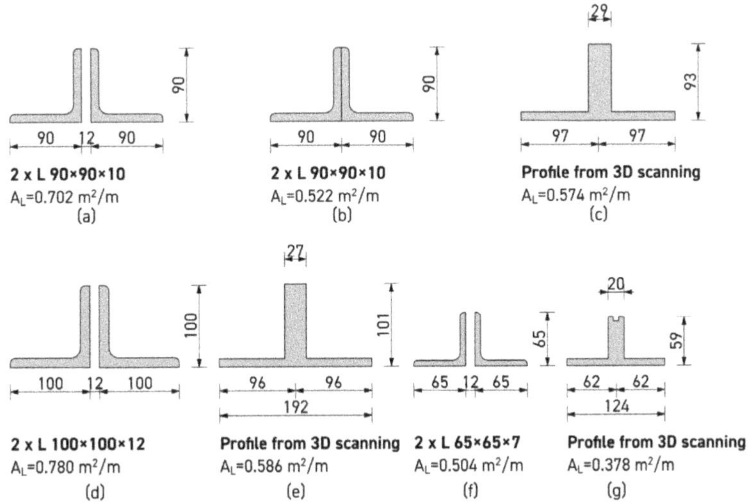

Figure 7. In reality, all the profiles are spaced apart (**a**,**d**,**f**), but in the 3D model built after the 3D scan, the profiles were connected (**c**,**e**,**g**). (**a**) The actual section of the bottom chord of the truss composed of two spaced L-type profiles with equal angles (L90 × 90 × 10). The actual section has a coverage area of 0.702 m^2/m. (**b**) Two connected L-type profiles. (**c**) The section of the bottom chord modeled after the 3D scan. (**d**) The actual section of the top chord (L100 × 100 × 12). (**e**) The section of the top chord modeled after the 3D scan. (**f**) The actual section of the webs (L65 × 65 × 7). (**g**) The section of the webs modeled after the 3D scan.

As for the top chord of the truss, it is composed of two L-type profiles with equal angles and a section of 100 × 100 × 12 mm that is spaced by 12 m (Figure 7d), but it was modeled as a T-type profile of 192 × 101 mm (Figure 7e). Between the two cross-sections, there is a difference in coverage area of 33.11%. The same situation was observed regarding the modeling of the webs (vertical and diagonal elements) of the trusses. These were modeled as a 124 × 59 mm T-type profile (Figure 7g), although, in reality, they are made from two L-type profiles with a 65 × 65 × 7 mm section that are spaced apart (Figure 7f). Between the two cross-sections, there is a difference in the covered area of 33.33%. The cross-sections of these profiles are illustrated in Figure 7, and the dimensions and coverage areas are centralized in Table 2.

Table 2. Cross-section and coverage areas of real profiles and 3D Scan Model profiles.

Truss Element	Actual Section	3D Model Section	Actual Coverage Area AL (m^2/m)	3D Model Coverage Area AL (m^2/m)	Difference
Bottom chord	2 × 90 × 90 × 10	194 × 93 × 29	0.702	0.574	+22.3%
Top chord	2 × 100 × 100 × 12	192 × 101 × 27	0.780	0.586	+33.11%
Webs	2 × 65 × 65 × 7	124 × 59 × 20	0.504	0.378	+33.33%

After performing this analysis, the 3D model was corrected in ArchiCAD by replacing the profiles that were identified incorrectly using the point cloud with the real profiles identified in the standard steel profile catalogues, using data collected through the traditional survey.

This aspect is clearly visible in Figure 8b,c. Figure 8b illustrates a detail highlighting how the steel profiles were modeled in the first 3D model, which was made using only the information extracted from the 3D scanner data, and Figure 8c shows the same detail extracted from the second 3D model that was modeled using the traditional survey data. It can be seen that the elements composed of two metal profiles were modeled as a single profile.

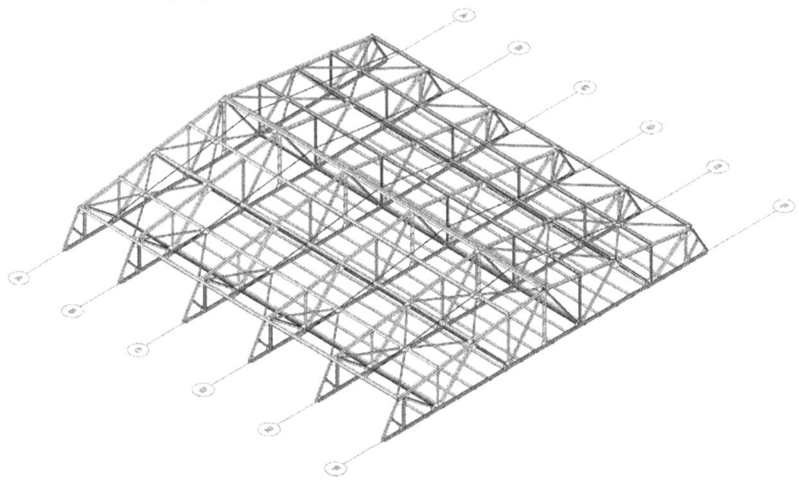

(**a**)

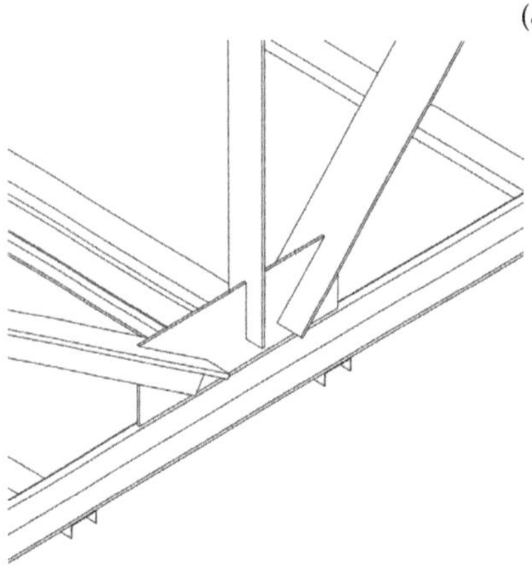

(**b**)

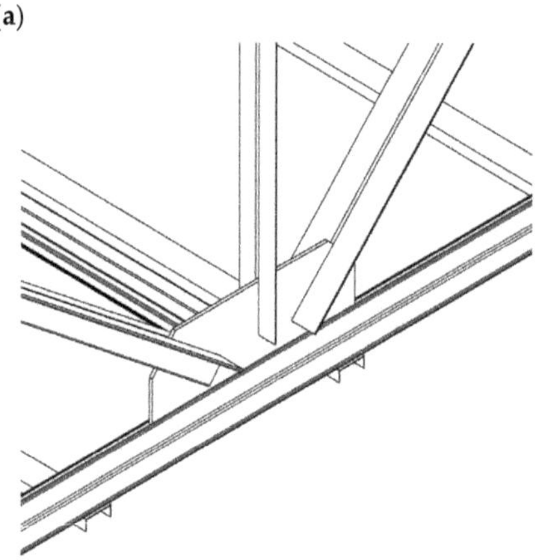

(**c**)

Figure 8. (**a**) Axonometric drawing of the second 3D model made in ArchiCAD 26 software. (**b**) Axonometric drawing with a detail of the metal profile joints extracted from the 3D model made using only 3D scan data. (**c**) Axonometric drawing with a detail of the metal profile joints extracted from the 3D model made using 3D scan and traditional survey data. It can be seen that the elements composed of two metal profiles were modeled as a single profile.

During the creation of the 3D model, the steel profiles were organized on different layers because we wanted the automatically calculated areas to be detailed by profile type to be able to analyze the results from different points of view, and also to more easily identify possible errors.

Based on the calculations made automatically by the BIM software on this 3D model, the surface that had to be covered with intumescent paint resulted in being 667.02 sqm. The actual area is therefore 57.72 sqm (9.47%) larger than the area automatically calculated by the BIM software using the 3D model made after the 3D scan. To validate these results, manual calculations were carried out. From this 3D model, the 2D drawings were extracted and exported in .dwg format so that the civil engineers could manually extract the lengths and sections of the profiles and then search for their coverage areas in the steel profile catalogues.

Manual calculations were carried out in a few hours and determined that the area that had to be covered with intumescent paint was 669.65 sqm. This area is 60.35 sqm (9.90%) larger than the automatically calculated area using the 3D scan model but only 2.63 sqm (0.39%) larger than the automatically calculated area based on the corrected 3D model. The fact that there is an insignificant difference between the two areas calculated using the corrected 3D model validates the results obtained automatically from the BIM software. The difference between the manually and automatically calculated area occurred due to certain approximations that were made in the manual calculation regarding the length of certain elements that were not represented in true size in the 2D drawings. In order to facilitate the understanding of the results obtained, all data are centralized in Table 3.

Table 3. Table centralizing the results obtained. The last two columns show the differences between the area calculated using the initial 3D model that was made using the 3D scanning survey and the area calculated using the second 3D model that was made via the manual survey.

3D Model Used for Calculation	Area Calculation Method	Calculated Area	Difference (sqm)	Difference (%)
3D scanning survey 3D model	Automatically calculated area	609.30 sqm	-	-
3D scanning + manual survey 3D model	Automatically calculated Area	667.02 sqm	57.72 sqm	+9.47%
	Manually calculated area	669.65 sqm	60.35 sqm	+9.90%

After analyzing these results, it can be seen that, by combining the data acquired by TLS with the traditional acquired data, an accurate 3D model can be produced in the BIM software, which facilitated both the automatic generation of accurate coverage areas and the production of the 2D drawings necessary for the accurate manual calculation of the areas.

In this case, the correct area is larger than the area calculated initially, and this means that the budgeted fireproofing cost and fireproofing materials ordered for this process would have been insufficient. In these situations, no material waste is generated because there is no surplus material that turns into waste, but non-physical waste is generated because of cost and time overruns. Cost overruns mean that the actual cost of fireproofing is higher than the approved budget for this work, and that an additional budget needs to be approved to complete the work. Time overruns are caused by the fact that, for public investments, approvals for budget supplements involve a time-consuming bureaucratic process, and fireproofing works should have been stopped until the necessary budget was approved.

4. Discussion

Terrestrial laser scanning (TLS) offers several advantages over traditional surveying methods when it comes to working time. Here are some key points highlighting how TLS can be more time-efficient compared to certain conventional surveying techniques:

- ➢ Rapid data acquisition and reduced field time: TLS can quickly capture a large amount of data in a relatively short period. Traditional surveying methods, such as manual measurements or total station surveys, may take significantly longer to cover the same area. This efficiency is particularly advantageous for projects with tight schedules. As was mentioned before, the whole scanning process took place in about 58 min, with this interval permitting the scanning of both the roof section and the Aula Magna Hall, with the timespan being significantly lower than that needed for traditional measurement techniques.

➢ Simultaneous data capture: TLS can capture data from multiple angles simultaneously, enabling a comprehensive view of the surveyed area in a single scan. This is in contrast to traditional methods, where each point or feature might need to be measured individually, leading to a more time-consuming process. Moreover, the scanner can capture plain 3D point clouds (as seen in Figure 9) but also colored details of the real-world environment, resulting in photorealistic 3D models (as can be observed in Figure 10b).

➢ Versatility in environments: TLS is highly versatile and well-suited for various environments, including complex or challenging terrains. Traditional survey methods may encounter difficulties in accessing certain areas or require additional time and effort to overcome obstacles. As can be observed in Figure 10b, the narrow beams and the void underneath the measured area would have made for a lengthy and unsafe operation for data acquisition using traditional methods.

➢ Real-time visualization combined with faster processing of data: TLS systems often provide real-time visualization of the scanned data. Surveyors can immediately assess the quality and coverage of the data, allowing for on-the-fly adjustments and ensuring that critical areas are adequately captured, without the need for rework. While post-processing is required for TLS data, advancements in software and processing algorithms have significantly reduced the time needed to generate usable results. Traditional survey methods may involve longer data processing times, especially for large datasets. The technique used in this application allowed for a rapid assessment of registration precision on the field, eliminating the need to check the precision afterwards and, in case of mismatches, to redo the data acquisition process, thus further enhancing the time and cost efficiency.

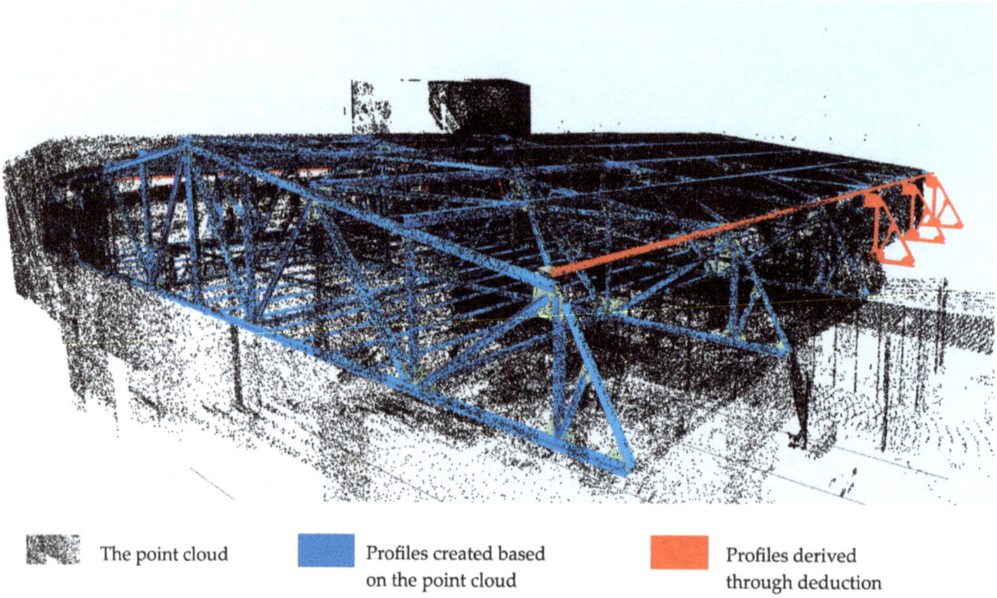

Figure 9. The black points represent the point cloud from the 3D scan. With blue are represented the steel truss elements that could be modeled from the point cloud, while with red are highlighted the elements that were modelled by deduction.

(**a**) (**b**)

Figure 10. Detailed area. (**a**) Representation of a metallic profile in cross-section from a 3D point cloud. (**b**) Overview photo of the spatial structure.

Due to the position of the scanning stations, the large cross-section of the profiles, and the 12 mm distance between them, the laser beam could not reach the space between the profiles, and therefore the shadowing effect occurred (Figure 10a), which led to the incorrect identification of the profiles in the first phase. The shadowing effect could have been eliminated by using a much larger number of scanning stations and by positioning some stations in line with the profile gaps, but this was not possible due to the roof structure configuration. At the same time, this strategy would have led to an increase in scanning costs and time, as well as to a significant increase in the point cloud. This would have made the modeling process more difficult, as it would have required more computing resources. Since the roof trusses are identical and there was safe access to one of these trusses, it was observed that, by manual surveying, the profiles could be correctly identified and that, therefore, the negative effects produced by shadowing could be compensated more efficiently with less resource consumption.

As there is access to only one of the roof trusses, surveying the other elements could not have been performed safely, as there is an increased risk of falling from a great height through the suspended plasterboard ceiling of the auditorium. By using TLS, it was possible to measure inaccessible areas safely. On the other hand, a small part of the roof structure could not be scanned using TLS because it was partially masked on the sides by walls delimiting some spaces that were too small for us to use the 3D scanning equipment on. The elements that could not be 3D scanned are highlighted in red in Figure 9. Following measurements using traditional techniques, it was found that those elements had the same dimensions and configuration as the elements that were 3D scanned and therefore could be easily modeled in the BIM software.

The existence of a 3D scan in the process of 3D modeling, particularly when using Autodesk Revit, has a significant impact on various aspects of the project. Below are some key areas where the impact is most notable:

Enhanced accuracy and detail: The 3D scanning process captures detailed and accurate measurements of the physical space or structure. When these data are imported into Revit, they provide a precise foundation for the modeling process, reducing the likelihood of errors that might occur when measurements are taken manually or estimated.

Time efficiency: Using 3D scans as a starting point in Revit accelerates the modeling process. It eliminates the initial phase of creating the basic structure from scratch, allowing designers to focus on refining and adding details to the already established base model [44].

Identification of complex elements: The scanned data help in identifying complex structural elements, especially in intricate areas like roof framing. This level of detail aids in creating more accurate and functional 3D models, as seen with the identification and 2D rendering of specific profiles and the delineation of unclear areas.

Challenges in data interpretation: While 3D scans provide detailed information, interpreting these data accurately in Revit can be challenging. As noted, ambiguities in the scan, such as blurred regions, can lead to misinterpretations in the model, such as confusing ventilation ducts with other structural elements. In 3D modeling, a notable challenge encountered was the differentiation of the profiles; specifically, two L-shaped profiles next to each other looked like T-shaped ones in many portions of the scan for reasons that are easy to understand. In the sections created through the 3D cloud for profiles, they appeared as in the image from Figure 10a. This ambiguity in profile classification precluded the effective use of beam tools in the modeling process. This ambiguity led to errors regarding areas generated solely based on the 3D model derived from the scanning. During the modeling process, it was noted the existence of L-shaped profiles, which were modeled in a later phase, and areas were corrected.

Data limitations: In order to obtain optimal data, the profiles had to be measured and verified manually, in situ. Photographic information is very useful for identifying different materials, but it should also be focused on the details if physical presence in situ of the modeler is not possible. Figure 10b exemplifies photographic information from which no details of the structure can be extracted; thus, profiles generated from the 3D scan, as in Figure 10a, remain unresolved.

Collaboration enhancement: The ability to export the model in IFC and DWG formats from Revit enhances collaboration. These formats are widely accepted and enable different stakeholders, even those using different software, to access, review, and collaborate on the project. The original 3D model made in Revit could be imported into ArchiCAD, using the .ifc format, without any problems. Thus, the 3D model could be easily corrected in a few hours in another program, by another person, because a large part of the 3D model could be kept, and only the elements that were initially wrongly identified were replaced.

Quantitative analysis and reporting: The precision of 3D scans can streamline quantitative analyses, such as quantity takeoffs. However, ambiguities in the scan can lead to errors in these reports, emphasizing the need for careful review and interpretation of the scanned data.

Automatic, fast, and accurate area calculation: Because BIM software has certain functions that automatically calculate the areas of different elements, it helped us to obtain the total areas in a few minutes after the completion of the modeling process. This aspect demonstrates that the use of BIM contributes significantly to easing the process of obtaining steel truss cover areas, as well as to obtaining more accurate bills of quantities. The fact that there is an insignificant difference of only 2.63 sqm (0.39%) between the area calculated automatically in the BIM software (667.02 sqm) and the area calculated manually (669.65 sqm) demonstrates that, by using BIM software, it is possible to obtain very accurate truss coverage areas automatically.

Miscalculation of the coverage areas of steel fireproofing trusses can produce waste in the following ways:

- When the calculated coverage area is larger than the actual coverage area, physical waste will result because excess material will be ordered and finally disposed of in the landfill.
- When the calculated coverage area is less than the actual coverage area, as could have happened in the case of the study presented in this paper, non-physical waste will result because not enough material is ordered, and therefore the budget allocated for fireproofing would not be sufficient. For public investments, the approval of the additional budget needed to complete the fireproofing works would have taken a long time and would have led to delays in the fireproofing works and ultimately to cost and time overruns.

- The steel truss elements described in this paper require a single coat of intumescent paint because the maximum coating thickness is 0.47 mm. It should be mentioned that, under certain conditions, depending on the cross-section of the profiles, fire resistance, and design temperature, the coating thickness may exceed 6 mm, and this involves applying the coating in multiple layers. Thus, the multiplication of coating layers leads to a multiplication of the waste produced.

The initial 3D model made only after the 3D scan was incorrect mainly because the information extracted from the point cloud was not sufficient to understand the composite profile configuration of the steel trusses because of the shadowing effect. In order to eliminate the risk of major errors it is necessary that the surveying of trusses or other complex steel structures be carried out using multiple surveying techniques and that the modeling process be informed by additional data (hand sketches, photographs, and video). For this purpose, the survey, 3D modeling, and calculation of the cover areas of steel trusses or steel structures that are made of composite profiles should be carried out according to the methodology shown in Figure 11. This methodology involves all the data collected from the site, whether collected using 3D scanning or traditional techniques, in order to inform the process of producing a single 3D model.

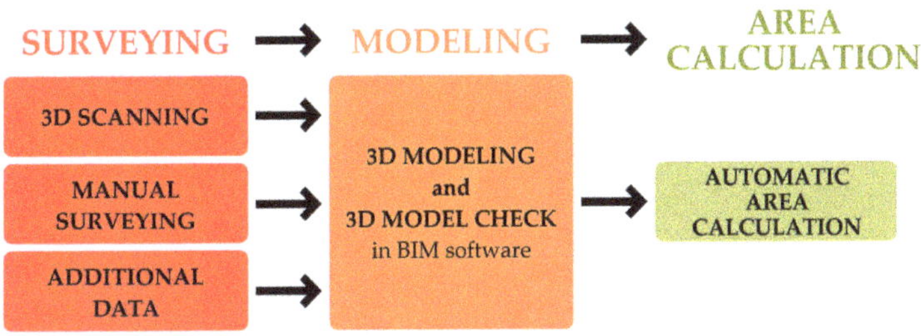

Figure 11. Graphic illustration of the proposed methodology for quantity surveying using 3D scanning, traditional surveying techniques, and additional data (hand sketches, photos, and videos).

5. Conclusions

The Aula Magna Hall at the University of Oradea, a key feature of the "SMART Campus—University of Oradea" project, showcases the university's dedication to preserving heritage while embracing modernization and technological advancement.

The present project offers several pertinent conclusions and recommendations regarding the interpretation of point cloud scanning and its modeling using BIM software, such as Revit or ArchiCAD, in a harsh and highly complex environment. The main challenge that was identified for the automatic 3D modeling was the shadowing effect, which was caused by the fact that the laser wave cannot reach the narrow space between the profiles, and this can lead to the misinterpretation of the different sections of elements that have a cross-section composed of several profiles. The proposed solution recommends using multiple scanning stations and a combination of data acquisition methods, including traditional techniques and photography, overcoming the limitations of TLS scanning and generating precise and comprehensive 3D models for a better assessment of the construction site. Understanding the configuration of the steel truss is important, requiring site visits or visual materials for modelers in the absence of direct access. Calculation errors in composite trusses can lead to an underestimation of the materials needed for fireproofing, especially when multiple layers of intumescent paint are required. Thus, BIM programs become essential for efficient data collection and error prevention that could lead to waste.

While most of the research conducted with the aim of proving that the 3D scanning and BIM software's have the ability to reduce the physical waste [45–48], the innovative point in this research unveiled that, by using a mixture of survey techniques, including traditional survey and 3D scanning systems, we are able to avoid a more undesirable type of waste—the non-physical waste, which will led to a time-consuming bureaucratic process to obtain the needed additional budget for the fireproofing process, generating cost and time overruns.

While other research studies are related to different types of trusses, such as wood [49], this research innovates the way that the 3D point cloud and 3D modeling of steel trusses should be forged in order to automatically generate accurate coverage areas and to reduce the influence of blunders with the purpose of avoiding physical and/or non-physical waste. Hence, this method represents a significant advancement in optimizing 3D modeling in the architecture, engineering, and construction (AEC) sector.

Even though the initial automated modeling was generated inaccurately, this model created the necessary premises for verification of complex structural elements to be correctly identified. Thus, it resulted in the development of an innovative and accurate methodology so that a multidisciplinary team composed of surveyors, architects, and civil engineers are able the enhance their collaboration to generate an accurate and sustainable 3D model that is usable for multiple and highly complex analyses.

The research in this paper demonstrates that the use of 3D scanning and BIM contributes significantly to the reduction of non-physical waste that can result in the fireproofing process of steel trusses and helps yield a more precise budget, while taking into account the necessary recommendations presented in the article.

Author Contributions: Conceptualization, C.S., A.-H.P., N.-S.S. and S.N.; methodology, C.S., A.-H.P., I.-G.Z., A.-M.D., N.-S.S. and S.N.; software, A.-H.P., I.-G.Z., A.-P.F., A.S.B. and S.N.; validation, C.S., A.-P.F. and A.S.B.; formal analysis, C.S., A.-H.P., I.-G.Z., A.-M.D. and S.N.; investigation, C.S., A.-H.P., I.-G.Z., A.-M.D., A.-P.F., N.-S.S., A.S.B. and S.N.; resources, A.S.B.; data curation, N.-S.S.; writing—original draft preparation, C.S., A.-H.P., I.-G.Z., A.-M.D., A.-P.F., N.-S.S., A.S.B. and S.N.; writing—review and editing, C.S., N.-S.S., I.-G.Z. and S.N.; visualization, A.-P.F.; supervision, C.S., A.-H.P. and N.-S.S.; project administration, A.-H.P., I.-G.Z., A.-M.D. and A.-P.F.; funding acquisition, A.S.B. and S.N. All authors have read and agreed to the published version of the manuscript.

Funding: This research was funded by the University of Oradea.

Institutional Review Board Statement: Not applicable.

Informed Consent Statement: Not applicable.

Data Availability Statement: The data presented in this study are available on request from the corresponding author. The data are not publicly available due to privacy restriction.

Acknowledgments: We gratefully acknowledge the anonymous reviewers and the editor for their thoughtful comments and suggestions, which contributed significantly to improving the quality of the manuscript. Also, we want to thank the local Trimble distributor, Liviu Gherman, and Sergiu Ceausu for all their support.

Conflicts of Interest: The authors declare no conflict of interest.

References

1. McKinsey Company. The Construction Productivity Imperative. Available online: https://www.mckinsey.com/capabilities/operations/our-insights/the-construction-productivity-imperative (accessed on 11 January 2024).
2. Coates, S. BIM Implementation Strategy Framework for Small Architectural Practices. 2020. Available online: https://salford-repository.worktribe.com/output/1429393 (accessed on 11 January 2024).
3. Aziz, R.F.; Hafez, S.M. Applying lean thinking in construction and performance improvement. *Alex. Eng. J.* **2013**, *52*, 679–695. [CrossRef]
4. Nagapan, S.; Rahman, I.A.; Asmi, A. Factors contributing to physical and non-physical waste generation in construction industry. *Int. J. Adv. Appl. Sci.* **2012**, *1*, 1–10. [CrossRef]
5. Nguyen, T.A.; Nguyen, P.T.; Do, S.T. Application of BIM and 3D Laser Scanning for Quantity Management in Construction Projects. *Adv. Civ. Eng.* **2020**, *2020*, 8839923. [CrossRef]
6. Lucherini, A.; Maluk, C. Intumescent coatings used for the fire-safe design of steel structures: A review. *J. Constr. Steel Res.* **2019**, *162*, 105712. [CrossRef]
7. Wang, K.; Le, H. The Development of Cement-Based, Intumescent and Geopolymer Fire-Retardation Coatings for Metal Structures: A Review. *Coatings* **2023**, *13*, 495. [CrossRef]

8. Inerhunwa, I.; Wang, Y.C.; Su, M. Reliability analysis of intumescent coating protected steel members under the standard fire condition. *Fire Saf. J.* **2019**, *104*, 43–56. [CrossRef]
9. Waste Statistics. Available online: https://ec.europa.eu/eurostat/statistics-explained/index.php?title=Waste_statistics#Total_waste_generation (accessed on 11 July 2023).
10. SmartWaste. Available online: https://www.bresmartsite.com/sustainable-construction-methods-2/ (accessed on 14 July 2023).
11. Plebankiewicz, E.; Wieczorek, D. Prediction of Cost Overrun Risk in Construction Projects. *Sustainability* **2020**, *12*, 9341. [CrossRef]
12. Keung, C.C.; Yiu, T.W.; Feng, Z. Building Information Modelling in quantity surveying practices: Current state and future challenges in Hong Kong. *J. Inf. Technol. Constr.* **2022**, *27*, 1060–1074. [CrossRef]
13. Harmening, C.; Neuner, H. A spatio-temporal deformation model for laser scanning point clouds. *J. Geod.* **2020**, *94*, 26. [CrossRef]
14. Skrzypczak, I.; Oleniacz, G.; Leśniak, A.; Zima, K.; Mrówczyńska, M.; Kazak, J.K. Scan-to-BIM method in construction: Assessment of the 3D buildings model accuracy in terms inventory measurements. *Build. Res. Inf.* **2022**, *50*, 859–880. [CrossRef]
15. Trembecka, A.; Droj, G.; Kwartnik-Pruc, A. Management of Municipal Housing Resources By Local Authorities in the Municipalities of Krakow (Poland) and Oradea (Romania)—A Comparative Study. *Real Estate Manag. Valuat.* **2023**, *31*, 23–35. [CrossRef]
16. Droj, G.; Droj, L.; Badea, A.-C.; Dragomir, P.I. GIS-Based Urban Traffic Assessment in a Historical European City under the Influence of Infrastructure Works and COVID-19. *Appl. Sci.* **2023**, *13*, 1355. [CrossRef]
17. Kalyan, T.S.; Zadeh, P.A.; Staub-French, S.; Froese, T.M. Construction Quality Assessment Using 3D as-built Models Generated with Project Tango. *Procedia Eng.* **2016**, *145*, 1416–1423. [CrossRef]
18. Apollo, M.; Jakubiak, M.; Nistor, S.; Lewińska, P.; Krawczyk, A.; Borowski, L.; Specht, M.; Krzykowska-Piotrowska, K.; Marchel, Ł.; Pęska-Siwik, A.; et al. Geodata in science—A review of selected scientific fields. *Acta Sci. Polonorum. Form. Circumiectus* **2023**, *22*, 17–40. [CrossRef]
19. Lin, Y.-C.; Chen, Y.-P.; Huang, W.-T.; Hong, C.-C. Development of BIM Execution Plan for BIM Model Management during the Pre-Operation Phase: A Case Study. *Buildings* **2016**, *6*, 8. [CrossRef]
20. Billi, D.; Croce, V.; Bevilacqua, M.G.; Caroti, G.; Pasqualetti, A.; Piemonte, A.; Russo, M. Machine Learning and Deep Learning for the Built Heritage Analysis: Laser Scanning and UAV-Based Surveying Applications on a Complex Spatial Grid Structure. *Remote Sens.* **2023**, *15*, 1961. [CrossRef]
21. Russo, M.; Russo, V. Geometric analysis of a space grid structure by an integrated 3D survey approach. *Int. Arch. Photogramm. Remote Sens. Spat. Inf. Sci.* **2022**, *46*, 465–472. [CrossRef]
22. Almukhtar, A.; Saeed, Z.O.; Abanda, H.; Tah, J.H.M. Reality Capture of Buildings Using 3D Laser Scanners. *CivilEng* **2021**, *2*, 214–235. [CrossRef]
23. Ham, Y.; Golparvar-Fard, M. Mapping actual thermal properties to building elements in gbXML-based BIM for reliable building energy performance modeling. *Autom. Constr.* **2015**, *49*, 214–224. [CrossRef]
24. Sanhudo, L.; Ramos, N.M.M.; Martins, J.P.; Almeida, R.M.S.F.; Barreira, E.; Simões, M.L.; Cardoso, V. A framework for in-situ geometric data acquisition using laser scanning for BIM modelling. *J. Build. Eng.* **2020**, *28*, 101073. [CrossRef]
25. Bieda, A.; Balawejder, M.; Warchoł, A.; Bydłosz, J.; Kolodiy, P.; Pukanska, K. Use of 3D technology in underground tourism: Example of Rzeszow (Poland) and Lviv (Ukraine). *Acta Montan. Slovaca* **2021**, *26*, 205–221.
26. Sacks, R.; Eastman, C.; Lee, G.; Teicholz, P.; BIM Handbook. *A Guide to Building Information Modeling for Owners, Designers, Engineers, Contractors, and Facility Managers*, 3rd ed.; John Wiley & Sons, Inc.: Hoboken, NJ, USA, 2018; p. 688.
27. Bieda, A.; Bydłosz, J.; Parzych, P.; Pukanská, K.; Wójciak, E. 3D technologies as the future of spatial planning: The example of Krakow. *Geomat. Environ. Eng.* **2020**, *14*, 15–33. [CrossRef]
28. Hamma-adama, M.; Kouider, T. Comparative Analysis of BIM Adoption Efforts by Developed Countries as Precedent for New Adopter Countries. *Curr. J. Appl. Sci. Technol.* **2019**, *36*, 1–15. [CrossRef]
29. Zhan, Z.; Tang, Y.; Wang, C.; Yap, J.B.H.; Lim, Y.S. System Dynamics Outlook on BIM and LEAN Interaction in Construction Quantity Surveying. *Iran. J. Sci. Technol. Trans. Civ. Eng.* **2022**, *46*, 3947–3962. [CrossRef]
30. Olsen, D.; Taylor, J.M. Quantity Take-Off Using Building Information Modeling (BIM), and Its Limiting Factors. *Procedia Eng.* **2017**, *196*, 1098–1105. [CrossRef]
31. Uusitalo, P.; Seppänen, O.; Lappalainen, E.; Peltokorpi, A.; Olivieri, H. Applying Level of Detail in a BIM-Based Project: An Overall Process for Lean Design Management. *Buildings* **2019**, *9*, 109. [CrossRef]
32. Marzouk, M.; Thabet, R. A BIM-Based Tool for Assessing Sustainability in Buildings Using the Green Pyramid Rating System. *Buildings* **2023**, *13*, 1274. [CrossRef]
33. Franco, J.; Mahdi, F.; Abaza, H. Using building information modeling (BIM) for estimating and scheduling, adoption barriers. *Univers. J. Manag.* **2015**, *3*, 376–384. [CrossRef]
34. Doumbouya, L.; Gao, G.; Guan, C. Adoption of the Building Information Modeling (BIM) for construction project effectiveness: The review of BIM benefits. *Am. J. Civ. Eng. Archit.* **2016**, *4*, 74–79.
35. Khosakitchalert, C.; Yabuki, N.; Fukuda, T. Improving the accuracy of BIM-based quantity takeoff for compound elements. *Autom. Constr.* **2019**, *106*, 102891. [CrossRef]
36. Paşca, M. *Oradea 1900: Un Ghid de Arhitectură*; Argonaut: Cluj-Napoca, Romania, 2019.
37. Buda, A.S.; Frenț, A.-P.; Nistor, S.; Suba, N.-S. Possibilities of Registering Natural Underground Cavities into the National Integrated Information System of Cadastre and Land Registry. *J. Appl. Eng. Sci.* **2023**, *13*, 53–56. [CrossRef]

38. Herman, G.V.; Caciora, T.; Ilies, D.C.; Ilies, A.; Deac, A.; Sturza, A.; Sonko, S.M.; Suba, N.S.; Nistor, S. 3D Modeling of the Cultural Heritage: Between Opportunity and Necessity. *J. Appl. Eng. Sci.* **2020**, *10*, 27–30. [CrossRef]
39. Mukupa, W.; Roberts, G.W.; Hancock, C.M.; Al-Manasir, K. A review of the use of terrestrial laser scanning application for change detection and deformation monitoring of structures. *Surv. Rev.* **2017**, *49*, 99–116. [CrossRef]
40. Gordon, S.J.; Lichti, D.D. Modeling Terrestrial Laser Scanner Data for Precise Structural Deformation Measurement. *J. Surv. Eng.* **2007**, *133*, 72–80. [CrossRef]
41. Trimble X7. Available online: https://geospatial.trimble.com/en/products/hardware/trimble-x7 (accessed on 4 September 2023).
42. *IEC 60825-1:2014*; Safety of Laser Products—Part 1: Equipment Classification and Requirements. International Electrotechnical Commission: Geneva, Switzerland, 2014.
43. Danial, C.E.; Mahmoud, A.H.A.; Tawfik, M.Y. Methodology for retrofitting energy in existing office buildings using building information modelling programs. *Ain Shams Eng. J.* **2023**, *14*, 102175. [CrossRef]
44. Syed, E.U.; Manzoor, K.M. Analysis and design of buildings using Revit and ETABS software. *Mater. Today Proc.* **2022**, *65*, 1478–1485. [CrossRef]
45. Guerra, B.C.; Leite, F.; Faust, K.M. 4D-BIM to enhance construction waste reuse and recycle planning: Case studies on concrete and drywall waste streams. *Waste Manag.* **2020**, *116*, 79–90. [CrossRef]
46. Won, J.; Cheng, J.C.P. Identifying potential opportunities of building information modeling for construction and demolition waste management and minimization. *Autom. Constr.* **2017**, *79*, 3–18. [CrossRef]
47. Hu, X.; Zhou, Y.; Vanhullebusch, S.; Mestdagh, R.; Cui, Z.; Li, J. Smart building demolition and waste management frame with image-to-BIM. *J. Build. Eng.* **2022**, *49*, 104058. [CrossRef]
48. Han, D.; Kalantari, M.; Rajabifard, A. Building Information Modeling (BIM) for Construction and Demolition Waste Management in Australia: A Research Agenda. *Sustainability* **2021**, *13*, 12983. [CrossRef]
49. Massafra, A.; Prati, D.; Predari, G.; Gulli, R. Wooden Truss Analysis, Preservation Strategies, and Digital Documentation through Parametric 3D Modeling and HBIM Workflow. *Sustainability* **2020**, *12*, 4975. [CrossRef]

sustainability MDPI

Article

Intervention Works Conducted to Ensure the Stability of a Slope: A Sustainability Study

Mircea Raul Tudorica [1,2,*] and Corneliu Ioan Bob [3]

1 Department of Civil Engineering, Faculty of Constructions, Cadaster and Architecture, University of Oradea, 410058 Oradea, Romania
2 Doctoral School of Engineering Sciences, "Polytechnic" University of Timisoara, 300006 Timisoara, Romania
3 Department of Civil Engineering and Installations, Faculty of Civil Engineering, "Polytechnic" University of Timisoara, 300233 Timisoara, Romania; corneliu.bob@gmail.com
* Correspondence: tudorica.mircea@gmail.com

Abstract: Challenges related to sustainability arise in all areas of human activity, but with a significant impact on the environment considering that the construction industry is held accountable for nearly one-third of the world's final energy consumption. The aim of this paper is to assess through the use of the Bob–Dencsak specific model a sustainable slope design taking into account environmental, economic, and safety variables. Thus, analysis was performed on four intervention works, two versions of reinforced concrete retaining walls and two versions of reinforced soil with a biaxial geogrid, which ensure the stability of a slope that serves as a base for an access road to an ecological landfill located in Alba County, Romania. The study's analysis points out that reinforced soil retaining walls are far more sustainable, providing the best sustainability indices, which is also supported by the impact of geogrids compared to reinforced concrete, thus resulting in the finding that reinforced concrete is less sustainable, achieving increases of up to 23% for embodied energy and 66% of CO_2 emissions in the atmosphere. Finally, the paper provides recommendations for future research on the sustainability assessment of slopes, with the intention of reducing environmental damage, while keeping costs to a minimum.

Keywords: sustainability; safety factor; embodied energy; GHG gas emissions; retaining wall; reinforced concrete; environmental protection; geogrids; soil improvement

Citation: Tudorica, M.R.; Bob, C.I. Intervention Works Conducted to Ensure the Stability of a Slope: A Sustainability Study. *Sustainability* **2024**, *16*, 1544. https://doi.org/10.3390/su16041544

Academic Editor: Jianjun Ma

Received: 31 December 2023
Revised: 2 February 2024
Accepted: 10 February 2024
Published: 12 February 2024

1. Introduction

The contextual meaning of the frequently used definition of sustainable development appears to be "The development that meets the needs of the present without compromising the ability of future generations to meet their own needs" [1,2]. The notion of sustainability, encompassing environmental, economic, and social aspects, is a significant topic in many different industries. The construction industry, as one of the most important industries in both developed and developing countries, has had a significant impact on various aspects of the environment, economy, and society. Sustainable construction has been introduced as a technique to assess the different stages of construction in recent years, in terms of social, economic, and environmental dimensions, also referred to as the three bottom lines (TBL) [3]. Energy demand and consumption have increased rapidly in recent years as a result of humankind's ever-increasing needs in the economy, industries, and agriculture [4]. Currently, a substantial amount of energy and material consumption is attributable to the construction industry [5]. As a consequence, the European Union has developed an interest in construction and its energy efficiency through innovative approaches, concentrating on present and future trends and concerns. Europe's priorities for the upcoming years until 2050 will be the decarbonization of the building industry and climate neutrality. This shift is highly intriguing as well, and it appears that a true revolution has started in this direction [6–8]. In response to the expanding demand for both urban and rural infrastructure development, more investigations are necessary to ensure the safety of numerous civil constructions, especially embankments and highways [9]. The life cycle assessment (LCA) method is currently used worldwide by many nations and organizations to assess and examine the energy use and environmental impact of diverse projects [10]. This approach was initially focused primarily on the fields of building construction, and relevant practices and research on roads and railways, as well as the layers beneath them, have only started in the last ten years [11,12]. Using life cycle assessment (LCA), Chang et al. investigated the distribution of carbon emissions in a segment of the California high-speed railway [13]. Improving the life cycle assessment approach by including the cost, Chan A. made a comparative analysis of three types of road surfaces: new, reconstruction, and regeneration, where he found that while a cement road surface has the highest greenhouse gas emissions, it consumes less energy than an asphalt road surface [14].

Because geotechnical engineering acts as an integrator, bringing together different civil engineering sub-disciplines, its significance has grown over time. In the field of geotechnical engineering, landslides are a major problem that can cause catastrophic consequences like infrastructure damage and potential casualties. Apart from these, there are also those which have suffered severe functional and structural damage as a result of earthquakes, phenomenon that has become increasingly common in recent years around the world [15]. Any mitigation plan for this issue, which impedes development efforts, must start with an assessment of these geologic risks [16]. Masses of rocks, soil material, or muddy flows that slide down a slope due to modifications in the slope's natural stability are known as landslides [17]. The slope stability behavior is influenced both by internal factors, such as the physical–mechanical properties of the material like friction angle and cohesion, as well as by external factors, where we can include the amount of rainfall and seismic activity. In this direction, Saptono and Rezky conducted a case study in Southeast Sulawesi, Indonesia, to analyze the sensitivity of embankment slopes, using the coefficient of variation (CV) approach, mainly following the most important physical parameters, namely the internal friction angle and cohesion. The study's findings provide data showing that the internal friction angle has the greatest impact on the stability of embankment slopes and highlights the fact that for the highest value of the variation coefficient CV, there is a serious risk of producing an avalanche [18]. Another relevant case study for geotechnical engineering was conducted in Algeria, where Boubazine et al. investigated the occurrence of landslides in the Tarzoust region, based on geophysical approaches. Using Vertical Electric Soundings (VES) and the Seismic Refraction Method (SRM) for underground exploration, as well as Electrical Resistivity Tomography (ERT), it was proven that the combination of geological and geotechnical data with geophysical deterministic methods can help engineers and decision-makers in land management. In addition, this approach offers recommendations consisting of topographical, inclinometric, and piezometric monitoring to track landslides and the effectiveness of soil reinforcement measures [19]. In areas that are impacted by this phenomenon, preventing social, economic, and sustainable vulnerabilities requires an efficient and secure slope stabilization execution. Therefore, due to the complexity of the slope stability analysis, but also due to the lack of research in this field, more thorough analyses are required to measure several different parameters such as the ground's volumetric weight, the elasticity modulus of the earth's geological layers, or the slope's geometrical configuration, together with the addition of drains, vertical columns, retaining walls, and reinforcements [20].

The importance of the landslide-related effects of climate change was highlighted in this context by policymakers, scientists, designers, and engineers. Our society has new opportunities for dealing with the global energy crisis through the sustainable design of a large-scale civil engineering project like slope stability and landslide management. These resources provide a practical response to the environmental problems and the world's energy requirements. Even though slope stability is essential for maintaining public safety and protecting the infrastructure, it frequently has disastrous results, underlining the importance of creating long-lasting and efficient methods to reduce the risks related to landslides [21]. In the meantime, the environment is unintentionally destroyed by construction activities, leading to the formation of numerous engineering slopes [22]. In addition to causing landslides and other natural disasters, these can also have an impact on the effectiveness and safety of constructions. Furthermore, as a result of the digging operation, a significant amount of soil subdivisions could migrate to the topsoil, reducing biodiversity, upsetting the ecological balance, and negatively impacting the long-term development of the local economy [23–25]. In this manner, Shen et al. describe the technologies that are frequently used in China, which combine soil improvement with bioremediation procedures. Even if the ecological restoration process is now highly mechanized, there are various problems that need to be studied further such as ecological restoration plans which are not designed with the local geographic conditions, the assessment of the ecological restoration sometimes being unclear due to a lack of quantitative data, restored slopes not being adequately monitored over the long term, and their environmental protection being ignored occasionally in the construction sector in an effort to increase profits. The authors also include a summary of the advantages and an assessment of their social impact [26]. Lastly, environmental restoration can help to improve the ecosystem and biodiversity's ability to function throughout addition to reducing landslides, soil erosion, and other local geological natural disasters [27,28]. A frequent situation is represented by the existence of expansive soils with a high concentration of hydrophilic mineral parts, like illite and montmorillonite, which are extremely sensitive to changes in water content. Their volume also fluctuates as a result of the variation in water quantity [29,30]. Since the expansive soil slopes are constantly expanding and contracting due to the action of the wet–dry cycles, cracks will develop on their surface. The resilience of slope soil will be reduced as more rainfall filters in, resulting in the development of shallow slope collapses [31,32]. Maintaining the long-term equilibrium of an expansive soil requires a slope support system that is highly sustainable [33]. To increase the flexible support structure's capacity to withstand, recover after, and react to the collapse of an expansive soil slope, systematic analysis and investigations are required to determine its structural stability in the future. Zhang et al. concluded that according to the life cycle evaluation analysis, the flexible support system uses approximately 50% less resources and energy and emits 10 times less carbon than the rigid support system based on the results of their research in this area [34]. Frischknecht et al. conducted an environmental assessment of the two types of retaining walls and compared the principles of the reinforced concrete retaining walls and those strengthened using geosynthetic materials. The analysis was performed on a slope of 3 m high and 1 m wide, which revealed that the environmental effect of the slope could be decreased by using geosynthetic-reinforced retaining [35].

In the specialized literature specific to the field of civil engineering, there are numerous standards dealing with structural safety, as well as various studies and models for assessing their sustainability, most of which are specific to new constructions and are developed during the design phase. Achieving the desired level of ensuring a sustainable environment on a global scale requires taking the best decisions to protect the environment through the rational and productive use of economic resources, all while meeting society's current needs without affecting future generations who will benefit from them directly. In this sense, the researchers' attention must be directed to the old existing structures, which may or may not present some structural, aesthetic, or energetic vulnerabilities due to the age of the materials and equipment used, in order to meet the needs of the present in terms of their safety and exploitation. However, special attention needs to be given to the land under the structure in question, both in the case of new or old structures, known in geotechnical engineering as the foundation soil. This natural resource is indispensable for both structures, being the most ancient building material, to which the choice of its resistance and stability characteristics is not an option, only their improvement through various mechanical or chemical technological processes but which inevitably result in higher costs. In the last decade, with the understanding worldwide that it is vital in all fields of activity to find and apply effective solutions to reduce emissions, the concept has been extended to the total elimination of embodied energy consumption and greenhouse gas emissions into the atmosphere resulting from the consumption of building materials across all industries, starting with tracking the manufacturing process, transportation, and the equipment used for putting them into operation, and real interest has started to appear for the sustainability study of the soil layers in the construction field for railways, tunnels, dams, roads, and highways. Nevertheless, there is still a big gap in the specialized literature that directly targets the sustainability of soil foundations and the possible intervention works that must be conducted on them.

This paper carries out a sustainability investigation focused on four intervention works to ensure the stability of a slope. The analysis was carried out applying the Bob–Dencsak specific model, which presents a series of advantages such as the method's focus on all three factors associated with sustainability, having a wide range of applications and consisting only of quantifiable parameters. The main purpose is to compare different solutions, in order to determine which is most efficient from a sustainable perspective. Thus, two intervention works have been explored which involved soil reinforcement with a geogrid in the configuration of the slope 2:3 and 1:1, where the sustainability index was obtained as $SI_1 = 0.920$, for the first mentioned case, and $SI_3 = 0.951$ for the second case, and another two intervention works of reinforced concrete: a retaining wall with a height of 2.50 m situated at the base of the slope, for which a sustainability index $SI_2 = 0.779$ was obtained; and the second retaining wall of 6.40 m in height, which shows the lowest value of the sustainability index at $SI_4 = 0.573$. Following the final values in the slope sustainability analysis, we can assert that the reinforced soil retaining walls obtained the highest sustainability scores, being much more sustainable than the ones using reinforced concrete. This can be highlighted by comparing the reinforced soil configuration with a slope of 2:3 and that of the 2.5 m reinforced concrete retaining wall, where approximately the same amount of filling material was used. As a consequence, it turned out that the reinforced concrete's embodied energy is only 2.69 times, while the GHG gas emissions are 7.40 times higher than those generated by geogrids, resulting in an 18% more sustainable solution than the version with a reinforced concrete retaining wall.

2. Materials and Methods

2.1. The Assessement of the Slope Safety Factor

The safety factor F_S of a slope is defined as the "ratio between the actual soil's shear strength value and the lowest possible shear strength value needed to avoid failure" or the rate that must be decreased in soil shear strength to push a slope toward collapse [36].

$$F_S = \frac{\tau_f}{\tau} = \frac{\sigma \times tg\varphi + c}{\sigma \times tg\varphi_m + c_m} \tag{1}$$

where F_S signifies the slope stability factor, τ_f signifies the ground's available shear strength, τ signifies the required or mobilized shear strength, σ signifies the normal stress, φ, c signifies the soil's shear properties, φ_m, c_m signifies the mobilized shear characteristics, φ signifies the soil particle's coefficient of frictions, and c signifies the soil's cohesiveness.

However, the development of methods focused on the stability of the slope surface has been restricted by the lack of information on soil shear strength characteristics and their relationships with other soil properties [37]. The soil's shear capacity is the highest level of shear stresses that soil can withstand without collapsing, and it is determined based on the characteristics of cohesiveness and the internal friction angle between soil particles. Even though most modern technologies are used worldwide, it is impossible to guarantee slope safety in every situation. Furthermore, design and sizing standards have been established. These align strength and effectiveness guarantees with "safety" margins regarded as "comfortable" by experts in the field. Methods based on the concept of ultimate equilibrium assume a known sliding surface (real or possible) and admit $F_s = 1$ over the entire sliding surface. These methods are not based on a mathematical foundation, and their biggest deficiency is that it presumes an incipient failure. The method is used even in the case of stable slopes, with $F_s > 1$, which leads to situations that obviously do not reflect the reality in the field [38].

The limit equilibrium techniques that Fellenius introduced in 1927 have resulted in significant improvements, which presume that resistance follows Coulomb's formula along the sliding line, splits the sliding soil volume included within the circular arc into slices, and assesses its equilibrium by reducing the forces and moments to zero. Since then, comparable approaches have also been established, which covers Janbu in 1954, Bishop in 1955, Simplified Bishop in 1960, Morgenstern and Price in 1965, Spencer in 1967, Simplified Janbu in 1973, and Sarma in 1973 [39]. Slope failure is far more complicated than the limit equilibrium approach has been able to simulate. In reality, failure does not occur concurrently along a single distinct normal surface, but rather a localized failure gradually expands over a larger failure surface. With the exception of strictly structural slope failures, like those governed by a discontinuity in a delicate rock mass, internal deformation is a crucial factor in the progression of these failures [40].

Methods using finite elements, abbreviated as FEM, are crucial for resolving stress–strain issues, especially in situations involving the interaction of soil–structure and slope stability [41]. In FEM, the structure and performance of geotechnical materials is analyzed by an elastoplastic simulation based on the Mohr–Coulomb failure criteria. It makes sense, therefore, that its application in the context of civil works safety be taken into account. Nevertheless, the analysis is not direct due to the distinctive nature of slope issues under unsaturated conditions in which suction has a major impact, where extra care must be taken to accurately replicate certain details [42]. The basic procedures in FEM include the discretization process, choosing approximations for functions, equation derivation, collecting element properties to form universal equations, and primary quantity calculation (e.g., displacements) and secondary calculation (e.g., stresses). Discretization is the process of breaking down a continuous material into a system of comparable small individual components (also known as finite elements), where each element is examined and handled separately. Physical properties or constitutive characteristics are assigned to each element, and matrices for the assembly's rigidity are generated [43]. FEM is a numerical method for estimating limit value solutions for a variety of partial differential equations. Theoretically, it fulfills every prerequisite needed for a comprehensive resolution of a slope stability issue [44].

2.2. *Sustainability Assessment Models*

Establishing sustainability performance can be carried out using a variety of models. Among the most widely recognized that are always being developed are the Building Research Establishment Environmental Assessment Method (BREEAM) and the Leadership in Energy and Environmental Design (LEED) [45].

The evolving and re-scoping of an understanding of what constitutes sustainable construction is reflected in the progress and constant improvement of different performance rating systems [46,47]. Worldwide, several types of complex models for determining a building's sustainability are offered in the specialized literature. They include a range of parameters from multiple perspectives that impact the sustainability research. The total number of parameters for every dimension, the importance of each dimension's proportion in the final result, and how they are classified from a sustainability perspective are shown in Table 1 [48].

Table 1. Establishing classification from the perspective of sustainability in construction.

Sustainability Model	Ecological Dimension	Economic Dimension	Social Dimension	Construction Classification
UK 1990 (59) BREEAM	(100%) 59	-	-	Insufficient, points <30 Good enough, points 30–85 Very good, points >85
USA 1993 (57) LEED	(100%) 57	-	-	Bronze, points 40–49 Silver, points 50–59 Gold, points 60–79 Platinum, points >85
Japan 2001 (80) CASBEE	(70%) 56	-	(30%) 24	C Class, grades <0.5 B- Class, grades 0.5–1 B+ Class, grades 1–1.5 A Class, grades 1.5–3 S Class, grades >3
International 1996 (14–122) SBTool Model	48%	24%	24%	Acceptable, score <1 Good, score 1–3 Excellent, score >3
CEN TC350 (51)	(33.3%) 16	(33.3%) 17	(33.3%) 18	A score of 100 points is the maximum. The classification based on the obtained score
Romania 2010 (45) Bob–Dencsak	(40%) 21	(30%) 11	(30%) 13	Very good, points >80 (>4) Good, points 60–80 (3–4) Acceptable, points 40–60 (2–3) Insufficient, points <40 (<2)

In many cases, these global models indicate certain disadvantages:

- The models do not take into account all three aspects of sustainability;
- They have a large number of parameters, some of which are difficult or impossible to quantify;
- The instruments are primarily designed for complete buildings, though they can be used, albeit with some difficulty for other kinds of construction projects and tasks.

To avoid the previously mentioned disadvantages which characterize global sustainability models, some specific models were proposed and applied, with the purpose of serving engineers in assessing the sustainability of certain particular construction works. The most significant advantages of these specific models are:

- They deal with all three aspects of sustainability;
- A wide range of application;
- They consist only of quantifiable parameters.

The main purpose of the specific models is to compare different solutions, in order to determine which is the most efficient from a sustainable point of view. Thus, a relative value is obtained for each solution, which is compared to the ideal value [49]. A similar approach has been proposed by Ding [50] and Diaz–Balteiro and Romero [51], but there are certain difficulties with using the models.

2.3. *Bob-Dencsak Specific Sustainability Model*

Based on fundamental mathematical formulas, the specific model logically combines the results of the parameters that were quantified to achieve a sustainability index SI.

$$SI = S_{env} + S_{eco} + S_{soc} \tag{2}$$

$$S_{env} = \sum_{i=1}^{n} \alpha_i \times \frac{P^R_{i,env}}{P_{i,env}} \quad (3)$$

$$S_{eco} = \sum_{i=1}^{n} \beta_i \times \frac{P^R_{i,eco}}{P_{i,eco}} \quad (4)$$

$$S_{soc} = \sum_{i=1}^{n} \gamma_i \times \frac{P^R_{i,soc}}{P_{i,soc}} \quad (5)$$

where: SI represents the sustainability index, S_{env}, S_{eco}, S_{soc} represent the sustainability indexes to the social, economic, and environmental aspects, α_i, β_i, γ_i represent how each parameter in the environmental, economic, and social dimensions is rated, $P^R_{i,env}$, $P^R_{i,eco}$, $P^R_{i,soc}$ represent the reference value for each parameter, and $P_{i,env}$, $P_{i,eco}$, $P_{i,soc}$ represent the calculated values for each parameter.

If two or more solutions are compared, the values obtained as references can be regarded as the optimal values from each parameter; when conducting a self-assessment, the best practices that are currently available are used as standards. For those circumstances, where a parameter's higher value is thought to be more sustainable, in Equations (3)–(5) the parameters in the ratio of the reference value and the calculated one will be reversed. The final result of the developed specific model is the sustainability index SI, with a dimensionless value between 0 and 1, in which 0 represents the worst value and 1 the best value [52].

3. Case Study

3.1. A Brief Description of the Geographic Location

The examined objective is situated to the north-east of the municipality of Alba Iulia, at about 20 km in the north of Galda de Jos Village. The area in question has a polygonal shape, with a total surface of approx. 22 ha out of which 7 ha will be occupied by Cell 1. The ground surface records level differences from 277 m to 320 m, with a gentle slope of 1:8 ... 1:10 in the direction southwest–northeast. Before the works started, the investigated settlement in its natural form did not show instability phenomena affecting the analyzed perimeter or the slopes from the settlement vicinity. The settlement is crossed from SW to NE, respectively, on the lines of the greatest slopes by ravines with depths of 0.5 to 2 m and slopes of 1:1 ... 1:1.5 with a high potential for losing local stability. The geotechnical study involved performing 10 bore-holes within the site, as shown in Figure 1.

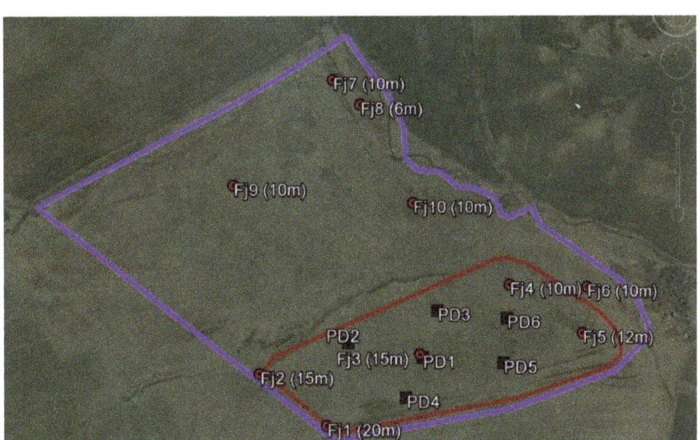

Figure 1. Bore-holes' arrangement within the site.

The geotechnical investigation revealed that the foundation soil is composed of a package of cohesive materials like clay—sandy loam, brown-yellow color, in a state of consistency from plastic to hard, located beneath a layer of vegetable soil varying in thickness from 10 to 30 cm.

Slope stability analysis was performed with the Geostru Slope application, with multiple tactics concerning the shape of failure surfaces, using the circular surface (the simplest shape). To avoid the situation of ultimate equilibrium, an acceptable safety level of 1.50 was proposed. The step search was set to 30, with a number of 30 strips in order to have a reasonable time period for the stability analysis. The partial coefficients for soil geotechnical parameters were considered at 1.25 for the angle tangent of internal friction and for the effective cohesion, respectively, 1.40 for the undrained cohesion.

The characteristics of the soil foundation and the adjacent soil layers were both introduced in the program according to the analyzed transverse profile, located next to the borehole Fj6, as can be seen in Table 2 [53].

Table 2. Characteristics of the slope analyzed.

Layer No.	c (kN/m^2)	∅ (deg)	G (kN/m^3)	G_S (kN/m^3)
1	33.5	11.62	18.63	20.59
2	0	30	19.93	21.13
3	56.60	15.30	20.53	23.76

where c represents the cohesion, ∅ represents the friction angle, G represents the specific weight, and G_S represents the saturated specific weight.

3.2. *The Intervention Works Analysed on the Slope*

The stability factor's analysis was conducted using the computational application with imposed surfaces in Geostru Slope, which is based on the Finite Element Method (FEM).

The geometric configurations of the intervention works analyzed (two retaining walls made of reinforced concrete and two soil reinforcement with a biaxial geogrid) in the sustainability study are presented in Figure 2.

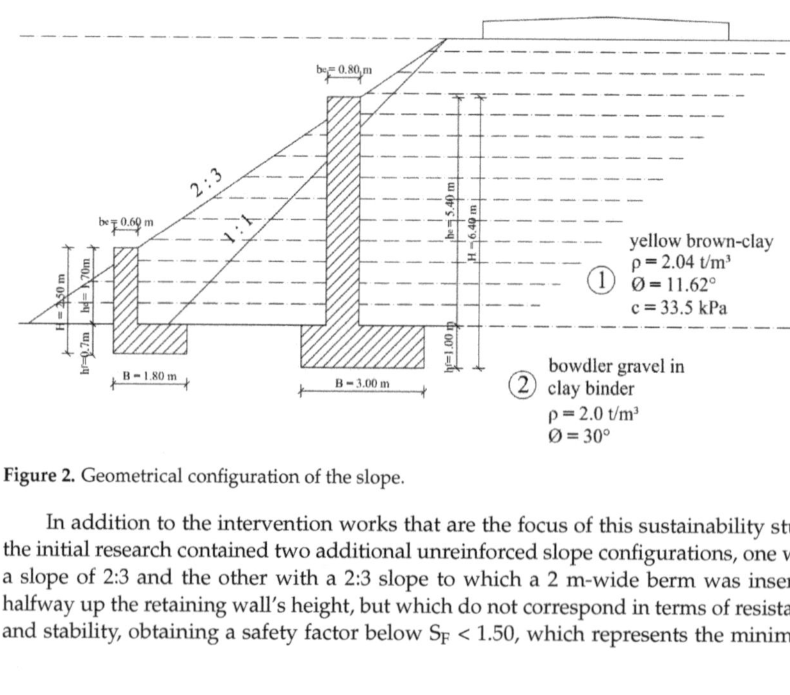

Figure 2. Geometrical configuration of the slope.

In addition to the intervention works that are the focus of this sustainability study, the initial research contained two additional unreinforced slope configurations, one with a slope of 2:3 and the other with a 2:3 slope to which a 2 m-wide berm was inserted halfway up the retaining wall's height, but which do not correspond in terms of resistance and stability, obtaining a safety factor below $S_F < 1.50$, which represents the minimum accepted value. The safety factor's assessment is not the subject of this article; it is very well detailed in [53]. To evaluate the sustainability indices, we will analyze the following intervention works:

1. Reinforced soil, inclination of slope 2:3;
2. Retaining wall H = 2.50 m;
3. Reinforced soil, inclination of slope 1:1;
4. Retaining wall H = 6.40 m.

The technical data and the cost of the materials used in the intervention works, as well as the coefficients regarding the embodied energy and gas emissions in terms of manufacturing, transport, and equipment or machines used in the construction process, are given in Table 3.

Table 3. Materials, technical information, and coefficients utilized in the sustainability study.

Material Used in the Intervention Works	Material Density (kg/m^3)	Embodied Energy Coefficient EE (MJ/kg)	GHG Emissions Coefficient EC-CO_2 (kgCO_2/kg)	Material Cost (EURO/m^3)
Stabilized filler material	2240	0.082	0.0048	15
Concrete C20/25	2400	0.91	0.131	100
Steel bar ∅12 and ∅14	7850	29.20	2.59	9812.5
Biaxial geogrid polypropylene	900	101.30	4.10	10,385

The quantities of materials required for performing the intervention works were estimated for a section of 5 m in length, following the profile with the highest level difference, measuring nearly eight meters.

Thus, the analysis started with the retaining wall made of reinforced soil with geogrilles in the geometric configuration with the smoothest slope of 2:3, where it is necessary to bring and compact 424 m^3 of local soil. Biaxial geogrids with a specific weight of 0.284 kg/m^2 were used in both reinforced soil configurations, being arranged in layers along the entire height of the retaining wall, at distances of 50 cm between them, consolidating each layer on a 50 m^2 surface. The required quantity was established taking into account the design length of 10 m, which was supplemented with the overlapped length required to secure the geogrid to the layer above over a distance of one meter.

The reinforced concrete retaining walls were designed with the C20/25 concrete class and an S500 reinforcement mark. The retaining wall with a height of H = 2.50 m, located at the base of the slope, is characterized by an elevation width b_e = 60 cm, foundation height h_f = 70 cm, and foundation width B = 1.80 m. The retaining wall's elevation and foundation are reinforced with ∅12/10 cm bars on both directions, forming closed edges both in the longitudinal and cross sections, obtaining 745.75 kg of iron for the analyzed intervention work.

In the last intervention work, we have a retaining wall with a height of H = 6.40 m located very close to the ecological landfill's access road, which has the following geometric dimensions: elevation width b_e = 80 cm, foundation height h_f = 1.00 m, and foundation width B = 3.00 m. And in this case, closed edges are formed on both sections from bars ∅14/10, obtaining a total steel amount of 2512 kg.

With the exception of the filler material that was brought to the site from a distance of maximum 10 km away, all materials were delivered to the site from the nearest local construction supplies warehouse, positioned 30 km away. All materials were delivered in trucks that could transport between 3.5 and 20 tons, which have the following coefficients: embodied energy EE = 4.60 MJ/tkm and GHG gas emissions EC-CO_{2e} = 0.28 kgCO_2/tkm.

The initial embodied energy, which depends on the embodied energy (EE) in terms of manufacturing materials, transport, and equipment or machines used in the construction process (En), is calculated with Formula (6).

$$\mathrm{En} = \mathrm{EE} \times \mathrm{m} \quad (6)$$

The GHG gas emissions which resulted from construction materials in terms of manufacturing materials, transport, and equipment or machines used in the construction process (G) are calculated with Formula (7).

$$\mathrm{G} = \mathrm{CO}_{2\mathrm{eq}} \times \mathrm{m} \quad (7)$$

4. Results and Discussions

The results of the sustainability study were obtained using the Bob–Dencsak specific model, thus calculating all the parameters in question. The ecological dimension is represented by the consumption of embodied energies (En) and the total amount GHG gas emissions (G) in the process of manufacturing and transporting the materials used in the intervention works. In the results of the calculation, these factors are given equal weight, accounting together for 40% of the sustainability indices' value. The economic dimension of sustainability is expressed through the labor (W) and material costs (C) required to complete these interventions works, which also represent 40% of the final result, divided equally among the parameters within the dimension. The safety factor (S_F) expresses the social dimension of sustainability, assigning 20% of the final value of each intervention work within the sustainability study.

The quantities of materials used for each intervention work, as well as the data regarding the environmental dimension through the embodied energies and gas emissions from each material, are presented in Table 4.

Table 4. The quantities of materials used and the embodied energy and GHG emissions for each intervention work.

Intervention Work	Name of Material	Quantity Volume (m^3)	Energy (MJ)	GHG Emission ($kgCO_2$)
1. Reinforced soil, inclination of slope 2:3	Filler Material	424	121,569.28	7218.18
	Biaxial Geogrid	0.23	21,230.97	859.89
2. Retaining wall, H = 2.50 m	Filler Material	412	118,128.64	7013.89
	Concrete C20/25	11.70	29,427.84	3914.35
	Steel Bar	0.12	27,636.40	2447.70
3. Reinforced soil, inclination of slope 1:1	Filler Material	366	104,939.52	6230.79
	Biaxial Geogrid	0.23	21,230.97	859.89
4. Retaining wall, H = 6.40 m	Filler Material	315	90,316.80	5362.56
	Concrete C20/25	36.60	92,056.32	12,244.90
	Steel Bar	0.33	76,000.09	6731.16

Based on the previous quantities of materials used in the sustainability study, the economic dimension regarding the cost of materials and the cost of labor is presented in Table 5.

The final values of the coefficients which were used in the sustainability analysis of each individual intervention work are presented in Table 6.

Table 5. The cost of the materials and labor involved in the sustainability analysis.

Intervention Work	Name of Material	Quantity Volume (m^3)	Material Cost (EURO)	Labor (man × h)
1. Reinforced soil, inclination of slope 2:3	Filler Material	424	6360	100
	Biaxial Geogrid	0.23	2389	128
2. Retaining wall, H = 2.50 m	Filler Material	412	6180	97
	Concrete C20/25	11.70	1170	16
	Steel Bar	0.12	1178	160
3. Reinforced soil, inclination of slope 1:1	Filler Material	366	5490	85
	Biaxial Geogrid	0.23	2389	128
4. Retaining wall, H = 6.40 m	Filler Material	315	4725	75
	Concrete C20/25	36.60	3660	50
	Steel Bar	0.33	3239	320

Table 6. The coefficients values used in the sustainability analysis.

Intervention Work	Environmental		Economic		Social
	Energy (MJ)	GHG ($kgCO_2$)	Material (EURO)	Labor (man × h)	Safety Factor
1. Reinforced soil, inclination of slope 2:3	142,800.25	8078.07	8749	228	2.18
2. Retaining wall, H = 2.50 m	175,192.88	13,375.94	8528	273	2.05
3. Reinforced soil, inclination of slope 1:1	126,170.49	7090.68	7879	213	1.65
4. Retaining wall, H = 6.40 m	258,373.21	24,338.62	11,624	445	2.01

The sustainability index for each intervention work is calculated with the formula:

$$SI = 0.2\frac{En^R}{En} + 0.2\frac{G^R}{G} + 0.2\frac{C^R}{C} + 0.2\frac{W^R}{W} 0.2\frac{S_F}{S_F^R} \quad (8)$$

where the reference values are: En^R = 126170.49 MJ, G^R = 7090.68 $kgCO_2$, C^R = EUR 7879, W^R = 213 man × h, and $S_F^R = 2.18$.

Thus, after performing the calculations, the following sustainability index values were obtained: SI_1 = 0.920 for the reinforced soil, with a slope inclination of 2:3, SI_2 = 0.779 for the retaining wall with the height of 2.50 m, SI_3 = 0.951 for the reinforced soil, with a slope inclination of 1:1, and SI_4 = 0.573 for the retaining wall with the height of 6.40 m.

The initial objective of this project was to ensure the resistance and stability of a slope that serves as an access road to an ecological landfill, where the intervention works involve the use of local ground that is used as a filling material to support the road. Thus, the first intervention analyzed was a retaining wall with unreinforced soil in the most stable geometric configuration, with a slope of 2:3, where the safety factor was found to be less than the minimum acceptable by the current regulations. In order to obtain an acceptable safety factor, the base of the slope was increased by adding a berm of 2 m wide at the midpoint of the retaining wall's height, but also with an unfavorable result in terms of the resistance and stability of the slope, obtaining a value that is less than the 1.50 minimum acceptable value, which is a mandatory requirement. Each of the analyzed variants fulfill the condition of resistance and stability, trying to achieve a high value for the safety factor, a fact that plays a major role in raising the social dimension.

In the version with the retaining wall reinforced with biaxial geogrids, with a slope of 2:3, a safety factor value of $F_S = 2.18$ was obtained, which was used as the social criterion's reference value when analyzing the sustainability of the intervention works. The 2.50 m-high reinforced concrete retaining wall placed at the base of the slope reduces its base by approximately 2 m, which leads to a greater storage capacity, but it strengthens the slope by increasing the base's stiffness, obtaining a satisfactory safety factor $F_S = 2.05$, which is comparable to the reinforced soil with a slope of 2:3. Due to the retaining wall's small dimensions, as well as the difference in filling material required to perform the intervention work, the total cost of the materials is lower than in the case of reinforced soil with a slope of 2:3, but taking into account all the parameters it offers a 18% lower sustainability index.

Investigating an additional decrease in the slope's base, which directly implies an overall reduction in the slope's stiffness due to its lack of massiveness, the geometric configuration of the reinforced soil retaining wall with a slope of 1:1 was analyzed, where a lower safety factor value $F_S = 1.65$ resulted. Although it has the lowest stability factor of all the configurations that were examined, this intervention work provides the reference values for the economic and ecological dimensions because it requires 58 m^3 less filling material than the reinforced soil retaining wall with a 2:3 slope configuration.

The geometric configuration of the slope with the smallest base, which provides an adequate total stiffness due to the large amount of concrete and steel bars used, is the intervention work of the reinforced concrete retaining wall with a height of 6.40 m, which offers a more than acceptable stability factor of $F_S = 2.01$. Due to the large volume of reinforced concrete, it recorded the highest values for the embodied energy En = 258373.21 MJ and GHG gas emissions G = 24338.62 $kgCO_2$, as well as the cost of materials and labor.

5. Conclusions

After carrying out the sustainability study of the four intervention works, it has been clearly observed that the best value $SI_3 = 0.951$ was obtained for the reinforced soil with a biaxial geogrid, with a slope of 1:1. Based on the parameters' values that were determined for this intervention work, it is important to draw attention to the optimal balance between the materials' energy consumption and gas emissions as well as their costs, including labor, as these represent the study's reference values. All of this supports the result, including the fact that the slope's geometrical configuration is stable, even if it recorded the lowest value of the safety factor, which is 10% higher than the minimum accepted value.

The second option is represented by the reinforced soil retaining wall with a slope of 2:3, with a sustainability index value of $SI_2 = 0.920$. Compared to the version with a slope of 1:1, there were increases in energy consumption of 13.5% and in the amount of gas released into the atmosphere of about 14% due to the filling material that must be brought additionally in order to achieve the slope geometrical configuration.

The third intervention work option from the perspective of sustainability, with an index value of $SI_3 = 0.779$, is represented by the reinforced concrete retaining wall with a height of 2.50 m placed at the slope base. Following the ecological dimension, if we make a comparison with the version of reinforced soil with a slope of 2:3, there are significant increases of 23% in energy consumption, respectively 66% for GHG gas emissions released into the atmosphere. This is clearly underlined by the material values in terms of embodied energy and GHG gas emissions, where for the geogrid the following values were obtained En = 21230.97 MJ and G = 859.88 $kgCO_2$ and respectively for reinforced concrete En = 57064.24 MJ and G = 6362.05 $kgCO_2$.

The lowest value of the sustainability index $SI_4 = 0.573$ was obtained by the reinforced concrete retaining wall with a height of 6.40 m. This result is highlighted by the values obtained by all of the studied parameters, specific ecological and economic dimensions, due to the large volume of reinforced concrete required, which unavoidably raises the total cost of the intervention work. Even though the storage capacity of the ecological landfill was not a criterion, for further research it should be noted that this intervention work provides the smallest base of the slope which offers the biggest storage capacity. This aspect can also be taken into account in the case of intervention works to stabilize slopes that have a limited base for various reasons, such as the presence of railroads nearby, or any type of structure, or even just the simple existence of flowing water.

As a final conclusion, we can state that compared to the retaining walls made of reinforced concrete, it can be clearly seen that the reinforced soil intervention works obtained the highest scores from the sustainability point of view. This points out the fact that using geogrids for soil reinforcement is a much more efficient solution from an ecological perspective, following energy consumption and gas emissions, but also from an economic aspect, analyzing both the cost of the materials used and the labor, compared to the case of reinforced concrete as a construction material. Ensuring environmental sustainability is an admirable activity that civil engineers should definitely perform, not only in the preservation or renovation of existing structures but also in the design of future infrastructure. What engineers design and build today will have a long-term impact on the environment and society.

Author Contributions: Conceptualization, M.R.T.; Methodology, M.R.T. and C.I.B.; Validation, M.R.T.; Formal analysis, M.R.T.; Investigation, M.R.T.; Data curation, M.R.T.; Writing—original draft, M.R.T.; Writing—review & editing, M.R.T.; Supervision, M.R.T. and C.I.B. All authors have read and agreed to the published version of the manuscript.

Funding: This research received no external funding.

Institutional Review Board Statement: Not applicable.

Informed Consent Statement: Not applicable.

Data Availability Statement: Data are available from the first author. The literature data are in accordance with the references inserted.

Conflicts of Interest: The authors declare no conflicts of interest.

References

1. United Nations. *Report of the World Commission on Environment and Development: Our Common Future*; United Nations: New York, NY, USA, 1987.
2. Vieira Nunhes, T.; Viviani Garcia, E.; Espuny, M.; Homem de Mello Santos, V.; Isaksson, R.; José de Oliveira, O. Where to go with Corporate Sustainability? Opening paths for sustainable businesses through the collaboration between universities, governments, and organizations. *Sustainability* **2021**, *13*, 1492. [CrossRef]
3. Oladazimi, A.; Mansour, S.; Hosseinijou, S.A.; Majdfaghihi, M.H. Sustainability Identification of Steel and Concrete Construction Frames with Respect to Triple Bottom Line. *Buildings* **2021**, *11*, 565. [CrossRef]
4. Butera, F.M. Climatic change and the built environment. *Adv. Build. Energy Res.* **2011**, *4*, 45–75. [CrossRef]
5. Akadiri, P.O.; Chinyio, E.A.; Olomolaiye, P.O. Design of A Sustainable Building: A Conceptual Framework for Implementing Sustainability in the Building Sector. *Buildings* **2012**, *2*, 126–152. [CrossRef]
6. Prada, M.; Prada, I.F.; Cristea, M.; Popescu, D.E.; Bungău, C.; Aleya, L.; Bungău, C.C. New solutions to reduce greenhouse gas emissions through energy efficiency of buildings of special importance—Hospitals. *Sci. Total Environ.* **2020**, *718*, 137446. [CrossRef] [PubMed]
7. Energy Roadmap. 2050. Available online: https://ec.europa.eu/energy/sites/ener/files/documents/2012_energy_roadmap_2050_en_0.pdf (accessed on 13 August 2023).
8. Prada, M.; Popescu, D.; Bungau, C. Building Education, Source of Energy Saving in Romania. In Proceedings of the 15th National Technical-Scientific Conference on Modern Technologies for the 3rd Millennium, Oradea, Romania, 27–28 November 2015; pp. 157–162.
9. Wang, L.; Wu, C.; Gu, X.; Liu, H.; Mei, G.; Zhang, W. Probabilistic stability analysis of earth dam slope under transient seepage using multivariate adaptive regression splines. *Bull. Eng. Geol. Environ.* **2020**, *79*, 2763–2775. [CrossRef]
10. Murat, K.; Omer, T. Ecologically based hybrid life cycle analysis of continuously reinforced concrete and hot-mix asphalt pavement. *Transp. Res. Part D Transp. Environ.* **2012**, *17*, 86–90.
11. Chinyere, D.; Caroline, H. Evaluation of low-impact modular housing using energy optimization and life cycle analysis. *Energy Ecol. Environ.* **2019**, *4*, 286–299.

12. Huang, Y.; Roger, B.; Oliver, H. Development of a life cycle assessment tool for construction and maintenance of asphalt pavements. *J. Clean. Prod.* **2009**, *17*, 283–296. [CrossRef]
13. Chang, B.; Kendalla, A. Life Cycle Greenhouse Gas Assessment of Infrastructure for California's High-speed Rail System. *Transp. Res. Part D Transp. Environ.* **2011**, *16*, 429–434. [CrossRef]
14. Chan, A. Economic and Environmental Evaluations of Life Cycle Cost Analysis Practice: A Case Study of Michigan DOT Pavement Projects. Mater's Thesis, Mater of Science Thesis in Natural Resource and Environment, University of Michigan, Ann Arbor, MI, USA, 2007.
15. Achour, Y.; Pourghasemi, H.R. How do machine learning techniques help in increasing accuracy of landslide susceptibility maps? *Geosci. Front.* **2020**, *11*, 871–883. [CrossRef]
16. Thennavan, E.; Ganapathy, G.P.; Chandrasekaran, S.; Rajawat, A. Probabilistic rainfall thresholds for shallow landslides initiation—A case study from The Nilgiris district, Western Ghats, India. *Int. J. Disaster Risk Manag.* **2020**, *2*, 1–14. [CrossRef]
17. Karim, Z.; Hadji, R.; Hamed, Y. GIS-based approaches for the landslide susceptibility prediction in Setif Region (NE Algeria). *Geotech. Geol. Eng.* **2019**, *37*, 359–374. [CrossRef]
18. Saptono, S.; Rezky, D.M. Sensitivity analysis of nickel haul road embankment slopes using the coefficient of variation approach. *Min. Miner. Depos.* **2022**, *16*, 48–53. [CrossRef]
19. Boubazine, L.; Boumazbeur, A.; Hadji, R.; Fares, K. Slope failure characterization: A joint multi-geophysical and geotechnical analysis, case study of Babor Mountains range, NE Algeria. *Min. Miner. Depos.* **2022**, *16*, 65–70. [CrossRef]
20. Sui, Z.F.; Yuan, W.; Yi, W.; Yang, W. Stability analysis of ecological slopes based on a 3D finite element model. *Adv. Mater. Sci. Eng.* **2021**, *2021*, 3785943. [CrossRef]
21. Tuskan, Y.; Basari, E. Evaluation of Sustainable Slope Stability with Anti-Slide Piles Using an Integrated AHP-VIKOR Methodology. *Sustainability* **2023**, *15*, 12075. [CrossRef]
22. Gu, W.; Cui, W.J.; Xu, J.; Shao, Q. Study on the industrialization of ecological restoration for surfaces destroyed by engineering. *Adv. Sci. Technol. Water Resour.* **2008**, *28*, 66–70.
23. Guan, Y.J.; Kang, R.P.; Liu, J.G. Evolution of the field of ecological restoration over the last three decades: A bibliometric analysis. *Restor. Ecol.* **2019**, *27*, 647–660. [CrossRef]
24. Tang, B.; Song, F.M.; Huang, L.; Xu, J.X.; Zhao, L. Typical Case Analysis of Ecological Restoration for Surfaces Destroyed by Engineering. *Chin. Landsc. Archit.* **2017**, *33*, 25–29.
25. Bo, Q.G.; Li, X.M.; Zhang, J.; Wang, X.M.; Bi, T. The Importance of Protection and Utilization of Topsoil Resources in Highway Construction to the Ecological Restoration of Engineering Wounds-Taking the TJSTFH-1 Section of Jingxin Expressway as an Example. *Traffic Saf. Environ. Prot.* **2014**, *22*, 13–17.
26. Shen, Y.; Li, Q.; Pei, X.; Wei, R.; Yang, B.; Lei, N.; Zhang, X.; Yin, D.; Wang, S.; Tao, Q. Ecological Restoration of Engineering Slopes in China—A Review. *Sustainability* **2023**, *15*, 5354. [CrossRef]
27. Peng, X. Ecosystem Restoration, Conservation, and Diversity under Climate Response. In Proceedings of the 3rd International Conference on Global Economy and Business Management (GEBM 2021), Suzhou, China, 25 December 2021; Wuhan Zhicheng Times Cultural Development Co., Ltd.: Wuhan, China, 2021; pp. 79–82.
28. Zhou, Y. Some Understanding about Soil-Vegetation and Slope Eco-Engineering. *J. Mt. Sci.* **1999**, *03*, 33–38.
29. Ng, C.W.W.; Zhan, L.-T.; Bao, C.-G.; Fredlund, D.G.; Gong, B.-W. Performance of an unsaturated expansive soil slope subjected to artificial rainfall infiltration. *Géotechnique* **2003**, *53*, 143–157. [CrossRef]
30. Tiwari, N.; Satyam, N.; Sharma, M. Micro-mechanical performance evaluation of expansive soil biotreated with indigenous bacteria using MICP method. *Sci. Rep.* **2021**, *11*, 10324. [CrossRef] [PubMed]
31. Puppala, A.; Congress, S.; Banerjee, A. Research advancements in expansive soil characterization, stabilization and geoinfrastructure monitoring. *Front. Geotech. Eng.* **2019**, 15–29.
32. Chehade, H.; Dias, D.; Sadek, M.; Jenck, O.; Chehade, F.H. Seismic analysis of geosynthetic-reinforced retaining wall in cohesive soils. *Geotext. Geomembr.* **2019**, *47*, 315–332. [CrossRef]
33. Tiwari, N.; Satyam, N.; Puppala, A.J. Strength and durability assessment of expansive soil stabilized with recycled ash and natural fibers. *Transp. Geotech.* **2021**, *29*, 100556. [CrossRef]
34. Zhang, R.; Tang, P.; Lan, T.; Liu, Z.; Ling, S. Resilient and Sustainability Analysis of Flexible Supporting Structure of Expansive Soil Slope. *Sustainability* **2022**, *14*, 12813. [CrossRef]
35. Frischknecht, R.; Büsser-Knöpfel, S.; Itten, R. Comparative Life Cycle Assessment of Geosynthetics versus Concrete Retaining Wall. In Proceedings of the 18th International Conference on Soil Mechanics and Geotechnical Engineering, Paris, France, 2–6 September 2013; pp. 1979–1982.
36. Duncan, J.M. State of the art: Limit equilibrium and finite-element analysis of slopes. *J. Geotech. Eng.* **1996**, *122*, 577–596. [CrossRef]
37. Rocha, W.W.; Dias Junior, M.S.; Lima, J.M.; Miranda, E.E.V.; Silva, A.R. Resistência ao cisalhamento e grau de intemperismo de cinco solos na região de Lavras (MG). *Rev. Bras. Ciência Do Solo* **2002**, *26*, 297–303. [CrossRef]
38. Ghowdhury, R.; Flentje, P.; Bhattacharya, G. Geotechnical Slope Analysis. Available online: https://www.researchgate.net/publication/267420172_Geotechnical_Slope_Analysis (accessed on 11 November 2023).
39. Steward, T.; Sivakugan, M.; Shukla, S.K.; Das, B.M. Taylor's slope stability charts revisited. *Int. J. Geomech.* **2011**, *11*, 1943–1956. [CrossRef]

40. Suhatril, M.; Osman, N.; Azura Sari, P.; Shariati, M.; Marto, A. Significance of Surface Eco-Protection Techniques for Cohesive Soils Slope in Selangor, Malaysia. *Geotech. Geol. Eng.* **2019**, *37*, 2007–2014. [CrossRef]
41. Cala, M.; Flisiak, J. Slope Stability Analysis with FLAC and Limit Equilibrium Methods. Available online: https://home.agh.edu.pl/~cala/papers/2003_3.pdf (accessed on 21 November 2023).
42. Johari, A.; Gholampour, A. A practical approach for reliability analysis of unsaturated slope by conditional random finite element method. *Comput. Geotech.* **2019**, *102*, 79–91. [CrossRef]
43. Johari, A.; Talebi, A. Stochastic Analysis of Rainfall-Induced Slope Instability and Steady-State Seepage Flow Using Random Finite-Element Method. *Int. J. Geomech.* **2019**, *19*, 1943–1956. [CrossRef]
44. Kamchoom, V.; Leung, A.K. Hydro-mechanical reinforcements of live poles to slope stability. *Soils Found.* **2018**, *58*, 1423–1434. [CrossRef]
45. Díaz-López, C.; Carpio, M.; Martín-Morales, M.; Zamorano, M. Analysis of the scientific evolution of sustainable building assessment methods. *Sustain. Cities Soc.* **2019**, *49*, 101610. [CrossRef]
46. Doan, D.T.; Ghaffarianhoseini, A.; Naismith, N.; Zhang, T.; Ghaffarianhoseini, A.; Tookey, J. A critical comparison of green building rating systems. *Build. Environ.* **2017**, *123*, 243–260. [CrossRef]
47. Shan, M.; Hwang, B. Gang Green building rating systems: Global reviews of practices and research efforts. *Sustain. Cities Soc.* **2018**, *39*, 172–180. [CrossRef]
48. Tudorica, M.R.; Toadere, M.T.; Bob, C.I. The Sustainability Study Done for a Consolidation Work on a Historical Building. *Sustainability* **2023**, *15*, 15285. [CrossRef]
49. Bob, C. *Durabilitatea si Sustenabilitatea Constructiilor*; Editura Politehnica: Timisoara, Romania, 2020; pp. 84–85.
50. Ding, G.K.C. Sustainable construction: The role of environmental assessment tools. *J. Environ. Manag.* **2007**, *86*, 451–464. [CrossRef] [PubMed]
51. Diaz-Balteiro, L.; Romero, C. In search of a natural systems sustainability index. *Ecol. Econ.* **2004**, *49*, 401–405. [CrossRef]
52. Bob, C. A New Approach of a Specific Sustainability Model. *J. Civ. Environ. Eng.* **2018**, *8*, 1000314. [CrossRef]
53. Tudorica, M.; Golea, L. Comparative Study of Slope Staility. In Proceedings of the 15th National Technical-Scientific Conference on Modern Technologies for the 3rd Millennium, Oradea, Romania, 27–28 November 2015.

sustainability

Article

Modular Autonomous Vehicles' Application in Public Transport Networks: Conceptual Analysis on Airport Connection

Ioan-Tudor Oargă , Bogdan Ovidiu Varga , Dan Moldovanu , Horațiu Cărăușan * and Gabriel Prunean

Department of Automotive Engineering and Transports, Technical University of Cluj-Napoca, 103-105 B-dul Muncii Street, 400641 Cluj-Napoca, Romania; tudor.oarga@auto.utcluj.ro (I.-T.O.); bogdan.varga@auto.utcluj.ro (B.O.V.); dan.moldovanu@auto.utcluj.ro (D.M.); gabriel.prunean@auto.utcluj.ro (G.P.)
* Correspondence: horatiu.carausan@auto.utcluj.ro

Abstract: Increased efficiency and optimized operation of transport networks represent two of the main topics of interest when discussing modern road vehicle solutions. Taking steps towards more sustainable options, manufacturers of road vehicles are looking into advanced technologies that allow vehicles to run more efficiently and take advantage of all the available data on the road. When looking at public transportation applications, trends point in the direction of using varied types of vehicles that can carry people around. The intermodality of these types of vehicles represents the most optimized way of traveling, combining the fast and secure characteristics of airplanes and trains with the flexibility of last-mile options, such as taxis, buses, or trams. This paper discusses the aspects of implementing a modular autonomous vehicle (MAV) solution for the last-mile part of travel routes, connecting key points of a city, such as an airport or a train station, to other key locations in the city, such as the city center, important facilities, or marginal neighborhoods.

Keywords: modular autonomous vehicles; public transport; energy efficiency; airport connection

Citation: Oargă, I.-T.; Varga, B.O.; Moldovanu, D.; Cărăușan, H.; Prunean, G. Modular Autonomous Vehicles' Application in Public Transport Networks: Conceptual Analysis on Airport Connection. *Sustainability* **2024**, *16*, 1512. https://doi.org/10.3390/su16041512

Academic Editors: Marc A. Rosen and Armando Cartenì

Received: 22 December 2023
Revised: 19 January 2024
Accepted: 6 February 2024
Published: 10 February 2024

1. Introduction

1.1. Background

The rapid pace of technological evolution in recent decades made its mark in various industries, allowing for difficult solutions to become mainstream and easier to implement, leading to discoveries and improvements in all types of domains. As a downside of this worldwide spread of accessible solutions, pollution stands out as one of the most destructive consequences to the environment [1]. All industries that find themselves in a position of harming the environment by polluting, exploiting limited resources, or crowding certain locations are acting to optimize the way they operate and, in the case of pollution, reduce their carbon footprint in the atmosphere.

One of the main players interested in reducing carbon emissions is the transportation industry, the effects of which have been studied for decades. The World Health Organization classifies emissions coming from diesel-powered vehicles as carcinogenic, following studies from 2010 [2] and 2012 [3]. As similar consequences are found for all fossil fuel-powered vehicles, the industry has started to study the alternatives when it comes to powertrain solutions, with battery electric vehicles, fuel cell vehicles, and hydrogen vehicles coming into the picture [4,5]. As optimization is key when talking about reducing the negative effects on the planet, developments of automated systems that allow assisted and autonomous driving offer solutions for reducing operating costs of commercial and passenger vehicles, minimizing energy consumption of all sorts, and offering a safer and more reliable way of transportation.

The benefits of similar automated systems are to be found in domains that use them for a longer period of time, such as aeronautics and railroad industries. Here, human intervention has been reduced or removed completely from controlling airplanes, trains, and subways, resulting in safer exploitation of the equipment, increasing the lifecycle, and reducing operational costs. In the past decade, intelligent vehicles (IVs) have benefited from the development of artificial intelligence (AI) to make use of the available technologies and data in order to bring new applications to the public domain, such as border transport and urban taxis [6].

In the direction of evolving these automated systems, the Society of Automotive Engineers (SAE) defines six specific levels of driving automation that a vehicle can achieve, based on features and functions that it can offer. The first three levels (SAE Level 0–SAE Level 2) imply that the driver is still responsible for all driving duties and the vehicle does not present self-driving capabilities. In these levels, driver support features are present, such as automatic emergency braking (SAE Level 0), adaptive cruise control (SAE Level 1), and adaptive cruise control together with lane centering (SAE Level 2). The next three levels (SAE Level 3–SAE Level 5) describe self-driving vehicles. Driver interventions are expected only for SAE Level 3, as vehicles in this class are intended to serve as traffic jam chauffeurs, while at higher speeds, the automated systems are not fully operational. SAE Level 4 and SAE Level 5 applications represent vehicles that can drive autonomously, e.g., local driverless taxis. The difference between the two levels is that the latter can drive the vehicle under all conditions, whereas the other one can operate safely under limited conditions [7].

Self-driving systems and advanced driver-assistance systems (ADAS) use complete sets of sensors, communication drivers, and information processing algorithms to ensure autonomous transportation from one location to another, ranging from V2I communication drivers, GPS, inertial motion units (IMUs), cameras, LiDAR, and ultrasonic sensors, as T. K. Chan and C. S. Chin highlight in their review [8]. In China, detailed research studies and development programs facilitate clear target points and roadmaps for the industry regarding intelligent and connected vehicles (ICVs), as Q. Xu et al. interpret in their 2022 work [9].

1.2. State of the Art of Autonomous Vehicles

In applications of passenger transportation, such as public transport networks, autonomous vehicles (AVs) of various kinds are starting to make their appearance [10–12], ranging from small shuttles with reduced carrying capacity, such as the units implemented in the Horizon Europe ULTIMO project [13], to full-sized buses, like the specimen presented by CAVForth in Scotland this year [14]. As the need for autonomous vehicles with the sole purpose of transporting passengers exists in various locations, with or without the optimal conditions to operate such vehicles, the more popular solutions that are adopted and implemented for public use are represented by the shuttle buses. Being smaller in dimension and with a limited carrying capacity of up to fifteen persons, applications are often found to be in restrained areas and controlled environments with less traffic, such as campuses, industrial areas, and airports. In 2020, upon a comprehensive review of autonomous shuttle buses as a solution for public transport, C. Iclodean presented the active fleets of autonomous shuttle buses all over the world, summing up to 55 applications [15].

Progressing from the concept of autonomous vehicles, several applications in public transport networks can greatly benefit from the implementation of modular autonomous vehicles (MAVs), similar to other domains that already use such concepts: farming, freight transportation, and warehouses. By combining the power and capacity of multiple autonomous vehicles in shuttle configurations, it is possible to form a train of vehicles that execute the same transportation task and run on the same route until each of the modules breaks apart the connection to the leader and starts performing its own transportation task on a different route. In such a way, a fleet of a limited number of vehicles can cover a larger pool of requests for different routes. Each vehicle can be treated as a module in a connected fleet, which, if managed in a proper manner, can optimize the operation tasks with the purpose of reducing waiting times for the passengers, taking care of all driving-related tasks in order to obtain a safer trip, and using the available energy in the most optimized way possible.

Modular autonomous vehicles represent a current sustainable transport solution that has the potential to revolutionize local public transport networks, especially airport connections. The main advantages of MAVs compared to the classic transport system are energy efficiency, the comfort and safety of the passengers, the reliability of the vehicles, the flexibility of the reconfiguration of the routes and the operating schedule, respectively, and the possibilities of integrating any public people transport network.

Another important aspect resulting from the integration of MAVs into a public passenger transport system is the research and development side for the concept of autonomous driving, especially due to feedback from a large target group of people, which opens up the possibility of operating on mixed routes (airport—closed circuit, urban routes—roads open to public traffic, respectively) and their integration into various IT platforms for ticketing and GPS tracking that serve public people transport needs.

Correlated with European and international legislation regarding the implementation of autonomous vehicles on roads open to public traffic, before testing the operation of an autonomous vehicle in real conditions, it is necessary to carry out simulations on a virtual model of this vehicle, covering real application scenarios in a virtual environment similar to the real environment, highlighting any possible deviations from real scenarios ("The manufacturer of the autonomous vehicle must evaluate the functional safety of the autonomous driving system using a number of test scenarios that include false negative and false positive ones. Simulation method may be used, subject to their validation by the approval authorities/technical services in accordance with the procedure for virtual testing in Directive 2007/46/EC or Regulation 858/2018").

Hence, there is a need to develop virtual models of autonomous MAVs configured based on the real characteristics of the main shuttle bus models spread across Europe: Navya, EasyMile, Auve Tech, 2GetThere, e.Go, etc.

Using advanced scheduling and optimization methods, a carefully developed management system can take into account all types of aspects, from available energy for each module and route requests to traffic jams and other hazardous events. By using intelligent algorithms and accessing data over a large period of time (e.g., one year in order to cover all possibilities of transport requests, crowded seasons, and hazards), an advanced system can predict the necessary capacities and ensure all modules are ready from all perspectives to act accordingly and complete the tasks without or with limited human intervention. Such systems may also track the usage and wear of each module and schedule any mandatory or needed service work and inspections.

1.3. Objective and Contributions

The goals of the research project hOListic Green Airport (OLGA) are increasing the energy efficiency in airport-related activities, on both landside and airside, along with reducing pollutant emissions for persons traveling to and from the airport. Focusing on these goals, our team proposes a mobility-as-a-service (MaaS) solution, which provides efficient and easy-to-use travel from the aircraft to key locations inside the city or residential areas. This uses the existing infrastructure of passenger transport on the landside and airside of the International Airport of Cluj-Napoca to bring persons to their accommodation or points of interest (e.g., venues, central squares, office buildings, or technology and industrial parks).

The correlation between aircraft arrivals at the airport and public transport departures from the airport into the urban area is calculated based on the total number of incoming passengers into the terminal, arrival times, which are changing dynamically, and public transport requests. As the majority of the total number of passengers choose to use public transportation, fleet resources are allocated in accordance with travel demand. Our proposal of using modular vehicles comes in support of this dynamic resource allocation strategy, allowing an optimized assignment of transport tasks and keeping the overall energy consumption of the fleet to a minimum.

A comparison to the existing transportation solution (i.e., battery electric buses) is performed in order to identify the advantages of the MAV solution in an already existing scenario. The first physical implementation target of our proposed system is to run modular autonomous vehicles during the nighttime, ensuring safe travel, optimal waiting times for passengers, and reduced operational costs of the equipment.

To analyze the benefits of implementing an MAV fleet in an existing public transport network, a specific case is taken into consideration: the public transport routes from the city of Cluj-Napoca, Romania, linking key locations of the city to the "Avram Iancu" Cluj International Airport. These locations represent extremities of the urban area, such as the southern end of the city and the western end, which is located in the opposite direction to the airport, located in the eastern part of Cluj-Napoca, as well as a high-interest area: the bus terminal. These specific routes are picked due to their demand and capability to serve as a vital component of the intermodality concept, having the role of last-mile personal transport. In this idea, a person coming into the city via airplane or train (considering the train stop near the airport terminal) can use public transport to reach their destination, whether that be a bus terminal that connects the metropolitan area or other small cities nearby or an accommodation in one of the main neighborhoods of the city. This proposed solution may be put into use in a nonstop operating schedule or only at night when the conventional public transport routes are limited to just a small number. The second option also benefits the limitations of autonomous systems of SAE Level 4, which thrive in a less crowded environment with controlled routes.

2. Current Solution

The requests for transport tasks and covering of the routes are directly linked to the population, people flow through the airport, key places in the city, events (e.g., concerts, summits, cultural or sports events, manifestations, meetings), and more. As an overview of the situation, Table 1 presents some of the relevant factors defining the needs taken into consideration when developing a public transport network.

Table 1. Public transport network overview [16].

Parameter	Value
Population	286.598
Number of travelers using public transport (annually)	76.918
Number of available routes	55
Number of bus stops	305
Number of available buses	246

Looking at the current solution that is implemented in the public transport network, the linkage between the bus terminal, which is located north of the city center, is served by a combination of two bus routes during the day, which continue further to other parts of the city, as presented in Figure 1. The routes linking the airport to the bus terminal are marked in purple and pink on the city map.

However, during the nighttime, these routes are limited and connections from the airport to the bus terminal are made with full-sized buses, which often travel more than half-empty through their entire route. In this application, energy consumption is compromised in order to honor regular routes.

Figure 1 presents the current planned route linking the airport to the most western part of the city, passing through the city center and multiple neighborhoods. This is composed by combining the connection from the airport to the city center (marked in purple), together with the illustrated bus line marked in red.

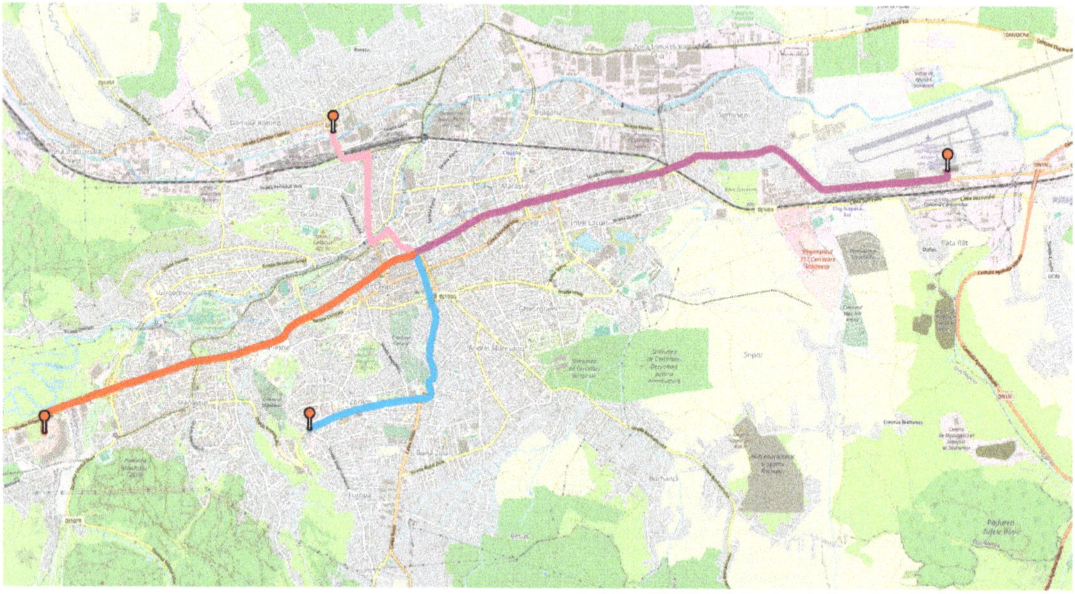

Figure 1. Public transport routes from the airport to Cluj-Napoca city center (purple), bus terminal (pink), southern area (blue), and western area (red) of the city.

While still going halfway through the city center, the routes connecting the airport to the southern neighborhoods steer away from the most crowded areas and experience smaller requests than those that head west, where a nearby locality with 52.735 inhabitants lies. Although the route through the southern area of the city leads to important objectives in the metropolitan area, such as industrial parks (Turda, Câmpia Turzii), the public transport routes do not reach those locations, as they are further than 30 km from the city. This route is presented in Figure 1, marked in blue.

Studying the available data leads to the conclusion that not all routes are optimized in a meaningful way, often having the same outcome: buses travel more than half empty to the end of the line in order to satisfy a reduced request from travelers. With the scope of reducing the time buses are riding at less than 50% capacity, a feasible solution that allows modification of the configuration of the vehicles is to be studied. The main application that can satisfy these criteria is the implementation of modular vehicles along all specified routes.

3. Materials and Methods—MAV Solution

The present chapter is split into two subsections—firstly, the simulation scenario is defined, containing the digitization of driving routes, and secondly, the simulation subjects are modeled (i.e., electric buses and autonomous modules).

3.1. Definition and Digitization of Travel Routes

In order to present the modular autonomous vehicles' implementation in a more accessible approach, the routes of the existing public transport network presented in the previous chapter are broken down into relevant segments, which correspond to parts of the course on which the modules (i.e., individual autonomous shuttles) are riding together, connected to one another, and parts that are covered by one or more vehicles, disconnected from the main train of vehicles.

For this scenario, three separate classes are considered:

- ABT—modules that ride from the airport to the bus terminal, in the northern vicinity of the city center;

- ASC—modules that ride from the airport to the southern area of the city, passing through some of the main neighborhoods;
- AWC—modules that ride from the airport to the western area of the city, passing through different main neighborhoods than the ones ASC is passing through.

By picking only these three routes, the majority of main neighborhoods, which are the oldest, most developed, and most populated ones, are covered in one way or another. This does not mean that the presented connections satisfy all the transport requests in these locations, as there is a large area to cover, making it virtually impossible to accomplish using only three routes.

As the airport is situated in this case on the outside of the city, near the Eastern border of the urban area, the marked route in Figure 2 shows the starting point for all routes, which is the International Airport bus terminal (marked as "Airport").

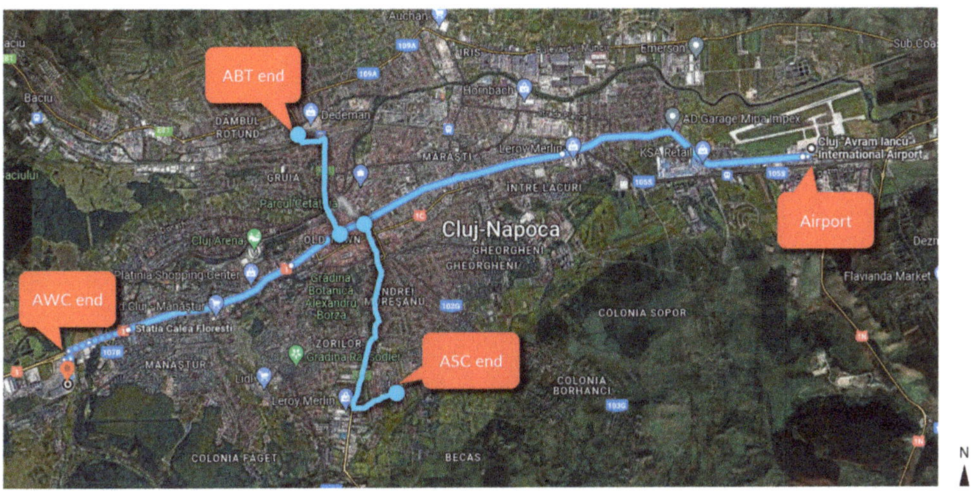

Figure 2. Planned routes for MAV solution simulation.

The ends of each line or route are highlighted in the same figure as follows:

- ABT end—end of the line for the route from the airport to the bus terminal;
- ASC end—end of the line for the route from the airport to the southern area of Cluj;
- AWC end—end of the line for the route from the airport to the western area of Cluj.

The above-mentioned modules are considered to be designed in such a fashion that they are allowed to physically connect to each other and form a train of vehicles, with the modules that detach last from the formation leading the pack. Note that each designated module, like AWC, ASC, or ABT, is presented as a singular unit here for simplification. Depending on the circumstances, the module might be formed by multiple units, leading to an increased capacity for each individual route. Figure 3 illustrates the vehicle train order, as the modules ride together until one of them needs to detach and follow their own designated route.

Along the route, there are so-called "detaching points", where modules disconnect from the pack and start operating on their own. These locations might differ over time depending on various hazards, events, traffic jams, or restrictions and are not necessarily the same as "attaching points", which are to be considered the locations on the map where these modules come back together and form a pack traveling to the International Airport bus terminal. It is not mandatory that the same modules must pair on the return routes, as the segments that they cover are not the same length or do not have the same duration of travel. In addition, on the return route, there might be fewer restrictions or lower passenger capacity, leading to suboptimal operation of the fleet if the total capacity of the module train is similar to the cases of conventional buses—less than 50%.

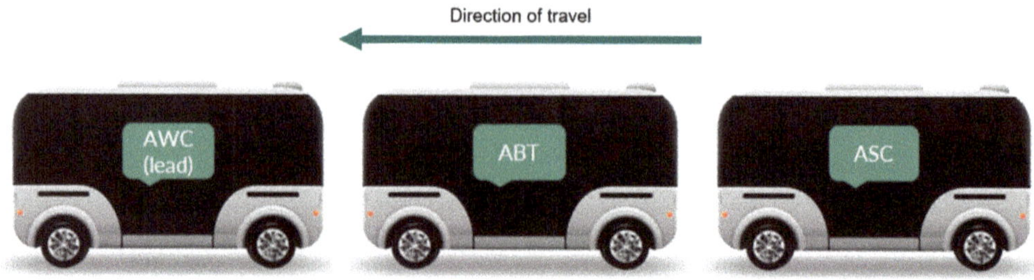

Figure 3. Illustration of a road train of vehicles.

Detaching points are noted on the map shown in Figure 4 as "ABT out" and "ASC out". There is no "AWC out" point in this case, as the AWC module is leading the formation and continues its journey to the end of the line—AWC end.

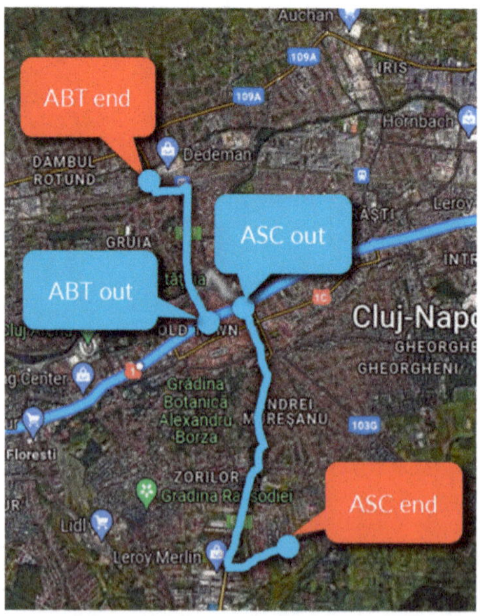

Figure 4. Detaching points for each module.

The order of attaching and detaching of the train module, for exemplification purposes, considers the AWC component as a leading module. Table 2 shows the progress of each module while covering their respective routes and fully executing each task.

Table 2. MAV train configuration along the routes.

Start End	A [1] ASC Out	ASC Out ABT Out	ASC Out ASC End	ABT Out AWC End	ABT Out ABT End
AWC	L [2]	L		L	
ABT					L
ASC			L		

[1] airport; [2] MAV train leader; train configurations are highlighted in grey color.

To expand on the concept definition and obtain data for studying the differences between the conventional way of transportation and the MAV solution, IPG CarMaker is used to model and simulate the public transportation tasks that need to be executed in order to cover the specified routes.

IPG CarMaker 12.0.1 is a simulation software developed by IPG Automotive GmbH, which is used in the industry as a support application for developing and testing automobiles and light-duty vehicles. Throughout the catalog, some solutions offered by IPG fit with this paper's direction of analysis—autonomous vehicles. Being an open platform, it can be integrated into other software, leading to complex data acquisition and processing, as well as advanced modeling of vehicles, traffic characteristics, and driving courses.

Based on available market solutions such as EasyMile EZ10 or Navya Arma autonomous shuttle buses, a module is defined in CarMaker as an AV with a maximum capacity of 900 kg (approximately twelve people). Road trains containing more modular vehicles are then defined by putting modules together. Defining parameters of the modules are added up to obtain the specification list of the train, holding two or four modules together.

By acquiring relevant information (distance, trajectories, elevation, etc.) from Google Earth and Google Maps in GPX (GPS Exchange Format) and KML (Keyhole Markup Language) files of the routes linking the city with the airport, accurate representations of the courses can be loaded into the CarMaker software. The defined routes are accurate representations of the existing public transport routes in the city, as described in the second chapter, having the starting point as the International Airport bus terminal, leading to the city center. From the city center, one route is continuing its course to the western area of Cluj-Napoca, whereas the other one is linking the city center to the city bus terminal. Along the routes, the vehicles drive entirely on public roads with unrestricted access and the same speed limit: 50 km/h for all vehicles, except heavy-duty vehicles, which have a speed limit of 30 km/h. At intersections, their speed limit is 30 km/h. All bus stops are placed on the right-hand side of the road, on the first lane, with some having pull-in space. A small portion of 1.9 km contains a dedicated lane for which the bus stops are placed inside the lane with no pull-in space. The dedicated lane is accessible for public transport vehicles, taxis, and cyclists.

On the above-mentioned courses, specific maneuvers are defined in the application (e.g., driving on the first lane, not cutting corners), as well as driving behavior regarding speed regulation and accelerations (e.g., rapid acceleration until reaching the velocity thresholds, tolerances around the target traveling speed). Using these details, transport solutions can be configured precisely to fit the needs of individual routes.

3.2. Definition and Modeling of Vehicles

This subsection shows the modeling of the two presented solutions: conventional electric buses and modular autonomous vehicles. From the modeling perspective, the electric bus is defined to be approximately the same weight and capacity as a road train containing four separate modules. Table 3 shows the specifications of the bus model parameters, based on the Solaris Urbino 9 LE electric bus [17,18].

Table 3. Electric bus model parameters.

Parameter	Value
Body Mass	11.098 [kg]
Maximum Mechanical Power	160 [kW]
Maximum Torque	1.400 [Nm]
Battery Capacity	124 [kWh]
Idle Voltage	600 [V]
Passenger Capacity	50
Load (front/center/rear)	1.200/1.200/1.200 [kg]

By considering a passenger capacity of twelve for one autonomous module, in order to match the electric bus capacity, four modules are coupled together at the beginning of the simulation until the first detaching point along the route. MAV solution modeling implies the parametrization of one module in IPG CarMaker and then expands the specifications to match an entire road train of four vehicles. In Table 4, modeling parameters for one autonomous module are listed based on specifications of autonomous shuttle buses, such as Navya Arma and EasyMile EZ10.

Table 4. Autonomous module model parameters.

Parameter	Value
Body Mass	2.130 [kg]
Maximum Mechanical Power	16 [kW]
Maximum Torque	41.25 [Nm]
Battery Capacity	10 [kWh]
Idle Voltage	48 [V]
Passenger Capacity	12
Load (front/center/rear)	300/300/300 [kg]

From the above specifications, it results that, for a total passenger capacity similar to that of a Solaris Urbino 9 LE electric, a combination of four autonomous modules with a total weight load of 3.600 kg is modeled based on the parameters in Table 5.

Table 5. Road train of four autonomous modules' model parameters.

Parameter	Value
Total Body Mass	8.520 [kg]
Total Maximum Mechanical Power	64 [kW]
Total Maximum Torque	165 [Nm]
Total Battery Capacity	10 [kWh]
Idle Voltage	48 [V]
Passenger Capacity	48
Total Load (leader, follower #1, #2, #3)	900/900/900/900 [kg]

For each of the two solutions, simulations are made on each defined segment of the road, with an equally distributed passenger load of 900 kg per module (front, center, and rear of each module) at the start of the simulation scenarios. By following the algorithm of pairing and unpairing the modules defined in Table 2, passenger loads, total body mass, and powertrain-specific parameters are altered on each disconnection of modules, simulating people boarding and exiting the vehicle and modules detaching from the main road train and forming smaller couples of autonomous modules to execute their specific transport task. Traffic characteristics are not considered, only road-specific variables, such as elevation changes and curves. The maximum traveling velocity is set to 30 km/h, while the driving behavior of the electric bus driver and self-driving systems is set to neutral—keeping a safe distance, using less aggressive acceleration and braking profiles, steering lightly, not cutting corners, and keeping in the designated traveling lane (first lane) all the time. The goal of the simulated trip is to be as safe, cautious, and comfortable as possible.

4. Results

The considered routes cover a distance of 14.2 km from the International Airport to the most western area of Cluj-Napoca and another of 10.5 km from the International Airport to the bus terminal. By applying the same concept on different routes, which may present even longer dedicated bus lanes and common road sectors, the impact of MAVs on energy efficiency increases. When traffic conditions are also taken into account, traveling in dedicated lanes improves the overall energy consumption and efficiency of the vehicles, reducing standing times and additional braking–accelerating actions while navigating through traffic. In the studied case, all dedicated lanes are also prioritized through the city center, allowing an easy flow of public transportation in and out of the most central area of the city. The traffic lights are synchronized in such a way that the vehicles in dedicated lanes are not held in traffic and go through intersections when there are no other vehicles driving through them. This significantly reduces the waiting times of passengers at bus stops and improves the overall consumption of resources of the vehicles.

To better understand the overall configuration of the analyzed public transport routes, common road sectors, dedicated bus lanes, and the total length of the routes are described in Figure 5.

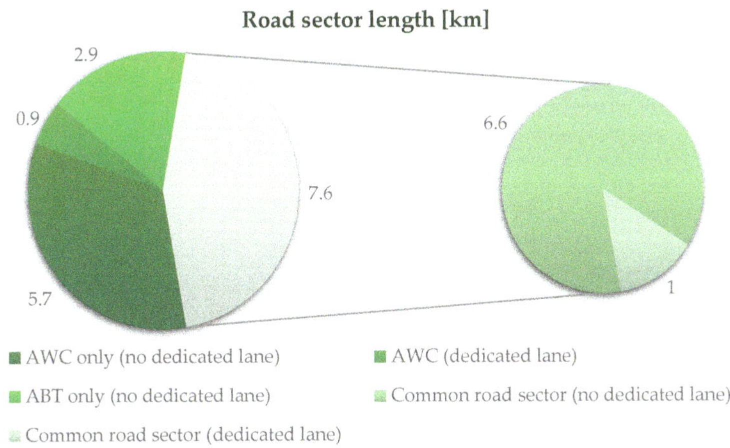

Figure 5. Road sector distribution.

Using the conventional solution simulation results as the benchmark, the outcome of the modular autonomous vehicle simulations shows significant improvement in overall battery management efficiency. As multiple electric buses are needed in order to cover multiple routes within the urban area, the cumulative energy consumption of those buses exceeds the total consumption of all the modules used in the MAV setup. Even though the electric buses are not riding at full passenger loading capacity, the energy consumption differences between a half-empty bus and a fully loaded bus do not match the energy consumption of a road vehicle train. Starting from 85% state of charge (SoC) for both test cases, Figure 6 shows a comparison between the final state of charge of the vehicles after transportation task execution for the buses (conventional solution) and MAVs (proposed solution).

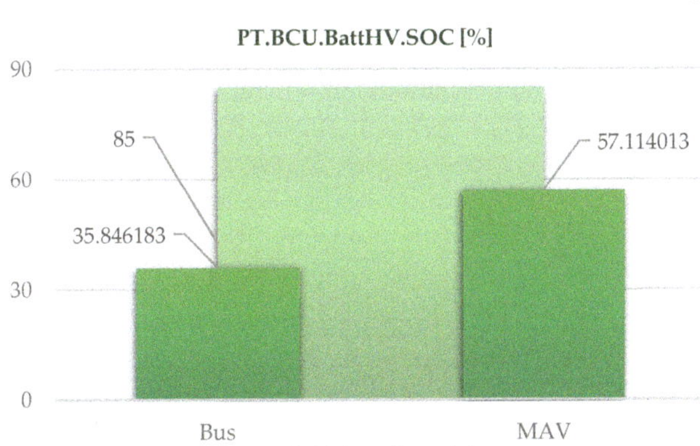

Figure 6. State of charge comparison.

The results point to a significant improvement in the remaining state of charge, measuring 59% more energy stored in the MAV battery pack while carrying the same number of passengers in the same road conditions. Evaluating the absolute energy consumption, the two electric buses that are needed to cover the two routes are 40.67% less efficient than the modular system configuration, as shown in Figure 7.

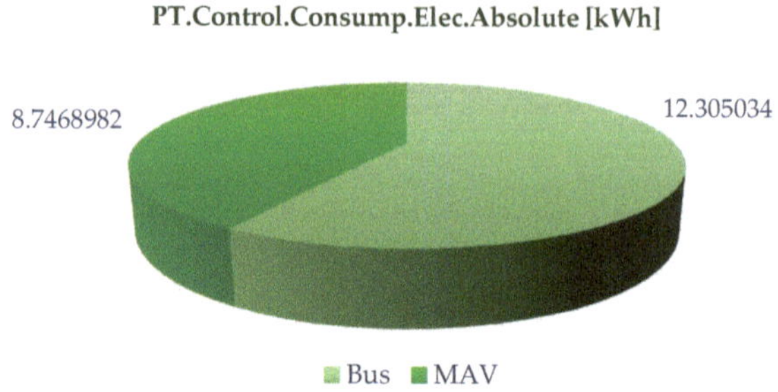

Figure 7. Absolute energy consumption comparison.

By considering the passenger loads of the two compared solutions, fifty people for the electric bus and forty-eight people for the modular setup, the total absolute energy consumption is translated per passenger, as illustrated in Figure 8.

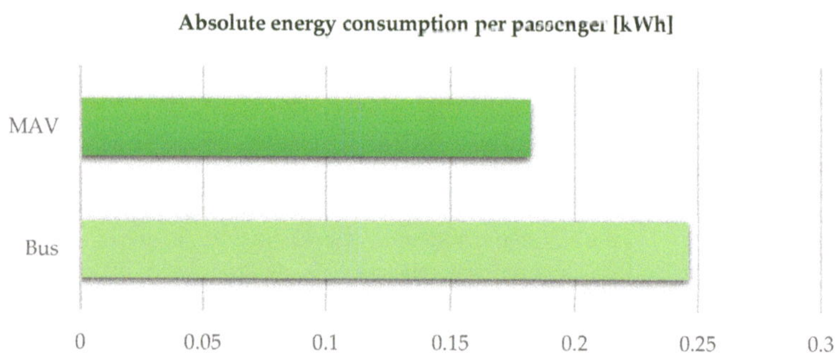

Figure 8. Absolute energy consumption per passenger comparison.

These evaluation criteria on electrical energy consumption confirm even further the advantages of energy efficiency of a modular autonomous vehicle configuration for two or more distinct routes through the city with common sectors along the course.

5. Discussion

As a first step in tackling the issue of public transport network optimization and the implementation of green solutions that lead to more sustainable transportation in the future, the electrification of bus fleets is the way to go. Even though the costs of starting and maintaining such applications are high, local pollution is demonstrated to be reduced in the case of the same city, Cluj-Napoca, by 668.45 tons of CO_2 and 5.618 tons of NO_x each year [19].

Moving forward within this trend and following the latest and most advanced solutions on the market, the autonomous shuttles, especially in a modified form of modular autonomous vehicles, show a feasible solution for an existing network in an old city with narrow streets and existing infrastructure that presents limited possibilities of expanding and highly expensive options to repurpose urban areas and streets.

As autonomous vehicle trends are expanding, even more manufacturers, researchers, and officials turn their attention to this domain. A legal framework, although sometimes not existing or existing in a very limited form, is starting to show recognition, understand the necessity of such applications, and move forward with creating laws, regulations, and specifications that allow others to research and develop with the scope of obtaining safer, more cost-efficient, and optimized road networks [20].

Investigation from A. Ansariyar and M. Tahmasebi concluded that connected and autonomous vehicles reduce delay times in urban traffic by dynamically changing the traveling routes, directly impacting the congestion rates and urban traffic quality. As the market penetration rate (MPR) of connected vehicles (CVs) rises, the delay times are improved significantly, especially when passing the 50 percent mark [21].

Considering the concept of MAVs implemented in an existing transport network, for the case study of the city of Cluj-Napoca, Romania, there are clear advantages that suggest an implementation of such sort will lead to an optimal system that allows a reduction in the wear of equipment, operational costs, and energy consumption and an improvement in the overall efficiency of the public transport network while maintaining or reducing waiting times for the passengers and providing a safer environment for inhabitants of the city.

By evaluating aspects of energy consumption and the overall efficiency of the fleet in scenarios with multiple transport routes, the key benefit point of modular vehicles is represented by the ability to share and restructure equipment resources based on travel demand. Although driving on the same path with a similar weight load, the energy consumption on a conceptual level is similar to a battery electric bus. When referring to multiple routes, the main advantage of MAVs is that they directly influence the overall energy consumption of the fleet, as illustrated in the above simulation results. The use of language models for route optimization may highly impact MAV solutions and ICVs as a whole, starting from studies in urban route optimization, as Y. Liu et al. reported in 2023 [22].

The disadvantages of this kind of application are the same worldwide: there are clear concerns about the safety and reliability of the systems when interacting with populated areas, and there is no clear legal framework to be put in place at the moment, as these solutions are starting to operate in limited areas. As time advances and solutions become safer, more dependable, and more adaptive, there is a certainty that an increasing number of cities, industrial areas, and university and sports campuses will benefit from the advantages of MAVs' integration into existing solutions.

In a broader context, solutions such as the one presented in this paper may greatly benefit from the implementation of artificial neural networks (ANNs) in various applications. As W. Zhu et al. found when they studied the dynamic prediction of traffic incidents, deep learning algorithms prove to add value to ICV applications, providing accurate and quick information on incident duration for operators of transport equipment and vehicles, as well as for travelers [23]. Machine learning applications prove to be beneficial for traffic flow predictions, traffic management, and data generation, leading to a better-trained decision-making algorithm, ultimately increasing the comfort, safety, and user-friendliness of automated and intelligent transport systems, as X. Qu et al. envisioned in 2023 [24].

Author Contributions: Conceptualization, I.-T.O. and B.O.V.; methodology, I.-T.O.; software, I.-T.O., H.C. and G.P.; validation, I.-T.O., H.C. and G.P.; formal analysis, B.O.V. and D.M.; investigation, I.-T.O.; resources, I.-T.O. and B.O.V.; data curation, I.-T.O.; writing—original draft preparation, I.-T.O.; writing—review and editing, I.-T.O.; visualization, I.-T.O.; supervision, B.O.V. and D.M.; project administration, B.O.V. and D.M.; funding acquisition, B.O.V. All authors have read and agreed to the published version of the manuscript.

Funding: This research was funded by European Union's Horizon 2020 research and innovation program under grant agreement No. 101036871, project hOListic Green Airport (OLGA).

Institutional Review Board Statement: Not applicable.

Informed Consent Statement: Not applicable.

Data Availability Statement: The data presented in this study are available on request from the corresponding author.

Conflicts of Interest: The authors declare no conflict of interest.

References

1. Ukaogo, P.O.; Ewuzie, U.; Onwuka, C. Environmental pollution: Causes, effects, and the remedies. In *Microorganisms for Sustainable Environment and Health*; Elsevier: Amsterdam, The Netherlands, 2020; pp. 419–429.
2. Olsson, A.C.; Gustavsson, P.; Kromhout, H.; Peters, S.; Vermeulen, R.; Brüske, I.; Pesch, B.; Siemiatycki, J.; Pintos, J.; Brüning, T.; et al. Exposure to Diesel Motor Exhaust and Lung Cancer Risk in a Pooled Analysis from Case-Control Studies in Europe and Canada. *Am. J. Respir. Crit. Care Med.* **2010**, *183*, 941–948. [CrossRef] [PubMed]
3. Kelland, K. *Diesel Exhaust Fumes Can. Cause Cancer, WHO Says*; Reuters Agency: London, UK, 2012.
4. Bethoux, O. Hydrogen Fuel Cell Road Vehicles: State of the Art and Perspectives. *Energies* **2020**, *13*, 5843. [CrossRef]
5. Beckers, C.J.J.; Besselink, I.J.M.; Nijmeijer, H. The State-of-the-Art of Battery Electric City Buses. In Proceedings of the 34th International Electric Vehicle Symposium and Exhibition, Nanjing, China, 3–6 June 2021.
6. Chen, L.; Li, Y.; Huang, C.; Li, B.; Xing, Y.; Tian, D.; Li, L.; Hu, Z.; Na, X.; Li, Z.; et al. Milestones in Autonomous Driving and Intelligent Vehicles: Survey of Surveys. *IEEE Trans. Intell. Veh.* **2023**, *8*, 1046–1056. [CrossRef]
7. *Society of Automotive Engineers SAE Levels of Driving Automation™ Refined for Clarity and International Audience*; SAE International: Warrendale, PA, USA, 3 May 2021. Available online: https://www.sae.org/blog/sae-j3016-update (accessed on 1 March 2022).
8. Chan, T.K.; Chin, C.S. Review of Autonomous Intelligent Vehicles for Urban Driving and Parking. *Electronics* **2021**, *10*, 1021. [CrossRef]
9. Xu, Q.; Li, K.; Wang, J.; Yuan, Q.; Yuan, Y.; Chu, W. The status, challenges, and trends: An interpretation of technology roadmap of intelligent and connected vehicles in China (2020). *J. Intell. Connect. Veh.* **2022**, *5*, 1–7. [CrossRef]
10. Lanxon I Rode the World's First Autonomous Public Bus Service. Cnet, 18 May 2023. Available online: https://www.cnet.com/tech/i-rode-the-worlds-first-autonomous-public-bus-service/ (accessed on 7 October 2023).
11. Hall, S. *Driverless Buses Are Arriving Soon in These 3 European Cities*; World Economic Forum: Cologny, Switzerland, 27 January 2023. Available online: https://www.weforum.org/agenda/2023/01/autonomous-buses-geneva-project/ (accessed on 7 October 2023).
12. Goldbach; Sickmann, J.; Pitz, T.; Zimasa, T. Towards autonomous public transportation: Attitudes and intentions of the local population. *Transp. Res. Interdiscip. Perspect.* **2022**, *13*, 100504. [CrossRef]
13. ULTIMO Overview. Available online: https://ultimo-he.eu/about-ultimo/#overview (accessed on 7 October 2023).
14. CAVForth Ambitious and Complex CAVForth Autonomous Bus Service Launches in Scotland Controlled by Fusion Processing's Automated Drive System. 11 May 2023. Available online: https://www.cavforth.com/ambitious-and-complex-cavforth-autonomous-bus-service-launches-in-scotland-controlled-by-fusion-processings-automated-drive-system/ (accessed on 7 October 2023).
15. Iclodean, C.; Cordos, N.; Varga, B.O. Autonomous Shuttle Bus for Public Transportation: A Review. *Energies* **2020**, *13*, 2917. [CrossRef]
16. Cluj-Napoca CTP Cluj-Napoca Website. Available online: https://ctpcj.ro/index.php/ro/ (accessed on 10 October 2023).
17. Solaris Bus & Coach sp. z o.o. Zero Emissions Powertrains—Product Catalogue 2023. Available online: https://www.solarisbus.com/public/assets/content/pojazdy/Katalogi_2023/EN_Zeroemisyjne_1920_x_1080.pdf (accessed on 5 December 2023).
18. Solaris Bus & Coach sp. z o.o. Solaris Urbino 9 LE Electric—Technical Details. Available online: https://www.solarisbus.com/public/assets/Biuro_prasowe/2021_09_30_premiera_U9_LE_electric/Technical_data_Urbino_9_LE_electric.pdf (accessed on 5 December 2023).
19. Varga, B.O.; Mariasiu, F.; Miclea, C.D.; Szabo, I.; Sirca, A.; Nicolae, V. Direct and Indirect Environmental Aspects of an Electric Bus Fleet Under Service. *Energies* **2020**, *13*, 336. [CrossRef]
20. The European Parliament and The Council of the European Union Regulation (EU) 2019/2144 of the European Parliament and of the Council of 27 November 2019. 27 November 2019. Available online: https://eur-lex.europa.eu/eli/reg/2019/2144/oj (accessed on 10 December 2023).
21. Ansariyar, A.; Tahmasebi, M. Investigating the effects of gradual deployment of market penetration rates (MPR) of connected vehicles on delay time and fuel consumption. *J. Intell. Connect. Veh.* **2022**, *5*, 188–198. [CrossRef]
22. Liu, Y.; Wu, F.; Liu, Z.; Wang, K.; Wang, F.; Qu, X. Can language models be used for real-world urban-delivery route optimization? *Innov.* **2023**, *4*, 100520. [CrossRef] [PubMed]
23. Zhu, W.; Wu, J.; Fu, T.; Wang, J.; Zhang, J.; Shangguan, Q. Dynamic prediction of traffic incident duration on urban expressways: A deep learning approach based on LSTM and MLP. *J. Intell. Connect. Veh.* **2021**, *4*, 80–91. [CrossRef]
24. Qu, X.; Lin, H.; Liu, Y. Envisioning the future of transportation: Inspiration of ChatGPT and large models. *Commun. Transp. Res.* **2023**, *3*, 100103. [CrossRef]

sustainability

Article

Financial Perspectives on Human Capital: Building Sustainable HR Strategies

Mirabela-Constanța Matei [1,*], Leonard-Călin Abrudan [2] and Maria-Madela Abrudan [1]

[1] Department of Management-Marketing, Faculty of Economic Sciences, University of Oradea, 410087 Oradea, Romania; mabrudan@uoradea.ro
[2] Department of International Business, Faculty of Economic Sciences, University of Oradea, 410087 Oradea, Romania; labrudan@uoradea.ro
* Correspondence: mmatei@uoradea.ro

Abstract: This paper examines the challenges surrounding sustainable human resources management (HRM), particularly in the context of budget constraints that often lead to the reduction of employee development investments. Our research focuses on developing a comprehensive model that integrates financial management tools into HRM strategies, ensuring the prioritization of sustainable practices. Through a systematic analysis of existing knowledge, we propose a solution-oriented approach that supports the financial substantiation of investments in employee development. This study addresses key research questions, emphasizing the adaptation of corporate finance tools to meet HR's specific requirements. Our research not only identifies challenges but, more importantly, offers solutions by presenting a model that empowers organizations to align financial goals with HR development objectives. The results of our research aim to formulate a pragmatic and inventive model, offering a systematic framework for assessing the financial feasibility of initiatives in human resources development. Our model offers a practical framework for assessing the financial feasibility of HR development initiatives, facilitating informed decision-making and the promotion of sustainable HRM practices.

Keywords: employee development; sustainable HRM; financial tools; human resources; investment strategy

Citation: Matei, M.-C.; Abrudan, L.-C.; Abrudan, M.-M. Financial Perspectives on Human Capital: Building Sustainable HR Strategies. *Sustainability* **2024**, *16*, 1441. https://doi.org/10.3390/su16041441

Academic Editor: Assunta Di Vaio

Received: 20 November 2023
Revised: 18 January 2024
Accepted: 2 February 2024
Published: 8 February 2024

1. Introduction

This paper explores the complex field of sustainable human resources management (HRM) with a specific focus on integrating financial management tools into HRM strategies. This study proposes a novel and practical framework that systematically prioritizes sustainable practices within HRM. Our analysis of existing knowledge leads to a solution-oriented approach that substantiates the financial justification for robust investments in employee development initiatives. This research addresses essential questions surrounding the adaptation of corporate finance tools to meet the evolving requirements of HR. The scope extends to the systematic examination of the financial feasibility of HR development initiatives, with the proposed model offering practical insights for informed decision-making within organizations. Furthermore, this paper aims to highlight the individuality of its findings due to the lack of comparable publications on the utilization of financial tools in HRM.

The primary objective of this research is to develop a theoretical framework for integrating financial analysis into human resources development (HRD) from a sustainable HRM perspective, with a focus on proposing relevant financial tools. In this paper, we first conducted a review of the existing literature related to sustainable HRM, HRD, financial analysis, and their intersections. We opted for presenting the literature review chronologically to illustrate the evolution of the research in this field of study. This paper organizes the existing literature based on the sequence of when the studies were conducted or when key concepts emerged to provide a clear picture of how the field of sustainable HRM has evolved, starting with earlier works and progressing to more recent research. The second part of the paper elaborates on some financial tools that could be used to substantiate the investments in human resources development (HRD) as an important part of sustainable HRM.

The novelty of our paper proposing financial tools to substantiate investments in human resources development (HRD) lies in its ability to bridge the gap between theoretical concepts and practical application. The paper brings together two critical areas of organizational management: HRD and financial analysis. It provides a conceptual framework for integrating financial analysis into HRD, which is a novel approach and helps organizations align their human capital investments with financial goals. The paper also offers a holistic view of HRM by considering the financial aspect. This perspective allows organizations to see HRD as an investment with potential returns rather than just a cost, which is a paradigm shift from traditional HR practices.

Sustainable HRM focuses on fostering socially and environmentally responsible HR practices within organizations. Sustainable HRM is concerned not only with social and environmental sustainability but also with financial sustainability. When integrated with HRD, it promotes the development of employees' skills, knowledge, and abilities in ways that align with sustainability goals. This alignment creates a workforce better equipped to drive sustainability initiatives. Financial analysis, on the other hand, provides a quantitative lens through which organizations can evaluate the impact of HRD efforts on their financial performance. By employing financial tools, organizations can assess the economic benefits and costs associated with HRD programs, thereby substantiating the value of sustainable HRM practices. These intersections enable organizations to not only enhance their sustainability efforts but also justify their investments in human development by demonstrating how such investments contribute to long-term financial viability and competitiveness. As such, the convergence of sustainable HRM, HRD, and financial analysis represents a strategic and holistic approach to achieving both sustainability and financial objectives within organizations.

Managers have a key role in maintaining the balance between various aspects of the activity of the company. To maintain the company on a sustainable growing trend, it is necessary to satisfy the needs of major stakeholders. While in this paper we will refer to only five categories (listed here in random order), shareholders, employees, suppliers, government, and customers, there are other important stakeholders which should not be overlooked when developing the overall business strategy: environment, future generations, and so on [1].

A balance must be maintained between all these categories of stakeholders, and because of the divergence in their interests, the endeavor is even more difficult. Figure 1 illustrates what this balance could look like. The level of satisfaction was arbitrarily considered to be the same size for each stakeholder, and this explains the relative perfection of the image. In practice, reality does not correspond to the same pentagon form; however, the wider and more perfect the image, the better the company stands from the point of view of its sustainable growing trend. Of course, the importance of these stakeholders can fluctuate depending on the circumstances, but the idea is for a balance to be pursued and maintained.

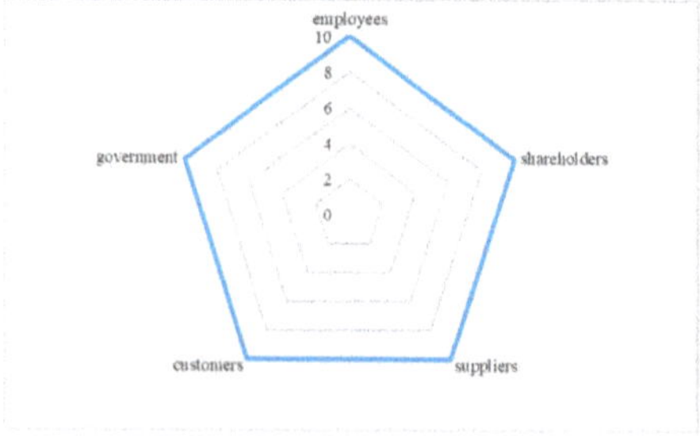

Figure 1. The stakeholders' balance pentagon. Source: own elaboration.

2. Theoretical Foundation

When considering stakeholders and their levels of satisfaction, it is important to be aware of the diverse nature and perceptions of their satisfaction. Simplistically, shareholders are typically pleased by higher dividends, employees strive for larger paychecks, governments expect companies to pay taxes, suppliers are satisfied when their debts are settled upon maturity, and customers are pleased with a good price–quality ratio.

All these stakeholders' requirements are considered under normal circumstances, but there are many other aspects that need to be addressed. In recent years, customers have diversified their preferences and now consider factors beyond just the price–quality ratio when measuring their satisfaction. Many customers are seeking products that are produced sustainably, with greater respect for the environment and socially responsible actions. Shareholders may also have other dimensions of their overall interests. They may focus on maximizing the value of the company and analyzing differences in stock prices at various times. Additionally, they may be more inclined towards ESG-based (Environmental, Social, and Governance) investment strategies rather than strictly focusing on maximizing shareholder value (MSV).

Employees are discussed last among all stakeholders in this paper because it specifically elaborates on this particular category of stakeholders. Additionally, employees expect more than just a monthly paycheck from their employers. They value good working conditions, which encompass both the physical space where employees carry out their regular activities and the work environment, including relationships with colleagues, career-related opportunities, and more. When people apply for a job, they typically assess the requirements presented in the job description. In doing so, they may identify certain incentives or statements that can influence their decision-making process. Cultural aspects are not negligible in terms of job perception [2,3]. Kamoche identifies some differences in the way employees could relate to their job depending on the cultural factors: "corporate welfarism in Cadbury Schweppes, the 'caring company' reputation of Johnson&Johnson, or the sense of belonging and job security associated with lifetime employment in some Japanese firms" [4].

On the other hand, a sustainable approach to doing business is preferred, and that includes identifying sustainable ways of tackling the needs of stakeholders. Employees are one of the major stakeholders usually considered. Of all the characteristics of sustainable HRM, we will mainly focus on care of employees and employee development.

The main purpose of this paper is to demonstrate that the need for HR development, as a sustainable approach in HRM, can be analyzed and supported using financial management tools. Our intention is to advocate for the identification of sustainable methods to support employees in their skill development. This research was conducted in two stages. In the first stage, a bibliometric review was performed to generate a comprehensive overview of existing knowledge in the field of sustainable HRM, which is presented in the theoretical background section. In the second stage, we present several financial management tools that can be utilized to support HR development initiatives.

To establish the coverage for this topic, an analysis on relevant articles in the field of sustainable HRM was conducted. The Web of Science Core collection was used because it offers a wide collection of scientific resources and the interest in our topic can be dynamically analyzed. When searching WoS for articles on "sustainable human resources management", more than 8700 articles were found for the 1992–2022 time frame. However, the interest in this topic has grown steadily since 2014. One interesting aspect in many of these articles is the scaling-down of ESG criteria—Environment, Social, Governance—to a size which seems fitting for a company: the environmental, social, and economic vision of the company's performance. This paper categorizes the available literature by the chronological order of study conduct and the emergence of pivotal concepts. This approach aims to offer a comprehensive overview of the evolution of the sustainable human resource management (HRM) field, commencing with earlier works and advancing to contemporary research.

This image of a company's performance was best captured by Elkington's triple bottom line concept [5]. This means that the company's activity should not only be seen in terms of revenues-minus-costs as a single bottom line, but it should be expanded to a more developed approach with a broad emphasis on the social and environmental dimensions of the full process. First, Elkington revealed a negative triad, inspired by thinkers such as Commoner, Goldsmith, and Ehrlich, whose interest was circumscribed to mostly environmental issues. The negative triad and its poisonous results that Elkington pointed out were "the combined trajectories of population growth, industrial pollution, and ecosystem destruction [which] threatened to undermine the future, whether capitalist or communist" [5]. This assertion was made in the author's early years (late 1960s), when communist regimes were powerful contenders to capitalism, and, while right-wing thinkers labelled Commoner, Goldsmith, and Ehrlich as "watermelon jibe—green on the outside, red on the inside", Elkington considers the thinkers were from a "much wider political spectrum".

One of the first mentions of HRM in terms of sustainability was that of Brotherton, Woolfenden, and Himmetoglu (1994), who were preoccupied with studies in the field of tourism. The authors identified that one of the problems in Turkish tourism was the quality of human resources in the hospitality industry. They were worried about the sustainability of Turkish tourism growth in the following period (after 1994), despite the "spectacular increase in the size of Turkish tourism" [6]. However, the authors focused only on the "shortage of skilled staff", the workplace conditions not being considered.

Kamoche [4] argues that the interaction between organizational routines, good practices, and human resources (HR) policies is generating HR competencies with certain strategic value going beyond the existing (at that moment) conditions. Barney and Wright (1998) assert that human resources should be regarded as a significant stakeholder and potential source of sustainable competitive advantage within an organization. To analyze the behavior of the HR function across different companies, the authors introduce the VRIO framework, which stands for "value, rareness, imitability, and organization." According to Barney and Wright, although managers may acknowledge employees as the most important asset in theory, their decisions and actions often do not align with this perspective. They also emphasize the relative ease of imitating many processes within a company. To address this challenge, the authors suggest that managers should focus their efforts on developing and nurturing unique characteristics of the firm's human resources that are not easily replicable by competitors [7].

The proposed solution involves delving into the distinctive qualities of each company to identify elements that can provide strategic advantages that are difficult for competitors to imitate. By understanding and leveraging these unique attributes, organizations can differentiate themselves and create sustainable competitive advantages in the marketplace. This approach moves beyond simply acknowledging the value of human resources and emphasizes the importance of actively developing and cultivating specific traits that set the organization apart from its rivals. An interesting statement was made by Lee and Miller (1999)—"People matter". Even if the authors studied mostly Korean companies, the statement and the proposed commitment to employees could be a mantra for every company, adapted, of course, to its cultural background [8]. Lee and Miller brought into the discussion the association between an organization's commitment to its employees (OCE) and the return on assets (ROA) indicator (otherwise a powerful and meaningful indicator). Usually, human resources as an asset of a company should respond as positively as possible to an analysis of their financial performance, but the authors have found that the previously mentioned association between the two notions is rather weak in some cases. They also found that the interaction of OCE with "the dedicated pursuit of Porter's (1980) strategies for achieving competitive advantage" is influencing ROA "strongly and positively" [8]. An enhanced level of OCE in a company leads to more effectively executed strategies, with one condition—that these strategies must be "dedicated, [. . .] intensive and thorough, positioning strategies" [8].

Daily and Huang (2001) discuss the relationship between two different major stakeholders—the environment and human resources—as a means to gain a competitive advantage over their competitors. According to the authors, implementing Environmental Management Systems (EMS) can be a lengthy and costly process. Daily and Huang identify numerous HRM-related instruments that could serve as key elements capable of positively influencing the implementation of an EMS. These instruments include "top management support, environmental training, employee empowerment, teamwork, and reward systems" [9]. By arguing that balancing these instruments with the cost and time duration of, for example, ISO 14001 guidelines, the authors demonstrate that developing environmentally oriented human resources could be a beneficial approach.

In another article focused on gaining competitive advantage by shaping human resources, Aragón-Sánchez et al. (2003) underline the importance of training and underline the difficulties in quantifying the outcomes of the training process. Their study found evidence of "significant relationships between training and performance" [10].

Jabbour and Santos identify the link between human resources and environmental management in a company [11] and, further, places HRM in a central role in pursuing the sustainable character of an organization [12]. In short, the two authors consider the environmental and human resources aspects as significantly related to the sustainability of companies.

In another article, Kramar (2013) discusses whether sustainable HRM will be the next HR approach. The article positions sustainable HRM as part of a broader strategic HRM framework, which encompasses the integration of sustainability and HRM. It also identifies various meanings associated with the relationship between the two within the broader strategic approach. The article recognizes organizational outcomes that extend beyond financial measures, presenting a more comprehensive perspective compared to MSV (maximizing shareholder value)-based strategies. Additionally, it introduces the moral aspect inherent in the entire HRM process [13].

Figurska and Sokół, in their research on Polish organizations, highlight the strategic nature of knowledge resources and their management. They propose benchmarking as a method to enhance the effectiveness and optimization of knowledge management processes [14].

Another area of interest for researchers is the relationship between managers (i.e., employers) and employees, examined through the lens of "workplace mutuality" [15]. In their article, Dobbins and Dundon promote a theory consisting of three elements—"system, society, and dominance"—and utilize this framework to examine the notion of "management-labor workplace partnerships" as a chimera. Through case study research conducted in organizations in Ireland, the authors discuss the dynamics between management and labor in the context of "neoliberal work regimes".

In a 2015 article, Doering et al. made a comparison between various types of capitalism according to their sustainable character. They were trying to answer the question of whether there are any sustainable types of capitalism by studying a large steel company, which operates both in Germany and in Brazil: the authors studied a company that was facing relatively new constraints regarding the sustainable development requirements in "Germany—as an example of a coordinated market economy (CME)—as well as in Brazil—this time an example of a hierarchical market economy (HME)" [16]. Their article demonstrates that different institutional contexts could generate different operational strategies regarding the green transition of the steel industry.

In an article discussing the efforts made by authorities, Rodriguez et al. (2017) consider that changes that happened in the work dynamic require more procedures, policies, and other actions from responsible institutions. The authors revealed an interesting facet of regulations and argue that "regulation sits at the center of competing economic and social demands, which are seen as both complementary and irreconcilable, and its complexity needs to be theorized and empirically mapped", which cannot be disputed [17].

Stankevičiūtė and Stankevičiūtė (2017) discuss smart power, a term borrowed from political science, as a useful tool in pursuing the sustainable character of HRM. In their article, the authors describe smart power as a combined tool (using both soft and hard power) to manage "work-life balance, management of employees' relations, and stress management". The premises are an aging society and health issues among employees. The authors also consider sustainable HRM to be an appropriate tool for "enhancing organization's profit, minimizing "ecological footprint" and reducing the harm on employees" [18].

In their research, Wepfer et al. (2017) examined the increasingly blurred boundaries between work and personal life among approximately 2000 employees. They discovered that many individuals struggle to establish a well-defined balance between these two domains and often find them intertwined [19]. The proliferation of new technologies, which enable constant connectivity and communication, has played a significant role in facilitating this merging of work and personal life. The study highlighted that employees, driven by the ease of access to work-related communication through technology, tend to work longer hours and with greater intensity. They frequently engage with work-related tasks even during their designated time off. This constant engagement and inability to disconnect from work activities can impede the availability of sufficient time for recovery activities, ultimately leading to adverse effects on their overall well-being. The findings shed light on the negative implications of the blurring boundaries between work and personal life. This study emphasizes the need for individuals and organizations to be aware of these challenges and strive to establish strategies and practices that promote a healthier work–life balance. This research underscores the importance of managing technology use and implementing policies that allow employees to have dedicated time for rest and recovery, ultimately supporting their well-being and overall quality of life.

In a research study conducted by Baum in 2018, the author proposes a set of concerted actions to be undertaken by local and national authorities to establish a framework that fosters developments in the quantity and quality of the tourism-involved workforce [20]. The research emphasizes the significance of the tourism industry and the role it plays in local and national economies. To support the growth and sustainability of this sector, Baum suggests that policymakers and authorities need to take proactive measures. Baum argues that local and national authorities can create an enabling environment that supports the development of a skilled and sustainable tourism workforce. This, in turn, contributes to the growth and competitiveness of the tourism industry and ensures its long-term success.

In a study conducted by Blašková et al. in 2018, the researchers explored the sustainable nature of motivation and its impact on key processes related to the development of human potential. The findings of their research revealed a strong correlation between the level of motivation and the quality of key processes such as leadership, appraisal, communication, and the creation of an atmosphere of trust [21]. The study highlights the importance of motivation in fostering an environment that supports the development of human potential within an organization. The researchers suggest that when employees are highly motivated, it positively influences the effectiveness of key processes that are essential for their development and growth.

In a study conducted by Stankevičiūtė and Stankevičiūtė in 2018, the authors address the issue of a lack of clarity in the attributes of sustainable human resource management (HRM). The authors argue that the traditional view of employees as resources to be consumed rather than developed is no longer appropriate. They propose a set of characteristics that define sustainable HRM, emphasizing the need for a shift towards a more holistic and sustainable approach: "long-term orientation, care of employees, care of environment, profitability, employee participation and social dialogue, employee development, external partnership, flexibility, compliance beyond labour regulations, employee cooperation, fairness, and equality" [22]. By delineating these characteristics, Stankevičiūtė and Stankevičiūtė offer a broader understanding of sustainable HRM that goes beyond traditional HR practices. They emphasize the need for organizations to adopt a comprehensive and integrated approach to HRM that aligns with the principles of sustainability, taking into account the well-being of employees, the environment, and the long-term success of the organization.

The research conducted by Zaid, Jaaron, and Talib Bon in 2018 examines the relationship between sustainable human resource management (Sustainable HRM) and sustainable supply chain management. They find a linkage between these two concepts and suggest that they share a connection that has a direct impact on the triple bottom line (TBL), which encompasses the environmental and social performance of companies. The study highlights that sustainable HRM practices, which focus on integrating sustainability principles into human resource management, can contribute to the sustainability of supply chain management. Sustainable supply chain management aims to incorporate environmental and social considerations into the management of supply chains to minimize negative impacts and create sustainable value [23].

The research conducted by Gutiérrez Crocco and Martin in 2019 focuses on the role of the union–management relationship and the process of resolving work conflicts in a sustainable manner. The authors explore how collaboration and effective communication between labor unions and management can contribute to sustainable outcomes within organizations. By fostering a cooperative approach to addressing work conflicts and finding mutually beneficial solutions, organizations can promote sustainability in their operations [24]. In a similar vein, Chams and García-Blandón (2019) highlight the growing awareness among companies regarding the importance of social, ethical, and ecological objectives. They argue that many organizations are now actively pursuing the 17 Sustainable Development Goals (SDGs) set forth by the United Nations. The authors suggest that implementing sustainable HRM practices can enhance the capability of human resources to understand and align their thinking with these SDGs. By incorporating sustainability principles into HRM practices, organizations can facilitate the achievement of SDGs more effectively [25].

Hitka et al. (2019) study the quality of employees and their level of engagement. The authors highlight the importance of improvement in the leaders' ability to be innovative and to "manage, motivate, and encourage other employees", leading in this way to comparative advantages for companies. The outcomes of their research showed a strong and direct relationship between career aspirations and the level of education [26].

Richards (2020) considers that sustainable HRM should put employees at the center of the whole process. The author thinks that this approach goes "far beyond" mainstream HRM [27], while Van Buren III pleads for employees' inclusion to establish a "pluralist perspective" and to pursue "social sustainability outside of the employment context" [28]. Along with employees, Nilsson and Nilsson (2021) introduce "first-line managers, trade union representatives, and HR practitioners" for an "extended work life and employability" [29]. Many people want to work even when in their older age, and it could be feasible to find a framework to better accommodate them. Vraňaková et al. (2021) studied the objective temporal coexistence of different generational groups at the level of almost every company. The authors consider that sustainable HRM should consider "the importance of age management pillars" [30]. The same age-related problem is revealed by Meidutė-Kavaliauskienė et al., who consider the aging population to be one of the major impediments in the way of the SDGs' implementation [31]. The relevance of the aging populations is related to certain goals among the 17 SDGs: inequality, poverty, health, and well-being. According to the authors, inequality must be seen both as between countries and within them.

Medina-Garrido et al. (2019) studied the balance between work and personal life in a sample of Spanish banks [32]. Strenitzerová and Achimský (2019), on the other hand, are interested in employee satisfaction and loyalty in the Slovak postal sector [33]. Di Fabio and Saklofske (2019) discuss personality traits and emotional intelligence in relational management towards sustainable development [34]. Chillakuri and Vanka (2020) reveal the positive effect of perceived organizational support on employees' health status and well-being within high-performance work systems [35]. In the same field, Durand et al. bring to light a sustainable return to work in the case of aging employees who are experiencing a work disability [36], while Radvila and Šilingienė (2020) are preoccupied by the remuneration system in a sustainable HRM context in a study based on a sample of Lithuanian enterprises [37].

In addition, two papers that provide examples of systematic literature reviews for sustainable HRM were identified. Macke and Genari (2018) analyzed "the state-of-the-art" of sustainable HRM using the Scopus database [38]. The authors grouped the 115 scientific papers from the period 2001–2018 into four categories. The first category is dedicated to sustainable leadership, which reveals the existing power at the individual and group level embedded in its principles, processes, practices, and organizational values. The second category demonstrates the relationship that exists among human resource management, environmental sustainability, and organizational performance. The third category includes tensions and paradoxes that exist between sustainability and HRM. These tensions are determined by the dichotomy present in the sustainability versus profitability discourse. The fourth category includes articles on the connection between sustainable HRM and the stakeholders of the company, especially those involved in social aspects. The second paper, by Kainzbauer and Rungruang (2019), studied scientific papers over a larger time period, from 1982 to 2019. In their bibliometric review, the authors mapped all articles, key articles, and journals and identified four Schools of Thought in this field: three focused on sustainability and one focused on general and mainstream HRM [39].

One of the most influential scholars in the field of sustainable HRM is Susan E. Jackson, with numerous citations to her work. As of 2023, according to Google Scholar, she has been cited alone and in co-authorship more than 122,000 times. Jackson's research covers various aspects of sustainable HRM, including burnout-related issues [40–42] and team and organizational diversity [43–45]. Jackson has also expressed concern about the "greening of strategic HRM scholarship" [46]. Furthermore, Jackson and her colleagues have addressed international HRM from the perspective of global talent management in their article [47]. In her comprehensive book titled "Managing Human Resources," Jackson discusses various HRM issues, with a significant emphasis on sustainable HRM [46].

The link between sustainability and HRM has also been a topic of investigation, with research exploring the role of HRM in fostering economically, socially, and ecologically sustainable organizations. Ehnert et al. (2014) delve into this subject, highlighting the importance of HRM in pursuing sustainability objectives [48]. This aligns with the concept of the triple bottom line, which encompasses environmental, social, and economic performance. Furthermore, the relationship between social sustainability and the quality of working life has also been examined. This research delves into how HRM practices can contribute to social sustainability by promoting a positive work environment, employee well-being, and work–life balance. The aim is to create organizations that not only achieve economic success but also prioritize the welfare of employees and society.

3. Research Design

The research design for this paper is centered around the exploration and analysis of theoretical concepts and ideas related to the concept of sustainable HRM. This study aims to develop a comprehensive theoretical framework that elucidates the intricate interplay between financial tools and sustainable HRM. While empirical data collection is not the primary focus, the research design employs a deductive approach, drawing on the existing literature and conceptual analysis to build and test theoretical propositions.

As stated, the aim of this paper is to demonstrate that the need for HR development can be analyzed and supported using financial management tools. Our research objective is to design a model that can be utilized for the financial justification of HR development initiatives. The proposed model consists of two parts: one based on a feed-before approach, and the other relying on traditional feedback analysis. It considers investment in HR development to be a project, similar to other economic projects within the company. The proposed model adapts specific corporate finance tools to the unique characteristics of HR development, which is a key component of sustainable HRM.

Our intention is to advocate for the identification of sustainable methods to support employees in their skills development. The research was conducted in two stages. In the first stage, a bibliometric review was performed to generate a comprehensive overview of existing knowledge in the field of sustainable human resource management (SHRM). In the second stage, we presented several financial management tools that can be utilized to support HR development initiatives. Our paper revolves around two primary research questions:

RQ1: Is Sustainable HRM sufficiently substantiated to be considered a stand-alone/legitimate approach?

RQ2: How can HR development, as a main characteristic of sustainable HRM, be transposed into long-term strategy?

The research follows a deductive approach, where existing theories and empirical findings in the fields of sustainable HRM are synthesized and applied to the context of human resources development. The deductive approach involves the formulation of theoretical propositions derived from established theories and concepts. Our research relies primarily on scholarly articles, books, and reports from disciplines such as finance and management.

To address the research questions, the following steps were undertaken:

- Literature Review and Content Analysis: A comprehensive review of the literature in the field of sustainable HRM was conducted, utilizing the Web of Science Core Collection and selecting relevant articles for analysis. This review aimed to gather existing knowledge and theories related to sustainable HRM.
- Critical Analysis of Existing Approaches: The selected literature was critically analyzed to evaluate the different approaches and theoretical perspectives within sustainable HRM. This analysis aimed to assess the level of substantiation and legitimacy of sustainable HRM as a stand-alone approach.
- Analysis of the Connection between HR Development and Sustainable HRM: The relationship between HR development and sustainable HRM was explored and examined. This analysis sought to understand the integration of HR development as a central element within long-term sustainable HRM strategies.
- Designing a Model for Substantiating HR Development Investments: Based on the findings from the literature review and analysis, a model was developed to assist in substantiating investments in HR development as a sustainable line of action in HRM. This model aimed to provide a framework for evaluating the financial aspects of HR development initiatives.

Assumptions and Limitations

It is assumed that the existing literature adequately represents the diverse approaches on sustainable HRM and human resources development. However, limitations include the potential simplification of complex HRD issues due to the theoretical nature of the study and the absence of direct empirical data collection. It should be noted that the articles selected for the literature review and content analysis were chosen for illustrative purposes. The main scientific database used was the Web of Science Core Collection. The primary focus was not only on analyzing the existing literature but also critically evaluating the various approaches from different theoretical perspectives.

4. Results and Discussion

The alignment of sustainable HRM with broader sustainability goals reflects a paradigm shift in organizational values. Companies increasingly recognize that sustainable practices in human resource management not only contribute to environmental and social responsibility but also enhance overall organizational performance. The integration of financial tools into HRM strategies, as emphasized in sustainable HRM, solidifies its legitimacy. The acknowledgment that investing in human capital development is not only a socially responsible act but also a financially sound decision demonstrates the approach's practical viability.

To answer RQ1: Is sustainable HRM sufficiently substantiated to be considered a stand-alone/legitimate approach? a literature review was conducted. Based on the critical analysis of the scientific literature in the field of sustainable HRM, conducted in Section 2 of the paper, several conclusions can be drawn:

- Sustainable HRM lacks a solid theoretical background and a consolidated approach. Similar to stakeholder theory, it lacks a clear normative framework, and often relies on good practices rather than well-defined regulations or instruments to guide decision-making.
- HR managers often do not view employees as developable assets, and instead focus on utilizing them without making significant efforts to support their skill development or tap into their true potential.
- Sustainable HRM encompasses a broader vision than mainstream HRM, considering outcomes beyond just financial results. The adoption of concepts like the triple bottom line (economic, social, and environmental performance) can significantly improve business practices.
- Placing employees at the center of sustainable HRM and recognizing them as major stakeholders can increase awareness and prioritize their well-being and development.
- HR practitioners should strive to create a unique HR system that focuses on attracting and retaining talent, while minimizing the possibility of replication by competitors. Developing a unique system that aligns with sustainability principles can provide a solid competitive advantage.

Overall, these conclusions highlight the need for further development and refinement of sustainable HRM, as well as the importance of integrating sustainability principles into HR practices.

4.1. Financial Support of Sustainable Human Resources Management

When it comes to financially supporting the decision-making process in a company, particularly in the field of HR, an interesting opinion arises from Wright, Dunford, and Snell. They assert that during prosperous times, companies readily justify expenditures on training, staffing, rewards, and employee involvement systems. However, when faced with financial difficulties, these HR systems become vulnerable to early cutbacks [49]. This perspective is influenced by the resource-based view (RBV) of the company, which suggests that the key instruments for achieving a competitive advantage should be sought within the company, rather than externally. This RBV strategy can be applied to various types of resources, including HR. Thus, HR can be considered as part of a use–develop–repeat process, leading to a potential competitive advantage. Furthermore, the authors highlight that HR is involved in both strategy design and implementation, emphasizing the need for the development of HRM. This underscores the importance of HR managers/practitioners taking a broader interest in creating HR strategies characterized by distinctiveness. This perspective emphasizes the significance of providing financial support for the decision-making process, particularly in HR. It suggests that HR should be recognized as an internal resource that contributes to competitive advantage. HR managers/practitioners should actively engage in developing unique HR strategies that align with the organization's overall strategy.

In terms of utilizing the human asset, it is important to consider the atypical nature of labor supply. Specifically, attention should be given to the relationship between wage levels and the number of hours employees are willing to work (see Figure 2). It has been observed that as wages increase, there reaches a point where the labor supply curve starts to bend back, resulting in a decrease in the number of hours employees are willing to work. This phenomenon suggests that there is a threshold beyond which higher wages do not necessarily lead to increased labor supply. As wages continue to rise, individuals may choose to work fewer hours due to factors such as increased leisure time, a desire for work–life balance, or alternative opportunities outside of work. This implies that the relationship between wages and labor supply is not linear, but rather exhibits diminishing returns. Understanding this non-linear relationship is crucial for employers and policymakers when making decisions related to wage levels and workforce planning. By recognizing the atypical nature of labor supply and considering the complexities of employee preferences, organizations can make more informed decisions regarding wage structures and work arrangements that effectively meet the needs and motivations of their workforce.

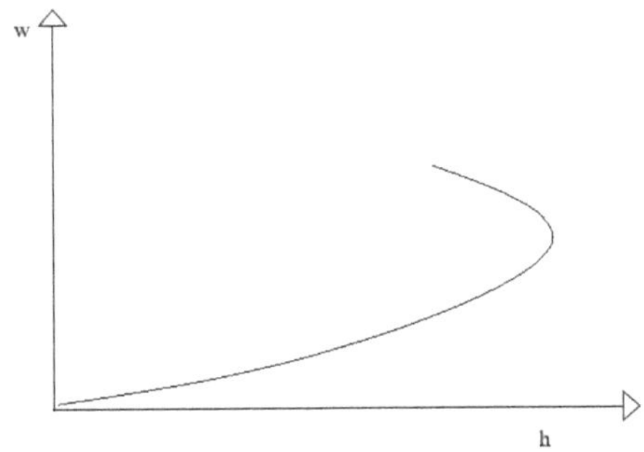

Figure 2. Labor supply, where w stands for wage and h for number of working hours. Source: own elaboration.

The substitution effect between money and leisure time is indeed reversing its direction, with leisure time becoming more valuable than monetary compensation. As a result, employees are inclined to reduce the number of hours they are willing to work. This shift in preference indicates that the allure of extrinsic financial motivation is diminishing, prompting HR managers and practitioners to focus more on intrinsic motivation. While extrinsic motivation may be more straightforward to address and implement in HR strategies, it is essential to recognize that easier does not necessarily equate to better. Numerous studies and arguments highlight the effects of extrinsic motivation on performance and creativity, as well as the existing gap between theory and practice in this area. Scientific evidence supports the notion that relying solely on extrinsic motivators may not lead to optimal outcomes. Studies by Zhang and Liu [50], Locke and Schattke [51], and Auger and Woodman [52] provide insights into the complex relationship between extrinsic motivation and employee performance and creativity. These works shed light on the limitations of extrinsic motivators and emphasize the importance of intrinsic factors, such as autonomy, mastery, and purpose, in driving employee engagement and productivity.

When it comes to leveraging intrinsic work motivation, HR specialists have various means at their disposal to harness the benefits associated with this type of motivation. One crucial aspect to consider is the impact of workplace conditions on employees' intrinsic motivation. Research by Matei and Abrudan (2016) highlights the significance of workplace conditions in influencing employee motivation. The study emphasizes that different employees place importance on various aspects of the work environment. Factors such as physical comfort, safety, supportive relationships, opportunities for growth and development, and meaningful work contribute to creating an environment that nurtures intrinsic motivation [53].

With both extrinsic and intrinsic implications, workplace conditions encompass at least two dimensions: physical conditions and psychologically subjective attributes. In the context of sustainable HRM, these attributes hold significant importance, and HR managers would demonstrate astuteness by taking actions to address them. While physical conditions are relatively straightforward, the psychologically subjective attributes of the workplace raise several issues that need to be considered. However, even if the physical conditions of the workplace seem to be a sine qua non, their meaning is not negligible. As Maslow (1943) pointed out, the situation in which individuals operate must be taken into account, but it alone cannot fully explain behavior [54]. Translated into the context of this paper, this implies that a specific workplace, devoid of people, has no impact on the behavior of individuals who work there. People's behavior is a motivational response to situational factors and motivation can serve as a nexus between the way in which the workplace presents itself and individual behavior. Humans can experience satisfaction or dissatisfaction with the physical conditions of their workplace, and their subsequent behavior can reflect their level of motivation in carrying out their activities. Situated within the intrinsic dimension of motivation, the workplace and its physical conditions can serve as a valuable tool for managers to enhance employee performance. However, a challenge arises from the fact that motivation is deeply personal and subjective, which can influence the decision-making process.

Considering the importance of workplace conditions, we paid attention to an article published by Astvik, Welander, and Larsson (2019), which focuses on work conditions in social services in Sweden and other European countries and how these conditions impact employees' willingness to stay in their organizations. The research was conducted using a quantitative analysis based on the results of two online questionnaires. The findings indicate that certain work conditions have an influence on employees' willingness to remain in their current organization. These conditions include low levels of conflicting demands and quantitative demands, high levels of openness and human resource orientation within the organization, and a high perception of service quality [55].

Authenticity, which is associated with job satisfaction and work engagement, is another characteristic of the workplace that HR managers should take into consideration [56].

Workplace actors, including colleagues and middle and top management, play a crucial role in shaping the organizational culture. Sustainable HRM actions are necessary in this regard to address issues such as discrimination and workplace bullying. For instance, Thorpe et al. (2014) examined the status and involvement of women in the fisheries sector in Sierra Leone. Despite fishing being predominantly seen as a male occupation, the authors found that women have a significant presence, particularly in postharvest stages. The authors emphasize the importance of gender awareness in this specific workplace [57]. In another article, Charlesworth and Macdonald (2015) discuss policy and regulatory developments in Australia that impact the working conditions of over 5.4 million women. These changes include the establishment of minimum labor standards and anti-discrimination protection for women workers. The authors argue that a national gender equality policy framework is crucial in ensuring fair treatment and opportunities for women [58].

4.2. Developing a Framework to Financially Support HR Development

Developing a framework to financially support human resources (HR) development involves creating a structured approach to allocate resources for the growth and enhancement of HR capabilities within an organization. By developing a comprehensive financial support framework, organizations can ensure that their HR professionals have the resources they need to stay current, contribute effectively to organizational success, and drive continuous improvement in HR practices. The acquisition and processing of data on a large scale is important to enhance organizational efficiency and provide access and understanding to necessary resources [59]. One way HR professionals could identify gaps in employee skills and knowledge is by using data analytics. This information can be used to tailor learning and development programs, ensuring that employees acquire the skills needed for both current and future roles, promoting sustainability in talent development.

To address RQ2: How can HR development, a key component of sustainable HRM, be integrated into long-term strategy? we examined and adapted concepts and tools from corporate finance. HR development is frequently linked with expenses, which raises questions about financial considerations. Training sessions and courses, although they enhance the quality of HR, necessitate a cost–benefit analysis. Engaging employees in their own skill development entails various implications. Decisions regarding whether to temporarily remove employees from their regular tasks to attend courses or to balance their involvement in ongoing processes while participating in after-work training sessions all involve financial costs.

In the first case, when employees are removed from their work, replacements are needed, and these replacements must be remunerated. Keeping employees at work full-time and sending them for additional training after hours can potentially affect their well-being and, subsequently, lead to decreased productivity, which can have negative financial consequences for the firm's revenues.

To facilitate informed decision-making in HR development, this paper proposes a model that enables managers to identify the most suitable approach. The model consists of two parts: a feed-before approach and a traditional feedback analysis. Both approaches treat HR development investment as a project similar to other economic projects within the company. These approaches incorporate relevant tools from corporate finance, such as net present value (NPV) and future value (FV), adapting them to their specific requirements. While other methods like internal rate of return (IRR) or profitability index (PI) can also be valuable in identifying the optimal option, this paper aligns with Brigham's (1982) viewpoint that the NPV method is the most reliable and recommended approach [60].

(a) Appling the net present value (NPV) method

It would indeed be valuable to estimate the future value of the impact of HR development on the firm's revenues. The first adaptation made is that instead of a risk-free rate assimilated to the deposit interest rate in a bank, the average rate of return for the analyzed firm could be considered as an investment alternative. When financial managers and HR managers evaluate the opportunity of investing in HR development, they can utilize the net present value (NPV) method to assess its feasibility. Viewing HR development as a project allows for prioritization in comparison to the company's regular activities. If the development of human capital can yield a higher return than the typical business operations, managers should give priority to this initiative. The focus is first on the positive nature of NPV and then on the magnitude of NPV. If the NPV is negative, the discussion becomes irrelevant. However, if the NPV is positive, the decision depends on the size of the NPV. Determining an appropriate significance threshold for the NPV value is crucial to give importance to the discussion. Estimating the future value of the impact on the company's revenues is challenging as it depends on the nature of the company's activities and the scale and quality of the HR segment being developed. Additionally, there may be various other factors influencing the impact's size, but the significance of the two factors mentioned earlier should overshadow them.

$$\mathrm{NPV} = \mathrm{PV} - \mathrm{IC} \tag{1}$$

where:

NPV—net present value.

PV—present value of the future increase in the firm's revenues.

IC—implementation cost for the HR development method—i.e., the amount that must be paid for the training session or for the course.

Further, the present value of the future impact of the investment must be calculated.

$$\mathrm{PV} = \frac{\mathrm{FI}}{(1+\mathrm{r})^{\mathrm{n}}} \tag{2}$$

where:

PV—the present value of the future impact of the investment.

FI—the future impact of the investment.

r—the average profit rate at the firm's level. It can be the arithmetic average, the mean, or the weighted average of the profit rates the company is recording.

n—the number of years in which the impact should happen.

Once the calculations are completed, the NPV for the HR development project can be determined and used to evaluate the feasibility of the chosen method. It is important to acknowledge that the main limitation of this tool lies in the challenge of accurately calculating the future impact of the investment. However, there are companies for which this difficulty may be mitigated. The essence of this approach is its feed-before nature, which assists in identifying the appropriate HR development method. Nevertheless, it is crucial for every company to incorporate references to this type of impact in its future activity projections. By including considerations for HR development impact, companies can enhance their strategic planning and decision-making processes.

(b) Using the future value (FV) tool

Another valuable instrument (feedback-type) for conducting a cost–benefit analysis is calculating the future value of the amount invested in HR development. This instrument allows for analysis over various time intervals, such as one year, two years, and so on. Two main approaches can be employed in this stage.

The first approach involves comparing the calculated future value with the increase in the firm's revenues. By calculating the difference between these two values, one can obtain a clear understanding of the profitability of the HR development investment.

The second approach entails assessing the future value through the break-even point of the calculated amount. In some cases, this analysis may yield a result in terms of the number of products that need to be sold to recoup the investment. However, a potential issue arises if the analysis is correlated with market share reports, revealing that the projected number of products cannot be sold within the existing framework.

It is important to carefully consider these findings and evaluate the feasibility of the HR development investment based on the projected future value and its alignment with market conditions and potential sales.

Translated in equations, this last paragraph appears as follows:

$$FV = IC \times (1 + r)^n \tag{3}$$

in which

FV is the future value of the amount paid as a fee for a training session or course.

The other notations remain as in the previous example, as future impact, used in the following lines, will be written as FI.

The analysis continues by calculating the ratios between the two indicators. Comparing the results with the value 1 could provide valuable feedback for the whole HR development operation.

$FI/FV > 1$—means that the investment was properly chosen.

$FI/FV < 1$—means that the investment should have not been made.

At the end of this feedback analysis, the timeframe in which such an investment could be profitable should be found.

4.3. Practical Implications

This paper proposes the integration of financial tools into human resource management (HRM) strategies, with a specific focus on sustainable practices. The findings underscore the significance of adopting a comprehensive model that not only prioritizes sustainable HRM but also substantiates the financial justification for investments in employee development initiatives.

The implications of this research extend beyond theoretical considerations, offering practical insights for organizations navigating the challenges of managing human capital within budget constraints. By introducing a systematic framework for evaluating the financial viability of HR development initiatives, the paper empowers decision-makers to align organizational goals with sustainable HRM practices. The proposed model represents an asset for strategic decision-making, helping HR professionals to bridge the gap between financial considerations and sustainable HRM practices.

By offering a solution-oriented approach, this paper supports organizations in optimizing their investments in employee development. This is crucial for enhancing workforce capabilities, fostering talent retention, and ultimately contributing to improved organizational performance.

The paper serves as a guide for organizations seeking to embed sustainability into their HRM practices. By integrating financial tools into HR strategies, organizations can not only achieve economic efficiency but also contribute to environmental and social sustainability through the development and retention of a skilled workforce.

This paper's focus on sustainable HRM practices aligns with broader CSR objectives. Organizations can utilize the insights from the research to integrate socially responsible practices into their HRM, showcasing a commitment to employee well-being and development as part of their broader corporate citizenship.

5. Conclusions

The aim of this paper is to analyze the existing knowledge in the field of sustainable human resource management (HRM) and address the main inconsistencies through an interconnected approach. This approach supports the implementation of sustainable HRM strategies while also proposing the use of financial management tools to support and substantiate investments in human resource development.

The paper emphasizes the importance of a use–develop–repeat process, which, if properly designed, can lead to achieving a comparative advantage in pursuing sustainability in the workplace. Two main approaches regarding HR are proposed in this paper. The first approach focuses on the utilization of HR and discusses both intrinsic and extrinsic motivations, as well as other methods to enhance the appropriate utilization of HR resources.

The second approach is dedicated to the development process of sustainable HRM and introduces tools that can be used to establish the financial feasibility of this process. These tools provide a means to assess the financial viability of investing in HR development initiatives within the context of sustainability.

Overall, the paper aims to address the gaps and inconsistencies in sustainable HRM by integrating financial management tools into the decision-making process. By considering the utilization and development of HR from a sustainable perspective, organizations can enhance their competitiveness and achieve sustainable outcomes in the long run.

This paper provides a theoretical foundation for organizations to make more informed and data-driven decisions regarding HRD investments. This aligns with sustainable HRM principles by ensuring that resources are allocated efficiently to initiatives that contribute positively to both human development and organizational sustainability.

Our paper proposes the integration of financial analysis into sustainable HRM, addressing a gap in the literature. By doing so, it extends the scope of HRM beyond traditional HR functions, making it more strategically aligned with organizational financial goals. The proposed financial tools offer a means to economically justify HRD investments. This contributes to the sustainability of HRM by demonstrating the financial returns and benefits of human capital development, making a strong case for continued investment in employee growth and well-being. The paper argues the importance of aligning HRD efforts with the organization's sustainability goals. The proposed tools help HRM practitioners ensure that HRD investments contribute positively to all dimensions of sustainability. The theoretical framework and proposed financial tools create opportunities for further research in the field. Empirical studies that focus on validating the effectiveness of these financial tools in diverse organizational contexts are needed to advance the field of sustainable HRM. Sustainability is an evolving and crucial aspect of modern business and integrating it with HRD and finance creates a forward-looking approach.

Using the proposed tools, HR managers and practitioners can effectively substantiate their decisions regarding HR development. However, implementing such a process may encounter challenges, as some employees may resist change or struggle to meet the required performance levels. This highlights the importance of skillful HR managers who can identify suitable employees, nurture their talents, retain them within the company, and achieve a favorable cost–benefit ratio. Moreover, HR managers must develop a unique approach that cannot be easily replicated by competitors. This is crucial for realizing a comparative advantage and maintaining a competitive edge in the market.

By employing the adapted NPV or FV methods to establish the feasibility of HR development, the quality of sustainable HRM in a company can be enhanced. This, in turn, enables companies to grow in a sustainable manner. The main objective of this research was to design a model for the financial substantiation of HR development, adapting corporate finance tools to the specific needs of HR development within the context of sustainable HRM. The proposed model has significant practical implications.

However, it is important to acknowledge the limitations of this study. Future research should aim to test the proposed model in practice and make adjustments if necessary. Testing the model would not only contribute to further research but also reveal any limitations or challenges that may arise in its implementation. Additionally, it is crucial to consider the long-term effects of implementing such a model, which are beyond the scope of this study. Furthermore, the willingness to use the model for substantiating HR development investments may vary due to various factors such as company size, national and organizational culture, personal beliefs, and more. Data regarding this aspect are limited and warrants further investigation.

Even though this paper is mainly theoretical, it has the potential to guide practical applications in real-world organizations. By proposing financial tools, our paper lays the foundation for future practical implementations and empirical research, which can validate the effectiveness of the proposed tools.

Author Contributions: Conceptualization, L.-C.A. and M.-C.M.; methodology, M.-C.M.; formal analysis, L.-C.A. and M.-M.A.; data curation, L.-C.A.; writing—original draft preparation, L.-C.A. and M.-C.M.; writing—review and editing, M.-C.M. and M.-M.A. All authors have read and agreed to the published version of the manuscript.

Funding: This research received no external funding.

Institutional Review Board Statement: Not applicable.

Informed Consent Statement: Not applicable.

Data Availability Statement: Data are contained within the article.

Conflicts of Interest: The authors declare no conflicts of interest.

References

1. Abrudan, L.; Matei, M.; Abrudan, M. Towards Sustainable Finance: Conceptualizing Future Generations as Stakeholders. *Sustainability* **2023**, *13*, 13717. [CrossRef]
2. Matei, M.; Abrudan, M. Cultural dimensions and work motivation in the European Union. *Ann. Univ. Oradea Econ. Sci. Ser.* **2013**, *22*, 1511–1519.
3. Prince, N.; Prince, J.; Kabst, R. National culture and incentives: Are incentive practices always good? *J. World Bus.* **2020**, *55*, 101075. [CrossRef]
4. Kamoche, K. Strategic Human Resource Management within a Resource-Capability View of the Firm. *J. Manag. Stud.* **1996**, *33*, 213–233. [CrossRef]
5. Elkington, J. *Cannibals with Forks: The Triple Bottom Line of 21st Century Business*; Capstone Publishing Limited: Oxford, UK, 1998.
6. Brotherton, B.; Woolfenden, G.; Himmetoglu, B. Developing Human-Resources for Turkeys Tourism Industry in the 1990s. *Tour. Manag.* **1994**, *15*, 109–116. [CrossRef]
7. Barney, J.B.; Wright, P.M. On becoming a strategic partner: The role of human resources in gaining competitive advantage. *Hum. Resour. Manag.* **1998**, *37*, 31–46. [CrossRef]
8. Lee, J.; Miller, D. People matter: Commitment to employees, strategy and performance in Korean firms. *Strateg. Manag. J.* **1999**, *20*, 579–593. [CrossRef]
9. Daily, B.F.; Huang, S. Achieving sustainability through attention to human resource factors in environmental management. *Int. J. Oper. Prod. Manag.* **2001**, *21*, 1539–1552. [CrossRef]
10. Aragón-Sánchez, A.; Barba-Aragón, I.; Sanz-Valle, R. Effects of training on business results. *Int. J. Hum. Resour. Manag.* **2003**, *14*, 956–980. [CrossRef]

11. Jabbour, C.J.C.; Santos, F.C.A. Relationships between human resource dimensions and environmental management in companies: Proposal of a model. *J. Clean. Prod.* **2008**, *16*, 51–58. [CrossRef]
12. Jabbour, C.J.C.; Santos, F.C.A. The central role of human resource management in the search for sustainable organizations. *Int. J. Hum. Resour. Manag.* **2008**, *19*, 2133–2154. [CrossRef]
13. Kramar, R. Beyond strategic human resource management: Is sustainable human resource management the next approach? *Int. J. Hum. Resour. Manag.* **2013**, *25*, 1069–1089. [CrossRef]
14. Figurska, I.; Sokół, A. Optimization of Knowledge Management Processes through Benchmarking in Organizations. *Mediterr. J. Soc. Sci.* **2014**, *5*, 235–246. [CrossRef]
15. Dobbins, T.; Dundon, T. The Chimera of Sustainable Labour-Management Partnership. *Br. J. Manag.* **2015**, *28*, 519–533. [CrossRef]
16. Doering, H.; Evans, C.; Stroud, D. Sustainable Varieties of Capitalism? The Greening of Steel Work in Brazil and Germany. *Relat. Ind.* **2015**, *70*, 621–644. [CrossRef]
17. Rodriguez, J.K.; Johnstone, S.; Procter, S. Regulation of work and employment: Advances, tensions and future directions in research in international and comparative HRM. *Int. J. Hum. Resour. Manag.* **2017**, *28*, 2957–2982. [CrossRef]
18. Savanevičienė, A.; Stankevičiūtė, Z. Smart Power as A Pathway for Employing Sustainable Human Resource Management. *Inz. Ekon. Eng. Econ.* **2017**, *28*, 198–206. [CrossRef]
19. Wepfer, A.G.; Allen, T.D.; Brauchli, R.; Jenny, G.J.; Bauer, G.F. Work-Life Boundaries and Well-Being: Does Work-to-Life Integration Impair Well-Being through Lack of Recovery? *J. Bus. Psychol.* **2017**, *33*, 727–740. [CrossRef]
20. Baum, T. Sustainable human resource management as a driver in tourism policy and planning: A serious sin of omission? *J. Sustain. Tour.* **2018**, *26*, 873–889. [CrossRef]
21. Blašková, M.; Figurska, I.; Adamoniene, R.; Poláčková, K.; Blaško, R. Responsible Decision making for Sustainable Motivation. *Sustainability* **2018**, *10*, 3393. [CrossRef]
22. Stankevičiūtė, Ž.; Savanevičienė, A. Designing Sustainable HRM: The Core Characteristics of Emerging Field. *Sustainability* **2018**, *10*, 4798. [CrossRef]
23. Zaid, A.A.; Jaaron, A.A.M.; Talib Bon, A. The impact of green human resource management and green supply chain management practices on sustainable performance: An empirical study. *J. Clean. Prod.* **2018**, *204*, 965–979. [CrossRef]
24. Gutiérrez Crocco, F.; Martin, A. Towards a sustainable HRM in Latin America? Union-management relationship in Chile. *Empl. Relat.* 2019; *ahead-of-print*. [CrossRef]
25. Chams, N.; García-Blandón, J. On the importance of sustainable human resource management for the adoption of sustainable development goals. *Resour. Conserv. Recycl.* **2019**, *141*, 109–122. [CrossRef]
26. Hitka, M.; Kucharčíková, A.; Štarchoň, P.; Balážová, Ž.; Lukáč, M.; Stacho, Z. Knowledge and Human Capital as Sustainable Competitive Advantage in Human Resource Management. *Sustainability* **2019**, *11*, 4985. [CrossRef]
27. Richards, J. Putting employees at the centre of sustainable HRM: A review, map and research agenda. *Empl. Relat.* 2020; *ahead-of-print*. [CrossRef]
28. Van Buren, H.J., III. The value of including employees: A pluralist perspective on sustainable HRM. *Empl. Relat.* 2020; *ahead-of-print*. [CrossRef]
29. Nilsson, K.; Nilsson, E. Organisational Measures and Strategies for a Healthy and Sustainable Extended Working Life and Employability—A Deductive Content Analysis with Data Including Employees, First Line Managers, Trade Union Representatives and HR-Practitioners. *Int. J. Environ. Res. Public Health* **2021**, *18*, 5626. [CrossRef]
30. Vraňaková, N.; Gyurák Babeľová, Z.; Chlpeková, A. Sustainable Human Resource Management and Generational Diversity: The Importance of the Age Management Pillars. *Sustainability* **2021**, *13*, 8496. [CrossRef]
31. Meidutė-Kavaliauskienė, I.; Dudzevičiūtė, G.; Maknickienė, N.; Vasilis Vasiliauskas, A. The relation between aging of population and sustainable development of EU countries. *Entrep. Sustain. Issues* **2020**, *7*, 2026–2042. [CrossRef]
32. Medina-Garrido, J.A.; Biedma-Ferrer, J.M.; Ramos-Rodríguez, A.R. Moderating effects of gender and family responsibilities on the relations between work–family policies and job performance. *Int. J. Hum. Resour. Manag.* **2019**, *32*, 1006–1037. [CrossRef]
33. Strenitzerová, M.; Achimský, K. Employee Satisfaction and Loyalty as a Part of Sustainable Human Resource Management in Postal Sector. *Sustainability* **2019**, *11*, 4591. [CrossRef]
34. Di Fabio, A.; Saklofske, D. Positive Relational Management for Sustainable Development: Beyond Personality Traits—The Contribution of Emotional Intelligence. *Sustainability* **2019**, *11*, 330. [CrossRef]
35. Chillakuri, B.; Vanka, S. Understanding the effects of perceived organizational support and high-performance work systems on health harm through sustainable HRM lens: A moderated mediated examination. *Empl. Relat.* 2020; *ahead-of-print*. [CrossRef]
36. Durand, M.J.; Coutu, M.F.; Tremblay, D.; Sylvain, C.; Gouin, M.M.; Bilodeau, K.; Kirouac, L.; Coté, D. Insights into the Sustainable Return to Work of Aging Workers with a Work Disability: An Interpretative Description Study. *J. Occup. Rehabil.* **2020**, *31*, 92–106. [CrossRef] [PubMed]
37. Radvila, G.; Šilingienė, V. Designing Remuneration Systems of Organizations for Sustainable HRM: The Core Characteristics of an Emerging Field. *Int. J. Hum. Resour. Stud.* **2020**, *10*, 252. [CrossRef]
38. Macke, J.; Genari, D. Systematic Literature Review on Sustainable Human Resource Management. *J. Clean. Prod.* **2018**, *208*, 806–815. [CrossRef]
39. Kainzbauer, A.; Rungruang, P. Science Mapping the Knowledge Base on Sustainable Human Resource Management, 1982–2019. *Sustainability* **2019**, *11*, 3938. [CrossRef]

40. Maslach, C.; Jackson, S.E. The measurement of experienced burnout. *J. Organ. Behav.* **1981**, *2*, 99–113. [CrossRef]
41. Maslach, C.; Jackson, S.E. The role of sex and family variables in burnout. *Sex Roles* **1985**, *12*, 837–851. [CrossRef]
42. Jackson, S.E.; Turner, J.A.; Brief, A.P. Correlates of burnout among public service lawyers. *J. Organ. Behav.* **1987**, *8*, 339–349. [CrossRef]
43. Jackson, S.E. Research on Work Team Diversity: Progress and Promise. *Perform. Improv. Q.* **2008**, *12*, 200–203. [CrossRef]
44. Jackson, S.E.; Seo, J. The greening of strategic HRM scholarship. *Organ. Manag. J.* **2010**, *7*, 278–290. [CrossRef]
45. Chung, Y.; Jackson, S.E. Co-worker trust and knowledge creation: A multilevel analysis. *J. Trust Res.* **2011**, *1*, 65–83. [CrossRef]
46. Jackson, S.E.; Schuller, R.S.; Werner, S. *Managing Human Resources*; Oxford University Press: New York, NY, USA, 2018.
47. Schuler, R.S.; Jackson, S.E.; Tarique, I. Global talent management and global talent challenges: Strategic opportunities for IHRM. *J. World Bus.* **2011**, *46*, 506–516. [CrossRef]
48. Ehnert, I.; Harry, W.; Zink, K.J. (Eds.) *Sustainability and Human*; Springer: Berlin/Heidelberg, Germany, 2014. [CrossRef]
49. Wright, P.M.; Dunford, B.B.; Snell, S.A. Human resources and the resource based view of the firm. *J. Manag.* **2001**, *27*, 701–721. [CrossRef]
50. Zhang, Y.; Liu, S. Balancing employees' extrinsic requirements and intrinsic motivation: A paradoxical leader behaviour perspective. *Eur. Manag. J.* **2022**, *40*, 127–136. [CrossRef]
51. Locke, E.; Schattke, K. Intrinsic and Extrinsic Motivation: Time for Expansion and Clarification. *Motiv. Sci.* **2019**, *5*, 277–290. [CrossRef]
52. Auger, P.; Woodman, R. Creativity and Intrinsic Motivation: Exploring a Complex Relationship. *J. Appl. Behav. Sci.* **2016**, *52*, 342–366. [CrossRef]
53. Matei, M.; Abrudan, M. Adapting Herzberg's Two Factor Theory to the Cultural Context of Romania. *Procedia Soc. Behav. Sci.* **2016**, *221*, 95–104. [CrossRef]
54. Maslow, A.H. A theory of human motivation. *Psychol. Rev.* **1943**, *50*, 370–396. [CrossRef]
55. Astvik, W.; Welander, J.; Larsson, R. Reasons for Staying: A Longitudinal Study of Work Conditions Predicting Social Workers' Willingness to Stay in Their Organisation. *Br. J. Soc. Work* **2019**, *50*, 1382–1400. [CrossRef]
56. Van den Bosch, R.; Taris, T.W. Authenticity at Work: Development and Validation of an Individual Authenticity Measure at Work. *J. Happiness Stud.* **2013**, *15*, 1–18. [CrossRef]
57. Thorpe, A.; Pouw, N.; Baio, A.; Sandi, R.; Ndomahina, E.T.; Lebbie, T. Fishing Na Everybody Business": Women's Work and Gender Relations in Sierra Leone's Fisheries. *Fem. Econ.* **2014**, *20*, 53–77. [CrossRef]
58. Charlesworth, S.; Macdonald, F. Women, work and industrial relations in Australia in 2014. *J. Ind. Relat.* **2015**, *57*, 366–382. [CrossRef]
59. Sobolu, R.; Stanca, L.; Bodog, S. Automated Recognition Systems: Theoretical and Practical Implementation of Active Learning for Extracting Knowledge in Image-based Transfer Learning of Living Organisms. *Int. J. Comput. Commun. Control* **2023**, *18*, 5728. [CrossRef]
60. Brigham, E.F. *Financial Management Theory and Practice*; The Dryden Press: New York, NY, USA, 1982.

sustainability

MDPI

Article

Influence of Environmental Perception on Place Attachment in Romanian Rural Areas

Dragos Darabaneanu *, Daniela Maci and Ionut Mihai Oprea

Department of Sociology and Social Work, Faculty of Social-Humanistic Sciences, University of Oradea, Str. Universitatii, No. 6, Campus II, 410087 Oradea, Romania; dmaci@uoradea.ro (D.M.); ioprea@uoradea.ro (I.M.O.)
* Correspondence: ddarabaneanu@uoradea.ro or darabaneanu@gmail.com

Abstract: This study analyzes aspects of place attachment in rural areas, as an element of social stability that determines attitudinal and behavioral patterns within a harmonious relationship between human beings and the environment. A higher level of place attachment generates efficient behavior patterns for the improvement of the problems caused by pollution and the degradation of natural environments. In the second section, we set out to measure the forms of manifestation of place attachment in rural areas and to identify effective strategies that can contribute to increasing the intensity of this phenomenon. We set out a study of the causality between environmental perception and place attachment. We carried out an investigation based on a questionnaire to determine the forms of manifestation of place attachment and environmental perception. We tested a statistical model to confirm or not the determining relationship between the two social phenomena. Our study also offers an original interpretation of environmental perception and explains the degree of intensity with which this phenomenon is felt at the individual level. The practical importance of this study lies in the fact that it offers a strategy proven by sociological analysis, which can be applied to stimulate an increase in intensity of the manifestation of feelings of place attachment, which ultimately leads to the spread of pro-environmental attitudinal and behavioral patterns.

Keywords: pro-environment behavior; place attachment; rural communities; collective attitudes; social inclusion; behavioral education; ecosystems conservation

Citation: Darabaneanu, D.; Maci, D.; Oprea, I.M. Influence of Environmental Perception on Place Attachment in Romanian Rural Areas. *Sustainability* **2024**, *16*, 1106. https://doi.org/10.3390/su16031106

Academic Editor: Marc A. Rosen

Received: 22 December 2023
Revised: 25 January 2024
Accepted: 26 January 2024
Published: 28 January 2024

1. Introduction

In this study, we set out to investigate the relationship between place attachment and environmental perception in the Romanian countryside, taking into account visibility and concerns about the home and the environment defined from a plurality of perspectives, and also the historical development of interest in this topic. The literature review reveals, on the one hand, the abundance of types of relationships between the two phrases, and, on the other hand, the need to understand, establish, and formulate the meaning of the terms for the investigation of any type of relationship [1–23]. A brief history of the semantics of the terms and how they have been delimited and (re)signified becomes necessary. This can only be achieved philosophically, because all these terms existed and were explored in this field before the emergence of other fields such as sociology, psychology, urbanism, etc. [2–5,7,8,11,24–34]. In the second part of the study, we apply the research methodology to measure the intensity with which environmental perceptions are manifested at the level of the rural population [35,36]. The general premise is that people's interest in the environment is a necessary condition in the context of place attachment and pro-environmental behavior. The secondary premise is that there is a semantic impoverishment of how people relate to place and environment as the world becomes more technical and urbanized. However, this terminological technicality renames the same content with other names.

After the literature review, there follows an archeology of vocabulary (analysis, operationalization, and hermeneutics of terms) and we discuss the multiplicity of meanings.

The use of these methods stems from the problematization and history of terms that belong primarily to philosophy. Philosophy is the field that first formulated, delineated, and problematized these terms before the emergence of the social sciences and their interest in these issues. For the association between the first two terms, place and attachment, the interdependent relationship between place as home or community and self as self-identity is taken into account [1–5]. For the association between the environment and perception, the intentionality that shifts the focus from one term to another is taken into account [6].

The instrument used for the research methodology was a questionnaire model with closed questions that contained a set of Likert scales for measuring attachment to the residential environment and the degree of satisfaction related to the surrounding environment. This was applied to rural subjects. The variables used were landscapes of the home area, agriculture, the place of residence, one's own household, neighbors, and the street, and also relationships with the health system, public institutions, the village hall, school, relatives, and colleagues. The use of certain items in developing the place attachment measurement scale is based on the model by Scannell and Gifford, who proposed a three-dimensional framework of place attachment (personal, psychological, and place dimensions) after analyzing a diversity of studies [8].

The relevance of our study lies in an original interpretation of environmental perception and the relationship between it and place attachment and how we try to show and explain the level of intensity that each individual feels at the level of the phenomenon of environmental perception. Last but not least, the practical relevance of this study is that the influence of the elements that describe environmental perception must start from environmental elements with which people are in permanent contact, because these elements are more accessible and have the highest degree of repeatability in daily behavior, as proven by the methodology used. Therefore, the possibility of applying the strategy proposed in this study can open up further research that involves sharing models of pro-environmental attitudes and behavior.

2. Literature Review

The extent of development of place attachment in the past 40 years on the one hand, and, especially in the 1980s, definitions of place attachment related to relation with neighborhoods, affective aspects between people and the environment, and human behavior, on the other hand, have led to a diversity of definitions of place attachment [7]. Scannell and Gifford synthesized the multidimensionality of this term in which meanings are "synthesized into a three-dimensional, person–process–place organizing framework" given the semantic multiplicity of place attachment [8]. This tripartite model was often used in further studies because the three dimensions (personal, psychological, and place) cover the definition of place attachment. When considering the personal dimension of place attachment, it is necessary to consider the multitude of individual and collective meanings related to places. The connection between the person and sense of place, cognition, and behavior is focused on the psychological process [8,9]. The importance of this dimension derives from the inclusion of human beings and the way they relate to the place. The understanding of how individuals and the groups to which they belong relate to a place takes into account the importance of the nature of psychological interactions in environments that mean something important for people, such as "the psychological dimension includes the affective, cognitive, and behavioral components of attachment" [8]. Therefore, place attachment cannot be defined apart from social interactions and the psychological components that occur at the level of these interactions. In this context, therefore, "spatial level, specificity, and the prominence of social or physical elements" are the characteristics of place attachment that provide place dimension, namely the social, symbolic, and physical aspects of place [9].

On the other hand, from investigation of the term "place attachment" derives concepts such as dwelling identity, community identity, and regional identity, in whose conceptual delimitations social or environmental factors are considered as demographic qualities of residents and interpretive residential affiliations, social participation in the local community, and patterns of intercommunity spatial activity [1]. In the relationship between place and attachment, the way in which the person relates to the place and to the symbolism and affectivity of the place such as a house, home, residence, location, etc., matters. Therefore, the way in which "self is situated in the social-spatial environment" is important for place identity [1]. In this relationship between place and attachment, both the houses and the term of dwelling that opens towards the environment are important. The dwelling means both the way the person relates to their own house and the relationship between person and the exterior of the house, namely with the environment and nature. However, while there are numerous studies "that have attempted to assess the restorative properties of nature and/or urban green space", only a few of those studies have considered the inclusion of humans in the natural environment [10]. This horizontal spatial openness is problematic when people need to delineate a place in a natural environment, because they can only do so by limiting or delimiting a place. This delimitation takes into account visual boundaries ("a portion of land which the eye can comprehend at a glance" [11] (p. 2) that are in an interdependent relationship with "three-dimensional perception of the landscape and, therefore, visibility" [2]. The visibility of place depends on the eye that looks at the place and how the place becomes symbolized either by feelings or by pragmatic functions, etc. The place becomes visible (visible in a certain way) through the way people perceive it. Place perception opens towards attachment. The look discovers three spatial ranges (house, neighborhood, and city) which measure the social and physical levels of place attachment, taking into account the physical and social dimensions [12]. Dang and Weiss started from a diversity of meanings, perspectives, and indicators of place attachment and conducted research in which they quantified all studies investigating place attachment between 2010 and 2021; the studies are indexed in Web of Science and ProQuest. One of the authors' conclusions is that there are different results within the subdimensions used. They synthesized these subdimensions that measuring place attachment into dependence, place identity, place affect, and place bonding [7].

The indicators used and the referential field depend on the perspective approach from which term "place attachment" is viewed. The psychological perspective mainly uses the terminology of behavioral intention to explain "the development of place attachment with a particular behavioral intention". Most of the relevant psychological studies consider the relationship between place attachment and pro-environmental behavior intentions, also providing empirical evidence [7]. There is a sociological point of view according to which place becomes a location for social interactions or is valued as a symbol for a social group. Social relationships are included in attachment formation, and place attachment "is a significant positive predictor of social norms" [7,9]. The ways in which sociological indicators are classified, categorized, and used lead to the investigation of the diversity of relationships. For example, Fried conducted a study that shows that even though place attachment is a characteristic feature of life in many poor, ethnic, or immigrant communities, the development of sense of spatial identity for members of these communities is a critical component [9]. Therefore, the diversity of the reports and relationships between place and attachment or between place attachment and other indicators may be better understood depending on the definition of groups and communities as well as the indicators for measuring place attachment.

Place attachment has been considered from the point of view of how people relate to their own identity and to the identity of communities/societies, because people are attached to a certain place where they (re)find their own identity in an environment. When the environment is considered as landscape, landscape identity becomes possible because there is an interrelationship between people's own identity and the environment. Defining landscape identity is achieved through several activities such as landscape protection, management, and planning. Also, the definition is achieved by evaluation of landscape character from three aspects—the physical, the visual, and the image, and last but not least, through "the unique psycho-sociological perception of a place defined in a spatial-cultural space" [2–4,14]. All of these can help with the conceptual delineations of this term, as there are issues with boundaries and how they can be determined [2]. Residential environment derives from landscape identity. One of Scannell and Gifford's hypotheses is that those who are place-attached have an identity relationship with the environment and those whose relationships are not based on an identity principle do not have an intense sense of place attachment. The two authors show that the principles that emerge from identity related to residential attachment are continuity, self-esteem, self-efficacy, and distinctiveness [8]. Given the relationship between place attachment and the environment, it appears that place attachment "was primarily coined in the context of environmental psychology" [7], but we have already seen that this kind of relationship is more nuanced, at least in philosophy. On the other hand, in this relationship between place attachment and environment there is the concept of sense of place defined as "beliefs about the relationship between self and place" and "feelings toward the place; and the behavioral exclusivity of the place in relation to alternatives" [15].

From the point of view of human geography, the relationship with the natural environment predominates, in which the relationship between the person and the physical environment opens up to place attachment as a universal phenomenon [9]. Edward Relph, professor of geography at the University of Toronto, developed a measurement scale that relates the physical characteristics of place to the connection between people and social environment, on the one hand, and to subjective perceptions of place of origin, on the other hand [16]. The notion of environment is a general perspective of the physical spaces developed by geography. Scientists quickly understood that the environment has consistency only as an observed reality, so a dual perspective was formed in studies in this field. Thus, phenomena and concepts from the sciences of nature were combined with phenomena and notions from the social sciences [17,18]. The stake of research activities about the environment is rendered by human actions, including the way in which human activities have influenced nature. This stake is all the greater as it has revealed the emergence of global phenomena due to human activities which endanger the quality of the environment.

People's perceptions differ about the importance of aspects that explain the environment [19]. In their study conducted in Australia, Brown and Raymond divided the population into two, residents and visitors, and the authors obtained different results about environmental perception for the two groups. Therefore, the functionality of daily activities and individual perspectives on place are criteria for personal reporting relating to the environment. The quality of jobs and their numbers depend on the economic performance of an area [19].

The implications of the phenomena of place attachment and environmental perception go beyond issues of environmental quality. These phenomena significantly determine people's well-being and quality of life [20]. In 1963, a study was carried out that showed that even if residents had good or reasonable reasons to move to another place, when they had to move, they experienced major problems because they had strong place attachment. Persistence of regret or nostalgia exist long after people are relocated [21]. These emotional states have a strong influence on quality of life and feelings of well-being. Individuals' subjective perceptions of their own lifestyle will be disturbed, even if many other daily demands and needs are met, because there is a satisfaction related to the sense of place attachment. Rollero and Piccoli conducted a study based on a sample of 443 subjects. These subjects were first-year students and were chosen because they had changed their home and expressed feelings and emotional states generated by changes and uncertainties related to place attachment [22]. A significant correlation between place attachment and social well-being resulted, and the dimension was operationalized using a social well-being scale. This was composed of five items measuring five dimensions: social integration, social acceptance, social contribution, social actualization, and social coherence [22,23]. Perception of well-being and quality of life are the mandatory attributes for ensuring an individual's emotional balance. These characteristics are highlighted by measuring the different types of relationships subjects have with the social networks and social institutions they interact with every day. Therefore, desirable characteristics have a strong sociological component, as can be seen from the mentioned studies.

3. Methodology

The general premise of this study is that people's interest in the environment is a prerequisite in the context of place attachment. Hence the implicit premise that as the world becomes more technical and, implicitly, urbanized, there is a semantic impoverishment of the way human beings relate to home and place, and their perception of the environment is. However, this terminological technicalization only renames the same content by other names. Therefore, the section dedicated to philosophy becomes necessary to show, on the one hand, the pre-existence of the terms used by sociology, and on the other hand, that the variables used in this study are included in the semantic richness of philosophy.

Therefore, for the association between the first two terms, place and attachment, the interdependent relationship between place as home or community and self as self-identity is taken into account [1–5]. For the association between the environment and perception, the intentionality that shifts the focus from one term to another is taken into account [6].

In second part of this study, we apply the research methodology to measure the intensity with which environmental perceptions are manifested at the level of the rural population [35,36]. The use of certain items in developing the place attachment measurement scale is based on the model proposed by Scannell and Gifford. Those authors proposed a three-dimensional framework of place attachment (person, psychological, and place dimensions), after analyzing a diversity of studies [8]. The phenomenon of place attachment has an individual and a group dimension at the same time. The group perspective has a strong incidence on the personal level because people need social integration. Any form of adherence to social values and norms is a confirmation of social integration. People, consciously or not, tend to adhere to currents of opinion specific to the social environment they belong to, because social integration allows access to available forms of social support. Social support is a phenomenon that is strongly linked to quality of life.

The measurement scales for place attachment and environment perceptions were constructed according to the Likert model. In the case of place attachment we used 12 variables with scores from 1 to 10, and in the case of environment perception we used 10 variables measured with scales of four degrees of intensity. The development of the scales was based on the Social Identity Model of Collective Action (SIMCA), which is a measure of collective action developed from a meta-analysis of over 180 studies investigating predictors of collective actions [35].

The instrument used for the research methodology was a questionnaire model with closed questions that contained a set of Likert scales for measuring attachment to the residential environment and the degree of satisfaction related to the surrounding environment. This was applied to rural subjects. The survey operators administered 1576 questionnaires to as many subjects aged over 18, who were familiar with the reality of rural areas in north-western Romania. The variables used were landscape of the home area, agriculture, place of residence, one's own household, neighbors, the street, and also relationships with the health system, public institutions, the village hall, school, relatives, and colleagues.

4. Philosophical Meanings Used in Place Attachment Definitions

Dang and Weiss confirm that place attachment has been studied in a diversity of scientific disciplines such as environmental psychology, human geography, and sociology, but also in other research area as business and management, risk and crisis, urban planning, leisure, hospitality, and tourism. Terms such as community attachment, sense of community, place identity, place dependence, and sense of place are used in descriptions of place attachment. However, the authors observe that the meanings "are not easy to differentiate and the concepts partly overlap" [7]. Our observation is that all these meanings were already defined and had been used by philosophy since Greek antiquity. So, it is important to take a glance at the philosophical vocabulary to understand, on one hand, the conceptual boundaries and clarification of meanings and, on the other hand, which meanings of place have been used by the other sciences. Last but not least, it is necessary to understand the transition that affected the technicalities of this vocabulary. Similar to the observation "common to all definitions is that place attachments refer to the relationship between individuals and their environment" [7], there is a philosophical presence because there is a vocabulary that describes and explains the relationships between people, place, and environment. If we take into account the etymologies and primary meanings of the ancient Greek terms including such verbs as oiken, naiein, and demein, and such nouns as domos or doma, ethos, hestia, horos, peras, etc. [24], five primary meanings of place attachment can be separated:

(a) The first meaning is that all the essential things of a human being's life are carried on under the order of the oikos and ensure the creation of memory as a durable link to the place where human life unfolded. The focus is on memory—a memory of affectivity—that opens towards the transcendent because the place of the human being (a place where the ancestors lived and where the descendants will live) is also a place of the gods. If it does not take into account the transcendent, the oikos meaning overlaps with that of place attachment [8].

(b) The second meaning derives from the first; there is a divine meaning of dwelling (the gods protect the house), and the link between human beings and gods is accomplished by ancestors and the dwelling of gods (naos—temple). Attachment is accomplished through a horizontal relationship of forming memory and identity as an individual, as family, and as community, and through a vertical relationship sacralizing of the place where we live. Inside of these appear environment and behavior. Human behavior is regulated by the relationship with gods, and the place—home and temple—means the environment as a world [24].

(c) The third meaning of place attachment states that the link with place is ensured through the imprint on the place by dwelling, because the primary meaning of ethos was the habits of the house. This could be a meaning of place identity [2–4,14].

(d) The fourth meaning of place attachment and environmental behavior has a unifying role, in which a central place is symbolized as center that finally attracts, gathers, and unifies the family and the household gods (Hestia, the protective goddess of the home).

(e) The fifth meaning of place attachment and environmental behavior implies that if we live within the border/limit (peras as limit crossing, horos as the visible territorial limit and hyper-oria as the territory located beyond the border), then we keep place attachment within that environment [24], but if we transgress this limit, the loss of the oikos, hestia, and naos occurs, that is, the loss of all previous meanings. If we go beyond the borders, we lose the attachment. In this case, environment influences place attachment. From these three meanings of limit, we can identify that the visible limit that we perceive with eye shows us the place as territory [2,11].

If for the Romans the meanings remain almost unchanged (each place has its own god—genius loci), in Christianity the primary sense of the naos (temple/the God's house) has been reconfigured. The church becomes a place outside the place where we live daily (house), but retains its centrality within the community. On the other hand, a major change is that the only real place becomes that of the divine presence, that is, the soul in a personal sense, especially in the first centuries of Christianity. Place matters only as a place where contemplation of the created world and of the creator is accomplished or possible. If geography is not symbolically invested, it loses its meaning [25]. On the other hand, the place of the church as place to find yourself, or to return to yourself, or to leave yourself in the care of God, etc., is the axis mundi of any rural community (and up to a point, urban communities as well). Practically, Greco–Roman and Christian semantic diversity is found in almost all areas until modernity. In a certain sense, Heidegger can be considered both the last thinker who named, thought about, and formulated the problem of the meaning of place through a relationship with the semantics of the ancient Greek language, as well as the first opener of new ways of thinking about place and dwelling, not only in philosophy but also in other fields. For Heidegger, there are boundaries that separate the dwelling place from the rest of the space and that have impregnated/permeated the interiority qualitatively, both horizontally and vertically [5].

Perhaps it is precisely here that an ontological mutation (the limitation that delimits the domestic space and the limitation that delimits the public space) takes place that will later be taken over by sociologists. However, perception has a central place in the careful description of lived experience. We can change the elements of our visual field so that stable things appear [6]. From this moment, technicality of terms emerges depending on the autonomy of the diversity of fields and the perspective that each field has on the terms.

The Heideggerian phenomenological approach is re-signified by Schultz, who explains how space is converted into place. Place has only the meaning of house/public building, which is understood through the relationship between context (putting the place in a context) and dwelling. Schultz aims to slide towards utility and efficiency and explains the transition from space to house referring to aspects such as infrastructure, facilities, etc. [26]. Thus, Schultz's theory only applies to the urban. However, for Harvey, the home is that place where we live accidentally/by chance (a building that protects us from bad weather) and is a private space that ensures privacy and a retreat from the intrusive gaze of others. The habitation is related to location because there is an importance of emplacement, infrastructure, etc. that reflects the social status of the inhabitants [27].

The indicators that measure wellness and the quality of individual life appear once the habitation begins to be defined and problematized according to several criteria such as emplacement, thermal efficiency, pollution level, public institutions in the vicinity of house, etc. The ontological mutation was achieved when the standard of living and individual prosperity was measured/quantified by means of indicators that apply to the way the inhabitants live and not to the way the human beings themselves live. New terms such as housing quality index appeared. For example, Ranci consider quality of living to be a key element of quality of life, and for Trudel quality of living is the basic dimension of quality of life [28] (p. 46). The importance of a classification in relation to the indicators used as statistical variables eliminated human beings and gods from the topic of place. The house can become home only after a period that is long enough to ensure the emergence of the sense of belonging, the sense of identity that is the place attachment [29]. On the other hand, other terms appear, such as sustainable development, which in 1991 became the main concept in the U.N.O. document entitled Caring for the Earth in 1991. Since 1992, the Sustainable Development Commission of Rio de Janeiro has analyzed, informed, and set strategy. Authors such as Backer, who introduced the term behavior setting (any behavior is formed according to the spatial setting in which it manifests) [30], and participants at the Stockholm Conference in 1970 prepared the ground for the debates that take place today. On the other hand, Braud observes that the environmental problem that has been regulated by the whole community becomes a political problem when those who hold the power take over the regulations and the criteria of these regulations [31] (p. 11). So, in this relationship between place attachment and environmental behavior, new indicators established by experts in fields such as ecology, sociology, urbanism, etc., are added.

In premodernity, both the village (rural) and the city (urban) used the same vocabulary to define the boundaries of place or the world as intramuros or extramuros. Vernant shows that the way the ancient Greeks conceived nature/world and religion influenced the way people organized both the interior space of the house and the exterior space, the space of the polis. The ancient Greeks introduced the term agora as a public space (debating), public square, and city center [32]. The symbolism of the center (axis mundi) as a common place that connects and unifies exists from antiquity until today, especially in countryside. The agora, the church, and the public square become places where the whole community meets, communicates, and participates together, in communion with God. Rural residents have a sense of belonging to the place. This place is delimited as the center (the hearth of the village), as a horizontal limit (boundary/frontier), and by a verticality mediated by the church. All these give people a sense of belonging to the world as a whole (God's world). The preservation of one's own identity can only be achieved in the place where people are themselves (home) and to which they belong (home, village, community, church, world).

In the countryside, limits are very important, in the sense of the old terms horos, hyper-oria, and peras. Preserving one's own identity can only be achieved in the place where one is oneself (the house with all the elements such as family, environment, etc.) and to which one belongs (the village, the community, the church with all the elements such as neighbors, agriculture, landscape, relationships with others and institutions, etc.). With the migration of the peasant to the city, some of these meanings are lost. Oiko-nomia (economics) became the simple accounting of household goods before becoming the science of today [24]. Basically, the essential meanings of the terms nostos and oikos have been lost.

Are these meanings still preserved in the countryside? Stahl shows that there is a wide variety of worldviews from village to village and from person to person. There is a diversity of elements as a result of previous generations, and the humanization of the landscape where people are born is a tradition, namely, a part of contemporary history is kept alive. Fields such as demography, sociology, anthropology, history of mentalities, etc., show us a past that lives in current forms [33] (pp. 118, 257–258), namely, the classifications and categories that the social sciences use and which have gradually changed from the discovery of the peasant's point of view as different from that of the city dweller [34] to the current re-signification of the peasant as a human being who lives harmoniously in the environment and respects nature. We aim to discover which of the meanings that place attachment once had still exist today in the rural environment and what is the influence of environmental perception.

5. Materials and Methods

The research instrument used for this study was a closed-question questionnaire design, which included a set of Likert scales to measure attachment to the residential environment and satisfaction with the environment. The questionnaire was administered to rural subjects and are constructed in accordance with the perceptual sensitivities and everyday concerns theoretically attributed to residents of the target area. The survey operators administered 1576 questionnaires to as many subjects aged over 18, who were familiar with the reality of rural areas in north-western Romania. The selection of respondents involved multi-stage sampling, the study being based on three stages. In the first stage, a random sample of localities was taken. The second stage involved applying the random step method in the field for the selection of households, and in the last stage, the selection of subjects who responded to the questionnaires, according to age and gender category. The sampling error was approximated to $+/-2.7\%$ with a probability of 95%.

The measurement scales for place attachment and environmental perceptions were constructed according to the Likert model. In the case of place attachment we used 12 variables with scores from 1 to 10, and in the case of environmental perception we used 10 variables measured with scales of 4 degrees of intensity. The development of the scales was based on the Social Identity Model of Collective Action (SIMCA), which is a measure of collective action developed from a meta-analysis of over 180 studies investigating predictors of collective actions [35]. This methodological model is based on three key variables: emotional reactions to injustice, efficacy, and identification. Studies show that emotional appraisals can heighten people's willingness to engage in collective action [36]. In this respect, it should be noted that the model for interpreting forms of place attachment in rural areas has the ultimate aim of determining pro-environmental behavior, which is materialized in everyday life as a form of manifestation of collective action. Place attachment is a multi-dimensional concept with person, psychological process, and place dimensions [8]. This study focuses on aspects related to place dimensions. Of the three dimensions, this is the most accessible in terms of the possibility of intervention and influence, with the objective of generating pro-environmental behaviors. Place attachment is a phenomenon strongly correlated with feelings of well-being and, therefore, the measurement scale we used in our study took into account the social well-being scale [37]. Thus, the variables describing the intensity of place attachment can be grouped into three categories: city attachment, neighborhood attachment, and attachment to public institutions.

We attempted to identify environmental perceptions as predictors of place attachment through a model that combines the variables considered in the literature. We based our hypotheses on the results of previous studies: Hypothesis 1: Subjects with a higher intensity of environmental perception propose a more positive image of the place where they live, because place attachment represents the emotional connection to a physical area [38]. Hypothesis 2: Representations of the environment are based on social experiences and acquire different values, and they are also a deep expression of the person's subjectivity [39]. Hypothesis 3: There is a causal relationship between environmental perception and place attachment that manifests itself as a directly proportional relationship.

6. Results

Based on factor analysis, we aimed to reduce the 12 variables used to describe the phenomenon of place attachment. We generated a factor characterized by a common variance (at the level of the 12 initial factors) of 44.7% of the sample, which represents approximately 700 respondents (Figure 1).

Component	Initial Eigenvalues		
	Total	% of Variance	Cumulative %
1	5.374	44.780	44.780
2	1.538	12.817	57.597
3	0.952	7.937	65.534
4	0.737	6.139	71.673
5	0.687	5.723	77.396
6	0.561	4.673	82.068
7	0.462	3.853	85.921
8	0.423	3.521	89.443
9	0.378	3.146	92.589
10	0.359	2.991	95.580
11	0.295	2.461	98.041
12	0.235	1.959	100.000

Extraction method: principal component analysis.

Component Matrix [a]	1
The landscape of the home area	0.776
Agricultural landscape	0.751
The landscape of the place of residence	0.735
Relationship with the health system	0.709
The landscape of your own household	0.670
Neighborhood landscape, street landscape	0.664
Relationship with the school	0.663
Relations with relatives	0.652
Relations with colleagues	0.619
Relations with other public institutions	0.607
Relationships with friends	0.597
Relationship with the town hall	0.548

Extraction method: principal component analysis.
[a]. 1 component extracted.

Figure 1. Total variance explained.

The questions referred to the quality perceived by the subjects regarding the different aspects that characterize place attachment. The minimum score is 1 and represents the greatest distance from the ideal situation, and 10 is the maximum value and represents the ideal situation that the subject perceives in relation to the different characteristics by which we defined place attachment.

Table 1 highlights the scores resulting from respondents' ratings of the characteristics that define place attachment. We also set out to measure the intensity of environmental perceptions in the rural population, based on the idea that people's interest in the environment is a prerequisite in the context of place attachment and pro-environmental behavior. Indeed, it can be expected that there are people who do not show sufficient interest in environmental issues, who do not understand, do not appreciate, and do not emphasize the characteristics of the environment as an important part in determining the quality of daily existence. This category may develop a place attachment based mainly on the phenomenon of socialization, but it will be difficult to manifest pro-environmental attitudes and behavior if their relationship with the environment is characterized by indifference rather than active involvement. Environmental perceptions are therefore the foundation of pro-environmental behavior, and we aim to describe how this phenomenon manifests itself in rural areas. On the other hand, we think that it is significant to construct an explanatory scheme about the influence of environmental perceptions on place attachment, since these two phenomena are mutually conditional.

Table 1. Place attachment variables.

	N	Minimum	Maximum	Mean
Relations with colleagues	1550	1	10	8.74
Relationships with friends	1551	1	10	8.68
The landscape of your own household	1555	1	10	8.39
The relationship with the town hall	1473	1	10	7.81
Agricultural landscape	1546	1	10	7.74
The landscape of the home area	1547	1	10	7.42
Relations with relatives	1551	1	10	7.35
Neighborhood landscape, street landscape	1551	1	10	7.09
Relationship with the health system	1551	1	10	7.04
Relationship with the school	1549	1	10	6.97
Relations with other public institutions	1477	1	10	6.93
Relationship with other aspects of daily life	1531	1	10	6.81
The landscape of the place of residence	1555	1	10	6.75

Using factor analysis again, the 10 variables measuring environmental perception were synthesized into three factors (Table 2). Here we can see that the variables clustered around the third factor represent close everyday visual and interactional aspects with an almost permanent and constant presence in everyday experiences. The first factor refers to environmental elements with which the individual interacts frequently. The second factor is made up of variables that characterize things that the individual interacts with rarely that are contextually more removed from everyday experience, and the third factor is made up of environmental elements with which man is in permanent contact. We can therefore say that environmental perception is a social phenomenon described by three factors that describe the frequency of the individual's interaction with environmental elements: factor 1—contextual level of interaction with environmental elements; factor 2—everyday level of interaction; and factor 3—permanent level of contact with environmental elements.

Table 2. Environmental perception component.

Rotated Component Matrix [a]	Component		
	1	2	3
things from the house	0.220	0.216	0.665
the yard of the household	0.592	0.072	0.525
the street you live on	0.108	0.331	0.712
own garden	0.006	0.475	0.655
the center of the village	0.641	0.055	0.302
hay or pasture	0.759	0.333	0.044
stables and places for animals	0.843	0.204	0.011
meadow/hill/mountain/lake/river/swamp/forest	0.278	0.697	0.050
shop/factory/wind farm/hydroelectric dam	0.035	0.696	0.173
exploitation of raw materials	0.164	0.757	0.036

Extraction method: principal component analysis. Rotation method: varimax with Kaiser normalization. [a] Rotation converged in eight iterations.

Furthermore, in Table 2, through factor analysis, we show that the variables measuring environmental perception are grouped into three categories according to the significance of how they correlate with each other. The first category consists of four variables: things from the house; the yard of the household; the street you live on; the center of the village. The second factor consists of the variables: own garden; hay or pasture; stables and places for animals. The third factor contains the variables meadow/hill/mountain/lake/river/swamp/forest, shop/factory/wind farm/hydroelectric dam, and exploitation of raw materials. Analyzing this situation, we found that the homogenizing element of the factors is the intensity of interaction of the subjects with the elements of the environment. Thus, if we look at the composition of factor 3 (Table 3), it includes the variables things from the house, the yard of the household, the street you live on, and the center of the village, which refer to elements of the environment with a high level of accessibility. Subjects come into contact with these elements very frequently. The conclusion is that spatial proximity, which determines the frequency of interactions between humans and environmental elements, is the main homogenizing element of environmental perception. The three factors that capture and measure environmental perception are named accordingly: environmental elements with permanent contact, environmental elements with daily contact, and environmental elements with occasional contact.

Table 3. Model summary regression of influence of environmental perception on place attachment.

Model	R	R Square	Adjusted R Square	Std. Error of the Estimate
1	0.775 [a]	0.600	0.596	1.15516
2	0.841 [b]	0.708	0.702	0.99215
3	0.878 [c]	0.771	0.764	0.88346

[a] Predictors: (constant), environmental elements with permanent contact. [b] Predictors: (constant), environmental elements with permanent contact, environmental elements with daily contact. [c] Predictors: (constant), environmental elements with permanent contact, environmental elements with daily contact, environmental elements with occasional contact.

Figure 2 shows the intensity with which the environment is perceived from the perspective of the three factors that explain this phenomenon. We can see that the intensity of perception of environmental elements is directly proportional to the frequency of interactions; in other words, the more interactions there are, the greater is the intensity of environmental perception. The elements of environmental perception most intensely manifested are those with permanent contact. In total, 44% of subjects stated that the things from the house, the yard of the household, the street they live on, and their own garden are the elements of environment that they notice to a very high degree. Only 24% of the population gave very much notice to the aspects meadow/hill/mountain/lake/river/swamp/forest, shop/factory/wind farm/hydroelectric dam, or exploitation of raw materials.

The coefficient of determination (R square) in Table 3 shows that each of the three regression models causes significant variation in the place attachment phenomenon. Environmental perception elements with permanent contact cause variation in place attachment in 60% of the study population. Elements with permanent contact together with elements with everyday contact cause a variation in place attachment in 70.8% of the population. Finally, if we add the occasional interaction elements and obtain the total of environmental perception, there is significant variation in place attachment in 77.1% of the study population.

Table 4 allows us to analyze the regression equations and make predictions about the evolution of feelings of place attachment. If we disregard the significance level of the multiple regression models, all three models have a significance level less than 0.1, which means that they have a strongly significant linear regression. Model 3 is the complete model, which includes all elements of environment perception. We therefore choose to discuss the prediction that this model allows. The regression equation is: place attachment = 2.743 + 0.457 (environmental elements with occasional contact) + 0.527 (environmental elements with daily contact) + 0.733 (environmental elements with permanent contact). Place attachment emotion changes by one unit when environmental perception changes by 1.717 units. In other words, any increase or decrease in environmental perception by 1.717 units will result in an increase (decrease) in place attachment by 1 unit. If we look at the three elements of environmental perception, we can see that each element increases the effect on the phenomenon of place attachment. Each element has a directly proportional influence on place attachment. The influence can be estimated for each of the three models of environmental perception. The prediction of the influence of the elements of environment perception on place attachment must start from the environmental elements with permanent contact, because these elements are the most accessible, with the highest degree of repeatability in everyday behavior.

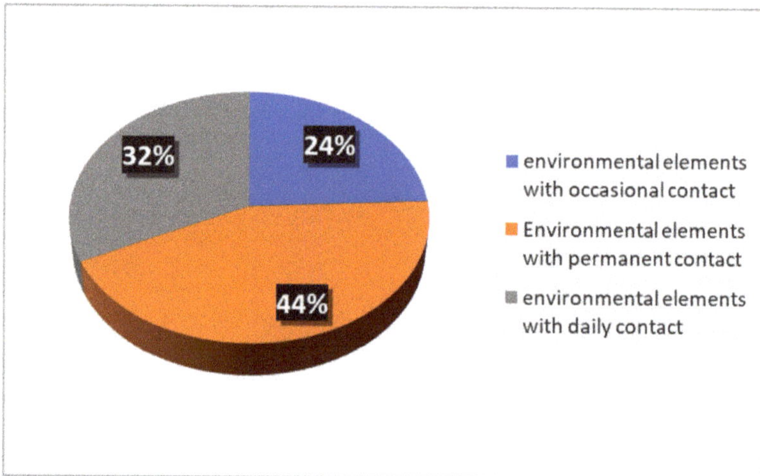

Figure 2. Elements of environmental perception.

Table 4. Coefficients table of the influence of environmental perception on place attachment.

Model		Unstandardized Coefficients B	Unstandardized Coefficients Std. Error	Standardized Coefficients Beta	*t*	Sig.
1	(Constant)	4.192	0.288		14.555	0.000
	Environmental elements with permanent contact	1.224	0.100	0.775	12.252	0.000
2	(Constant)	3.457	0.276		12.541	0.000
	Environmental elements with occasional contact	0.673	0.125	0.426	5.370	0.000
	Environmental elements with daily contact	0.812	0.134	0.479	6.047	0.000
3	(Constant)	20.743	0.281		9.745	0.000
	Environmental elements with occasional contact	0.457	0.119	0.289	3.836	0.000
	Environmental elements with daily contact	0.527	0.132	0.311	3.997	0.000
	Environmental elements with occasional contact	0.733	0.141	0.379	5.182	0.000

Dependent variable: Q6PA (place attachment).

7. Discussion

First, we propose some more precise details about the role of philosophical analysis in the design of this study. First of all, we would like to point out that philosophical analysis is a methodological approach that derives from the field of knowledge developed by various authors under the name of "philosophy of social science" [40–42]. Specifying the philosophy of social science to our research subject, we highlight some ideas:

1. Philosophical analysis covers important issues that could not be described by statistical analysis. One such situation is that place attachment analysis raises an ontological dispute in that this phenomenon exists outside the symbolic or cultural meanings that actors manifest. In simple terms, people are usually not aware of feelings like place attachment or environmental perception, they do not think about them, and they do not use them in conversations. Yet, they exist and their consequences are decisive for behavioral patterns such as those based on pro-environmental attitudes.
2. Sociological research covers a limited fraction of a very large phenomenon in terms of its social implications. Philosophical analysis supports sociological analysis. Thus, the etymological perspective offers knowledge at the historical level; philosophical analysis brings into question the role of church and faith in crystallizing place attachment; Heideggerian phenomenology emphasizes the emergence of meaning, as space is becoming place and house is becoming home; last but not least, philosophical analysis argues the authors' choice to limit this sociological study to rural communities.
3. From a methodological point of view, philosophical analysis contributed to the creation of the theoretical entities that we have used in the research. Philosophy of social sciences introduces an empiricist approach to social reality that promotes the construction of theoretical entities by means of "useful fictions" that subsequently allow scientific prediction by virtue of mathematical content. This type of approach is called "instrumentalism" [43], and is precisely what has happened in this study by relating philosophical analysis to sociological interpretation. Theoretical entities were defined by the items on the scales and then placed in relation as independent variables that determine place attachment as a dependent variable.

Given that "a concept has a specific defining attributes because of its role in the theory", a brief foray into the field of philosophy became necessary because in the methodological literature, these philosophical questions are often called problems of "construct validity". The main concept that we used in this study had many attributes and meanings, but all of them are kept under other names in sociology. Last but not least, "real features of the world correspond to the theoretical concepts or constructs, and valid surveys (or other tests) can measure them" [44].

The phenomenon of place attachment has an individual and a group dimension at the same time. The group perspective has a strong incidence on the personal level because people need social integration. Any form of adherence to social values and norms is a confirmation of social integration. People, consciously or not, tend to adhere to currents of opinion specific to the social environment they belong to, because social integration allows access to available forms of social support. Social support is a phenomenon that is strongly linked to quality of life [45]. The quality of personal life is, most often, one of the powerful factors that give meaning to human life, even if a kind of mercantilism is observed. This is suggested by the meanings that rural people give to the term environment. Most of them consider that the environment is "place around us", which is a general perspective with a high degree of impartiality and can be seen as representative of the relationship between population and nature in the targeted social area [46].

If we calculate the answers given for all 12 variables in relation to place attachment, we obtain the percentage results presented in Table 5, which highlight the intensity with which the phenomenon of place attachment is manifested among the inhabitants of rural areas in the north-western part of Romania. The total number of respondents differs and is smaller than the number of subjects who answered each variable separately, because in the case of the data in Table 5, only subjects who answered all questions related to place attachment at the same time were considered.

It is important to note that 10% of the population manifest a lack of a sense of place attachment, while 26.1% of the population manifest characteristics that suggest a strong attachment to the area in which they live. People in rural areas are more likely to be attached to the area where they live, and this characteristic is a very good context for promoting pro-environmental behavior. However, there is a need to raise awareness of environmental needs, and then the chances are good that the majority of the population will make efforts to conserve the environment. At the same time, other social forces and phenomena will also have an impact on this behavioral pattern. As a rule, when it comes to collective issues that are part of a social environment's agenda of priorities, the influence of the majority's views is also taken up by those who have different perspectives or are indifferent. Adherence to majority opinion is a strategy of social integration and is often mandatory, especially for issues that are intensely perceived by the public. Pollution issues have been and are intensely debated in the public arena. In addition to public analyses, marketing arguments such as the pollution standard for cars or green certificates have also emerged. So, failure to meet standardized pollution reduction performance can lead to additional costs. In this context, environmental cleanliness issues are willingly or unwillingly becoming part of people's daily concerns.

Table 5. Place attachment results.

	Frequency	Valid Percent	Cumulative Percent
soft place attachment	125	10.0	10.0
medium place attachment	803	63.9	73.9
strong place attachment	328	26.1	100.0
Total	1256	100.0	

Even if the focus of this research is ultimately on pro-environmental behavior, we believe that this topic needs to be investigated from related perspectives, primarily because of the notoriety of pollution problems and their consequences. This notoriety generates an intense phenomenon of social desirability among the population, especially since the positions towards environmental problems, which we find almost permanently in the public space, are unidirectional and cannot be otherwise since the problems generated by pollution cannot be seen in a positive sense. In this context, investigations into the phenomenon of place attachment have a double functional role. Firstly, place attachment is a barometer for the individual's willingness to manifest pro-environmental behavioral patterns and, secondly, it has the role of avoiding systematic errors generated by the phenomenon of social desirability.

Of the variables describing environmental perception (Figure 3), we note that three were listed by 80% of respondents. These are the things from the house, the yard of the household, and one's own garden. It can be said that these three variables describe the closest areas of the environment, which appear most frequently in the action areas of the respondents. Environmental perception is a social manifestation that can be found in four distinct forms: the environment as external object, as representation of self, as embodiment of value, and as arena for action [47]. The three aspects of the environment perceived with high intensity by the subjects, can be classified simultaneously in two categories: as representation of self and as arena for action. The frequency with which the individual interacts with the environmental element does not necessarily represent an advantage in relation to increased intensity of perception. On the contrary, frequent interaction inhibits attention and induces a sense of habituation that generates a tendency to carelessness and a perception of the natural as ordinary and banal. The literature shows that increased attention to environmental elements with which the individual interacts all the time is an indicator of social isolation [48]. It is not the same when we think of environmental elements as forms of representation of the self or as environmental tools in carrying out one's own activities. The conclusion is that the increased intensity we see in the manifestation of environmental perception has deeply subjective explanations. More concretely, environmental perception manifests itself in the context of elements that can be used as a representation of the self but also as a pragmatic form of providing a suitable environment for the performance of everyday activities. This is not unimportant, because such information can decisively guide strategies to increase the intensity with which environmental perceptions are manifested in order to induce pro-environmental behavior.

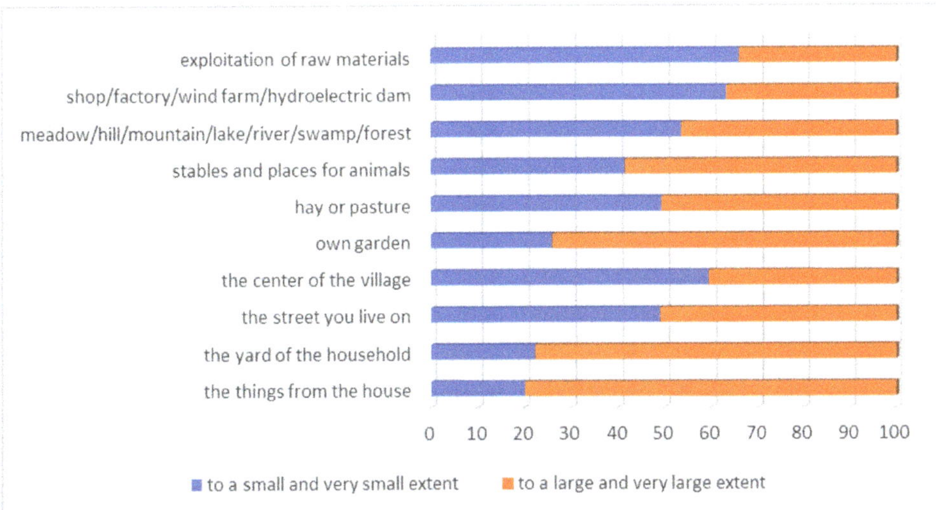

Figure 3. Intensity of occurrence of variables defining environmental perception.

8. Conclusions

Therefore, we consider that, in the rural environment, the meanings of the primary philosophical terms have been preserved in sociology, even if they have other names. All these meanings are maintained within the variables that we used in the research section.

Analyzing and understanding the forms of manifestation of the phenomenon of place attachment is an important issue for contemporary societies. The need to monitor and stimulate the growth of place attachment is stronger today than in the past. This is due to the problems faced by mankind in relation to the environment and pollution. Inadequate care of areas occupied and frequented by humans also results in pollution of nature, and this leads to major imbalances in ecosystems, affecting or destroying flora and fauna and generating extreme climatic phenomena associated with global warming. From a social point of view, all this inevitably leads to a lower quality of life in general.

The effectiveness of knowing the forms of manifestation of place attachment and subsequently developing productive strategies to stimulate its spread and intensity is based on identifying elements of everyday life that facilitate intervention in this phenomenon. Our study is oriented towards environmental perception as a general form that facilitates the interaction between man and the world around him. Also, environmental perception is an area of everyday perception that plays an important role in well-being and in the adoption of positive attitudinal patterns towards people, towards institutional environments, and towards everyday activities.

As stated in the literature review, there are differences between rural and urban environments in terms of the relationship between people and the environment. The pace of life, social symbols, and even the attitudinal patterns that people adopt in relation to the world are things that occur differently in rural environments compared with the urban space [49]. First of all, referring to urban environments, the rules of interaction with elements of the environment are stricter. This is largely due to the fact that elements of the urban environment are often the result of public investment. Quality assessment in public investments has sustainability as an indicator of performance, so public institutions impose behavioral patterns that generate sustainability. On the other hand, in terms of surface area of environmental zones, rural inhabitants influence and interact with larger areas. These areas are characterized as natural environmental zones. The measurement of the forms of manifestation of place attachment in rural social environments is based on this consideration of areas of interaction and influence on the part of the inhabitants.

Environmental perception is a basic form of interaction between man and nature, a way in which individuals define themselves in relation to the world around them. The hypothesis that environmental perceptions exert a determining force in relation to place attachment has been confirmed. The ten variables through which we described environment perception were structured into three categories according to the ways in which they correlate with each other. The factor analysis resulted in three categories describing environmental perception: environmental elements with permanent contact, environmental elements with daily contact, and environmental elements with occasional contact.

There is a deterministic relationship between environmental perception and place attachment, a relationship that was tested by regression analysis. According to the results of this study, it is possible to accurately predict the intensity of increase or decrease in the feeling of place attachment as a function of the measured variation in the intensity of environmental perception. For example, if we consider the category of the population that shows a low level of place attachment, we will be able to increase this level through concrete interventions in environmental perception. So, if the intensity of environment perception increases by 1.7 units (which concretely means an increased interest in the environment), according to the regression analysis we will obtain an increase in the intensity of place attachment by one unit. Based on these analyses, we are able to provide a strategy to improve pro-environmental behavior through feelings of place attachment, which is modified by directly intervening in the interest that people living in rural environments show for the elements of the environment they interact with in everyday life.

Author Contributions: Conceptualization, D.D., D.M. and I.M.O.; methodology, D.D.; software, D.D.; validation, D.D., D.M. and I.M.O.; formal analysis, D.D. and D.M.; investigation, D.D., D.M. and I.M.O.; resources, D.D., D.M. and I.M.O.; writing—original draft preparation, D.D. and D.M.; writing—review and editing, D.D., D.M. and I.M.O.; visualization, D.M. and I.M.O.; project administration, D.D., D.M. and I.M.O. All authors have read and agreed to the published version of the manuscript.

Funding: This research received no external funding.

Institutional Review Board Statement: The study did not require ethical approval.

Informed Consent Statement: Informed consent was obtained from all subjects involved in the study.

Data Availability Statement: The data are not publicly available due to privacy considerations.

Conflicts of Interest: The authors declare no conflicts of interest.

References

1. Lee, C.; Hummon, D.M. A Place to Call Home: Identification with Dwelling, Community, and Region. *Sociol. Q.* **1993**, *34*, 111–131.
2. Penko Seidl, N.; Šmid Hribar, M.; Hudoklin, J.; Pipan, T.; Golobič, M. Defining Landscapes, and Their Importance for National Identity—A Case Study from Slovenia. *Sustainability* **2021**, *13*, 6475. [CrossRef]
3. McCormack, A.; O'Leary, T. Development and Application of Landscape Assessment Guidelines in Ireland: Case Studies Using Forestry and Wind Farm Developments. In *Countryside Planning*; Routledge: London, UK, 2012; pp. 158–172.
4. Stobbelaar, D.J.; Pedroli, B. Perspectives on Landscape Identity: A Conceptual Challenge. *Landsc. Res.* **2011**, *36*, 321–339. [CrossRef]
5. Heidegger, M. *Originea operei de Artă, Traducere de Gabriel Liiceanu si Thomas Kleininger*; Editura Univers: București, Romania, 1983.
6. Cosmescu, A. Teoria percepției la A. Spir și E. Husserl: Similitudini și diferențe. *Rev. Filos.* **2016**, *2*, 142–151.
7. Dang, L.; Weiss, J. Evidence on a Relationship between Place Attachment and Behavioral Intentions between 2010 and 2021: A Systematic Literature Review. *Sustainability* **2021**, *13*, 13138. [CrossRef]
8. Scannell, L.; Gifford, R. Defining place attachment: A tripartite organizing framework. *J. Environ. Psychol.* **2010**, *30*, 1–10. [CrossRef]
9. Mjahed Hammami, S.; Abdulrahman Al Moosa, H. Place Attachment in Land Use Changes: A Phenomenological Investigation in Residents' Lived Experiences with a Renewable Energy Project Development. *Sustainability* **2021**, *13*, 8856. [CrossRef]

10. Neale, C.; Lopez, S.; Roe, J. Psychological Restoration and the Effect of People in Nature and Urban Scenes: A Laboratory Experiment. *Sustainability* **2021**, *13*, 6464. [CrossRef]
11. Jackson, J.B. *Discovering the Vernacular Landscape*; Yale University Press: London, UK, 1984.
12. Hidalgo, M.C.; Hernández, B. Place attachment: Conceptual and empirical questions. *J. Environ. Psychol.* **2001**, *21*, 273–281. [CrossRef]
13. Fried, M. Continuities and Discontinuities of Place. *J. Environ. Psychol.* **2000**, *20*, 193–205. [CrossRef]
14. The European Landscape Convention (Florence, 2000). Available online: https://www.coe.int/en/web/landscape/the-european-landscape-convention (accessed on 30 November 2023).
15. Jorgensen, B.; Stedman, R. Sense of Place as an Attitude: Lakeshore Owners Attitudes Toward Their Properties. *J. Environ. Psychol.* **2001**, *21*, 233–248. [CrossRef]
16. Relph, E. *Place and Placelessness*; Pion Limited: London, UK, 1976.
17. Sack, R.D. *Homo Geographics: A Framework for Action, Awareness and Moral Concern*; John Hopkins University Press: London, UK, 1997.
18. Patterson, M.E.; Williams, D.R. Maintaining research traditions on place: Diversity of thought and scientific progress. *J. Environ. Psychol.* **2005**, *25*, 361–380. [CrossRef]
19. Brown, G.; Raymond, C. The relationship between place attachment and landscape values: Toward mapping place attachment. *Appl. Geogr.* **2007**, *27*, 89–111. [CrossRef]
20. Moser, G. Quality of life and sustainability: Toward person 415 environment congruity. *J. Environ. Psychol.* **2009**, *29*, 351–357. [CrossRef]
21. Brown, B.B.; Perkins, D.D. Disruptions in place attachment. In *Place Attachment*; Altman, I., Low, M., Eds.; Plenum Press: New York, NY, USA, 1992.
22. Rollero, C.; De Piccoli, N. Does place attachment affect social well-being? *Rev. Eur. Psychol. Appliquée* **2010**, *60*, 233–238. [CrossRef]
23. Keyes, C.L.M. Social well-being. *Soc. Psychol. Q.* **1998**, *61*, 121–140. [CrossRef]
24. Liiceanu, G. Repere pentru o hermeneutica a locuirii. *Rev. Secolul 20 Loc–Locuire Poluare* **1999**, *1*, 37–43.
25. Manolescu, A. Peisaje modelatoare. *Rev. Secolul 20 Loc–Locuire Poluare* **1999**, *1*, 23–34.
26. Norberg-Schulz, C. *Habiter: Vers une Architecture Figurative*; Electa Editrice: Milano, Italy; Paris, France, 1985.
27. Harvey, D. (Ed.) *Society, the City, and the Space–Economy of Urbanism*; Association of American Geographers: Washington, DC, USA, 1972.
28. Vâlceanu, D.G.; Zulaica, L. Indicele calității locuirii—Instrument de evaluare a calității locuirii urbane. *Urban. Arhit. Construcții* **2012**, *3*, 45–56.
29. ten Brinke, S.S. Imagining, Appropriating and Silencing: Street Working Chil-dren's Strategies of Home-Making in Public and Private Space. *Childhoods Today* **2014**, *8*, 1–23.
30. Barker, R.G. *Ecological Psychology: Concepts and Methods for Studying the Environment of Human Behavior*; Stanford University Press: Palo Alto, CA, USA, 1968.
31. Braud, P. (Ed.) *La Science Politique*; Presses Universitaires de France: Paris, France, 1990.
32. Vernant, J.P. (Ed.) *Mit şi Gândire în Grecia Antică*; Meridiane Publishing: București, Romania, 1995.
33. Rostás, Z. (Ed.) *Monografia ca Utopie. Interviuri cu Henri H. Stahl*; Paideia Publishing House: București, Romania, 2000.
34. Mihăilescu, V. (Ed.) *Antropologie. Cinci Introduceri*; Polirom: Iași, Romania, 2007.
35. Van Zomeren, M.; Postmes, T.; Spears, R. Toward an integrative model of collective action: A quantitative research synthesis of three sociopsychological perspectives. *Psychol. Bull.* **2008**, *134*, 504–535. [CrossRef]
36. Mackie, D.; Devos, T.; Smith, E. Intergroup emotions: Explaining offensive action tendencies in an intergroup context. *J. Personal. Soc. Psychol.* **2000**, *79*, 602–616. [CrossRef]
37. Tartaglia, S. A comparison between theoretical and naïve definitions of community. In *Community Participation and Empowerment*; Hindsworth, M.F., Lang, T.B., Eds.; Nova Science Publishers, Inc.: Hauppauge, NY, USA, 2009.
38. Felonneau, M.L. Love and loathing of the city: Urbanophilia and urbanophobia, topological identity and perceived incivilities. *J. Environ. Psychol.* **2004**, *24*, 43–52. [CrossRef]
39. Ryff, C.D.; Keyes, C.L.M. The structure of psychological well-being revisited. *J. Personal. Soc. Psychol.* **1995**, *69*, 719. [CrossRef]
40. Rosenberg, A. *Philosophy of Social Science*; Westview Press: Boulder, CO, USA, 1988; Volume 2.
41. Potter, G. *The Philosophy of Social Science: New Perspectives*; Taylor & Francis: Abingdon, UK, 2016.
42. Little, D. *New Directions in the Philosophy of Social Science*; Rowman & Littlefield: Lanham, MD, USA, 2016.
43. Benton, T.; Craib, I. *Philosophy of Social Science: The Philosophical Foundations of Social Thought*; Bloomsbury Publishing: London, UK, 2023.
44. Risjord, M. *Philosophy of Social Science: A Contemporary Introduction*; Taylor & Francis: Abingdon, UK, 2022.
45. Marian, M.; Dărăbăneanu, D.; Chirodea, F.; Țoca, C.V. Analysis of social support as an argument for the sustainable construction of the European Community space. *Sustainability* **2022**, *14*, 7448. [CrossRef]
46. Dincă, I.; Dărăbăneanu, D.; Oprea, I.M. Collective and Social Representations on Nature and Environment: Social Psychology Investigation in Rural Areas. *Land* **2021**, *10*, 1385. [CrossRef]
47. Ittelson, W.H. Environmental perception and urban experience. *Environ. Behav.* **1978**, *10*, 193–213. [CrossRef]

48. Schweiker, M.; Ampatzi, E.; Andargie, M.S.; Andersen, R.K.; Azar, E.; Barthelmes, V.M.; Berger, C.; Bourikas, L.; Carlucci, S.; Chinazzo, G.; et al. Review of multi-domain approaches to indoor environ-mental perception and behaviour. *Build. Environ.* **2020**, *176*, 106804. [CrossRef]
49. Bădescu, I.; Buruiană, C.; Şişeştean, G.; Abraham, D. *Ţăranii şi Noua Europă*; Editura Mica Valahie: București, Romania, 2003.

sustainability

Article

The Relationship between Sustainable Economic Growth, R&D Expenditures and Employment: A Regional Perspective for the North-West Development Region of Romania

Daniel Badulescu [1], Darie Gavrilut [1], Ramona Simut [1], Simona-Aurelia Bodog [2], Daniela Zapodeanu [3], Constantin-Vasile Toca [4] and Alina Badulescu [1,*]

1 Department of Economics and Business, Faculty of Economic Sciences, University of Oradea, 410087 Oradea, Romania; dbadulescu@uoradea.ro (D.B.); gavrilut.darie@didactic.uoradea.ro (D.G.); rsimut@uoradea.ro (R.S.)
2 Department of Management and Marketing, Faculty of Economic Sciences, University of Oradea, 410087 Oradea, Romania; simona.bodog@didactic.uoradea.ro
3 Department of Finance and Accountancy, Faculty of Economic Sciences, University of Oradea, 410087 Oradea, Romania; daniela.zapodeanu@didactic.uoradea.ro
4 Department of International Relations and European Studies, University of Oradea, 410087 Oradea, Romania; ctoca@uoradea.ro
* Correspondence: abadulescu@uoradea.ro

Citation: Badulescu, D.; Gavrilut, D.; Simut, R.; Bodog, S.-A.; Zapodeanu, D.; Toca, C.-V.; Badulescu, A. The Relationship between Sustainable Economic Growth, R&D Expenditures and Employment: A Regional Perspective for the North-West Development Region of Romania. *Sustainability* **2024**, *16*, 760. https://doi.org/10.3390/su16020760

Academic Editor: Grigorios L. Kyriakopoulos

Received: 20 December 2023
Revised: 12 January 2024
Accepted: 14 January 2024
Published: 16 January 2024

Abstract: The role of research and development (R&D) in economic growth has been intensively promoted by scholars and policy-makers of the last decades, emphasizing its impact on technological innovation, intensive and sustainable growth and economic revival. The R&D sector is considered a main driver in the fight against chronic underdevelopment, regional disparities, isolation and lack of socio-economic perspectives. Although the steady economic growth of Romania in the last 15 years continues to converge with the European Union's average, the regional disparities persisted and even deepened, the country still being considered a modest innovator, and resources allocated to research and innovation are far below the European level. In this paper, we aim to identify the existence, direction and duration of the relationship between economic growth, expenditure and employment in the R&D sector. We applied the Johansen cointegration test, the VECM model and Granger causality both at the county and component region levels during the 1995–2021 period. The results of our research reveal the consistency of these bidirectional relationships at the regional and sub-regional levels, especially in the long run. We also emphasize the importance of economic growth in supporting public and private efforts for R&D: the regions that can allocate more resources to research, development and innovation (RD&I) will benefit from the more reliable and long-run-oriented economic growth.

Keywords: regional economic development; research and development expenditures; innovation; Romania; North-West Region

1. Introduction

Understanding the factors that determine economic growth has been a major and constant concern of scholars and policy-makers. Since the end of the 1980s, numerous theoretical and empirical studies have emphasized the role of technology as an essential factor for national and regional economic growth, based on seminal contributions through the New Growth Theory. The role of technological change in propelling economic growth [1], endogenous growth models [2,3], and technological diffusion and convergence [4] emphasize that the primary force driving growth is closely tied to technology. This can occur through its incorporation into labor [5], the expansion of the quantity of intermediate available goods [3], or an enhancement in the quality of the relationship between these inputs [6,7].

Implicitly, the support and adequate public and private funding of research and development (R&D) activities, together with an educated human capital, will drive and spread technological innovations, increase the competitiveness of companies, contribute to the modernization of economic structures, and will lead to the growth and prosperity of countries, regions and local communities. However, the growth of technological capital can also be achieved, apart from innovation, through imitation [8] or technological transfer [9]. Many developing countries, with an insufficient stock of knowledge, low-qualified human capital, and deficient institutional structures, have resorted to this second option to reduce the gaps and speed up economic development.

Despite the complexities, many states engage in R&D to foster domestic growth, produce competitive goods through independent efforts, decrease reliance on uncompetitive and unsustainable domestic resources [10], and stimulate competition and entrepreneurial spirit, etc. [11]. Reducing the technological gap requires the allocation of considerable resources and deep institutional transformations, but it is often noticed that either as a result of different allocations or due to other factors, regional development is not homogeneous.

Additionally, the rationale of our study is based on the fact that the regions of Central and Eastern Europe (including the regions of Romania) experience non-uniform rates of economic growth, which are more pronounced compared to other regions of the EU [12]. Thus, the benefits of development and modernization are unequally distributed in territorial, social and sectoral terms, thus aggravating urban–rural disparities and the decline of peripheral areas, leading to reduced investment in deprived areas that are further drained by the exodus of human capital to more prosperous countries and regions [10–12]. Investments in R&D could mitigate these imbalances, underlining the importance of choosing appropriate indicators to measure R&D efforts and their effects on economic growth [13]. Although there are several indicators that can capture the relationship between research, development, innovation and economic growth, namely *number of patents filed and trademarks, turnover from active R&D local units, number of projects for R&D activities*, it is only the *total expenditure on R&D* and *individuals employed in the R&D sector* that are available for a long enough time period on a national, regional (NUTS2) or sub-regional (NUTS3) level.

In this paper, we are interested in finding, for the case of Romania and some territorial administrative units (region and counties), if there is a positive correlation between R&D efforts (expressed by total expenditure on R&D and individuals employed in the R&D sector) and economic growth (Gross Domestic Product—GDP). The novelty of this study resides in enriching the national and regional perspectives, which have already been studied in the recent literature, with a sub-regional (county/NUTS 3) approach. The presence (or absence) and strength of this relationship can offer insights for decision-makers at various levels to incentivize R&D to stimulate growth. Alternatively, it may prompt a reassessment and adjustment of policies to bolster innovation and technological progress, supporting both regional and national economic prosperity.

The paper's structure is as follows: after this introduction, we provide a review of key contributions to the relationship between R&D expenditure and (regional) economic growth. Following that, we present the context of R&D and economic development in the Nord-West Region of Romania and selected counties, discuss the research methodology, present the main results, and conclude with a discussion. We end with conclusions and highlight the study's contribution to scientific knowledge on this topic, along with its main limitations.

2. Theoretical Overview

2.1. R&D and Economic Growth

Research and development support has a positive impact on economic growth [14,15], creating productive external effects [16] and emphasizing sustainable innovation impacts and spillover effects at the regional level [17,18]. According to Goel et al. [19], public R&D expenditures allocated at the central level are essential for economic growth, the causality being direct and positive in the long term [20], or even bidirectional [21], simultaneously driving the diffusion of innovation on a regional and sectoral scale and enhancing entrepreneurial initiatives [22].

While many recent studies establish a direct causal link between increased expenditure on R&D, innovation, education, and sustainable economic growth, closing gaps between countries and regions [23], numerous studies temper these optimistic claims and even question their existence or effectiveness.

Lichtenberg [24] and Park [25] admit the positive role of R&D expenditures in private sector growth, but their effect seems insignificant in the public sector [26], while other scholars, analyzing data from developed countries, are cautious about the existence of an obvious relationship between economic growth and R&D expenditures [27], or innovation [28,29] or, simply, the direction of causality between the variables cannot be determined [30]. The relationship between R&D expenditures and economic growth in developing countries is unconvincing [31], the effects are much smaller and uneven compared to developed countries [32], or weak in the short run but strong in the long run [33]. Even in OECD countries, although the R&D expenditures have a positive and significant effect on economic growth, the magnitude and duration of the effects are variable in the medium and long terms [21].

Understanding the motivations and sources of R&D in relation to regional economic growth is subject to two major explanations—the substitution and complementarity effects, but often, the real world shows that the two explanations are not mutually exclusive [34,35]. Public funding for R&D appears to be essential in increasing the R&D capabilities of firms and regions [36], but it does not greatly influence the commercialization and diffusion of innovations. The motivations of companies to get involved in research and innovation are, primarily, to increase profits and strengthen their market position [37], and this somewhat contradicts the objectives of governmental funding for R&D—to stimulate regional growth, reduce disparities and have a more "collective distribution" of the benefits [38–40].

Research in fundamental fields, which is perceived as riskier and less likely to produce practical results, is typically avoided by companies and instead becomes the responsibility of public authorities [41]. On the other hand, applied research with innovative potential is also of interest to the private sector, which suggests complementarity rather than substitutability; specifically, public spending in R&D complements private investment pre- and post-economic crisis [42]. However, it is controversial whether the increased public involvement in R&D will drive and encourage more private spending in R&D [35]. Somewhat, firms improve their own technological and innovation capabilities starting from the results of public investments in R&D [43], influencing productivity growth which, in turn, will be reflected in a greater share of production and GDP at sectoral and regional levels [41], or, in other words, the regions susceptible to (and equipped for) innovation are "capable of transforming a larger share of their own R&D into innovation and economic activity" [44] (p. 82).

2.2. R&D in the European Union and Romania

In the European Union (EU), member countries and regions have enacted public policies to support R&D, aiming for efficient public spending and the promotion of investments in innovation [45]. These policies also target private sector growth, employment [36], and the dynamism of small and medium-sized companies, utilizing local resources effectively. The EU's Regional Policy (Cohesion Policy—CP) has significantly backed public and private R&D investment, particularly in lagging regions, aiming to enhance regional competitiveness and diminish economic disparities. In the recent programming periods (from 2007 to the present), 25% to 30% of PC funds were allocated to R&D [40].

Celli et al. [39] find that CP has had a positive and significant impact on regional economic growth in the EU over the last 15 years, confirming the previous results of Becker et al. [46] and Pellegrini et al. [47], but also, they find that many of the regions that invested additional funds in R&D did not experience higher economic growth compared to those regions that did not. Thus, Celli et al. [39] consider that increasing R&D expenditures in lagging regions does not have a significant effect on economic growth in the short term, recalling here the negative or divergent influence of the lack of skilled labor, a critical number of firms generating ambitions and entrepreneurial spirit, institutional deficiencies and insufficient support for entrepreneurship [48,49], or in essence, the reduced and ineffective absorption capacity of funds dedicated to research, innovation and development (RD&I) [50]. The literature also shows that consistent investments in R&D do not have the same effect in less-developed regions [39]. On the one hand, this is explained by the lack of favorable circumstances (the ability to move from technological progress to innovation, the agglomeration of innovative firms, the quality of human capital, and the existence of innovative value chains) [45]; on the other hand, policies that are trying to compensate these disadvantages by providing significant incentives for R&D could overestimate the limited absorptive capacity of underdeveloped regions. This could eventually generate negative effects by discouraging economic recovery and growth [31–33,39]. Pop Silaghi et al. [26] argue that the impact of private R&D spending on economic growth in Central and Eastern European countries is much higher than the impact of public spending, and existing regional economic structures tend to influence this perception [51].

Due to the importance of the topic and the challenges and opportunities that followed Romania's accession to the European Union in 2007, the relationship between R&D, innovation and regional economic growth found a consistent reflection in the Romanian economic literature. Research has shown that, in conditions where other comparative advantages decrease, focus on innovation becomes inevitable [52–54]. However, hindered by low R&D funding, weak national–regional coordination of technology transfer support programs [55], and challenges in predictability and innovation diffusion [56], the contribution of RD&I to national and regional economic growth is unsatisfactory. The limited share of innovative companies and volatile innovation performance, concentrated in specific sectors [57] or spatial "pockets of excellence" [58], further underscores the lack of proportionality between R&D financing efforts and economic impact [52].

According to Dachin and Postoiu [59], innovation (supported by investments in R&D, education, and technology transfer) still does not generate regional economic development in Romania. However, at the sub-regional level (NUTS 3), a direct relation can be identified between investments in R&D and the GDP/capita level. Goschin [60] (p. 33) supports, however, the positive effect of R&D expenditures on regional economic growth in the period 1995–2010, and "the existence of stable regional characteristics that influence the economic growth patterns". Therefore, when formulating regional development policies, it is crucial to consider peculiarities and structural features [58,60], the specialization of local RD&I systems, university involvement in technology transfer, support for innovative entrepreneurship, and co-financing the acquisition of advanced technologies by companies [61]. Additionally, enhancing the integration of centers of excellence in research and innovation into international networks and value chains can transform them into service centers for other regions [58], thus aiding in the alleviation of regional disparities.

2.3. The Case of Romania in a Regional Approach

In 2020, the World Bank classified Romania (for the first time) as a country with a high income level [62]. Regarding GDP/capita (measured in purchasing power standard—PPS), Romania has been converging with the EU average, increasing from 44% in 2007 to 72% in 2020. However, persistent regional disparities, with the ratio between GDP/capita in the most prosperous and poorest regions of Romania being almost three times higher [63], remain a challenge.

In terms of investing in RD&I, Romania ranks at the bottom of the European Union. The RD&I system in Romania faces chronic challenges characterized by insufficient funding, fragmentation, and unpredictability [64]. Closing the gap in Romania's R&D system compared to the EU may take anywhere from 5 to 25 years, depending on the indicator [65].

As per the 2019 European Innovation Scoreboard (EIS), Romania ranks at the lowest position within the EU concerning innovation performance, having a country score of 20, and being considered as an emerging innovator. For example, the first three countries in this ranking (Denmark, Sweden and Finland) have scores of above 130 points each [66]. For EU average levels, between 2015 and 2022 we can notice an increase in innovation performance of 9.9%; however, in Romania, performance has exhibited a downward trend in 2016–2018, and 2022. Notably, the most pronounced decline in performance in 2022 was observed in indicators such as Innovation Expenditures per Employee and Collaborative Initiatives among Innovative SMEs. These findings highlight a concerning pattern of decreasing performance in specific aspects of innovation, signifying potential challenges or areas requiring focused attention and strategic intervention. Ever since 2015, the most significant decreases in terms of innovation-related indicators refer to environment-related technologies, doctorate graduates and non-R&D innovation expenditures [67].

According to the NUTS classification for 2021, Romania is divided into 4 NUTS 1 macro-regions, NUTS 2 regions, and 42 NUTS 3 counties (including Bucharest) [68] (see Figure 1 below).

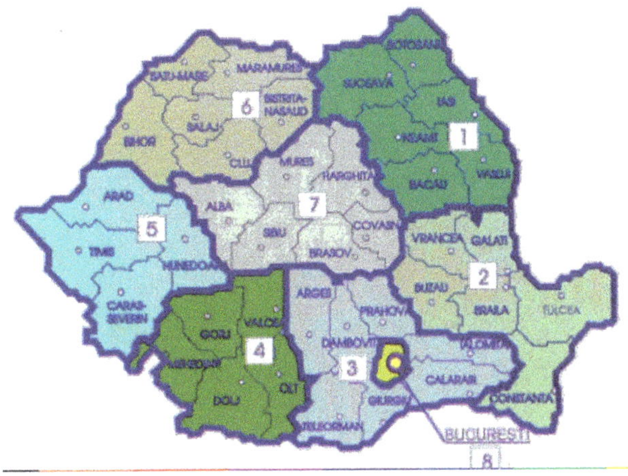

Figure 1. Romania' NUTS 2 region and component counties. Legend: 1—North-East Region, 2—South-East Region, 3—South Muntenia Region, 4—South West Oltenia Region, 5—West Region, 6—North-West Region, 7—Center Region, 8—Bucharest–Ilfov Region. Source: [69].

The strongest region throughout Romania (in terms of GDP/capita) is the capital region, namely the Bucharest–Ilfov Region, with over 25,400 EUR/capita, followed by the West Region (11,200 EUR/capita), Centre Region (10,600 EUR/capita) and North-West Region (10,400 EUR/capita) [70]. The innovation scores for each region are the following: Bucharest–Ilfov (59.5 points; rank within the EU: 199), North-East (35.8; 230), North-West (34.5; 232), West (32.6; 234) Center (25.7; 236), South (23.0; 237), South-West (19.8; 238), South-East (18.9; 239) [67].

In our analysis, we will look at the North-West Development Region (NW Region) and two (out of six) constituent counties, namely Cluj County and Bihor County (see Figure 1). Regarding Cluj County, out of a total county population of 741,000 inhabitants, the capital county, namely Cluj–Napoca, has a population of 307,000 individuals [70], it being the 2nd-largest city in Romania. Cluj–Napoca is by far one of the largest and most respected academic, cultural, industrial, and business centers in Romania, one of the main IT clusters in Romania. Regarding Bihor County (551,000 inhabitants), the county's capital city, Oradea is the 10th-largest city in Romania (221,000 inhabitants), and the second-biggest city of the NW Region, after Cluj–Napoca [70].

3. Data Analysis and Research Methodology

3.1. Data Analysis

In order to investigate the relationship between economic growth (expressed in current prices—million RON) (GDP) and the total expenditure on R&D (expressed in current prices—million RON) (EXP), and also the relationship between the economic growth (GDP) and the individuals employed in the R&D sector (full-time equivalent) (EMP), in Romania, in the NW Region, and in the two constituent counties (namely Cluj County and Bihor County), this analysis applies the Johansen cointegration test, the VECM model and Granger causality. The data utilized in this study have been sourced from the National Institute of Statistics (Romania) [70]. Given the limited availability and accessibility of data, this research focuses on the timeframe from 1995 to 2021. Table 1 presents the descriptive statistics for the dataset in Romania, the NW Region, Cluj County and Bihor County.

Table 1. The descriptive statistics of the variables—GDP, total expenditure on R&D (EXP) and individuals employed in the R&D sector (EMP).

Variables	Minimum	Maximum	Mean	Median	Std. Dev.	Skewness	Kurtosis
			Romania				
GDP	7648.9	957,554	411,069.7	425,691.1	310,750.5	0.218027	−1.20984
EXP	577,148	9,528,718	3,278,001	2,786,830	2,086,427	1.444748	2.458032
EMP	26,171	60,939	36,003.36	32,507	9916.696	1.735055	1.718173
			NW Region				
GDP	911.1	162,865.8	49,891.85	51,229	41,306.6	0.850533	0.707835
EXP	23,113	386,870	204,135.4	204,056	100,169.5	0.007063	−0.85062
EMP	1757	3919	2451.16	2352	554.2362	1.402258	1.863705
			Cluj County				
GDP	266.9	66,861.55	18,912.54	18,265.7	16,965.97	1.067688	1.144931
EXP	16,400	332,665	165,766.9	176,353	82,750.57	−0.19194	−0.83136
EMP	1119	2436	1738.84	1790	388.459	−0.06118	−0.74113
			Bihor County				
GDP	214.6	31,106.52	10,306.54	11,807.1	7841.948	0.629216	0.397997
EXP	574	76,026	10,121.72	5956	16,711.5	3.299811	11.14963
EMP	22	840	254.12	172	227.8661	0.964202	0.103342

Note: Std. Dev—standard deviation. Source: own elaboration based on data provided by the Romanian National Institute of Statistics [70].

According to descriptive statistics, EXP and EMP variables in Cluj County exhibit negative skewness, while in all other cases, the variables show positive skewness. In terms of kurtosis, most variables demonstrate low kurtosis, except for the EXP variable in Bihor County. To test for the presence of outliers, we used the interquartile range (IQR) and applied winsorized estimators to replace extreme values with percentiles, in particular trimmed minimum and maximum values. The application of logarithmic transformation to the variables was employed to address issues related to asymmetry and achieve a more symmetrical distribution of the data. The outcomes revealed a decrease in both variance and asymmetry. To facilitate interpretation and address heteroscedasticity issues, we transformed the variables using natural logarithms, denoted as follows: LGDP—natural logarithm of GDP, LEXP—natural logarithm of total expenditure on R&D and economic growth, and LEMP—natural logarithm of individuals employed in the R&D sector (full-time equivalent).

3.2. Methodology and Research Hypotheses

In examining the time series, we have started from the hypothesis that the observed series are stationary. According to the literature, the majority of time series fail to meet the condition of stationarity, attributed to the numerous changes that occur in the economic environment. To assess the relationship and the causality among the three picked variables in Romania, the NW Region, and two constituent counties, the initial step involved examining the stationarity of all variables. Following this, cointegration tests were conducted to identify a long-run equilibrium relationship among the variables. The third step involved the testing of causality between these variables.

The analysis starts by investigating the stationarity of the following variables: economic growth (LGDP), total expenditure on R&D (LEXP) and individuals employed in the R&D sector (LEMP). In order to test the stationarity, we have used the Augmented Dickey–Fuller test [71]. The null hypothesis of the ADF test is the presence of a unit root, indicating non-stationarity. The alternative hypothesis is the absence of a unit root, implying stationarity. The ADF test involves estimating an autoregressive model of the form:

$$\Delta Y_t = \delta Y_{t-1} + \sum_{j=1}^{k} \gamma_j \Delta Y_{t-j} + \alpha + \beta t + u_t \quad (1)$$

where: Y_{t-1}—the lagged dependent variable, ΔY_{t-j}—the lagged differenced values; u_t is the error term; the coefficients δ, γ, α, and β are the initially estimated coefficients.

To assess the existence of cointegration between the selected variables, we will employ the method introduced by Johansen and Juselius [69], which involves testing the null hypothesis of non-cointegration. Starting from this procedure, we will use the following two tests: the Maximum Eigenvalue test and the likelihood ratio (LR) tests, specifically the Trace test (LRtr) [72,73]. As stated by Engle and Granger [74], when there is cointegration between time series, a long-run effect exists, preventing the time series from diverging.

To explore the long-term equilibrium relationship among the LGDP, LEXP and LEMP variables, we will employ the Johansen cointegration test. Following the results obtained from the Johansen–Juselius cointegration tests, we can determine the suitability levels by applying the vector error correction model (VECM). Following the normalization of the cointegrating vector in order to estimate the error correction model (ECM) of the dynamic structure, we obtain [73]:

$$y_t = \alpha_1 + \beta_1 x_t + \varepsilon_t \quad (2)$$

where: y_t—is the dependent variable, x_t—is the independent variable, α_1, β_1—the coefficients, ε_t—the aleatory variable. The error correction term can be obtained from the abovementioned equation, providing valuable insights into the adjustment process of the variables over time (Equation (1)):

$$EC_t = y_t - \alpha_1 - \beta_1 x_t \quad (3)$$

Thus, the equation becomes:

$$\Delta y_t = a_0 + \delta EC_{t-1} + \sum a_{1j} \Delta y_{t-j} + \sum a_{2j} \Delta x_{t-j} + u_t \quad (4)$$

A negative and significant coefficient of the EC shows that any short-term association between the dependent and independent variables will yield a consistent long-run relationship between them [75].

Based on the literature review, i.e., GDP, EXP and EMP, we have formulated and examined the following six hypotheses in the case of Romania, the NW Region, Cluj County and Bihor County:

H1a: *A long-run equilibrium defines the relationship between economic growth (GDP) and total expenditure on R&D (EXP).*

H1b: *A long-run equilibrium defines the relationship between economic growth (GDP) and individuals employed in the R&D sector (EMP).*

H2a: *A unidirectional relationship exists between economic growth (GDP) and total expenditure on R&D (EXP).*

H2b: *A unidirectional relationship exists between economic growth (GDP) and individuals employed in the R&D sector (EMP).*

H3a: *A bidirectional relationship exists between economic growth (GDP) and total expenditure on R&D (EXP).*

H3b: *A bidirectional relationship exists between economic growth (GDP) and individuals employed in the R&D sector (EMP).*

4. Results and Discussion

In most instances, economic variables tend to be non-stationary. Therefore, our initial step involves examining the stationarity of the variables under consideration. In Table 2, we have presented the results of the Augmented Dickey–Fuller (ADF) test for the three analyzed variables.

Table 2. Stationarity test of data series.

	LGDP (*p*-Value)		LEXP (*p*-Value)		LEMP (*p*-Value)	
	Level	**First Diff**	**Level**	**First Diff**	**Level**	**First Diff**
Romania I(1)	−0.749 (0.869)	−1.705 (0.082)	−2.741 (0.230)	4.778 (0.004)	−1.575 (0.772)	−3.855 (0.031)
NW Region I(1)	−1.225 (0.938)	−3.194 (0.002)	−2.775 (0.219)	−4.725 (0.005)	−2.054 (0.543)	−5.065 (0.002)
Cluj County I(1)	−1.242 (0.940)	−3.081 (0.003)	−1.027 (0.914)	−5.418 (0.000)	−0.344 (0.549)	−6.259 (0.000)
Bihor County I(1)	−1.149 (0.930)	−3.112 (0.003)	−0.032 (0.662)	−6.570 (0.000)	−1.285 (0.177)	−7.556 (0.000)

Note: *p*-values are in ( ) and the optimal lag length is determined based on Akaike, Schwarz and Hannan–Quinn information criteria. Source: own elaboration using Eviews 12.

The ADF test results indicate that, following the first difference, the null hypothesis is rejected for nearly all variables, considering a significance level of p-value < 0.05. However, in the case of Romania's LGDP variable, the null hypothesis is only rejected when considering a significance level of p-value < 0.10 after the first difference. Therefore, it can be inferred that all variables exhibit a unit root. Since the time series become stationary after the first difference, it is established that these are integrated of order one, I(1). Given that, in our case, all the variables share the same order of integration I(1) at the 10% significance level, the condition to be cointegrated is respected. To evaluate the presence of cointegration among the chosen variables, we will utilize the approach introduced by Johansen and Juselius.

The results (Table 3) show the rejection of the null hypothesis suggesting a no-cointegration relationship among the variables at the 5% significance level. Thus, we accept the alternative hypothesis that suggests the presence of at least one cointegration equation between the LGDP and LEXP, respectively, between the LGDP and LEMP in the Romanian economy, the NW Region, and the Cluj and Bihor Counties. The two tests used in the cointegration analysis suggest that the set of time series has an error correction representation that reflects the long-run adjustment mechanism.

Table 3. Cointegration test by Johansen and Juselius.

Series	Hypothesized No. of CE(s)		Trace Statistic	Max-Eigen Statistic	Critical Value (0.05)	
					Trace	Max-Eigen
Romania (total)						
LGDP/LEXP	None	H_0: r = 0	41.93 **	34.66 **	25.87	19.38
	At most 1	H_0: r ≤ 1	6.82	6.82	12.51	12.51
LGDP/LEMP	None	H_0: r = 0	43.96 **	38.76 **	20.26	15.89
	At most 1	H_0: r ≤ 1	5.20	5.20	9.16	9.16
NW Region						
LGDP/LEXP	None	H_0: r = 0	30.73 **	21.28 **	25.87	19.38
	At most 1	H_0: r ≤ 1	9.45	9.45	12.51	12.51
LGDP/LEMP	None	H_0: r = 0	39.36 **	33.16 **	25.87	19.38
	At most 1	H_0: r ≤ 1	6.19	6.19	12.51	12.51
Cluj County						
LGDP/LEXP	None	H_0: r = 0	32.79 **	22.68 **	25.87	19.38
	At most 1	H_0: r ≤ 1	10.10	10.10	12.51	12.51
LGDP/LEMP	None	H_0: r = 0	37.10 **	32.37 **	25.87	19.38
	At most 1	H_0: r ≤ 1	4.72	4.72	12.51	12.51
Bihor County						
LGDP/LEXP	None	H_0: r = 0	29.53 **	19.40 **	25.87	19.38
	At most 1	H_0: r ≤ 1	10.13	10.13	12.51	12.51
LGDP/LEMP	None	H_0: r = 0	20.91 **	18.70 **	15.49	14.26
	At most 1	H_0: r ≤ 1	2.20	2.20	3.84	3.84

Note: ** denotes significance (at the 5% significance level); r denotes the number of cointegrated vectors. Source: own elaboration using Eviews 12.

Considering the results obtained after testing the cointegration between the variables, we can state that long-run relationships exist between them. Therefore, in order to examine the relationship between these variables, we can apply a vector error correction model (VECM). The long-run relationship between economic growth (LGDP) and total expenditure on R&D (LEXP), respectively between economic growth (GDP) and individuals employed in the R&D sector (EMP) for one cointegrating vector, is presented in Table 4.

Based on the findings provided in Table 4, we can argue that in the case of Romania, the causal effect of the LGDP on the LEXP and on the LEMP variables is significant in the long run. This conclusion is supported by the statistically significant nature of the estimated adjusted coefficients (p-value > 0.05). More than that, the negative sign of these coefficients indicates that a long-run equilibrium characterizes the relationship between the LGDP and LEXP variables, respectively, between the LGDP and LEMP. Regarding the long-run causal effect of the LEXP and the LEMP variables on the LGDP, the presence of negative values suggests a long-run equilibrium in the relationship between these variables. In the case of the causal effect of the LEXP on the LGDP, the causality is significant in the long run only for the 10% significance level, as the value of the t-statistics is greater than 1.3160, while in the case of the causal effect of the LEMP on the LGDP, significance in the long run is observed at both the 5% and 1% significance levels. At the sample level, the short-run coefficients suggest convergence from LGDP to LEXP and, likewise, from LEMP to LGDP. Therefore, at the sample level, we can state that if the GDP at Romania's level increases by one unit, the total expenditure on R&D at the Romanian level will increase by 0.026 units, while if the number of individuals employed in the R&D sector increases by one unit, the GDP will increase by 0.127 units. In terms of the short-run causal results of LGDP on the LEMP and of LEXP on the LGDP, the results suggest divergence and non-significant coefficients. As such, the results confirm hypotheses H1a and H1b in the case of Romania.

Table 4. The results of VECM estimation by the OLS method.

Causality Direction	Error Correction Term (ECT) [t-Statistics]	Short-Run Coefficient [t-Statistics]	Lag Coefficient [t-Statistics]	R-Squared	F-Statistic
Romania					
LEXP → LGDP	−0.033 * [−1.380]	−0.808 [−0.706]	−0.047 [−0.200]	0.103	0.708
LGDP → LEXP	−0.167 *** [−7.729]	0.026 [0.822]	−0.279 ** [−1.804]	0.878	45.924 ***
LEMP → LGDP	−0.087 *** [−2.216]	0.127 [0.853]	0.194 [0.965]	0.377	3.835
LGDP → LEMP	−0.125 *** [−7.700]	−0.199 [−0.945]	−0.277 ** [−1.819]	0.867	65.723 ***
NW Region					
LEXP → LGDP	−0.379 *** [−2.478]	0.977 * [1.430]	0.114 [0.599]	0.258	2.207
LGDP → LEXP	−0.154 *** [−4.944]	0.075 * [1.669]	0.010 [0.062]	0.784	23.060 **
LEMP → LGDP	−0.082 [−1.022]	0.432 * [1.725]	0.057 [0.289]	0.155	1.163
LGDP → LEMP	−0.171 *** [−4.677]	0.296 ** [1.868]	0.097 [0.488]	0.755	19.609 **
Cluj County					
LEXP → LGDP	−0.356 *** [−2.062]	−0.617 [−0.970]	0.002 [0.013]	0.208	1.664
LGDP → LEXP	−0.174 *** [−5.459]	0.100 *** [2.271]	−0.063 [−0.409]	0.814	27.837 **
LEMP → LGDP	−0.017 [−0.580]	−0.149 [−0.408]	−0.244 [−1.224]	0.176	1.356
LGDP → LEMP	−0.123 *** [−3.866]	0.057 [0.471]	−0.044 [−0.197]	0.690	14.116 **
Bihor County					
LEXP → LGDP	−0.291 ** [−1.820]	1.471 [0.713]	−0.232 [−1.070]	0.269	2.332
LGDP → LEXP	−0.202 *** [−4.942]	0.009 [0.583]	0.061 [0.365]	0.759	19.984 **
LEMP → LGDP	−0.093 [−1.038]	0.181 [0.172]	−0.494 *** [−2.495]	0.342	3.301
LGDP → LEMP	−0.175 *** [−4.783]	0.029 [0.876]	0.039 [0.225]	0.751	19.116 **

Note: *, **, *** denote significance at the 10%, 5% and 1% confidence levels. Source: authors' estimates using Eviews 12.

The causal effect of the LGDP on the LEXP and on the LEMP variables in the case of the NW Region is significant in the long run, with the estimated adjusted coefficients demonstrating statistical significance. This is evident as the t-statistics value is greater than 1.708 for the 5% significance level and 2.060 for the 1% significance level. Also, the negative sign of these coefficients implies the existence of a long-run equilibrium in the association between the LGDP and LEXP variables, respectively between the LGDP and LEMP. Regarding the long-run causal effect of the LEXP and the LEMP variable on the LGDP, we can also observe negative values that presuppose a long-run equilibrium in their relationship. The causal effect of the LEMP on the LGDP is not statistically significant in the long run, with t-statistics below 1.3160. On the other hand, the causality of LEXP on LGDP is significant in the long run at both the 5% and 1% significance levels. The results confirm hypotheses H1a and H1b only partially in the case of the NW Region. The short-run coefficients indicate convergence from LGDP to the LEXP, from LGDP to the LEMP, and also from the LEMP to LGDP, respectively from LEXP to LGDP. Thus, the results confirm hypotheses H2a and H2b in the case of the NW Region.

In the case of Cluj County, the causal effect of the LGDP on the LEXP and on the LEMP variables is significant in the long run, with the estimated adjusted coefficients demonstrating statistical significance. This is evident as the value of t-statistics is greater than 1.708 for the 5% significance level and 2.060 for the 1% significance level. The negative sign of these coefficients shows that a long-run equilibrium defines the relationship between the LGDP and LEXP variables, as well as between the LGDP and LEMP. Regarding the long-run causal effect of the LEXP and the LEMP variable on the LGDP, we have also observed negative values, which presuppose a long-run equilibrium in their relationship. The causal effect of the LEMP on the LGDP is not significant in the long run as the value of t-statistics is lower than 1.3160 for the 10% significance level. In the case of the causal effect of the LEXP on the LGDP, the causality is significant in the long-run for the 5% and 1% significance levels. Therefore, the results confirm hypotheses H1a and H1b only partially in the case of Cluj County. Convergence from LGDP to the LEXP is suggested by the short-run coefficients. Therefore, we can state that if GDP at Cluj County's level increases by one unit, the total expenditure on R&D at Cluj County's level will increase by 0.10 units. Thus, the results confirm only the H2a hypothesis.

Regarding Bihor County, we can argue that the long-run causal impact of LGDP on both the LEXP and LEMP variables is substantial, with the estimated adjusted coefficients being statistically significant. The relationship between the LGDP and LEXP variables, as well as between the LGDP and LEMP, is defined by a long-term equilibrium, as indicated by the negative sign of these coefficients. Regarding the long-run causal effect of the LEXP and the LEMP variables on the LGDP, negative values are also observed, presupposing a long-run equilibrium characterizing the relationship between these variables. As concerns the causal effect of the LEMP on the LGDP, the causality is not significant in the long run as the value of t-statistics is lower than 1.3160 for the 10% significance level, while in the case of the causal effect of the LEXP on the LGDP, the causality is significant in the long run for the 5% and 1% significance levels. Therefore, the results confirm hypothesis H1a, whereas it confirms H1b only partially in the case of Bihor County. The short-run coefficients indicate convergence from the LGDP to the LEXP, from LGDP to the LEMP, from LEXP to the LDP and from the LEMP to the LGDP, but only at the sample level. Thus, the second hypothesis cannot be confirmed. Regarding R-squared, in the case of Bihor County, we notice that the highest intensity of the correlation is found in the case of models that have as an independent variable the GDP (R-squared = 0.75). Thus, we can say that at the sample level, GDP influences to a large extent both the total expenditure on R&D and the individuals employed in the R&D sector. We could not say the same thing if GDP were considered as the dependent variable. Thus, according to the results, we can say that GDP influences these two variables to a much greater extent than these two influence GDP.

We have also checked the quality of the residual, namely: the homoscedasticity, the autocorrelation, and the normal distribution. Table 5 presents the test values and p-values associated with the three hypotheses on the residuals listed above.

Table 5. Residual Analysis.

Models Based on Causality Direction	Null Hypothesis—The Errors Are Homoscedastic ARCH LM (*p*-Value)	Null Hypothesis—The Errors Are Independent Breusch Godfrey LM (*p*-Value)	Null Hypothesis—The Errors Are Normally Distributed Jarque-Bera (*p*-Value)
Romania			
LEXP → LGDP	18.95 (0.39)	6.92 (0.13)	0.36 (0.83)
LGDP → LEXP	18.95 (0.39)	6.92 (0.13)	0.24 (0.88)
LEMP → LGDP	23.15 (0.18)	4.37 (0.35)	0.73 (0.69)
LGDP → LEMP	21.91 (0.23)	6.69 (0.15)	0.23 (0.89)
NW Region			
LEXP → LGDP	20.15 (0.32)	3.21 (0.52)	3.64 (0.16)
LGDP → LEXP	14.78 (0.67)	4.31 (0.36)	2.72 (0.25)
LEMP → LGDP	15.44 (0.63)	3.80 (0.43)	0.31 (0.85)
LGDP → LEMP	15.44 (0.63)	3.80 (0.43)	0.86 (0.64)
Cluj County			
LEXP → LGDP	18.56 (0.41)	3.17 (0.52)	3.55 (0.16)
LGDP → LEXP	10.87 (0.89)	3.52 (0.47)	2.32 (0.31)
LEMP → LGDP	24.48 (0.13)	11.56 (0.02)	0.91 (0.63)
LGDP → LEMP	24.48 (0.13)	11.56 (0.02)	0.93 (0.62)
Bihor County			
LEXP → LGDP	14.28 (0.71)	6.31 (0.17)	0.97 (0.61)
LGDP → LEXP	9.51 (0.94)	2.19 (0.70)	0.61 (0.73)
LEMP → LGDP	11.68 (0.86)	4.89 (0.29)	1.57 (0.45)
LGDP → LEMP	11.68 (0.86)	4.89 (0.29)	1.58 (0.45)

Source: authors' estimates using Eviews 12.

According to the results presented in Table 5, we can affirm that the null hypothesis is accepted for all three residual tests. This implies that the validity of the representation of residuals in all four estimated models for Romania, the NW Region and Bihor County is confirmed, as the associated probabilities exceed the 5% threshold (p-value > 0.05). Consequently, the accurate representation of the residuals in the estimated models is verified. In the case of Cluj County, the probabilities associated with the values of the first test and the third test are higher than the 5% threshold, so the null hypothesis is accepted, while in the case of the autocorrelation, the values of this test are higher than the 5% threshold, but only in the case of the first two models (LGDP and LEXP). Thus, as regards the LGDP and LEXP variables, it can be affirmed that the null hypothesis is accepted for all three residual tests, confirming the accurate representation of the residuals in the estimated models.

The cointegration test results indicate the presence of a long-run, stable equilibrium among total expenditure on R&D and economic growth, respectively, among individuals employed in the R&D sector and economic growth. This implies a causal relationship between the variables, including GDP and R&D sector variables (LEXP and LEMP), in at least one direction. In order to find this causality, we employed the Granger causality test. Figure 2 illustrates both the long-run Granger causality from the exogenous variables to endogenous variable and the one-way short-run Granger causality.

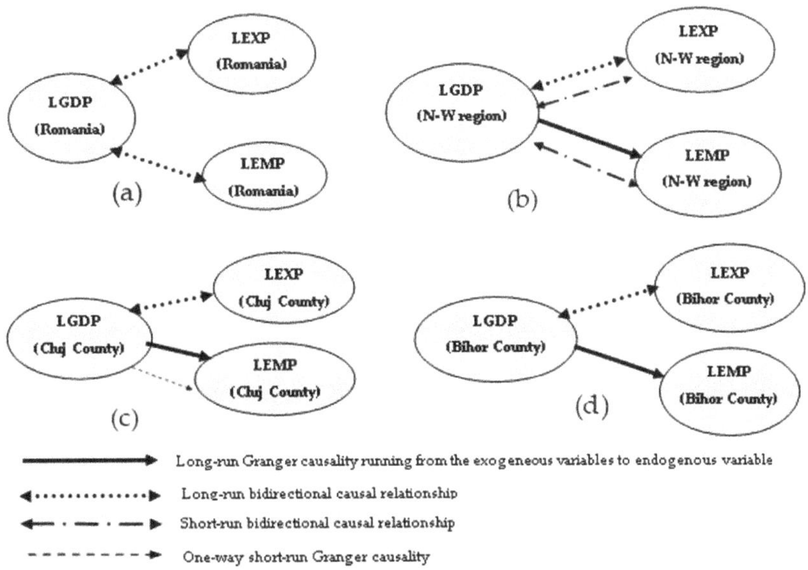

Figure 2. The Granger causality relationship between economic growth (GDP), total expenditure on R&D (EXP) and individuals employed in the R&D sector (EMP) in Romania (**a**), in the NW Region (**b**), in Cluj County (**c**) and in Bihor County (**d**). Source: authors' contribution.

The findings reveal the existence of a long-run bidirectional Granger causality between Romania's total expenditure on R&D (R&D) and economic growth, respectively, between Romania's individuals employed in the R&D sector and economic growth. Therefore, Romania's economic growth is the cause of Granger's development of the Romanian R&D sector, i.e., national economic growth can promote the long-run development of the total expenditure on R&D and of the individuals employed in the R&D sector. We also can state that, in the long run, Romania's total expenditure on R&D and the individuals employed in the R&D sector in Romania are the cause of Granger's development of Romania's economic growth. The results confirm the first two hypotheses, H1a and H1b, and the last two hypotheses, H3a and H3b.

According to Figure 2, we can argue that in the NW Region, there is a long-run unidirectional Granger causality between total expenditure on R&D and economic growth. According to Figure 2, we can argue that in the NW Region, there is a long-run unidirectional Granger causality only from economic growth to individuals employed in the R&D sector and a long-run bidirectional Granger causality between Romania's NW Region's total expenditure on R&D and economic growth. Also, we find evidence of a long-run bidirectional Granger causality between Romania's NW Region's total expenditure on R&D and economic growth. In the short run, we have identified a bidirectional causal relationship between Romania's NW Region's total expenditure on R&D and economic growth, respectively, between Romania's NW Region's individuals employed in the R&D sector and economic growth. The results confirm hypotheses H1a, H2b—only from GDP to individuals employed in the R&D sector, H3a and H3b.

Regarding Cluj County, the results indicate the existence of a long-run bidirectional causal relationship between the GDP and the total expenditure on R&D. Also, we have identified a long-run unidirectional Granger causality between the individuals employed in the R&D sector in Cluj County and the economic growth of this county. Thus, we can conclude that the economic growth of Cluj County is the cause of Granger's development of the Cluj County R&D sector, i.e., the economic growth can promote the long-run development of total expenditure on R&D and individuals employed in the R&D sector. However, the existence of the long-run influence of individuals employed in the R&D sector on economic growth is doubtful. In the short run, we have identified a unidirectional causal relationship between GDP and individuals employed in the R&D sector. The results thus confirm hypothesis H3a, while hypothesis H3b cannot be confirmed.

Regarding Bihor County, it is notable that a long-run bidirectional Granger causality was found only in the case of GDP and the total expenditure on R&D, while in the case of individuals employed in the R&D sector, we have found a long-run unidirectional Granger causality from GDP to this indicator. Thus, the economic growth of Bihor County is the cause of Granger's development of the Bihor County R&D sector. What is more, we can affirm that the total expenditure on R&D from Bihor County is the cause of Granger's development of Bihor County GDP but only in the long run. The results confirm hypothesis H3a.

As can be seen from Table 6, several hypotheses were confirmed. In general, the influence of GDP growth on total expenses and, respectively, on employment in the R&D sector is stronger than the effect of these two variables on GDP, in the case of Romania, the Cluj and Bihor Counties. As such, we can assume that regions with higher or better-performing GDP levels may attract more public funds and private investments in RD&I compared to other regions, where GDP levels are low. However, the influence that these two variables (total R&D expenditures and employment) have on regional development is also noteworthy, and regions already advanced and equipped for research and innovation will go further and attract more human and material resources than other less-developed regions.

Table 6. Results of testing the hypotheses.

Hypothesis	Romania	NW Region	Cluj County	Bihor County
H1a	confirmed	confirmed	confirmed	confirmed
H1b	Confirmed	partially confirmed	partially confirmed	partially confirmed
H2a	not confirmed	not confirmed	not confirmed	not confirmed
H2b	not confirmed	partially confirmed	partially confirmed	partially confirmed
H3a	confirmed	confirmed	confirmed	confirmed
H3b	confirmed	confirmed	not confirmed	not confirmed

Source: authors' elaboration.

The main limitations of the research consist in the incomplete number of factors influencing the relationship between the R&D sector and GDP, as well as the specific focus of findings, limited to a single Romanian region and two component counties. Another limitation of the study may be that it overshadows the role of other factors that shape economic growth (socio-economic policies, sectoral particularities, variations in economic structures, innovation and technology diffusion, entrepreneurship).

Another limitation of our study is its focus on the efforts dedicated to innovation (i.e., expenditure and employment in R&D activities) and not the actual results generated by these efforts (for instance, the number of patents filed and trademarks, turnover of innovative local units, the number of projects for R&D activities, etc.). Therefore, R&D expenditure does not capture the actual introduction of new products, services or processes, i.e., "expenditures will not necessarily produce outcomes in terms of output" [13] (p.18) and can hide or roughly approximate other inputs, for example: product design, market analysis, initial staff training, experiments. However, the indicators investigated by us are suitable for the analysis of long-term series and therefore useful for identifying a relationship with the relevant macroeconomic indicators, such as the evolution of GDP. In other words, the output indicators (patents and trademarks, turnover, etc.) are more suitable for medium-term analyses, at the micro- and meso-economic levels, but with more limited relevance for long time series.

Moreover, our analysis could not consider the level of autonomy in directing public R&D expenditure at the county and regional levels, given that the development regions in Romania are not administrative units and do not have formal powers for RD&I. Therefore, in future studies, we intend to extrapolate these results to other time periods or regional and national contexts but also other R&D-related indicators that were not addressed in the present investigation.

5. Conclusions and Implications

For over three decades, experts have emphasized the vital role of technological development, especially in RD&I, in fostering economic growth. The goals of national economic growth are closely linked to regional development, particularly in less-resource-endowed, peripheral, or isolated areas. Scholars have identified challenges such as a shortage of skilled labor, insufficient innovative firms, a lack of entrepreneurial spirit, insufficient infrastructural investments or unfavorable market dynamics, along with the inefficient use of RD&I funds, hindering regional prosperity and individual well-being.

Our research considered the analysis of the relationship between economic growth (expressed by GDP) and expenditure and employment in the R&D sector during the 1995–2021 period, running the Johansen cointegration test, the VECM model and Granger causality at the level of Romania and a component region. For a better understanding of these connections, we also deepened the analysis at the level of two counties in this region, engaged in innovative and sustainable technological development in the last two decades.

Our research findings support the existence of a long-run bidirectional Granger causality between Romania's total expenditure on R&D and employment in this sector, and economic growth, respectively. We can also state that, in the long term, Romania's total R&D expenditures and employment directly influence the country's economic growth.

At the regional level, the economic growth registered in the NW Region is the cause of the Granger development of the total R&D expenditures in the long run. Reciprocally, the NW Region's R&D expenditures support regional economic growth. In the short run, we also identified a bidirectional causality relationship between the NW Region's total R&D expenditure and economic growth. In the case of R&D employment in the NW Region, there is a short-run Granger causality relationship to economic growth but not vice versa.

For Cluj County, economic growth supports the long-run development of R&D expenditures and employment in this sector. However, the impact of R&D sector employment on economic growth is uncertain in the long run. Short-run causality exists between GDP and R&D sector employment. In Bihor County, there is a long-run bidirectional Granger causality between GDP and total R&D expenditures, while a long-run unidirectional causality exists between GDP and R&D sector employment. Total R&D expenditures drive the long-run development of Bihor County's GDP.

Our results are consistent with previous research on the favorable role of R&D on economic growth [24,25], stimulating innovation and economic recovery [28,29].

Previous studies [31–33] suggested that the effects of R&D spending on economic growth are weaker in developing countries; we observed that a similar pattern exists within a single county. Specifically, regions and counties that are better endowed in human and investment capital, and that are better prepared for innovation, tend to display a stronger, long-run link between R&D spending and employment and, respectively, economic growth.

Adapting RD&I policies to the national and regional specificities and comparative advantages can significantly contribute to economic growth. The relationships identified in our paper are, mainly, long-run relationships, and, consequently, our paper should act as a guide for both legislators and the business sector to multiply and potentiate the positive long-lasting effects of present-day decisions. The authorities must pay attention not only to the R&D allocation (from the state budget or from EU or private funds) but also to their intensity, duration and correct orientation towards technological and innovative sectors. Also, our study emphasizes the importance of the quality of governance and the increase in regional competencies, regarding public funding of research in fundamental fields, in parallel with the stimulation of private investments in applied research, profitable in the short term. Finally, theoretical studies and practical actions must consider the propagation at the national level of the results recorded in those regions that allocate greater public and private resources to research and innovation, the alignment of different R&D priorities and strategies at the national, sectoral and regional levels.

The novelty of this article resides in identifying the relationship between economic growth, expenditure and employment in the R&D sector in Romania, focusing on a specific development region (i.e., the Northwest Region) and two of its constituent counties. Additionally, we aimed to investigate whether the long-run or short-run relationship, as well as the causality identified at the national level in Romania, applies to the analyzed regions and, specifically, to the examined counties.

Author Contributions: Conceptualization, D.B. and D.G.; formal analysis, D.B., R.S., S.-A.B. and D.Z.; methodology, D.B., D.G., R.S. and A.B.; supervision, A.B.; writing—original draft, D.B., D.G., R.S., S.-A.B., D.Z. and C.-V.T.; writing—review and editing, D.G., D.B., R.S., S.-A.B., D.Z., C.-V.T. and A.B. All authors have read and agreed to the published version of the manuscript.

Funding: This research received no external funding.

Institutional Review Board Statement: Not applicable.

Informed Consent Statement: Not applicable.

Data Availability Statement: The data presented in this study are available on request from the corresponding author.

Acknowledgments: This work has been partially supported by the University of Oradea, within the Grants Competition "Scientific Research of Excellence Related to Priority Areas with Capitalization through Technology Transfer: INO—TRANSFER—UO—2nd Edition", Project Nos. 233/2022 and 234/2022.

Conflicts of Interest: The authors declare no conflicts of interest.

References

1. Solow, R. A Contribution to the Theory of Economic Growth. *Q. J. Econ.* **1956**, *70*, 65–94. [CrossRef]
2. Romer, P. Increasing Returns and Long-Run Growth. *J. Political Econ.* **1986**, *94*, 1002–1037. [CrossRef]
3. Romer, P. Endogenous Technological Change. *J. Political Econ.* **1990**, *98*, 71–102. [CrossRef]
4. Barro, R.; Sala-I-Martin, X. Technological Diffusion, Convergence and Growth. *J. Econ. Growth* **1997**, *2*, 1–26. [CrossRef]
5. Lucas, R. On the Mechanics of Economic Development. *J. Mon. Econ.* **1988**, *22*, 3–42. [CrossRef]
6. Grossman, G.; Helpman, E. *Innovation and Growth in the Global Economy*; MIT Press: Cambridge, MA, USA, 1991.
7. Aghion, P.; Howitt, P. A Model of Growth through Creative Destruction. *Econometrica* **1992**, *60*, 323–351. [CrossRef]
8. Segerstrom, P. Innovation, Imitation, and Economic Growth. *J. Political Econ.* **1991**, *99*, 807–827. [CrossRef]
9. Fernandez, N.; Martinez-San Roman, V.; Sanchez-Robles, B. R&D and Growth in the Spanish Regions, 1995-2007—An Empirical Approximation. *Int. J. Bus. Soc. Sci.* **2012**, *3*, 22–31. [CrossRef]
10. Ma, J.; Ahn, Y.-G.; Lee, M.-K. The Interactive Influence of Institutional Quality and Resource Dependence on Regional Economic Growth: Evidence from China's Resource-Based Provinces. *Sustainability* **2022**, *14*, 6173. [CrossRef]
11. Archibugi, D.; Filippetti, A.; Frenz, M. The Impact of the Economic Crisis on Innovation: Evidence from Europe. *Technol. Forecast. Soc.* **2013**, *80*, 1247–1260. [CrossRef]
12. Smętkowski, M. Regional Disparities in Central and Eastern European Countries: Trends, Drivers and Prospects. *Eur.-Asia Stud.* **2013**, *65*, 1529–1554. [CrossRef]
13. Taques, F.H.; López, M.G.; Basso, L.F.; Areal, N. Indicators used to measure service innovation and manufacturing innovation. *J. Innov. Knowl.* **2021**, *6*, 11–26. [CrossRef]
14. Wang, E. R&D Efficiency and Economic Performance: A cross Country Analysis Using the Stochastic Frontier Approach. *J. Policy Model.* **2007**, *29*, 345–360. [CrossRef]
15. Kuo, C.-C.; Yang, C.-H. Knowledge Capital and Spillover on Regional-Economic Growth: Evidence From China. China Economic Review. *China Econ. Rev.* **2008**, *19*, 594–604. [CrossRef]
16. Rodríguez-Pose, A.; Crescenzi, R. Research and Development, Spillovers, Innovation Systems, and the Genesis of Regional Growth in Europe. *Reg. Stud.* **2006**, *42*, 51–67. [CrossRef]
17. Sarpong, D.; Boakye, D.; Ofosu, G.; Botchie, D. The Three Pointers of Research and Development (R&D) for Growth-Boosting Sustainable Innovation System. *Technovation* **2003**, *122*, 102581. [CrossRef]

18. Wang, M.; Choi, B. An Analysis of the Impact of International R&D Spillovers and Technology Innovation in China. *Sustainability* **2023**, *15*, 1968. [CrossRef]
19. Goel, R.; Payne, J.; Ram, R. R&D Expenditures and US Economic Growth: A Disaagrgated Approach. *J. Pol. Model.* **2008**, *30*, 237–250. [CrossRef]
20. Bravo-Ortega, C.; Garcia Marin, A. R&D and Productivity: A Two Way Avenue? *World Dev.* **2011**, *39*, 1090–1107. [CrossRef]
21. Ildırar, M.; Özmen, M.; İşcan, E. The Effect of Research and Development Expenditures on Economic Growth: New Evidences. In Proceedings of the International Conference on Eurasian Economies, Kaposvár, Hungary, 29–31 August 2016.
22. Amaghouss, J.; Ibourk, A. Entrepreneurial Activities, Innovation and Economic Growth: The Role of Cyclical Factors. Evidence from OECD Countries for the Period 2001–2009. *Int. Bus. Res.* **2013**, *6*, 153–162. [CrossRef]
23. Woo, Y.; Kim, E.; Lim, J. The Impact of Education and R&D Investment on Regional Economic Growth. *Sustainability* **2017**, *9*, 676. [CrossRef]
24. Lichtenberg, F. *R&D Investment and International Productivity Differences*; NBER Working Paper No. 4161; NBER Working Paper Series; NBER: Cambridge, MA, USA, 1991. [CrossRef]
25. Park, W. International Spillovers of R&D Investment and OECD Economic Growth. *Econ. Inq.* **1995**, *33*, 571–591. [CrossRef]
26. Pop Silaghi, M.; Alexa, D.; Jude, C.; Litan, C. Do Business and Public Sector Research and Development Expenditures Contribute to Economic Growth in Central and Eastern European Countries? A Dynamic Panel Estimation. *Econ. Model.* **2014**, *36*, 108–109. [CrossRef]
27. Sylwester, K. R&D and Economic Growth. *Knowl. Technol. Policy* **2001**, *13*, 71–84. [CrossRef]
28. Ulku, H. *R&D, Innovation and Economic Growth: An Emprical Analysis*; IMF Working Paper, W/P/04/185; IMF: Washington, DC, USA, 2004.
29. Ulku, H. R&D, Innovation, and Growth: Evidence From Four Manufacturing Sectors in OECD Countries. *Oxf. Econ. Pap.* **2007**, *59*, 513–535. [CrossRef]
30. Goel, R.; Ram, R. Research and Development Expenditures and Economic Growth: A Cross-Country Study. *Econ. Dev. Cult. Chang.* **1994**, *42*, 403–411. [CrossRef]
31. Samimi, A.; Alerasoul, S. R&D and Economic Growth: New Evidence from Some Developing Countries. *Aust. J. Basic Appl. Sci.* **2009**, *3*, 3464–3469.
32. Inekwe, J. The Countribution of R&D Expenditures to Economic Growth in Developing Countries. *Soc. Indic. Res.* **2015**, *124*, 727–745. [CrossRef]
33. Gumus, E.; Celikay, F. R&D Expenditure and Economic Growth: New Empirical Evidence. *Margin J. Appl. Econ. Res.* **2015**, *9*, 205–217. [CrossRef]
34. Coccia, M. Public and Private R&D Investments as Complementary Inputs for Productivity growth. *Int. J. Technol. Policy Manag.* **2010**, *10*, 73–91. [CrossRef]
35. Salvekar, S. R&D Expenditure and Economic Performance: A Regional German Panel Analysis. Master's Thesis, Jönköping University, Jönköping, Sweden, 2020.
36. Beise, M.; Stahl, H. Public Research and Industrial Innovations in Germany. *Res. Policy* **1999**, *28*, 397–422. [CrossRef]
37. Harhoff, D. R&D Spillovers, Technological Proximity, and Productivity Growth—Evidence From German Panel Data. *Schmalenbach Bus. Rev.* **2000**, *52*, 238–260. [CrossRef]
38. Moutinho, R.; Au-Yong-Oliveira, M.; Coelho, A.; Pires Manso, J. The Role of Regional Innovation Systems (RIS) in Translating R&D Investments Into Economic and Employment Growth. *J. Technol. Manag. Innov.* **2015**, *10*, 9–23. [CrossRef]
39. Celli, V.; Cerqua, A.; Pellegrini, G. Does R&D Expenditure Boost Economic Growth in Lagging Regions? *Soc. Indic. Res.* **2021**. [CrossRef]
40. European Commission. Report from the Commission to the European Parliament, the Council, the European Economic and Social Committee and the Committee of the Regions. European Structural and Investment Funds 2014–2020, Brussels, Belgium. 2021. Available online: https://eur-lex.europa.eu/legal-content/EN/TXT/?uri=CELEX:52021DC0213 (accessed on 17 November 2023).
41. Guellec, D.; Van Pottelsberghe De La Potterie, B. The Impact of Public R&D Expenditure on Business R&D. *Econ. Innov. New Technol.* **2003**, *12*, 225–243. [CrossRef]
42. Rehman, N.U.; Hysa, E.; Mao, X. Does public R&D complement or crowd-out private R&D in pre and post economic crisis of 2008? *J. Appl. Econ.* **2020**, *23*, 349–371. [CrossRef]
43. David, P.; Hall, B.; Toole, A. Is Public R&D a Complement or Substitute for Private R&D? A Review of the Econometric Evidence. *Res. Policy* **2000**, *29*, 497–529. [CrossRef]
44. Rodríguez-Pose, A. Innovation Prone and Innovation Averse Societies. Economic Performance in Europe. *Growth Chang.* **1999**, *30*, 75–105. [CrossRef]
45. Bronzini, R.; Piselli, P. The Impact of R&D Subsidies on Firm Innovation. *Res. Policy* **2016**, *45*, 442–457. [CrossRef]
46. Becker, S.; Egger, P.; von Ehrlich, M. Going NUTS: The Effect of EU Structural Funds on Regional Performance. *J. Public Econ.* **2010**, *94*, 578–590. [CrossRef]
47. Pellegrini, G.; Terribile, F.; Tarola, O.; Muccigrosso, T.; Busillo, F. Measuring the Effects of European Regional Policy On Economic Growth: A Regression Discontinuity Approach. *Pap. Reg. Sci.* **2013**, *92*, 217–233. [CrossRef]

48. Fratesi, U.; Perucca, G. EU Regional Development Policy and Territorial Capital: A Systemic Approach. *Pap. Reg. Sci.* **2019**, *98*, 265–281. [CrossRef]
49. Kijek, T.; Kijek, A.; Matras-Bolibok, A. Club Convergence in R&D Expenditure across European Regions. *Sustainability* **2022**, *14*, 832. [CrossRef]
50. Becker, S.; Egger, P.; von Ehrlich, M. Absorptive Capacity and the Growth and Investment Effects of Regional Transfers: Regression Discontinuity Design with Heterogeneous Treatment Effects. *Am. Econ. J.-Econ. Policy* **2013**, *5*, 29–77. [CrossRef]
51. Bilbao-Osorio, B.; Rodriguez-Pose, A. From R&D to Innovation and Economic Growth in the EU. *Growth Chang.* **2004**, *35*, 434–455. [CrossRef]
52. Diukanova, O.; Chioncel, M. *A GDP Impact Evaluation of R&D Investments in Romania Using the CGE Model Rhomolo*; JRC Working Papers on Territorial Modelling and Analysis No. 10/2021; European Commission: Seville, Spain, 2021; JRC126690.
53. Ur Rehman, N.; Hysa, E. The effect of financial development and remittances on economic growth. *Cogent Econ. Financ.* **2021**, *9*, 1932060. [CrossRef]
54. Badulescu, D.; Cadar, O. Romania: Many Entrepreneurs But Few Innovators. *Ann. Univ. Oradea Econ. Sci.* **2016**, *XXV*, 465–475.
55. Serbanica, C.; Pupinis, M. *PSF Review of the Romanian R&I System—Background Report*; Publications Office of the European Union: Luxembourg, 2021; Available online: https://data.europa.eu/doi/10.2777/38334 (accessed on 19 December 2023).
56. Chioncel, M. Analysis of the Factors That Obstruct the Diffusion of Innovation in Romania. 2017. Available online: https://uefiscdi.gov.ro/resource-821136-analysis-of-the-factors-that-obstruct-the-diffusion-of-innovatio.pdf (accessed on 11 November 2023).
57. Diaconu, M.; Vilcu, A. Business Innovation Activity in Romania: The Main Trends and Weaknesses. *Procd. Soc. Behv.* **2018**, *238*, 157–166. [CrossRef]
58. Șerbănică, C. Territorial Innovation Patterns in Romania. Future Pathways for Smart Specialization. *Transylv. Rev. Adm. Sci.* **2021**, *62*, 153–175. [CrossRef]
59. Dachin, A.; Postoiu, C. Innovation and Regional Performance in Romania. *Theor. Appl. Econ.* **2015**, *2*, 55–64.
60. Goschin, Z. R&D as an Engine of Regional Economic Growth in Romania. *Rom. J. Reg. Sci.* **2014**, *8*, 24–37.
61. Herte, D.A.; Dianu, D.; Ciucos, M.; Badulescu, D.; Badulescu, A. SMEs and Innovation in The European Context. *J. Innov. Manag. Small Medium Enterp.* **2021**, 238722. [CrossRef]
62. The World Bank. Romania. Overwiew. 2023. Available online: https://www.worldbank.org/en/country/romania/overview (accessed on 10 November 2023).
63. Eurostat. Gross Domestic Product (GDP) at Current Market Prices by NUTS 3 Regions. 2023. Available online: https://ec.europa.eu/eurostat/databrowser/product/page/nama_10r_3gdp (accessed on 11 November 2023).
64. Chioncel, M.; Del Rio, J.C. *RIO Country Report 2017: Romania*; Publications Office of the European Union: Luxembourg, 2018; ISBN 978-92-79-81280-4. [CrossRef]
65. Goschin, Z.; Sandu, S.; Goschin, G.G. How Soon Could Romania Close the R&D Gaps against EU-28. *Proc. Econ. Financ.* **2015**, *22*, 160–167. [CrossRef]
66. Hollanders, H.; European Commission. *European Innovation Scoreboard 2019*; Publications Office of the European Union: Luxembourg, 2019.
67. European Commission; Directorate-General for Research and Innovation; Hollanders, H.; Es-Sadki, N. *Regional Innovation Scoreboard 2023—Regional Profiles Romania*; Publications Office of the European Union: Bruxelles, Belgium, 2023; Available online: https://data.europa.eu/doi/10.2777/70412 (accessed on 18 December 2023).
68. Eurostat, NUTS—Nomenclature of Territorial Units for Statistics. 2021. Available online: https://ec.europa.eu/eurostat/web/nuts/background (accessed on 20 November 2023).
69. Ministry of Regional Development and Housing (Romania). The Urban Dimension of Romania's Cohesion Policy Regional Operational Programme 2007–2013. Available online: https://slideplayer.com/slide/4762639/ (accessed on 20 November 2023).
70. National Institute of Statistics (Romania). Tempo Online. Domains. 2022. Available online: http://statistici.insse.ro:8077/tempo-online/#/pages/tables/insse-table (accessed on 2 November 2023).
71. Dickey, D.; Fuller, W. Distribution of the Estimators for Autoregressive Time Series with a Unit Root. *J. Am. Stat. Assoc.* **1979**, *74*, 427–431. [CrossRef]
72. Johansen, S.; Juselius, K. Maximum Likelihood Estimation and Inference on Cointegration—With Application to the Demand for Money. *Oxf. Bull. Econ. Stat.* **1990**, *52*, 169–210. [CrossRef]
73. Andrei, T.; Bourbonnais, R. *Econometrie*, 2nd ed.; Economica Publishing: Bucharest, Romania, 2017; ISBN 978-973-709-812-2.
74. Engle, R.; Granger, C. Co-Integration and Error Correction: Representation, Estimation and Testing. *Econometrica* **1987**, *52*, 251–276. [CrossRef]
75. Greene, W.H. *Econométrie*, 7th ed.; Pearson Education France: Paris, France, 2021; ISBN 9782744054402.

sustainability

Article

The Comet Assay as a Sustainable Method for Evaluating the Genotoxicity Caused by the Soluble Fraction Derived from Sewage Sludge on Diverse Cell Types, Including Lymphocytes, Coelomocytes and *Allium cepa* L. Cells

Monica Adriana Costea, Cristina Adriana Rosan *, Vasile Laslo, Eliza Agud *, Cornelia Purcarea and Simona Ioana Vicas

Faculty of Environmental Protection, University of Oradea, 26 Gen. Magheru Street, 410048 Oradea, Romania; mcostea@uoradea.ro (M.A.C.); vasile.laslo@uoradea.ro (V.L.); cpurcarea@uoradea.ro (C.P.); svicas@uoradea.ro (S.I.V.)
* Correspondence: crosan@uoradea.ro (C.A.R.); eliza.agud@uoradea.ro (E.A.)

Abstract: With the expansion of urban areas, the amount of sludge produced by sewage treatment plants is increasing, raising big problems regarding the reintroduction of this sludge into nature in order to fully solve the wastewater problem. The application of sludge to agricultural surfaces or degraded land is a controversial solution since, despite the well-known benefits, sludge can, in certain cases, represent a real threat to both human health and the environment, with long-term harmful effects. The present study evaluates the potential genotoxicity of sludge using the Comet Test and three cellular bioindicators (lymphocytes, coelomocytes, and *Allium cepa* L.) for its quantification. To perform the tests, the soluble fraction of the sludge was used at concentrations of 25%, 50%, 75%, and 100%, as well as a negative control (H_2O) and a positive control (H_2O_2). The Comet test indicated an increase in DNA damage among cells exposed for 4 h in the following order: coelomocytes, lymphocytes, and *Allium cepa* L. cells. Our results indicate that *Allium cepa* L. nuclei are more sensitive, with genotoxic effects being evident at concentrations as low as 25%. In coelomocytes, we recorded nuclear damage starting at a concentration of 75%. These results indicate the necessity of using multiple genotoxicity tests, combined in a test battery, to achieve a greater level of relevance. The concentration of the soluble fraction of the sludge has an inverse relationship with the auxin content in leaves and roots, suggesting varying levels of stress. The results of this study can contribute to the creation of a genotoxic profile of sewage sludge, facilitating decisions related to reducing its negative impact.

Keywords: Comet assay; DNA damage; sewage sludge; lymphocytes; coelomocytes; *Allium cepa* L.; bioindicators

Citation: Costea, M.A.; Rosan, C.A.; Laslo, V.; Agud, E.; Purcarea, C.; Vicas, S.I. The Comet Assay as a Sustainable Method for Evaluating the Genotoxicity Caused by the Soluble Fraction Derived from Sewage Sludge on Diverse Cell Types, Including Lymphocytes, Coelomocytes and *Allium cepa* L. Cells. *Sustainability* **2024**, *16*, 457. https://doi.org/10.3390/su16010457

Academic Editor: Agostina Chiavola

Received: 8 December 2023
Revised: 26 December 2023
Accepted: 2 January 2024
Published: 4 January 2024

1. Introduction

The amount of wastewater produced by urban agglomerations and industrial activities is constantly increasing, at the same time increasing the amount of sludge resulting from the purification of these waters [1]. The properties of sludge depend on the concentration of pollutants in wastewater and the specific treatments applied to it [2,3]. Sludge contains compounds with high agricultural value (organic matter, nitrogen, phosphorus, potassium, and, to a lesser extent, calcium, sulfur, and magnesium) [4,5]. When applied on agricultural surfaces, it can partially substitute for classic fertilizers [4,6,7], and, at the same time, it can improve the physical characteristics of the soil (density, porosity, permeability, water retention capacity, etc.) as well as its chemical characteristics (pH and cation exchange) [8]. Worldwide, the use of sludge in agriculture is considered a way of reducing environmental pollution and contributing to the circular economy, specifically via using sludge as a fertilizer and soil improver [9]. The amount of sewage sludge produced in Europe (EU27) in 2020 was estimated to be over 13 million tons of dry matter [10].

Even if the economic advantages of using sludge as fertilizer are obvious in the current energy context, the risks of its application should also be noted in the absence of proper management. These risks are related to the accumulation over time of potentially genotoxic compounds such as polycyclic aromatic hydrocarbons (PAH), polychlorinated biphenyls (PCB), organochlorine pesticides, and heavy metals [8,11,12], which can translocate into plants and then pass on throughout the food chain [13]. The presence of heavy metals in soil induces genotoxicity through different mechanisms: metal ions induce oxidative changes in free amino acids and proteins [14]; the binding of metal to the cell nucleus causes genotoxicity [15]; and metal-mediated oxygen induces genotoxicity in plants [16].

Some studies have reported on plant growth inhibition mediated by heavy-metal stress, especially in the early stages of plant growth [17,18], while others [19,20] have demonstrated favorable plant yield responses to sewage sludge application. A decrease in soil microbial biomass and enzymatic activities due to the application of sewage sludge was observed in [8,21]. In sewage sludge, Hg and Pb frequently appear, both being metals responsible for the generation of very toxic oxygen species, such as the superoxide radical (O^{2-}), the hydroxyl radical ($OH^{\bullet -}$), and hydrogen peroxide (H_2O_2) [18,22]. They act directly on membrane lipids and proteins, or indirectly, damaging the genome by degrading nitrogenous bases, generating single- and double-strand breaks, and inducing DNA–protein crosslinks [23]. The different types of DNA damage induced by heavy metals, through ROS activity, affect the stability of the genome by affecting replication and transcription [24–26].This fact leads to different physiological effects, including reduced protein synthesis and damage to the cell membrane and to proteins involved in photosynthesis that ultimately negatively affect plant growth and development.

The spread of mutagenic and genotoxic agents in aquatic ecosystems [27,28] has an adverse effect on aquatic animals, posing a significant risk to human health. The evaluation of genotoxic contamination in the aquatic environment can be conducted by using the Comet test and by using various aquatic organisms as biosensors [29]. The toxicity and bioavailability of metals in the aquatic environment are affected by various abiotic factors, including pH, water hardness, alkalinity, and the accumulation of humic substances. A significant correlation exists between these parameters and the accumulation of metals in the biota.

The heavy metals responsible for soil genotoxicity, when sewage sludge is applied for a long time, are arsenic (As), barium (Ba), cadmium (Cd), chromium (Cr), copper (Cu), lead (Pb), mercury (Hg), molybdenum (Mo), nickel (Ni), selenium (Se), and zinc (Zn) [19,30,31]. The mobility of heavy metals in soil is conditioned by some of the latter's characteristics, such as texture and pH, and by the organic complexation of the metals [32]. Increased mobility of heavy metals is associated with their increased absorption by plants or with their leaching into underground water [33].

The concentration of heavy metals (including As, Pb, Cd, and Cr) in plants varies depending on the species [34]. In one study, treatment with sewage sludge did not significantly affect the uptake of heavy metals by *Loropetalum chinense* var. *rubrum* but significantly affected heavy metal uptake by *Dendranthema morifolium* and *Viola tricolor* [35]. Henning et al. (2001) [36] reported that *Zea mays* plants grown on a mixture of soil and sludge showed high levels of Pb, Cu, and Zn in their tissues. Another study revealed that heavy-metal accumulation occurs in plants grown in soil mixed with sewage sludge [37]. The reductions in the concentrations of Pb, Cd, Hg, and Zn in soil mixed with sludge after harvesting *Solanum lycopersicum* plants were 39.38%, 47.93%, 6.18%, and 49.89%, respectively [38]. According to Emamverdian et al. (2015) [39], lead is easily absorbed by plants through the roots but is translocated only in small amounts, while cadmium, although not essential for plants, is easily absorbed and accumulates easily in plant tissues.

Results have demonstrated that by interfering with essential metabolic processes like photosynthesis, heavy metals cause a decrease in endogenous auxin levels, which, in turn, slows down shoot and root growth processes [40,41]. Exogenous auxin supplementation mitigates growth inhibition by enhancing heavy metal tolerance, as demonstrated by Bücker-Neto [42]. Several investigations have provided evidence for the significant contribution of auxins in the production of enzymatic and non-enzymatic antioxidants to combat nitro-oxidative stress caused by metalloids or other abiotic factors. According to Zhang [43], auxins are a class of phytohormones that play important roles in the life cycle of plants and form part of a hormonal signaling network, and they affect how plants react to stressors.

In the European Union, the use of sewage sludge in agriculture is regulated by Directive 86/278/CEE [44].

Before application, the chemical composition of sewage sludge should be determined [4]. The complexity of the corresponding mixture greatly increases the costs of determinations. The determination of the genotoxic potential of sewage sludge through relatively simple and cheap tests such as the Comet assay can provide important information about the genotoxic potential of sludge.

The in vitro assessment of the genotoxic effects of different agents using human cells allows for a good extrapolation to the state of human health [45], and conducting the same procedure using plant cells allows a more accurate assessment of the health of the environment [46–48]. Earthworms are good environmental indicators, as they are present in the soil in large numbers and respond to a variety of environmental and ecological factors, including alterations in soil chemistry as well as in forestry and agricultural practices [49,50]. Verschaeve and Gilles (1995) [51] used, for the first time, coelomocytes obtained from earthworms of the species *Eisenia fetida* for the detection of genotoxic compounds in soil, and, later, Salagovic et al. (1996) [52] proposed the use of the Comet assay and *coelomocytes* to monitor pollution in terrestrial ecosystems. Earthworm coelomocytes represent a very suitable bioindicator for pollution monitoring because they are exposed to genotoxicity through dermal uptake/diffusion and the ingestion of soil pollutants. Three main subpopulations of coelomocyte have been proposed, namely, eleocytes, granulocytes, and amoebocytes, but this distinction is still under debate due to the different stages of cell functioning [53].

Several studies have shown the a negative impact of sewage sludge on soil organisms, particularly regarding the physiology, activity, and diversity of earthworms [54–57]. On the other hand, other studies have shown that there is a good correlation between the results of bioassays performed on plant and animal cells [58].

The comet assay was first proposed by Ostling and Johanson in 1984 [59]. They demonstrated that under neutral conditions, DNA fragments from nuclei migrate to an anode. This version of the comet test detects only DNA double-strand breaks, which are characteristic effects of radiation and radiomimetic agents [60]. This technique was later improved by Singh et al. [61], who found that alkaline conditions increased the sensitivity, specificity, and reproducibility of this method. This comet assay variant is widely employed and capable of identifying DNA single-strand breaks, alkali-labile sites, and cross-linked DNA within individual cells. This test is based on the ability of DNA breaks to relax supercoiling, allowing DNA loops to extend from the nuclear core (nucleoid) under an electric field to form a comet-like tail [62]. It is used in human-monitoring studies as a biomarker of exposure to agents that cause DNA damage [63] and in ecotoxicological studies for a variety of sentinel organisms [64].

Based on the above, in this study, the possible genotoxic potential of treated sludge from an urban wastewater treatment plant was investigated using cells of different origins, namely, from humans (lymphocytes), animals (coelomocyte), and plants (*Allim cepa* L.), in order to identify which of them are more sensitive and can be used as a bioindicator of environmental pollution. In addition, the effect of different sludge concentrations on the phytohormone auxin content was evaluated in barley leaves and roots.

2. Materials and Methods

2.1. Chemical Analysis of Sludge and Acquisition of the Soluble Fraction from Biosolids

The sludge was obtained from the Oradea sewage treatment plant (Romania), and the chemical analyses conducted on it were carried out in the WESSLING Hungary Kft laboratory accredited with the number NAH-1-1398, based on the accredited methods shown in Table S1.

The soluble fraction of the sludge (SFS) was obtained according to the recommendations of DIN procedure 38414-S4 (1984): 5 g of biosolid sample dried for 24 h at 105 °C was mixed with 50 mL of deionized water, and the mixture was shaken for 24 h at 25 °C (Orbital Shaker OS 20 Biosan, Riga, Latvia). The mixture was separated via centrifugation at 4500 rpm (EBA 20, Andreas Hettich GmbH & Co. KG, Tuttlingen, Germany) for 30 min, and the suspension was then passed through a sterile 0.22 µm pore membrane filter (LABBOX, Labbox Labware, S.L Barcelona, Spain) to avoid microbiological contamination.

2.2. Isolation of Lymphocytes and Treatment with SFS

This task was performed according to the protocol described by Álvarez-Moya, C., and Reynoso-Silva, M., in 2015 [65], with minor modifications. A total of 100 µL heparinized blood + 400 µL RMPI 1640 (BIOWEST, Nuaillé, France) was gently mixed and deposited on 400 µL of Histopaque-1077 (SIGMA) without mixing the layers. This mixture was centrifuged for 20 min at 2000 rpm. At the interface between the 2 layers, a whitish layer of lymphocytes became visible. Using a micropipette tip (ISOLAB, Isolab Laborgeräte GmbH-Wertheim Deutschland, Germany), the layer of lymphocytes was collected and placed in a microcentrifuge tube with 500 µL of RMPI 1640 medium. After centrifugation for 10 min at 2000 rpm, the supernatant was removed, and the pellet was washed once more, after which it was taken up in 200 µL RMPI 1640. The lymphocytes (Figure 1a) were brought to a cell density of 2×10^5 (Bürker-Türk chamber, BRAND GMBH + CO. KG Wertheim, Germany) and exposed to the experimental variants (SFS 25%, 50%, 75%, and 100% and 200 µM of H_2O_2 (positive control) and H_2O (negative control) for 4 h.

2.3. Isolation of Coelomocytes and Treatment with SFS

Earthworms were purchased from a company specializing in their culture for fishing (S.C. Maggot Fish S.R.L. Oradea, Romania). They belonged to the species *Eisenia fetida* (Annelida, Oligochaeta). To obtain coelomocytes (Figure 1b), we used a non-invasive method described by Eyambe GS et al. in 1991 [66]. Before extrusion, earthworms were washed with tap water at room temperature and placed on a paper towel overnight to allow them to empty their gut contents. Groups of 3 worms (approx. 15 g) were placed for 2 min in a Petri dish containing an extrusion medium (2 mL/individual) consisting of saline solution (95% *v*/*v*) and absolute ethanol (5% *v*/*v*) (S.C. CHIMREACTIV, S.R.L., Bucuresti, Romania), supplemented with 2.5 mg/mL of EDTA (Amresco Chemicals) and 10 mg/mL of the mucolytic agent guaiacol glycerol (1:1) (Sigma-Aldrich, Merck KGaA, Darmstadt, Germany), adjusted to pH 7.3. The extruded cells were transferred to centrifuge tubes containing 6 mL of PBS (4 °C) (Roth), centrifuged at 8000 for 3 min, and washed again with 10 mL of PBS (4 °C). After washing, the cell density was determined using a Bürker-Türk counting chamber, and viability was determined via staining with trypan-blue 0.4% (Sigma-Aldrich, Merck KGaA, Darmstadt, Germany). The cell density was 10^5 cells/mL. A total of 150 µL of cell suspension was exposed for 4 h to each of the experimental variants tested (SFS 25%, 50%, 75%, and 100% and 200 µM of H_2O_2 (positive control) and H_2O (negative control)) for 4 h.

2.4. Isolation of Allium cepa L. *nuclei and Treatment with SFS*

Bulbs of *Allium cepa* L. var. Alba de Buzău purchased from a local market, identified in the Vegetables Laboratory of the Faculty of Environmental Protection, and of approximately equal size were placed on the tops of 800 mL jars, with the root meristem submerged in water. They were kept in the dark at room temperature. After 5 days, the roots emerging from the rhizogenic discs were about 4–5 cm long. The root discs together with the roots were detached from the bulb through a transverse section and then divided into 4 fragments (Figure 1c), which were immersed in the solutions with the experimental variants (SFS 25%, 50%, 75%, and 100% and 200 μM of H_2O_2 (positive control) and H_2O (negative control)) for 4 h. After exposure, the roots were washed and placed in a Petri dish with 400 mM of Tris-HCl buffer (AppliChem) at pH 7.5 (at an ice-cold temperature). The tips of roots measuring 1–1.5 cm long were fragmented with a fresh razor blade, and the isolated root nuclei were collected in 320 μL of 0.4 M Tris-HCl buffer (pH 7.5). This method ensures a low level of damage to the nuclei [67]. For all types of bioindicators, exposure to the different concentrations of SFS (soluble fraction of sludge) was performed for 4 h at 4 °C and in the dark.

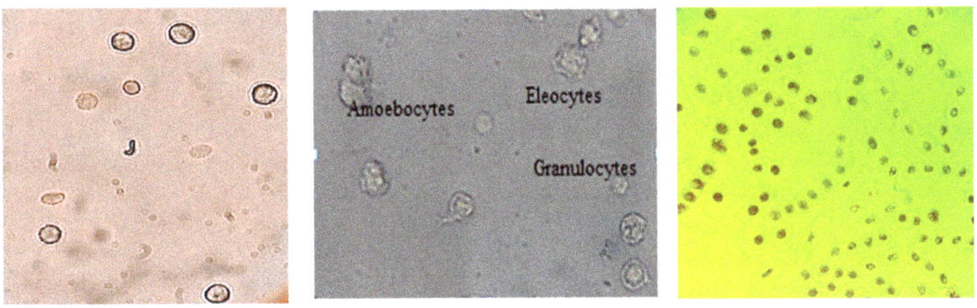

(**a**) (**b**) (**c**)

Figure 1. The biological material used in the Comet assay: (**a**) lymphocytes; (**b**) coelomocytes; (**c**) cells from the root tips of *Allium cepa* L.

2.5. The Comet Assay

2.5.1. Slide Preparation

Slide preparation was performed according to the method reported by Singh (1996) [68] with minor modifications. Each slide was pre-coated with 1% normal-melting-point agarose (NMPA) (Sigma-Aldrich) and completely dried at room temperature. A total of 50 μL of 0.8% low-melting-point agarose (LMPA) (Thermo Fisher Scientific Inc., Waltham, MA, USA) kept at 37 °C was mixed with 50 μL of nuclear suspension and placed on top of the agarose layer, and a lamella was placed on top. The slides thus prepared were placed on ice for 10 min for solidification, after which the lamella was removed, and the slide was used in the next lysis step.

2.5.2. Lysis

The lymphocyte lysis solution contained 2.5 M of NaCl, 10 mM of Tris, and 100 mM of Na-EDTA plus 1% sodium lauryl sarcosinate (Sigma-Aldrich, Merck KGaA, Darmstadt, Germany) at pH 10.0. Immediately before use, 1% Triton X-100 (Sigma Aldrich, Merck KGaA, Darmstadt, Germany) and 10% DMSO (dimethyl sulfoxide) (Sigma-Aldrich, Merck KGaA, Darmstadt, Germany) were added, with the role of capturing the radicals generated by hemoglobin iron. The lysis time was 1h, and lysis occurred at 40 °C and in the dark [69].

For lysis, slides with *Allium cepa* L. nuclei were immersed in freshly prepared, ice-cold lysis solution (1 M of NaCl and 30 mM of NaOH, 0.5% *w*/*v* SDS, pH 12.3) for 1 h at 40 °C in the dark [70].

Slides with coelomocytes were immersed in freshly prepared, ice-cold lysis solution (2.5 M of NaCl, 100 mM of EDTA, 10 mM of Tris, 10% DMSO, and 1% Triton X-100; pH 10.0) for 2 h [71].

2.5.3. Pre-Electrophoresis

The slides were placed in an electrophoresis unit filled with a freshly prepared alkaline buffer (300 mM of NaOH and 1 mM of Na EDTA) (Sigma-Aldrich) and placed at 40 °C. For lymphocytes, the despiralization time was 20 min; for coelomocytes, it was 20 min, and it was 30 min for *Allium cepa* L. nuclei [72].

2.5.4. Electrophoresis

A fresh, cold electrophoresis buffer (1 mM of Na_2EDTA and 300 mM of NaOH, pH 13) was prepared. Slides placed in an electrophoresis tank (SC PRECISA SRL, Sibiu, Romania) were exposed to 26 V (0.72 V/cm) at 300 mA for 25 min. After electrophoresis, the slides underwent three rinses using 400 mM of Tris (pH 7.5). They were then stained with 80 µL of ethidium bromide (20 µg/mL) (Amresco Inc., Cincinnati, OH, USA) for 10 min. Finally, the slides were washed with ice-cold water to eliminate the dye. All operations were carried out under red lights. Visualization was performed with a fluorescent microscope (BIOSYSTEMS Co., Ltd., Ulaanbaatar, Mongolia) equipped with an excitation filter of 515–560 nm, a barrier filter of 590 nm at a magnification of 200×, and a Microscope Digital Camera SP83. A minimum of 50 comets were analyzed per repetition using the software program CometScore2.00.38.

2.6. Determining 3-Indole Acetic Acid (IAA)

The *Hordeum vulgare* L. cv. Salamander seeds used originated from S.C. Totem SRL, Oradea, Romania, and were harvested in 2022. Following a washing procedure conducted using tap water, the items were disinfected by immersing them in a solution containing 0.15% mercuric chloride ($HgCl_2$) for 2 min. Subsequently, the seeds were rinsed 5 times in sterile distilled water.

Barley seedlings were obtained by germinating the seeds on an inert substrate moistened with solutions at the concentrations established in the experiment. After 14 days, root and leaf samples were collected in three repetitions, and the IAA content was determined. Data were analyzed in three independent replicates.

The procedure described by Goldschmidt et al. (1968) [73], with minor modifications, was used to obtain the auxin extract. Briefly, sterile plant tissue (160 mg) was cold-mortared with fine sand in 2 mL phosphate buffer (pH 6) and then centrifuged at 5000 rpm for 5 min. For the determination of IAA, 2.0 mL of Salkowski's reagent (1.0 mL of 0.5 M $FeCl_3$ (Fluka Chemie GmbH, Buchs, Switzerland) in 50 mL of 35% perchloric acid) was added to the supernatant (1.0 mL) (Alpha Chemika, Mumbai, India). Pink color development was measured after 1 h at 530 nm (Shimadzu UVmini 1240, Shimadzu Corporation Tokyo, Japan), and the results were expressed in µg/100 mg of fresh weight.

2.7. Statistical Analysis

All the experiments were carried out in triplicate, and the results indicate the mean ± standard deviation (SD). Statistical significance between treatment groups was evaluated using the one-way ANOVA test followed by Tukey's multiple comparison test, which were carried out using GraphPad Prism (version 8.01) software (GraphPad Software, Inc., La Jolla, CA, USA). A value of $p < 0.05$ was considered statistically significant. Different letters for each SFS treatment indicate statistically significant differences.

3. Results and Discussion

Certain sludges possess traits that define them as organic fertilizers owing to their significant concentrations of organic matter and essential macro-elements (such as nitrogen and phosphorus) that are essential for soil health and plant growth. However, an important barrier to using sludge as soil and plant fertilizer is its high concentration of heavy metals and other contaminants, which can be released into the soil and consequently negatively impact the development of cultivated plants. The physico-chemical parameters (pH and conductivity) and mineral content of sludge are presented in Table 1.

Table 1. The physico-chemical characteristics of sludge.

Elements Analyzed	u.m. *	Determined Values	Analysis Methods **
pH (25 °C)-SFS L/S:10/1	pH units	6	(1)
Conductivity (25 °C)-SFS L/S:10/1	µS/cm	1979	(2)
Arsenic (As)	mg/kg	1.18	(3)
Barium (Ba)	mg/kg	2.90	(3)
Cadmium (Cd)	mg/kg	<0.5	(3)
Chromium (Cr)	mg/kg	<0.5	(3)
Copper (Cu)	mg/kg	3.50	(3)
Mercury (Hg)	mg/kg	0.23	(4)
Molybdenum (Mo)	mg/kg	<1	(3)
Nickel (Ni)	mg/kg	<2	(3)
Lead (Pb)	mg/kg	<2	(3)
Selenium (Se)	mg/kg	<0.1	(5)
Antimony (Sb)	mg/kg	<0.1	(6)
Zinc (Zn)	mg/kg	18.9	(3)
Chloride	mg/kg	482	(7)
Fluoride	mg/kg	<100	(7)
Sulphate	mg/kg	109	(7)
Total dissolved solids (TDS)	mg/kg	16,560	(8)
Dissolved organic carbon (DOC)	mg/kg	620	(9)

* Unit of measurement. All the results are reported for dry matter. ** The methods are mentioned in Table S1.

The soluble fraction of the sludge was slightly acidic (pH = 6). The importance of sludge pH results from the fact that the solubility of heavy metals in a sludge sample is dependent on pH, which controls the bioavailability of metals that exist mostly in labile form. Concentrations of As, Ba, Cd, Cr, Cu, Hg, Mo, Ni, Pb, Se, Sb, and Zn in the soluble fraction of the sludge were lower than the limited values for heavy metals defined in Order 95/2005 and European Directive 86/278/EEC. Mercury exhibited a slight excess of 0.23 mg/kg, surpassing the national standard of 0.2 mg/kg.

The present study exploits the sensitivity of the comet test, an indicator of exposure to genotoxic agents, with reference to DNA damage in the presence of chemical contaminants from the soluble phase of sludge from urban water treatment plants. Several systems have been developed to evaluate genotoxicity using a wide range of biosensors (bacteria, human, plant, or animal cells). Several studies addressing this topic concluded that the response of most bioindicators is not strictly linked to a specific contaminant but is associated with the complex mixture of contaminants present in treated sewage sludge [8,74–76].

The parameters of the comets obtained are presented in Table 2. The values for the comet tail length (µm), tail DNA (%), and olive tail moment (OTM) are described.

Table 2. Main parameters in comets obtained by exposing 3 cell types to the soluble phase of sewage sludge.

Cell Bioindicators	Tail Length (µm)	Tail DNA %	OTM * (µm)
Lymphocytes			
H_2O	2.34 ± 1.11^{c}	10.95 ± 6.36^{c}	0.41 ± 0.14^{c}
25	3.45 ± 1.38^{c}	16.42 ± 8.33^{c}	0.62 ± 0.11^{c}
50	37.96 ± 3.13^{b}	43.77 ± 3.71^{b}	5.18 ± 0.70^{b}
75	40.32 ± 4.01^{ab}	76.49 ± 7.81^{a}	6.91 ± 1.36^{b}
100	44.61 ± 0.54^{a}	80.72 ± 2.69^{a}	7.37 ± 1.33^{b}
H_2O_2	44.94 ± 2.08^{a}	82.92 ± 2.85^{a}	14.46 ± 0.82^{a}

Table 2. *Cont.*

Cell Bioindicators	Tail Length (μm)	Tail DNA %	OTM * (μm)
Coelomocytes			
H_2O	2.70 ± 0.39 [c]	7.8562 ± 0.88 [c]	0.2071 ± 0.04 [c]
25%	2.80 ± 0.22 [c]	8.25 ± 0.30 [c]	0.23 ± 0.02 [c]
50%	2.84 ± 0.15 [c]	8.27 ± 0.08 [c]	0.24 ± 0.01 [c]
75%	14.25 ± 1.48 [b]	16.80 ± 1.23 [b]	2.36 ± 0.25 [b]
100%	26.94 ± 4.53 [a]	18.01 ± 2.60 [ab]	5.22 ± 1.50 [a]
H_2O_2	30.08 ± 4.16 [a]	20.98 ± 3.143a	6.63 ± 1.51 [a]
Allium cepa L.			
H_2O	0.18 ± 0.15 [d]	7.75 ± 5.29 [c]	0.79 ± 0.15 [c]
25%	30.55 ± 5.91 [c]	72.34 ± 14.88 [b]	4.67 ± 0.68 [b]
50%	39.59 ± 3.87 [b]	75.02 ± 10.31 [b]	5.31 ± 0.27 [b]
75%	43.00 ± 3.34 [a]	77.91 ± 7.74 [b]	5.31 ± 0.55 [b]
100%	44.61 ± 6.35 [a]	90.39 ± 7.24 [a]	5.64 ± 0.41 [b]
H_2O_2	45.11 ± 1.65 [a]	98.39 ± 0.75 [a]	6.45 ± 0.42 [a]

The values are expressed as the mean ± SD of experiments conducted in triplicate (n = 3). Tukey's multiple comparison test was conducted to identify significant differences. Distinct lowercase letters denote a statistically significant difference ($p < 0.05$) in each column, for each cell type, and according to the treatments. Lymphocytes s, coelomocytes, and *Allium cepa* L. cells were treated with concentrations of 25, 50, 75, and 100% SFS; H_2O_2 was the positive control, and H_2O served as the negative control. * The olive tail moment (OTM) was used to assess DNA damage. The OTM is calculated as a product of two factors: the percentage of tail DNA (Tail DNA %) and the distance between the intensity centroids (centers of gravity) of the head and tail along the comet's x-axis.

The lymphocytes indicated the genotoxicity of the soluble phase of the biosolid with respect to the tail-length parameter, starting at a concentration of 50% (Figure 2). The degree of DNA damage, for the significance threshold ($p < 0.05$), at a concentration of 25% was not significant compared to that of the negative control ($p = 0.96863$) but was significantly lower compared to the level of damage recorded at the concentrations of 75%, 100%, and compared to the positive control. For the OTM descriptor, we recorded significant differences between the positive control and all other evaluated concentrations and an insignificant difference between the negative control and the 25% concentration. The *Allium cepa* L. nuclei exposed to the positive control (H_2O_2, 200 μM) showed a higher level of DNA damage, illustrated by increased comet length and OTM values, while the nuclei treated with the negative control (H_2O) were intact, round, and did not show signs of DNA fragmentation or significant increases in the length of the comet or OTM (Figure 2). The results showed that the nuclei used as bioassay, exposed to different concentrations of the soluble phase of the biosolid, reacted in a specific way depending on their origin. Nuclei in *Allium cepa* L., with regard to the significance threshold employed ($p < 0.05$), indicated genotoxicity starting at a concentration of 25% (Figure 2). Compared to the negative control, the differences recorded at this concentration are significant in terms of comet tail length (+30.41), the % of DNA in the tail (+64.59), and OMT (+3.88). The length of the comets increased with the increase in the concentrations of the soluble phase of the biosolid; however, the differences between the concentrations of 50%, 75%, and 100% were not significant, nor were they significant when compared to the positive control. The differences in terms of the amount of DNA in the tails of the comets were significantly higher in the positive control compared to the 25% ($p = 0.00063$), 50% ($p = 0.00288$), and 75% ($p = 0.01279$) variants and insignificant compared to the 100% variant ($p = 0.71014$). The values determined for the OTM parameter indicate significant differences between the positive control and the 25% ($p = 0.00005$), 50% ($p = 0.02366$), and 75% ($p = 0.02453$) variants and insignificant compared to the 100% concentration ($p = 0.19940$).

The higher sensitivity of plant cell nuclei to genotoxic agents can be associated with the higher DNA content (33 pg for onion diploid cells and 6 pg for human diploid cells). The results suggested [77,78] that compared to plant cells, animal cells may have an apparatus that is more efficient in repairing damaged DNA.

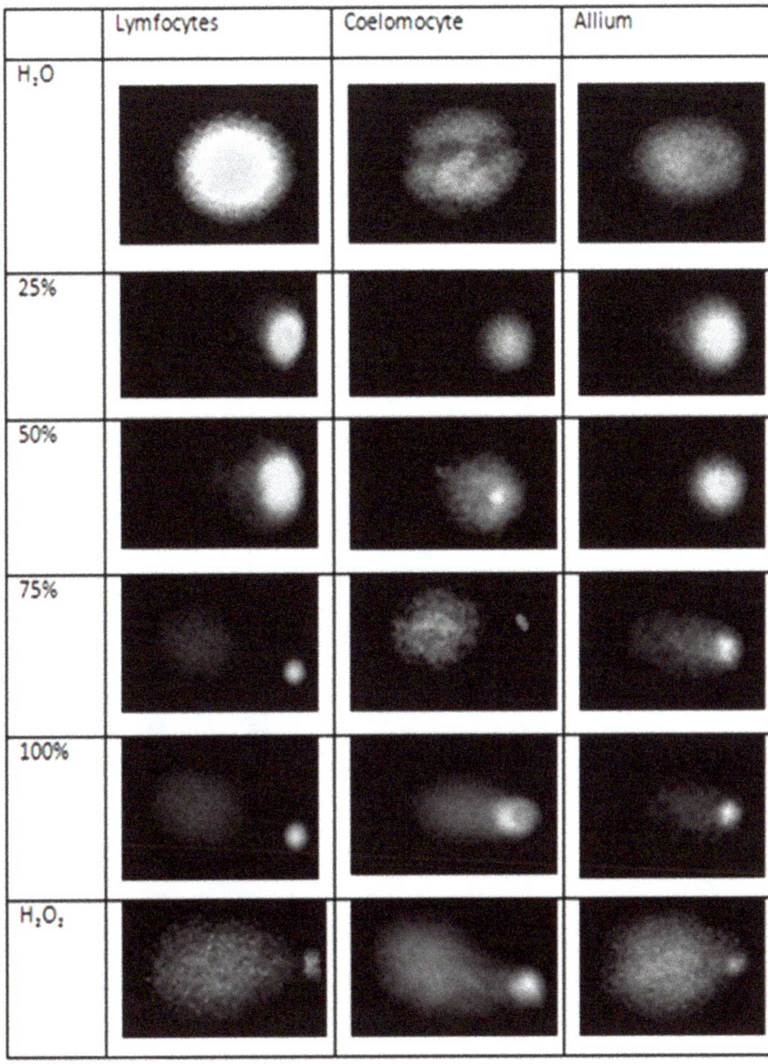

Figure 2. Morphology of DNA comets in relation to the tested bioindicator and SFS concentration. H_2O—negative control; 25%, 50%, 75%, and 100%—the concentrations of SFS with which the leucocytes, coelomocytes, and *Allium cepa* L cells were treated; H_2O_2—positive control.

The results of the Comet coelomocyte test showed that for all three analyzed parameters, the differences between the negative control and the concentrations of 25% and 50% were insignificant. This fact suggests a greater stability of the genetic material to the genotoxic factors present in the sewage sludge. Several studies have revealed mechanisms for reducing DNA damage in coelomocytes by eliminating damaged cells via apoptosis (programmed cell death), stopping them from becoming mutant cells [79] via DNA repair [80] or through the existence of distinct adaptive strategies (epigenetic or genetic) in response to exposure to a genotoxic environment [81]. The average length of the comet tails, % of DNA in the tail, and OTM determined in the three types of cells (leucocytes, coelomocytes, and *Allium cepa* L.) exposed to different concentrations of the soluble phase of the biosolid are shown in Figure 3a–c. The existence of a clear relationship between the dose applied and the response obtained as well as the higher stability of coelomocytes with respect to the genotoxicity of the soluble phase were observed. It is important to mention that our genotoxicity study includes all coelomocytes in the population, without distinguishing between subpopulations.

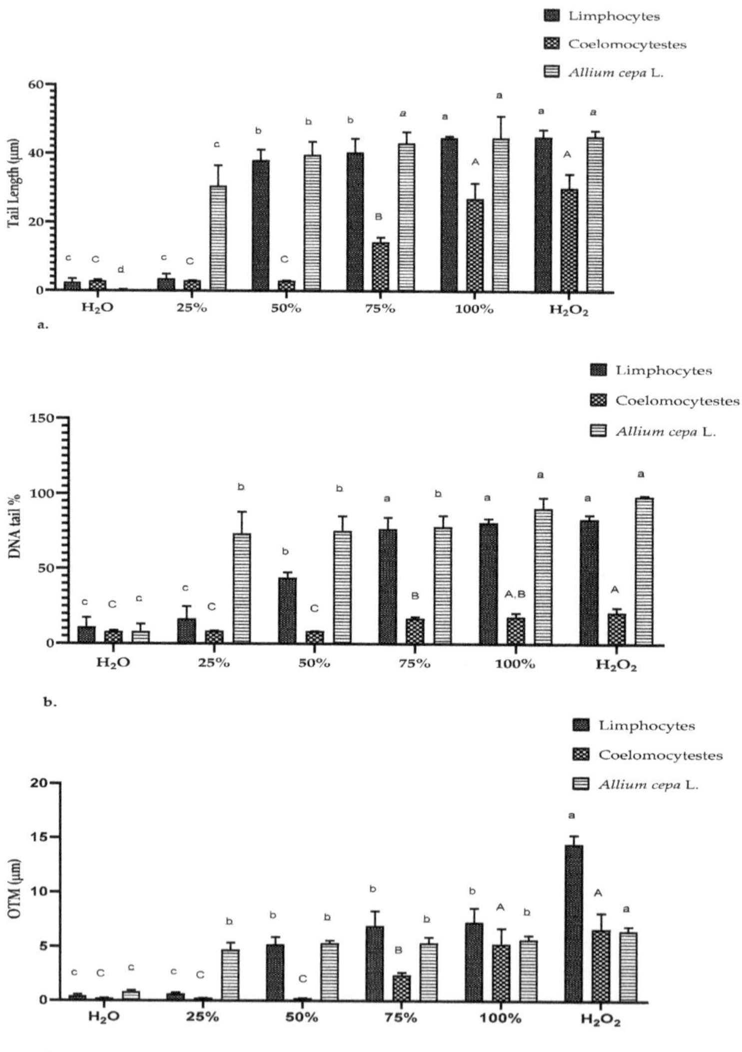

Figure 3. Evaluation of DNA damage using the Comet assay. (**a**) Mean tail length values. (**b**) Mean values of DNA tail %. (**c**) Olive tail moment mean values (OTM). The error bars depict the standard deviation of the mean values obtained from three repeated tests. Tukey's multiple comparison test was conducted to identify significant differences. Significant differences ($p < 0.05$) in the DNA tail of the experimental variants are indicated by different small letters. Significant differences in the experimental variants' DNA lengths are indicated by different capital letters. Significant differences in the OTM values between the experimental variants are indicated by different lowercase letters that are underlined. The experimental variants included a negative control using H_2O, and different concentrations (25%, 50%, 75%, and 100%) of SFS were used to treat leucocytes, coelomocytes, and *Allium cepa* L. cells. H_2O_2 was used as a positive control.

Heavy metals prevent the growth of roots, causing changes in water balance and in their absorption of nutrients [82] and thus affecting the growth of shoots and the accumulation of biomass in the above-ground parts of a plant. The stress induced by heavy metals in plants causes a decrease in endogenous levels of auxins affecting their growth and development [42]. For example, arsenic (As) was found to be able to alter the levels of three auxins (IAA, NAA, and IBA) in Brassica juncea [83]. Other studies demonstrated that short-term exposure of barley root tips to cadmium disrupted IAA homeostasis [84] or that it suppressed root elongation in Arabidopsis [85]. Upon entering the soil, heavy metal ions become involved in a competitive process with essential nutrient cations, namely, a competition to bind and be absorbed at the surface of the roots. After entering the plant cell, they exert their cytotoxic and genotoxic effects by disrupting the protein structure through attacking their thiol groups [18]. Stimulation of ROS production induced by heavy metals inflicts oxidative damage on cellular macromolecules and the photosynthetic apparatus, leading to a decrease in membrane stability and photosynthetic yield. Assimilative pigment production is also compromised, hormonal and nutritional imbalances are established, and the inhibition of DNA replication, gene expression, and cell division occurs [86]. To reduce and/or avoid heavy metal toxicity, plant roots use several strategies: avoidance of heavy metal uptake, the regulation of heavy metal toxicity through transporters, the activation of antioxidant mechanisms, the sequestration of heavy metals in vacuoles, and the synthesis and deposition of callose [87].

The effects of stress induced by pollutants in SFS on the auxin content in the roots and leaves of barley (*Hordeum vulgare* L. var. Salamandra) are shown in Figure 4. Compared to the control roots, treatment with 25% SFS reduced the content of endogenous IAA from 5.78 to 4.16 μg IAA equivalent/mL, and in the treatment with 100%, this content was reduced to 1.37 μg IAA equivalent/mL. In the case of leaves, the endogenous auxin decrease was from 3.36 μg IAA equivalent/mL for the control (H_2O) to 0.97 μg IAA equivalent/mL at the 100% concentration. Our observations are consistent with studies on sorghum [88,89], Pteris vittata and rice [43,90], and rice and Brassica juncea [91,92].

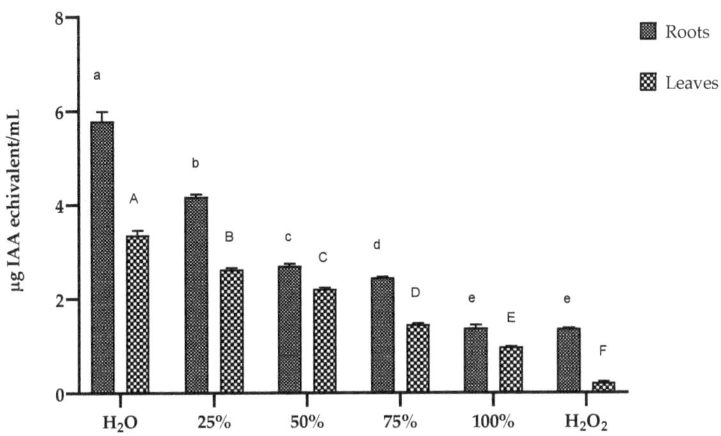

Figure 4. Effects of stress induced by heavy metals in SFS on auxin content in roots and leaves of barley (*Hordeum vulgare* L. var. Salamandra). The error bars depict the standard deviation of the mean values obtained from three repeated tests. Tukey's multiple comparison test was performed to determine significant differences ($p < 0.05$). The lowercase letters indicate significant differences between the root samples, while the uppercase letters denote statistical significance between the leaf samples based on the treatments applied. The treatments included a negative control, for which H_2O was used, and different concentrations (25%, 50%, 75%, and 100%) of SFS, while H_2O_2 was used as the positive control.

Studies on the genotoxicity of sewage sludge have shown different results (Table 3), probably due both to the different assay systems that were used and differences in the characteristics of the sludge due to the treatment technology applied [11,44,93]. Our results suggest the possibility of evaluating the sensitivity of particular tests, as integrated in a test battery, to accurately determine the genotoxic effects on relevant organisms.

Table 3. A summary of research confirming the incidence of genotoxic damage resulting from exposure to sewage sludge to different cell types.

Cell Types	Test/Assay	Findings	Reference
Allium cepa L.	Chromosomal aberrations, Micronucleus test	- Chromosomal aberration and micronucleus in root meristem cells; - A genotoxic effect was observed; - Inhibition of seed germination occurs when heavy metals are present in concentrations that are below the legal standards set in Brazil.	[37]
	Chromosomal aberrations, Micronucleus test	- Mitotic and chromosomal abnormalities (chromosome adherence chromosomal bridge, cell polyploidy, micronucleus); - Increase in phytotoxicity, cytotoxicity, genotoxicity, and/or mutagenicity.	[37]
	Chromosomal aberrations	- Chromosomal aberration in root meristem cells; - Genotoxic effects were observed for 10% sludge leachate that contains Cr, Cu, Ni, and Pb.	[94]
Lymphocytes	Comet assay	- There was no observed increase in the rate of DNA damage in peripheral blood leukocytes as a result of sewage sludge exposure; - There was not an increase in the levels of DNA damage.	[95]
Human Peripheral Blood	Comet assay	- Significant differences in the levels of lead (Pb) and cadmium (Cd) were noted in the blood of sewage workers; - Significantly increased levels of DNA damage were also detected in the groups that were exposed.	[96]
Lymphocytes	Comet assay	- No evidence was found indicating that channeling workers are more exposed to genotoxic substances compared to other workers; - No increase in DNA damage was observed among the sewage workers examined.	[97]
Lymphocytes	DNA diffusion assay, Micronucleus test, Comet assay	- Exposure to untreated waste leachate for a duration of 6 to 24 h resulted in significantly higher values of comet assay parameters in comparison to the other groups that were tested; - The lymphocytes examined were found to be cytogenotoxic when exposed to sludge leachate.	[98]
Eisenia fetida (coelomocytes)	Comet assay	- Earthworms exposed to the two highest concentrations (20 and 40%) showed a significant increase in DNA damage compared to the corresponding control group. Additionally, there was a decrease in the number of coelomocytes.	[80]

Table 3. *Cont.*

Cell Types	Test/Assay	Findings	Reference
Eisenia fetida	Comet assay	- Evaluating soil mutagenicity by measuring oxidative DNA damage caused by earthworms was used as a bio-monitoring technique; - This method is exclusively reliable for bio-accumulating metals, such as cadmium, but not for non-bio-accumulating metals, such as nickel.	[99]
Pheretima posthuma (earthworms)	Comet assay RAPD-PCR *	- As the concentration of sewage sludge increased, the amount of fragmentation in DNA also increased; - Sewage sludge induced DNA damage and inhibited the activities of esterase and phosphatase.	[100]
Eisenia fetida	Allium bioassay, Chromosomal aberration, Micronucleus, Mitotic index	- A decrease in chromosomal aberrations as well as an increase in root length and mitotic index in the final vermicompost were observed; - Earthworms have the ability to accumulate heavy metals, and the intestinal microflora and chloragocytic cells of earthworms have the ability to detoxify toxic substances.	[101]

* RAPD-PCR—Random Amplified Polymerase DNA Polymerase Chain Reaction.

However, there are other relevant aspects that need to be considered. Correa et al. [102] showed that the sewage sludge analyzed in their study presented genetic toxicity and mutagenicity in the context in which all physico-chemical parameters stay within the limits established by internal standards. The apparent absence of lymphocyte DNA damage in situations in which such damage should be obvious may be due to the selection of inadequate target cells Amresco [97]. The genotoxicity of sewage sludge applied on agricultural terrain may be diminished locally by the activity of earthworms, which can accumulate and detoxify heavy metals via their intestinal microflora and their chloragogen cells [101]. There is a certain difficulty in comparing results obtained via the Comet test [60] due to the lack of a standard procedure for the application of the test (processing duration, the comet parameters considered representative, sorting method, etc.).

4. Conclusions

The present study focuses on the impact of genotoxic agents found in sewage sludge on three trophic levels: human lymphocytes, plant cells, and coelomocytes. All the tested bioindicators demonstrated the existence of genotoxicity, with varying degrees of sensitivity. Our findings indicate that it is necessary to employ multiple tests, combined in a test battery, in order to properly evaluate genotoxic effects on relevant bioindicators. This approach is necessary for multiple significant reasons: (i) the diverse mixture of substances in sludge makes it challenging to identify a specific genotoxic factor within a broader context, such as a heavy metal; (ii) the acceptable levels of substances in sludge may not be relevant in terms of their genotoxic effects, as indicated in the existing literature and our own observations; (iii) the biosolid investigated herein (sewage sludge) contains a significant biotic component, and its metabolic processes are part of a complex network of interactions that can either enhance or diminish a specific effect. In addition, as the concentration of sludge increased, a decrease in the auxin's concentration was noted in both leaves and roots. Prior to utilizing sludge as a fertilizer and soil enhancer, it is necessary to develop an in-depth strategy that considers the sustainable management of soil's productive capacity. Before implementing a program of fertilization with sewage sludge, it is necessary to investigate the potential toxicity of treated sewage sludge. The results presented here can contribute to the compilation of a toxicological profile of sludge and to the completion of a database for the assessment of possible risks, aiding in making fully informed technological decisions.

Supplementary Materials: The following supporting information can be downloaded at https://www.mdpi.com/article/10.3390/su16010457/s1. Table S1. The standards and methods used for the chemical analysis of sludge.

Author Contributions: Conceptualization, M.A.C. and V.L.; methodology, M.A.C.; validation, C.A.R., E.A. and M.A.C.; formal analysis, C.P. and V.L.; investigation, C.A.R. and E.A.; resources, V.L.; writing—original draft preparation, V.L.; writing—review and editing, C.P. and S.I.V.; supervision, V.L.; supervision, C.A.R. and E.A.; project administration, S.I.V. All authors have read and agreed to the published version of the manuscript.

Funding: The APC was funded by University of Oradea.

Institutional Review Board Statement: This study was conducted in accordance with the Declaration of Helsinki and approved by the Ethics Committee of UNIVERSITY OF ORADEA, Faculty of Medicine and Pharmacy (29 November 2023).

Informed Consent Statement: Informed consent was obtained from the subjects involved in the study.

Data Availability Statement: Data are contained within the article and supplementary materials.

Acknowledgments: The authors acknowledge S.C. Maggot Fish S.R.L., Oradea, Romania for donating the earthworms used in this study.

Conflicts of Interest: The authors declare no conflict of interest.

References

1. Rusănescu, C.O.; Rusănescu, M.; Voicu, G.; Paraschiv, G.; Biriș, S.Ș.; Popescu, I.N. The Recovery of Vermicompost Sewage Sludge in Agriculture. *Agronomy* **2022**, *12*, 2653. [CrossRef]
2. Arthur Andersen Environment Risk Consulting Department. Disposal and Recycling Routes for Sewage Sludge. In *Part 3—Scientific and Technical Report*; Office for Official Publications of the European Communities: Luxembourg, 2001; pp. 151–2001.
3. Lima, M.; Mattos, C.; Vieira, P.; Almeida, L. Geração de Lodo de Esgoto e Seu Potencial Como Fonte de Matéria Orgânica Para a Agricultura. In *Manual de uso Agrícola e Disposição do Lodo de Esgoto Para o Estado do Espirito Santo*; INCAPER: Vitoria, Brazil, 2011.
4. Singh, R.P.; Agrawal, M. Potential Benefits and Risks of Land Application of Sewage Sludge. *Waste Manag.* **2008**, *28*, 347–358. [CrossRef] [PubMed]
5. Oleszczuk, P.; Malara, A.; Jośko, I.; Lesiuk, A. The Phytotoxicity Changes of Sewage Sludge-Amended Soils. *Water Air Soil. Pollut.* **2012**, *223*, 4937–4948. [CrossRef] [PubMed]
6. Yurievna Selivanovskaya, S.; Zinnatovna Latypova, V.; Aleksandrovna Artamonova, L. Use of Sewage Sludge Compost as the Restoration Agent on the Degraded Soil of Tatarstan. *J. Environ. Sci. Health Part. A* **2003**, *38*, 1549–1556. [CrossRef] [PubMed]
7. Nascimento, C.W.A.; Barros, D.A.S.; Melo, E.E.C.; Oliveira, A.B. Alterações Químicas Em Solos e Crescimento de Milho e Feijoeiro Após Aplicação de Lodo de Esgoto. *Rev. Bras. Ciênc. Solo* **2004**, *28*, 385–392. [CrossRef]
8. de Lourdes Marzo Solano, M.; de Lima, P.L.; Luvizutto, J.F.; Silva, P.R.; de Aragão Umbuzeiro, G.; de Camargo, J.L. In Vivo Genotoxicity Evaluation of a Treated Urban Sewage Sludge Sample. *Mutat. Res./Genet. Toxicol. Environ. Mutagen.* **2009**, *676*, 69–73. [CrossRef]
9. Sabbahi, S.; Ben Ayed, L.; Trad, M.; Berndtsson, R.; Karanis, P. Parasitological Assessment of Sewage Sludge Samples for Potential Agricultural Reuse in Tunisia. *Int. J. Environ. Res. Public. Health* **2022**, *19*, 1657. [CrossRef]
10. Collivignarelli, M.C.; Abbà, A.; Benigna, I. The Reuse of Biosolids on Agricultural Land: Critical Issues and Perspective. *Water Environ. Res.* **2020**, *92*, 11–25. [CrossRef]
11. Chenon, P.; Gauthier, L.; Loubières, P.; Séverac, A.; Delpoux, M. Evaluation of the Genotoxic and Teratogenic Potential of a Municipal Sludge and Sludge-Amended Soil Using the Amphibian *Xenopus laevis* and the Tobacco: *Nicotiana tabacum* L. var. *xanthi Dulieu. Sci. Total Environ.* **2003**, *301*, 139–150. [CrossRef]
12. Rank, J.; Nielsen, M.H. Genotoxicity Testing of Wastewater Sludge Using the *Allium cepa* Anaphase-Telophase Chromosome Aberration Assay. *Mutat. Res./Genet. Toxicol. Environ. Mutagen.* **1998**, *418*, 113–119. [CrossRef]
13. Grotto, D.; Carneiro, M.F.H.; Sauer, E.; Garcia, S.C.; De Melo, W.J.; Barbosa, F. Evaluation of Biochemical and Redox Parameters in Rats Fed with Corn Grown in Soil Amended with Urban Sewage Sludge. *Ecotoxicol. Environ. Saf.* **2013**, *95*, 188–194. [CrossRef] [PubMed]
14. Stadtman, E.R. Oxidation of free amino acids and amino acid residues in proteins by radiolysis and by metal-catalyzed reactions. *Annu. Rev. Biochem.* **1993**, *62*, 797–821. [CrossRef]
15. Kasprzak, K.S. Possible Role of Oxidative Damage in Metal-Induced Carcinogenesis. *Cancer Investig.* **1995**, *13*, 411–430. [CrossRef] [PubMed]
16. Cunningham, R.P. DNA Repair: Caretakers of the Genome? *Curr. Biol.* **1997**, *7*, R576–R579. [CrossRef] [PubMed]
17. Li, W.-X.; Chen, T.-B.; Huang, Z.-C.; Lei, M.; Liao, X.-Y. Effect of Arsenic on Chloroplast Ultrastructure and Calcium Distribution in Arsenic Hyperaccumulator *Pteris vittata* L. *Chemosphere* **2006**, *62*, 803–809. [CrossRef] [PubMed]

18. Dutta, S.; Mitra, M.; Agarwal, P.; Mahapatra, K.; De, S.; Sett, U.; Roy, S. Oxidative and Genotoxic Damages in Plants in Response to Heavy Metal Stress and Maintenance of Genome Stability. *Plant Signal Behav.* **2018**, *13*, e1460048. [CrossRef] [PubMed]
19. Amin, A.W.; Sherif, F.K.; El-Atar, H.; Ez-Eldin, H. Residual Effect of Sewage Sludge on Soil and Several Yield Parameters of *Zea mays*. *Res. J. Environ. Toxicol.* **2009**, *3*, 86–93. [CrossRef]
20. Marzougui, N.; Ounalli, N.; Sabbahi, S.; Fezzani, T.; Abidi, F.; Jebari, S.; Melki, S.; Berndtsson, R.; Oueslati, W. How Can Sewage Sludge Use in Sustainable Tunisian Agriculture Be Increased? *Sustainability* **2022**, *14*, 13722. [CrossRef]
21. Kızılkaya, R.; Bayraklı, B. Effects of N-Enriched Sewage Sludge on Soil Enzyme Activities. *Appl. Soil. Ecol.* **2005**, *30*, 192–202. [CrossRef]
22. Gao, Y.; Guo, Y.-K.; Lin, S.-H.; Fang, Y.-Y.; Bai, J.-G. Hydrogen Peroxide Pretreatment Alters the Activity of Antioxidant Enzymes and Protects Chloroplast Ultrastructure in Heat-Stressed Cucumber Leaves. *Sci. Hortic.* **2010**, *126*, 20–26. [CrossRef]
23. Gjorgieva, D.; Kadifkova Panovska, T.; Ruskovska, T.; Bačeva, K.; Stafilov, T. Influence of Heavy Metal Stress on Antioxidant Status and DNA Damage in *Urtica Dioica*. *BioMed Res. Int.* **2013**, *2013*, 276417. [CrossRef] [PubMed]
24. Britt, A.B. Molecular Genetics of DNA Repair in Higher Plants. *Trends Plant Sci.* **1999**, *4*, 20–25. [CrossRef] [PubMed]
25. Lin, A.-J.; Zhang, X.-H.; Chen, M.-M.; Cao, Q. Oxidative Stress and DNA Damages Induced by Cadmium Accumulation. *J. Environ. Sci.* **2007**, *19*, 596–602. [CrossRef] [PubMed]
26. Siddiqui, S. DNA Damage in Cicer Plant Grown on Soil Polluted with Heavy Metals. *J. King Saud. Univ. Sci.* **2015**, *27*, 217–223. [CrossRef]
27. Das, A.; Parbat, D.; Shome, A.; Manna, U. Sustainable Biomimicked Oil/Water Wettability That Performs Under Severe Challenges. *ACS Sustain. Chem. Eng.* **2019**, *7*, 11350–11359. [CrossRef]
28. Das, A.; Naskar, S.; Dhar, M.; Manna, U. Rapid and Scalable Synthesis of a Vanillin-Based Organogelator and Its Durable Composite for a Comprehensive Remediation of Crude-Oil Spillages. *ACS Appl. Mater. Interfaces* **2021**, *13*, 46803–46812. [CrossRef]
29. Jiang, N.; Naz, S.; Ma, Y.; Ullah, Q.; Khan, M.Z.; Wang, J.; Lu, X.; Luosang, D.-Z.; Tabassum, S.; Chatha, A.M.M.; et al. An Overview of Comet Assay Application for Detecting DNA Damage in Aquatic Animals. *Agriculture* **2023**, *13*, 623. [CrossRef]
30. Amin, A.W. Evaluation of the Genotoxicity of Residual Repeated Applications of Sewage Sludge on M_2 Meiocytes of Zea Plants. *Res. J. Environ. Toxicol.* **2011**, *5*, 235–250. [CrossRef]
31. Gawdzik, J.; Gawdzik, B. Mobility of Heavy Metals in Municipal Sewage Sludge from Different Throughput Sewage Treatment Plants. *Pol. J. Environ. Stud.* **2012**, *21*, 1603–1611.
32. McBride, M.B. Toxic Metal Accumulation from Agricultural Use of Sludge: Are USEPA Regulations Protective? *J. Environ. Qual.* **1995**, *24*, 5–18. [CrossRef]
33. Udom, B.E.; Mbagwu, J.S.C.; Adesodun, J.K.; Agbim, N.N. Distributions of Zinc, Copper, Cadmium and Lead in a Tropical Ultisol after Long-Term Disposal of Sewage Sludge. *Environ. Int.* **2004**, *30*, 467–470. [CrossRef] [PubMed]
34. Mateuca, R.; Lombaert, N.; Aka, P.V.; Decordier, I.; Kirsch-Volders, M. Chromosomal Changes: Induction, Detection Methods and Applicability in Human Biomonitoring. *Biochimie* **2006**, *88*, 1515–1531. [CrossRef] [PubMed]
35. Dai, J.; Chen, L.; Zhao, J.; Ma, N. Characteristics of Sewage Sludge and Distribution of Heavy Metal in Plants with Amendment of Sewage Sludge. *J. Environ. Sci.* **2006**, *18*, 1094–1100. [CrossRef] [PubMed]
36. Henning, B.; Snyman, H.; Aveling, T. Plant-Soil Interactions of Sludge-Borne Heavy Metals and the Effect on Maize (*Zea mays* L.) Seedling Growth. *WSA* **2004**, *27*, 71–78. [CrossRef]
37. Corrêa Martins, M.N.; de Souza, V.V.; Souza, T. da S. Genotoxic and Mutagenic Effects of Sewage Sludge on Higher Plants. *Ecotoxicol. Environ. Saf.* **2016**, *124*, 489–496. [CrossRef] [PubMed]
38. Krishna, V.; Pandey, A.K.; Gupta, P.K. Heavy Metals in Sludge Produced from UASB Treatment Plant at Mirzapur, India. *Pollution* **2021**, *7*, 607–616. [CrossRef]
39. Emamverdian, A.; Ding, Y.; Mokhberdoran, F.; Xie, Y. Heavy Metal Stress and Some Mechanisms of Plant Defense Response. *Sci. World J.* **2015**, *2015*, 756120. [CrossRef]
40. Hąc-Wydro, K.; Sroka, A.; Jabłońska, K. The Impact of Auxins Used in Assisted Phytoextraction of Metals from the Contaminated Environment on the Alterations Caused by Lead(II) Ions in the Organization of Model Lipid Membranes. *Colloids Surf. B Biointerfaces* **2016**, *143*, 124–130. [CrossRef]
41. Rodriguez, E.; Santos, C.; Azevedo, R.; Moutinho-Pereira, J.; Correia, C.; Dias, M.C. Chromium (VI) Induces Toxicity at Different Photosynthetic Levels in Pea. *Plant Physiol. Biochem.* **2012**, *53*, 94–100. [CrossRef]
42. Bücker-Neto, L.; Paiva, A.L.S.; Machado, R.D.; Arenhart, R.A.; Margis-Pinheiro, M. Interactions between Plant Hormones and Heavy Metals Responses. *Genet. Mol. Biol.* **2017**, *40*, 373–386. [CrossRef]
43. Zhao, F.; Han, M.; Zhang, S.; Wang, K.; Zhang, C.; Liu, T.; Liu, W. Hydrogen Peroxide-Mediated Growth of the Root System Occurs via Auxin Signaling Modification and Variations in the Expression of Cell-Cycle Genes in Rice Seedlings Exposed to Cadmium Stress. *J. Integr. Plant Biol.* **2012**, *54*, 991–1006. [CrossRef] [PubMed]
44. Schowanek, D.; Carr, R.; David, H.; Douben, P.; Hall, J.; Kirchmann, H.; Patria, L.; Sequi, P.; Smith, S.; Webb, S. A Risk-Based Methodology for Deriving Quality Standards for Organic Contaminants in Sewage Sludge for Use in Agriculture—Conceptual Framework. *Regul. Toxicol. Pharmacol.* **2004**, *40*, 227–251. [CrossRef] [PubMed]

45. Azqueta, A.; Ladeira, C.; Giovannelli, L.; Boutet-Robinet, E.; Bonassi, S.; Neri, M.; Gajski, G.; Duthie, S.; Bo', C.D.; Riso, P.; et al. Application of the Comet Assay in Human Biomonitoring: An hCOMET Perspective. *Mutat. Res./Rev. Mutat. Res.* **2020**, *783*, 108288. [CrossRef] [PubMed]
46. Fomin, A.; Hafner, C. Evaluation of Genotoxicity of Emissions from Municipal Waste Incinerators with Tradescantia-Micronucleus Bioassay (Trad-MCN). *Mutat. Res.* **1998**, *414*, 139–148. [CrossRef] [PubMed]
47. Grant, W.F. Higher Plant Assays for the Detection of Genotoxicity in Air Polluted Environments. *Ecosyst. Health* **1998**, *4*, 210–229. [CrossRef]
48. Martins, M.; Costa, P.M. The Comet Assay in Environmental Risk Assessment of Marine Pollutants: Applications, Assets and Handicaps of Surveying Genotoxicity in Non-Model Organisms. *Mutagenesis* **2015**, *30*, 89–106. [CrossRef]
49. Tondoh, J.E.; Monin, L.M.; Tiho, S.; Csuzdi, C. Can Earthworms Be Used as Bio-Indicators of Land-Use Perturbations in Semi-Deciduous Forest? *Biol. Fertil. Soils* **2007**, *43*, 585–592. [CrossRef]
50. Peres, G.; Cluzeau, D.; Ferrand, C.; Peron, D. Earthworms Used as Indicators of Agricultural Managements. In *BIO-BIO Project. Biodiversity-Bioindication to Evaluate Soil Health*; Cenci, R.M., Sena, F., Eds.; 2006; Volume 22245, pp. 107–115; ISBN 92-79-02011-0. Available online: https://esdac.jrc.ec.europa.eu/ESDB_Archive/eusoils_docs/other/EUR22245.pdf (accessed on 7 December 2023).
51. Verschaeve, L.; Gilles, J. Single Cell Gel Electrophoresis Assay in the Earthworm for the Detection of Genotoxic Compounds in Soils. *Bull. Environ. Contam. Toxicol.* **1995**, *54*, 112–119. [CrossRef]
52. Salagovic, J.; Gilles, J.; Verschaeve, L.; Kalina, I. The Comet Assay for the Detection of Genotoxic Damage in the Earthworms: A Promising Tool for Assessing the Biological Hazards of Polluted Sites. *Folia Biol.* **1996**, *42*, 17–21.
53. Adamowicz, A. Morphology and Ultrastructure of the Earthworm *Dendrobaena veneta* (Lumbricidae) Coelomocytes. *Tissue Cell* **2005**, *37*, 125–133. [CrossRef]
54. Liu, W.; Li, P.J.; Qi, X.M.; Zhou, Q.X.; Zheng, L.; Sun, T.H.; Yang, Y.S. DNA Changes in Barley (*Hordeum vulgare*) Seedlings Induced by Cadmium Pollution Using RAPD Analysis. *Chemosphere* **2005**, *61*, 158–167. [CrossRef] [PubMed]
55. Lemtiri, A.; Liénard, A.; Alabi, T.; Brostaux, Y.; Cluzeau, D.; Francis, F.; Colinet, G. Earthworms *Eisenia fetida* Affect the Uptake of Heavy Metals by Plants *Vicia faba* and *Zea mays* in Metal-Contaminated Soils. *Appl. Soil. Ecol.* **2016**, *104*, 67–78. [CrossRef]
56. Karimi, F.; Rahimi, G.; Kolahchi, Z. Interaction Effects of Salinity, Sewage Sludge, and Earthworms on the Fractionations of Zn and Cu, and the Metals Uptake by the Earthworms in a Zn- and Cu-Contaminated Calcareous Soil. *Environ. Sci. Pollut. Res.* **2020**, *27*, 10565–10580. [CrossRef] [PubMed]
57. He, X.; Zhang, Y.; Shen, M.; Zeng, G.; Zhou, M.; Li, M. Effect of Vermicomposting on Concentration and Speciation of Heavy Metals in Sewage Sludge with Additive Materials. *Bioresour. Technol.* **2016**, *218*, 867–873. [CrossRef] [PubMed]
58. Poli, P.; Buschini, A.; Restivo, F.M.; Ficarelli, A.; Cassoni, F.; Ferrero, I.; Rossi, C. Comet Assay Application in Environmental Monitoring: DNA Damage in Human Leukocytes and Plant Cells in Comparison with Bacterial and Yeast Tests. *Mutagenesis* **1999**, *14*, 547–556. [CrossRef] [PubMed]
59. Ostling, O.; Johanson, K.J. Microelectrophoretic Study of Radiation-Induced DNA Damages in Individual Mammalian Cells. *Biochem. Biophys. Res. Commun.* **1984**, *123*, 291–298. [CrossRef]
60. Faust, F.; Kassie, F.; Knasmüller, S.; Boedecker, R.H.; Mann, M.; Mersch-Sundermann, V. The Use of the Alkaline Comet Assay with Lymphocytes in Human Biomonitoring Studies. *Mutat. Res./Rev. Mutat. Res.* **2004**, *566*, 209–229. [CrossRef]
61. Singh, N.P.; McCoy, M.T.; Tice, R.R.; Schneider, E.L. A Simple Technique for Quantitation of Low Levels of DNA Damage in Individual Cells. *Exp. Cell Res.* **1988**, *175*, 184–191. [CrossRef]
62. Azqueta, A.; Muruzabal, D.; Boutet-Robinet, E.; Milic, M.; Dusinska, M.; Brunborg, G.; Møller, P.; Collins, A.R. Technical Recommendations to Perform the Alkaline Standard and Enzyme-Modified Comet Assay in Human Biomonitoring Studies. *Mutat. Res./Genet. Toxicol. Environ. Mutagen.* **2019**, *843*, 24–32. [CrossRef]
63. Speit, G.; Vasquez, M.; Hartmann, A. The Comet Assay as an Indicator Test for Germ Cell Genotoxicity. *Mutat. Res./Rev. Mutat. Res.* **2009**, *681*, 3–12. [CrossRef]
64. Cordelli, E.; Bignami, M.; Pacchierotti, F. Comet Assay: A Versatile but Complex Tool in Genotoxicity Testing. *Toxicol. Res.* **2021**, *10*, 68–78. [CrossRef] [PubMed]
65. Alvarez-Moya, C. Use of Comet Assay in Human Lymphocytes as a Molecular Biomarker for Simultaneous Monitoring of Genetic Damage and Genotoxicity in Residents Who Lived Nearby the Santiago River, Mexico, in 2012. *Glob. J. Biotechnol. Biomater. Sci.* **2015**, *1*, 4–8. [CrossRef]
66. Eyambe, G.S.; Goven, A.J.; Fitzpatrick, L.C.; Venables, B.J.; Cooper, E.L. A Non-Invasive Technique for Sequential Collection of Earthworm (*Lumbricus Terrestris*) Leukocytes during Subchronic Immunotoxicity Studies. *Lab. Anim.* **1991**, *25*, 61–67. [CrossRef] [PubMed]
67. Gichner, T.; Mühlfeldová, Z. Induced DNA Damage Measured by the Comet Assay in 10 Weed Species. *Biol. Plant.* **2002**, *45*, 509–516. [CrossRef]
68. Singh, N.P. Microgel Electrophoresis of DNA from Individual Cells. In *Technologies for Detection of DNA Damage and Mutations*; Pfeifer, G.P., Ed.; Springer: Boston, MA, USA, 1996; pp. 3–24; ISBN 978-1-4899-0303-7.
69. Juchimiuk, J.; Gnys, A.; Maluszynska, J. DNA Damage Induced by Mutagens in Plant and Human Cell Nuclei in Acellular Comet Assay. *Folia Histochem. Cytobiol.* **2006**, *44*, 127–131.

70. Bandyopadhyay, A.; Mukherjee, A. Sensitivity of Allium and Nicotiana in Cellular and Acellular Comet Assays to Assess Differential Genotoxicity of Direct and Indirect Acting Mutagens. *Ecotoxicol. Environ. Saf.* **2011**, *74*, 860–865. [CrossRef] [PubMed]
71. Rajaguru, P.; Suba, S.; Palanivel, M.; Kalaiselvi, K. Genotoxicity of a Polluted River System Measured Using the Alkaline Comet Assay on Fish and Earthworm Tissues. *Environ. Mol. Mutagen.* **2003**, *41*, 85–91. [CrossRef] [PubMed]
72. Liu, W.; Zhu, L.-S.; Wang, J.; Wang, J.-H.; Xie, H.; Song, Y. Assessment of the Genotoxicity of Endosulfan in Earthworm and White Clover Plants Using the Comet Assay. *Arch. Environ. Contam. Toxicol.* **2009**, *56*, 742–746. [CrossRef]
73. Goldschmidt, E.E.; Goren, R.; Monselise, S.P. The IAA-Oxidase System of Citrus Roots. *Planta* **1966**, *72*, 213–222. [CrossRef]
74. Júnior, E.O.D.C.; Pereira, B.B.; Morelli, S. Monitoring Genotoxicity Potential in the Mumbuca Stream, Minas Gerais, Brazil. *J. Toxicol. Environ. Health Part. A* **2015**, *78*, 1277–1287. [CrossRef]
75. Fang, W.; Qi, G.; Wei, Y.; Kosson, D.S.; Van Der Sloot, H.A.; Liu, J. Leaching Characteristic of Toxic Trace Elements in Soils Amended by Sewage Sludge Compost: A Comparison of Field and Laboratory Investigations. *Environ. Pollut.* **2018**, *237*, 244–252. [CrossRef] [PubMed]
76. Siqueira, W.N.D.; França, E.J.D.; Pereira, D.R.; Lima, M.D.V.; Silva, H.A.M.F.; Araújo, H.D.A.D.; Sá, J.L.F.; Melo, A.M.M.D.A. Study of Genotoxic and Cytotoxic Effects after Acute and Chronic Exposures to Industrial Sewage Sludge on Biomphalaria Glabrata Hemocytes. *Chemosphere* **2020**, *249*, 126218. [CrossRef]
77. Navarrete, M.H.; Carrera, P.; de Miguel, M. Consuelo de la Torre A Fast Comet Assay Variant for Solid Tissue Cells. The Assessment of DNA Damage in Higher Plants. *Mutat. Res./Genet. Toxicol. Environ. Mutagen.* **1997**, *389*, 271–277. [CrossRef] [PubMed]
78. de Lima, R.; Oliveira Feitosa, L.; Rodrigues Maruyama, C.; Abreu Barga, M.; Yamawaki, P.C.; Vieira, I.J.; Teixeira, E.M.; Corrêa, A.C.; Caparelli Mattoso, L.H.; Fernandes Fraceto, L. Evaluation of the Genotoxicity of Cellulose Nanofibers. *Int. J. Nanomed.* **2012**, *7*, 3555. [CrossRef] [PubMed]
79. Evan, G.; Littlewood, T. A Matter of Life and Cell Death. *Science* **1998**, *281*, 1317–1322. [CrossRef] [PubMed]
80. Bonnard, M.; Devin, S.; Leyval, C.; Morel, J.-L.; Vasseur, P. The Influence of Thermal Desorption on Genotoxicity of Multipolluted Soil. *Ecotoxicol. Environ. Saf.* **2010**, *73*, 955–960. [CrossRef] [PubMed]
81. Kille, P.; Andre, J.; Anderson, C.; Ang, H.N.; Bruford, M.W.; Bundy, J.G.; Donnelly, R.; Hodson, M.E.; Juma, G.; Lahive, E.; et al. DNA Sequence Variation and Methylation in an Arsenic Tolerant Earthworm Population. *Soil. Biol. Biochem.* **2013**, *57*, 524–532. [CrossRef]
82. González-Moscoso, M.; Martínez-Villegas, N.; Cadenas-Pliego, G.; Juárez-Maldonado, A. Effect of Silicon Nanoparticles on Tomato Plants Exposed to Two Forms of Inorganic Arsenic. *Agronomy* **2022**, *12*, 2366. [CrossRef]
83. Srivastava, S.; Srivastava, A.K.; Suprasanna, P.; D'Souza, S.F. Identification and Profiling of Arsenic Stress-Induced microRNAs in *Brassica juncea*. *J. Exp. Bot.* **2013**, *64*, 303–315. [CrossRef]
84. Zelinová, V.; Alemayehu, A.; Bočová, B.; Huttová, J.; Tamás, L. Cadmium-Induced Reactive Oxygen Species Generation, Changes in Morphogenic Responses and Activity of Some Enzymes in Barley Root Tip Are Regulated by Auxin. *Biologia* **2015**, *70*, 356–364. [CrossRef]
85. Bennett, T.; Sieberer, T.; Willett, B.; Booker, J.; Luschnig, C.; Leyser, O. The Arabidopsis MAX Pathway Controls Shoot Branching by Regulating Auxin Transport. *Curr. Biol.* **2006**, *16*, 553–563. [CrossRef] [PubMed]
86. Singh, N.; Ma, L.Q.; Vu, J.C.; Raj, A. Effects of Arsenic on Nitrate Metabolism in Arsenic Hyperaccumulating and Non-Hyperaccumulating Ferns. *Environ. Pollut.* **2009**, *157*, 2300–2305. [CrossRef] [PubMed]
87. Riaz, M.; Kamran, M.; Rizwan, M.; Ali, S.; Wang, X. Foliar Application of Silica Sol Alleviates Boron Toxicity in Rice (*Oryza sativa*) Seedlings. *J. Hazard. Mater.* **2022**, *423*, 127175. [CrossRef] [PubMed]
88. Stobart, A.K.; Griffiths, W.T.; Ameen-Bukhari, I.; Sherwood, R.P. The Effect of Cd^{2+} on the Biosynthesis of Chlorophyll in Leaves of Barley. *Physiol. Plant.* **1985**, *63*, 293–298. [CrossRef]
89. Zhan, Y.; Zhang, C.; Zheng, Q.; Huang, Z.; Yu, C. Cadmium Stress Inhibits the Growth of Primary Roots by Interfering Auxin Homeostasis in Sorghum Bicolor Seedlings. *J. Plant Biol.* **2017**, *60*, 593–603. [CrossRef]
90. Balestri, M.; Ceccarini, A.; Forino, L.M.C.; Zelko, I.; Martinka, M.; Lux, A.; Ruffini Castiglione, M. Cadmium Uptake, Localization and Stress-Induced Morphogenic Response in the Fern *Pteris vittata*. *Planta* **2014**, *239*, 1055–1064. [CrossRef] [PubMed]
91. Shah, K.; Kumar, R.G.; Verma, S.; Dubey, R.S. Effect of Cadmium on Lipid Peroxidation, Superoxide Anion Generation and Activities of Antioxidant Enzymes in Growing Rice Seedlings. *Plant Sci.* **2001**, *161*, 1135–1144. [CrossRef]
92. Verma, K.; Shekhawat, G.S.; Sharma, A.; Mehta, S.K.; Sharma, V. Cadmium Induced Oxidative Stress and Changes in Soluble and Ionically Bound Cell Wall Peroxidase Activities in Roots of Seedling and 3–4 Leaf Stage Plants of *Brassica juncea* (L.) Czern. *Plant Cell Rep.* **2008**, *27*, 1261–1269. [CrossRef]
93. Gustavsson, L.; Engwall, M. Genotoxic Activity of Nitroarene-Contaminated Industrial Sludge Following Large-Scale Treatment in Aerated and Non-Aerated Sacs. *Sci. Total Environ.* **2006**, *367*, 694–703. [CrossRef]
94. Srivastava, R.; Kumar, D.; Gupta, S.K. Bioremediation of Municipal Sludge by Vermitechnology and Toxicity Assessment by *Allium cepa*. *Bioresour. Technol.* **2005**, *96*, 1867–1871. [CrossRef]
95. Silva, P.R.P.; Barbisan, L.F.; Dagli, M.L.Z.; Saldiva, P.H.N. Sewage Sludge Does Not Induce Genotoxicity and Carcinogenesis. *Genet. Mol. Biol.* **2012**, *35*, 657–663. [CrossRef]

96. Basu, R.; Talapatra, S.N.; Mukhopadhyay, A.; Roy Goswami, M.; Ray, S.S.; Chakrabarti, P.; Ram, S.S.; Sudarshan, M.; Chakraborty, A.; Dasgupta, A.; et al. Genotoxicity Study with Special Reference to Comet Test in the Blood Cells of Workers Exposed to Sewage Water. *Adv. Toxicol.* **2014**, *2014*, 251812. [CrossRef]
97. Friis, L.; Vaghef, H.; Edling, C.; Hellman, B. No Increased DNA Damage in Peripheral Lymphocytes of Sewage Workers as Evaluated by Alkaline Single Cell Gel Electrophoresis. *Occup. Environ. Med.* **1997**, *54*, 494–498. [CrossRef] [PubMed]
98. Gajski, G.; Oreščanin, V.; Garaj-Vrhovac, V. Cytogenotoxicity of Sewage Sludge Leachate before and after Calcium Oxide-Based Solidification in Human Lymphocytes. *Ecotoxicol. Environ. Saf.* **2011**, *74*, 1408–1415. [CrossRef] [PubMed]
99. Hirano, T.; Tamae, K. Earthworms and Soil Pollutants. *Sensors* **2011**, *11*, 11157–11167. [CrossRef] [PubMed]
100. Khalid, H.; Kashif Zahoor, M.; Riaz, D.; Arshad, M.; Yaqoob, R.; Ranian, K. Sewage Sludge-Induced Effect on Growth, Enzyme Inhibition, and Genotoxicity Can Be Ameliorated Using Wheat Straw and Biochar in *Pheretima posthuma* Earthworms. *Front. Environ. Sci.* **2022**, *10*, 888394. [CrossRef]
101. Bhat, S.A.; Singh, J.; Singh, K.; Vig, A.P. Genotoxicity Monitoring of Industrial Wastes Using Plant Bioassays and Management through Vermitechnology: A Review. *Agric. Nat. Resour.* **2017**, *51*, 325–337. [CrossRef]
102. Corrêa Martins, M.N.; Souza, V.V.D.; Silva Souza, T.D. Cytotoxic, Genotoxic and Mutagenic Effects of Sewage Sludge on *Allium cepa*. *Chemosphere* **2016**, *148*, 481–486. [CrossRef]

sustainability

MDPI

Article

Modified Social Group Optimization to Solve the Problem of Economic Emission Dispatch with the Incorporation of Wind Power

Dinu Calin Secui [1], Cristina Hora [1,*], Codruta Bendea [1], Monica Liana Secui [2], Gabriel Bendea [1] and Florin Ciprian Dan [1]

1 Department of Energy Engineering, Faculty of Energy Engineering and Industrial Management, University of Oradea, 410058 Oradea, Romania; gbendea@uoradea.ro (G.B.); florin.dan@uoradea.ro (F.C.D.)

2 Department of Psychology, Faculty of Social-Humanistic Sciences, University of Oradea, 410058 Oradea, Romania; msecui@uoradea.ro

* Correspondence: chora@uoradea.ro

Abstract: Economic dispatch, emission dispatch, or their combination (EcD, EmD, EED) are essential issues in power systems optimization that focus on optimizing the efficient and sustainable use of energy resources to meet power demand. A new algorithm is proposed in this article to solve the dispatch problems with/without considering wind units. It is based on the Social Group Optimization (SGO) algorithm, but some features related to the selection and update of heuristics used to generate new solutions are changed. By applying the highly disruptive polynomial operator (HDP) and by generating sequences of random and chaotic numbers, the perturbation of the vectors composing the heuristics is achieved in our Modified Social Group Optimization (MSGO). Its effectiveness was investigated in 10-unit and 40-unit power systems, considering valve-point effects, transmission line losses, and inclusion of wind-based sources, implemented in four case studies. The results obtained for the 10-unit system indicate a very good MSGO performance, in terms of cost and emissions. The average cost reduction of MSGO compared to SGO is 368.1 \$/h, 416.7 \$/h, and 525.0 \$/h for the 40-unit systems. The inclusion of wind units leads to 10% reduction in cost and 45% in emissions. Our modifications to MSGO lead to better convergence and higher-quality solutions than SGO or other competing algorithms.

Keywords: economic/emission dispatch; wind power; logistic map; highly disruptive polynomial operator; optimization

Citation: Secui, D.C.; Hora, C.; Bendea, C.; Secui, M.L.; Bendea, G.; Dan, F.C. Modified Social Group Optimization to Solve the Problem of Economic Emission Dispatch with the Incorporation of Wind Power. *Sustainability* **2024**, *16*, 397. https://doi.org/10.3390/su16010397

Academic Editor: Yagang Zhang

Received: 21 September 2023
Revised: 3 November 2023
Accepted: 6 November 2023
Published: 2 January 2024

1. Introduction

The power generation sector has been undergoing continuous development in recent years, with a focus on diversification of energy sources and production technologies, but also on efficiency and constant innovation. Decision makers aim through regulations and policies to create a competitive and attractive framework for investors while ensuring the sustainable development of the energy system [1]. In order for electricity producers and companies to remain in a competitive market, it is necessary for them to operate with the lowest possible costs and to use sources/technologies with the lowest possible environmental impact. One way to optimize operating costs while taking emissions into account is to solve the economic emission dispatch (EED) problem [2]. The goal of the EED problem is to determine the optimal operating mode of energy sources to minimize the two objectives—cost and emissions—considering a given power demand, as well as the operating limits of the generating units [3]. If the EED problem aims only at cost optimization (without considering emissions), then this is called the economic dispatch problem (EcD) [4]. If EED only aims to optimize emissions (without considering costs), then it is called the emission dispatch (EmD) problem [5]. For a more complete approach, both the economic aspects and the level of emissions released into the atmosphere must be taken into account, which implies solving the EED problem.

Initially, the solution to EcD, EmD, and EED problems focused mainly on conventional unit power plants (firing coal, gas, etc.), as they had a very large share of the power system structure. However, last year's warnings about the environmental impact of these types of power plants led to an increased use of both high-efficiency co-generation units (CU) (that simultaneously produce both electricity and heat), and renewable energy power generation systems (RESPSs) to steer the electricity generation sector towards sustainable development [6,7]. The EED problem that includes CU aims at the optimal scheduling (in terms of cost and emissions) of power-only units, heat-only units, and their combination: co-generation units [8]. It is a similar situation for EcD [9] or EmD [8] problems, with both aiming to manage all categories of units operating in the power system: power-only, heat-only, and CU units. For systems that include CU units, in addition to costs and constraints specific to power-only units, the EED optimization model must also consider the fuel costs related to heat-only and CU units, power-heat dependencies, and the produced-demand heat balance [10]. In the case of including RESPSs in the power system, an important option is the use of wind energy, which can be converted into electricity without producing greenhouse gases. Wind energy is considered a renewable and clean energy source, having only a low secondary impact on the environment due to the manufacturing process of the equipment and its transport. Increasing the share of wind energy in the energy mix has a positive impact on the quality of the environment and reduces dependence on conventional sources. Since wind is an uncontrollable and variable source of energy, the energy production from wind units is influenced by weather conditions and wind speed. Thus, the operation of power systems that include wind units must also take into account the unpredictable variations in the power of these units [4].

The integration of wind units into EcD, EmD, or EED problems brings challenges related to wind speed distribution, mathematical optimization models, and solution algorithms. One of the most common parametric distributions used in wind speed modeling is the Weibull distribution [11–13], but other distributions can be considered, such as [14] Gamma, inverse Gamma, Gaussian, Burr, Halphen, etc.

In the following, some articles are presented which propose mathematical models and solution algorithms for solving EcD, EmD, or EED problems considering the uncertainties caused by unpredictable wind speed fluctuations. Thus, in [4], an optimization model is presented that includes the costs related to the overestimation and underestimation of the wind power. The case study considers a Weibull distribution of the wind speed, and it shows that the optimal solutions to the EcD problem can be influenced by factors associated with the overestimation and underestimation of wind power. Based on a linear relationship between wind speed and wind power, and considering a Weibull distribution for wind speed, in [11], an optimization model is developed for the minimization of emissions, in which the uncertainty of the wind power is included in the constraints of the model. In [12], both the cost of emissions and the cost of overestimation and underestimation of wind power are included in the bi-objective EED problem. Starting from the classical formulation of EcD and EmD problems, in [2], the objective functions related to costs and emissions are extended with terms that include the uncertainty of the wind power. The resulting models are tested on a system consisting of two conventional units and two wind units, considering a mixed Gamma-Weibull distribution for the wind speed. To solve the EED problem with the inclusion of wind power, in [15], the Honey Bee Mating Optimizer (HBMO) is applied, where a so-called 2m-point model is used to estimate the uncertainty of the wind power. In [16], a new evolutionary technique called Lightning Flash Algorithm (LFA) is proposed to solve the bi-objective EED problem by considering different levels of wind power penetration and multiple fuel sources. The LFA technique is efficient for the EED problem, having a good convergence and achieving lower costs and emissions than other algorithms. A robust and efficient optimization model for dispatching wind-thermal power under uncertainties is developed in [1], taking into account robustness, economics, and environmental aspects. In [17], another robust model for the optimization of the EcD problem is presented. It is based on the identification of a set of discrete bad scenarios, where the objective function aims at minimizing all the penalties associated with the bad scenarios.

The EcD, EmD, or EED problems that include or do not include wind power uncertainty are non-convex with multiple local optima and nonlinear constraints [12,15,16] requiring advanced solution algorithms.

For the study of both systems (the ones comprising conventional-only units and the ones comprising thermal-wind units), several algorithms are presented in the following lines. Considering the first category of systems, in [18], an Improved Class Topper Optimization (ICTO) has been developed by including in the classical CTO three new concepts: adaptive improvement factor, adaptive acceleration coefficient, and chaos local search, which help to enhance the exploration and exploitation capability of the ICTO. The ICTO is tested on five systems with conventional units, and it performs better, in terms of cost and emissions, than the original CTO and several other competing algorithms. A quasi-oppositional-based political optimizer is used, in [19], to solve the bi-objective EED problem considering valve-point loading effect, transmission line losses, and other constraints related to conventional units. Recently, two new population-based algorithms (Criminal Search Optimization Algorithm (CSOA) [20] and Kho-Kho optimization algorithm (KKO) [21]) have been developed and applied to solve the EED problem. Both algorithms (CSOA and KKO) show good exploitation and exploration capabilities, and they can be used to solve complex problems in various domains. Another option for solving the EED problem is presented in [22], where two metaheuristic algorithms, Exchange Market Algorithm (EMA) and Adaptive Inertia Weight Particle Swarm Optimization (AIWPSO), are combined to obtain a new algorithm with improved global and local search abilities. The constraints of the EED problem are maintained using the multiple constraints ranking technique. The best compromise solution (BCS) obtained using EMA-AIWPSO dominates the BCSs obtained using other algorithms (such as KKO, ISA, or GSA). Also, several multi-objective algorithms have been proposed for solving the EED problem, such as multi-objective SSA [23], multi-objective cultural algorithm [24], NSGA-III algorithm [25], or multi-objective quasi-oppositional TLBO [26], each of which uses a basic metaheuristic algorithm (SSA, CA, GA, or TLBO) that is endowed with the Pareto-dominance principle to generate successive Pareto fronts. The multi-objective algorithms mentioned [23–26] have been tested on various power systems with conventional units, their performance being superior to other recognized multi-objective techniques (MODE, NSGA II, or SPEA-2).

Over the time, when trying to solve the EED problem, various techniques have been used to optimally program the units of the thermal-wind systems. Thus, in [6], the techniques of weighted goal programming and the progressive bounded constraint method are combined to generate a set of Pareto solutions that are efficient in the cost–emission space, and then extract the best compromise solution. For instance, four cases are analyzed to quantify and demonstrate the benefits of including wind units in the structure of a power system comprising only thermal units. In [12], the Artificial Bee Colony (ABC) algorithm is strengthened by including the best solution in the update equations, and in [27], chaos is inserted into the sine–cosine algorithm to obtain better-quality solutions to the EED problem. Additionally, in [13]—where wind units are considered—a number of eight metaheuristic algorithms (Flower Pollination Algorithm (FPA), Mine Blast Algorithm (MBA), Backtracking Search Algorithm (BSA), Symbiotic Organisms Search (SOS), Ant Lion Optimizer (ALO), Moth-Flame Optimization (MFO), Stochastic Fractal Search (SFS), and Lightning Search Algorithm (LSA)) are applied to identify the best solutions to the EcD, EmD, or EED problem. The FPA and BSA algorithms provide the best scheduling of thermal/wind units for cost and/or emission minimization. To study the behavior of systems in which thermal, wind, and solar units operate, in [28], the Dragonfly Algorithm (DA) is proposed for the EcD problem. Uncertainties related to the power generated by using wind and solar energy are modeled using the 2-m point estimation technique. The

DA algorithm outperforms other algorithms, such as Crow Search Algorithm (CSA), Ant Lion Optimizer (ALO), oppositional RCCRO, Biogeography-Based Optimization (BBO), PSO, and GA, in terms of cost and execution time.

SGO is a metaheuristic algorithm proposed relatively recently by Satapathy SC (2016) [29], which is based on the fact that a social group of individuals has a greater ability to solve a real-life problem than a single individual. For some mathematical functions, SGO is more efficient than other well-known metaheuristic algorithms (such as [29]: DE, PSO, ABC, GA, TLBO, etc.), being an easy-to-implement algorithm, having two main phases (improving phase and acquiring phase) and a single specific parameter. However, for some mathematical functions or applications, it can provide a low performance due to an imbalance between exploration and exploitation, which ultimately leads to getting stuck in a local optimum [30]. Also, SGO may suffer due to the reduced diversity of the population [31]. To overcome the mentioned shortcomings and to obtain better quality solutions, there have been attempts to improve the performance of SGO by different strategies, such as modifying the relations for updating the solutions [30], inertia weight strategies [32], and hybridizations with other algorithms [31].

In this paper, some changes have been made to the original SGO algorithm, aiming to increase efficiency, resulting in the MSGO algorithm. In order to improve the exploration–exploitation balance, the commutation condition between the equations for updating the solutions in the acquiring phase was changed. Also, to increase the diversity of the population, a logistic map and a highly disruptive polynomial operator were inserted.

The main contributions of this paper are:

- We propose a modified version of SGO (called MSGO) in which the way of updating and adapting the individuals in the social group is changed by inserting chaos and an HDP operator (in the original SGO only uniformly randomly generated number sequences are used). The operators associated with chaos and HDP aim at increasing the efficiency of the MSGO algorithm by reducing the number of close solutions and overcoming some drawbacks related to slow convergence. To the best of the authors' knowledge the HDP operator has never been used to improve the performance of the SGO algorithm.
- Implementation of MSGO to solve EcD, EmD, and EED problems with or without consideration of wind units.
- Conducting experiments to evaluate and statistically compare the effectiveness of MSGO with SGO and other well-known algorithms (or their varieties) for thermal or wind-thermal power systems of different sizes and characteristics.

2. Statement of the EcD, EmD, and EED Problems

2.1. Statement of the EcD Problem

The EcD problem for a power system that includes thermal and wind units aims to determine the power outputs of these generating units so that the operating cost of the entire system is minimized, and a number of technical constraints are met. For the formulation of the mathematical optimization model, we consider a power system comprising Nt thermal units and Nw wind units, and the total power demand (P_D) of the consumers in the system is considered known and constant for the period of analysis. The variables to be optimized are continuos, being represented by the output power vectors of the thermal units $PT = [PT_1, PT_2, \ldots, PT_i, \ldots, PT_{Nt}]$ and of wind units $PW = [PW_1, PW_2, \ldots, PW_j, \ldots, PW_{Nw}]$.

The objective function $C(PT, PW)$ is represented by two components; one related to the fuel cost of the thermal units, $C^T(PT)$, and the other related to the operating cost of the wind units, $C^W(PW)$ [4]:

$$C(PT, PW) = C^T(PT) + C^W(PW) \quad (1)$$

$$C^T(PT) = \sum_{i=1}^{Nt} C_i^T(PT_i) \tag{2}$$

$$C^W(PW) = \sum_{j=1}^{Nw} C_j^W(PW_j) = \sum_{j=1}^{Nw} \left\{ c_j^d \cdot PW_j + c_j^o \cdot E_j^o(PW_j) + c_j^u \cdot E_j^u(PW_j) \right\} \tag{3}$$

where, $C^T{}_i(PT_i)$ is the fuel cost corresponding to thermal unit i, and $C^W{}_j(PW_j)$ is the operating cost corresponding to the wind unit j.

In this paper, the relationship between the fuel cost $C^T{}_i(PT_i)$ and output power PT_i of a thermal unit i is modeled via a non-convex function consisting of a quadratic and a sinusoidal term [33]:

$$C_i^T(PT_i) = a_i PT_i^2 + b_i PT_i + c_i + |e_i sin(f_i(PT_{min,i} - PT_i))|,\ i = 1, 2, \ldots, Nt \tag{4}$$

The operating cost of wind unit j, $C^W{}_j(PW_j)$, consists of three terms [4]: the direct operating cost of unit j ($c_j^d \cdot PW_j$), the cost corresponding to the overestimation of the wind power ($c_j^o \cdot E_j^o(PW_j)$), and the cost corresponding the underestimation the wind power ($c_j^u \cdot E_j^u(PW_j)$).

The wind power overestimation for unit j occurs if the estimated wind power for this unit PW_j is higher than the available power represented by a random variable W_j. In this situation, the difference between the two powers is covered by a reserve source and implies the reserve cost $c_j^o \cdot E_j^o(PW_j)$. In the opposite situation of the underestimation (if PW_j is less than W_j), a part of the available power of unit j remains unused, which implies a penalty cost $c_j^u \cdot E_j^u(PW_j)$. Since W_j is a random variable, the powers $E_j^o(PW_j)$ and $E_j^u(PW_j)$ will include the uncertainty of the wind power, and their calculation method is presented below.

For the calculation of the average powers $E_j^o(PW_j)$ and $E_j^u(PW_j)$, it is considered that the wind speed is modeled by a Weibull distribution whose probability density function (pdf) is given by relation (5) [11]:

$$f_V(v) = \frac{k}{c}\left(\frac{v}{c}\right)^{k-1} exp\left[-\left(\frac{v}{c}\right)^k\right] \tag{5}$$

where, V is the random variable wind speed, v denotes the wind speed (a value of V), and $f_V(v)$ is the pdf of the variable V, k is the shape parameter, and c is the scale parameter.

Also, between the random variable wind power (W) and the random variable wind speed (V), we consider a linear relationship expressed as follows [11]:

$$W = \begin{cases} 0, & if\ V < v_{in} or\ V > v_{out} \\ \frac{(V - v_{in})PWr}{v_r - v_{in}}, & if\ v_{in} \le V \le v_r \\ PWr, & if\ v_r \le V \le v_{out} \end{cases} \tag{6}$$

The use of the linear model (6) requires knowledge of three limit speeds: cut-in wind speed (v_{in}), rated wind speed (v_r), and cut-out wind speed (v_{out}). PWr is the rated output power of the wind unit, which corresponds to the speed v_r.

The pdf for wind power (W) over the continuous interval (0, PWr) has the expression:

$$f_W(w) = \frac{k \cdot v_{in} \cdot R}{c \cdot PWr}\left(\frac{v_{in}}{c} \cdot \left(1 + \frac{w \cdot R}{PWr}\right)\right)^{k-1} exp\left[-\left(\frac{v_{in}}{c} \cdot \left(1 + \frac{w \cdot R}{PWr}\right)\right)^k\right] \tag{7}$$

where $R = (v_r - v_{in})/v_{in}$. Given relation (6) and the Weibull distribution of the wind speed, the event probabilities $W = 0$ and $W = PWr$ are calculated using relations [4]:

$$Prob(W = 0) = Prob(V < v_{in}) + Prob(V > v_{out}) = 1 - exp[-(v_{in}/c)^k] + exp[-(v_{out}/c)^k] \tag{8}$$

$$Prob(W = PWr) = Prob(v_r < V < v_{out}) = exp[-(v_r/c)^k] - exp[-(v_{out}/c)^k] \quad (9)$$

We must say that, in general, the distributions of the variables V and W, as well as the pdfs derived from them, differ depending on the location of the wind turbine units. Thus, for wind unit j, the average powers associated with the overestimation and the underestimation of the wind power are mathematically expressed as follows [4]:

$$E_j^o(PW_j) = \int_0^{PW_j} (PW_j - w) f_{W_j}(w) dw + (PW_j - 0) \cdot \text{Prob}(W_j = 0) \quad (10)$$

$$E_j^u(PW_j) = \int_{PW_j}^{PWr_j} (w - PW_j) f_{W_j}(w) dw + (PWr_j - PW_j) \cdot \text{Prob}(W_j = PWr_j) \quad (11)$$

where $f_{W_j}(w)$ is probability density function for the random variable W_j having the form given by relation (7), while w are the values of the continuous random variable W_j. The discrete probabilities $Prob(W_j = 0)$ and $Prob(W_j = PW_{rj})$, for unit j, are calculated using relations (8) and (9), where PWr_j represents rated wind power for unit j.

In Table A1 are presented the steps to evaluate the cost related to wind power, as well as a calculation example for a wind unit.

The feasible space of EcD problem solutions is limited by the following constraints [4,12]:

1. The thermal units i must be operated between the minimum capacity ($PT_{min,i}$) and the maximum capacity ($PT_{max,i}$):

$$PT_{min,i} \leq PT_i \leq PT_{max,i}, i = 1, 2, \ldots, Nt \quad (12a)$$

If the power of the thermal unit i in the previous hour is known or specified (PT_i^0), then the operating limits are given by the constraint:

$$Max\left(PT_{min,i}, PT_i^0 - DR_i\right) \leq PT_i \leq Min\left(PT_{max,i}, PT_i^0 + UR_i\right), i = 1, 2, \ldots, Nt \quad (12b)$$

where DR_i and UR_i are the down-ramp and up-ramp limits of the unit i.

2. The wind units j must be operated between the minimum ($PW_{min,j}$) and maximum ($PW_{max,j}$) capacity:

$$PW_{min,j} \leq PW_j \leq PW_{max,j}, j = 1, 2, \ldots, Nw \quad (13)$$

3. The actual powers generated by the thermal and wind units must cover the power consumed in the system:

$$\sum_{i=1}^{Nt} PT_i + \sum_{j=1}^{Nw} PW_j - P_L - P_D = 0 \quad (14)$$

where P_D is the load demand of the system, and P_L represents the transmission line losses, which can be determined considering the constant B coefficients formula:

$$P_L = \sum_{i=1}^{Nt+Nw} \sum_{j=1}^{Nt+Nw} PTW_i \cdot B_{ij} \cdot PTW_j + \sum_{i=1}^{Nt+Nw} B_{0i} \cdot PTW_i + B_{00} \quad (15)$$

where B_{ij} is an element of the loss coefficients matrix, B_{0i} is i element of the loss coefficients vector, and B_{00} is the loss coefficient constant; $PTW = [PTW_1, PTW_2, \ldots, PTW_i, \ldots, PTW_{Nt+Nw}]$ is a vector that combines the powers of thermal PT and wind PW units, while PTW_i represents the ith component of the PTW vector.

2.2. Statement of the EmD Problem

The EmD problem has a high level of similarity with the EcD problem, aiming to determine the *PT* and *PW* vectors, so that the emissions released into the atmosphere is minimal while maintaining some technical constrains at the units and system level. Thus, in the EmD problem, the variables to be optimized (*PT* and *PW* vectors), as well as the constraint relations (12)–(15), are identical to those in the EcD problem. The objective function is represented by the total amount of emissions released into the atmosphere, mathematically defined by the following relation [2]:

$$E(PT,PW) = E^T(PT) + E^W(PW) \tag{16}$$

$$E^T(PT) = \sum_{i=1}^{Nt} E_i^T(PT_i) \tag{17}$$

$$E^W(PW) = \sum_{j=1}^{Nw} E_j^W(PW_j) = \sum_{j=1}^{Nw} \left\{ e_j^o \cdot E_j^o(PW_j) + e_j^u \cdot E_j^u(PW_j) \right\} \tag{18}$$

where $E_i^T(PT_i)$ and $E^T(PT)$ represent the pollutant emissions released into the atmosphere due to the operation of thermal unit *i*, respective of all thermal units in the analyzed power system. $E_j^W(PW_j)$ and $E^W(PW)$ are the pollutant emissions produced due to the need to use other thermal units to cover the uncertainty of the availability of wind unit *j*, respective of all wind units considered. Average powers, $E_j^o(PW_j)$ and $E_j^u(PW_j)$, are calculated with relations (10) and (11), which include the uncertainty of wind power in the estimation of pollutant emissions.

The term $e_j^o \cdot E_j^o(PW_j)$ represents the emissions released into the environment due to the need to use some thermal units in the system to cover the difference between the scheduled power of the wind unit PW_j (power that cannot be realized due to the unavailability of the wind resource) and its available power W_j (the case of overestimation of wind power). The term $e_j^u \cdot E_j^u(PW_j)$ represents the emissions released into the environment by other thermal units due to the non-use of the full available power of wind unit *j* (the case of underestimation of wind power). The quantity of emissions $E_i^T(PT_i)$ released into the atmosphere by a thermal unit *i* can be defined by relation [2]:

$$E_i^T(PT_i) = \sum_{i=1}^{Nt} (\gamma_i + \beta_i PT_i + \alpha_i PT_i^2 + \delta_i exp(\lambda_i PT_i)) \tag{19}$$

2.3. Statement of the EED Problem

The EED problem is similar to the EcD and EmD problems, in that the variables to be optimized and the constraints of the problem are the same, but the objective function is different. In the EED problem, the objective function Φ(*PT*, *PW*) can be formed by the weighted and normalized summation of the cost *C*(*PT*, *PW*) and emissions *E*(*PT*, *PW*) objectives [34]:

$$\Phi(PT,PW) = \omega \frac{C(PT,PW) - C_{min}}{C_{max} - C_{min}} + (1-\omega)\frac{E(PT,PW) - E_{min}}{E_{max} - E_{min}} \tag{20}$$

where C_{min}, C_{max} are the minimum and maximum costs corresponding to the function *C*(*PT*, *PW*); E_{min}, E_{max} are the minimum and maximum emissions corresponding to the function *E*(*PT*, *PW*); ω and (1 − ω) are weighting factors associated with the normalized cost and emissions objectives, $0 \leq \omega \leq 1$.

To use relation (20) it is necessary to calculate the minimum (C_{min} and E_{min}) and maximum (C_{max} and E_{max}) values for the two objectives. The determination of the minimum values (C_{min}, E_{min}) is done by first solving the EcD and EmD problems formulated in Sections 2.1 and 2.2. The maximum values (C_{max}, E_{max}) are determined while considering that the EcD and EmD problems have opposite tendencies [35]. Thus, the solution for which a minimum cost is obtained in the EcD problem will be considered to result in maximum emissions, and vice versa.

In current practice, system operators or decision makers look for a single solution that takes into account both objectives, which is generally the best compromise solution. By solving the EED problem for different values of the ω factor, one can estimate the set of non-dominant solutions that form the discrete Pareto front [5] (from which the BCS between the two objectives is extracted). To extract the BCS from the set of solutions belonging to the Pareto front, a fuzzy approach is used [5]. Thus, the solutions in the Pareto front are ordered according to one of the objectives, then for each objective i and non-dominant solution s, a fuzzy membership function $\mu_{i,s}$ is assigned. This is defined by relation [5]: $\mu_{i,s} = (f_{i,max} - f_{i,s})/(f_{i,max} - f_{i,min})$, $i = \{1, 2\}$ and $s = \{1, 2,\ldots, P\}$, where $f_{i,min}$ and $f_{i,max}$ are the minimum and maximum value of the ith objective function; $f_{i,s}$ is the value of the objective function corresponding to solution s and objective i; P is the number of non-dominated solutions from the Pareto front. In order to demonstrate the merit of all the objectives corresponding to solution s, the normalized membership function μ_S^* is calculated [5]:

$$\mu_S^* = \frac{\sum_{i=1}^{2} \mu_{i,s}}{\sum_{s=1}^{P} \sum_{i=1}^{2} \mu_{i,s}}, \; s = 1, 2, \ldots, P \tag{21}$$

The BCS corresponds to the maximum value of the index μ_S^*: $\mu_{max} = Max(\mu_S^*, s = 1, 2,\ldots, P)$.

3. The Modified SGO Algorithm

3.1. Classic SGO

The SGO is a metaheuristic, population-based algorithm that is inspired by human social group behavior and is used for solving complex problems [29]. In the SGO, the population consists of a social group of N individuals $X_i = [x_{1,i}, x_{2,i},\ldots, x_{j,i},\ldots, x_{n,i}]_{|\, i = 1,2,\ldots, N}$ interacting and exchanging knowledge with each other, each individual representing a solution X_i to the problem. The characteristics of the individuals represent the components $x_{j,i}$, $j = 1, 2,\ldots, n$ (n is the number of characteristics of an individual, which equals the size of the problem to be solved) of the solutions X_i, and the ability of an individual to find a solution to the problem is measured by the fitness function $f_i = f(X_i)$. The SGO algorithm has three phases: initialization phase, improving phase, and acquiring phase. In the initialization phase, each component $x_{j,i}$, $j = 1, 2,\ldots, n$ of the solution X_i, is randomly generated between the minimum ($x_{min,j}$) and maximum ($x_{max,j}$) limits:

$$x_{j,i} = x_{min,j} + r_1(x_{max,j} - x_{min,j}) \tag{22}$$

where r_1 is a random number uniformly distributed in the range (0, 1).

In the improving phase, each individual X_i, $i = 1, 2,\ldots, N$ of the group seeks to improve their traits $x_{j,I}$ by interacting with the best individual of the group at that moment, $X^{best} = [x_1^{best}, x_2^{best},\ldots, x_j^{best},\ldots x_n^{best}]$. Thus, the new traits ($x_{j,i}^{new}$) of the individual X_i^{new} are determined with relation [29]:

$$x^{new}_{j,i} = c\, x_{j,i} + r_2\, (x_j^{best} - x_{j,i}) \tag{23}$$

where c is the self-introspection parameter with values between 0 and 1, and r_2 is a uniformly generated random number in the range (0, 1). If the new vector X_i^{new} is better, then it is retained; otherwise, the old solution is kept.

In the acquiring phase, each individual X_i aims to improve its traits (level of knowledge) through mutual, cumulative interactions with both the best individual X^{best} and an individual randomly r ($r \neq i$) chosen from the population, $X_r = [x_{1,r}, x_{2,r},\ldots, x_{j,r},\ldots, x_{n,r}]$. Depending on the quality of the individuals X_i and X_r, the new traits ($x_{j,i}^{new}$) of the individual X_i^{new} are determined as follows [29]:

If $f(X_i) < f(X_r)$ Then

$$x_{j,i}^{new} = x_{j,i} + r_3\,(x_{j,i} - x_{j,r}) + r_4\,(x_j^{best} - x_{j,i}) \quad (24a)$$

Else

$$x_{j,i}^{new} = x_{j,i} + r_3\,(x_{j,r} - x_{j,i}) + r_4\,(x_j^{best} - x_{j,i}) \quad (24b)$$

where, r_3 and r_4 denote independent arbitrary value in the (0, 1).

The solutions obtained using the SGO algorithm are iteratively improved using the updating relations from the improving and acquiring phases, until the maximum number of iterations t_{max} is reached. A pseudo-code of the SGO algorithm is presented in detail in [29].

3.2. Modified SGO (MSGO)

The proposed MSGO algorithm includes the same phases as the original SGO algorithm, but some features of SGO have been modified to increase the efficiency of the MSGO algorithm. Two of these changes, presented below, aim at inserting two operators (one chaotic type and the other being highly disruptive polynomial (HDP)) in the structure of the solutions update relations to improve MSGO's ability to escape from local minima and obtain better quality solutions.

Chaotic operator: Chaos, due to its properties (unpredictability, non-periodic, ergodicity, non-converging, pseudo-randomness etc. [18]), has been successfully incorporated into the structure of various optimization algorithms (such as adaptive sparrow search algorithm [36] and moth–flame optimizer [37]) to cover a number of shortcomings related to low population diversity, slow convergence of the optimization process, stagnation in a local area etc. In this paper, chaos is included in the MSGO algorithm using a chaotic sequence (cx_p) generated by Logistic map [37] which can be effective in solving some EcD problems [18]. The Logistic map is defined by relation [18]:

$$cx_{p+1} = 4\,cx_p\,(1 - cx_p),\ cx_p \in (0, 1) \quad (25)$$

where $\{cx_p\}_p = 1\ldots\infty$ represents the chaotic sequence generated by the Logistic map, at the pth iteration. The initial conditions are $cx_0 \in (0, 1)$ and $cx_0 \neq \{0.0, 0.25, 0.5, 0.75, 1\}$.

HDP operator: In case of more complex mathematical functions, SGO cannot provide an efficient search in the whole solutions space by relying only on randomly generated sequences through numbers ($r1$–$r4$) or on randomly extracting a solution from the population (X_r). As a result, in order to increase the exploration and exploitation capacity of the MSGO algorithm, as well as the chances of exceeding local minima, a perturbation based on the HDP operator [38] is introduced into MSGO. Mathematically, the HDP operator is defined by relation [38]:

$$\delta = \begin{cases} \delta_H = \left[2\cdot ru + (1 - 2\cdot ru)(1-\delta_1)^{\eta+1}\right]^{\frac{1}{\eta+1}-1} & \textit{If } ru \le \alpha \\ \delta_L = 1 - \left[2(1 - ru) + 2(ru - 0.5)(1-\delta_2)^{\eta+1}\right]^{\frac{1}{\eta+1}} & \textit{otherwise} \end{cases} \quad (26)$$

where, ru is a random number uniformly generated in the range [0, 1]; LB_i and UB_i indicate the lower and upper boundary of the variables $x_{j,i}$; η is the index for the polynomial operator; α is a control parameter that is considered to have a value of 0.5 [39] (in this paper, the values of the parameter α are determined by experimental trials in the range (0, 1)); δ_H is the equation that alternates moderate values with high values for δ, while δ_L is the equation that generates low values for δ; δ_1 and δ_2 are calculated, based on the LB_i and UB_i imposed boundaries for the variable $x_{j,i}$, with the relations $\delta_1 = (x_{j,i} - LB_i)/(UB_i - LB_i)$ and $\delta_2 = (UB_i - x_{j,i})/(UB_i - LB_i)$.

In general, the HDP operator is used as a mutation operator that intervenes in altering some $x_{j,i}$ values of a X_i solution with a certain predetermined probability. In MSGO, the HDP operator is not used as a mutation operator, but as an operator applied to modify each variable $x_{j,i}$ of a solution X_i (practically, the mutation probability is considered equal to 1). Thus, depending on the HDP parameters (η, ru, δ_1, α), the δ values generated by (26) may be high, which could cause some of the $x_{j,i}$ components of some X_i solutions to reach the limits of the search space relatively quickly (the δ values could generate excessively large search steps). To avoid this problem, the δ values generated by (26) will be attenuated/smoothed according to the current iteration t, if δ exceeds a certain limit value (δ_{max}):

$$\delta_a = \begin{cases} \delta \cdot \left(1 - \frac{t}{t_{max}}\right) & \text{If } \delta \geq \delta_{max} \\ \delta & \text{otherwise} \end{cases} \tag{27}$$

Figure 1 shows the first 10,000 values of δ generated by relation (26), considering the following settings for the HDP operator parameters: $\delta_1 = rnd(1)$, $\delta_2 = 1 - \delta_1$; $\eta = 4$, $\alpha = 0.5$, $\delta_{max} = 1.5$. The δ_H values generated by (26) under the condition $ru \leq \alpha = 0.5$ are marked in blue, and the δ_L values in red. Since the δ_H values are much higher than δ_L, in order to have a clearer picture, the δ_L values have been detailed in a separate graph on the interval (1, 10,000). It is seen that the δ_H equation alternates moderate values and high ones, while the δ_L equation generates only low values. Figure 2 shows the attenuated δ_a values obtained using (27). Initially, the δ_a values are higher (being able to ensure an efficient search in the whole solutions space), then with the increase in the number of simulations they attenuate, the search step is reduced, making the transition from exploration to exploitation, and at the end of the process the values are equal to or lower than the δ_{max} limit (which facilitates the exploitation phase).

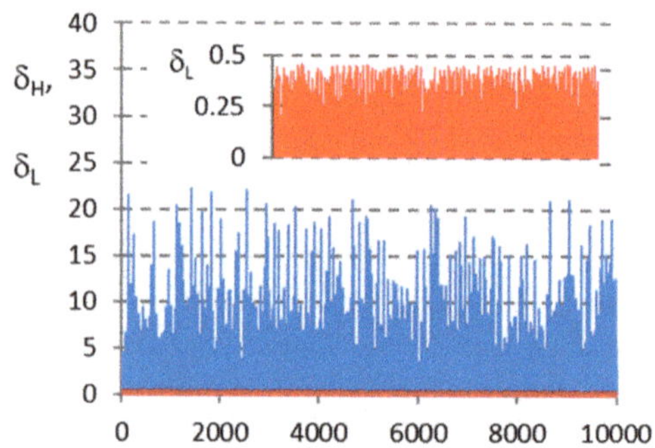

Figure 1. HDP values generated for δ_H and δ_L.

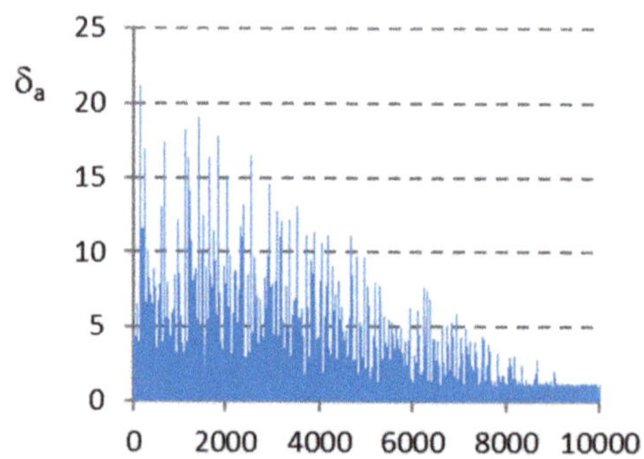

Figure 2. The attenuated values obtained for δ_a.

The changes implemented in MSGO compared to SGO are presented below:

a. The improving phase of MSGO is similar to that of SGO but the sequences of numbers r_2 randomly generated in SGO by (23) are replaced by sequences of numbers generated by the attenuated DHP operator δ_a by relation (27) in the form:

$$x_{j,i}^{new} = x_{j,i} + \delta_a \left(x_j^{best} - x_{j,i} \right) \quad (28)$$

b. The acquiring phase of MSGO is similar to that of SGO, except for the following three modifications:
    b1. The sequences of numbers r_3 randomly generated in SGO by (24a) are replaced by sequences of numbers generated by the attenuated DHP operator δ_a by relation (27).
    b2. The sequences of numbers r_3 randomly generated in SGO by (24b) are replaced by chaotic sequences (cx) generated by the Logistic map with relation (25).
    b3. In SGO, switching between relations (24a) and (24b) is performed by considering the fitness of the competitive solutions X_i and X_r, based on the condition $f(X_i) < f(X_r)$. In MSGO, this condition is replaced by a random one, having the form $rnd(1) < \beta$, where $rnd(1)$ is a uniformly generated random number in the range (0, 1); β is a value determined by experimental trials.

The three changes (b1)–(b3) made in MSGO are embodied in relations (29a) and (29b):
If $rnd(1) < \beta$ Then

$$x_{j,i}{}^{new} = x_{j,i} + \delta_a \, (x_{j,i} - x_{j,r}) + r_4 \, (x_j{}^{best} - x_{j,i}) \quad (29a)$$

Else

$$x_{j,i}{}^{new} = x_{j,i} + cx \, (x_{j,r} - x_{j,i}) + r_4 \, (x_j{}^{best} - x_{j,i}) \quad (29b)$$

In Algorithm 1, the calculation steps for applying the MSGO algorithm are presented in detail, with the modifications (b1)–(b3) mentioned above.

Algorithm 1: MSGO algorithm

{Initialization phase}
Initialize the iterations (t = 0), t is counter of iterations;
Initialize the solutions X_i, $i = 1, 2, \ldots, N$ using relation (22);
Evaluate the initial solutions and identify the best X^{best} solution;
repeat
    $t = t + 1$
    For i = 1 To N Do {*improving phase*}
    For j = 1 To n Do
    Generate a value δ of the HDP operator using (26);
    Determine δ_a using relation (27);
    Update the components $x_{j,i}$ using relation (28), obtaining the new solution $X_i{}^{new}$; End For j
If the new solution $X_i{}^{new}$ is better, then it is retained; otherwise, the old solution is maintained;
Find the best X^{best} solution from the population; End For i
For i = 1 To N Do {acquiring phase}
Randomly select a solution X_r, $r \in \{1, 2, \ldots, N\}$, $r \neq i$;
If $rnd(1) < \beta$ Then
For j = 1 To n Do
    Generate a value δ of the HDP operator using (26);
    Determine δ_a using relation (27);
    Update $x_{j,i}$ using (29a), obtaining $X_i{}^{new}$; End For j; End If
Else
For j = 1 To n Do
    Generate a chaotic value cx using (25);
    Update $x_{j,i}$ using (29b), obtaining $X_i{}^{new}$; End For j; End Else
If the new solution $X_i{}^{new}$ is better, then it is retained; otherwise, the old solution is maintained;
Find the best X^{best} solution from the population; End For i
Until $t \geq t_{max}$ {*stopping criterion*}
The best solution X^{best} and the fitness $f(X^{best})$ are retained.

The main steps of the MSGO algorithm are summarized in a flowchart shown in Figure 3.

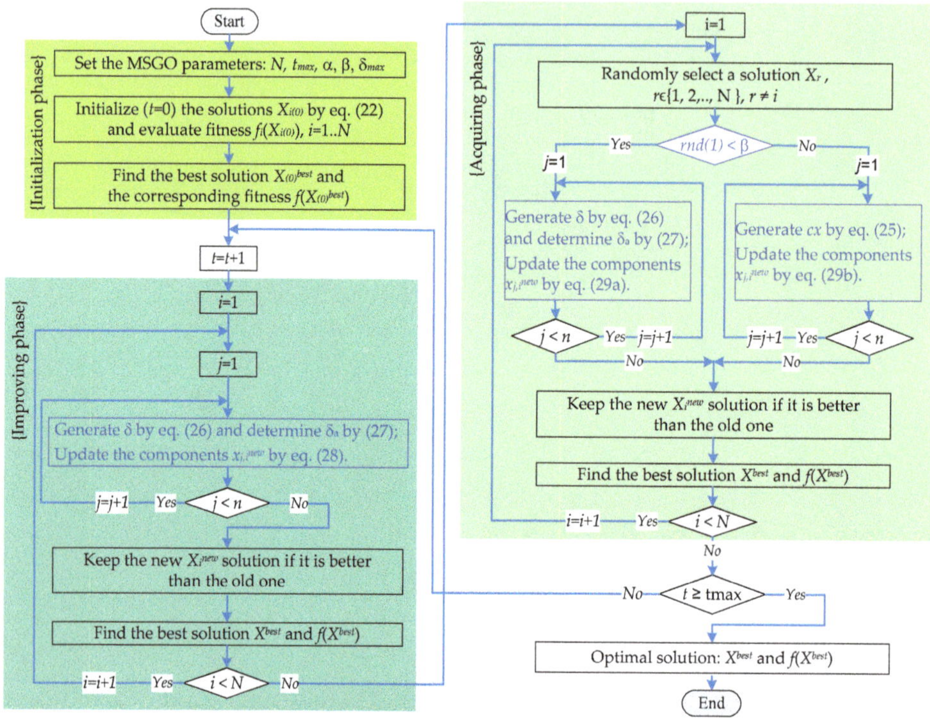

Figure 3. Flowchart of the MSGO algorithm.

4. Implementing the MSGO for the EcD, EmD, or EED Problems

This section shows how to implement the MSGO algorithm for solving EcD, EmD, or EED problems. The application of MSGO for solving EcD, EmD, or EED problems is similar, the difference being given by the type of objective function to be minimized: cost $C(PT, PW)$ defined by (1) for EcD problems, emission $E(PT, PW)$ defined by (16) for EmD problems or $\phi(PT, PW)$ defined by (20) for EED problems. For the implementation of the MSGO algorithm, a solution i is considered to be represented by the vector $PTW_i = [PTW_{1,i}, PTW_{2,i}, \ldots, PTW_{j,i}, \ldots, PTW_{Nt+Nw,i}] \mid_{i=1,2,\ldots N}$ which includes the powers of thermal units ($PTW_{j,i} \mid_{j=1,2,\ldots Nt}$) and wind units ($PTW_{j,i} \mid_{j=Nt+1, Nt+2,\ldots, Nt+Nw}$). If the system contains only thermal units, then the vector PTW_i consists only of the powers generated by these units. We also denote $PTW^{best(t)}$ as the best solution obtained using MSGO up to iteration t.

4.1. Stages of MSGO Implementation for the EcD, EmD, or EED Problem

The MSGO algorithm applied to solve the EcD, Em, or EED problems formulated in Section 2 includes the following steps:

Step 1: Specify test system input data for the EcD, EmD, or EED problem: the number of thermal (Nt) and wind (Nw) units; cost coefficients for the thermal units (a, b, c, e, f); parameters (c, k) to the Weibull distribution; the characteristic values of the wind speed (v_{in}, v_r, v_{out}); the rated output power of the wind units (PWr); active power limits for thermal and wind units (PT_{min}, PT_{max}; PW_{min}, PW_{max}); the cost coefficients (c^d, c^o, c^u) and emissions coefficients (e^o, e^u) to the wind units; the B-loss coefficients (B_{ij}, B_{0i}, B_{00}); load demand (P_D) and accuracy ε.

Step 2: Set the parameters of the MSGO algorithm: N and t_{max};

Step 3: Random initialization of a population of N solutions, represented by the vectors $PTW_{i\,|\,i\,=\,1,2,\ldots N}$

3.1: $t = 0$;

3.2: Chaotic sequence cx are initialized using the logistic map;

3.3: Randomly generate an initial population with N solutions ($PTW_i^{(0)}\,|_{\,i\,=\,1,2,\ldots,N}$) using relation (22). Each solution respects the constraints defined by relations (12)–(14); The constraint (14) is handled by a heuristic procedure CHM presented in Section 4.2;

3.4: Evaluate the initial solutions $PTW_i^{(0)}\,|_{\,i\,=\,1,2,\ldots,N}$ using relation (1), (16) or (20) associated with the EcD, EmD, or EED problems;

3.5: Find the best initial solution $PTW^{best(0)}$ and the objective function associated with the addressed problem (EcD, EmD, or EED).

For $t = 1$ Do t_{max}

Step 4: Update solutions $PTW_i^{(t)}\,|_{\,i\,=\,1,2,\ldots,N}$ in the improving phase.

4.1: For $i = 1$ To N Do

4.2: For $j = 1$ to n Do

4.3: Generate a value δ of the HDP operator using (26);

4.4: Determine δ_a using relation (27);

4.5: Updating the components $PTW_{j,i}^{(t)}$ using relation (28), obtaining the new components $PTW_{j,i}^{new(t)}$, at iteration t;

4.6: Checking the inequality constraints (12) and (13): if the $PTW_{j,i}^{new(t)}$ power is outside the limits, then CHM from Section 4.2 is applied; End For j;

4.7: Checking the equality constraint (14): the new solution $PTW_i^{new(t)}$ is adjusted using the equality constraint handling mechanism from Section 4.2;

4.8: Evaluate the new solution $PTW_i^{new(t)}$ using relation (1), (16) or (20) depending on the addressed problem (EcD, EmD or EED): If the new solution $PTW_i^{new(t)}$ is better, then it is retained; otherwise, the old solution is maintained;

4.9: Update the best solution $PTW^{best(t)}$; End For i;

Step 5: Update solutions $PTW_i^{(t)}\,|_{\,i\,=\,1,2,\ldots,N}$ in the acquiring phase.

5.1: For $i = 1$ To N Do

5.2: Randomly select a solution $PTW_r^{(t)}$, $r\epsilon\{1, 2,\ldots, N\}$, $r \neq i$;

5.3: If $rnd(1) < \beta$ Then

5.4: For $j = 1$ to n Do

5.5: Generate a δ value of the HDP operator using (26);

5.6: Determine δ_a using relation (27);

5.7: Updating $PTW_{j,i}^{(t)}$ using (29a), getting the new components $PTW_{j,i}^{new(t)}$ of the new solutions $PTW_i^{new(t)}$;

5.8: Apply the procedure for handling inequality constraints (12) and (13) from Section 4.2; End For j; End If

5.9: Else

5.10: For $j = 1$ to n Do

5.11: Generate a chaotic value cx by (25);

5.12: Updating $PTW_{j,i}^{(t)}$ by (29b), obtaining $PTW_{j,i}^{new(t)}$;

5.13: Apply CHM from Section 4.2 for constraints (12) and (13); End For j; End Else

5.14: Checking the equality constraint (14): the new solution $PTW_i^{new(t)}$ is adjusted using the equality constraint handling mechanism from Section 4.2;

5.15: Evaluate the new solution $PTW_i^{new(t)}$ using relation (1), (16) or (20) depending on the addressed problem (EcD, EmD, or EED): if the new solution $PTW_i^{new(t)}$ is better, then it is retained; otherwise the old solution is maintained;

5.16: Update the best solution $PTW^{new(t)}$; End For i;

Step 6: Stop the process: the calculation process is stopped when the maximum number of iterations (t_{max}) is reached. {End For t}
Step 7: Memorize the best solution: The best solution PTW^{best} and the objective function associated with the problem addressed (EcD, EmD, or EED).

4.2. Constraints Handling Mechanism (CHM)

The CHM refers to the adjusting method of the solutions that do not satisfy the inequality and equality constraints of the EcD, EmD, or EED problems defined by relations (12)–(14). In the case of inequality constraints, defined by relations (12) and (13), if the power of a thermal or wind unit is outside the imposed operating limits (PT_{min} or PT_{max}, and PW_{min} or PW_{max}, respectively), then the power of this unit is set with the exceeded limit [40]. To handle the equality constraint, defined by (14), we apply a CHM procedure presented in [41]. The CHM(*PTW*) procedure has a single parameter, represented by the power vector *PTW*, and finally it returns a feasible solution that respects relation (14) with the imposed error ε. The CHM(*PTW*) is called whenever a *PTW* vector is updated in the initializing, the improving, or the acquiring phases of the MSGO or SGO algorithms.

5. Case Studies

The efficiency of the MSGO algorithm was tested on medium-sized (10-unit) and large (40-unit) systems, having different characteristics, such as operating limits of the units, transmission line losses, valve-point effect, wind power. According to the characteristics of the systems, four cases were studied, described below.

Case 1 (C_1): This case analyses a 10-unit system taking into account the transmission line losses. The power demand is 2000 MW. The cost coefficients (a_i, b_i, c_i, e_i, f_i, i = 1, 2,. . ., 10), B-loss coefficients (B_{ij}), and emission coefficients (α_i, β_i, γ_i, η_i, δ_i) are taken from [42].
Case 2 (C_2): The second case is a system having 40 units that considers the valve-point effects. The system is analyzed without considering transmission line losses. The power demand is 10,500 MW. The cost coefficients are considered from [33] and emission coefficients are provided by [43].
Case 3 (C_3): The case C_3 is a system with 40 units similar to case C_2 (the cost and emission coefficients are identical to those in case C_2 [33,43]), but transmission losses are considered, as well. The power demand is 10,500 MW. The B-loss coefficients are taken from [44].
Case 4 (C_4): A 40-unit system derived from case C_3 by replacing the first two thermal units (PT_1 and PT_2) with the two wind units (PW_1 and PW_2). Each wind unit has a nominal power of 550 MW, and the minimum and maximum capacities are $PW_{min,1} = PW_{min,2} = 0$, $PW_{max,1} = PW_{max,2}$ = 550 MW. The power demand is P_D = 10,500 MW. We consider that the wind speed has a Weibull distribution. The shape and scale parameters (k, c) corresponding to the sites of the two wind units have the values [13]: $k_1 = 1.5$; $c_1 = 15$; $k_2 = 1.5$; $c_2 = 15$. Other wind-related characteristics have the following values [13]: cut-in wind speed (v_{in} = 5 m/s), rated wind speed (v_r = 15 m/s), cut-out wind speed (v_{out} = 45 m/s), the cost coefficients associated with underestimation ($c_1^u = c_2^u = 5$) and overestimation ($c_1^o = c_2^o = 5$). The cost and emission coefficients for the thermal units (P_3–P_{40}) and the B-loss coefficients are identical to those mentioned in C_3 case, being taken from [44]. The characteristics of the thermal units and the B-loss coefficients for the 10-unit and 40-unit systems are presented in Appendix A, Tables A2–A5.

Each of the cases mentioned in Table 1 (C_1–C_4) were studied based on the mathematical optimization models (presented in Section 2) of the problems EcD (cases C_{1a}–C_{4a}), EmD (cases C_{1b}–C_{4b}), and EED (cases C_{1c}–C_{4c}). The SGO and MSGO algorithms perform 50 independent runs for each case study, retaining four statistical indicators: Best cost/emission (B), Average cost/emission (A), Worst cost/emission (W), and the standard deviation (SD). Algorithms SGO and MSGO have been implemented in MathCAD and run on a PC with an Intel i5 processor, 2.2 GHz CPU and 4 GB of RAM.

5.1. Setting the Parameters

Adjusting the parameters of metaheuristic algorithms is an essential task with multiple positive effects, such as improved performance, adaptation of the algorithms to a specific problem, better execution times, and stability, etc. The proposed MSGO algorithm has three specific parameters (α, β, and δ_{max}). A procedure based on the design of an experimental plan is used to set the specific parameters. They are successively set (for example, in the following order α, β, and δ_{max}), starting from a set of initial values [45]. Thus, one parameter is varied, and the others are fixed either with the initial values or with the already set values. This procedure allows the evaluation of the MSGO algorithm performance through testing different combinations of the interest parameter values, followed by the selection of the best combination based on a predetermined criterion. The test values considered for setting the specific parameters are α = {0, 0.25, 0.5, 0.75, 1}, β = {0, 0.25, 0.5, 0.75, 1}, and δ_{max} = {1.2, 1.5, 2}. The initial values are $\alpha = 0$, $\beta = 0$, and $\delta_{max} = 1.2$. The parameter selection criterion is based on Average cost/emission obtained from 25 runs for each case study. Note that the experimental analysis does not guarantee the best values for the parameters of MSGO algorithm. However, the results obtained using MSGO following this selection process show that the algorithm parameters were reasonably set Table 1 shows the parameters of the MSGO algorithm (N, t_{max}, α, β, δ_{max}) for all the analyzed cases.

Table 1. Characteristics of the analyzed systems and values set for the SGO and MSGO parameters.

Cases	Type of Problem	SGO and MSGO				MSGO Parameters			P_L	VPE	Wind
		n	N	t_{max}	NE	α	β	δ_{max}			
C_{1a}	EcD	10	15	30	900	0.75	0.25	1.2	√	√	-
C_{1b}	EmD	10	15	30	900	0.75	0.25	1.2	√	√	-
C_{1c}	EED	10	15	30	900	0.75	0.25	1.2	-	√	-
C_{2a}	EcD	40	50	350	35,000	0.05	0.5	1.2	-	√	-
C_{2b}	EmD	40	50	350	35,000	0.05	0.5	1.2	-	√	-
C_{2c}	EED	40	50	350	35,000	0.05	0.5	1.2	√	√	-
C_{3a}	EcD	40	50	350	35,000	0.05	0.5	1.5	√	√	-
C_{3b}	EmD	40	50	200	20,000	0.05	0.5	1.5	√	√	-
C_{3c}	EED	40	50	350	35,000	0.05	0.5	1.5	√	√	-
C_{4a}	EcD-Wind	40	50	350	35,000	0.05	0.5	1.5	√	√	√
C_{4b}	EmD-Wind	40	50	200	20,000	0.05	0.25	1.5	√	√	√
C_{4c}	EED-Wind	40	50	350	35,000	0.05	0.5	1.5	√	√	√

In the case of the SGO algorithm, the values of the common parameters (N and t_{max}) are the same as the values set for the MSGO algorithm in each case analyzed (these are shown in Table 1). The SGO algorithm has only one specific parameter c, the value of which being set to the one recommended in [29] ($c = 0.2$). For the fair comparison of the SGO and MSGO algorithms, the number of evaluations of the objective functions (NE) is considered equal in all analyzed cases (C_1–C_4), which is mentioned in Table 1.

5.2. Results for EcD (Cases C_{1a}–C_{4a}) and EmD (Cases C_{1b}–C_{4b}) Problems

The best solutions obtained using SGO and MSGO: The optimal scheduling of the thermal units for the cases C_{1a}, C_{1b} (10 units), and C_{2a}, C_{2b} (40 units) obtained using the MSGO and SGO algorithms are presented in Tables 2 and 3, respectively.

In the cases C_{1a} and C_{1b}, the system being of relatively small sizes, the MSGO and SGO algorithms find approximately the same optimal operating solution, and as a result, the derived quantities (Cost, Emission, P_L) will be approximately the same. However, mathematically, MSGO performs slightly better than SGO. Thus, in the case of C_{1a} the best costs obtained using MSGO and SGO are $\text{Cost}_{C1a(MSGO)}$ = 111,497.6301 \$/h and $Cost_{C1a(SGO)}$ = 111,497.6302 \$/h, respectively. For the C_{1b} case, the best emissions found by MSGO and SGO are $\text{Emission}_{C1b(MSGO)}$ = 3932.2432 lb/h and $\text{Emission}_{C1b(SGO)}$ = 3932.2433 lb/h, respectively. Also, in the case of C_{2b} (Table 3), the best solutions (P_1–P_{40}, Cost, Emission) found by MSGO and SGO are very close (both algorithms have the ability to identify quality solutions): $\text{Emission}_{C2b(MSGO)}$ = 176,682.26363 t/h and $\text{Emission}_{C2b(SGO)}$ = 176,682.26364 t/h, respectively.

Table 2. Generation schedule obtained using SGO and MSGO for cases C_{1a}, C_{1b}, and C_{1c} (P_D = 2000 MW).

Outputs Algorithms	Best Cost (C_{1a})		Best Emission (C_{1b})		BCS (C_{1c})	
	SGO	MSGO	SGO	MSGO	SGO	MSGO
PT_1 (MW)	55	55	55	55	55	55
PT_2 (MW)	80	80	80	80	79.9919138	80
PT_3 (MW)	106.9541594	106.9339816	81.1395049	81.13389758	84.9307435	86.1178873
PT_4 (MW)	100.5724243	100.5785797	81.3521442	81.365596459	84.4692580	84.4754324
PT_5 (MW)	81.4829314	81.5030072	160	160	132.5861458	132.0947090
PT_6 (MW)	83.0293259	83.0232077	240	240	148.8844943	150.1405606
PT_7 (MW)	299.9999999	300	294.5013582	294.50918872	298.3856220	300
PT_8 (MW)	340	340	297.2767111	297.26088151	317.3109360	318.6699829
PT_9 (MW)	470	470	396.7327820	396.76630122	436.6060140	437.5797908
PT_{10} (MW)	470	470	395.5926758	395.55896593	446.4576771	440.4116407
Cost C^T (PT) ($/h)	111,497.6302221	111,497.6301430	116,412.4961603	116,412.5597722	112,884.4848042	112,913.8454551
Emission E^T (PT) (lb/h)	4572.2503628	4572.1759096	3932.2432876	3932.2432406	4177.6632977	4173.0883561
P_L (MW)	87.0388508	87.03878611	81.59518555	81.594840685	84.62280525	84.49001346
P_D (MW)	2000	2000	2000	2000	2000	2000
ΔP * (MW)	-9.9×10^{-6}	-9.8×10^{-6}	-9.3×10^{-6}	-9.2×10^{-6}	-8.3×10^{-7}	-9.8×10^{-6}
Time (s)	0.17	0.18	0.17	0.18	0.18	0.19

* ΔP represents the accuracy with which the power balance is satisfied: $\Delta P = \Sigma PT_i - P_L - P_D$.

Table 3. Generation schedule obtained using SGO and MSGO for cases C_{2a}, C_{2b}, and C_{2c} (P_D = 10,500 MW).

Outputs Algorithms	Best Cost (C_{2a})		Best Emission (C_{2b})		BCS (C_{2c})	
	SGO	MSGO	SGO	MSGO	SGO	MSGO
PT_1 (MW)	110.922179	111.716634	114.000000	114.000000	113.999966	110.800341
PT_2 (MW)	111.243322	110.860214	114.000000	114.000000	113.999796	110.800220
PT_3 (MW)	119.999923	97.403591	120.000000	120.000000	120.000000	119.999963
PT_4 (MW)	179.733198	179.743937	169.368013	169.367866	179.733101	179.733075
PT_5 (MW)	87.968669	87.780202	97.000000	97.000000	97.000000	87.801870
PT_6 (MW)	139.999997	139.999967	124.257317	124.257125	140.000000	139.999941
PT_7 (MW)	299.999977	259.611600	299.711092	299.711096	299.999999	299.999869
PT_8 (MW)	284.599970	284.603778	297.914825	297.914414	284.599650	284.599715
PT_9 (MW)	284.600310	284.603707	297.259861	297.260410	284.599652	284.599803
PT_{10} (MW)	130.000055	130.000000	130.000000	130.000000	204.799826	130.000011
PT_{11} (MW)	168.799938	94.001033	298.409765	298.409555	243.599650	318.397097
PT_{12} (MW)	94.000100	168.802785	298.026091	298.026393	318.399211	318.396537
PT_{13} (MW)	125.000068	214.762696	433.557450	433.557827	394.279369	394.279333
PT_{14} (MW)	304.519644	394.279917	421.727984	421.729503	394.279373	394.279355
PT_{15} (MW)	394.279363	304.520136	422.779051	422.780633	394.279370	394.279352
PT_{16} (MW)	394.279396	394.281435	422.779145	422.779765	394.279371	394.279339
PT_{17} (MW)	489.279362	489.279715	439.413302	439.412095	489.279356	489.279199
PT_{18} (MW)	489.279374	489.277936	439.402981	439.402147	489.279365	489.278987
PT_{19} (MW)	511.279452	511.286046	439.413322	439.413830	472.436489	421.519583
PT_{20} (MW)	511.279469	511.288520	439.413375	439.412524	421.519581	421.519572
PT_{21} (MW)	523.279375	523.282365	439.446230	439.446547	433.519581	433.519577
PT_{22} (MW)	523.279470	523.288232	439.447404	439.446608	433.519581	433.519606
PT_{23} (MW)	523.279716	523.286753	439.771529	439.772178	433.519581	433.519604
PT_{24} (MW)	523.279630	523.298896	439.771899	439.771421	433.519585	433.519606
PT_{25} (MW)	523.279479	523.282145	440.111752	440.112274	433.519584	433.519649
PT_{26} (MW)	523.279413	523.285836	440.111656	440.111780	433.519583	433.519592
PT_{27} (MW)	10.000142	10.001290	28.994136	28.993709	10.000009	10.000005
PT_{28} (MW)	10.000000	10.000158	28.994289	28.993427	10.000000	10.000098
PT_{29} (MW)	10.000004	10.000029	28.993716	28.994114	10.000000	10.000015
PT_{30} (MW)	87.979665	96.047579	97.000000	97.000000	97.000000	96.999848
PT_{31} (MW)	189.999986	189.999914	172.332024	172.331705	190.000000	189.999825
PT_{32} (MW)	190.000000	190.000000	172.332025	172.331635	189.999867	189.999975
PT_{33} (MW)	190.000000	189.999740	172.331486	172.331960	189.999915	189.999989
PT_{34} (MW)	199.999981	164.840604	200.000000	200.000000	200.000000	199.999986
PT_{35} (MW)	199.999969	199.999540	200.000000	200.000000	200.000000	199.999994
PT_{36} (MW)	199.999996	199.999900	200.000000	200.000000	200.000000	199.999976
PT_{37} (MW)	110.000000	109.999989	100.839198	100.838331	109.999999	109.999997
PT_{38} (MW)	109.999997	109.999981	100.838254	100.838389	110.000000	109.999850
PT_{39} (MW)	110.000000	109.999727	100.838245	100.838200	110.000000	109.999659
PT_{40} (MW)	511.279414	511.283472	439.412572	439.412526	421.519580	488.039989
Cost C^T *(PT)* (\$/h)	121,509.82092	121,426.70390	129,995.28508	129,995.30108	125,526.34268	125,434.46554
Emission E^T *(PT)* (t/h)	359,251.81848	356,231.07314	176,682.26364	176,682.26363	201,944.58419	200,613.67971
ΔP * (MW)	-1.7×10^{-11}	1.5×10^{-11}	-9.9×10^{-6}	-9.9×10^{-6}	-8.7×10^{-6}	5.1×10^{-7}
Time (s)	7.64	8.03	7.34	7.65	9.28	9.97

* ΔP represents the accuracy with which the power balance is satisfied: $\Delta P = \Sigma PT_i - P_L - P_D$.

On the contrary, in the case of C_{2a} (where the functions that model the cost have a higher complexity and determine a larger number of local minima), the MSGO algorithm can identify a much better solution than SGO ($Cost_{C2a(MSGO)}$ = 121,426.7039 \$/h < $Cost_{C2a(SGO)}$ = 121,509.8209 \$/h).

Table 4 shows the best solutions in terms of cost and emissions obtained using the MSGO algorithm for the cases C_{3a}, C_{3b} (without wind), C_{4a}, and C_{4b} (with wind).

Table 4. Generation schedule obtained using MSGO for cases C_{3a}, C_{3b}, and C_{3c} (with loss of power and without wind) and C_{4a}, C_{4b}, C_{4c} (with loss of power and wind).

Outputs Cases	Best Cost		Best Emission		BCS	
	Case C_{3a} (Without Wind)	Case C_{4a} (With Wind)	Case C_{3b} (Without Wind)	Case C_{4b} (With Wind)	Case C_{3c} (Without Wind)	Case C_{4c} (With Wind)
PT_1/PW_1 (MW)	114.000000	549.9999977	114.000000	550.000000	114.0000000	549.9997828
PT_2/PW_2 (MW)	113.999999	549.9999998	114.000000	550.000000	114.0000000	549.9997818
PT_3 (MW)	120.000000	97.39994714	120.000000	120.000000	120.0000000	119.9988116
PT_4 (MW)	189.999995	179.7331053	190.000000	175.313561	182.8929206	179.7374211
PT_5 (MW)	96.999999	87.79995033	97.000000	97.000000	96.99999982	96.99883835
PT_6 (MW)	140.000000	68.00000364	132.951212	119.769488	139.9999995	105.4022758
PT_7 (MW)	300.000000	259.5996859	300.000000	300.000000	300.0000000	299.9999742
PT_8 (MW)	300.000000	284.5996597	300.000000	299.999999	299.9999998	285.7135868
PT_9 (MW)	299.999998	284.5997061	300.000000	300.000000	299.9999995	287.9519754
PT_{10} (MW)	279.599683	204.79983	270.805053	161.263885	279.5996268	204.8086989
PT_{11} (MW)	168.799860	94.00001916	322.733043	297.366373	318.3994557	243.6012566
PT_{12} (MW)	94.000003	94.00000226	315.129176	289.219449	318.3993251	243.6003501
PT_{13} (MW)	484.039161	304.519587	480.860016	441.638969	484.0391469	394.2851366
PT_{14} (MW)	484.039166	304.5195816	475.905463	431.552565	484.0391695	394.2799802
PT_{15} (MW)	484.039164	394.2793758	476.271511	433.133288	484.0391541	394.2841752
PT_{16} (MW)	484.039178	484.039163	480.455232	439.072444	484.0391647	484.0292251
PT_{17} (MW)	489.279372	489.2793773	471.830727	438.041844	489.2793744	489.2692227
PT_{18} (MW)	489.279372	489.2793735	461.973780	427.916684	399.5196438	399.5209877
PT_{19} (MW)	511.279600	511.2793699	483.440794	446.749851	510.840761	506.0067276
PT_{20} (MW)	511.279490	511.2793799	483.258780	446.772552	510.666052	421.5364187
PT_{21} (MW)	526.732209	523.2793949	483.252001	447.336714	511.7221827	433.5530376
PT_{22} (MW)	550.000000	523.2793745	486.947376	452.005128	516.1579618	514.2353651
PT_{23} (MW)	523.279384	523.2793957	472.040407	438.402949	433.5204637	433.5329212
PT_{24} (MW)	523.279383	523.2793722	462.227523	428.365213	433.5195802	433.5268561
PT_{25} (MW)	524.239856	523.2793935	483.754300	447.367210	511.6249809	433.5342317
PT_{26} (MW)	523.815577	523.2793731	483.575006	447.380310	511.4580636	433.7489257
PT_{27} (MW)	10.000020	10.00000931	67.543932	33.650576	15.28948883	10.05540017
PT_{28} (MW)	10.000113	10.00000064	72.653376	36.778266	17.79565013	10.24418652
PT_{29} (MW)	10.000007	10.00000189	54.008031	28.253062	10.36768925	10.0022597
PT_{30} (MW)	87.800155	87.79991386	97.000000	97.000000	96.99999662	89.80426078
PT_{31} (MW)	190.000000	190.0000000	190.000000	175.455547	190.0000000	189.998884
PT_{32} (MW)	190.000000	189.9999989	190.000000	175.467999	190.0000000	189.9955403
PT_{33} (MW)	190.000000	189.9999997	190.000000	175.691210	190.0000000	189.9978565
PT_{34} (MW)	200.000000	195.3472481	200.000000	200.000000	200.0000000	199.9994215
PT_{35} (MW)	199.999999	164.7998866	200.000000	200.000000	200.0000000	199.9994151
PT_{36} (MW)	164.799870	164.7998371	200.000000	200.000000	199.9999989	199.998299
PT_{37} (MW)	110.000000	109.9999997	110.000000	103.209409	110.0000000	110.0000000
PT_{38} (MW)	109.999999	109.9999992	110.000000	103.216803	110.0000000	109.9997758
PT_{39} (MW)	110.000000	109.9999985	110.000000	103.385396	110.0000000	109.9962344
PT_{40} (MW)	550.000000	511.2794082	486.926373	451.971343	511.2792996	509.5675741

Table 4. *Cont.*

Outputs Cases	Best Cost		Best Emission		BCS	
	Case C_{3a} (Without Wind)	Case C_{4a} (With Wind)	Case C_{3b} (Without Wind)	Case C_{4b} (With Wind)	Case C_{3c} (Without Wind)	Case C_{4c} (With Wind)
Cost $C(PT,PW)$ ($/h)	136,454.33693	123,161.88666	147,526.93169	133,213.62916	139,049.0921	126,645.13420
Emission $E(PT,PW)$ (t/h)	501,366.97888	375,873.82903	347,578.49057	193,311.54070	388,020.2454	239,155.73184
P_L (MW)	958.6206217	936.7097306	1040.543121	1009.748098	1000.489155	962.815076
ΔP (MW)	-9.7×10^{-6}	-9.8×10^{-6}	-9.9×10^{-6}	-9.7×10^{-6}	-6.1×10^{-6}	-4.1×10^{-6}
Time (s)	28.61	30.09	16.34	17.37	30.07	32.96

ΔP represents the accuracy with which the power balance is satisfied: $\Delta P = (\Sigma PT_i + \Sigma PT_i) - P_L - P_D$.

The solutions for C_{4a} and C_{4b} cases indicate that the two wind units (PW_1, PW_2) are scheduled to operate at full capacity (550 MW) in both the EcD and EmD problems. Thus, according to the statement in case C_4, the wind units will replace two thermal units PT_1, PT_2 (that operate at maximum capacity $PT_1 \approx PT_2 \approx 114$ MW in C_{3a} and C_{3b} cases), and will also reduce the operating power level of other thermal units (PT_3–PT_{40}). Based on the EcD and EmD models, it can be seen that if the wind units operate at full capacity it results in an E^u

power very close to zero ($E_1^u(PW_1 = 550) \approx E_2^u(PW_2) \approx 0\ MW$), so their corresponding costs and emissions will be close to zero. In contrast, the average powers associated with overestimation will tend towards maximum values $\left(E_1^o(PW_1 = 550) \approx E_2^o(PW_2) \approx 230.7635\ MW\right)$, while their corresponding costs and emissions are 2 × 1153.817 \$/h, and 2 × 1153.817 t/h. The risks taken by the system operator when planning the wind units to operate at values close to the maximum capacity are covered by the reduction of costs and emissions from the thermal units (due to the reduction of the PT_3–PT_{40} powers in $C_{4a}(C_{4b})$ cases compared to $C_{3a}(C_{3b})$).

Comparing the solutions for the cases with wind and those without wind, it can be seen that the inclusion of the wind units (PW_1 and PW_2) results in a substantial reduction in cost (from $\text{Cost}_{\text{C3a(MSGO)}}$ = 136,454.33693 \$/h to $\text{Cost}_{\text{C4a(MSGO)}}$ = 123,161.88666 \$/h, representing 9.74%), and emissions (from $\text{Emission}_{\text{C3b(MSGO)}}$ = 347,578.49052 t/h to $\text{Emission}_{\text{C4b(MSGO)}}$ = 193,311.54070 t/h representing 44.38%). Also, power losses are reduced from 958.62 MW in the case of C_{3a} to 936.70 MW in the case of C_{4a} (2.28%) and from 1040.54 MW in the case of C_{3b} to 1009.74 MW in the case of C_{4b} (2.96%), respectively.

Comparison of MSGO algorithm with other algorithms: In Tables 5–12 the values of the statistical items (B, A, W, and SD) are presented. They are obtained using different algorithms for EcD problem (cases C_{1a}–C_{4a}) and EmD problem (C_{1b}–C_{4b}), respectively.

Analyzing the values from the mentioned tables, it can be seen that in all studied cases the MSGO algorithm obtains statistical indicators as good or better than the competing algorithms mentioned in these tables. The exception is the Worst cost item for the algorithms Jaya-SML [46] (case C_{2a}—Table 6), ORCCRO [47] (case C_{3a}—Table 7), DE (case C_{4a}—Table 8), and the SD item for the algorithms CSS [48] (cases C_{3a}—Table 7), DE, and SCA (cases C_{4a}—Table 8).

The MSGO algorithm shows a better performance than well-known algorithms, such as DE [49], RCCRO [35] (in cases C_{1a}—Table 5 and C_{1b}—Table 9), GA [50], DE [42], ABC [51], PSO [52], TLBO [53], SCA [54], CSO [55] (cases C_{2a}—Table 6), DE [26], TLBO [26], BSA [56] (cases C_{2b}—Table 10), ACS [57], BBO [47], ORCCRO [47], CSS [48], IMO [58] (cases C_{3a}—Table 7), PSO, DE, SCA (cases C_{3b}—Table 11, C_{4a}—Table 8, C_{4b}—Table 12). Also, the performance of MSGO is superior to other metaheuristic algorithms obtained using various procedures (such as modification of solution update relations, inclusion of chaos and/or opposite solutions, hybridizations, etc.): GQPSO [59], CSCA [27], QOPO [19] (in cases C_{1a}—Table 5), MIMO [58], HSCA [60], CTLBO [53], TLABC [61] (cases C_{2a}—Table 6), GAAPI [44], SDE [62], HPSO-DE [63], MIMO [58] (cases C_{3a}—Table 7).

Table 5. Values for the statistical items obtained using different algorithms for the cases C_{1a}—*Best Cost* (10 units with losses, 2000 MW).

Algorithm	Best Cost (\$/h)	Average Cost (\$/h)	Worst Cost (\$/h)	SD (\$/h)	Cost Saving * (\$/h)
DE [49]	111,500	-	-	-	-
TLBO [26]	111,500	-	-	-	-
QOTLBO [26]	111,498	-	-	-	-
QPSO [59]	119,005.3030	121,621.7556	122,144.8454	372	10,124.12
GQPSO [59]	112,429.7444	113,102.4627	113,327.0680	256	1604.83
RCCRO [35]	111,497.6319	-	-	-	-
BSA [56]	111,497.6308	-	-	-	-
CSCA [27]	111,497.6307	-	-	-	-
QOPO [19]	111,892.4096	-	-	-	-
SGO	111,497.6302	111,497.7362	111,502.7703	7.27×10^{-1}	0.10
MSGO	111,497.6301	111,497.6302	111,497.6304	6.25×10^{-5}	0

* Indicates the cost saving through MSGO compared to other algorithms using the *Average Cost* item.

Table 6. Values of the statistical items obtained using different algorithms for the cases C_{2a}—*Best Cost* (40 units without losses, 10,500 MW).

Algorithm	Best Cost ($/h)	Average Cost ($/h)	Worst Cost ($/h)	SD ($/h)	Cost Saving ($/h)
MIMO [58]	122,758.7	124,621.8	126,059.2	866.20	2964.84
FSS-IPSO2 [64]	122,535.56	125,025.86	127,401.23	1134.43	3368.90
GA [50]	121,996.4	122,919.77	123,807.97	320.31	1262.81
HSCA [60]	121,983.5	-	-	-	-
NGWO [65]	121,881.81	122,787.77	-	-	1130.81
DE [42]	121,840	-	-	-	-
PSO [52]	122,588.5093	123,544.88	124,733.67	-	1887.92
TLBO [53]	124,517.27	126,581.56	128,207.06	1060	4924.60
CTLBO [53]	121,553.83	121,790.23	122,116.18	150	133.27
SMA [66]	121,658.6656	-	-	-	-
L-SHADE [67]	121,543.43	122,105.39	122,983.68	-	448.43
S-Jaya [68]	121,517.6513	121,948.42	122,283.83	193.57	291.46
SCA [54]	121,506.58	121,857.90	122,056.15	347.26	200.94
ABC [51]	121,479.6467	121,984.24	122,137.42	–	327.28
Jaya-SML [46]	121,476.3977	121,689.07	122,039.87	147.89	32.11
TLABC [61]	121,468.3847	121,739.4406	122,192.3263	160.88	82.48
CSO [55]	121,465.99	121,988.48	122,781.75	275.92	331.52
IJaya [68]	121,454.3785	121,770.32	122,109.01	173.70	113.36
ESCSDO10 [69]	121,626.97	122,351.7	123,128.9	412.2976	694.74
SDO [69]	121,750.2	122,460.1	123,222.7	405.019	803.14
SGO	121,509.8209	122,025.1179	123,527.6187	380	368.16
MSGO	121,426.7039	121,656.9571	122,048.2807	143	0

Table 7. Values of the statistical items obtained using different algorithms for the cases C_{3a}—*Best Cost* (40 units with losses and without wind, 10,500 MW).

Algorithm	Best Cost ($/h)	Average Cost ($/h)	Worst Cost ($/h)	SD ($/h)	Cost Saving ($/h)
GAAPI [44]	139,864.96	-	-	-	-
SDE [62]	138,157.46	-	-		-
ACS [57]	137,413.73	-	-	-	-
BBO [47]	137,026.82	137,116.58	137,587.8200	-	382.64
ORCCRO [47]	136,855.1900	136,855.1900	136,855.1900	-	121.25
HPSO-DE [63]	136,835.0021	-	-	-	-
CSS [48]	136,679.0228	136,993.6115	137,447.4131	171.26	259.67
MIMO [58]	137,034.2000	138,472.9000	140,124.3000	752.74	1738.96
IMO [58]	138,789.6000	140,486.3000	142,106.7000	765.71	3752.36
SGO	136,510.7626	137,150.7361	138,082.7271	415	416.80
MSGO	136,454.6072	136,733.9409	137,185.4625	174	0

Table 8. Values of the statistical items obtained using different algorithms for the cases C_{4a}—*Best Cost* (40 units with losses and with wind, 10,500 MW).

Algorithm	Best Cost ($/h)	Average Cost ($/h)	Worst Cost ($/h)	SD ($/h)	Cost Saving ($/h)
PSO	123,607.9479	124,438.1644	125,509.7725	463.89	877.02
DE	123,804.0394	123,962.8486	124,160.1466	78.42	401.70
SCA	125,895.2706	126,196.7393	126,479.6091	132.44	2635.60
SGO	123,289.7874	124,086.1871	125,789.8060	532.90	525.04
MSGO	123,161.8867	123,561.1438	124,239.9876	212.44	0

Table 9. Values of the statistical items obtained using different algorithms for the cases C_{1b}—*Best Emission* (10 units with losses, 2000 MW).

Algorithm	Best Emission (lb/h)	Average Emission (lb/h)	Worst Emission (lb/h)	SD (lb/h)	Emission Reduction * (lb/h)
DE [49]	3923.40 **	-	-	-	-
QPSO [59]	4032.3875	4041.9171	4058.3615	8.06	109.6
GQPSO [59]	4011.9244	4032.9320	4042.1878	7.55	100.6
RCCRO [35]	3932.243269	-	-	-	-
BSA [56]	3932.243269	-	-	-	-
NSGA-III [25]	3932.5	-	-	-	-
SGO	3932.243252	3932.2484	3932.2821	9.05×10^{-3}	≈0
MSGO	3932.243240	3932.2432	3932.2433	2.16×10^{-5}	0

* Indicates the emisssion reduction through MSGO compared to other algorithms using the *Average Cost* item; ** in the case of the DE algorithm, the correct value is 3932.41728 lb/h.

Table 10. Values of the statistical items obtained using different algorithms for the cases C_{2b}—*Best Emission* (40 units without losses, 10,500 MW).

Algorithm	Best Emission (t/h)	Average Emission (t/h)	Worst Emission (t/h)	SD (t/h)	Emission Reduction (t/h)
MBFA [43]	176,682.269	-	-	-	-
DE [26]	176,683.3	-	-	-	-
TLBO [26]	176,683.5	-	-	-	-
QOTLBO [26]	176,682.5	-	-	-	-
BSA [56]	176,682.2646	-	-	-	-
SGO	176,682.26364	176,682.26380	176,682.2646	2.04×10^{-4}	≈0
MSGO	176,682.26363	176,682.26376	176,682.2641	9.80×10^{-5}	0

Table 11. Values of the statistical items obtained using different algorithms for the cases C_{3b}—*Best Issue* (40 units with losses and without wind, 10,500 MW).

Algorithm	Best Emission (t/h)	Average Emission (t/h)	Worst Emission (t/h)	SD (t/h)	Emission Reduction (t/h)
PSO	347,578.65776	347,581.8140	347,594.6001	3.39	3.3
DE	347,877.89873	348,120.8380	348,591.0764	135	542.3
SCA	364,849.12062	372,931.5027	380,287.1626	3880	25,353.0
SGO	347,578.49061	347,578.4912	347,578.4922	3.57×10^{-4}	≈0
MSGO	347,578.49057	347,578.4910	347,578.4919	2.88×10^{-4}	0

Table 12. Values of the statistical items obtained using different algorithms for the cases C_{4b}—*Best Issue* (40 units with losses and with wind, 10,500 MW).

Algorithm	Best Emission (t/h)	Average Emission (t/h)	Worst Emission (t/h)	SD (t/h)	Emission Reduction (t/h)
PSO	193,313.7047	193,331.5342	193,373.1017	13.59	19.9
DE	193,953.9668	194,532.0465	195,040.0923	246.50	1220.5
SCA	210,484.9674	218,736.9094	227,559.3490	3971.80	25,425.3
SGO	193,311.54075	193,311.5414	193,311.5437	4.88×10^{-4}	≈0
MSGO	193,311.54071	193,311.5410	193,311.5415	2.02×10^{-4}	0

The stability of the MSGO algorithm is very good (SD item value being below 2.1×10^{-4}) for all cases on the EmD problem (C_{1b}–C_{4b}, Tables 9–12) and relatively good for the EcD problem (C_{1a}–C_{4a}, Tables 5–8).

We specify that the specific parameters of the PSO, DE, and SCA algorithms were set by performing several experiments in which the number of evaluations is considered to be the same as for the SGO and MSGO algorithms (shown in Table 1). Thus, the settings performed for these algorithms applied in the cases C_{3b}, C_{4a}, and C_{4b} are the following:

- For the PSO algorithm ($N = 50$, $t_{max} = 400$, $c_1 = c_2 = 2$, $w_{min} = 0.3$, $w_{max} = 0.9$ in cases C_{3b} and C_{4b}, respectively, $N = 50$, $t_{max} = 700$, $c_1 = c_2 = 2$, $w_{min} = 0.3$, $w_{max} = 0.9$ in the case of C_{4a}; where c_1 and c_2 are acceleration coefficients, w_{min} and w_{max} are the initial and final inertial weights).

- For the DE algorithm (N = 50, t_{max} = 400, CR = 0.2, F = 0.4 in cases C_{3b} and C_{4b}, respectively, N = 50, t_{max} = 700, CR = 0.1, F = 0.8 in the case of C_{4a}; where, CR is the crossover rate, and F is the scaling factor).
- For the SCA algorithm (N = 50, t_{max} = 400, a = 1 in cases C_{3b} and C_{4b}, respectively, N = 50, t_{max} = 700, a = 1 in the case of C_{4a}).

The comparison of the MSGO and SGO algorithms: Tables 5–12 present the values of the statistical items (B, A, W, SD) obtained using the SGO and MSGO algorithms for all the study cases (C_{1a}–C_{4a} and C_{1b}–C_{4b}, respectively). In all situations, the statistical items obtained using MSGO are higher than those obtained using SGO (except for the cases mentioned in the previous paragraph), but there are also cases where the differences between the SGO and MSGO algorithms are small (in particular, the cases C_{1b}–C_{4b} associated with EmD problem, to which case C_{1a} can be added). For this reason, the SGO and MSGO algorithms are compared using the non-parametric Wilcoxon statistical test, considering a significance-level of 1%. Also, 50 values/algorithm were simulated to compare SGO and MSGO, resulting in 50 pairs of values that are compared using the Wilcoxon test.

Table 13 shows the statistical item p-value after applying the Wilcoxon test for the comparison of SGO and MSGO in all cases studied. For the EcD (C_{1a}–C_{4a}) and EmD (C_{1b}–C_{4b}) problems, the p-value is less than 0.01 (except for cases C_{2b} and C_{3b}), which indicates that MSGO is statistically significantly better than the SGO algorithm (getting six wins out of a possible eight).

Table 13. Comparison of SGO and MSGO by the Wilcoxon test.

Cases	R^-	R^+	p-Values	Winner
C_{1a}	3.00	26.44	0.000	MSGO
C_{1b}	1.00	26.00	0.000	MSGO
C_{2a}	12.50	27.27	0.000	MSGO
C_{2b}	22.11	29.82	0.858	-
C_{3a}	12.67	28.32	0.000	MSGO
C_{3b}	21.95	27.87	0.055	-
C_{4a}	11.67	28.54	0.000	MSGO
C_{4b}	13.54	29.70	0.000	MSGO

R^- and R^+ represent the mean of the negative and positive ranks, respectively.

In order to investigate the ability of MSGO to identify a better solution compared to SGO, we will evaluate the cost savings, as well as the emission reduction for cases C_{1a}–C_{4a} and C_{1b}–C_{4b}. Thus, for C_{1b}–C_{4b} (Tables 9–12) the MSGO and SGO algorithms perform very well at solving the EmD problem. As a result, the emission reduction achieved by MSGO compared to SGO is insignificant for this type of problem. The situation is similar in the EcD problem, case C_{1a} (Table 5), when both algorithms perform very well on the small 10-unit system. However, the cost savings and emission reduction may be important when comparing MSGO with other competing algorithms. Thus, for C_{1a} case (Table 5), analyzing the Average cost item, the average cost reduction is 10,124.12 $/h compared to QPSO [59]. Also, in the EmD problem (Tables 9–12), based on the Average emission item, the average hourly emission reduction is 109.6 lb/h (compared to QPSO [59], in case C_{1b}), 542.3 t/h (relative to DE, in case C_{3b}), and 1220.5 t/h (relative to DE, in C_{4b}). Following the results in Tables 6–8, the cost savings achieved by MSGO compared to SGO is significant: 368.16 $/h (for case C_{2a}), 416.80 $/h (case C_{3a}), and 525.04 $/h (case C_{4a}). These cost reductions are maintained or increased when comparing MSGO with other algorithms presented in Tables 6–8.

Convergence process: Figures 4 and 5 show the convergence characteristics in terms of cost and emissions obtained using the MSGO algorithm for all studied cases C_{1a}–C_{4a} and C_{1b}–C_{4b}. In all analyzed situations, convergence is fast in the first iterations (approximately 15% of the maximum number of iterations), and in the last iterations the convergence is slow, the iterative process being stabilized.

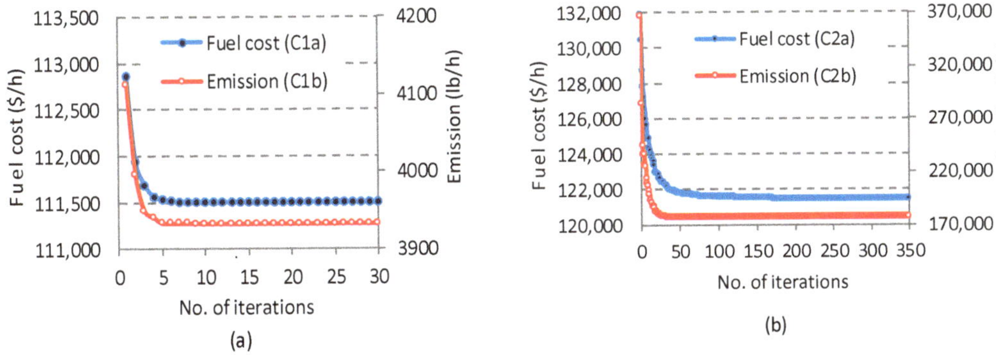

Figure 4. Cost and emission convergence characteristics obtained using MSGO: (**a**) for cases C_{1a}–C_{1b}; (**b**) for cases C_{2a}–C_{2b}.

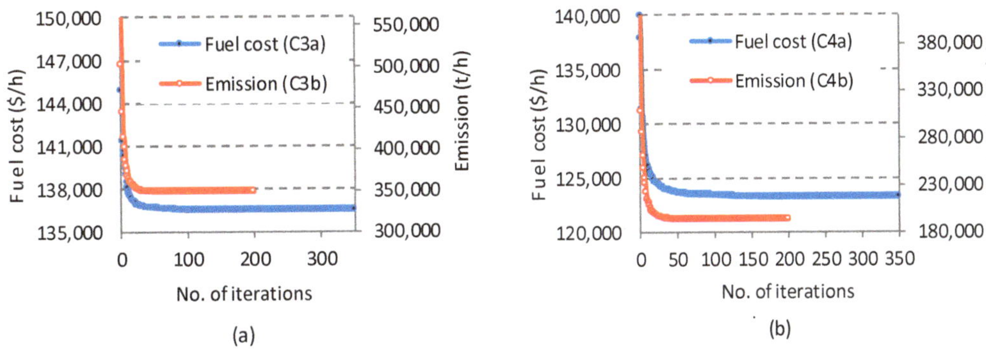

Figure 5. Cost and emission convergence characteristics obtained using MSGO: (**a**) for cases C_{3a}–C_{3b}; (**b**) for cases C_{4a}–C_{4b}.

In the case of emission convergence characteristics (C_{1b}–C_{4b}), the iterative process stabilizes much faster than in the case of cost characteristics (C_{1a}–C_{4a}). Thus, in the case of the EmD problem (C_{1b}–C_{4b}), the quasi-stable process starts after about 20–30% of the maximum number of iterations, and in the EcD problem after approximately 40–50% of the maximum number of iterations. We mention that the stable values towards which the convergence characteristics of the MSGO algorithm tend are mentioned for each case in Tables 5–12.

The accuracy and computation time*:* The accuracy (ΔP) of the computations refers to the inequality and equality constraints of the optimization model presented in Section 2 and can be seen for the best solutions presented in Tables 2–4. It is observed that the constraints are respected, the size of ΔP being less than 10^{-5} MW for all cases studied. Also, for each case (C_{1a}–C_{4a} and C_{1b}–C_{4b}), the average execution time (Time) obtained using the MSGO algorithm is indicated in Tables 2–4. In cases C_{1a} and C_{1b}, the execution time is very good (under 0.2 s); for cases C_{2a} and C_{2b} it is higher (under 10 s) due to the larger size of the analyzed system (40-unit). In the cases C_{3a}–C_{4a} and C_{3b}–C_{4b}, due to the large size of the analyzed systems (40-unit) and the need to calculate power losses, the average execution time goes up to 33 s.

5.3. Results for EED Problem (Cases C_{1c}–C_{4c})

The EED problem considers the cases C_{1c}–C_{4c}, in which the single-objective function ϕ defined by relation (20) is minimized. In these cases, it is of interest to determine the Pareto front solutions, as well as the best compromise solution (BCS) obtained using the SGO and MSGO algorithms.

To estimate the Pareto front, the weighting factor ω in relation (20) is varied between 0 and 1 with a step of 0.1 (in the end 11 points are obtained in the Cost–Emission objective plan). However, in order to obtain a Pareto front as uniform as possible and to show graphically that the solutions obtained using MSGO dominate the solutions obtained using other algorithms, another 12 points have been added to the 11 points for the case C_{1c} (the added points correspond to the values ω = {0.25, 0.35, 0.41, 0.42, 0.43, 0.435, 0.436, 0.44, 0.441, 0.445, 0.47, 0.55}) and 3 another points in the case of C_{2c} (ω = {0.51, 0.52, 0.55}), respectively. In the case of C_{3c} and C_{4c}, the number of points considered is 11.

Figures 6–8 show the Pareto fronts obtained using the SGO and MSGO algorithms for the cases C_{1c}–C_{4c}. Also, the best compromise solutions obtained using SGO, MSGO, and other competing algorithms applied in the literature are indicated.

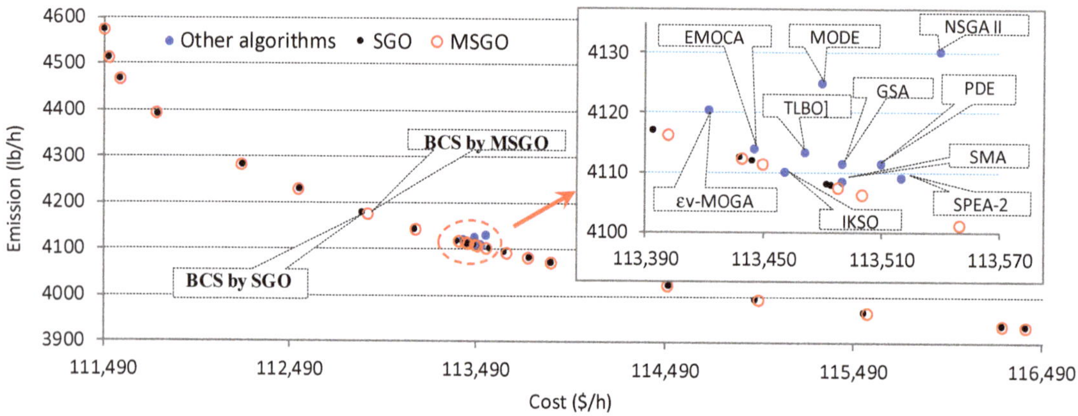

Figure 6. Pareto fronts and BCSs obtained using SGO, MSGO, and other algorithms for case C_{1c}.

Best compromise solution by SGO and MSGO: The BCSs obtained using the SGO and MSGO algorithms are identified using a fuzzy-based mechanism presented in Section 2.3. The results obtained through this mechanism indicate that BCS corresponds to a weighting factor equal to $\omega = 0.5$, both for SGO and MSGO.

As a result, in Table 14, we present for each case C_{1c}–C_{4c}, the values for Cost, Emission, and item μ_{max} (determined according to the methodology in Section 2.3) corresponding to the best compromise solution determined by SGO and MSGO.

Comparison of solutions from the Pareto front of MSGO with BCSs obtained using other algorithms: In the case of C_{1c}, in order to test the ability of MSGO to identify solutions from the Pareto front, it was compared with BCSs obtained using a group of algorithms (denoted G1) presented in the literature G1 = {EMOCA [24], MODE [42], NSGA II [42], TLBO [26], GSA [70], PDE [42], SMA [71], εv-MOGA [49], SPES-2 [42]}. To highlight the Pareto front solutions obtained using MSGO and BCSs obtained using competing algorithms in the G1 group, in Figure 6, a detail of the Cost–Emission plan is shown. In this detail of Figure 6, it can be visually observed that the Pareto front solutions obtained using MSGO (marked with red circles) dominate, in terms of Cost and Emission, all BCSs obtained using competing algorithms G1 (marked with blue circles). Thus, the points marked with red circles (obtained using MSGO) will remove from the Pareto front of MSGO all the points marked by blue circles (obtained using competing G1 algorithms). The exception is the IKSO algorithm [72], which obtains a non-dominated solution using the solutions in the Pareto front of the MSGO (a discussion on IKSO [72] will be made below).

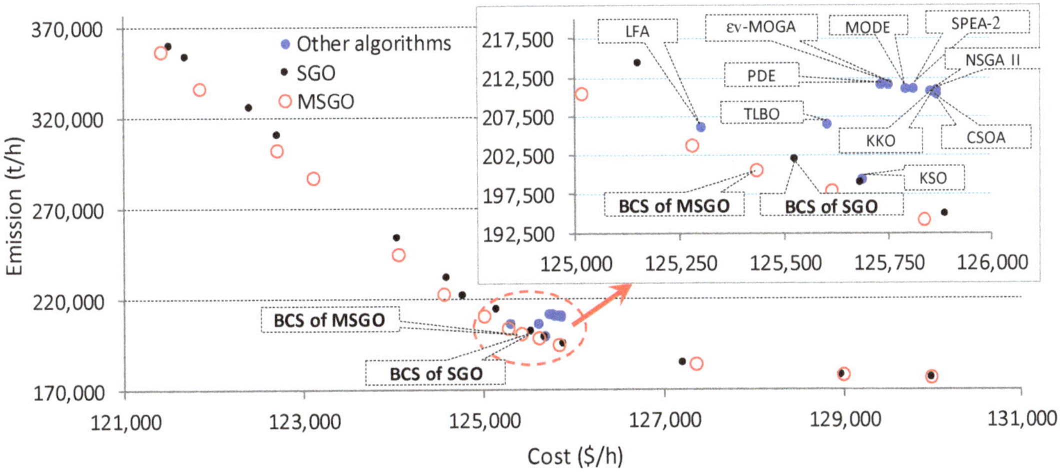

Figure 7. Pareto fronts and BCSs obtained using SGO and MSGO for case C_{2c}.

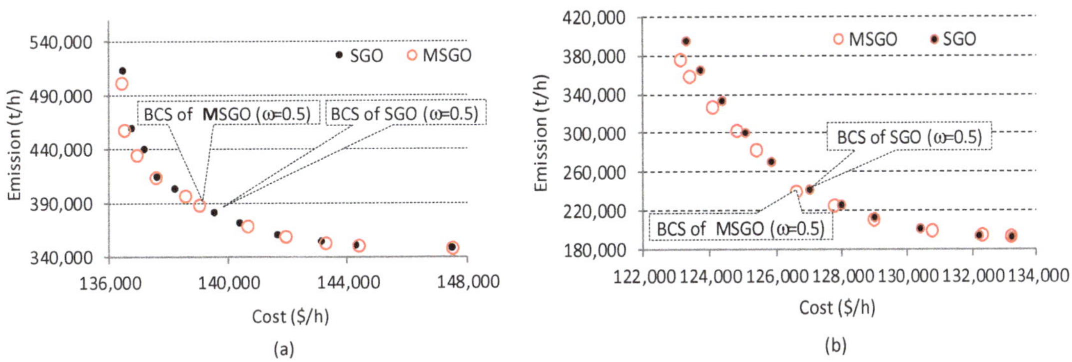

Figure 8. Pareto fronts and BCSs obtained using SGO and MSGO: (**a**) for case C_{3c}; (**b**) for case C_{4c}.

Table 14. The values corresponding to cost and emission obtained using SGO and MSGO for the cases C_{1c}–C_{4c}.

Items Cases	SGO Algorithm (ω = 0.5)			MSGO Algorithm (ω = 0.5)			Cost/Emission Reduced by MSGO **	
	Cost ($/h)	Emission (t/h)	μ_{max}	Cost ($/h)	Emission (t/h)	μ_{max}	Cost Saving ($/h)	Emission Reduction (t/h)
C_{1c}	112,884.4848	4177.6633	0.04903	112,913.8455	4173.0883	0.04908	−29.36	4.57
C_{2c}*	125,526.3426	201,944.5841	0.08106	125,434.4655	200,613.6797	0.08017	91.88	1330.90
C_{3c}	139,588.2247	380,601.7819	0.10265	139,049.0921	388,020.2454	0.10244	539.13	−7418.46
C_{4c}*	127,033.6502	242,252.893	0.10342	126,645.1342	239,155.7318	0.10661	388.52	3097.16

* MSGO's BCS dominates SGO's BCS in terms of *Cost* and *Emission*; ** MSGO Cost savings and emission reduction comparing to SGO, for each case study BSCs.

In case of C_{2c}, in Figure 7, a detail of the Cost–Emission plan is shown highlighting the Pareto front solutions obtained using MSGO, and BCSs obtained through several algorithms presented in the literature, denoted G2, G2 = {LFA [16], MODE [42], SPEA-2 [42], εv-MOGA [49], PDE [42], NSGA II [22], TLBO [26], CSOA [20], KKO [21], KSO [22]}. From the image it can be visually observed that the Pareto front solutions obtained using MSGO (marked with red circles) dominate, in terms of Cost and Emission, the BCSs obtained using the G2 group algorithms (marked with blue circles).

Comparison of BCSs obtained using MSGO and SGO: To compare BCSs obtained using the SGO and MSGO algorithms, we will include in the Pareto front of MSGO the best compromise solution identified for the SGO algorithm. Similarly, in the case of C_{1c}, the BCS identified using the IKSO algorithm [72] will be included in the Pareto front of MSGO. Note that BCSs obtained using the SGO algorithm (cases C_{1c}–C_{4c}) or IKSO [72] (in the case of C_{1c}) must be non-dominated solutions in the MSGO Pareto front. Thus, MSGO will have a Pareto front increased by one value (in cases C_{3c}, corresponding to the BCS of SGO), respectively by two values (in cases C_{1c}, corresponding to the BCSs of SGO and IKSO [72]). In cases C_{2c} and C_{4c}, the BCS of MSGO dominates the BCS of SGO in terms of Cost and Emission (see Table 14). The values from these extended Pareto fronts (with one or two other values) are subject to the methodology for establishing the BCS presented in Section 2.3, and the results of interest are given in Table 15. Table 15 shows the values of the item μ_{max} obtained using the SGO, MSGO, and IKSO algorithms (we specify that the IKSO algorithm [72] is used only in the case of C_{1c}). From Table 15, it can be seen that, for the cases C_{1c} and C_{3c}, the maximum value of the item μ corresponds to the MSGO algorithm (the weighting factor being for each case ω = 0.5). As a result, in all cases (C_{1c}–C_{4c}), the MSGO algorithm can identify a better compromise solution than SGO. Figures 5–8 show the Pareto fronts obtained using SGO and MSGO and indicate the BCSs identified using these algorithms.

Table 15. Values of item μ_{max} obtained using SGO, MSGO and IKSO for cases C_{1c}–C_{4c}.

Case	Case C_{1c}	Case C_{2c}	Case C_{3c}	Case C_{4c}
IKSO [72]	0.044284	–	–	–
SGO (ω = 0.5)	0.044676	MSGO's BCS dominates SGO's BCS, from Table 14 (MSGO winner)	0.092898	MSGO's BCS dominates SGO's BCS, from Table 14 (MSGO winner)
MSGO (ω = 0.5)	μ_{max} = **0.044715 (MSGO winner)**		μ_{max} = **0.092926 (MSGO winner)**	

– the IKSO algorithm [72] is not used in the C_{2c}–C_{4c} cases. The bold values correspond to the winning algorithm.

Comparison of SGO and MSGO Pareto fronts: Two metrics are used to evaluate and compare the quality of the Pareto fronts obtained using the SGO and MSGO algorithms: C-metric [73] and hyper-volume [74]. The C-metric, denoted C(S_1, S_2) and applied to the non-dominated solution sets S_1 and S_2, indicates the percentage of solutions from the S_2 set dominated by the solutions from S_1. For example, if C(S_1, S_2) = 100%, all solutions in S_2 are dominated by those in S_1. Currently, C(S_1, S_2) $\neq$ C(S_2, S_1) is required to calculate both metrics C(S_1, S_2) and C(S_2, S_1). The hyper-volume metric measures the volume (area in the case of two objectives) that is dominated by the Pareto front and is located below a predetermined reference point. Typically, higher values of the hyper-volume metric are associated with a better-quality Pareto front.

Table 16 shows, for each case, the values of the C-metric and hyper-volume metrics corresponding to the solution sets obtained using the SGO and MSGO algorithms. From Table 16 it can be seen that the C-metric (SGO, MSGO) = 0, which indicates that the solutions in the Pareto front obtained using SGO do not dominate any solution in the Pareto front obtained using the MSGO algorithm. Instead, the inverse C-metric (MSGO, SGO) indicates the percentages of SGO solutions dominated by MSGO solutions. Also, the hyper-volume metric shows higher values in the case of MSGO compared to SGO. Considering the values obtained using the two metrics, we estimate that MSGO can obtain a Pareto front of better quality than SGO.

Table 16. Values of the metrics for assessing the Pareto fronts of SGO and MSGO.

Item Cases	Hyper Volume		C-Metric (%)	
	SGO	MSGO	C(SGO, MSGO)	C(MSGO, SGO)
Case C_{1c}	0.9710	0.9712	0	4.55
Case C_{2c}	0.9775	0.9836	0	42.86
Case C_{3c}	1.0887	1.0942	0	27.27
Case C_{4c}	1.0007	1.0172	0	54.54

When analyzing Table 4 from the EED problem point of view—case C_{4c}—it shows that the BCS is obtained when the wind units operate at maximum capacity ($PW_1 \approx PW_2 \approx 550$ MW). When comparing the BCS-including wind (case C_{4c}) with the BCS-without wind (case C_{3c}), it results in lower costs (by 8.9%) and emissions (by 38.3%) for case C_{4c}. Table 14 shows Cost savings and Emission reduction by MSGO compared to SGO for BCSs obtained in cases C_{1c}–C_{4c}. Positive values indicate that MSGO achieves better cost or lower emissions than SGO, and negative values indicate the opposite situation.

Average execution time: Tables 2–4 show the values of the average execution time (Time) obtained using the MSGO algorithm for each case (C_{1c}–C_{4c}) of the EED problem. It can be seen that the values obtained for the EED problem are slightly higher than for the EcD or EmD problems.

6. Conclusions

In this article, a modified SGO is proposed in which several terms in the solution update relations are disturbed by including chaotic sequences generated by the Logistic map and sequences generated by the HDP operator. Also, switching between solution update relations (in the acquiring phase) is achieved by introducing a random condition instead of a condition based on the value of the fitness function of the competing solutions. MSGO has been successfully tested for solving EcD, EmD, and EED problems for medium and large power systems.

The effectiveness of MSGO was tested on four cases for each type of problem (EcD, EmD, and EED). For all case studies targeting EcD, and EmD problems, MSGO obtained solutions of better or equal quality than well-known algorithms (such as: DE, ABC, PSO, SCA, etc.) or improved varieties (except for the cases mentioned in Section 5.2). Also, the convergence process was fast and the stability of the MSGO algorithm evaluated by the SD item was very good in cases C_{1b}–C_{4b} and relatively good in the cases C_{1a}–C_{4a}. The statistical items (B, A, W and SD) and the results of the Wilcoxon test show that MSGO is more efficient than SGO in six cases (out of a possible eight) and equally good in two cases (C_{2b} and C_{3b} in Table 13), indicating that the changes made in MSGO had a positive effect. In the case of the EED problem, MSGO was able to extract a better compromise solution than SGO or other algorithms for all studied cases C_{1c}–C_{4c}. Moreover, the values of the hyper-volume and C-metric indicate that MSGO can obtain a Pareto front of superior quality compared to SGO. It should be mentioned that the inclusion of wind sources (in the case of C_{4a}/C_{4b} compared to C_{3a}/C_{3b}) brings benefits both in terms of cost reduction (about 10%) and polluting emissions (about 45%). The average computation time of MSGO and SGO algorithms is similar, and the efficiency of both algorithms is good in all case studies. By applying economic and/or emissions dispatching using MSGO, electricity producers can make informed decisions to operate a sustainable energy system.

Author Contributions: Conceptualization, D.C.S., C.H. and M.L.S.; software data curation, D.C.S.; investigation, D.C.S., C.H., M.L.S., G.B., C.B. and F.C.D.; writing—original draft preparation, D.C.S., M.L.S. and C.B.; writing—review and editing, C.H., G.B., C.B. and F.C.D. All authors have read and agreed to the published version of the manuscript.

Funding: This research was funded by the University of Oradea within the framework of the grant competition "Scientific Research of Excellence Related to Priority Areas with Capitalization through Technology Transfer: INO-TRANSFER-UO-2nd Edition" projects no. 239/2022 and no. 236/2022.

Institutional Review Board Statement: Not applicable.

Informed Consent Statement: Not applicable.

Data Availability Statement: Data are contained within the article.

Acknowledgments: The research undertaken was made possible by the equal scientific involvement of all the authors concerned.

Conflicts of Interest: The authors declare no conflicts of interest.

Nomenclature

a_i, b_i, c_i, e_i, and f_i	Fuel cost coefficients of thermal unit i;
B_{ij}, B_{0i}, B_{00}	Loss coefficients;
c	Self-introspection parameter from SGO;
c_j^d, c_j^o, c_j^u	Direct, reserve, and penalty cost coefficient for unit j;
$C(PT,PW)$	Total fuel cost;
e_j^o, e_j^u	Reserve, and penalty emission coefficient for unit j;
$E_j^o\left(PW_j\right)$, $E_j^u\left(PW_j\right)$	Mean powers associated with over and underestimation W_j for unit j;
$E(PT,PW)$	Total emission;
$f(X^{best})$	Objective function associated to X^{best};
$f_i = f(X_i), f(X_r)$	Objective function associated to X_i and X_r solutions;
k, c	Shape, and scale parameters of Weibull distribution;
n	Problem dimension;
N	Population size;
Nt, Nw	The number of thermal and wind units;
o, u	Superscript symbols attached to some quantities reflecting overestimation, and underestimation of the available wind power;
P_D	The total power demand;
P_L	Transmission line losses;
PT, PW	The output power vectors of the thermal and wind units;
PT_i	Power of the thermal unit i;
$PT_{min,i}$, $PT_{max,i}$	Real minimum and maximum power of the thermal unit i;
PW_j	Scheduled wind power of the wind unit j;
PWr	Rated power of the wind unit;
ru, δ	Parameters specific to the HDP operator;
t	Current iteration;
t_{max}	Maximum number of iterations
v_{in}, v_r, v_{out}	Cut-in, rated, and cut-out wind speeds of the wind unit;
W_j	Random variable that represents available wind power for unit j;
X_i, X_r	n-dimensional vectors associated with solutions i and r;
X_i^{new}	New solution vector X_i;
X^{best}	The vector of the best solution;
$x_{j,i}$	j^{th} component of solution X_i;
$x_{j,i}^{new}$	j^{th} component of solution X_i^{new}
x_j^{best}	j^{th} component of solution X^{best}
α, β, δ_{max}	Specific parameters of the MSGO algorithm;
α_j, β_j, γ_j, δ_j, and λ_j	Emission coefficients of the thermal unit i;
ω	Weighting factor;
$\{cx_p\}$	Chaotic sequence;
$\exp(\bullet)$	exponential function;
$f_W(\bullet)$	Probability density function (pdf) of random variable W;
$Prob(\bullet)$	The probability of the event;
min, max	Indicate the minimum and maximum limits of some variables/functions;
Min, Max	Mathematical functions that determine the minimum and maximum value in a set.

Abbreviations

ABC: Artificial bee colony algorithm; ACS: Artificial cooperative search; BBO: Biogeography-based optimization; BCS: Best compromise solution; BSA: Backtracking search optimization; CA: Cultural algorithm; CSCA: Chaotic SCA; CSO: Cuckoo search optimization; CSOA: Criminal search optimization algorithm; CSS: Charged system search; CTLBO: Chaotic TLBO; CTO: Class topper optimization; DE: Differential evolution; EcD: Economic dispatch; EED: Economic emission dispatch; EmD: Emission dispatch; EMA: Exchange market algorithm; EMOCA: Enhanced multi-objective cultural algorithm; εv-MOGA: Epsilon-multi-objective GA; ESCSDO10: Eagle-strategy supply–demand based optimization algorithm with chaotic map ten; FSS-IPSO2: Floating search space with improved PSO; GA: Genetic algorithm; GAAPI: GA with ant colony optimization; GQPSO: Gaussian quantum-behaved PSO; GSA: Gravitational search algorithm; GWO: Grey wolf optimization; HDP: Highly disruptive polynomial; HPSO-DE: Hybrid PSO with DE; HSCA: Hybrid SCA; IJaya: Improved Jaya algorithm; IKSO: Improve kernel search optimization; IMO: Ion motion optimization; ISA: Interior search algorithm; Jaya-SML: Jaya algorithm with SAMP and Lévy flights; KKO: Kho-Kho optimization algorithm; KSO: Kernel search optimization; LFA: Lightning flash algorithm; LPSR: Linear population size reduction; L-SHADE: LPSR with success-history based adaptive DE; MBFA: Modified bacterial foraging algorithm; MIMO: Modified IMO; MODE: Multi-objective DE; MSGO: Modified SGO; NGWO: Novel grey wolf optimization; NSGA II/III: Non-dominated sorting genetic algorithm-version II/III; ORCCRO: Oppositional real coded chemical reaction optimization; PDE: Pareto differential evolution; pdf: Probability density function; PSO: Particle swarm optimization; QOPO: Quasi-oppositional-based political optimizer; QOTLBO: Quasi-oppositional TLBO; QPSO: Quantum PSO; RCCRO: Real coded chemical reaction optimization; SAMP: Self-adaptive multi-population; SCA: Sine–cosine algorithm; SDE: Shuffled differential evolution; SDO: Supply–demand optimization; SGO: Social group optimization; S-Jaya: Jaya with self-adaptive po-pulation mechanism; SMA: Slime mould algorithm; SPES-2: Strength pareto evolutionary algorithm 2; SSA: Squirrel search algorithm; TLABC: Teaching-learning-based ABC; TLBO: Teaching learning-based-optimization; VPE: Valve-point effects; Wind: Indicates the inclusion of wind power in EcD, EmD, or EED problems.

Appendix A

Table A1. Assessment of the cost related to wind power: steps and numerical example.

Step	Description and Exemplification **To Evaluate the Cost Related to the Wind Power, the Main Steps That Have to Be Done Are Presented, as Well as How Each of Them Is Applied for a Specific Case of a Wind Unit:**
1	Set the input data for a wind unit: the shape (k = 1.5) and scale (c = 15) parameters; rated output power of the wind unit (PWr = 550 MW); cut-in wind speed (v_{in} = 5 m/s), rated wind speed (v_r = 15 m/s), cut-out wind speed (v_{out} = 45 m/s); direct cost coefficient (c^d = 0), reserve cost coefficient (c^o = 5), penalty cost coefficient (c^u = 5). Because during the MSGO optimization process, the variable "scheduled wind power" (PW) vary between the limits (PW_{min} = 0, PW_{max} = 550) MW, the calculation is performed for an arbitrarily chosen value, for example PW = 400 MW
2	Set the expression of the pdf $f_W(w)$ related to the wind power (W) using relation (7): $f_W(w) = \frac{1.5 \cdot 5 \cdot 2}{15 \cdot 550}\left(\frac{5}{15} \cdot \left(1 + \frac{w \cdot 2}{550}\right)\right)^{1.5-1} exp\left[-\left(\frac{5}{15} \cdot \left(1 + \frac{w \cdot 2}{550}\right)\right)^{1.5}\right] = \frac{1}{550}\left(\frac{1}{3} \cdot \left(1 + \frac{w}{275}\right)\right)^{0.5} exp\left[-\left(\frac{1}{3} \cdot \left(1 + \frac{w}{275}\right)\right)^{1.5}\right],$ where, $R = (15 - 5)/5 = 2$
3	Calculate the discrete probabilities $Prob(W = 0)$, and $Prob(W = PWr)$, using relations (8) and (9): $Prob(W = 0) = 1 - exp[-(v_{in}/c)^k] + exp[-(v_{out}/c)^k] = 1 - \exp[(-(5/15)^{1.5}] + \exp[-(45/15)^{1.5}] = 0.180602$ $Prob(W = PWr) = Prob(W = 550) = exp[-(v_r/c)^k] - exp[-(v_{out}/c)^k] = \exp[-(15/15)^{1.5}] - \exp[-(45/15)^{1.5}] = 0.362342$
4	Calculate the average power E^o(PW = 400) and E^u(PW = 400) according to (10) and (11): $E^o(PW = 400) = \int_0^{400} (400 - w) f_W(w) dw + (400 - 0) \cdot 0.180602 = 143.087\text{MW}$ $E^u(PW = 400) = \int_{400}^{550} (w - 550) f_W(w) dw + (550 - 400) \cdot 0.362342 = 45.944\text{MW}$ where $f_W(w)$ is presented in Step 2 and the integrals may by directly calculated in mathcad.
5	Calculate the cost $C^W(PW)$ related to the wind power PW, using (3): $C^W(PW) = \sum_{j=1}^{Nw=1} C_j^W(PW_j) = c^d \cdot PW + c^o \cdot E^o(PW = 400) + c^u \cdot E^u(PW = 400) = 0 \cdot 400 + 5 \cdot 143.087 + 5 \cdot 45.944 = 945.153\ \$/\text{h}$

Table A2. Thermal units' characteristics for the 10-unit test system.

Unit	Power Limits		Fuel Cost Coefficients					Emission Coefficients				
	$P_{min,i}$	$P_{max,i}$	a_i	b_i	c_i	e_i	f_i	α_i	β_i	γ_i	δ_i	λ_i
	MW	MW	$/MW²h	$/MWh	$/h	$/h	rad/MW	lb/MW²h	lb/MWh	lb/h	lb/h	1/MW
1	10	55	0.12951	40.5407	1000.403	33	0.0174	0.04702	−3.9864	360.0012	0.25475	0.01234
2	20	80	0.10908	39.5804	950.606	25	0.0178	0.04652	−3.9524	350.0056	0.25475	0.01234
3	47	120	0.12511	36.5104	900.705	32	0.0162	0.04652	−3.9023	330.0056	0.25163	0.01215
4	20	130	0.12111	39.5104	800.705	30	0.0168	0.04652	−3.9023	330.0056	0.25163	0.01215
5	50	160	0.15247	38.5390	756.799	30	0.0148	0.00420	0.3277	13.8593	0.24970	0.01200
6	70	240	0.10587	46.1592	451.325	20	0.0163	0.00420	0.3277	13.8593	0.24970	0.01200
7	60	300	0.03546	38.3055	1243.531	20	0.0152	0.00680	−0.5455	40.2669	0.24800	0.01290
8	70	340	0.02803	40.3965	1049.998	30	0.0128	0.00680	−0.5455	40.2669	0.24990	0.01203
9	135	470	0.02111	36.3278	1658.569	60	0.0136	0.00460	−0.5112	42.8955	0.25470	0.01234
10	150	470	0.01799	38.2704	1356.659	40	0.0141	0.00460	−0.5112	42.8955	0.25470	0.01234

Table A3. The B-loss coefficients ($[B_{ij}]$, $[B_{0i}]$, and B_{00}) for 10-unit system.

		1	2	3	4	5	6	7	8	9	10
	1	4.9×10^{-5}	1.4×10^{-5}	1.5×10^{-5}	1.5×10^{-5}	1.6×10^{-5}	1.7×10^{-5}	1.7×10^{-5}	1.8×10^{-5}	1.9×10^{-5}	2.0×10^{-5}
	2	1.4×10^{-5}	4.5×10^{-5}	1.6×10^{-5}	1.6×10^{-5}	1.7×10^{-5}	1.5×10^{-5}	1.5×10^{-5}	1.6×10^{-5}	1.8×10^{-5}	1.8×10^{-5}
	3	1.5×10^{-5}	1.6×10^{-5}	3.9×10^{-5}	1.0×10^{-5}	1.2×10^{-5}	1.2×10^{-5}	1.4×10^{-5}	1.4×10^{-5}	1.6×10^{-5}	1.6×10^{-5}
	4	1.5×10^{-5}	1.6×10^{-5}	1.0×10^{-5}	4.0×10^{-5}	1.4×10^{-5}	1.0×10^{-5}	1.1×10^{-5}	1.2×10^{-5}	1.4×10^{-5}	1.5×10^{-5}
$[B_{ij}]_{10\times10}=$	5	1.6×10^{-5}	1.7×10^{-5}	1.2×10^{-5}	1.4×10^{-5}	3.5×10^{-5}	1.1×10^{-5}	1.3×10^{-5}	1.3×10^{-5}	1.5×10^{-5}	1.6×10^{-5}
	6	1.7×10^{-5}	1.5×10^{-5}	1.2×10^{-5}	1.0×10^{-5}	1.1×10^{-5}	3.6×10^{-5}	1.2×10^{-5}	1.2×10^{-5}	1.4×10^{-5}	1.5×10^{-5}
	7	1.7×10^{-5}	1.5×10^{-5}	1.4×10^{-5}	1.1×10^{-5}	1.3×10^{-5}	1.2×10^{-5}	3.8×10^{-5}	1.6×10^{-5}	1.6×10^{-5}	1.8×10^{-5}
	8	1.8×10^{-5}	1.6×10^{-5}	1.4×10^{-5}	1.2×10^{-5}	1.3×10^{-5}	1.2×10^{-5}	1.6×10^{-5}	4.0×10^{-5}	1.5×10^{-5}	1.6×10^{-5}
	9	1.9×10^{-5}	1.8×10^{-5}	1.6×10^{-5}	1.4×10^{-5}	1.5×10^{-5}	1.4×10^{-5}	1.6×10^{-5}	1.5×10^{-5}	4.2×10^{-5}	1.9×10^{-5}
	10	2.0×10^{-5}	1.8×10^{-5}	1.6×10^{-5}	1.5×10^{-5}	1.6×10^{-5}	1.5×10^{-5}	1.8×10^{-5}	1.6×10^{-5}	1.9×10^{-5}	4.4×10^{-5}
$[B_{0i}]_{1\times10}=$		0	0	0	0	0	0	0	0	0	0
$B_{00}=$		0									

Table A4. Thermal units' characteristics for the 40-unit test system.

Unit	Power Limits		Fuel Cost Coefficients					Emission Coefficients				
	$P_{min,i}$	$P_{max,i}$	a_i	b_i	c_i	e_i	f_i	α_i	β_i	γ_i	δ_i	λ_i
	MW	MW	$/MW²h	$/MWh	$/h	$/h	rad/MW	t/MW²h	t/MWh	t/h	t/h	1/MW
1	36	114	0.0069	6.73	94.705	100	0.084	0.048	−2.22	60	1.31	0.0569
2	36	114	0.0069	6.73	94.705	100	0.084	0.048	−2.22	60	1.31	0.0569
3	60	120	0.02028	7.07	309.54	100	0.084	0.0762	−2.36	100	1.31	0.0569
4	80	190	0.00942	8.18	369.03	150	0.063	0.054	−3.14	120	0.9142	0.0454
5	47	97	0.0114	5.35	148.89	120	0.077	0.085	−1.89	50	0.9936	0.0406
6	68	140	0.01142	8.05	222.33	100	0.084	0.0854	−3.08	80	1.31	0.0569
7	110	300	0.00357	8.03	287.71	200	0.042	0.0242	−3.06	100	0.6550	0.02846
8	135	300	0.00492	6.99	391.98	200	0.042	0.031	−2.32	130	0.6550	0.02846
9	135	300	0.00573	6.6	455.76	200	0.042	0.0335	−2.11	150	0.6550	0.02846
10	130	300	0.00605	12.9	722.82	200	0.042	0.425	−4.34	280	0.6550	0.02846
11	94	375	0.00515	12.9	635.2	200	0.042	0.0322	−4.34	220	0.6550	0.02846
12	94	375	0.00569	12.8	654.69	200	0.042	0.0338	−4.28	225	0.6550	0.02846
13	125	500	0.00421	12.5	913.4	300	0.035	0.0296	−4.18	300	0.5035	0.02075
14	125	500	0.00752	8.84	1760.4	300	0.035	0.0512	−3.34	520	0.5035	0.02075
15	125	500	0.00708	9.15	1728.3	300	0.035	0.0496	−3.55	510	0.5035	0.02075
16	125	500	0.00708	9.15	1728.3	300	0.035	0.0496	−3.55	510	0.5035	0.02075
17	220	500	0.00313	7.97	647.85	300	0.035	0.0151	−2.68	220	0.5035	0.02075
18	220	500	0.00313	7.95	649.69	300	0.035	0.0151	−2.66	222	0.5035	0.02075
19	242	550	0.00313	7.97	647.83	300	0.035	0.0151	−2.68	220	0.5035	0.02075
20	242	550	0.00313	7.97	647.81	300	0.035	0.0151	−2.68	220	0.5035	0.02075
21	254	550	0.00298	6.63	785.96	300	0.035	0.0145	−2.22	290	0.5035	0.02075
22	254	550	0.00298	6.63	785.96	300	0.035	0.0145	−2.22	285	0.5035	0.02075
23	254	550	0.00284	6.66	794.53	300	0.035	0.0138	−2.26	295	0.5035	0.02075
24	254	550	0.00284	6.66	794.53	300	0.035	0.0138	−2.26	295	0.5035	0.02075

Table A4. *Cont.*

Unit	Power Limits		Fuel Cost Coefficients					Emission Coefficients				
	$P_{min,i}$	$P_{max,i}$	a_i	b_i	c_i	e_i	f_i	α_i	β_i	γ_i	δ_i	λ_i
	MW	MW	\$/MW^{2}h	\$/MWh	\$/h	\$/h	rad/MW	t/MW^{2}h	t/MWh	t/h	t/h	1/MW
25	254	550	0.00277	7.1	801.32	300	0.035	0.0132	−2.42	310	0.5035	0.02075
26	254	550	0.00277	7.1	801.32	300	0.035	0.0132	−2.42	310	0.5035	0.02075
27	10	150	0.52124	3.33	1055.1	120	0.077	1.842	−1.11	360	0.9936	0.0406
28	10	150	0.52124	3.33	1055.1	120	0.077	1.842	−1.11	360	0.9936	0.0406
29	10	150	0.52124	3.33	1055.1	120	0.077	1.842	−1.11	360	0.9936	0.0406
30	47	97	0.0114	5.35	148.89	120	0.077	0.085	−1.89	50	0.9936	0.0406
31	60	190	0.0016	6.43	222.92	150	0.063	0.0121	−2.08	80	0.9142	0.0454
32	60	190	0.0016	6.43	222.92	150	0.063	0.0121	−2.08	80	0.9142	0.0454
33	60	190	0.0016	6.43	222.92	150	0.063	0.0121	−2.08	80	0.9142	0.0454
34	90	200	0.0001	8.95	107.87	200	0.042	0.0012	−3.48	65	0.6550	0.02846
35	90	200	0.0001	8.62	116.58	200	0.042	0.0012	−3.24	70	0.6550	0.02846
36	90	200	0.0001	8.62	116.58	200	0.042	0.0012	−3.24	70	0.6550	0.02846
37	25	110	0.0161	5.88	307.45	80	0.098	0.095	−1.98	100	1.42	0.0677
38	25	110	0.0161	5.88	307.45	80	0.098	0.095	−1.98	100	1.42	0.0677
39	25	110	0.0161	5.88	307.45	80	0.098	0.095	−1.98	100	1.42	0.0677
40	242	550	0.00313	7.97	647.83	300	0.035	0.0151	−2.68	220	0.5035	0.02075

Table A5. The B-loss coefficients ($[B_{ij}]$, $[B_{0i}]$, and B_{00}) for 40-unit system.

$([B_{ij}]_{40\times40}) \times 10^{-6} =$

	1	2	3	4	5	6	7	8	9	10	11	12	13	14	15	16	17	18	19	20	21	22	23	24	25	26	27	28	29	30	31	32	33	34	35	36	37	38	39	40
1	17	12	7	-1	-5	-2	17	12	7	-1	-5	-2	17	12	7	-1	-5	-2	17	12	7	-1	-5	-2	17	12	7	-1	-5	-2	17	12	7	-1	-5	-2	17	12	7	-1
2	12	14	9	1	-6	-1	12	14	9	1	-6	-1	12	14	9	1	-6	-1	12	14	9	1	-6	-1	12	14	9	1	-6	-1	12	14	9	1	-6	-1	12	14	9	1
3	7	9	31	0	-10	-6	7	9	31	0	-10	-6	7	9	31	0	-10	-6	7	9	31	0	-10	-6	7	9	31	0	-10	-6	7	9	31	0	-10	-6	7	9	31	0
4	-1	1	0	24	-6	-8	-1	1	0	24	-6	-8	-1	1	0	24	-6	-8	-1	1	0	24	-6	-8	-1	1	0	24	-6	-8	-1	1	0	24	-6	-8	-1	1	0	24
5	-5	-6	-10	-6	129	-2	-5	-6	-10	-6	129	-2	-5	-6	-10	-6	129	-2	-5	-6	-10	-6	129	-2	-5	-6	-10	-6	129	-2	-5	-6	-10	-6	129	-2	-5	-6	-10	-6
6	-2	-1	-6	-8	-2	150	-2	-1	-6	-8	-2	150	-2	-1	-6	-8	-2	150	-2	-1	-6	-8	-2	150	-2	-1	-6	-8	-2	150	-2	-1	-6	-8	-2	150	-2	-1	-6	-8
7	17	12	7	-1	-5	-2	17	12	7	-1	-5	-2	17	12	7	-1	-5	-2	17	12	7	-1	-5	-2	17	12	7	-1	-5	-2	17	12	7	-1	-5	-2	17	12	7	-1
8	12	14	9	1	-6	-1	12	14	9	1	-6	-1	12	14	9	1	-6	-1	12	14	9	1	-6	-1	12	14	9	1	-6	-1	12	14	9	1	-6	-1	12	14	9	1
9	7	9	31	0	-10	-6	7	9	31	0	-10	-6	7	9	31	0	-10	-6	7	9	31	0	-10	-6	7	9	31	0	-10	-6	7	9	31	0	-10	-6	7	9	31	0
10	-1	1	0	24	-6	-8	-1	1	0	24	-6	-8	-1	1	0	24	-6	-8	-1	1	0	24	-6	-8	-1	1	0	24	-6	-8	-1	1	0	24	-6	-8	-1	1	0	24
11	-5	-6	-10	-6	129	-2	-5	-6	-10	-6	129	-2	-5	-6	-10	-6	129	-2	-5	-6	-10	-6	129	-2	-5	-6	-10	-6	129	-2	-5	-6	-10	-6	129	-2	-5	-6	-10	-6
12	-2	-1	-6	-8	-2	150	-2	-1	-6	-8	-2	150	-2	-1	-6	-8	-2	150	-2	-1	-6	-8	-2	150	-2	-1	-6	-8	-2	150	-2	-1	-6	-8	-2	150	-2	-1	-6	-8
13	17	12	7	-1	-5	-2	17	12	7	-1	-5	-2	17	12	7	-1	-5	-2	17	12	7	-1	-5	-2	17	12	7	-1	-5	-2	17	12	7	-1	-5	-2	17	12	7	-1
14	12	14	9	1	-6	-1	12	14	9	1	-6	-1	12	14	9	1	-6	-1	12	14	9	1	-6	-1	12	14	9	1	-6	-1	12	14	9	1	-6	-1	12	14	9	1
15	7	9	31	0	-10	-6	7	9	31	0	-10	-6	7	9	31	0	-10	-6	7	9	31	0	-10	-6	7	9	31	0	-10	-6	7	9	31	0	-10	-6	7	9	31	0
16	-1	1	0	24	-6	-8	-1	1	0	24	-6	-8	-1	1	0	24	-6	-8	-1	1	0	24	-6	-8	-1	1	0	24	-6	-8	-1	1	0	24	-6	-8	-1	1	0	24
17	-5	-6	-10	-6	129	-2	-5	-6	-10	-6	129	-2	-5	-6	-10	-6	129	-2	-5	-6	-10	-6	129	-2	-5	-6	-10	-6	129	-2	-5	-6	-10	-6	129	-2	-5	-6	-10	-6
18	-2	-1	-6	-8	-2	150	-2	-1	-6	-8	-2	150	-2	-1	-6	-8	-2	150	-2	-1	-6	-8	-2	150	-2	-1	-6	-8	-2	150	-2	-1	-6	-8	-2	150	-2	-1	-6	-8
19	17	12	7	-1	-5	-2	17	12	7	-1	-5	-2	17	12	7	-1	-5	-2	17	12	7	-1	-5	-2	17	12	7	-1	-5	-2	17	12	7	-1	-5	-2	17	12	7	-1
20	12	14	9	1	-6	-1	12	14	9	1	-6	-1	12	14	9	1	-6	-1	12	14	9	1	-6	-1	12	14	9	1	-6	-1	12	14	9	1	-6	-1	12	14	9	1
21	7	9	31	0	-10	-6	7	9	31	0	-10	-6	7	9	31	0	-10	-6	7	9	31	0	-10	-6	7	9	31	0	-10	-6	7	9	31	0	-10	-6	7	9	31	0
22	-1	1	0	24	-6	-8	-1	1	0	24	-6	-8	-1	1	0	24	-6	-8	-1	1	0	24	-6	-8	-1	1	0	24	-6	-8	-1	1	0	24	-6	-8	-1	1	0	24
23	-5	-6	-10	-6	129	-2	-5	-6	-10	-6	129	-2	-5	-6	-10	-6	129	-2	-5	-6	-10	-6	129	-2	-5	-6	-10	-6	129	-2	-5	-6	-10	-6	129	-2	-5	-6	-10	-6
24	-2	-1	-6	-8	-2	150	-2	-1	-6	-8	-2	150	-2	-1	-6	-8	-2	150	-2	-1	-6	-8	-2	150	-2	-1	-6	-8	-2	150	-2	-1	-6	-8	-2	150	-2	-1	-6	-8
25	17	12	7	-1	-5	-2	17	12	7	-1	-5	-2	17	12	7	-1	-5	-2	17	12	7	-1	-5	-2	17	12	7	-1	-5	-2	17	12	7	-1	-5	-2	17	12	7	-1
26	12	14	9	1	-6	-1	12	14	9	1	-6	-1	12	14	9	1	-6	-1	12	14	9	1	-6	-1	12	14	9	1	-6	-1	12	14	9	1	-6	-1	12	14	9	1
27	7	9	31	0	-10	-6	7	9	31	0	-10	-6	7	9	31	0	-10	-6	7	9	31	0	-10	-6	7	9	31	0	-10	-6	7	9	31	0	-10	-6	7	9	31	0
28	-1	1	0	24	-6	-8	-1	1	0	24	-6	-8	-1	1	0	24	-6	-8	-1	1	0	24	-6	-8	-1	1	0	24	-6	-8	-1	1	0	24	-6	-8	-1	1	0	24
29	-5	-6	-10	-6	129	-2	-5	-6	-10	-6	129	-2	-5	-6	-10	-6	129	-2	-5	-6	-10	-6	129	-2	-5	-6	-10	-6	129	-2	-5	-6	-10	-6	129	-2	-5	-6	-10	-6
30	-2	-1	-6	-8	-2	150	-2	-1	-6	-8	-2	150	-2	-1	-6	-8	-2	150	-2	-1	-6	-8	-2	150	-2	-1	-6	-8	-2	150	-2	-1	-6	-8	-2	150	-2	-1	-6	-8
31	17	12	7	-1	-5	-2	17	12	7	-1	-5	-2	17	12	7	-1	-5	-2	17	12	7	-1	-5	-2	17	12	7	-1	-5	-2	17	12	7	-1	-5	-2	17	12	7	-1
32	12	14	9	1	-6	-1	12	14	9	1	-6	-1	12	14	9	1	-6	-1	12	14	9	1	-6	-1	12	14	9	1	-6	-1	12	14	9	1	-6	-1	12	14	9	1
33	7	9	31	0	-10	-6	7	9	31	0	-10	-6	7	9	31	0	-10	-6	7	9	31	0	-10	-6	7	9	31	0	-10	-6	7	9	31	0	-10	-6	7	9	31	0
34	-1	1	0	24	-6	-8	-1	1	0	24	-6	-8	-1	1	0	24	-6	-8	-1	1	0	24	-6	-8	-1	1	0	24	-6	-8	-1	1	0	24	-6	-8	-1	1	0	24
35	-5	-6	-10	-6	129	-2	-5	-6	-10	-6	129	-2	-5	-6	-10	-6	129	-2	-5	-6	-10	-6	129	-2	-5	-6	-10	-6	129	-2	-5	-6	-10	-6	129	-2	-5	-6	-10	-6
36	-2	-1	-6	-8	-2	150	-2	-1	-6	-8	-2	150	-2	-1	-6	-8	-2	150	-2	-1	-6	-8	-2	150	-2	-1	-6	-8	-2	150	-2	-1	-6	-8	-2	150	-2	-1	-6	-8
37	17	12	7	-1	-5	-2	17	12	7	-1	-5	-2	17	12	7	-1	-5	-2	17	12	7	-1	-5	-2	17	12	7	-1	-5	-2	17	12	7	-1	-5	-2	17	12	7	-1
38	12	14	9	1	-6	-1	12	14	9	1	-6	-1	12	14	9	1	-6	-1	12	14	9	1	-6	-1	12	14	9	1	-6	-1	12	14	9	1	-6	-1	12	14	9	1
39	7	9	31	0	-10	-6	7	9	31	0	-10	-6	7	9	31	0	-10	-6	7	9	31	0	-10	-6	7	9	31	0	-10	-6	7	9	31	0	-10	-6	7	9	31	0
40	-1	1	0	24	-6	-8	-1	1	0	24	-6	-8	-1	1	0	24	-6	-8	-1	1	0	24	-6	-8	-1	1	0	24	-6	-8	-1	1	0	24	-6	-8	-1	1	0	24

$([B_{0i}]) \times 10^{-4} =$

1	2	3	4	5	6	7	8	9	10	11	12	13	14	15	16	17	18	19	20
-3.908	-1.297	7.047	0.591	2.161	-6.635	-3.908	-1.297	7.047	0.591	2.161	-6.635	-3.908	-1.297	7.047	0.591	2.161	-6.635	-3.908	-1.297
21	22	23	24	25	26	27	28	29	30	31	32	33	34	35	36	37	38	39	40
7.047	0.591	2.161	-6.635	-3.908	-1.297	7.047	0.591	2.161	-6.635	-3.908	-1.297	7.047	0.591	2.161	-6.635	-3.908	-1.297	7.047	0.591

$B_{00} =$ 0.56

References

1. Jin, J.; Wen, Q.; Qiu, Y.; Cheng, S.; Guo, X. Distributed robust optimization for low-carbon dispatch of wind-thermal power under uncertainties. *Environ. Sci. Pollut. Res.* **2023**, *30*, 20980–20994. [CrossRef] [PubMed]
2. Jin, J.; Zhou, D.; Zhou, P.; Miao, Z. Environmental/economic power dispatch with wind power. *Renew. Energy* **2014**, *71*, 234–242. [CrossRef]
3. Nanda, J.; Kothari, D.P.; Linga Murthy, K.S. Economic emission load dispatch through goal programming techniques. *IEEE Trans. Energy Convers.* **1988**, *3*, 26–32. [CrossRef] [PubMed]
4. Hetzer, J.; Yu, D.C.; Bhattarai, K. An economic dispatch model incorporating wind power. *IEEE Trans. Energy Convers.* **2008**, *23*, 603–611. [CrossRef]
5. Dhillon, J.S.; Parti, S.C. Stochastic economic emission load dispatch. *Electr. Power Syst. Res.* **1993**, *26*, 179–186. [CrossRef]
6. Martins, A.C.S.; Balbo, A.R.; Jones, D.; Nepomuceno, L.; Soler, E.M.; Baptista, E.C. A Hybrid Multi-Criteria Methodology for Solving the Sustainable Dispatch Problem. *Sustainability* **2020**, *12*, 6780. [CrossRef]
7. Mahdy, A.; El-Sehiemy, R.; Shaheen, A.; Ginidi, A.; Elbarbary, Z.M.S. An improved artificial ecosystem algorithm for economic dispatch with combined heat and power units. *Appl. Sci.* **2022**, *12*, 11773. [CrossRef]
8. El-Sehiemy, R.; Shaheen, A.; Ginidi, A.; Elhosseini, M. A honey badger optimization for minimizing the pollutant environmental emissions-based economic dispatch model integrating combined heat and power units. *Energies* **2022**, *15*, 7603. [CrossRef]
9. Ginidi, A.; Elsayed, A.; Shaheen, A.; Elattar, E.; El-Sehiemy, R. An innovative hybrid heap-based and jellyfish search algorithm for combined heat and power economic dispatch in electrical grids. *Mathematics* **2021**, *9*, 2053. [CrossRef]
10. Shaheen, A.M.; Ginidi, A.R.; El-Sehiemy, R.A.; Ghoneim, S.S.M. Economic power and heat dispatch in cogeneration energy systems using manta ray foraging optimizer. *IEEE Access* **2020**, *8*, 208281–208295. [CrossRef]
11. Liu, X.; Xu, W. Minimum emission dispatch constrained by stochastic wind power availability and cost. *IEEE Trans. Power Syst.* **2010**, *25*, 1705–1713.
12. Jadhav, H.T.; Roy, R. Gbest guided artificial bee colony algorithm for environmental/economic dispatch considering wind power. *Expert Syst. Appl.* **2013**, *40*, 6385–6399. [CrossRef]
13. Dubey, H.M.; Pandit, M.; Panigrahi, B.K. An overview and comparative analysis of recent bio-inspired optimization techniques for wind integrated multi-objective power dispatch. *Swarm Evol. Comput.* **2018**, *38*, 12–34. [CrossRef]
14. Tsvetkova, O.; Ouarda, T.B.M.J. Use of the Halphen distribution family for mean wind speed estimation with application to Eastern Canada. *Energy Convers. Manag.* **2023**, *276*, 116502. [CrossRef]
15. Ghasemi, A.; Gheydi, M.; Golkar, M.J.; Eslami, M. Modeling of Wind/Environment/Economic Dispatch in power systemand solving via an online learning meta-heuristic method. *Appl. Soft Comput.* **2016**, *43*, 454–468. [CrossRef]
16. Kheshti, M.; Kang, X.; Li, J.; Regulski, P.; Terzija, V. Lightning Flash Algorithm for Solving Nonconvex Combined Emission Economic Dispatch with Generator Constraints. *IET Gener. Transm. Distrib.* **2018**, *12*, 104–116. [CrossRef]
17. Wang, B.; Zhang, P.; He, Y.; Wang, X.; Zhang, X. Scenario-oriented hybrid particle swarm optimization algorithm for robust economic dispatch of power system with wind power. *J. Syst. Eng. Electron.* **2022**, *33*, 1143–1150. [CrossRef]
18. Srivastava, A.; Das, D.K. An adaptive chaotic class topper optimization technique to solve economic load dispatch and emission economic dispatch problem in power system. *Soft Comput.* **2022**, *26*, 2913–2934. [CrossRef]
19. Basetti, V.; Rangarajan, S.S.; Shiva, C.K.; Pulluri, H.; Kumar, R.; Collins, R.E.; Senjyu, T. Economic Emission Load Dispatch Problem with Valve-Point Loading Using a Novel Quasi-Oppositional-Based Political Optimizer. *Electronics* **2021**, *10*, 2596. [CrossRef]
20. Srivastava, A.; Das, D.K. Criminal Search Optimization Algorithm: A Population-Based Meta-Heuristic Optimization Technique to Solve Real-World Optimization Problems. *Arab. J. Sci. Eng.* **2022**, *47*, 3551–3571. [CrossRef]
21. Srivastava, A.; Das, D.K. A new Kho-Kho optimization algorithm: An application to solve combined emission economic dispatch and combined heat and power economic dispatch problem. *Eng. Appl. Artif. Intell.* **2020**, *94*, 103763. [CrossRef]
22. Nourianfar, H.; Abdi, H. Environmental/Economic dispatch using a new hybridizing algorithm integrated with an effective constraint handling technique. *Sustainability* **2022**, *14*, 3173. [CrossRef]
23. Sakthivel, V.P.; Suman, M.; Sathya, P.D. Combined economic and emission power dispatch problems through multi-objective squirrel search algorithm. *Appl. Soft Comput. J.* **2021**, *100*, 106950. [CrossRef]
24. Zhang, R.; Zhou, J.; Moa, L.; Ouyang, S.; Liao, X. Economic environmental dispatch using an enhanced multi-objective cultural algorithm. *Electr. Power Syst. Res.* **2013**, *99*, 18–29. [CrossRef]
25. Bhesdadiya, R.H.; Trivedi, I.N.; Jangir, P.; Jangir, N.; Kumar, A. An NSGA-III algorithm for solving multi-objective economic/environmental dispatch problem. *Cogent Eng.* **2016**, *3*, 1269383. [CrossRef]
26. Roy, P.K.; Bhui, S. Multi-objective quasi-oppositional teaching learning based optimization for economic emission load dispatch problem. *Electr. Power Energy Syst.* **2013**, *53*, 937–948. [CrossRef]
27. Guesmi, T.; Farah, A.; Marouani, I.; Alshammari, B.; Abdallah, H.H. Chaotic sine–cosine algorithm for chance-constrained economic emission dispatch problem including wind energy. *IET Renew. Power Gener.* **2020**, *14*, 1808–1821. [CrossRef]
28. Das, D.; Bhattacharya, A.; Ray, R.N. Dragonfly Algorithm for solving probabilistic Economic Load Dispatch problems. *Neural Comput. Appl.* **2020**, *32*, 3029–3045. [CrossRef]
29. Satapathy, S.C.; Naik, A. Social group optimization (SGO): A new population evolutionary optimization technique. *Complex Intell. Syst.* **2016**, *2*, 173–203. [CrossRef]

30. Naik, A.; Satapathy, S.C.; Abraham, A. Modified social group optimization—A meta-heuristic algorithm to solve short-term hydrothermal scheduling. *Appl. Soft Comput.* **2020**, *95*, 106524. [CrossRef]
31. Reddy, A.K.V.K.; Narayana, K.V.L. A combinatorial social group whale optimization algorithm for numerical and engineering optimization problems. *Appl. Soft Comput.* **2021**, *99*, 106903.
32. Jena, J.J.; Satapathy, S.C. A new adaptive tuned social group optimization (SGO) algorithm with sigmoid-adaptive inertia weight for solving engineering design problems. In *Multimedia Tools and Applications*; Springer: Berlin/Heidelberg, Germany, 2021. [CrossRef]
33. Sinha, N.; Chakrabarti, R.; Chattopadhyay, P.K. Evolutionary programming techniques for economic load dispatch. *IEEE Trans Evol. Comput.* **2003**, *7*, 83–94. [CrossRef]
34. Rajasomashekar, S.; Aravindhababu, P. Biogeography based optimization technique for best compromise solution of economic emission dispatch. *Swarm Evol. Comput.* **2012**, *7*, 47–57. [CrossRef]
35. Bhattacharjee, K.; Bhattacharya, A.; Halder, S. Solution of economic emission load dispatch problems of power systems by real coded chemical reaction algorithm. *Int. J. Electr. Power Energy Syst.* **2014**, *59*, 176–187. [CrossRef]
36. Geng, J.Q.; Sun, X.M.; Wang, H.H.; Bu, X.H.; Liu, D.H.; Li, F.; Zhao, Z.W. A modified adaptive sparrow search algorithm based on chaotic reverse learning and spiral search for global optimization. In *Neural Computing and Applications*; Springer: Berlin/Heidelberg, Germany, 2023. [CrossRef]
37. Gupta, A.; Tiwari, D.; Kumar, V.; Rana, K.P.S.; Mirjalili, S. A chaos–infused moth–flame optimizer. *Arab. J. Sci. Eng.* **2022**, *47*, 10769–10809. [CrossRef]
38. Abed-alguni, B.H.; Paul, D.; Hammad, R. Improved salp swarm algorithm for solving single-objective continuous optimization problems. *Appl. Intell.* **2022**, *52*, 17217–17236. [CrossRef]
39. Alawad, N.A.; Abed-alguni, B.H. Discrete island-based cuckoo search with highly disruptive polynomial mutation and opposition-based learning strategy for scheduling of workflow applications in cloud environments. *Arab. J. Sci. Eng.* **2021**, *46*, 3213–3233. [CrossRef]
40. Gaing, Z.-L. Particle swarm optimization to solving the economic dispatch considering the generator constraints. *IEEE Trans. Power Syst.* **2003**, *18*, 1187–1195. [CrossRef]
41. Park, J.-B.; Jeong, Y.-W.; Shin, J.-R.; Lee, K.Y. An improved particle swarm optimization for nonconvex economic dispatch problems. *IEEE Trans. Power Syst.* **2010**, *25*, 155–166.
42. Basu, M. Economic environmental dispatch using multi-objective differential evolution. *Appl. Soft Comput.* **2011**, *11*, 2845–2853. [CrossRef]
43. Hota, P.K.; Barisal, A.K.; Chakrabarti, R. Economic emission load dispatch through fuzzy based bacterial foraging algorithm. *Electr. Power Energy Syst.* **2010**, *32*, 794–803. [CrossRef]
44. Ciornei, I.; Kyriakides, E. A GA–API solution for the economic dispatch of generation in power system operation. *IEEE Trans. Power Syst.* **2012**, *27*, 233–242. [CrossRef]
45. Lam, A.Y.S.; Li, V.O.K.; Yu, J.J.Q. Real-coded chemical reaction optimization. *IEEE Trans. Evol. Comput.* **2012**, *16*, 339–353. [CrossRef]
46. Yu, J.T.; Kim, C.H.; Wadood, A.; Khurshaid, T.; Rhee, S.B. Jaya algorithm with self-adaptive multi-population and Lévy flights for solving economic load dispatch problems. *IEEE Access* **2019**, *7*, 21372–21384. [CrossRef]
47. Bhattacharjee, K.; Bhattacharya, A.; Dey, S.H.N. Oppositional real coded chemical reaction optimization for different economic dispatch problems. *Int. J. Electr. Power Energy Syst.* **2014**, *55*, 378–391. [CrossRef]
48. Zakian, P.; Kaveh, A. Economic dispatch of power systems using an adaptive charged system search algorithm. *Appl. Soft Comput. J.* **2018**, *73*, 607–622. [CrossRef]
49. Afzalan, E.; Joorabian, M. Emission, reserve and economic load dispatch problem with non-smooth and non-convex cost functions using epsilon-multi-objective genetic algorithm variable. *Electr. Power Energy Syst.* **2013**, *52*, 55–67. [CrossRef]
50. Pothiya, S.; Ngamroo, I.; Kongprawechnon, W. Ant colony optimisation for economic dispatch problem with non-smooth cost functions. *Int. J. Electr. Power Energy Syst.* **2010**, *32*, 478–487. [CrossRef]
51. Labbi, Y.; Attous, D.; Mahdad, B. Artificial bee colony optimization for economic dispatch with valve point effect. *Front. Energy* **2014**, *8*, 449–458. [CrossRef]
52. Selvakumar, A.I.; Thanushkodi, K. Optimization using civilized swarm: Solution to economic dispatch with multiple minima. *Electr. Power Syst. Res.* **2009**, *79*, 8–16. [CrossRef]
53. He, X.; Rao, Y.; Huang, J. A novel algorithm for economic load dispatch of power Systems. *Neurocomputing* **2016**, *171*, 1454–1461. [CrossRef]
54. Alghamdi, A.S. Greedy sine-cosine non-hierarchical grey wolf optimizer for solving non-convex economic load dispatch problems. *Energies* **2022**, *15*, 3904. [CrossRef]
55. Yu, J.-T.; Kim, C.-H.; Rhee, S.-B. Clustering cuckoo search optimization for economic load dispatch problem. *Neural Comput. Appl.* **2020**, *32*, 16951–16969. [CrossRef]
56. Bhattacharjee, K.; Bhattacharya, A.; Dey, S.H. Backtracking search optimization based economic environmental power dispatch problems. *Int. J. Electr. Power Energy Syst.* **2015**, *73*, 830–842. [CrossRef]
57. Kaboli, S.H.A.; Alqallaf, A.K. Solving non-convex economic load dispatch problem via artificial cooperative search algorithm. *Expert Syst. Appl.* **2019**, *128*, 14–27. [CrossRef]

58. Kumar, M.; Dhillon, J.S. A conglomerated ion-motion and crisscross search optimizer for electric power load dispatch. *Appl. Soft Comput.* **2019**, *83*, 105641. [CrossRef]
59. Coelho, L.S. Gaussian quantum-behaved particle swarm optimization approaches for constrained engineering design problems. *Expert Syst. Appl.* **2010**, *37*, 1676–1683. [CrossRef]
60. Babar, M.I.; Ahmad, A.; Fayyaz, S. A hybrid sine cosine algorithm with SQP for solving convex and nonconvex economic dispatch problem. *Mehran Univ. Res. J. Eng. Technol.* **2020**, *39*, 31–46. [CrossRef]
61. Xu, S.; Xiong, G.; Mohamed, A.W.; Bouchekara, H.R.E.H. Forgetting velocity based improved comprehensive learning particle swarm optimization for non-convex economic dispatch problems with valve-point effects and multi-fuel options. *Energy* **2022**, *256*, 124511. [CrossRef]
62. Reddy, A.S.; Vaisakh, K. Shuffled differential evolution for large scale economic dispatch. *Electr. Power Syst. Res.* **2013**, *96*, 237–245. [CrossRef]
63. Naderi, E.; Azizivahed, A.; Narimani, H.; Fathi, M.; Narimani, M.R. A comprehensive study of practical economic dispatch problems by a new hybrid evolutionary algorithm. *Appl. Soft Comput.* **2017**, *61*, 1186–1206. [CrossRef]
64. Amiri, M.; Khanmohammadi, S.; Badamchizadeh, M.A. Floating search space: A new idea for efficient solving the economic and emission dispatch problem. *Energy* **2018**, *158*, 564–579. [CrossRef]
65. Xu, J.; Yan, F.; Yun, K.; Su, L.; Li, F.; Guan, J. Noninferior solution grey wolf optimizer with an independent local search mechanism for solving economic load dispatch problems. *Energies* **2019**, *12*, 2274. [CrossRef]
66. Kamboj, V.K.; Kumari, C.L.; Bath, S.K.; Prashar, D.; Rashid, M.; Alshamrani, S.S.; AlGhamdi, A.S. A cost-effective solution for non-convex economic load dispatch problems in power systems using slime mould algorithm. *Sustainability* **2022**, *14*, 2586. [CrossRef]
67. Hamdi, M.; Idomghar, L.; Chaoui, M.; Kachouri, A. An improved adaptive differential evolution optimizer for non-convex economic dispatch problems. *Appl. Soft Comput.* **2019**, *85*, 105868. [CrossRef]
68. Chen, C.; Zou, D.; Li, C. Improved Jaya algorithm for economic dispatch considering valve-point effect and multi-fuel options. *IEEE Access* **2020**, *8*, 84981–84995. [CrossRef]
69. Hassan, M.H.; Kamel, S.; Eid, A.; Nasrat, L.; Jurado, F.; Elnaggar, M.F. A developed eagle-strategy supply-demand optimizer for solving economic load dispatch problems. *Ain Shams Eng. J.* **2023**, *14*, 102083. [CrossRef]
70. Güvenç, U.; Sonmez, Y.; Duman, S. Combined economic and emission dispatch solution using gravitational search algorithm. *Sci. Iran. D Comput. Sci. Eng. Electr. Eng.* **2012**, *19*, 1754–1762. [CrossRef]
71. Ahmed, T.A.; Ebeed, M.; Refai, A.; Kamel, S. Solving combined economic and emission dispatch problem using the slime mould algorithm. *Sohag Eng. J.* **2021**, *1*, 62–70. [CrossRef]
72. Dong, R.; Chen, H.; Heidari, A.A.; Turabieh, H.; Mafarja, M.; Wang, S. Boosted kernel search: Framework, analysis and case studies on the economic emission dispatch problem. *Knowl. Based Syst.* **2021**, *233*, 107529. [CrossRef]
73. Zitzler, E.; Thiele, L. Multiobjective evolutionary algorithms: A comparative case study and the strength pareto approach. *IEEE Trans. Evol. Comput.* **1999**, *3*, 257–271. [CrossRef]
74. Cao, Y.; Smucker, B.J.; Robinson, T.J. On using the hypervolume indicator to compare Pareto fronts: Applications to multi-criteria optimal experimental design. *J. Stat. Plan. Inference* **2015**, *160*, 60–74. [CrossRef]

sustainability

MDPI

Article

European Union Tools for the Sustainable Development of Border Regions

Florentina Chirodea [1,*], Luminita Soproni [1,*] and Mihai Marian [2]

[1] Department of International Relations and European Studies, University of Oradea, 410087 Oradea, Romania
[2] Psychology Department, University of Oradea, 410087 Oradea, Romania; mmarian@uoradea.ro
* Correspondence: fchirodea@uoradea.ro (F.C.); lsoproni@uoradea.ro (L.S.)

Abstract: The European Union uses several instruments aimed at reducing disparities between regions and strengthening their competitiveness and sustainability. The border regions have a special relevance, given their position and characteristics, as well as the challenges faced by local actors. The European Commission has introduced the action "b-solutions" to tackle the specific border obstacles along European Union internal borders. The article aims to analyze the integration mechanisms at the micro-regional level, which is considered as a viable and sustainable solution for cross-border regions with resources which can be exploited to attract investment and generate wealth and well-being. The qualitative analysis method involved consulting thematic publications of the b-solution program and extracting data from the presentation sheets of accepted proposals published by the Association of European Border Regions. The collected data were processed according to criteria such as direction of action, types and causes of obstacles, solutions proposed by experts, etc. Addressing the legal and administrative obstacles that hamper cross-border flows proved to be a good initiative, with 120 cases selected. The solutions offered to the particular cases can also be replicated for other obstacles identified at the European Union's internal borders so that cooperation between border regions is intensified to the benefit of increasing European territorial, economic, and social cohesion.

Keywords: border region; cross-border partnership; sustainable development; b-solutions program

Citation: Chirodea, F.; Soproni, L.; Marian, M. European Union Tools for the Sustainable Development of Border Regions. *Sustainability* **2024**, *16*, 388. https://doi.org/10.3390/su16010388

Academic Editors: Constantin Bungău, Alina Badulescu, Dorina Camelia Ilieș, Cosmin Mihai Vesa and Delia Mirela Tit

Received: 2 November 2023
Revised: 16 November 2023
Accepted: 30 December 2023
Published: 31 December 2023

1. Introduction and Theoretic Background

Regionalization is an attractive option in the contemporary socio-economic landscape because it increases the capacity of states to attract foreign investments and allows for development through the use of common resources, which results in a reduction in certain adjustment costs and helps to overcome some political and administrative obstacles [1–5]. It thus supports the more facile integration of the countries involved into the global market.

Regionalization is defined differently by specialists depending on the relationship they establish between globalization and regionalization. Those that see regionalization as an opposed process to globalization consider that this removes the advantages of liberalization and allows regional partners to be privileged over others [6]. Other authors define a second perspective, which considers regionalization as a logical response to the effects created by globalization, as agreements between states are necessary for these to become stronger in international economic competition [1,7–9]. The group of specialists of the third line of analysis considers regionalization as a stimulus of globalization and therefore as a basic component of a deeper integration. The fact that states demonstrated readiness to participate in regional agreements proves that such agreements are complementary to the globalization process rather than attempts to replace it [10–13].

The new type of regionalism is a worldwide phenomenon that emerged in the 1980s as a result of the need to harmonize national policies with the view of achieving global economic integration [14]. It occurs in several areas of the world and continues the processes and mechanisms created by the old regionalism (manifested in the 1950s–1970s), which mainly aimed to reduce or eliminate barriers to trade. The process was facilitated by the negotiations within the Uruguay Round of the General Agreement on Tariffs and Trade (GATT). At the same time, the new regionalism defines and develops new cooperation mechanisms and structures (Figure 1). If we refer to the actors involved in the regional process, we find that the nation-states played a central role in the case of old regionalism, while new regionalism involves actions and interactions of several players, such as regional and local authorities and civil society alongside the actors from businesses and non-governmental bodies [15–18]. In this sense, the European Union is considered a model of new regionalism due to its characteristics, which are seen as essential for deeper integration [8,19–22].

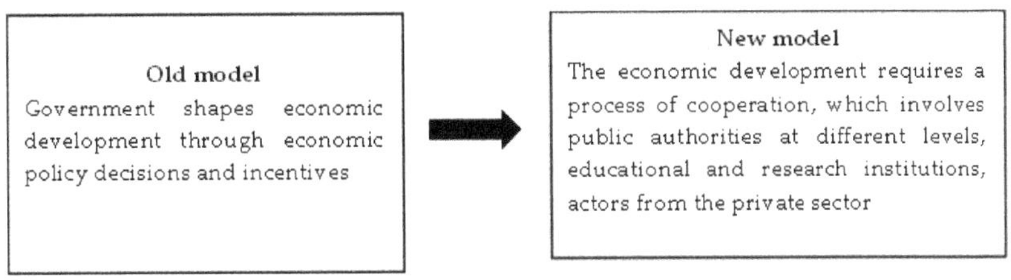

Figure 1. Shifting responsibilities for economic development. Source. Authors' own elaboration based on the theoretical background.

The new regionalism has two components. Its first direction concerns cross-border groupings established at the meso-regional (different states wishing to integrate their economic, political, or military activities) and macro-regional levels (trade blocs, regional organizations, or transcontinental networks). The decision of states to enter into a regional agreement is largely based on the balance between the costs and the economic benefits at stake. In addition, the size of a country influences both the decision concerning the association and its ability to use power in international relations [23]: small countries move towards regionalism because they believe that their relative position can be weakened if they remain isolated; at the same time, they aim to secure their position against the risks of disputes with major partners. On the other hand, large countries see regionalism as a way to increase their power and expand their influence in international negotiations.

If we look at the regional integration agreements made at the mezzo or macro-regional level [24–26], the problem that arises here relates to the fact that the main beneficiaries of these forms of integration are the developed states. Developing countries' governments therefore face both short-term and long-term challenges in increasing regional and global competitiveness, and traditional responses no longer offer viable solutions, hence the need for the second component of new regionalism: the micro-regional perspective.

In the context created by the mechanisms and processes of the new regionalism, the actions carried out at the sub-national level have become an integral part of the vocabulary related to development (especially in the case of the neo-liberal model in the northern hemisphere) [27,28], highlighting the relevance of the sub-national dimension of progress and development. Post-development theorists support the need for local objectives to be prioritized in the development of any policies, instead of including individuals and communities in programs and projects created for global purposes for which they do not have the necessary power. They envision and articulate development in different terms than the proponents of the global development concept [29,30], believing that actions with global scope and effects can be more effectively carried out at the local level. This theory is therefore about thinking and acting locally without excluding transnational alliances and networks [31,32].

This perspective fits into the theory of endogenous regional development [29,33,34], which is based on the internal development potential of the local community. The model prioritizes regional needs within national territory and capitalizes on territorial strategic advantages that provide a competitive position to the region. Thus, the region must be able to guarantee the autonomy of processes aiming at economic growth through its own resources and through the advantages offered by the local specificity. In addition, the endogenous growth model considers interactions between regions [35–37]. Thus, endogenous factors are considered drivers of regional development and growth and are enhanced by the economic, social, or cultural interconnections achieved between or among neighboring regions through cooperative actions.

At the level of the European Union, the need to overcome the obstacles generated by the processes of deepening economic and political integration has led to increased interest in the field of regional development policy. Cohesion policy, being a horizontal policy, addresses areas considered strategic by the European Commission (education, employment, energy, the environment, the single market, research, and innovation) by financing the territorial development programs proposed by the member states, which are implemented at regional level by local authorities. The EU has created a series of territorial cooperation instruments within its Cohesion Policy. The New Cohesion Policy (2021–2027) continues and complements the objectives pursued during previous periods [38,39].

Processes aimed at European integration have led to the transformation of border regions from peripheral areas into areas of growth and development. The changes generated by the European integration process (increased productivity, reduction of transaction costs, intensification of intra-European trade, and increase in the number of jobs) had both positive outcomes (increased cross-border interactions) and negative outcomes (reduction in the number of jobs in the field of customs administration) in border regions [40]. In the context of the development model described above, border areas are often described as "laboratories" of European integration and cross-border cooperation, as they are hotspots with intense cross-border interactions. They are regions where the advantages of the single market are visible and where new ideas and solutions can be tested for the first time and analyzed at a small scale [41–45].

The areas bounded by the internal borders of the EU, which are more or less open, represent interesting territories for researchers because three major changes can be analyzed here [46]: (1) the increase in cross-border trade and service flows, along with the increase in the international mobility of the workforce, as a result of the effects of European integration; (2) the expansion of transport networks, utilities and public services and the emergence of new models of economic activities, as a result of investments in transnational infrastructure; (3) strengthening cooperation between communities located on the two sides of the border and multiplying cross-border development initiatives, by standardizing legal and administrative procedures. Moreover, the border regions delimited by the old border that divided Western and Eastern Europe, also called "little Europe", seem to be the most suitable areas for analyzing opportunities for political, economic, cultural, environmental, and social welfare action [47].

The emergence and development of cooperative relations between border regions have been supported by a series of European Commission initiatives. Most of these have been transposed into EU legislation [48]. Since 1990, the most important instrument for the implementation of the European Union's Cohesion Policy in border areas has undoubtedly been the European Cross Border Cooperation program (Interreg), which aims to encourage economic growth in border areas, having among its objectives the acceleration of regional development [49]. The program was created to stimulate cooperation between institutions and communities located on both sides of the border, through the development of cross-border socio-economic centers with common development strategies (Euroregions) [50–52]. Interregional cooperation projects are included in the first pillar of the Interreg program, which aims to improve the exchange of experience and the sharing of common practices, as well as the preparation of action plans for the integration and implementation of good practices within regional development policies [42].

In this context, for the purposes of this work, we have formulated the following research hypothesis: the integration mechanisms at the micro-regional level (which form the second direction of the new regionalism) appear as a viable solution for cross-border states or regions that have resources which can be exploited in order to attract investment and generate wealth and well-being at their internal level.

The research questions subsumed to this objective are:

1. Is the b-solutions program an effective mechanism of European integration through the instruments offered to the beneficiary regions?
2. Have the obstacles identified in the selected proposals been eliminated by applying the solutions proposed by the experts?
3. Is the lack or poor functioning of institutional cooperation one of the important sources of obstacles?

The paper is organized in such a way as to provide a contextualization of the assumed research hypothesis by creating a theoretical framework that explains the premises and specificity of the new regionalism, seen as the foundation for the relevance of the sub-national dimension of progress and development (expressed starting from the theory of endogenous regional development). At the same time, regionalism is seen as context for the cross-border actions and programs analyzed. One section of the paper includes a review of the process that resulted in the financing of the b-solutions initiative by the European Commission, with a presentation of the objectives and directions of action related to the program. Cases associated with the process, as well as its results, are described in a separate section. The last part of the paper provides answers and validations/invalidations for the hypothesis and the research questions posed.

2. Materials and Methods

The study was conducted between 2021 and 2023, starting with the qualitative analysis of data made public by the Association of European Border Regions (AEBR). In the first part of the research, the three thematic publications developed by the AEBR in 2021 were consulted regarding obstacles and solutions to cross-border cooperation in the EU. In the second part of our research, focused on the fourth direction of the program (institutional cooperation), data were extracted from the presentation sheets of the proposals accepted for analysis, published in the annex to *B-solutions: Solving Border Obstacles. A Compendium of 43 Cases*, respectively in *B-solutions: Solving Border Obstacles. A Compendium 2020–2021*.

In addition, for an in-depth analysis of the effectiveness of the b-solutions initiative, we consulted two studies developed by groups of researchers led by Eduardo Medeiros, published in 2021 and 2023, which address the issue of the b-solutions program. The diagram shown in Figure 2 schematically shows the analysis process.

The collected data was processed according to a set of criteria such as: the direction of action proposed by the program; the types of obstacles (legal, administrative), the causes that led to the emergence of obstacles; the solutions proposed by experts (legislative, capacity building and administrative coordination, transversal solutions). In line with the standard criteria of exploratory qualitative analysis, we explained how the b-solutions initiative worked and its impact was. On the other hand, we have considered that the method can suggest possible relationships, causes, effects, or dynamic processes in our research area. Additionally, in the qualitative analysis we have targeted a set of keywords (e.g., direction for action, obstacles, causes, policy area, etc.) to facilitate the identification of relevant information.

For three directions of action of the program (cross-border public services, the labor market and education, the implementation of the Green Deal), we made a synthesis of the results published in the three publications of the AEBR mentioned above and referred briefly to the most important achievements in each field. For the fourth line of action, institutional cooperation, which is not the subject of an analysis by AEBR specialists or other researchers, we created a database for the 24 selected proposals, in which information was entered based on the following criteria: local actors who submitted the proposal; policy area; the indicated obstacle; type of obstacle (legal, administrative); causes of obstacles;

the proposed solutions. The data were analyzed qualitatively so that along with those presented synthetically, they can be used in a deep analysis of the b-solutions program.

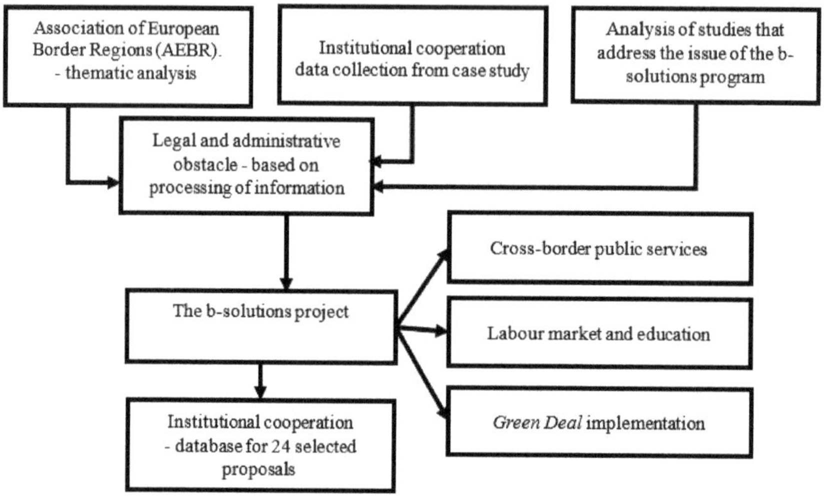

Figure 2. Data analysis mechanism used for the b-solutions initiative. Source. Authors' own elaboration.

3. Results

Since its launch in 2017, the b-solutions initiative, through the experts involved, has identified and promoted sustainable ways to reduce obstacles at the EU's internal borders, including neighboring EFTA countries. The program provided the opportunity to effectively test ways to overcome barriers and replicate solutions to achieve full cross-border cooperation [53]. Initially described as an EU pilot action through which 10 projects and 33 case studies were selected, where divergences between national and European legislation or incompatible administrative procedures are indicated, the initiative highlighted the European Commission's interest in "collecting practical, feasible, comprehensible, viable solutions, aimed at reducing the identified barriers" [54]. The thematic areas of the 10 pilot actions (labor market, health, public passenger transport, multilingualism, and institutional cooperation) and those in which the case studies are grouped (e-government, labor market, records and databases, health, information services, institutional cooperation, multilingualism, and transport) highlighted the fact that most cross-border obstacles are based on differences in the legislation of the member states, followed by incompatibility or duplication of administrative procedures. Their reduction or elimination requires actions that especially involve the public administrations at different decision-making levels in the member states but also other local actors. After the first phase of the b-solutions program, AEBR published a compendium with the cases analyzed and the resulting solutions, offering the opportunity to replicate them in the cases of other internal borders or introduce them into EU legislation [55].

In a complex analysis carried out in 2021, a group of researchers led by Eduardo Medeiros [56] showed that there is an unbalanced geographical distribution of pilot actions and cases accepted for analysis, with most projects (40%) coming from the Benelux and from the border of France with Germany. The explanation offered by the researchers refers to the cross-border institutional maturity, the intensity of cross-border flows, the deep integration of the regions in these areas, and the fact that the borders in northwest Europe are those with the most legal and administrative obstacles. Unlike the pilot projects, the 33 cases selected in the first phase of the project "are better distributed along the European borders", as Medeiros and his collaborators state (70% of the cases are concentrated in the border areas between the Benelux countries–France–Germany–Spain–Portugal), and the thematic area covered is considered to be much more relevant to the objectives of the program. In general, the result of this first phase consists in the identification of concrete solutions to reduce the various cross-border barriers in transport, trade, health, education, language, taxation, environment, cartage, etc., the implementation of the solutions not being a priority of the program. However, the actors involved in the development of pilot proposals and cases have started to take the first steps in involving different authorities or entities in a common process of reducing or eliminating obstacles [56].

The second, third, and fourth phases of the b-solutions program, carried out in the period from 2019–2021, meant the acceptance of another 47 cases for which AEBR experts offered solutions to reduce legislative and administrative obstacles. The Association's specialists have published a series of summaries of the data collected in the analysis sheets of projects and cases from the 4 years of implementation of the b-solution program, which cover only three of the four major directions of action of the program: cross-border public services; labor market and education; implementation of the Green Deal. A summary of the published results is presented synthetically in Table 1.

The data collected by AEBR experts for the field of cross-border public services highlighted the presence of obstacles of a legal and administrative nature in the following thematic areas: citizenship, justice, public security; civil protection, disaster management; communication and information of society; education; environment protection; health; social inclusion; labor market; spatial planning, culture; transport [57]. In the absence of common methodologies, collaboration protocols and coordination structures, the attempt of local authorities to offer or stimulate certain services is cumbersome or delayed, with consequences on the achievement of the European objective of being as close as possible to citizens in all regions [57].

Regarding the field of education and the labor market, the AEBR specialists say that due to the multitude of unclear rules that must be respected and the many existing administrative models in the Member States, coordination across internal borders is a complex process hampered by difficulties in areas such as social security; diplomas and certificates recognition; access to training and education; business opportunities; trade; and professional or educational status of non-EU nationals [58]. The solutions offered in this field are general and at the level of good practices, they must be customized according to the specific legislation and national practices by the competent actors in education and the labor market, with local public authorities having the task of applying common actions and to engage the opportunities of the single market [58].

For the implementation of the Green Deal strategy adopted by the EU in 2019, which had a major role in regional development [59], local actors from the regions on the internal borders of the EU have noticed difficulties in the following areas: clean, cheap and safe energy; clean and circular economy; energy-efficient buildings with low resource consumption; sustainable mobility; environmentally friendly food system; conservation of ecosystems and biodiversity; and zero pollution [60]. The collected data show that border regions have to deal with the same obstacles in carrying out sustainable common actions, and the signing of cross-border agreements between local or national authorities for the implementation of green projects is the most often offered solution. AEBR experts also recommend increasing access to training courses through the INTERREG program, harmonizing green initiatives and involving EGTC-type structures as potential promotors of cross-border initiatives aimed at sustainable development [60].

Table 1. Overview of AEBR publications that refer to the results of implementing the b-solution program for three directions of action.

Direction for Action	Obstacles Identified		Causes	Other Causes	Solutions Proposed		
	Legal	Administrative			Legal	To Strengthen Administrative Capacity and Coordination	Transversal
Cross-border public services	- Local and regional levels of government do not have legal powers to provide or operate services - The legislation does not allow the automatic recognition of diplomas issued by another EU member state - Different national rules governing services - Technical standards/requirements for the operation of services regulated differently at national level - The legislation regarding the provision of services does not take into account the cross-border dimension; - Conflicting transposition of EU legislation in cross-border regions.	- Long procedures for the recognition of diplomas - Absence of common administrative mechanisms to facilitate the operationalization of services - Lack of knowledge of the decision-makers regarding the enabling legislative framework to regulate certain services - Different approaches in data collection as a preliminary step towards the implementation of public services - The presence of many actors with different levels of administrative skills in certain fields.	- The failure to take into account the specificity of cross-border territories, which creates gaps where inconsistencies are generated	- Complex and burdensome administrative procedures for coordinating public services across national borders - Lack of horizontal coordination between responsible bodies - Lack of knowledge of the decision-makers regarding the legal or administrative framework - Incompatibility or lack of provisions at various legislative levels (sub-national, national, European).	- Changes in legislation - Revision or improvement of the existing one - Creating an ad hoc legislative framework.	- Creation of new coordination structures - The development of ad hoc conventions.	- Specific strategies for increasing coordination among the actors involved - Training courses to increase knowledge in new cooperation schemes - The complementarity of EU aid (e.g., INTERREG) - The establishment of cross-border structures - creating ad hoc plans - development of protocols and conventions.

Table 1. *Cont.*

Direction for Action	Obstacles Identified		Causes	Other Causes	Solutions Proposed		
	Legal	Administrative			Legal	To Strengthen Administrative Capacity and Coordination	Transversal
Labor market and education	- The EU framework contains general provisions, leaving room for interpretation - The provisions in force are not adapted to the complexity of the cross-border context - The national provisions regarding contracts, taxation and financing are not aligned - The regulations in force do not provide automatic recognition of diplomas/certificates - The provisions regulating new fields/working and living conditions are insufficient or outdated.	- The presence of complex/unclear bureaucratic procedures - The absence of common administrative mechanisms - The existence of different administrative protocols and approaches - Lack of knowledge as regards the frameworks in force that are already facilitators.	- Lack of provisions to regulate different issues - Lack of knowledge about the already existing framework that facilitates the provisions/tools offered by the legal/administrative framework in force - Divergent or inconsistent regulations at European, national or sub-national level.	- Complex and burdensome administrative procedures on both sides of the border - Lack of administrative coordination between competent actors from two or more neighboring countries.	- Amending or improving existing legislation at European level - Revising or updating the current provisions on one or both sides of the border - Creating an ad hoc legal framework.	- Creating ad hoc plans - Development of protocols and conventions.	

Table 1. *Cont.*

Direction for Action	Obstacles Identified		Causes	Other Causes	Solutions Proposed		
	Legal	Administrative			Legal	To Strengthen Administrative Capacity and Coordination	Transversal
Green Deal implementation	- Creation and management of common infrastructures - Divergent national rules on infrastructure design and construction approvals - Inconsistent legal powers of spatial planning - Different regulations for the necessary technical requirements; - Lack of revised legislation - Lack of specific provisions regarding the cross-border dimension - Absence of standardized European norms.	- Different references for spatial data needed for mapping and data collection - Different technical standards for environmental management criteria - The absence of an ad hoc cross-border structure or entity, responsible for the coordination of nature reserves - The absence of a common mechanism for regulating data exchange.	- Inconsistencies in the legal frameworks related to green policies in place in neighboring Member States - Member States' exclusive competence on certain matters that regulate actions, infrastructure and projects implementing the Green Deal - National laws have not transposed the most recent EU legal framework.	- Diverging practices - or technical features of specific actions hamper the completion of cross-border actions - The lack of horizontal. - cooperation among the stakeholders involved in a specific project or action.	- Amending and improving existing legislation at European level - Harmonization at the supranational level - Reviewing the current provisions on both sides of the border, - creating an ad hoc legal framework.	- Establishing a common management structure - Creating a single, unified command point - Harmonization of environmental management data sets, methods and technical standards.	- The creation of a specific consortium for relevant actors on both sides of the border - Information actions - The training of local actors engaged in specific projects.

Note: Own processing of the data presented by the AEBR in publications related to the implementation of the b-solution program [57–59].

For the fourth direction of action of the b-solutions initiative—institutional cooperation—AEBR has not yet developed a synthesis of the cases. Therefore, in what follows, we will carry out our own analysis of the information collected from the program's website. The importance of institutions for the success of reforms is widely recognized by the academic community, Dimitrios Zikos defining them "as systems of established and embedded social rules that structure social interactions. Institutional arrangements influence governance structures, shape the economy and affect public awareness and civic engagement". In this context, institutional cooperation is one of the tools often used to solve acute problems, in particular legal and administrative obstacles, at the internal borders of the EU. Zikos emphasizes the fact that this cooperation cannot exist outside of a framework of "formal rules", which generates a feeling of trust that the institutions are able to "provide expectations, stability, meaning essential to human existence, coordination, regularize life, support values and produce and protect interests" [61]. For their part, Ezers and Naglis-Liepa see institutional cooperation as a process carried out both horizontally and vertically, "opposed to competition", and demonstrate that its success influences regional development and economic progress. The two Latvian researchers emphasize that at least three of the principles of good governance "directly underpin the significance of institutional cooperation: openness—institutions have to be more open and actively communicate with one another; participation—the effectiveness of institutions is directly dependent on how successful the participation of the other ones is; coherence - cooperation among regional institutions have to be coordinated" [62]. It is also well known that the European Union has developed a wide spectrum of institutional cooperation mechanisms to achieve common objectives, which impose consultation rules and conflict management; provide predictability in decision-making processes; and impose certain obligations and constraints on different institutional actors [63]. The b-solution initiative contributes to the creation of such mechanisms, this time customized and much better anchored in the complex context of cross-border cooperation.

In the first phase of the program, 4 projects and 12 case studies were selected within the institutional cooperation direction of action, and in the period from 2020–2021, another 8 cases were analyzed by AEBR experts. The 24 selected proposals were registered in the following thematic areas: cross-border mobility; economy; sustainable management of rural areas; groundwater cross-border management; institutional cross-border cooperation; environmental management and circular economy; cross-border transport and traffic regulations; social and medical care; and protection and education of children.

The spatial distribution of the 24 proposals accepted for analysis in the direction of institutional cooperation, presented in Figure 3, is to some extent balanced in the EU space. It is observed, once again, to be a concentration of obstacles at the borders of the states that form the European core (Belgium, the Netherlands, Germany, France, and Luxembourg) but also a desire to eliminate difficulties on the part of the states of Central and Eastern Europe (Slovenia, Croatia, Hungary, Czech Republic, Poland, Lithuania, Latvia, Slovakia, Romania, and Bulgaria). We must also note the efforts of Spain and Portugal to make cooperation at their common border more efficient (see Figure 4).

The multitude and complexity of institutions from different levels of government; different institutional and legal cultures from one member state to another; the activity of cross-border actors (e.g., EGTC); the lack of bilateral agreements, ad hoc agreements, or unwritten customary procedures, customs control, and operations; and lack of coordination between stakeholders and actors are part of the factors that make it difficult to implement cross-border cooperation projects. The legal and administrative obstacles mentioned in the 24 selected proposals are presented in Table 2, which also highlights their distribution by identified thematic areas.

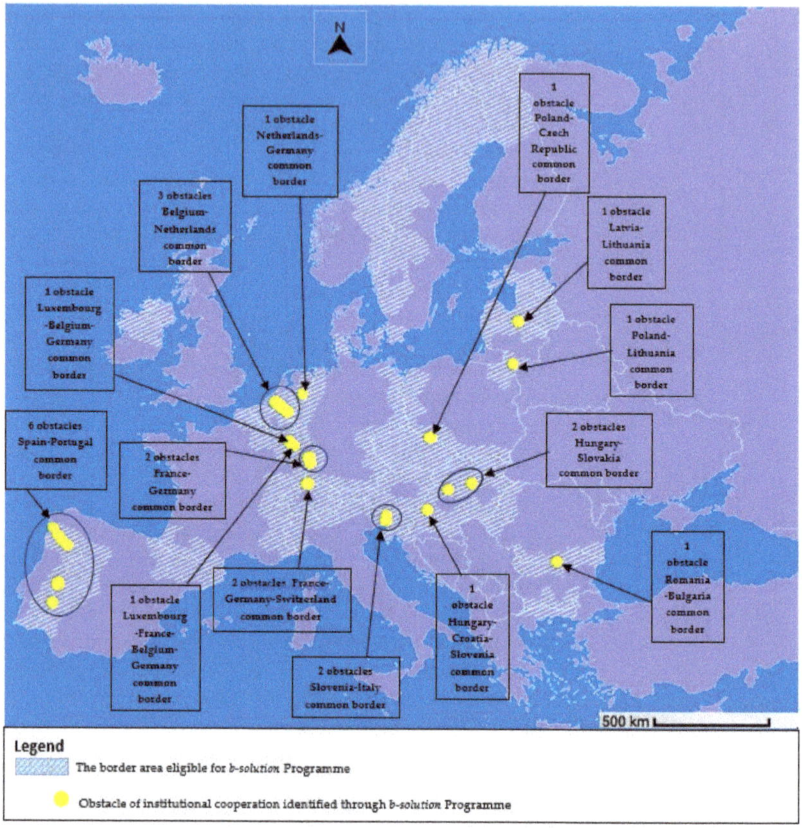

Figure 3. Distribution of institutional cooperation obstacles identified within the b-solutions program. Source: Our own elaboration, based on data found in Association of European Border Regions and European Union. *B-solutions: Solving Border Obstacles. A Compendium of 43 Cases*, Publications Office of the European Union: Luxembourg, Luxembourg, 2020 and Association of European Border Regions and European Union. *B-solutions: Solving Border Obstacles. A Compendium 2020–2021*, Publications Office of the European Union: Luxembourg, Luxembourg, 2021.

Table 2. Distribution of cross-border obstacles by thematic areas.

Policy Area	Common Obstacles
Cross-border mobility	Different, divergent, and complex national administrative procedures; lack of clarity in information; language gaps; lack of reciprocity; lack of established rules in the field of right of access in the enclave.
Economy	Complex national customs procedures that make difficult the interoperability, with consequences upon the cross-border commerce; different legal frameworks in tourism; different national laws on taxation in case of customs free zone.
Sustainable management of rural areas	Different administrative systems for the projects application and implementation.
Groundwater cross-border management	Harmonization of data and methodologies.

Table 2. *Cont.*

Policy Area	Common Obstacles
Institutional cross-border cooperation	Tax differences and non-harmonized procedures; lack of clarity in the application of existing national and European regulations; unsureness regarding the rights of the employee and the obligations of the employer(s); lack of a financial support for EGTCs and diverging national legal frameworks; lack of recognition of EGTCs as legal entities in the Member States; different legal, organizational and technical principles regarding standardization of spatial data; uncertainty in administration of a complex situation; divergence in the implementation of the Bologna standards.
Environmental management and circular economy	National regulations require many changes to adapt to the latest EU legal provisions, because of that—legal and administrative requirements have grown exponentially; the non-harmonized rules on public procurement.
Social and medical care	Application of the national law in relation to the European Regulations about emergency assistance, and health insurance registration and inadequate division of responsibilities in difficult case.
Cross-border transport and traffic regulation	Different traffic regulations and difficulties in the coordination regarding transportation of resources for industrial processes.
Children protection and education	Complexity of the national legal and administrative requirements regarding protecting the interests of minors abroad; lack of common procedures between administrative departments and social services.

Note: Own processing of the data presented by the AEBR in publications related to the implementation of the b-solution program [64,65].

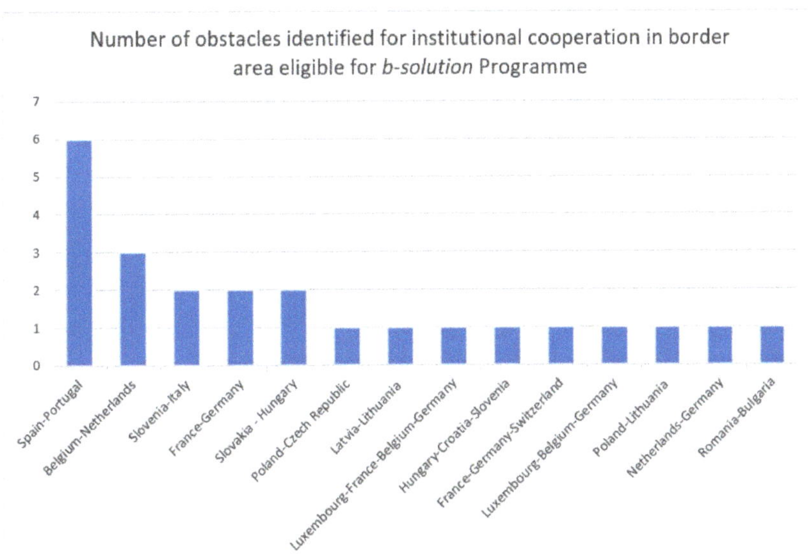

Figure 4. Number of obstacles identified for institutional cooperation in border area eligible for the b-solution program. Source: Our own elaboration, based on data from the Association of European Border Regions and European Union. *B-solutions: Solving Border Obstacles. A Compendium of 43 Cases*, Publications Office of the European Union: Luxembourg, Luxembourg, 2020 and Association of European Border Regions and European Union. *B-solutions: Solving Border Obstacles. A Compendium 2020–2021*, Publications Office of the European Union: Luxembourg, Luxembourg, 2021.

Analyzing the legal and administrative obstacles quantitatively, we observe the prevalence of those related to the different ways in which laws, regulations, and procedures are applied. At the same time, the differences between national legislative frameworks can lead to the difficulty or even the blocking of institutional cooperation among different local actors located on different sides of the borders. In addition, the transposition of EU directives into national legislation in a different way leads to confusion and a lack of clarity in the application of national or European norms, the consequence being the multitude of types of administrative obstacles reported in cross-border cooperation. On the other hand, these obstacles, whether legislative or administrative, are also based on the lack of institutional cooperation between the local authorities located on either side of the border or the national authorities that would generate coordinated procedures, coordination between stakeholders and actors, interoperability, digitization of public administrations, forms developed in at least three languages (one of which is widely used), ad hoc agreements, and usually unwritten procedures. The b-solution program, through its objectives, addresses these obstacles, and the solutions proposed by AEBR experts are in line with these objectives, as can be seen from Table 3.

Table 3. The solutions identified by AEBR experts for the 24 cases selected for institutional cooperation.

Objective of the b-Solution Program	Proposed Solutions		
	Legal	**Enhanced Administrative Capacity and Coordination**	**Cross-Cutting Solutions**
O1. Mitigate cross-border obstacles which are caused by a lack of coherence, inconsistencies or overlapping between legal provisions or administrative procedures on each side of the border but, particularly, because applicable European, national or regional/local legislation does not consider the specificity of cross-border interactions	- Adapting and amending national provisions to respond to the exceptional context of a cross-border area; - Taking into account the existing national legal framework for cross-border cooperation of enforcement authorities; - Application of European regulations.		- Lobbying administrations with the objective of simplifying procedures for citizens; - The organization of dedicated trainings in bilingual legal assistance centers to facilitate the simultaneous fulfilment of legal procurement obligations.
O2. Increase the understanding of every specific obstacle among key stakeholders on both sides of the border at local, regional, national and EU levels.	- Drawing up an action protocol between the actors involved with the support of legal experts.	- Cooperation between national governments and European institutions and organizations.	- Creating an inventory of administrative obstacles faced by border residents.
O3. Promote sustainable methods to solve cross-border obstacles through innovative proposals to inform further cross-border development and implementation by public authorities or through EU instruments.	- Modify the LEADER scheme to the entire EU area.	- Transfer the submission responsibility and accountability for cross-border LEADER projects to one managing authority.	- Formalizing cross-border cooperation by establishing a local version of the European EGTCs.
O4. Involve public bodies committed to jointly fostering, designing, and agreeing on feasible solutions to reduce cross-border barrier effects.	- Updating and adapting existing bilateral agreements to current needs.	- Establishment of bilateral agreement.	- Coordinate local development strategies.

Table 3. *Cont.*

Objective of the b-Solution Program	Proposed Solutions		
	Legal	Enhanced Administrative Capacity and Coordination	Cross-Cutting Solutions
O5. Stimulate an increased exchange of information and mutual engagement between the variety of administration levels in border areas to make possible the generation of joint initiatives involving multi-level governance across borders.		- The unification of digital platforms for multimodal logistics integration; - The IT specialists involved in the customs offices communicate closely to achieve the goals of interoperability and coordination.	- Information and knowledge sharing, structured consultation, organizing forums and events.
O6. Foster the replication of the solutions found.		- Harmonize the reference datasets, produce a digital terrain model.	- Design cross-border coordination points under the European Cross-Border Mechanism (ECBM)

Note: Own processing of the data presented by the AEBR in publications related to the implementation of the b-solution program [64,65].

4. Discussion

The new regionalism provides the appropriate framework for the fulfilment of the endogenous regional development model, as the actors involved in cooperation actions (states and regions) act strategically, pursuing the balance between the economic costs and the benefits at stake. Thus, new cross-border cooperation mechanisms and structures are generated and implicate the involvement of public authorities, educational and research institutions, and private sector companies on the playing board of regional competitiveness. The actions of local actors lead to the capitalization of territorial strategic advantages that provide a competitive position to the region. In this context, endogenous factors become drivers of regional development and growth, enhanced by the institutional, economic, social or cultural interconnections made between neighboring partner regions.

Cross-border areas have become laboratories of European integration and cohesion, as they are hot spots where multiple intense interactions are carried out. They are regions where the advantages of the single market can be observed and where new ideas and solutions can be analyzed on a small scale or be tested for the first time. The b-solutions program, through the general objective assumed, represents an effective way to promote sustainable solutions for legal or administrative obstacles that prevent or hinder cross-border cooperation. Thus, during its development, the program identified and promoted sustainable solutions to reduce legislative and administrative obstacles at the borders of the EU, including neighboring EFTA countries, providing the opportunity to effectively test ways to overcome obstacles and replicate solutions to achieve full cross-border cooperation.

Our study complements the three reports published by AEBR and other studies that address the subject with an analysis of the fourth line of action—institutional cooperation—thus contributing to the evaluation of the b-solution program. Therefore, the analysis of the data collected and processed has shown that our research objective and the starting hypotheses are mostly verified: integration mechanisms at the micro-regional level represent a viable solution for cross-border states or regions to ensure wealth and well-being within their territory.

In a special report published in 2021, the European Court of Auditors found that Interreg-type cooperation programs can only partially respond to cross-border challenges, and due to insufficient resources, it is necessary to direct funding where the added value is the highest. The Court found that in most cases, the cooperation between the partners was limited to the presentation of a joint proposal for the purpose of financing the investments. In addition, most of the cross-border challenges identified are generated by "legal obstacles, in relation to the legislative frameworks at the EU, national or regional level, the rest being administrative" [66].

In this context, financing the b-solutions initiative is a necessary decision for the proper functioning of integration processes at the micro-regional level. The way in which this mechanism was conceived led to the identification of numerous legal and administrative obstacles, for which AEBR experts were able to offer short-, medium-, or long-term solutions (customized solutions for each particular case). They confirm Research Question 1 (the b-solutions program is an effective mechanism of European integration through the instruments offered to the beneficiary regions) and refer to the following aspects: (1) modifying/updating/adapting national and European legislation to respond to the complex conditions in the border regions; (2) the adoption of an ad hoc legislative framework and the development of bilateral conventions/protocols; (3) establishment of common management structures or a single unified command point; (4) harmonization of methodologies and technical standards; (5) creation of consortia for relevant actors on both sides of the border; (6) carrying out information actions; and (7) training of local actors through projects financed by Interreg.

The European Court of Auditors estimates that if these solutions were implemented and 20% of the existing obstacles to cross-border cooperation were removed, border regions would register a 2% increase in GDP and create more than 1 million additional jobs [65]. Therefore, the effectiveness of the initiative is argued by the forecasted economic indicators, but the responsibility of the approaches belongs to all administrative levels (European, national, regional, and local), which partially confirms Research Question 2 (the obstacles noticed in the selected proposals were eliminated by applying the solutions proposed by experts).

In order to avoid cumbersome legislative procedures, AEBR experts proposed the signing of conventions, agreements, and protocols between local authorities on both sides of the borders. In addition to these short-term solutions, there are also possibilities that require an average implementation period as they aim to create common local management structures or unified command points. The application of all solutions depends, again, on the will of public authorities to attract investment for the creation of wealth and welfare in the border regions.

Another advantage of this initiative is the fact that by publishing the files of the accepted proposals and the results obtained in the period from 2018–2021, any solution proposed by the experts can be replicated to minimize a similar obstacle at the internal borders of the EU. The decision to continue the program in the 2021–2027 budget exercise demonstrates the usefulness of this mechanism, whose effectiveness can only be proven after these data are collected.

Local public actors are the ones that must apply both national and European legislation at sub-national levels and ensure, from an administrative point of view, that the procedures are effective, flexible, and easy to follow and that they do not overburden the daily lives of citizens. At the same time, development strategies and policies place them at the center of growth mechanisms that relate to living standards and life quality. In the case of actors living in border areas, these challenges are felt even more acutely due to the absence of administrative capacity, investments and the cross-border context in which they perform. Cross-border cooperation instruments provide numerous cooperation opportunities for local authorities on both sides of the European borders. However, the b-solutions program has highlighted the fact that efficient collaboration among authorities (at all levels) might generate "openness and dilution of barriers among communities and institutions" [67]. The limited number of selected proposals in the direction of institutional cooperation allows us to state that Research Question 3 (the absence of or the defective functioning of institutional cooperation is an importance source of generated obstacles) is confirmed to a lesser extent. Following the analysis of the typology of the registered obstacles, one can see that they fall within the same thematic areas as those observed within the first three directions of actions of the program. Moreover, as regards the particularized solutions provided by AEBR experts, it appears that all 90 matters share common factors that generate the most obstacles—lacks of information, communication, and cooperation between sub-national or national public authorities.

5. Conclusions

The b-solution program, with its current version b-solution 2.0, is a new tool through which the European Commission can increase cross-border cooperation at the internal borders of the EU. Tackling the legal and administrative obstacles that hamper cross-border flows and the daily lives of European citizens proved to be a good initiative, with 120 cases selected for analysis.

The solutions offered to the cases studied can also be replicated for other legal or administrative obstacles identified at the EU's internal borders so that cooperation between border regions is intensified and sustainable, to the benefit of increasing European territorial, economic, and social cohesion.

The main theoretical contribution of this study consists in formulating the hypothesis according to which the micro regional integration constitutes a favorable context for boosting up cross-border cooperation and provides the tools with the help of which the local actors from the regions close to the internal borders of the EU contribute to a deeper European integration.

The study highlights, through a qualitative analysis of the data published by AEBR, certain types of border obstacles that generate restrictions in micro-regional cooperation but also the solutions proposed to eliminate these blockages. The newest tool available to the local actors, the b-solutions program, validates through the proposed objectives and the results obtained the formulated research hypothesis, demonstrating that the elimination of legislative and administrative obstacles is one of the mechanisms that can support deeper European integration.

The proposed mechanism and solutions pave the way for cooperation also at the level of other border regions of the European Union. In this sense, we believe that the paper offers a model of analysis and work for regional public authorities looking for viable and sustainable solutions for the development of cross-border micro-regions.

From 2022, the b-solution initiative has continued with version 2.0 and will still remain in our evaluation for an even wider validation of our research hypothesis, especially since the eligibility area of the program has been extended to maritime border areas and common borders with countries involved in pre-accession assistance. It is also interesting to investigate how the solutions offered to solve legislative and administrative obstacles contribute to boosting up other cross-border cooperation programs (i.e., Interreg).

Author Contributions: Conceptualization, L.S.; methodology, M.M.; software, F.C.; validation, L.S., F.C. and M.M.; formal analysis, M.M.; investigation, F.C.; resources, L.S.; data curation, F.C.; writing—original draft preparation, L.S. and F.C.; writing—review and editing, M.M.; visualization, L.S., F.C., and M.M.; supervision, L.S., F.C. and M.M.; project administration, L.S.; funding acquisition, F.C. All authors have read and agreed to the published version of the manuscript.

Funding: This research received no external funding.

Institutional Review Board Statement: Not applicable.

Informed Consent Statement: Not applicable.

Data Availability Statement: The data presented in this study are available on request from the corresponding author. The data are not publicly available due to privacy considerations.

Conflicts of Interest: The authors declare no conflicts of interest.

References

1. Gilpin, R. *The Challenge of Global Capitalism: The World Economy in the 21st Century*; Princeton University Press: Princeton, NJ, USA, 2000.
2. Wunderlich, J.-U. *Regionalism, Globalisation and International Order: Europe and Southeast Asia*; Ashgate: London, UK, 2007.
3. Padoan, P.C. The Political Economy of New Regionalism and World Governance. In *European Union and New Regionalism. Regional Actors and Global Governance in a Post-Hegemonic Era*; Telo, M., Ed.; Ashgate: London, UK, 2007; pp. 37–54.
4. Fernández Jilberto, A.E.; Mommen, A. Globalization versus regionalization. In *Regionalization and Globalization in the Modern World Economy. Perspectives on the Third World and Transitional Economies*; Fernández Jilberto, A.E., Mommen, A., Eds.; Routledge: New York, NY, USA, 2002; pp. 1–26.
5. Mistry, P. New Regionalism and Economic Development. In *Theories of New Regionalism*; Söderbaum, F., Shaw, T., Eds.; Palgrave Macmillan: New York, NY, USA, 2003; pp. 117–139.
6. Bhagwati, J. Regionalism versus multilateralism. *World Econ.* **1992**, *15*, 535–556. [CrossRef]
7. Krugman, P. The Move Toward Free Trade Zones. *Econ. Rev.* **1991**, *76*, 5–25.
8. Higgott, R. The international political economy of regionalism: The Asia-Pacific and Europe compared. In *Regionalism and Global Economic Integration. Europe, Asia and the Americas*; Coleman, W., Underhill, G., Eds.; Routledge: London, UK, 2002; pp. 42–67.
9. Chase, K.A. *Trading Blocs. States, Firms, and Regions in the World Economy*; The University of Michigan Press: Ann Arbor, MI, USA, 2005.
10. Capello, R.; Fratesi, U.; Resmini, L. *Globalization and Regional Growth in Europe. Past Trends and Future Scenarios*; Springer: Berlin/Heidelberg, Germany, 2011.
11. Bowles, P. Regionalism and development after the global financial crises. In *New Regionalisms in the Global Political Economy: Theory and Cases*; Breslin, S., Hughes, W.C., Phillips, N., Rosamond, B., Eds.; Routledge: New York, NY, USA, 2002; pp. 81–103.
12. Yuzhu, W. New Regionalism Reshaping the Future of Globalization. *China Q. Int. Strateg. Stud.* **2020**, *6*, 249–265.
13. Hurrell, A. One World? Many Worlds? The Place of Regions in the Study of International Society. *Int. Aff.* **2007**, *83*, 127–146.
14. Hettne, B. The New Regionalism Revisited. In *Theories of New Regionalism*; Söderbaum, F., Shaw, T., Eds.; Palgrave Macmillan: New York, NY, USA, 2003; pp. 22–42.
15. Ethier, W. The new regionalism. *Econ. J.* **1998**, *108*, 1149–1161. [CrossRef]
16. Băhnăreanu, C.; Sarcinschi, A. *Procesul de Regionalizare a României şi Securitatea Naţională [Romania's Regionalization Process and National Security]*; "Carol I" National Defence University Publishing House: Bucharest, Romania, 2012.
17. Acharya, A. Regionalism and the emerging world order. Sovereignty, autonomy, identity. In *New Regionalisms in the Global Political Economy*; Breslin, S., Hughes, W.C., Phillips, N., Rosamond, B., Eds.; Routledge: New York, NY, USA, 2002; pp. 20–32.
18. Barbieri, G. Regionalism, globalism and complexity: A stimulus towards global IR? *TWQ J.* **2019**, *4*, 424–441. [CrossRef]
19. Anheier, H.; Juergensmeyer, M. *Encyclopedia of Global Studies*; Sage: Thousand Oaks, CA, USA, 2012.
20. Aggarwal, V.; Fogarty, E. Between Regionalism and Globalism: European Union Transregional and Inter-Regional Trade Strategies. In *European Union Trade Strategies: Between Globalism and Regionalism*; Aggarwal, V., Fogarty, E., Eds.; Palgrave: London, UK, 2004; pp. 1–40.
21. Lane, Y.-E. The European Union and NAFTA as Models of Regional Organisation. In *Global Trends and Regional Development*; Genov, N., Ed.; Routledge: New York, NY, USA, 2012; pp. 45–63.
22. Kolodko, W.G. Globalisation and Transformation: Illusions and Reality. *J. Emerg. Mark. Financ.* **2003**, *2*, 207–252. [CrossRef]
23. Hettne, B.; Söderbaum, F. Theorising the rise of regionness. In *New Regionalisms in the Global Political Economy*; Breslin, S., Hughes, W.C., Phillips, N., Rosamond, B., Eds.; Routledge: New York, NY, USA, 2002; pp. 33–47.
24. Molle, W. *Economia Integrării Europene [The Economy of European Integration]*; Epigraf: Chișinău, Moldova, 2009.
25. Gänzle, S. Experimentalist governance and EU macro-regional strategies: New dynamics in European regional and territorial cooperation? In *Regional Governance in the EU: Regions and the Future of Europe*; Abels, G., Battke, J., Eds.; Edward Elgar Publishing: Northampon, MA, USA, 2019; pp. 140–158.
26. Chirodea, F.; Șoproni, L.; Țoca, C.-V.; Czimre, K. (Eds.) *Regional Development at the Borders of the European Union*; "Tarii Crisurilor" Museum Publishing House: Oradea, Romania, 2019.
27. UNCTAD. Regional Integration and Foreign Direct Investment in Developing and Transition Economies (December 2012). Available online: http://unctad.org/meetings/en/Sessional%20Documents/ciimem4d2_en.pdf (accessed on 12 May 2023).
28. Torok, I.; Benedek, J.; Gomez-Zaldivar, M. Quatifying Subnational Economic Complexity: Evidence Form Romania. *Sustainability* **2022**, *14*, 10586. [CrossRef]
29. Willis, K. *Theories and Practices of Development*; Routledge: New York, NY, USA, 2005.
30. Esteva, G.A. *Critique of Development and Other Essays*; Routledge: New York, NY, USA, 2023.
31. Gibson-Graham, J.K. Forging post-development partnerships. Possibilities for local and regional development. In *Handbook of Local and Regional Development*; Pike, A., Rodríguez-Pose, A., Tomaney, J., Eds.; Routledge: New York, NY, USA, 2011; pp. 226–236.
32. Korosteleva, A.E.; Flockhart, T. Resilience in EU and international institutions: Redefining local ownership in a new global governance agenda. *Contemp. Sec. Pol.* **2020**, *41*, 153–175. [CrossRef]
33. Peet, R.; Hartwick, E. *Theories of Development. Contentions, Arguments, Alternatives*; The Guilford Press: New York, NY, USA, 2015.
34. Schumpeter, A.J. *The Theory of Economic Development*; Routledge: London, UK, 2021.

35. Martin, R.; Sunley, P. Slow Convergence? Post-Neoclassical Endogenous Growth Theory and Regional Development. In *ESRC Centre for Business Research*; Working Paper No. 44; University of Cambridge: Cambridge, UK, 1996; pp. 20–28.
36. Jones, I.C. Paul Romer: Ideas, Nonrivalry, and Endogenous Growth. *Scand. J. Econ.* **2019**, *121*, 859–883. [CrossRef]
37. Bhattacharjee, M.; Ghosh, D. On Dynamics of Economic Development in the Light of Endogenous Growth: A Theoretical and Empirical Investigation. In *Globalization, Income Distribution and Sustainable Development*; Das, R.C., Ed.; Emerald Publishing: Bingley, UK, 2022; pp. 21–30.
38. Pike, A.; Rodríguez-Pose, A.; Tomaney, J. *Local and Regional Development*; Routledge: New York, NY, USA, 2006.
39. Malizia, E.; Feser, E.; Renski, H.; Drucker, J. *Understanding Local Economic Development*, 2nd ed.; Routledge: New York, NY, USA, 2021.
40. Antonescu, D. Theoretical approaches of endogenous regional development. *Ann. Fac. Econ.* **2015**, *1*, 75–82.
41. European Commission. Regional Policy. Available online: https://ec.europa.eu/regional_policy/policy/how/priorities_en (accessed on 11 May 2023).
42. European Commission. *Communication from the Commission to the Council and the European Parliament. Boosting Growth and Cohesion in EU Border Regions*; Publications Office of the European Union: Brussels, Belgium, 2017.
43. Despiney Zochowska, A.B. Euroregions. Emerging New Forms of Cross-Border Cooperation. In *Borders and Border Regions in Europe: Changes, Challenges and Chances*; Lechevalier, A., Wielgohs, J., Eds.; Transcript Verlag, JSTOR: Bielefeld, Germany, 2013; pp. 71–94.
44. Tolstov, S. Theoretical and Methodological Aspects of Regionalism Problems of Interaction of Border Regions. *Probl. World Hist.* **2022**, *17*, 7–41. [CrossRef]
45. Makkonen, T.; Weidenfeld, A.; Williams, A.M. Cross-border regional innovation system integration: An analytical framework. *Tijdschr. Econ. Soc. Geogr.* **2017**, *108*, 805–820. [CrossRef]
46. Nijkamp, P. Moving Frontiers: A Local-Global Perspective. In *Developing Frontier Cities. Global Perspectives—Regional Contexts*; Lithwick, H., Gradus, Y., Eds.; Springer Science+Business Media: Dordrecht, The Netherlands, 2000; pp. 3–25.
47. Ruidisch, R. Territorial Cohesion and Border Areas. In *Borders and Border Regions in Europe: Changes, Challenges and Chances*; Lechevalier, A., Wielgohs, J., Eds.; Transcript Verlag, JSTOR: Bielefeld, Germany, 2013; pp. 95–110.
48. Czimre, K. *Cross-Border Co-Operation—Theory and Practice*; Debrecen University Press: Debrecen, Hungary, 2006.
49. European Commission. *Report from the Commission to the European Parliament, the Council, the European Economic and Social Committee and the Committee of the Regions*; EU Border Regions: Living labs of European integration; Publications Office of the European Union: Brussels, Belgium, 2021.
50. Interreg Europe Joint Secretariat. When Europe Cooperates, Regions Benefit. 30 Stories about the Results of Cooperation. October 2020. Available online: https://www.interregeurope.eu/sites/default/files/202107/30%20stories%20of%20results%20-%20full%20publication%20-%202020.pdf (accessed on 10 June 2023).
51. Wevers, T.H.; Voinea, L.C.; de Langen, F. Social Entrepreneurship as a Form of Cross-Border Cooperation: Complementarity in EU Border Regions. *Sustainability* **2020**, *12*, 8463. [CrossRef]
52. Wielgohs, J.; Lechevalier, A. Introduction. In *Borders and Border Regions in Europe: Changes, Challenges and Chances*; Lechevalier, A., Wielgohs, J., Eds.; Transcript Verlag, JSTOR: Bielefeld, Germany, 2013; pp. 9–18.
53. Demeterova, B.; Goodwin-Hawkins, B.; Fischer, T. Conceptualisations of Territorial Cohesion in Central European border regions. *Eur. Plan. Stud.* **2020**, *28*, 2287–2306. [CrossRef]
54. 2020 Update Evaluation of the Interreg Europe Programme. Final Report, July 2020. Available online: https://www.Interregeurope.eu/sites/default/files/2022-11/2020-07-28%20IR-E%20Update%20Evaluation%20-%20Final%20report.pdf (accessed on 5 June 2023).
55. Medeiros, E.; Guillermo Ramírez, M.; Brustia, G.; Dellagiacoma, A.C.; Mullan, A.C. Reducing border barriers for cross-border commuters in Europe via the EU *b-solutions* initiative. *Eur. Plan. Stud.* **2023**, *31*, 822–841. [CrossRef]
56. Medeiros, E.; Guillermo Ramírez, M.; Dellagiacoma, C.; Brustia, G. Will reducing border barriers via the EU's *b-solutions* lead towards greater European territorial integration? *Reg. Stud.* **2021**, *56*, 504–517. [CrossRef]
57. European Commission. *More and Better Cross-Border Public Services. Obstacles and Solutions to Cross-Border Cooperation in the EU*; Publication Office of the European Union: Luxembourg, 2021.
58. European Commission. *Vibrant Cross-Border Labour Markets. Obstacles and Solutions to Cross-Border Cooperation in the EU*; Publication Office of the European Union: Luxembourg, 2021.
59. Sandri, S.; Hussein, H.; Alshyab, N.; Sagatowski, J. The European Green Deal: Challenges and opportunities for Southern Mediterranean. *Mediterr. Politics* **2023**, 1–12. [CrossRef]
60. European Commission. *Border regions for the European Green Deal. Obstacles and Solutions to Cross-Border Cooperation in the EU*; Publication Office of the European Union: Luxembourg, 2021.
61. Zikos, D. Revisiting the Role of Institutions in Transformative Context: Institutional Change and Conflicts. *Sustainability* **2020**, *12*, 9036. [CrossRef]
62. Ezers, J.; Naglis-Liepa, K. The role of institutions in regional development. In Proceedings of the 2019 International Conference Economic Science for Rural Development, Jelgava, Latvia, 9–10 May 2019; No. 51, pp. 107–111.
63. Vanhoonacker, S.; Neuhold, C. Dynamics of institutional cooperation in the European Union: Dimensions and effects. An Introduction. *Eur. Integr. Online Pap.* **2015**, *19*, 1–15.

64. Association of European Border Regions. *B-Solutions: Solving Border Obstacles. A Compendium of 43 Cases*; Publications Office of the European Union: Luxembourg, 2020.
65. Association of European Border Regions. *B-Solutions: Solving Border Obstacles. A Compendium 2020–2021*; Publications Office of the European Union: Luxembourg, 2021.
66. European Court of Auditors. *Interreg Cooperation: The Potential of the European Union's Cross-Border Regions Has Not Yet Been Fully Unlocked*; Special Report 14; Publication Office of the European Union: Luxembourg, 2021.
67. Brie, M.; Mărcuț, M.; Polgar, I. Developing international cooperation capabilities to boost local development and social responsibility. Case study: Bihor County. *Civ. Sale* **2022**, *73*, 71–90.

sustainability

Article

Analysing the Impact of the Bleaching Process on Wet Spun Hemp Yarn Properties

Simona Tripa [1], Naz Kadınkız [2], Ayesha Kanwal [3], Muhammad Anwaar Nazeer [3,*], Ahsan Nazir [3], Florin Tripa [1] and Muhammet Uzun [2,*]

1 Department of Textiles, Leather and Industrial Management, Faculty of Energy Engineering and Industrial Management, University of Oradea, 410087 Oradea, Romania; simona.tripa@didactic.uoradea.ro (S.T.); tripa.florinmarin@student.uoradea.ro (F.T.)
2 Department of Textile Engineering, Faculty of Technology, Marmara University, Istanbul 34722, Türkiye; nazkadinkiz@gmail.com
3 School of Engineering and Technology, National Textile University, Faisalabad 37610, Pakistan; ahsan@ntu.edu.pk (A.N.)
* Correspondence: mnazeer13@ku.edu.tr (M.A.N.); m.uzun@marmara.edu.tr (M.U.)

Abstract: Historically, cotton has been regarded as a highly sustainable material; however, thorough research indicates otherwise. The increasing levels of pollution and the need to address climate change have led towards a global search for sustainable alternatives. Plants with comparable chemical compositions, such as hemp, are attracting growing attention. The cultivation of hemp can be done with sustainable methods, thereby making it a viable alternative to cotton. This study investigates the mechanical, physical, and dyeing properties of 100% wet-spun hemp yarn in its natural and bleached state with the objective of incorporating its use in both technical and traditional textiles. Although significant academic literature is available on the properties of cotton, there is a noticeable lack of literature based on wet-spun hemp. This research suggests that the bleaching process positively affects wet-spun hemp yarn, thus making it suitable for use by the textile industries in various applications.

Keywords: wet-spun hemp; bleached hemp yarn; sustainable alternatives

Citation: Tripa, S.; Kadınkız, N.; Kanwal, A.; Nazeer, M.A.; Nazir, A.; Tripa, F.; Uzun, M. Analysing the Impact of the Bleaching Process on Wet Spun Hemp Yarn Properties. *Sustainability* **2023**, *15*, 16894. https://doi.org/10.3390/su152416894

Academic Editor: Maria L. Auad

Received: 16 October 2023
Revised: 9 December 2023
Accepted: 12 December 2023
Published: 15 December 2023

1. Introduction

As a natural fibre, cotton was once considered the epitome of sustainability due to its biodegradable and compostable properties. Known for its incomparable functional and mechanical and physical properties, cotton is used in various applications and industries [1–3]. Comprehensive research has uncovered several drawbacks to the sustainability of cotton fibre, including its excessive use of land, water, pesticides, and its contamination of water. Although organic cotton doesn't use fertilizers, it cannot be viewed as a universal substitute for conventional cotton because it accounts for less than 1% of all cotton production [4]. The global search for a more sustainable alternative has been initiated due to rising pollution, efforts to combat climate change, and the textile industry's position as the second-most environmentally detrimental sector [5].

Plants with a comparable chemical composition, such as hemp, are being considered promising research subjects due to their lower land, water, and pesticide requirements [6,7]. Hemp fibre falls into the category of bast fibres [8]. It is derived from the hemp plant (Cannabis sativa). It is a member of the Cannabaceae family, the most well-known of which is the Cannabis genus, which consists of two main species: Cannabis sativa and Cannabis indica [9,10]. Each species has unique morphological characteristics, growth patterns, and chemical compositions. These plants can be differentiated by the presence of resinous structures primarily located in their stems, leaves, and flowers. These structures are responsible for the production of active compounds like tetrahydrocannabinol (THC) and cannabidiol (CBD). THC and CBD levels can vary between sativa and indica, with

Cannabis sativa plants having generally higher THC levels and Cannabis indica plants having generally higher CBD levels [11,12]. Due to the presence of such cannabinoids, the cultivation and use of hemp plants can be subjected to legal regulations in some countries, while being completely restricted in others [13–15].

Hemp cultivation can be done using environmentally sustainable practices which may result in positive effects on the soil. In particular, hemp has been observed to enhance the soil's nutrient and water retention capabilities, stimulate microbial activity, and enhance soil structure [16–18]. A water retting or dew retting process can be used to break down the parenchyma cells, after which the stalks are processed in decortication machines [19–21]. The process of retting hemp usually takes 10 to 20 days, which helps to separate fibres from the plant's stalk [22]. Furthermore, these fibres can be produced via two distinct methods: dry spinning and wet spinning [23,24]. For this study, yarn made of wet-spun fibres is used. During the wet spinning process, the fibres undergo a preliminary treatment of immersion in heated water tanks. This step softens the pectin, which facilitates the subsequent drawing and separation of the fibres. Additionally, this treatment helps the development of fibre ribbons which ultimately enhances the strength and durability of the fibres [25–27]. The fibres are then aligned in parallel and twisted into yarn. The yarn can be bleached, which substantially affects variables such as the colour, strength, and uniformity of the yarn. Various chemical processes, such as sodium sulphite, sodium bisulphite, sodium hydroxide with EDTA, boiling with sodium hydroxide after treatment with hydrochloric acid, and boiling in oxalic acid, are used for bleaching. The chemical composition of hemp fibre can be seen in Figure 1, which has been adapted from [28]. Hemp fibres are currently being used in composites, insulation, and building materials. Investigating the mechanical and physical properties as well as the dyeing properties of this fibre and its bleached variant produces valuable findings that can improve the utilisation of hemp in the textile sector for high-value applications.

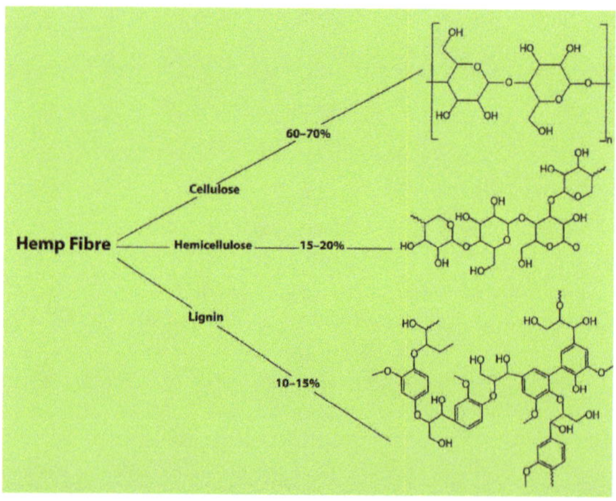

Figure 1. The chemical composition of hemp fibre.

The main objective of this study is to investigate the physical, mechanical, and dyeing properties of 100% wet-spun hemp yarn, both in its natural and bleached states, for use in technical and traditional textiles. It is necessary to analyse the mechanical and physical characteristics of the yarns in both their unbleached and bleached conditions to see the modifications in the structure as induced by the process of bleaching. Additionally, within the scope of the study, the natural and bleached variants of the yarns were dyed using reactive dyes, and a comprehensive analysis was conducted to examine the effects of the dyeing process. Hemp fibres are commonly characterized by their natural colouration, which normally ranges from beige to brown to light yellow. The variation in fibre colour is dependent on the colour of the fibres present in the plant stem, the age of the plant, and the post-harvest processing of the fibres. Based on the preferred hues of the textile industry, five potential yarn colours were selected which were then achieved by the reactive dyeing process and were studied. Although academic work is abundant on the properties of cotton, there is a lack of information when it comes to wet-spun hemp yarn.

2. Materials and Methods

2.1. Materials

Each test was performed by using 10 replicates of bleached and natural hemp. All tested parameters were conducted under standard test conditions at the temperature of 20 ± 2 °C and relative humidity of 65 ± 4%. The yarn replicates were conditioned for 24 h at standard atmospheric pressure with a relative humidity of 65 ± 2% and a temperature of 21 ± 1 °C before the dying process.

Natural and bleached twisted wet-spun hemp yarn was sourced from China. The abbreviations used for the types of yarn are given in Table 1.

Table 1. Sample abbreviation.

Sample Abbreviation	Sample Name
N-HEMP	Natural Wet-Spun Hemp
B-HEMP	Bleached Wet-Spun Hemp

Reactive dyes are used for this study due to their uniform colour distribution. The selected dyestuff complies with the OEKO-TEX and GOTS standards [29]. The list of dyestuffs used in this study is given in Table 2. Other ingredients in the dye solution include urea, caustic, and silicates.

Table 2. List of dyes.

Name of Dye	Trade Name of Dye
Reactive Yellow	Remazol Yellow SAM
Reactive Red	Remazol Red SAM
Reactive Blue	Remazol Blue SAM
	Remazol Navy RGB 150%
Reactive Black	Remazol Map Black NN

2.2. Methods

2.2.1. Yarn Count and Twist Test

The effect of bleach on the yarn count and twist was evaluated on a total of 10 replicates of each yarn. The yarn count test was conducted in accordance with ASTM D1059 standards [30] using a yarn count measuring device, while the twist determination test was carried out using a twist determination device. The experiments were conducted twice and the obtained data was averaged.

2.2.2. Yarn Evenness Test

To investigate the effect of bleach on yarn evenness, a yarn evenness level test was conducted using the Uster Tester 5. The measures include various fundamental characteristics that are crucial for the examination of yarn quality. These criteria include the unevenness percentage (%Um), the coefficient of variation, the mean value coefficient (%CVm), the index, thick places, thin places, neps, and the hairiness of the yarn. Each replicate was tested twice and the data was subjected to statistical analysis. The outcome was presented alongside measures of evenness and standard deviation. The surface appearance of yarn has also been determined using this test.

2.2.3. Surface and Morphological Structure

The impact of bleach on the structural properties and quality of yarn was assessed by observing the morphological characteristics of the fibre and yarn using a light microscope. The examination of the 10 replicates involved the analysis of the yarn diameter, the surface properties, and morphological data at various magnification levels. These images were processed using computer-aided image analysis software.

2.2.4. Yarn Cross-Section Fibre Count Test

The yarn cross-section fibre count determination test was conducted to evaluate the impact of bleach on the yarn's quality, performance, and durability by measuring the number of fibres it contains. To conduct this test, the yarn is cut across its width and is examined under a microscope to determine the fibre density.

2.2.5. Fourier-Transform Infrared Spectroscopy Test

The chemical composition of the yarns was analysed using a Thermo Scientific Nicolet iS50 FTIR spectrometer to detect the effects of bleach. The equipment was sourced from Massachusetts, USA.

2.2.6. Thermogravimetric Analysis

The thermal properties of wet-spun hemp yarn variants were examined through the thermogravimetric study (TGA). This study was conducted using the NETZSCH STA 449F3 Jupiter. The equipment was sourced from Selb, Germany. The acquired data was analysed in terms of thermal degradation behaviour and the profile of weight loss.

2.2.7. Strength Test

The strength of the yarn was determined through the Instron tensile strength test using the Instron universal testing machine. The 10 replicates were subjected to a constant rate of tension. The test was performed twice on each replicate, and the result was subjected to statistical analysis. The result was presented along with strength value and standard deviation. The tensile strength was again tested using a double-ply yarn to get more precise results with fewer discrepancies caused by variations in fibre thickness. The yarn was twisted at 300 turns per minute (TPM). The yarn twist test was then performed using the Saurer ring twisting machine to examine the increase in strength. The twisting machine was sourced from Arbon, Switzerland.

2.2.8. Strength Test of Dyed Yarn

The samples were subjected to reactive dyes for analysis. Table 3 presents the dye formulations used in the reactive dyeing process of the yarns. The dye was combined with hot water and stirred with a magnetic stirrer until it was fully dissolved. Urea was used as an immersion agent. The temperature was carefully controlled during the procedure. Caustic and silicate compounds were added. After achieving homogeneity, the dye solution was introduced into a vat to dye the yarn. Subsequently, the dyed yarn was carefully enclosed in an air-tight nylon bag for a standard waiting time of 20 h. The dyed yarn was technically washed for five minutes to remove any residual dye solution. The strength properties of the dyed yarn were then examined.

2.2.9. Colorimetric Data Analysis

ColorTools QC 2.4.3 software was used along with the Datacolor device for the Colorimetric Data Analysis. The L*, a*, and b* values of the dyed yarns were measured three times using a dual-position spectrophotometer under D65 daylight, F11 10, and A 10 light conditions. The primary objective of this test was to present quantifiable data that is not visible to the naked eye. This evaluation offered a foundational understanding of the colour characteristics of wet-spun hemp yarns.

Table 3. Dye recipe.

	%	Trade Name of Dye
Grey		
Reactive Yellow	0.06	Remazol Yellow SAM
Reactive Red	0.05	Remazol Red SAM
Reactive Blue	0.06	Remazol Blue SAM
Brown		
Reactive Yellow	0.5	Remazol Yellow SAM
Reactive Red	0.26	Remazol Red SAM
Reactive Blue	0.25	Remazol Blue SAM
Green		
Reactive Yellow	1.3	Remazol Yellow SAM
Reactive Red	0.5	Remazol Red SAM
Reactive Blue	1.0	Remazol Blue SAM
Navy Blue		
Reactive Yellow	0.6	Remazol Ultra Yellow RGB
Reactive Red	1	Remazol Ultra Red RGB
Reactive Blue	4	Remazol Navy RGB 150%
Black		
Reactive Black	6	Remazol Map Black NN

3. Results and Discussion

3.1. Yarn Count and Twist Test

To evaluate the effect of bleach on the yarn twist, the average twist values of the samples were measured. The findings in Table 4 suggest that bleaching had no impact on the yarn count but it did lead to an increase in the yarn twist. The twist value of hemp had an increase of 5–8%. This result can be attributed to the use of peroxide in the carding process which tightens the voids in the morphology of the fibres, resulting in greater twist values.

Table 4. Yarn count and twist values of N-HEMP and B-HEMP.

Sample Code	Yarn Count (Ne)	T/m
N-HEMP	16	337
B-HEMP	16	366

3.2. Yarn Evenness Test

The impact of bleach on the uniformity of the yarn is observed through the assessment of the yarn's evenness. The Um% specifies variation in the yarn thickness or diameter, with a lower value indicating more evenness. The CV% represents the variation in the yarn thickness, with lower values representing less variation. The index represents the overall evenness of the yarn, with a low index signifying improved evenness. The results in Table 5 suggest that the use of bleach has resulted in a partially positive outcome for the hemp yarn. It can be observed by the improved uniformity and a considerable reduction in the irregularities, such as in the thick and thin places of the yarn. However, the neps and hairiness have increased. This may be attributed to the partial modifications to the cellulose structure caused by the bleaching process, which led to alterations in the uniformity of the yarn structure.

Table 5. Yarn evenness values of N-HEMP and B-HEMP.

Sample	Um (%)	CVm (%)	Index (−)	Thin Places (−50%)	Thick Places (+50%)	Neps (+200%)	Hairiness (−)
N-HEMP	26.43	33.59	4.95	4814	2949	5596	1.87
B-HEMP	19.87	25.75	3.79	1104	2204	5614	2.62

3.3. Surface Appearance of the Yarn

The surface appearance of yarn is characterised by the fineness value of the yarn. The fineness values of the yarn indicate whether the fibres in the yarn are thin or thick. An abundance of thin places suggests a lack of homogeneity and weakness. Similarly, an abundance of thick places indicates irregularity in the yarn. In Table 6, bleached hemp shows considerably lowered minimum and maximum thickness values, which indicates that the yarn has fewer thin and thick places, specifying a more homogenous structure. The variation in the thickness of N-Hemp can be seen in Figure 2, while the variation in the thickness of B-HEMP can be seen in Figure 3. The sample can be seen in images A, B, and C in Figures 2 and 3 at different magnifications and places.

Table 6. Yarn surface thickness of N-HEMP and B-HEMP.

Thickness Value	N-Hemp	B-Hemp
Minimum (μm)	204.16	123.61
Maximum (μm)	373.19	267.61

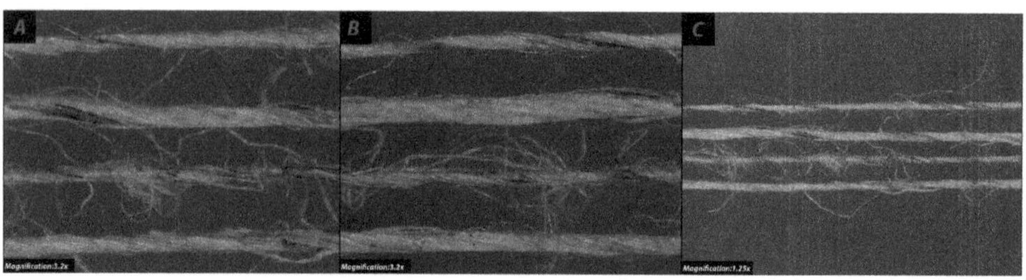

Figure 2. The thickness of N-HEMP under the microscope at different places and magnifications of 3.2× (**A**), 3.2× (**B**) and 1.25× (**C**).

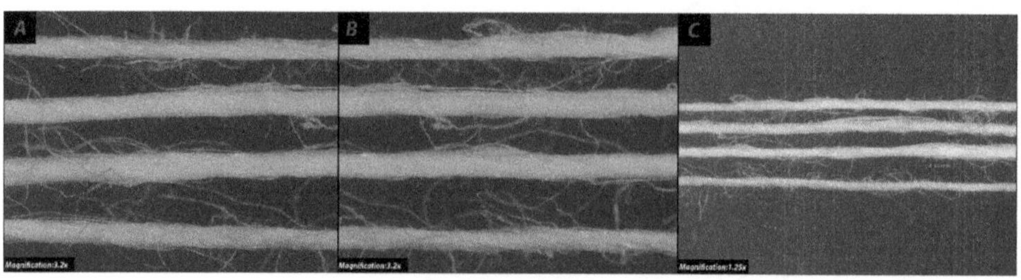

Figure 3. The thickness of B-HEMP under the microscope at different places and magnifications of 3.2× (**A**), 3.2× (**B**) and 1.25× (**C**).

3.4. Surface and Morphological Structure of the Yarn

By using a light microscope, the effect of bleach on the structural attributes and quality of the yarn was assessed. In the course of fibre development, there are specific stages at which growth ceases. At these specific points, referred to as nodes, fibres gather crucial nutrients and amorphous regions are formed. Figure 4 confirms when examining the samples under a microscope and using a compensator to bend light, it was noted that the nodal regions displayed colours that fall within the yellow and red spectrum at wavelengths ranging from 700 to 560 nm. In areas with an abundance of crystalline regions and continuous growth, the wavelengths of light ranged from 560 to 400 nm and visual observations indicated the presence of green, blue, and purple hues. Upon examination of the fibre diameters in the horizontal plane, it was obvious that the nodal regions exhibited greater thickness compared to the regions of uninterrupted growth. The strength loss is lower where the amorphous region is more dominant than the crystalline region. Therefore, it can be seen that fractures usually occur in the crystalline regions during strength tests. Moreover, in Figure 5 the microscopic examination of bleached yarns provided visual evidence of the impact of the bleach, signifying the presence of surface irregularities and chemical accumulation on the fibres.

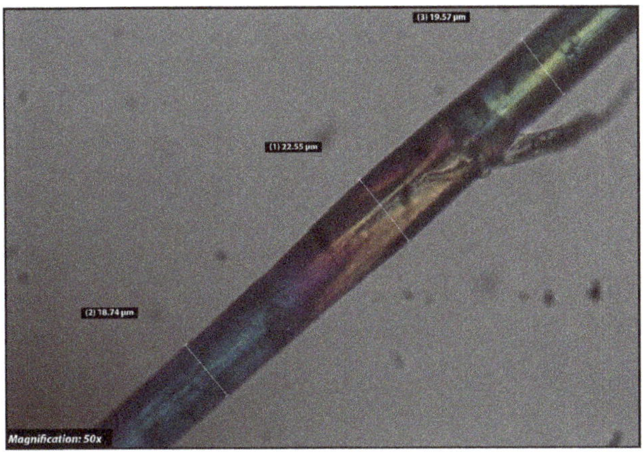

Figure 4. Amorphous and crystalline regions of N-HEMP fibre under the microscope.

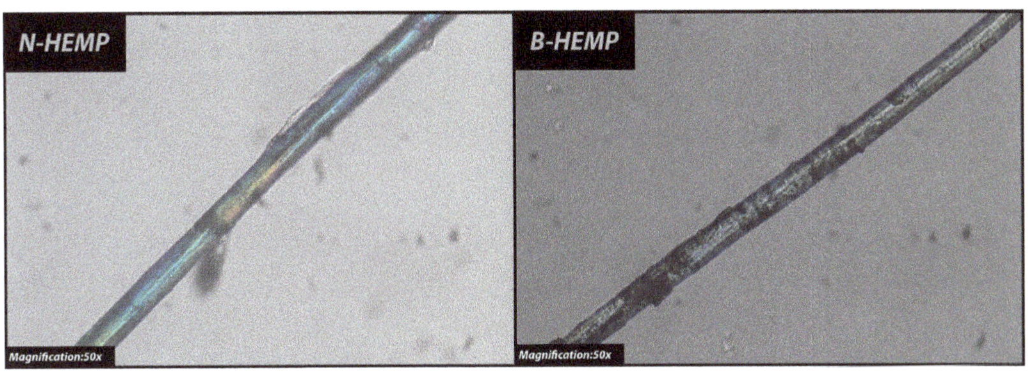

Figure 5. Comparison of N-HEMP and B-HEMP under a microscope.

3.5. Determination of Fiber Count in Yarn Cross-Section

The findings in Table 7 from the yarn cross-section fibre count determination test indicate that the measured total cross-sectional area of N-HEMP is 35,114.95 μm^2. Furthermore, the measurements for the diameters of the fibres are 18.24 μm, 16.23 μm, 24.43 μm, and 12.68 μm, respectively. The presence of different-sized fibres within the yarn has been observed. The average number of fibres per cross-sectional area is 132.50. Figure 6 contains the cross-sectional view of N-HEMP with fibre diameters. According to Table 8, the measured total cross-sectional area of B-HEMP is 56,944.08 μm^2 and it is noted that the fibres have varying diameters. The average number of fibres in the cross-section is 190.88. Based on the observed average fibre count of 190.88, it can be concluded that B-HEMP displays a higher fibre density. Figure 7 contains the cross-sectional view of B-HEMP with fibre diameters. The increased abundance of fibres inside the cross-sectional area may indicate a more compact yarn structure, potentially resulting in improved strength properties. However, the notable variations in diameters and the non-uniform dispersion of fibres suggest that the yarn is not homogeneous.

Table 7. N-HEMP cross-sectional fibre count.

Yarn Total Area	Fibre Diameter/Radius (μm)	Fibre Area (μm^2)	Average Fibre Count
35,114.95 μm^2	18.24–9.12	261.167	132.50
	16.23–8.115	206.779	
	24.43–12.215	468.507	
	12.68–6.34	126.214	
Average	17.895	265.5	

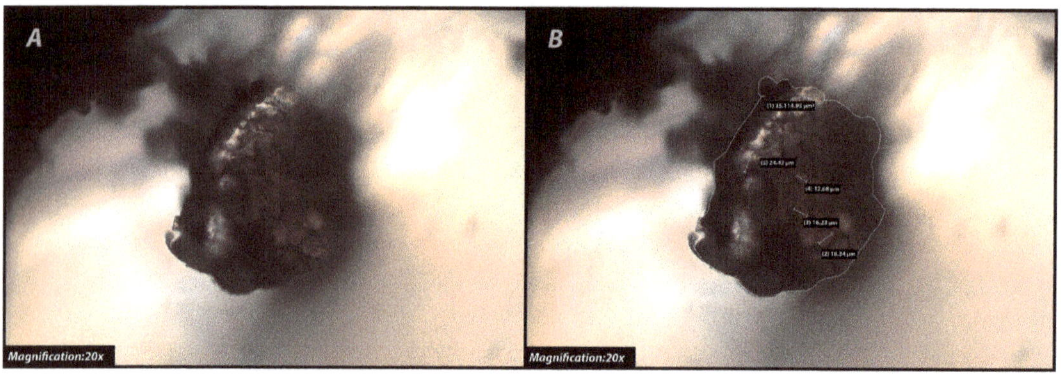

Figure 6. (**A**) N-HEMP cross-sectional view without the fibre diameters. (**B**) N-HEMP cross-sectional view with the fibre diameters.

Table 8. B-HEMP cross-sectional fibre count.

Yarn Total Area	Fibre Diameter/Radius (μm)	Fibre Area (μm^2)	Average Fibre Count
56,944.08 μm^2	20.6–10.3	333.122	190.881
	25.51–12.755	510.84	
	5.20–2.6		
	8.06–4.03	50.996	
Average		298.321	

3.6. Fourier-Transform Infrared Spectroscopy

FTIR analysis is used to determine the chemical bonds and components of the material. In this analysis, infrared light was used to examine how the material reacts in the infrared region (usually 4000 to 400 cm^{-1}). The results in Figure 8 indicate that there was no difference in wavelength range and intensity between N-HEMP and B-HEMP. This specifies that these yarns have the same chemical composition structurally.

3.7. Thermogravimetric Analysis

The thermogravimetric study revealed that the thermal decomposition of the N-HEMP initiates at 369.58 °C. At this point, the yarn begins a process of heat deterioration, resulting in a reduction of weight. The temperature at which the yarn shows the greatest weight loss is 513.25 °C. During this stage, the heat deterioration of the yarn becomes more rapid, resulting in additional reduced weight. The residual weight is determined to be 0.624%, representing the proportion of the yarn that remains after thermal degradation. On the other hand, the thermal decomposition of the B-HEMP initiates at 373.25 °C, which is slightly higher than N-HEMP. The temperature at which the highest weight loss occurred is 808.25 °C, which is also higher than the temperature observed for N-HEMP. The residual weight is 0.328%, which is lower compared to the residual weight observed in N-HEMP. Furthermore, this comparison of N-HEMP and B-HEMP is illustrated in Figure 9. The observed rise in the temperature at which initial weight loss occurs suggests that the bleaching process had a positive impact on the thermal stability of the yarn. This observation signifies that the bleached yarn experiences thermal deterioration at elevated temperatures, leading to a comparably substantial reduction in weight. The residual weight of B-HEMP is 0.328%, which suggests that the bleaching process altered the structural composition of the yarn, which led to a decreased number of organic components in the yarn.

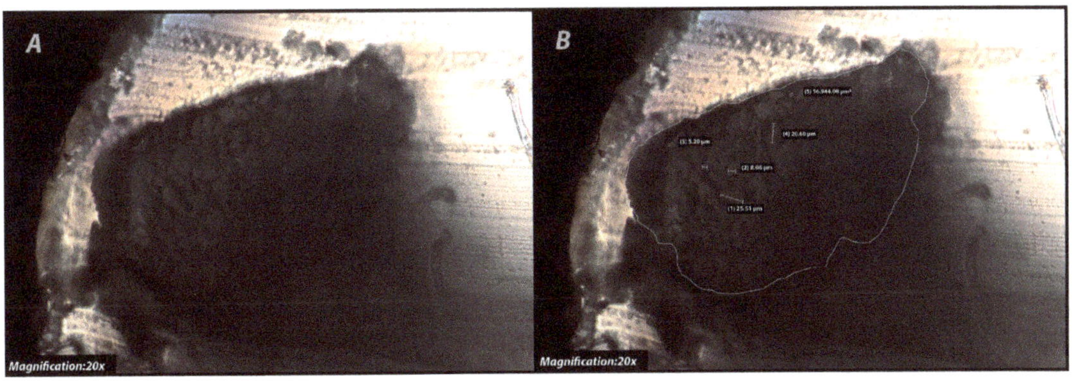

Figure 7. (**A**) B-HEMP cross-sectional view without the fibre diameters. (**B**) B-HEMP cross-sectional view with the fibre diameters.

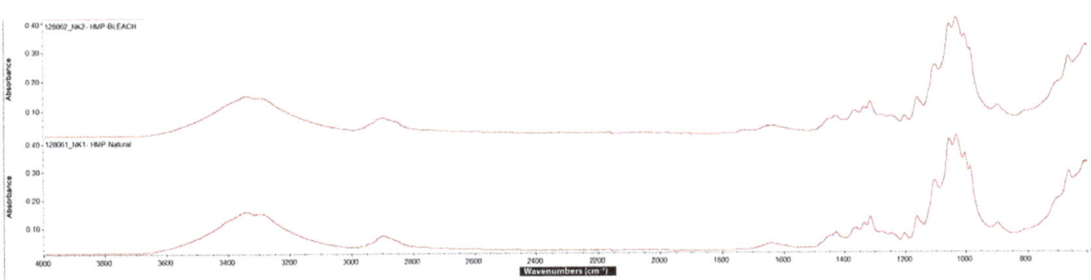

Figure 8. FTIR result of N-HEMP and B-HEMP.

3.8. Strength Test and the Examination of Breakage Points

The strength test results displayed in Table 9 indicate that the breaking strength, elongation, and tenacity values of hemp yarns display an increase following the bleaching process. This finding suggests that the bleaching process improved the mechanical characteristics, elasticity, and energy absorption of the yarn.

Different breakage points were detected throughout the strength tests conducted on both natural and bleached yarns as shown in Figure 10. During the examination under a microscope, a notched structure was seen at the breakage points. The presence of a notched structure in the cross-section of the yarns during the strength tests suggests that the breakage of the yarns occurred gradually. It was also noted that there were two distinct points of weakness at the breakage points. At one point, a complete breakage occurred instantly, while at the other point, there was a gradual weakening resulting in strength loss. This problem was attributed to inconsistencies in the yarn, resulting from variations in its thickness along the length of the yarn. Microscopic visual inspections were performed to determine the distribution of yarn thickness from a horizontal perspective, allowing the identification of thick and thin places within the yarn. The examination revealed varied thickness distributions and variations in fibre lengths.

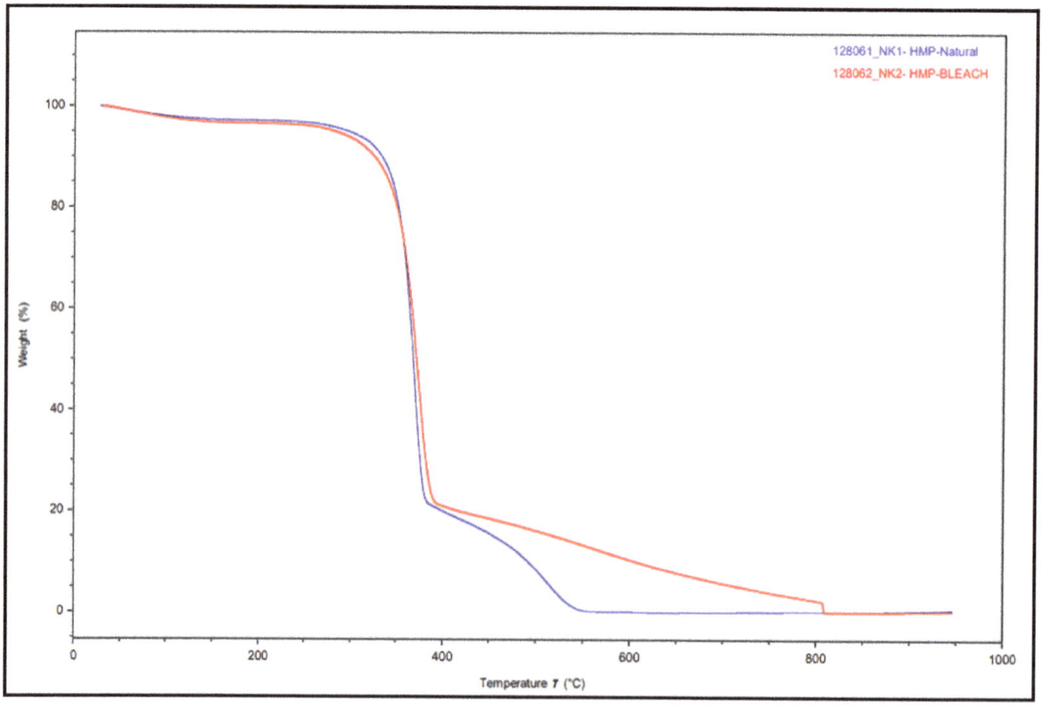

Figure 9. Comparison of TGA results of N-HEMP and B-HEMP.

Table 9. Strength values of samples.

Sample	Breaking Strength (N)	Elongation (%)	Tenacity (cN/tex)
N-HEMP	9.50	2.18	25.74
B-HEMP	12.62	2.43	34.19

To obtain more accurate results with fewer inconsistencies caused by variations in fibre thickness, the tensile strength test was repeated using a double-ply yarn. The process involved the double-plying of both 16-count natural and bleached hemp yarns using an Alma Saurer brand ring twisting machine, operating at a speed of 300 TPM. The twisting procedure led to a reduction in inconsistencies. The results in Table 10 indicate that double-plied twisted yarns display a more regular surface appearance with a reduction of the variation of thin and thick places. The thickness of the double-plied yarn samples can be seen in Figure 11 at different magnifications.

The bleached double-plied wet-spun hemp yarn exhibits superior maximum load and elongation at break values in comparison to the natural double-plied hemp yarn. However, the tenacity values show similarity. Based on the findings presented in Tables 11 and 12, it can be concluded that the bleached double-plied wet-spun hemp yarn shows a notable strength gain.

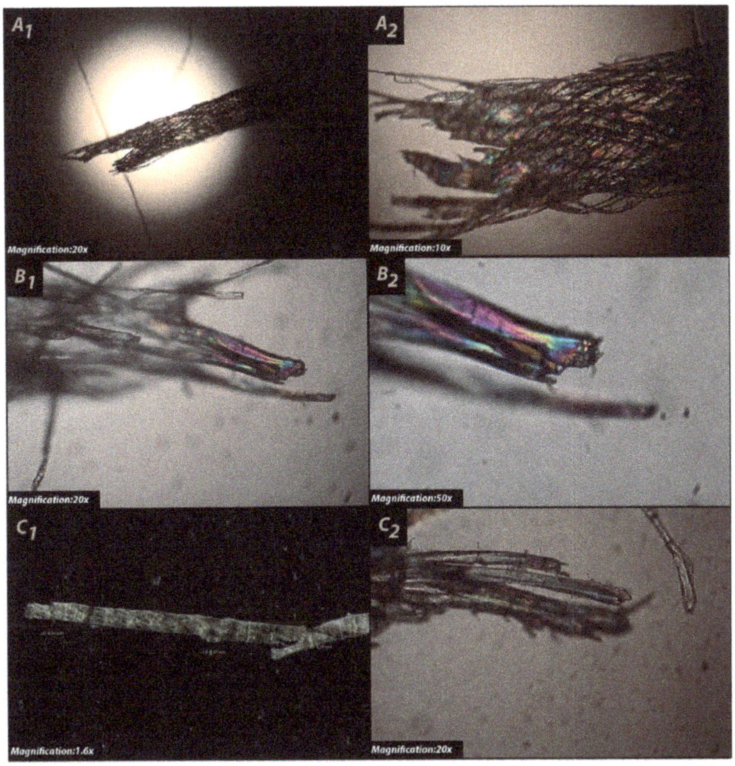

Figure 10. (A_1,A_2) Breakage points of N-HEMP. (B_1,B_2) Breakage points of B-HEMP fibre. (C_1) Breaking of B-HEMP during the strength test. (C_2) Breaking of B-HEMP fibre during the strength test.

Table 10. The surface thickness of the double-plied yarn.

Thickness Value	Natural Hemp Double—Ply Yarn	Bleached Hemp Double—Ply Yarn
Minimum (µm)	302.80	316.00
Maximum (µm)	352.00	415.60

According to Table 13, N-HEMP and B-HEMP with a yarn count of Ne 16 show different breaking strength and elongation properties. The average breaking strength value for N-HEMP is 1.41 kgf, while it is 1.28 kgf for B-HEMP. The standard deviation values are 0.15 and 0.09 respectively. On the other hand, the average elongation value of N-HEMP is 2.33%, whereas it is 2.45% for B-HEMP. In the case of 2-Ply yarns with a yarn count of Ne 16×2, the average breaking strength of N-HEMP is 1.21 kgf while it is 1.33 kgf for B-HEMP. The average elongation value of N-HEMP is 2.54% while is 3.02% for B-HEMP.

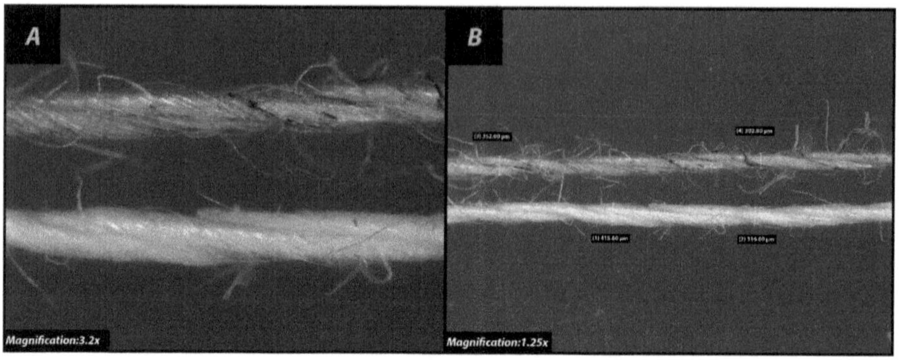

Figure 11. Thickness of double-plied samples under a microscope at different magnifications of 3.2× (**A**) and 1.25× (**B**).

Table 11. Strength values of natural double-plied hemp yarn.

	Maximum Load (N)	**Breaking Elongation (%)**	**Tenacity (%)**
Average	11.87	2.54	1.56
Standard Deviation	2.85	0.44	0.37
Minimum	7.79	1.84	1.03
Maximum	16.23	3.23	2.14

Table 12. Strength values of bleached double-plied hemp yarn.

	Maximum Load (N)	**Breaking Elongation (%)**	**Tenacity (%)**
Average	13.03	3.02	1.71
Standard Deviation	2.42	0.48	0.32
Minimum	7.87	2.24	1.04
Maximum	16.25	4.30	2.14

Table 13. Strength values of twisted and non-twisted samples.

	Yarn Count	**Breaking Strength (kgf)**		**Breaking Elongation (%)**	
		Average	**Standard Deviation**	**Average**	**Standard Deviation**
N-HEMP	Ne 16	1.41	0.15	2.33	0.20
B-HEMP	Ne 16	1.28	0.09	2.45	0.09
N-HEMP 2-Ply Twisted	Ne 16×2	1.21		2.54	
B-HEMP 2-Ply Twisted	Ne 16×2	1.33		3.02	

3.9. Strength Test of Dyed Yarn

The strength test of dyed N-HEMP revealed in Table 14, noticeable variations in strength characteristics across yarn of different colours, including maximum load, elongation at break, and stress under maximum load. Furthermore, it is worth noting that the coefficient of variation values fluctuated among yarns of different colours, thereby suggesting that certain colours possess more consistent strength properties. This may be due to multiple reasons such as the amount of the dye substance, the level of penetration of the dye, the size and structure of the molecule of the dye and the chemical interaction between the dye and the yarn. The grey-coloured N-HEMP showed the highest maximum load value among the other yarns, with a measurement of 1211.01 cN. Similarly, the grey colour exhibited the lowest coefficient of variation, suggesting that the strength qualities of grey yarns are of higher consistency. The stress and Strain graph of N-HEMP and B-HEMP dyed in grey colour is visualised in Figures 12 and 13.

The strength test results of dyed B-HEMP are presented in Table 15. The yarns with a brown colour showed the highest maximum load value of 1684.89 cN. The highest coefficient of variation among yarns of the same colour suggests a higher level of variability in their strength properties. Further observation revealed that dyed B-HEMP showed superior qualities in comparison to N-HEMP, specifically in the case of grey-coloured yarn. On the other hand, yarns of alternative colours exhibited poorer values in comparison to N-HEMP.

Table 14. Strength test values of dyed N-HEMP.

	Grey	Brown	Green	Navy Blue	Black
Maximum Load (cN)					
Coefficient of Variation	44.90	6.70	38.89	13.78	25.32
Maximum	1211.01	855.61	1105.71	934.58	947.75
Average	808.22	794.18	770.05	816.12	740.43
Standard Deviation	362.95	53.20	299.49	112.46	187.51
Elongation at Break (mm)					
Coefficient of Variation	28.74	7.01	29.50	24.64	14.46
Maximum	20.67	13.48	19.14	16.26	13.79
Average	14.25	12.51	13.36	12.70	11.43
Standard Deviation	4.09	0.88	9.94	3.13	1.65
Stress Under Maximum Load (%)					
Coefficient of Variation	28.28	11.55	26.80	6.40	17.82
Maximum	3.79	3.29	3.3	2.81	3.16
Average	3.91	3.02	2.73	2.64	2.60
Standard Deviation	0.82	0.35	0.73	0.17	0.46

3.10. Colorimetric Data Analysis

Colorimetric Data Analysis was conducted and the colour coordinates deviations through plane analysis were presented in Figures 14–18. While these deviations vary according to colour tones, no significant changes were observed in the samples.

The colour analysis was conducted under different light sources. The DL* values represent the difference in colour brightness, while the Da* and Db* values indicate colour differences along the green/red and blue/yellow axes, respectively. The DC* value represents the difference in chromaticity, and the Dh* value denotes the difference in colour tone. CMC dE value is used as a metric to calculate the overall colour difference. According to the results, there are some differences in the colour values of N-HEMP and B-HEMP under different light conditions.

For grey and brown samples, negligible differences in lightness (DL) were observed, while a significant difference in colour values Da, Db, Dh and Dc and the overall colour difference DE CMC were observed.

For green and navy-blue samples, a different trend was observed as compared to that observed for grey and brown samples. Here, the difference in colour coordinates for both samples was much lower. This was also evident in the overall colour difference having a maximum value of 0.95, which falls under the acceptable range in industrial colour evaluation practices.

The colour differences as described above were also observed during visual evaluation as depicted in Figure 19. These colour differences could be attributed to the difference in absorption and fixation tendency of the natural and bleached samples. However, the lower difference observed in Navy Blue and Green colours could be attributed to the high dye concentration used for these colours which could have accommodated the difference in dye absorption.

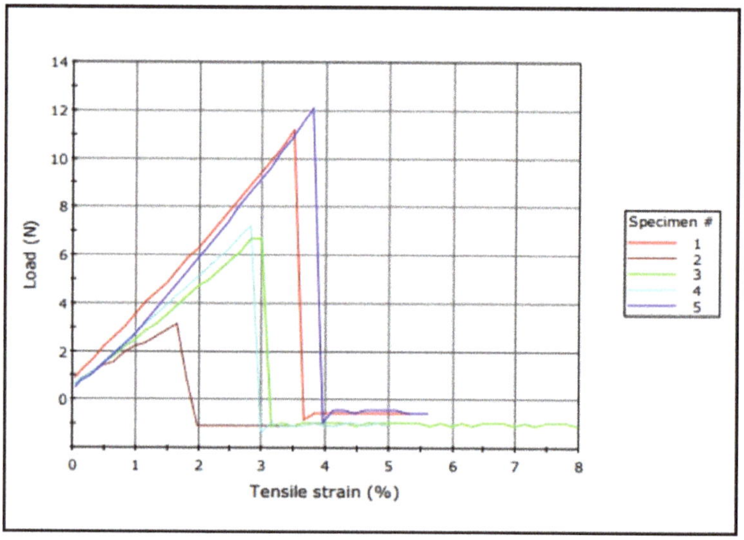

Figure 12. Stress-strain graph of N-HEMP dyed in Grey colour.

Table 15. Strength test values of dyed B-HEMP.

	Grey	Brown	Green	Navy Blue	Black
Maximum Load (cN)					
Coefficient of Variation	10.44	15.64	18.24	11.23	14.26
Maximum	1461.11	1684.89	1461.11	1474.28	1368.97
Average	1360.20	1443.57	1270.25	1307.55	1175.91
Standard Deviation	141.97	225.83	231.64	146.78	167.71
Elongation at Break (mm)					
Coefficient of Variation	6.91	12.98	16.37	8.97	25.21
Maximum	14.90	17.31	17.72	13.72	20.22
Average	13.81	15.35	15.15	13.01	15.83
Standard Deviation	0.96	1.99	2.48	1.7	3.99
Stress Under Maximum Load (%)					
Coefficient of Variation	4.27	4.53	12.42	6.87	7.80
Maximum	4.15	4.33	3.96	3.82	3.66
Average	3.97	4.21	3.56	3.60	3.38
Standard Deviation	0.17	0.19	0.44	0.25	0.26

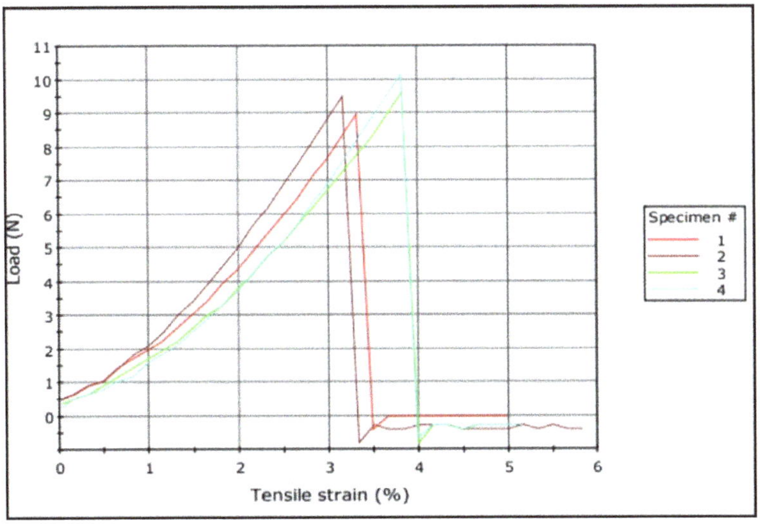

Figure 13. Stress/strain graph of B-HEMP dyed in Grey Colour.

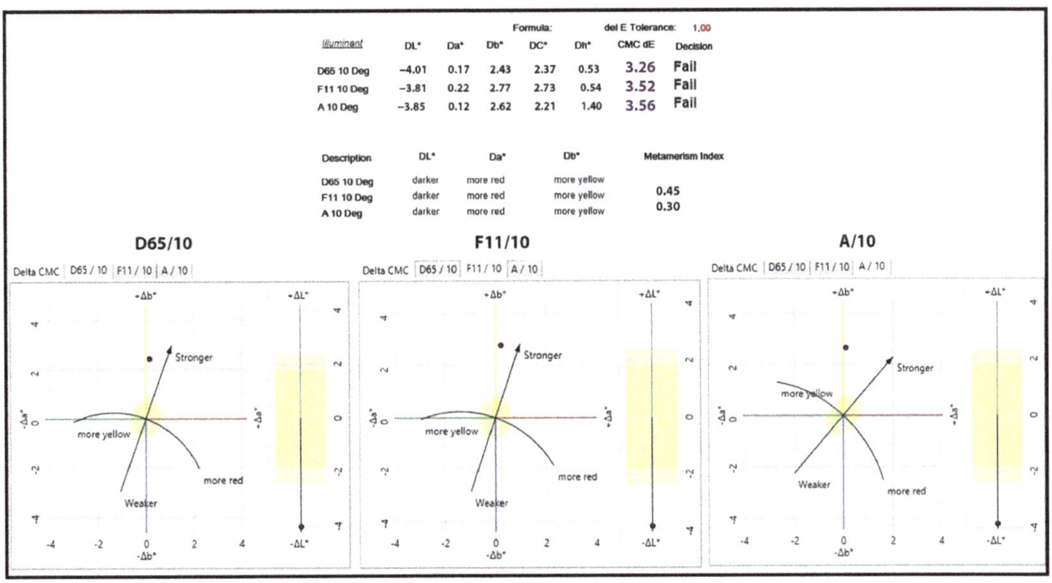

Figure 14. N-HEMP and B-HEMP in grey colour.

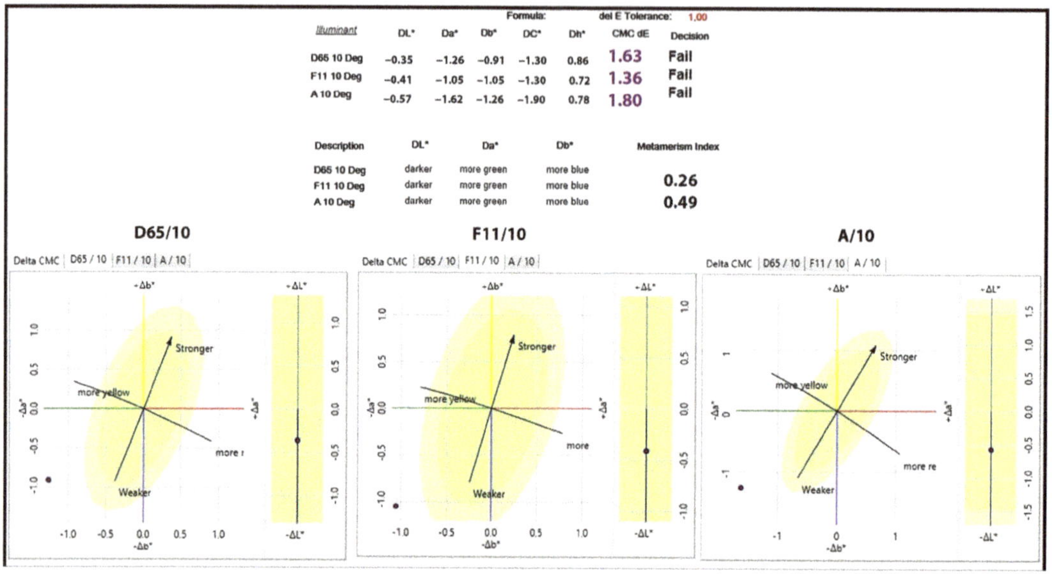

Figure 15. N-HEMP and B-HEMP in brown colour.

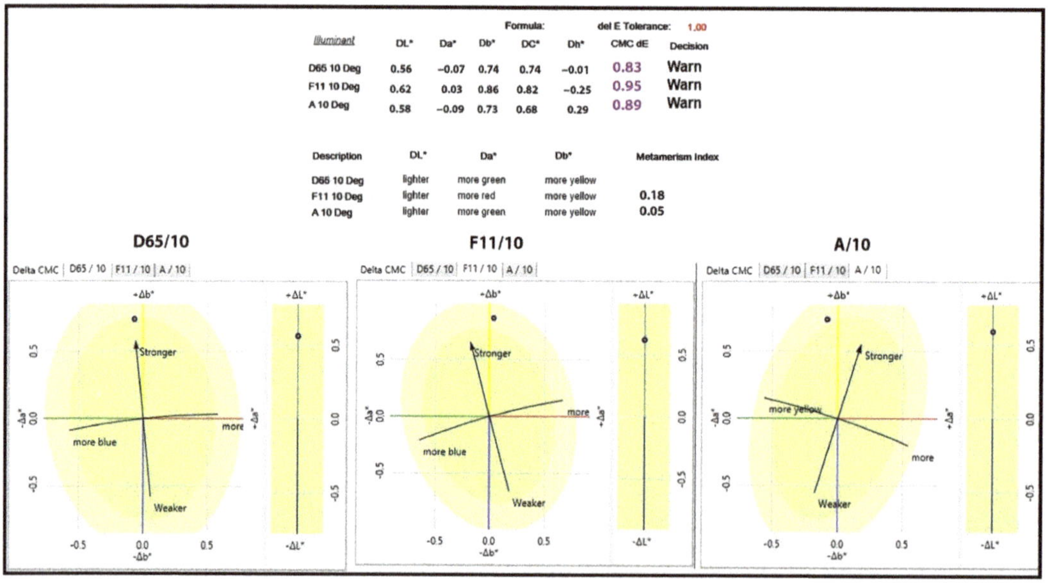

Figure 16. N-HEMP and B-HEMP in green colour.

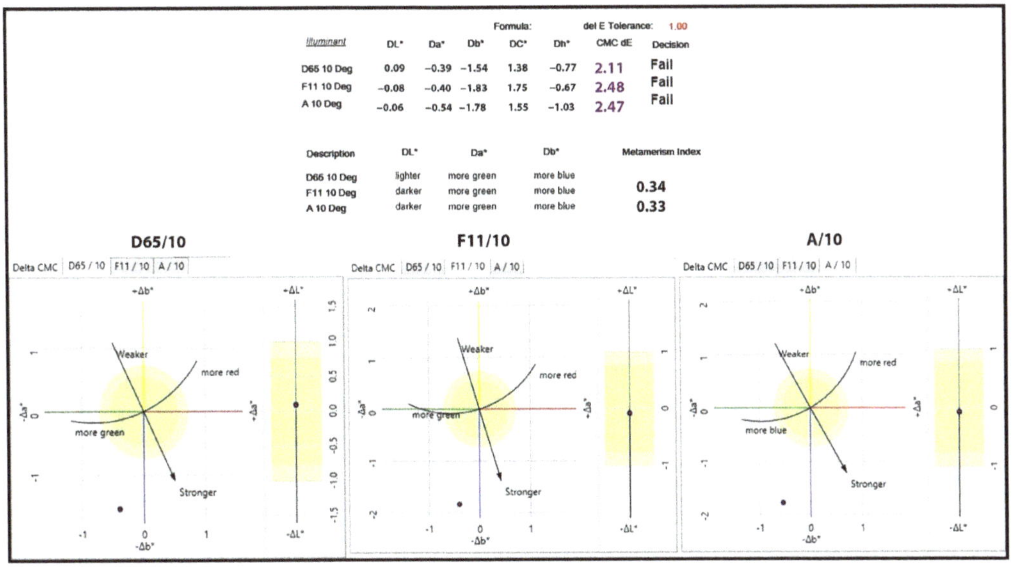

Figure 17. N-HEMP and B-HEMP in navy blue colour.

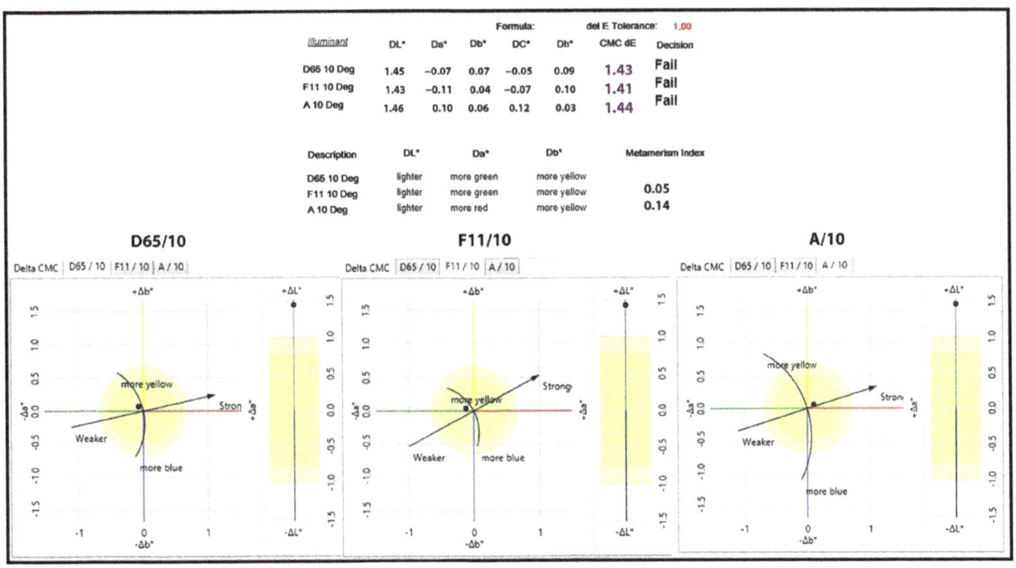

Figure 18. N-HEMP and B-HEMP in black colour.

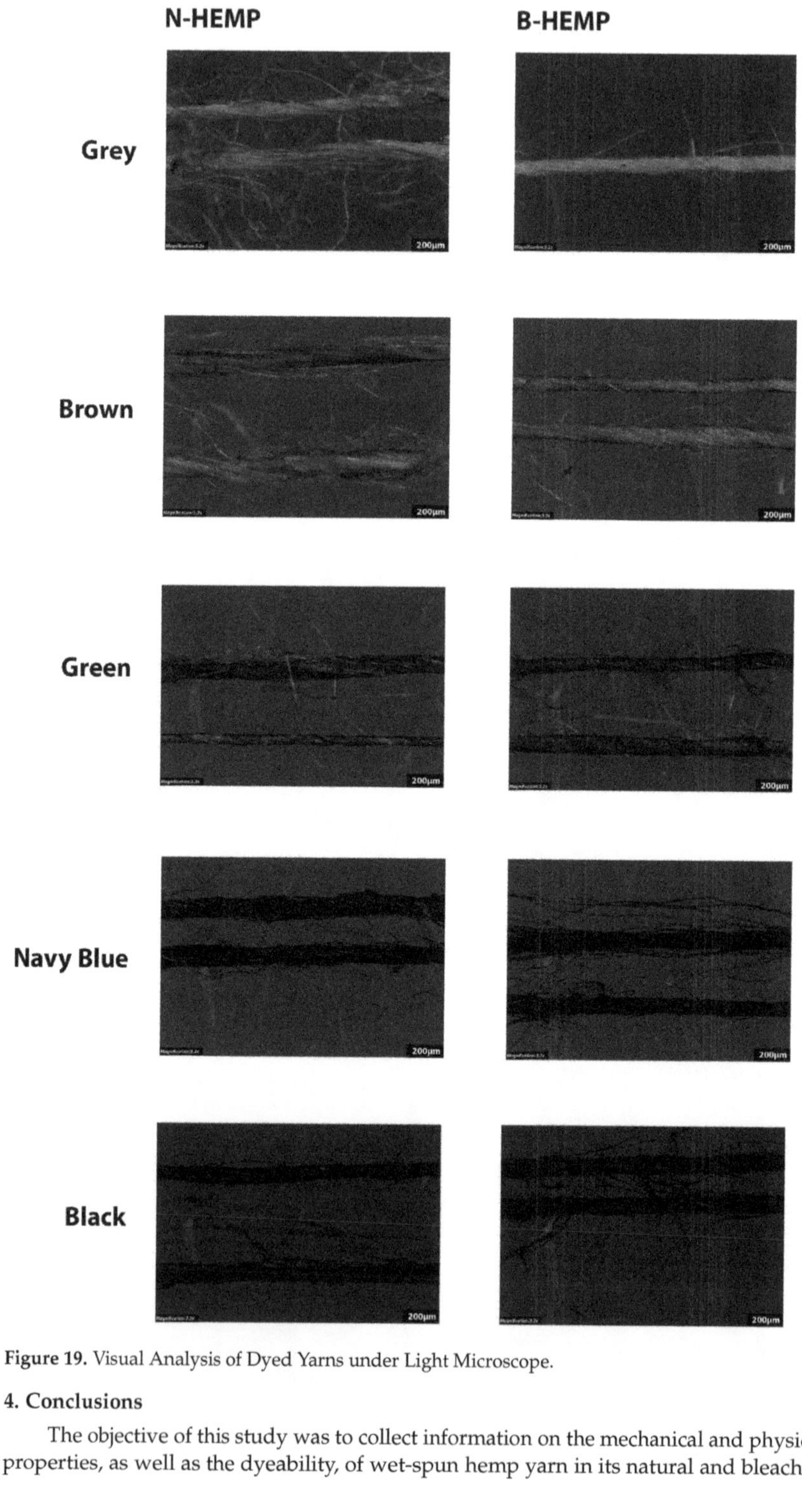

Figure 19. Visual Analysis of Dyed Yarns under Light Microscope.

4. Conclusions

The objective of this study was to collect information on the mechanical and physical properties, as well as the dyeability, of wet-spun hemp yarn in its natural and bleached states to facilitate its use in the textile industry. The results indicate that bleach did not affect the yarn count, but had a positive effect on the yarn twist, as its value increased. On the other hand, bleach had a partially positive effect on the yarn's evenness, as it enhanced overall evenness by reducing thick and thin places, however, the process increased the neps and hairiness. In B-HEMP, the surface appearance test revealed lower values of thick and thin places, indicating a more homogenous and uniform structure. Under a microscope, images of yarn samples revealed the presence of nodes and crystalline regions on the fibres, with nodes having a larger diameter than crystalline regions. On the bleached yarn, microscopic images also reveal surface irregularities and chemical accumulation. The results of the yarn cross-section fibre count determination test indicate that the B-HEMP has a higher fibre density, signifying enhanced yarn performance. During the FTIR test, no differences in wavelength range and intensity were detected, suggesting that the two materials have the same chemical composition. The initial weight loss temperature of B-HEMP was marginally higher than that of N-HEMP, as determined by thermogravimetric analysis. Furthermore, the residual weight is less than that of N-HEMP. This demonstrates that bleaching improved the thermal stability of the yarn. The results of strength tests indicate that the tensile strength, elongation, and tenacity of hemp yarns increased after the bleaching process. This finding suggests that the bleaching process improves yarn's mechanical properties, resulting in a more durable structure. Throughout the strength tests performed on both unbleached and bleached wet-spun hemp yarns, various types of breakage points were identified. There were two distinct areas of weakness in the yarn: one where complete breakage occurred instantly and the other where gradual weakening led to strength loss. Due to the irregularities in the yarn, the strength test was repeated, with the yarn twisted into a double-ply. These findings suggest that double-plied yarns have a more uniform surface appearance, with less variation of thin and thick places. In comparison to the natural double-plied wet-spun hemp yarn, the bleached double-plied wet-spun hemp yarn showed superior maximum load and elongation at break values. However, the tenacity values were comparable. The bleached wet-spun hemp yarn has an increased overall strength. The evaluation of the strength of the dyed yarn revealed that dyed B-HEMP are superior to N-HEMP, particularly in the case of grey-coloured yarn. In contrast to natural wet-spun hemp yarns, alternative-coloured yarns have lower quality. Under varied lighting conditions, the grey colour values of bleached wet-spun hemp yarn and natural wet-spun hemp yarn differ according to colorimetric data analysis. The test revealed notable variations in colour results when dyeing wet-spun natural and bleached hemp yarns in brown under different lighting conditions, resulting in considerable variations in tone. The findings indicate that it is possible to produce hemp yarns in a green hue with uniform colouration, as there is no visual difference in terms of colour perception. Distinctive variations in colour between navy blue N-HEMP and B-HEMP can be observed by the analysis of their respective colour values, indicating a visually noticeable colour difference. Noticeable differences can be observed in the black-dyed hemp yarns. In summary, it can be concluded that the process of bleaching has had a positive effect on the wet-spun hemp yarn.

5. Future Work

Due to the limited data available on wet-spun hemp yarn, there are some significant research gaps. For instance, to enhance the properties of the wet-spun hemp yarn, the wet-spinning parameters can be studied for optimisation. Additionally, wet-spun hemp can be blended with other natural fibres making it much more suitable for use in apparel and home textiles. Similarly, eco-friendly dyed wet-spun hemp yarn can be tested for colour fastness and other properties. The cost of production of hemp fibres can be compared to existing fibres to ensure the commercial viability for mass production. This can be combined with a market and consumer study. Furthermore, significant research can be conducted on the life cycle assessment of the wet-spun hemp yarn to evaluate its environmental impact.

Author Contributions: S.T.: Conceptualization, formal analysis, resources, visualization. N.K.: Data curation, writing—original draft preparation. A.K.: Data curation, writing—original draft preparation. M.A.N.: Conceptualization, validation, writing—review and editing. A.N.: Writing—review and editing, visualization. F.T.: Writing—original draft preparation. M.U.: Conceptualization, methodology, validation, resources, writing—review and editing, project administration. All authors have read and agreed to the published version of the manuscript.

Funding: This research was funded by the University of Oradea within the framework of the grants competition "Scientific Research of Excellence Related to Priority Areas with Capitalization through Technology Transfer: INO-TRANSFER-UO-2nd Edition", Projects no. 236/2022.

Institutional Review Board Statement: Not applicable.

Informed Consent Statement: Not applicable.

Data Availability Statement: The data presented in this study are available on request from the corresponding author.

Conflicts of Interest: The authors declare no conflict of interest.

References

1. Mwaikambo, L.Y. Review of the History, Properties and Application of Plant Fibres. *Afr. J. Sci. Technol.* **2006**, *7*, 120–133.
2. Mohanty, A.K.; Misra, M.; Drzal, L.T. Sustainable Bio-Composites from Renewable Resources: Opportunities and Challenges in the Green Materials World. *J. Polym. Environ.* **2002**, *10*, 19–26. [CrossRef]
3. Thyavihalli Girijappa, Y.G.; Mavinkere Rangappa, S.; Parameswaranpillai, J.; Siengchin, S. Natural Fibers as Sustainable and Renewable Resource for Development of Eco-Friendly Composites: A Comprehensive Review. *Front. Mater.* **2019**, *6*, 481024. [CrossRef]
4. Tausif, M.; Jabbar, A.; Naeem, M.S.; Basit, A.; Ahmad, F.; Cassidy, T. Cotton in the New Millennium: Advances, Economics, Perceptions and Problems. *Text. Progress* **2018**, *50*, 1–66. [CrossRef]
5. González-García, S.; Hospido, A.; Feijoo, G.; Moreira, M.T. Life Cycle Assessment of Raw Materials for Non-Wood Pulp Mills: Hemp and Flax. *Resour. Conserv. Recycl.* **2010**, *54*, 923–930. [CrossRef]
6. Van Der Werf, H.M.G. *Life Cycle Analysis of Field Production of Fibre Hemp, the Effect of Production Practices on Environmental Impacts*; Kluwer Academic Publishers: Boston, MA USA, 2004; Volume 140.
7. Kozłowski, R.M.; Mackiewicz Talarczyk, M. *Types, Properties and Factors Affecting Breeding and Cultivation*; Woodhead Pub.: Hamilton, UK, 2012; ISBN 978-1-84569-697-9. Available online: https://www.sciencedirect.com/book/9780128183984/handbook-of-natural-fibres (accessed on 8 December 2023).
8. Eyupoglu, S. Sustainable Plant-Based Natural Fibers. In *Sustainability in the Textile and Apparel Industries. Sustainable Textiles: Production, Processing, Manufacturing & Chemistry*; Springer: Cham, Switzerland, 2020; pp. 27–48. [CrossRef]
9. Visković, J.; Zheljazkov, V.D.; Sikora, V.; Noller, J.; Latković, D.; Ocamb, C.M.; Koren, A. Industrial Hemp (*Cannabis sativa* L.) Agronomy and Utilization: A Review. *Agronomy* **2023**, *13*, 931. [CrossRef]
10. Pollio, A. The Name of *Cannabis*: A Short Guide for Nonbotanists. *Cannabis Cannabinoid Res.* **2016**, *1*, 234–238. [CrossRef]
11. Parsons, J.L.; Martin, S.L.; James, T.; Golenia, G.; Boudko, E.A.; Hepworth, S.R. Polyploidization for the Genetic Improvement of *Cannabis sativa*. *Front. Plant Sci.* **2019**, *10*, 449166. [CrossRef]
12. Atakan, Z. Cannabis, a Complex Plant: Different Compounds and Different Effects on Individuals. *Ther. Adv. Psychopharmacol.* **2012**, *2*, 241. [CrossRef]
13. Fischedick, J.T.; Hazekamp, A.; Erkelens, T.; Choi, Y.H.; Verpoorte, R. Metabolic Fingerprinting of *Cannabis sativa* L., Cannabinoids and Terpenoids for Chemotaxonomic and Drug Standardization Purposes. *Phytochemistry* **2010**, *71*, 2058–2073. [CrossRef]
14. Small, E. Evolution and Classification of *Cannabis sativa* (Marijuana, Hemp) in Relation to Human Utilization. *Bot. Rev.* **2015**, *81*, 189–294. [CrossRef]
15. Stone, D. Cannabis, Pesticides and Conflicting Laws: The Dilemma for Legalized States and Implications for Public Health. *Regul. Toxicol. Pharmacol.* **2014**, *69*, 284–288. [CrossRef] [PubMed]
16. Amaducci, S.; Errani, M.; Venturi, G. Plant Population Effects on Fibre Hemp Morphology and Production. *J. Ind. Hemp* **2002**, *7*, 33–60. [CrossRef]
17. Amaducci, S.; Amaducci, M.T.; Benati, R.; Venturi, G. Crop Yield and Quality Parameters of Four Annual Fibre Crops (Hemp, Kenaf, Maize and Sorghum) in the North of Italy. *Ind. Crops Prod.* **2000**, *11*, 179–186. [CrossRef]
18. Duque Schumacher, A.G.; Pequito, S.; Pazour, J. Industrial Hemp Fiber: A Sustainable and Economical Alternative to Cotton. *J. Clean. Prod.* **2020**, *268*, 122180. [CrossRef]
19. Jankauskiene, Z.; Butkute, B.; Gruzdeviene, E.; Cesevičiene, J.; Fernando, A.L. Chemical Composition and Physical Properties of Dew- and Water-Retted Hemp Fibers. *Ind. Crops Prod.* **2015**, *75*, 206–211. [CrossRef]
20. Riddlestone, S.; Stott, E.; Blackburn, K.; Brighton, J. A Technical and Economic Feasibility Study of Green Decortication of Hemp Fibre for Textile Uses. *J. Ind. Hemp* **2006**, *11*, 25–55. [CrossRef]

21. Hobson, R.N.; Hepworth, D.G.; Bruce, D.M. Quality of Fibre Separated from Unretted Hemp Stems by Decortication. *J. Agric. Eng. Res.* **2001**, *78*, 153–158. [CrossRef]
22. Zhang, L.L.; Zhu, R.Y.; Chen, J.Y.; Chen, J.M.; Feng, X.X. Seawater-Retting Treatment of Hemp and Characterization of Bacterial Strains Involved in the Retting Process. *Process Biochem.* **2008**, *43*, 1195–1201. [CrossRef]
23. Zimniewska, M. Hemp Fibre Properties and Processing Target Textile: A Review. *Materials* **2022**, *15*, 1901. [CrossRef]
24. Horne, M.R.L. Bast Fibres: Hemp Cultivation and Production. In *Handbook of Natural Fibres*, 2nd ed.; Elsevier Inc.: Amsterdam, The Netherlands, 2020; Volume 1, pp. 163–196. ISBN 978-0-12820-666-9. Available online: https://www.researchgate.net/publication/289744018_Bast_fibres_Hemp_cultivation_and_production (accessed on 8 December 2023).
25. Idler, C.; Pecenka, R.; Fürll, C.; Gusovius, H.J. Wet Processing of Hemp: An Overview. *J. Nat. Fibers* **2011**, *8*, 59–80. [CrossRef]
26. Müssig, J.; Amaducci, S.; Bourmaud, A.; Beaugrand, J.; Shah, D.U. Transdisciplinary Top-down Review of Hemp Fibre Composites: From an Advanced Product Design to Crop Variety Selection. *Compos. Part C Open Access* **2020**, *2*, 100010. [CrossRef]
27. Dhondt, F.; Subramanian Senthilkannan, M. *Sustainable Textiles: Production, Processing, Manufacturing & Chemistry Hemp and Sustainability*; Springer: Berlin/Heidelberg, Germany, 2022.
28. Both, A.K.; Choudhry, D.; Cheung, C.L. Valorization of Hemp Fibers into Biocomposites via One-Step Pectin-Based Green Fabrication Process. *J. Appl. Polym. Sci.* **2023**, *140*, e53586. [CrossRef]
29. GOTS. Global Organic Textile Standard. Version 6.0. Available online: Https://Global-Standard.Org/Images/Resource-Library/Documents/Standard-and-Manual/Gots_version_6_0_en1.Pdf (accessed on 8 December 2023).
30. *D1059*; Standard Test Method for Yarn Number Based on Short-Length Specimens. ASTM: West Conshohocken, PA, USA, 2017.

sustainability

MDPI

Review

Current and Future Sustainability Traits of Digestive Endoscopy

Tiberia Ioana Ilias *, Cristian Sergiu Hocopan †, Roxana Brata † and Ovidiu Fratila

Department of Medical Disciplines, Faculty of Medicine and Pharmacy, University of Oradea, 410073 Oradea, Romania; hocopan.sergiucristian@student.uoradea.ro (C.S.H.); brata.roxanadaniela@didactic.uoradea.ro (R.B.); ovidiu.fratila@didactic.uoradea.ro (O.F.)
* Correspondence: ioana.ilias@didactic.uoradea.ro
† These authors contributed equally to this work.

Abstract: One of the most important parts of medical care is the endoscopy sector, like digestive endoscopy, which has gained extensive importance and is assumably going to increase in the future. We aimed to analyse and synthesize the impact of digestive endoscopy upon the environment and the possible measures that can be taken to minimize the negative effects of endoscopy related to environmental pollution and human health exposure. The means through which digestive endoscopy produces pollution have been analysed, considering the frame and the base of the last stage of a medical or pharmaceutical product. This research suggests a strategy for improving the impact of this sector on the sustainability of the healthcare system based on four pillars comprising the use of eco-friendly substances, materials, and devices, reducing the consumption of water and all possible devices and energy, reusing those components that can be safely reinserted in the endoscopic circuit and recycling everything that is possible. The conclusions highlight that there is a great need to take control of medical practice, admitting the impact that the healthcare system has on global warming and greenhouse gas emissions, acknowledging the limited assets and wealth of the planet, and applying standards and scales of sustainability that can lead to responsible services for patients.

Keywords: digestive endoscopy; sustainability; pollution; medical/pharmaceutical waste; last stage of a product

Citation: Ilias, T.I.; Hocopan, C.S.; Brata, R.; Fratila, O. Current and Future Sustainability Traits of Digestive Endoscopy. *Sustainability* **2023**, *15*, 15872. https://doi.org/10.3390/su152215872

Academic Editor: Elena Cristina Rada

Received: 6 September 2023
Revised: 9 November 2023
Accepted: 10 November 2023
Published: 12 November 2023

1. Introduction

Pollution appears when any form of substance or any type of energy is scattered into nature in a proportion that is bigger than the capacity of dispersion or storage of the environment. Therefore, pollution, whether created by natural pollutants or by human interventions, is obviously harmful not only to the planet Earth itself but also to all its inhabitants [1].

Environmental pollution remains the world's greatest problem facing humanity and the leading environmental cause of morbidity and mortality [2]. As is commonly known, there are four main types of pollution (i.e., noise, radioactive, biological, and chemical) in different environments (i.e., air, water, and soil). The main causes of pollution are as follows: burning fossil fuels, all types of industrial activities [3,4], waste destroying processes, mining activities [5], domestic sources [6], construction [7,8], microbial decaying processes [9], and agriculture [10].

Among these causes of pollution, as recent data suggest, the medical sector is proven to provoke an important part of the global emissions of greenhouse gases and air pollutants: 4.4% of greenhouse gases, 2.8% of harmful particulate matter (air particles), 3.4% of nitrogen oxides, and 3.6% of sulphur dioxide [11]. Therefore, hospitals and all the other medical care units are the first sources for generating sanitary waste, with high risks for populations and, consequently, imposing strict protocols for handling wastes, from generating them to the final treatment [12].

It is evident from the continuous provision of healthcare benefits to the populace that healthcare can inadvertently inflict harm through a cascading process, as delineated in Figure 1.

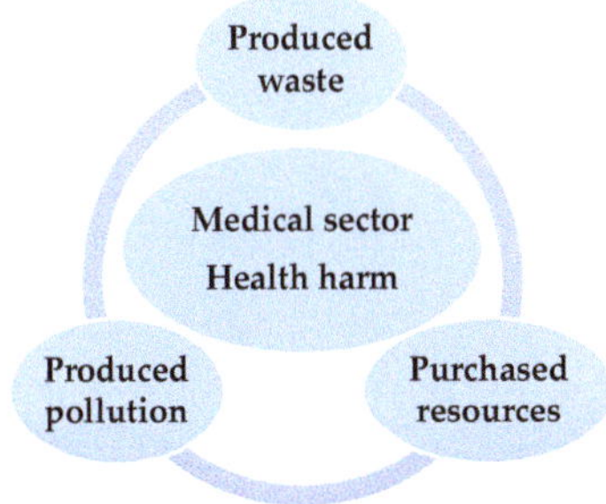

Figure 1. Mechanism of pollution in the heath sector.

One of the most important parts of medical care is the endoscopy sector, which has gained extensive importance lately due to a lot of factors such as real-time and optimal evaluation and diagnosis with minimally invasive procedures, the opportunity to collect tissue samples, and, more importantly, the ability to perform therapeutic manoeuvres with lower risks and less stress for the patient [13,14].

Almost all medical branches extensively use endoscopy procedures, considering:

- Gastroenterology (digestive endoscopy);
- Orthopaedic surgery (arthroscopy);
- Pulmonology and thoracic surgery (bronchoscopy and mediastinoscopy);
- Urology (cystoscopy and urethroscopy);
- Gynaecology (hysteroscopy);
- Various types of surgery (laparoscopy);
- Otolaryngology (laryngoscopy) [15].

Endoscopy is permanently advancing, and newer generations of endoscopes offer high-definition imaging and more and more therapeutic interventional possibilities in all the aforementioned medical specialties [16].

All over the world, due to the ever-increasing requirements in the field of health care, as well as the growing number of hospital units, clinics, laboratories, polyclinics, health centres, etc., imposed by the needs of the growing population, the variety and quantity of medical or pharmaceutical waste resulting from these health care activities have considerably increased [17]. These aforementioned wastes that result from the care of patients, through the provision of optimal health services, and through the promotion of health are themselves a serious threat not only to health but also to the environment in general [18].

According to a study carried out in several countries and published recently, in 2022, the generation rate of medical waste oscillated between 0.14 and 6.10 kg/bed/day. Of the total number of countries considered, approximately 25% selected medical waste, and 17% used the standard storage method for it. In addition, deficiencies found in the cases of some countries referred to the stages of collection, storage, transport, and transfer of these typical wastes, as well as the organization of the elimination of their management activities. In the same research, it was found that only a quarter of the investigated countries simultaneously applied three techniques for the elimination or treatment of sanitary waste (i.e., autoclaving, incineration, or storage), with 91% usually using incineration [19].

In this review paper, we aimed to analyse and summarize the current impact of digestive endoscopy on the environment through the waste it inevitably generates. Also, we considered the possible measures that can and should be taken in an attempt to minimize the negative effects of pollution caused by this medical field in an effort to create such a useful and sustainable sector. The idea of this study started with the desire to make digestive endoscopy as "ecological"/"green" as possible in an era where patients are becoming more and more dependent on the procedures offered by the health care system.

2. Last Stage of a Medical/Pharmaceutical Product

About 10 years ago, it was found in a sample of investigated countries that only a little more than half of the spaces dedicated to health care had systems and equipment considered necessary for the safe disposal of medical-pharmaceutical waste [20]. There is therefore a combination of factors that lead to the appearance of numerous problems regarding the approach to sanitary waste. Often, one of the most serious situations is the wrong use of incineration when waste of non-clinical origin is mixed with clinical (hazardous) waste, unnecessarily choosing internal incineration for the first of the mentioned categories of waste. The result is deficient and weak, leading to overcrowding of the installation, overloading of the incineration installation, and overloading of the personnel who handle it [21]. Biomedical waste is categorized into ten distinct groups (Figure 2), covering a wide array of materials, including human anatomical waste, animal waste, microbiological specimens, sharps, discarded medicines, soiled items, solid disposables, liquid waste, incineration ash, and chemical waste. These categories guide the scientific management and disposal of biomedical waste [22].

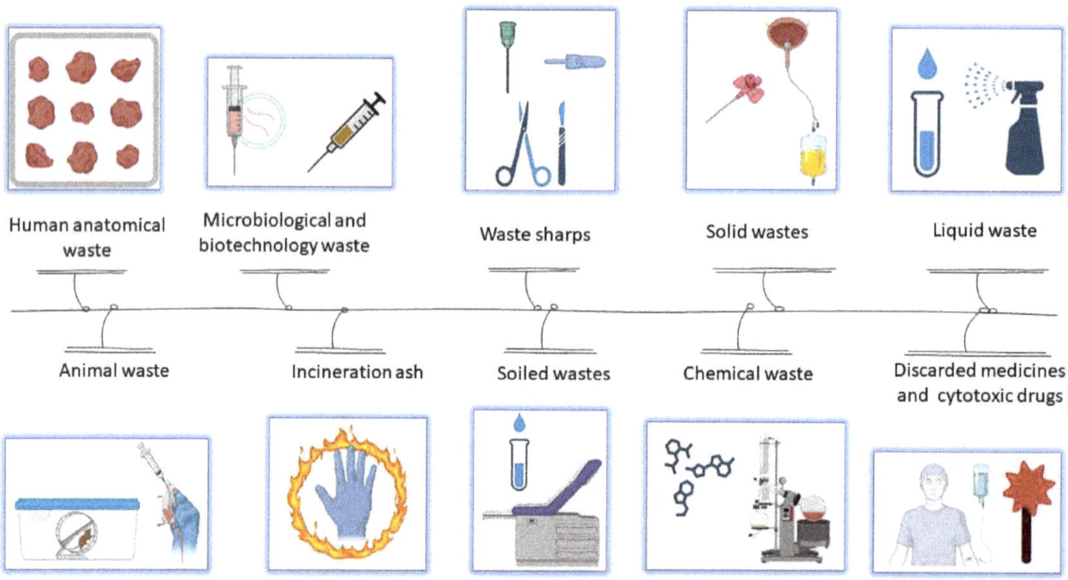

Figure 2. Biomedical waste categorization into ten key categories.

Most of the published research addresses the first three stages of the life of a medical-pharmaceutical product, which are in order as follows: design, manufacture, and use. The last stage in the life of any type of product is recycling or the management of the final waste, which is the result of the end of the respective product's life cycle. Therefore, the details of the approach to this last stage in the case of medical-pharmaceutical waste must be known, applied, and evaluated [23].

In this sense, an essential role is that of the medical staff, who must be educated in this sense and be fully aware of their own role in the optimal management of waste resulting from health care activities. They are the ones responsible for sorting waste exactly at the place of its generation. Non-clinical staff are often less aware of the importance of this selection, having at the same time less experience and knowledge in this area of waste treatment and segregation. Therefore, the medical assistance staff must know the ways to reduce the volume of waste, which implicitly leads to a reduction in their management costs as well as the operational efficiency of the respective sanitary unit. These desired goals can be achieved by implementing appropriate activities, appropriate measures, and even strict protocols that allow the correct flow of medical waste [24].

For implementing a recycling economy, the following stages were suggested:

- Elimination of waste generation;
- Reuse of waste;
- Emphasizing waste recycling;
- Energy recovery;
- Compliant disposal.

Respecting these stages, the application of those activities that result in the reduction of sanitary waste must be the responsibility of the front-line medical care staff [25]. In Romania, regarding the sanitary waste management activity, there are numerous deficiencies, mainly generated by the following:

- A lack of coherent, clear legislation with simple procedures, easily applicable by both citizens and medical and pharmaceutical units;
- High costs are incurred in the collection of sanitary waste from patients, pharmacies, and medical units;
- Personnel responsible for the initial selection of this waste face a lack of sufficient information.

However, some solutions have already been identified in the published literature:

- Effective and complex waste reduction policies and the establishment of clear responsibilities for the personnel who handle it;
- A continuous information system for nurses and pharmacists regarding the effective management of the resulting waste;
- Awareness of patients and health personnel regarding the importance of correct management of medical-pharmaceutical waste;
- Removing or reducing the financial impact of waste storage;
- Risk identification, etc. [26].

An unfortunate consequence of the improper use and disposal of medical and/or pharmaceutical products is their presence as waste in the environment (images of masks on all the beaches of the world and in the most inappropriate places have made the rounds of the planet after the pandemic). Moreover, due to extensive farming practices, wastes, including antibiotics and hormones, are often found in the excrement of domestic animals, such as cows and poultry [6,27], posing a direct threat to ecosystems and human health [9].

The National Health Service of England has issued a document developing the "*Principles for the disposal of pharmaceutical waste used in community health services*". Through it, coherent guidelines directly related to the correct disposal of sanitary waste are promoted. There are ways to select the pharmaceutical waste separately by class and ways of specific and correct colouring of the coded boxes and their corresponding seals when they are full, after which they will be assigned a transfer note. The mentioned document is an efficient, safe, and responsible example of managing a type of waste resulting from health care activity with minimized impact on the toxicity of chemical substances and reduced harmful exposure to the environment [28].

The Context of Waste and Packaging Materials in the Endoscopy Suite

The concept of "Green Endoscopy" involves implementing effective strategies to reduce waste and optimize the use of equipment and supplies in the endoscopy sector, benefiting both patients and the environment. Medical waste disposal is significantly costlier and has a larger environmental footprint in terms of carbon emissions compared to regular waste [29].

At the University of Michigan's Medical Procedure Unit, the management of waste was switched from reusable biopsy forceps to disposable ones due to cost savings. This economic choice, however, has led to a notable increase in waste generation, which must be managed properly to prevent environmental problems [30].

In recent years, there has been a shift from multi-use to single-use endoscopes and supplies. While this reduces infection risks, it raises concerns about increased costs, inventory management complexities, and waste production. Whether this shift is environmentally friendly and cost-effective remains a subject of debate [31–33].

Educating healthcare professionals about proper waste classification and disposal is crucial in creating a sustainable endoscopy unit. Research indicates that a significant percentage of endoscopy staff incorrectly dispose of accessories as regulated medical waste or sharps [34].

There are limited global data on waste generated by endoscopy services, but reports suggest that they are significant waste producers. In Italy, gastroenterology/endoscopy was the second-highest waste generator per procedure [35]. Research conducted in Japan, which assessed the quantity of waste generated in endoscopy facilities across three different hospitals, revealed an average range of 110.2 to 179.9 g of waste produced per procedure, with a predominant 92.9% portion categorized as infectious waste [30].

To address and alleviate the environmental impact associated with endoscopy procedures, a range of strategic measures can be implemented. Firstly, fostering enhanced clinical oversight and conducting internal audits are vital steps. These activities should be complemented with appropriate corrective actions, thereby ensuring that procedures are conducted in an environmentally responsible manner. Secondly, sustainable procurement practices are crucial. These encompass efforts to diminish emissions throughout the entire supply chain, emphasizing the importance of selecting eco-friendly products and suppliers who adhere to green standards [36].

The adoption of renewable energy sources to power hospitals and endoscopy units represents another essential facet of this endeavour. Transitioning to sustainable energy solutions can significantly reduce the carbon footprint associated with healthcare operations. Furthermore, healthcare institutions can leverage their anchor status to influence suppliers. By compelling these suppliers to disclose their carbon footprint and embrace sustainable practices, the healthcare sector can further promote environmentally responsible procurement [37].

Innovative medical alternatives can also contribute to a reduction in the environmental impact. Exploring less invasive procedures, such as the utilization of Cytosponge, offers eco-friendly alternatives to traditional methods. Materials used in endoscopy procedures can be replaced with compostable or recyclable plastics, diminishing the burden on landfills and encouraging sustainable waste management practices [36].

The emphasis on the use of recyclable equipment and the prioritization of multi-use and easily repairable devices are measures that can substantially decrease waste generation and resource consumption. Consideration of the entire lifecycle of procured items when making choices between single-use and reusable devices is essential. Evaluating the environmental impact throughout the product's life cycle can guide more eco-conscious decisions [38,39].

A reduction in the use of nitrous oxide, a potent greenhouse gas, and the maintenance of equipment to minimize gas leaks are essential steps in decreasing the ecological footprint of endoscopy procedures. Efficiency in the decontamination process is equally crucial. Streamlining the resources required for decontamination can contribute to a reduction in waste generation and energy consumption within the healthcare setting [40].

Collectively, these measures are designed to promote the transformation of endoscopy units into more environmentally sustainable entities while simultaneously reducing their ecological impact.

3. Means and Ways Digestive Endoscopy Produces Pollution

3.1. Air and Water Pollution

Air pollution is defined as the contamination of the indoor or outdoor environment by a mixture of any chemical, physical, or biological agent that modifies the natural atmosphere [41].

The indoor quality of the air and maintaining its proprieties [42] are matters of great importance for the health of every person who spends a lot of time in the same confined space, and also for interior objects, especially when they are of patrimony (books, paintings, furniture, etc.) [43]. Besides the toxic pollutants that arise from the construction, building materials, products, and installations used for the maintenance of a construction, etc. [7,8,44,45], or working in an environment that is constantly affected by air pollutants produced by different activities puts the workers or the people living in those buildings at high risk for pulmonary, cardiovascular, and other diseases. These air pollutants are mostly carbon dioxide (CO_2), volatile organic compounds (VOCs), particulate matter that has a diameter of <2.5 μm ($PM_{2.5}$), nitrogen dioxide (NO_2), carbon monoxide (CO), and ozone [46].

In this context, digestive endoscopy can contribute to these types of air pollution in multiple ways. First, to a smaller extent, there can be expulsions of gastrointestinal gas from the patients during the endoscopic procedure and coagulation of tissues through carbonation (haemostasis through thermal coagulation devices, endoscopic mucosal or submucosal resection procedures). Also, one of the most important parts of a lower gastrointestinal endoscopy is the insufflation of the bowel/digestive tract with environmental air, or CO_2, to assure space and visibility. CO_2 insufflation is known to do this with less distress (less abdominal pain) for the patient, but of course, it leaks out into the air. If it reaches a high concentration, it can obviously affect the personnel's health [47,48].

In Sweden, there are national occupational exposure limits (OELs) for CO_2. There are two commonly used OEL values: the level limit value (LLV) and the short-term exposure limit (STEL). LLV is the OEL value for exposure during a working day, normally eight hours (limit value (LLV) of 5000 ppm). STEL is the OEL value for a reference period of 15 min of exposure (highest accepted value 10,000 ppm) [49].

However, extensive reviews showed that linear physiological changes in the circulatory, cardiovascular, and autonomic systems become evident in the human body upon exposure to CO_2 at concentrations ranging from 500 to 5000 ppm. Therefore, is CO_2 insufflation safe for the air quality, or can it also contribute to the air pollution from the endoscopy unit? Research data are rather scarce regarding the CO_2 levels from digestive endoscopy units, as not many studies have been performed in this field. Nevertheless, we found an interesting small study from 2020 on laparoscopies in the operating rooms during 20 laparoscopic procedures using CO_2 insufflation. With the help of a gas detector (i.e., TM Dräger X-am 5600, Lübeck, Germany), they recorded point measurements of CO_2 concentrations during the surgeries. The CO_2 concentration during the surgeries was measured at 400–1100 ppm and never exceeded 22% of the LLV at 5000 ppm [49].

Second, and of course to a much bigger extent, the reprocessing of the endoscopes (cleaning and disinfection stages), which are of course repetitive after each procedure, produces chemically volatile vapours that can be inhaled and high amounts of contaminated water that need to be discarded into the sewer. This air pollution can potentially affect the health of medical personnel (nurses, doctors, etc.) that spend a lot of their time in the endoscopy unit and, to a lesser extent, the patients that undergo digestive endoscopy manoeuvres [47].

Moreover, there are no current standards for indoor air quality in the endoscopy unit, and no specific measurements are performed for air quality checks. Endoscope reprocessing is regulated by multiple local country guidelines, which clearly indicate all the compulsory stages and the chemical solutions that must be used [50].

3.2. Pollution Produced through the Reprocessing Sequence

According to the Spaulding Classification System [51], digestive endoscopes are devices that need high level disinfection (Table 1).

Table 1. Endoscope classifying according to Spaulding sorting system.

Spaulding Classification	Examples of Devices	Risk of Infection Transmission	Disinfection Level
Critical (enters tissues or vascular system)	Implants, scalpels, needles, other surgical instruments, etc.	High	Sterilization
Semi critical (touches mucous membranes)	Flexible endoscopes, endotracheal tubes	Medium	High-level
Noncritical (touches intact skin)	Stethoscopes, bed pans, etc.	Low	Intermediate or low

In general, as a consensus, most reprocessing guidelines recommend the following sequence: precleaning, cleaning, rinsing, disinfection, final rinsing, drying, and finally storage. Air and water pollution can appear during this process. Ideally, the reprocessing of the endoscopes has two components: a manual stage, which means that all the external and accessible internal components are exposed to a low-foaming, endoscope-compatible detergent (usually a nonenzymatic detergent is preferred), followed by the automatic disinfection, rinsing, and drying of all exposed parts of the endoscopes using specific chemicals or detergent [52].

It must be mentioned that although it is advisable that high-level disinfection be obtained using an automatic reprocessing or washing machine, it is still carried out manually in many units, meaning that the exposure of the medical stuff dedicated to this job is still high.

3.3. Chemicals Used in Disinfection Process of Digestive Endoscopes

Below, the most used disinfectants are described as follows:

a. Glutaraldehyde (2.4–3.5%), which is not expensive, and it is highly effective and readily available, with practically no damage to the endoscopes. Unfortunately, glutaraldehyde elicits adverse effects on individuals involved in its manipulation, and substantial reductions in atmospheric levels of glutaraldehyde have been recommended. Due to this major disadvantage, this agent was withdrawn from use in some countries. Also, its disposal is a concern, and it should not be directly emptied into the sewage system [51].

b. Orthophthal aldehyde (OPA) (0.55–0.60%) is a more stable alternative disinfectant that has a lower vapor pressure than glutaraldehyde, but it is more expensive. It has a barely perceptible odor. It is advisable that sprays, mists, and aerosols are not used during the use of OPA. All OPA solutions must be neutralized to inactivate the disinfectant before disposal into the sewer [53].

c. Peracetic acid is a highly effective disinfectant that may prove to be a suitable alternative to glutaraldehyde or OPA. It is considered a sustainable disinfectant because it decomposes in oxygen, water, and biodegradable acetic acid, thus not affecting the environment [54,55] (Table 2). Thus, it offers many sustainability advantages, like decomposition in environmentally friendly compounds, as mentioned above; toxic by-products are not generated during its use; and due to its potency, it is resource efficient [56].

d. Hydrogen peroxide (2–7.5%), also used in the terminal disinfection of endoscopes, is usually found in a dual formula that comprises vaporized hydrogen peroxide and ozone. It assures high-level disinfection of all types of digestive endoscopes, including duodenoscopes [57]. Hydrogen peroxide is largely considered eco-friendly as it decomposes in water and oxygen (Table 2). However, in high amounts, hydrogen peroxide can be toxic if ingested, inhaled, or through contact with the skin or eyes, or if it is evacuated into the water, especially for phytoplankton. The toxicity level of exposure varies with the duration and exposure dose [58].

e. Hypochlorous acid (HOCL) is basically a weak acid that results when chlorine is dissolved into water (Table 2). Due to this behaviour, it becomes clear that it represents no harm for the medical personnel and can be disposed of with no risk of producing toxic waste. Thus, hypochlorous acid 650–675 ppm is another potent disinfectant. It was declared by WHO, in 2021 during the Corona Virus pandemic, as the most potent and environmentally safe disinfectant available with a wide range of efficacy against many human pathogens, including the SARS-CoV-2 coronavirus, and it can also be used in digestive endoscopy [59]. The beneficial effects of HOCL as well as its safety for medical personnel and the environment depend on the purity of the solution and the avoidance of contaminating molecular species of aqueous chlorine (such as hypochlorite a.s.o.) [60]. HOCL can be degraded into an anion called hypochlorite (ClO^-). Usually, this compound can be combined with cations to form salts like sodium and calcium hypochlorite (NaClO). These hypochlorite solutions (i.e., bleach, as commonly known and widely used as a whitening, cleaning, and disinfectant agent) represent toxic compounds, which prompts the usage of personal protective equipment and special disposal measures, as they are an environmental hazard. If the manufacturing process is not properly performed according to strict regulations, HOCl products may lack stability in storage, lose part of their antiseptic efficacy, or even cause toxicity through contaminants that can be harmful to the environment and people [61,62].

Table 2. Types of digestive endoscopy disinfectants, their characteristics, and impact upon the environment.

Disinfectant	Advantages	Disadvantages
Glutaraldehyde (3.5%) *	Over 30 years of use in medical sector Excellent biocidal activity Cheap	Healthcare personnel exposure Air pollution Water pollution (requires neutralization)
Ortho-phal-aldehyde (0.55%)	Fast acting Excellent microbiocidal activity (superior to GA) Better material compatibility	More expensive Healthcare personnel exposure Air pollution Water pollution (may require neutralization before exposure)
Peracetic acid	Short time of action even at low temperatures Environmentally friendly Compatibility with many materials	More expensive Can corrode some types of material Unstable when diluted Serious eye and skin damage at high concentration
Hydrogen peroxide	Active against a wide range of microorganism No disposal issues	Material compatibility issues Health care personnel issues (excessive exposure may produce irreversible tissue damage, and vapours can severely affect the respiratory system)
Hypochlorous acid/hypochlorite	Cheap Efficient against many pathogens, including SarsCov2	If turned into hypochlorite, it can corrode some types of material Hypochlorite solutions can cause health care issues

* Adapted after: SGNA Practice Committee 2013–14 Guideline for Use of High-Level Disinfectants & Sterilant for Reprocessing Flex. Gastrointestinal Endoscopes.

Regarding all the above-mentioned disinfectants used in the reprocessing sequence of endoscopes, their impact upon the environment is summarized in Table 2 [63].

Depending on the chemical solutions used, the reprocessing sequences of the endoscopes are synthetized in Figure 3.

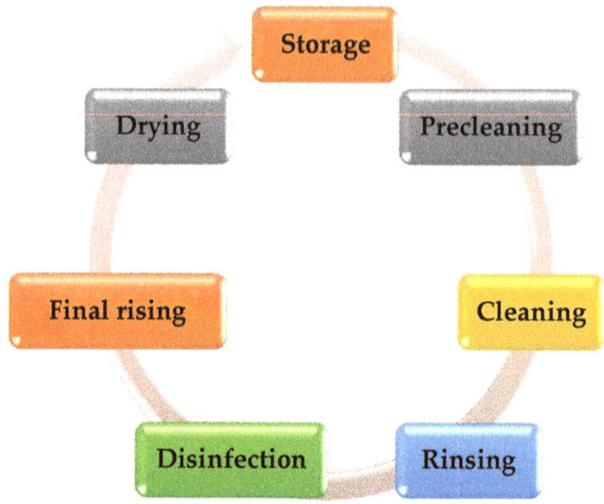

Figure 3. Disinfection scheme and stages of the endoscopes.

3.4. Waste through Disposable Materials and Instruments

Another important part regarding the sustainability of the endoscopy sector is the waste production before, during, and after the procedures take place. The endoscopy sector seems to be in 3rd place for waste production in hospitals [64]. According to recent sources, endoscopy is generating around 3.09 kg of waste per bed day, which represents a rather important addition to the environmental footprint [39]. It was estimated that a single endoscopy produces around 2 kg of waste from periprocedural medical and nonmedical materials and disposable devices [65]. Also, the study of Gayam et al. made an estimation regarding the production of plastic waste in endoscopy departments, and seemingly, about 13,500 tons of waste are made per year in high volume units that reach 40 endoscopies per one day [66].

The following sources of waste are produced in the endoscopy sector:

- Non contaminated/regular waste;
- Contaminated waste;
- Sharps;
- Recyclable waste.

Depending on this classification, these are the main sources of waste in each of the categories from above (Figure 4).

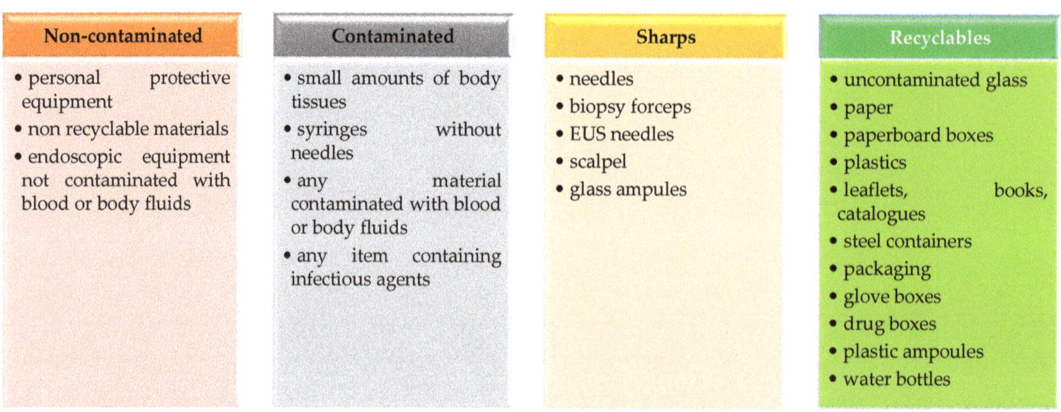

Figure 4. Waste categories in endoscopy sector.

If the types of materials are analysed, the majority of endoscopy materials and subsequently endoscopy waste are composed of plastic, followed by mixed materials (plastic and metals), only metals, and other materials (cotton, fabric, paper) as seen in the pictures below (Figure 5a,b).

(**a**) (**b**)

Figure 5. (**a**,**b**). Pictures from the authors' endoscopy unit with different types of waste (personal collection).

Published studies showed that plastic used in endoscopy represents the majority of disposable non-biohazard waste (54%) and personal protective equipment (PPE) was 8% of the disposable waste. Out of these materials, 48% of non-biohazard and 35% of all waste were candidates for recycling [65].

One of the most important sources of waste in hospitals and digestive endoscopy units, that increased dramatically during the COVID-19 pandemic, is PPE, leading to notable environmental implications [67,68].

Notably, ancillary disposable devices such as snares, needles, filters, clips, balloons, biopsy forceps, etc., employed during endoscopy are numerous. Most often, these items are single-use and made of plastic, thereby generating approximately 2 kg of waste per procedure [65].

Depending on each country and the legislation in place, the storage and disposal of medical waste can be more expensive than that of regular waste, as this process requires special containers and transport to incineration facilities as well as proper disposal at a landfill [30]. Therefore, the segregation of different types of waste must be compulsory due to recycling reasons versus more complex ways of waste disposal like incineration.

3.5. Single-Use Devices and Endoscopes versus Reusable Ones

This domain was always dominated by a constant shift between single-use medical devices and reusable ones. Debates are still ongoing, but we will try to summarize the current conclusions and facts in our work.

Comparable to the general non-medical trend of exchanging reusable with single-use products, the medical sector has also done this during the last years. This change was probably driven by the large access to plastics that are much cheaper and lighter in weight. However, the most important advantage that needs to be outlined in this setting is that single-use devices used in endoscopy guarantee for infection-free procedures. From single-use endoscopes to single-use secondary devices, the actual producing market is offering a high variety of products that are at hand. With the emergence of therapeutic digestive endoscopy, the need for sterilization, not only high-level disinfection, seemed obvious, so single-use devices rapidly became, where available, the preferred option among doctors (Figure 6).

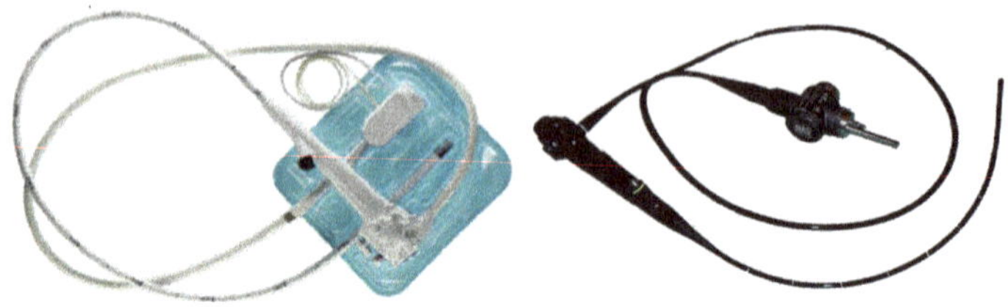

(a) (b)

Figure 6. Images of (**a**) single-use endoscope (personal collection), and (**b**) reusable endoscope (per CC BY-SA, [69]).

Regarding single-use endoscopes and compatible devices, we can speak of two types of sustainability: economic sustainability and environmental sustainability. Economical sustainability refers to the financial viability that interests any hospital facility. When it comes to endoscopy, it seems that single-use endoscopes are more financially beneficial for hospitals due to the avoidance of reprocessing costs and hospitalization costs secondary to endoscopy related infections.

There are many studies that focused on these financial elements, and they support the idea of a strategy based on single-use endoscopes and devices as being more cost efficient, with a saving amount between USD 124 and USD 261 per procedure [70,71]. When it comes to environmental sustainability of single-use endoscopes, it is clear that they are producing a substantial amount of waste [72].

Published results demonstrated that the total amount of waste produced by the reprocessing cycle of a reusable endoscope during its entire lifetime generates approximately 610 kg of waste, while single-use endoscopes (for the same number of procedures) generate 2520 kg of waste, accounting for a 4.1-fold rise in waste [65].

Disposable endoscopic accessories gained a lot of popularity during the last decade, due to the advantage of clarity in use, large variety, and comfort of sterility, but all this popularity is burdened by the waste disposal issue of these devices. Studies show that small units with low volume work, might shift their preferences towards these devices, whereas high volume units might consider evaluating these practices more attentively to obtain a sustainable course [73,74]. Current guidelines advise that single-use endoscopes should be restricted to select indications and environmental impact should be taken into account [29].

3.6. Other Sources of Increased Carbon Footprint from Digestive Endoscopy

Apart from the specific aspects of digestive endoscopy, like any other hospital department, endoscopy units also have some general ways to produce pollution like electricity, heating and cooling, paper use and printing, products transportation, etc., that need to be addressed when it comes to improving sustainability (Figure 7).

The British Guideline recommends, with the purpose of reducing auxiliary sources of pollution, the use of electronic documentation and dissemination to provide medical information for the patient and for colleague referrals in digital versions. Also, it recommends encouraging the patient to come to the endoscopy unit with his own personal reusable objects (mugs, water bottles, etc.) to reduce pollution caused by single-use objects. At the same time, it is advisable to pay great attention to any secondary source of energy consumption like computers, printers, etc., which should all be switched off when not in use [29].

Also of note, we need to mention the pollution caused by electricity, lighting, heating, and cooling systems that work almost non-stop in endoscopy units, which should also be addressed. Therefore, energy-efficient bulbs and motion-driven sensors should be used, and small adjustments in the degrees of temperature of heating and cooling systems should be taken into consideration for reducing energy expenditure. Studies show that the reduction or rise in the temperature by one degree in winter and summer can reduce energy costs by 5% [75].

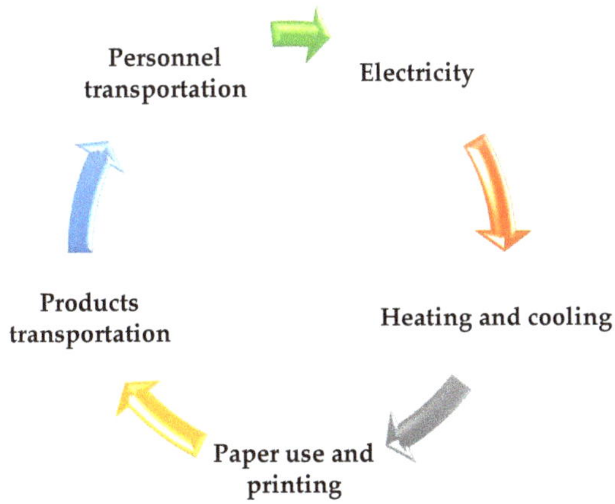

Figure 7. Other sources of pollution in digestive endoscopy.

4. What Can We Do to Increase Digestive Endoscopy Sustainability?

In an attempt to make endoscopy green, there are several statements that address this aspect, but there are also many unmet needs. Of course, the best way to make endoscopy more sustainable is to reduce the number of procedures performed on a single patient; this target can be achieved using more specialized triage by following specific guidelines that address screening strategies and clear indications of endoscopic procedures.

Unfortunately, it is clear, as studies are showing that most of the waste produced in the digestive endoscopy units is not handled properly and that most of the personnel (medical and auxiliary) are not sufficiently informed and trained about medical waste and disposal rules [34].

A good strategy to improve the impact of this sector upon the sustainability of the healthcare system must be based on four pillars, as schematized in Figure 8:

- Use eco-friendly substances, materials, and devices;
- Reduce the unnecessary consumption of water and all possible materials, devices, energy, etc.;
- Reuse those components that can be safely reinserted in the endoscopic circuit;
- Recycle everything that is possible.

Finally, our suggestions for improving the sustainability of the digestive endoscopy sector, which can also be considered future directions, are as follows:

- Raising perceptions among medical personnel and the auxiliary team about the risks and long-term implications of waste and pollution;
- Establishing clear standards for indoor air quality in the endoscopy unit and performing specific measurements for air quality checks, maybe even CO_2 monitoring devices;
- Performing all the disinfection stages of the endoscopes in a dedicated room or space with proper ventilation or even with air extraction devices;
- Making more efforts to reduce the waste quantity;
- Better understanding of sorting out the waste;
- Improving the standards of disposal practices;
- Stewardship towards safer and greener methods for the sterilization of medical devices (autoclaving, etc.) over incineration;
- Work education of the personnel regarding hazards associated with manoeuvring, storing, transporting, and processing wastes;
- Following guidelines;

- Promoting continuously wise resource distribution and safer practice.

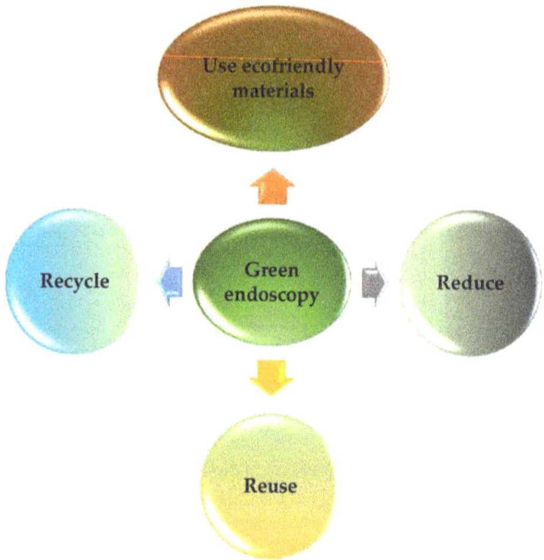

Figure 8. The 4 pillars of a sustainable digestive endoscopy.

Healthcare waste generates substantial costs annually [76] and is influenced by a range of factors specific to each hospital, including the hospital's size in terms of the number of beds, its classification (e.g., general hospital, specialized clinic), the range of medical services it provides, the volume of annual inpatients, the duration of patient stays, the quantity of scheduled surgeries, the presence of specialized units like Intensive Care Units, and the total workforce size [77,78].

Moreover, these factors, which impact healthcare waste generation, along with the financial implications, offer valuable insights for hospital management and governmental authorities. They can guide decision-makers in understanding what changes are necessary and what policies and action plans should be implemented to achieve environmental protection, reduce operational costs in healthcare facilities, and promote sustainable practices [79].

In the endeavour to achieve sustainability within the healthcare sector, a multifaceted approach is proposed. Firstly, a tailored and efficient healthcare waste management system should be established, accommodating the diverse characteristics of healthcare facilities. Secondly, comprehensive training programs for healthcare professionals are essential, focusing on both waste management and occupational safety. Thirdly, the implementation of a healthcare waste management policy and a customized Standard Operating Procedure ensures consistency and best practices. A critical evaluation of the existing legislation and policymaking is imperative to facilitate necessary revisions. The integration of cutting-edge medical waste treatment technologies and the promotion of recycling practices are pivotal for environmental responsibility. Furthermore, the development of standardized guidelines at national and international levels for healthcare waste management is essential. Lastly, concerted efforts are needed to minimize costs and risks associated with healthcare waste management, ensuring economic efficiency while upholding environmental and public health standards [80].

5. Conclusions

Gastrointestinal endoscopy is a domain of great importance and is, assumably, only going to increase in the next few years. Therefore, there is a great need to take control of our practice, admit the impact that the healthcare system has on global warming and greenhouse gas emissions, acknowledge the limited assets and wealth of the planet, and apply standards and scales of sustainability that can improve and make the services more responsible for the patients. The pollution print of the endoscopy sector is not fully established, but more and more work can be carried out in this field as it becomes clearer that there is a definite need to lower the pollution in this domain where a high volume of work and procedures are carried out.

Author Contributions: Conceptualization, T.I.I. and O.F.; validation, T.I.I., C.S.H. and R.B.; formal analysis, C.S.H.; investigation, T.I.I., C.S.H., R.B. and O.F.; resources, T.I.I., C.S.H. and R.B.; data curation, T.I.I.; writing—review and editing, T.I.I., C.S.H., R.B. and O.F.; visualization, R.B.; supervision, O.F.; project administration, T.I.I. All authors have read and agreed to the published version of the manuscript.

Funding: This research received no external funding.

Data Availability Statement: Information presented in this paper are supported by Web of Science Data base and by the references mentioned in the text.

Conflicts of Interest: The authors declare no conflict of interest.

References

1. Russell, V.S. Pollution: Concept and Definition. *Biol. Conserv.* **1974**, *6*, 157–161. [CrossRef]
2. Ukaogo, P.O.; Ewuzie, U.; Onwuka, C. V Environmental Pollution: Causes, Effects, and the Remedies. In *Microorganisms for Sustainable Environment and Health*; Chowdhary, P., Raj, A., Verma, D., Akhter, Y.B.T.-M., Eds.; Elsevier: Amsterdam, The Netherlands, 2020; pp. 419–429.
3. Perera, F. Pollution from Fossil-Fuel Combustion Is the Leading Environmental Threat to Global Pediatric Health and Equity: Solutions Exist. *Int. J. Environ. Res. Public Health* **2017**, *15*, 16. [CrossRef] [PubMed]
4. Fayomi, G.U.; Mini, S.E.; Fayomi, O.S.I.; Oyeleke, O.; Omole, D.O.; Akinwumi, I.I. Overview of Industrial Pollution Activities and Its Curbing Mechanisms. *IOP Conf. Ser. Mater. Sci. Eng.* **2019**, *640*, 012097. [CrossRef]
5. Matei, O.-R.; Dumitrescu Silaghi, L.; Dunca, E.-C.; Bungau, S.G.; Tit, D.M.; Mosteanu, D.-E.; Hodis, R. Study of Chemical Pollutants and Ecological Reconstruction Methods in the Tismana I Quarry, Rovinari Basin, Romania. *Sustainability* **2022**, *14*, 7160. [CrossRef]
6. Tit, D.M.; Bungau, S.G.; Nistor-Cseppento, D.C.; Copolovici, D.M.; Buhas, C.L. Disposal of Unused Medicines Resulting from Home Treatment in Romania. *J. Environ. Prot. Ecol.* **2016**, *17*, 1425–1433.
7. Prada, M.; Prada, I.F.; Cristea, M.; Popescu, D.E.; Bungău, C.; Aleya, L.; Bungău, C.C. New Solutions to Reduce Greenhouse Gas Emissions through Energy Efficiency of Buildings of Special Importance—Hospitals. *Sci. Total Environ.* **2020**, *718*, 137446. [CrossRef]
8. Bungau, C.C.; Bungau, T.; Prada, I.F.; Prada, M.F. Green Buildings as a Necessity for Sustainable Environment Development: Dilemmas and Challenges. *Sustainability* **2022**, *14*, 13121. [CrossRef]
9. Bungau, S.; Tit, D.M.; Behl, T.; Aleya, L.; Zaha, D.C. Aspects of Excessive Antibiotic Consumption and Environmental Influences Correlated with the Occurrence of Resistance to Antimicrobial Agents. *Curr. Opin. Environ. Sci. Heal.* **2021**, *19*, 100224. [CrossRef]
10. Behl, T.; Kaur, I.; Sehgal, A.; Singh, S.; Sharma, N.; Bhatia, S.; Al-Harrasi, A.; Bungau, S. The Dichotomy of Nanotechnology as the Cutting Edge of Agriculture: Nano-Farming as an Asset versus Nanotoxicity. *Chemosphere* **2022**, *288*, 132533. [CrossRef]
11. Capon, T.; Malik, A.; Pencheon, D.; Weisz, H.; Lenzen, M. Health Care Has a Huge Environmental Footprint, Which Then Harms Health. This Is a Matter of Ethics. Available online: https://theconversation.com/health-care-has-a-huge-environmental-footprint-which-then-harms-health-this-is-a-matter-of-ethics-142651 (accessed on 1 September 2023).
12. Padmanabhan, K.K.; Barik, D. Health Hazards of Medical Waste and Its Disposal. In *Energy from Toxic Organic Waste for Heat and Power Generation*; Woodhead Publishing: Cambridge, UK, 2019; pp. 99–118.
13. Moore, L.E. The Advantages and Disadvantages of Endoscopy. *Clin. Tech. Small Anim. Pract.* **2003**, *18*, 250–253. [CrossRef]
14. Baddeley, R.; de Santiago, E.R.; Maurice, J.; Siddhi, S.; Dhar, A.; Thomas-Gibson, S.; Hayee, B. Sustainability in Gastrointestinal Endoscopy. *Lancet. Gastroenterol. Hepatol.* **2022**, *7*, 9–12. [CrossRef] [PubMed]
15. Boese, A.; Wex, C.; Croner, R.; Liehr, U.B.; Wendler, J.J.; Weigt, J.; Walles, T.; Vorwerk, U.; Lohmann, C.H.; Friebe, M.; et al. Endoscopic Imaging Technology Today. *Diagnostics* **2022**, *12*, 1262. [CrossRef] [PubMed]
16. Gulati, S.; Patel, M.; Emmanuel, A.; Haji, A.; Hayee, B.; Neumann, H. The Future of Endoscopy: Advances in Endoscopic Image Innovations. *Dig. Endosc.* **2020**, *32*, 512–522. [CrossRef] [PubMed]
17. Almomani, H.; Obaidat, M.; Khazaleh, A.; Muneizel, O.; Afyouni, N.; Fayyad, S. Review of Medical Waste Management in Jordanian Health Care Organisations. *Br. J. Healthc. Manag.* **2019**, *25*, 1–8. [CrossRef]
18. Eren, E.; Tuzkaya, U.R. Occupational Health and Safety-Oriented Medical Waste Management: A Case Study of Istanbul. *Waste Manag. Res.* **2019**, *37*, 876–884. [CrossRef]
19. Fadaei, A. Comparison of Medical Waste Management Methods in Different Countries: A Systematic Review. *Rev. Environ. Health* **2023**, *38*, 339–348. [CrossRef] [PubMed]
20. World Health Organization. *Safe Management of Waste from Healthcare Activity*; WHO: Geneva, Switzerland, 2014; pp. 1–242.

21. Mbongwe, B.; Mmereki, B.T.; Magashula, A. Healthcare Waste Management: Current Practices in Selected Healthcare Facilities, Botswana. *Waste Manag.* **2008**, *28*, 226–233. [CrossRef]
22. Singh, Z.; Bhalwar, R.; Jayaram, J.; Tilak, V.W. An Introduction to Essentials of Bio-Medical Waste Management. *Med. J. Armed Forces India* **2001**, *57*, 144–147. [CrossRef]
23. Bungau, S.; Bungau, C.; Tit, D.M. Studies on the Last Stage of Product Lifecycle Management for a Pharmaceutical Product. *J. Environ. Prot. Ecol.* **2015**, *16*, 56–62.
24. Kwikiriza, S.; Stewart, A.G.; Mutahunga, B.; Dobson, A.E.; Wilkinson, E. A Whole Systems Approach to Hospital Waste Management in Rural Uganda. *Front. Public Heal* **2019**, *7*, 136. [CrossRef]
25. Preparation of National Health-Care Waste Management Plans in Sub-Saharan Countries Guidance Manual Secretariat of the Basel Convention and World Health Organization. Available online: https://apps.who.int/iris/bitstream/handle/10665/43118/924154662X.pdf?sequence=1 (accessed on 2 September 2023).
26. Bungau, S.G.; Suciu, R.N.; Bumbu, A.G.; Cioca, G.; Tit, D.M. Study on Hospital Waste Management in Medical Rehabilitation Clinical Hospital, Baile Felix. *J. Environ. Prot. Ecol.* **2015**, *16*, 980–987.
27. Pyhälä, M.; Zandaryaa, S.; Andresmaa, E.; Korsjukov, M.; Reisner, R.; Susanne Boutrup, D.; Päivi Fjäder, F.; Mehtonen, J.; Äystö, L.; Hein, A.; et al. Pharmaceuticals in the Aquatic Environment of the Baltic Sea Region: A Status Report. UNESCO: Paris, France, 2017.
28. Chisholm, J.M.; Zamani, R.; Negm, A.M.; Said, N.; Abdel Daiem, M.M.; Dibaj, M.; Akrami, M. Sustainable Waste Management of Medical Waste in African Developing Countries: A Narrative Review. *Waste Manag. Res.* **2021**, *39*, 1149–1163. [CrossRef] [PubMed]
29. Sebastian, S.; Dhar, A.; Baddeley, R.; Donnelly, L.; Haddock, R.; Arasaradnam, R.; Coulter, A.; Disney, B.R.; Griffiths, H.; Healey, C.; et al. Green Endoscopy: British Society of Gastroenterology (BSG), Joint Accreditation Group (JAG) and Centre for Sustainable Health (CSH) Joint Consensus on Practical Measures for Environmental Sustainability in Endoscopy. *Gut* **2023**, *72*, 12–26. [PubMed]
30. de Melo, S.W.; Taylor, G.L.; Kao, J.Y. Packaging and Waste in the Endoscopy Suite. *Tech. Innov. Gastrointest. Endosc.* **2021**, *23*, 371–375. [CrossRef]
31. Choi, H.H.; Cho, Y.-S. Endoscope Reprocessing: Update on Controversial Issues. *Clin. Endosc.* **2015**, *48*, 356–360. [CrossRef]
32. Larsen, S.; Russell, R.V.; Ockert, L.K.; Spanos, S.; Travis, H.S.; Ehlers, L.H.; Mærkedahl, A. Rate and Impact of Duodenoscope Contamination: A Systematic Review and Meta-Analysis. *EClinicalMedicine* **2020**, *25*, 100451. [CrossRef]
33. Brown, G.; Ong, A.; Juliebø-Jones, P.; Davis, N.F.; Skolarikos, A.; Somani, B. Single-Use Ureteroscopy and Environmental Footprint: Review of Current Evidence. *Curr. Urol. Rep.* **2023**, *24*, 281–285. [CrossRef]
34. Agrawal, D.; Shoup, V.; Montgomery, A.; Wosik, J.; Rockey, D.C. Disposal of Endoscopic Accessories After Use: Do We Know and Do We Care? *Gastroenterol. Nurs.* **2017**, *40*, 13–18. [CrossRef]
35. Vaccari, M.; Tudor, T.; Perteghella, A. Costs Associated with the Management of Waste from Healthcare Facilities: An Analysis at National and Site Level. *Waste Manag. Res.* **2018**, *36*, 39–47. [CrossRef]
36. Haddock, R.; Gopfert, A.; van Hove, M.; Stableforth, W. The Case for Sustainable Endoscopy as a Professional Priority. *Tech. Innov. Gastrointest. Endosc.* **2021**, *23*, 337–343. [CrossRef]
37. Vaziri, S.M.; Rezaee, B.; Monirian, M.A. Utilizing Renewable Energy Sources Efficiently in Hospitals Using Demand Dispatch. *Renew. Energy* **2020**, *151*, 551–562. [CrossRef]
38. Coelho, P.M.; Corona, B.; ten Klooster, R.; Worrell, E. Sustainability of Reusable Packaging–Current Situation and Trends. *Resour. Conserv. Recycl. X* **2020**, *6*, 100037. [CrossRef]
39. Donnelly Endoscopy, L. Green Endoscopy: Practical Implementation. *Frontline Gastroenterol.* **2022**, *13*, 7–12. [CrossRef] [PubMed]
40. Ratcliff, A.; Burns, C.; Gwinnutt, C.L. The Contribution of Medical Nitrous Oxide to the Greenhouse Effect. *Health Trends* **1991**, *23*, 119–120.
41. World Health Organization. Regional Office for Europe. Evolution of WHO Air Quality Guidelines: Past, Present and Future. Available online: https://apps.who.int/iris/handle/10665/341912 (accessed on 2 September 2023).
42. Nechifor, A.C.; Cotorcea, S.; Bungău, C.; Albu, P.C.; Pașcu, D.; Oprea, O.; Grosu, A.R.; Pîrțac, A.; Nechifor, G. Removing of the Sulfur Compounds by Impregnated Polypropylene Fibers with Silver Nanoparticles-Cellulose Derivatives for Air Odor Correction. *Membranes* **2021**, *11*, 256. [CrossRef]
43. Glevitzky, M.; Aleya, L.; Vică, M.L.; Dumitrel, G.A.; Avram, M.; Tit, D.M.; Popa, M.; Popa, V.C.; Behl, T.; Bungau, S. Assessing the Microbiological Contamination along with Environmental Factors of Old Books in the 1490-Founded Bistrița Monastery, Romania. *Environ. Sci. Pollut. Res.* **2021**, *28*, 8743–8757. [CrossRef]
44. Bungau, C.C.; Hanga Prada, F.I.; Bungau, T.; Bungau, C.; Bendea, G.; Prada, M.F. Web of Science Scientometrics on the Energy Efficiency of Buildings to Support Sustainable Construction Policies. *Sustainability* **2023**, *15*, 8772. [CrossRef]
45. Bungău, C.C.; Prada, I.F.; Prada, M.; Bungău, C. Design and Operation of Constructions: A Healthy Living Environment-Parametric Studies and New Solutions. *Sustainability* **2019**, *11*, 6824. [CrossRef]
46. Bang, C.S.; Lee, K.; Choi, J.H.; Soh, J.S.; Hong, J.Y.; Baik, G.H.; Kim, D.J. Ambient Air Pollution in Gastrointestinal Endoscopy Unit; Rationale and Design of a Prospective Study. *Medicine* **2018**, *97*, e13600. [CrossRef]
47. Bang, C.S.; Lee, K.; Yang, Y.J.; Baik, G.H. Ambient Air Pollution in Gastrointestinal Endoscopy Unit. *Surg. Endosc.* **2020**, *34*, 3795–3804. [CrossRef]

48. Sivananthan, A.; Glover, B.; Ayaru, L.; Patel, K.; Darzi, A.; Patel, N. The Evolution of Lower Gastrointestinal Endoscopy: Where Are We Now? *Ther. Adv. Gastrointest. Endosc.* **2020**, *13*, 2631774520979591. [CrossRef] [PubMed]
49. Af Petersens, M.; Andersson Fenger-Krog, F.; Jakobsson, J.G. Workplace Exposure to Carbon Dioxide during Routine Laparoscopy—Is It Safe? *F1000Research* **2020**, *9*, 571. [CrossRef] [PubMed]
50. Speer, T.; Alfa, M.; Jones, D.; Vickery, K.; Griffiths, H.; Sáenz, R.; LeMair, A. WGO Guideline-Endoscope Disinfection Update. *J. Clin. Gastroenterol.* **2023**, *57*, 1–9. [CrossRef] [PubMed]
51. Lichtenstein, D.; Alfa, M.J. Cleaning and Disinfecting Gastrointestinal Endoscopy Equipment. In *Clinical Gastrointestinal Endoscopy.*, 3rd, ed.; Chandrasekhara, V., Elmunzer, B.J., Khashab, M.A., Muthusamy, V.R.B.T.-C.G.E., Eds.; Elsevier: Philadelphia, PA, USA, 2019; pp. 32–50.e5.
52. Endoscope Disinfection Update: A Guide to Resource-Sensitive Reprocessing. Available online: https://www.worldgastroenterology.org/UserFiles/file/guidelines/endoscope-disinfection-english-2019.pdf (accessed on 1 September 2023).
53. Rutala, W.A.; Weber, D.J. Disinfection, Sterilization, and Control of Hospital Waste. In *Mandell, Douglas, and Bennett's Principles and Practice of Infectious Diseases*; Elsevier Health Sciences: New York, NY, USA, 2015; pp. 3294–3309.e4.
54. Wang, Y.-W.; Liao, M.-S.; Shu, C.-M. Thermal Hazards of a Green Antimicrobial Peracetic Acid Combining DSC Calorimeter with Thermal Analysis Equations. *J. Therm. Anal. Calorim.* **2015**, *119*, 2257–2267. [CrossRef]
55. Carrasco, G.; Urrestarazu, M. Green Chemistry in Protected Horticulture: The Use of Peroxyacetic Acid as a Sustainable Strategy. *Int. J. Mol. Sci.* **2010**, *11*, 1999–2009. [CrossRef] [PubMed]
56. McDonnell, G.; Russell, A.D. Antiseptics and Disinfectants: Activity, Action, and Resistance. *Clin. Microbiol. Rev.* **1999**, *12*, 147–179. [CrossRef]
57. Molloy-Simard, V.; Lemyre, J.L.; Martel, K.; Catalone, B.J. Elevating the Standard of Endoscope Processing: Terminal Sterilization of Duodenoscopes Using a Hydrogen Peroxide–Ozone Sterilizer. *Am. J. Infect. Control* **2019**, *47*, 243–250. [CrossRef]
58. Mahaseth, T.; Kuzminov, A. Potentiation of Hydrogen Peroxide Toxicity: From Catalase Inhibition to Stable DNA-Iron Complexes. *Mutat. Res. Rev. Mutat. Res.* **2017**, *773*, 274–281. [CrossRef]
59. World Health Organization. Cleaning and Disinfection of Environmental Surfaces in the Context of COVID-19. Available online: https://www.who.int/publications/i/item/cleaning-and-disinfection-of-environmental-surfaces-inthe-context-of-covid-19 (accessed on 3 September 2023).
60. Busch, M.; Simic, N.; Ahlberg, E. Exploring the Mechanism of Hypochlorous Acid Decomposition in Aqueous Solutions. *Phys. Chem. Chem. Phys.* **2019**, *21*, 19342–19348. [CrossRef]
61. World Health Organization. Applications for the 23rd WHO Expert Committee on Selection and Use of Essential Medicines Available for Public Review and Comment. Available online: https://www.who.int/news/item/19-04-2021-applications-for-the-23rd-who-expert-committee-on-selection-and-use-of-essential-medicines-available-for-public-review-and-comment (accessed on 1 September 2023).
62. World Health Organization. Hypochlorous Acid (HOCl) for Disinfection, Antisepsis, and Wound Care in Core Categories 15.1, 15.2, and 13. Available online: https://cdn.who.int/media/docs/default-source/essential-medicines/2021-eml-expert-committee/applications-for-addition-of-new-medicines/a.18_hypochlorous-acid.pdf?sfvrsn=35222172_4 (accessed on 4 September 2023).
63. Society of Gastroenterology Nurses and Associates. Guideline for Use of High-Level Disinfectants & Sterilants in the Gastroenterology Setting. Available online: https://www.sgna.org/portals/0/hld__final.pdf (accessed on 2 September 2023).
64. Peery, A.F.; Crockett, S.D.; Murphy, C.C.; Lund, J.L.; Dellon, E.S.; Williams, J.L.; Jensen, E.T.; Shaheen, N.J.; Barritt, A.S.; Lieber, S.R.; et al. Burden and Cost of Gastrointestinal, Liver, and Pancreatic Diseases in the United States: Update 2018. *Gastroenterology* **2019**, *156*, 254–272.e11. [CrossRef]
65. Namburar, S.; von Renteln, D.; Damianos, J.; Bradish, L.; Barrett, J.; Aguilera-Fish, A.; Cushman-Roisin, B.; Pohl, H. Estimating the Environmental Impact of Disposable Endoscopic Equipment and Endoscopes. *Gut* **2022**, *71*, 1326–1331. [CrossRef] [PubMed]
66. Gayam, S. Environmental Impact of Endoscopy: "Scope" of the Problem. *Am. J. Gastroenterol.* **2020**, *115*, 1931–1932. [CrossRef] [PubMed]
67. Rizan, C.; Reed, M.; Bhutta, M.F. Environmental Impact of Personal Protective Equipment Distributed for Use by Health and Social Care Services in England in the First Six Months of the COVID-19 Pandemic. *J. R. Soc. Med.* **2021**, *114*, 250–263. [CrossRef] [PubMed]
68. Bortoluzzi, F.; Sorge, A.; Vassallo, R.; Montalbano, L.M.; Monica, F.; La Mura, S.; Canova, D.; Checchin, D.; Fedeli, P.; Marmo, R.; et al. Sustainability in Gastroenterology and Digestive Endoscopy: Position Paper from the Italian Association of Hospital Gastroenterologists and Digestive Endoscopists (AIGO). *Dig. Liver Dis.* **2022**, *54*, 1623–1629. [CrossRef] [PubMed]
69. Endoscopes Evolve to Meet Increased Safety Demands. Available online: https://connectorsupplier.com/endoscopes-evolve-to-meet-increased-safety-demands/ (accessed on 3 September 2023).
70. Pang, S.; England, R.W.; Solomon, A.; Hong, K.; Singh, H. Single-Use versus Reusable Endoscopes for Percutaneous Biliary Endoscopy with Lithotripsy: Technical Metrics, Clinical Outcomes, and Cost Comparison. *J. Vasc. Interv. Radiol.* **2022**, *33*, 420–426. [CrossRef] [PubMed]
71. Mouritsen, J.M.; Ehlers, L.; Kovaleva, J.; Ahmad, I.; El-Boghdadly, K. A Systematic Review and Cost Effectiveness Analysis of Reusable vs. Single-Use Flexible Bronchoscopes. *Anaesthesia* **2020**, *75*, 529–540. [CrossRef]
72. Agrawal, D.; Tang, Z. Sustainability of Single-Use Endoscopes. *Tech. Innov. Gastrointest. Endosc.* **2021**, *23*, 353–362. [CrossRef]
73. Petersen, B.T. Advantages of Disposable Endoscopic Accessories. *Gastrointest. Endosc. Clin. N. Am.* **2000**, *10*, 341–348. [CrossRef]

74. Wolfsen, H.C. Advantages of Reusable Accessories. *Gastrointest. Endosc. Clin. N. Am.* **2000**, *10*, 349–359. [CrossRef]
75. Schroeder, K.; Thompson, T.D.B.; Frith, K.; Pencheon, D. *Sustainable Healthcare*; Wiley-Blackwell: Hoboken, NJ, USA, 2012; ISBN 978-0470656716.
76. Shrank, W.H.; Rogstad, T.L.; Parekh, N. Waste in the US Health Care System: Estimated Costs and Potential for Savings. *JAMA* **2019**, *322*, 1501–1509. [CrossRef]
77. Komilis, D.; Katsafaros, N.; Vassilopoulos, P. Hazardous Medical Waste Generation in Greece: Case Studies from Medical Facilities in Attica and from a Small Insular Hospital. *Waste Manag. Res.* **2011**, *29*, 807–814. [PubMed]
78. Bdour, A.; Altrabsheh, B.; Hadadin, N.; Al-Shareif, M. Assessment of Medical Wastes Management Practice: A Case Study of the Northern Part of Jordan. *Waste Manag.* **2007**, *27*, 746–759. [PubMed]
79. Sepetis, A.; Zaza, P.N.; Rizos, F.; Bagos, P.G. Identifying and Predicting Healthcare Waste Management Costs for an Optimal Sustainable Management System: Evidence from the Greek Public Sector. *Int. J. Environ. Res. Public Health* **2022**, *19*, 9821. [PubMed]
80. Lee, S.M.; Lee, D. Effective Medical Waste Management for Sustainable Green Healthcare. *Int. J. Environ. Res. Public Health* **2022**, *19*, 14820.

sustainability MDPI

Article

How Gender, Culture, and Economy Influence Field of Study Preferences in Higher Education: Exploring Gender Gaps in STEM, AHSS, and Medicine among International Students

Corina Florina Tătar [1], Marius Ioan Tătar [2,*], János Pénzes [3] and George W. White [4]

1 Department of Geography, Tourism and Territorial Planning, University of Oradea, Str. Universității nr. 1, 410087 Oradea, Romania; corina.tatar@didactic.uoradea.ro
2 Department of Political Science and Communication Sciences, University of Oradea, Str. Universității nr. 1, Campus II, C55 Building, 410087 Oradea, Romania
3 Department of Social Geography and Regional Development, University of Debrecen, Egyetem tér 1, 4032 Debrecen, Hungary; penzes.janos@science.unideb.hu
4 Department of Geography and Geospatial Sciences, South Dakota State University, 1025 Medary Ave, Brookings, SD 57006, USA; george.white@sdstate.edu
* Correspondence: mtatar@uoradea.ro; Tel.: +40-770-210-960

Abstract: International female and male students' segregation per academic fields of study designates an important challenge for educational equity, diversity, and gender equality in tertiary education institutions worldwide. This study probes the determinants of study field choice among 984 students from 57 countries who enrolled at the University of Oradea, Romania, during 2022–2023. By incorporating gender approaches and concepts within broader economic and cultural theories, we utilized the bivariate analysis and multinominal regression models to scrutinize how students' preferences for STEM (science, technology, engineering, and mathematics), AHSS (arts, humanities, and social sciences), or medicine are influenced by their gender, and the more general cultural and economic attributes of their home country. Our findings enrich the knowledge and understanding of gendered patterns of academic study field choice, providing a cross-cultural and integrative viewpoint that enables us to set forth recommendations to bridge higher education gender gaps.

Keywords: gender equality; gender gap; higher education; field of study; international students; STEM; AHSS; medicine; Romania

Citation: Tătar, C.F.; Tătar, M.I.; Pénzes, J.; White, G.W. How Gender, Culture, and Economy Influence Field of Study Preferences in Higher Education: Exploring Gender Gaps in STEM, AHSS, and Medicine among International Students. *Sustainability* **2023**, *15*, 15820. https://doi.org/10.3390/su152215820

Academic Editor: Tarah Wright

Received: 2 October 2023
Revised: 3 November 2023
Accepted: 8 November 2023
Published: 10 November 2023

1. Introduction

Gender differences concerning international students' academic choices are a focal interest for researchers, educators, and decision-makers. The students' choice to follow a certain study field strongly influences both their educational and professional outcomes as well as the human capital distribution and skills in society [1–5]. Nevertheless, the study field choice is not student-gender neutral as men and women tend to significantly differ in terms of their preferences and enrolments per different academic specializations [6–9]. More precisely, men are more prone to choose science, technology, engineering, and mathematics, whereas women show a higher probability of choosing arts, humanities, and social sciences, or medicine [10–17]. Although women are a majority among the high-education enrolled students at a global level, field study segregation remains a critical factor that accounts for the prevalence of wage inequalities among men and women in the labor market [18–24].

The specialized literature reveals more explanations for gender imparity in different fields and academic disciplines [23,25,26], but empirical findings undermine their explanatory power [27]. Some of these explanations focus on microlevel factor effects, such as cognitive skills specific to each gender, perceived self-efficacy in various study fields, gender role socialization, and gender beliefs and stereotypes [13,28–30]; for a review, see [31]. On one hand, these factors might only account for a small part of the study field choice variation, and on the other hand, they have inconsistent effects in different countries and contexts [15,32,33]. Other explanations underline macrolevel factors such as educational system structures, job market opportunities, and the incentives for different study fields, as well as the cultural and institutional contexts which shape gender norms and expectations [8,13,34,35]. Rather often, these factors have heterogeneous effects on the study field choice or interact with the microlevel factors in multiple and contradictory ways [16,34,35].

In addition, the academic literature mainly analyzed domestic college students' gendered preferences for fields of study [10,11]. Nevertheless, research on international students might deliver valuable opportunities for understanding how gender and broader contextual factors impact the academic choices of persons educated in different socioeconomic and cultural milieux from their country of origin. Moreover, the study of international students is essential as they represent a heterogeneous growing population in global higher education [36], and their choices may be impacted by different factors than the ones usually affecting domestic students [37]. International students also face specific issues and opportunities regarding their professional and academic routes, such as the adjustments to a new culture and educational system, as well as the chance of developing intercultural and linguistic skills that can influence their contribution to the global society and economy [38]. Henceforth, studying international students' gendered field of study choice can provide valuable information for addressing the gender gap in higher education and the inequalities in the global labor market [39].

In the current study, we adopt an integrative approach that combines micro-individual and contextual factors to analyze the academic choice of the international students who study in Romania at the University of Oradea. Romania provides a fascinating milieu to examine higher education gender disparities as it faced noteworthy social and economic changes from the fall of communism in 1989, which affected both the educational system and the labor market [40–42]. Furthermore, Romania has the highest ratio of female to male graduates (0.80) in science, technology, engineering, and mathematics (STEM) among the European Union countries (EU average is only 0.52) [43], a fact that questions some of the common assumptions related to higher education gender segregation. To examine the gender patterns in different study fields, we use a dataset that includes 984 international students enrolled within the University of Oradea in the academic year 2022–2023. These students come from different cultural and socioeconomic settings, namely 57 countries of different world regions.

The purpose of this study is to examine the extent to which international students' gender, in combination with their home country's development level and gender equality index, impacts their study field choice at the University of Oradea, Romania. More precisely, we aim to explore how these individual and contextual factors match international students' preference for one of the three primary study fields: science, technology, engineering, and mathematics (STEM); arts, humanities, and social sciences (AHSS), and medicine. Our general argument is that international students' academic choice patterns are context-dependent, reflecting the interaction between gender, culture, and economic development in different countries. In the following sections of this study, we are operationalizing the main argument into more specific hypotheses, which we are evaluating using statistical analysis on a dataset concerning international students provided by the University of Oradea and completed with contextual data concerning the students' countries of origin.

Besides the introduction, we organized this paper into five sections. First, we critically assess the main cultural and economic perspectives, focusing on factors that affect the study field choice, and then we formulate the relevant hypotheses of each theoretical approach. Second, we discuss the used data and methods for empirically testing our hypotheses. Third, we show the results of the bivariate and multinominal regression results. Fourth, we discuss and interpret our study results, considering the theoretical expectations highlighted in the section concerning the literature review. In the final section, we summarize this research's main contributions and implications and suggest specific ways to develop it further.

2. Literature Review and Hypotheses

Despite the increasing number of women who pursue higher education, gender inequality in academia remains an omnipresent and persistent phenomenon [26,31,44–51]. The decision process of choosing a field of study is complex and influenced by many circumstances. Understanding the factors that impact international students' field of study choice is crucial for academic institutions and decision-makers to shape educational policies and support systems more effectively. The theoretical framework that lays the ground of this research integrates micro and macro perspectives to explain the academic gender patterns, more precisely, the field of study choice.

Gender theories refer to diverse approaches that analyze how gender is built, performed, and replicated in different social contexts. Generally, constructivist perspectives oppose the essentialist opinion according to which gender is a natural attribute, innate and fixed, derived from the biological sex. On the other hand, they claim that gender is a fluid and dynamic category shaped by social interactions, norms, and expectations [52,53]. The followers of these perspectives also examine how gender intersects with other social categories, such as class, race, ethnicity, sexuality, and nationality, to generate multiple and diverse shapes of identity and inequality [54,55]. Gender theories are relevant for understanding the field of study choice as they focus on how individuals attach meanings and values to different academic disciplines. According to these theories, the expectations and pressures from the relevant ones, such as parents, professors, and role models, influence students' preferences and decisions on college paths [13,56–58]. For instance, Eagly and Wood [59] claim that individuals learn gender roles through socialization. Subsequently, these internalized gender roles impact the educational and career paths that students decide to follow. For instance, female students may be encouraged to pursue careers in fields perceived as traditionally feminine, such as education or medical assistance.

On the other hand, male students may be encouraged to pursue careers in disciplines conventionally considered masculine, such as engineering and informatics. The stereotype threat theory can also explain the gender gap in the academic field choice. Steele and Aronson [60] show that people can have worse results at different tasks when they are aware of the negative stereotypes concerning their group. If women are aware of the stereotype that they are not good at mathematics or science, this can undermine their potential and performance in fields such as STEM. Based on these theories, we formulate two hypotheses connecting international students' gender to their academic study field:

H1A. *Female international students are more likely to choose AHSS over STEM as their field of study than male international students.*

H1B. *Female international students are more likely to choose medicine over STEM as their field of study than male international students.*

Dominant gender norms and expectations in international students' home countries can also impact their preference for particular study fields. Socialization theories underline the role of culture in shaping individuals' values, beliefs, attitudes, and behaviors. In this context, culture refers to the shared symbols, meanings, norms, and practices that constitute the collective identity and vision of the world of a particular group or society. Cultural perspectives assume that individuals' social background marks their mindset, affecting their educational choices [61,62]. Thus, cultural approaches are relevant for understanding the study field choice. They suggest that students' preferences and decisions are influenced by the prevailing values and norms of their country of origin, as well as by the cultural diversity and integration in the host country [34,35,63]. According to this approach, gender roles and the associated stereotypes are not natural or inevitable. Instead, they are learned and consolidated through social interactions within diverse institutions and arenas such as family, media, the education system, and the labor market.

In addition, gender-related social norms and expectations are not fixed but dynamic and evolving phenomena. For example, Inglehart and Norris [64] claim that the shift from industrialized societies to postindustrialized ones is culturally associated with the evolution from materialist to postmaterialist values such as self-expression, autonomy, and equality. These changes further led to significant gains regarding gender equality in the public and workplace spheres [64]. Thus, according to modernization theories, higher gender equality reflects a higher level of postmaterialist values in a society, which can stimulate more diverse and flexible study field choices both for men and women. On the one hand, the socioeconomic and cultural changes substantiated by these theories have provided more opportunities for individuals' educational choices.

On the other hand, the educational path of people who live in countries with lower gender equality remains relatively constrained by gender stereotypes and expectations that limit individual options. Nevertheless, gendered patterns concerning the fields of study in higher education endure even in advanced postindustrial societies. For example, in a comparative study of 44 countries, Charles and Bradley [63] show that segregation by gender per study field is not an outcome of traditionalism or backwardness but rather a product of the modernization of society, which favors gender identities which in their turn lead to gendered academic choice patterns. The authors find that gender segregation by fields of study is higher in societies with more equalitarian contexts concerning gender in which the gender-related essentialist ideology and self-expression value systems create opportunities and incentives for the "gendered selves" expression [63]. Thus, based on this theoretical framework, we can argue that international students from more gender-equalitarian societies have a higher propensity to choose AHSS or medicine over STEM, as they have more freedom and diversity in their educational choices. Furthermore, they are motivated by their interests and hobbies rather than social norms and pressures. Based on the cultural approaches highlighted above, we formulate two hypotheses concerning the relationship between the gender equality levels in international students' country of origin and their field of study preferences:

H2A. *International students coming from countries with higher levels of gender equality are more likely to choose AHSS over STEM as their field of study compared with international students coming from countries with lower gender equality.*

H2B. *International students coming from countries with higher levels of gender equality are more likely to choose medicine over STEM as their field of study compared with international students coming from countries with lower gender equality.*

International students can choose disciplines relevant to their home country's economic development priorities, preparing for the challenges and opportunities from their region of origin. The human capital theory is an economic perspective of education as an investment. Thus, students perceive the skills and knowledge achieved in college as a means to improve productivity and gains in the labor market. The human capital theory assumes that individuals are rational actors who make educational choices based on different study fields' expected costs and benefits [65,66]. Thus, examining the economic development levels of international students' countries of origin may be relevant for understanding their field of study choice. This approach assumes that the individuals' preferences and decisions are influenced by the opportunities and rewards in the labor market associated with different study fields as well as the assumed risks and uncertainties concerning employment in a specific job [45,67,68]. Therefore, these theories would predict that students from developing economies are more likely to study disciplines that grant a higher income and status in their home country or abroad [67], which may favor medicine over STEM. Moreover, lower economic development levels in students' home countries could also limit the availability and quality of education and career-related opportunities both for men and women. Thus, weaker outcomes of STEM education and skills regarding income, employment, and social status can discourage students from developing countries from choosing STEM over medicine or AHSS.

On the other hand, international students from economies in transition may be less tempted to choose disciplines requiring higher tuition fees or a longer period of academic education [68]. Thus, they may be less likely to choose medicine over STEM than students from developed economies. More than that, there can be a lower demand and supply for medical skills in economies in transition. The public health sector in these economies may be relatively underdeveloped, underfunded, and understaffed, resulting in low quality and accessibility of health services, low salaries and incentives for those employed in healthcare, and high morbidity and mortality rates [69–71]. These factors may spawn lower expected results for education in medicine. Therefore, students from economies in transition may be less likely to choose medicine as a study field as they can perceive it as less valuable or full of satisfaction in terms of human capital accumulation and the outcomes in the labor market.

Furthermore, international students from countries with higher development levels may be more likely to choose AHSS over STEM than those from countries with transition economies. This hypothesis matches Bell's [72] postindustrialization theory, which claims that socioeconomic development favors the passage from industrialist to postindustrialist societies characterized by the growth of the services, information, and knowledge sectors. According to this theory, a higher development level implies a greater demand for more relevant and applicable skills in the AHSS fields, such as creativity, critical thinking, and communication. Based on the human capital theories highlighted above, we formulate four hypotheses linking international students' choice for a field of study and the economic development levels of their home country:

H3A. *International students from developing economies are more likely to choose AHSS over STEM as their field of study than international students from developed economies.*

H3B. *International students from developing economies are more likely to choose medicine over STEM as their field of study than international students from developed economies.*

H3C. *International students from economies in transition are less likely to choose AHSS over STEM as their field of study than international students from developed economies.*

H3D. *International students from economies in transition are less likely to choose medicine over STEM as their field of study than international students from developed economies.*

We empirically assessed these hypotheses using a multinomial logistic regression model having the study field as a dependent variable and the international students' gender, the gender equality index, and the economic development levels of their home country as independent variables. The following section describes the empirical data sources and the analysis methods used.

3. Data and Methods

3.1. Data Sources

The data for this study come from two sources. First, the University of Oradea administrative service has provided us with a database of the 984 individual international students enrolled in the academic year 2022–2023 [73]. Second, we have completed the dataset with country-level information concerning gender equality and economic development in the international students' home country (see Supplementary Materials). The dataset contains individual-level variables about gender, nationality, the field of study, study level (BSc/BA, MSc/MA, or Ph.D.), and the study year of the international students from

57 countries. We have compiled the country-level data from the online platforms of the Global Gender Gap Report 2022, elaborated by the World Economic Forum [74] and World Economic Situation and Prospects 2022–Statistical Annex Report by the United Nations–The Social and Economic Affairs Department [75]. The Gender Gap Index, renamed in our analysis as the Gender Equality Index, provides a composite country-level measure of the gender-based parity levels in four key areas: economic participation and opportunities, educational attainment, health and survival, and political empowerment. Regarding the economic-related data gathering, we used the UN report concerning the World Economic Situation and Prospects from 2022, which ranks the countries' development levels into three categories: developing economies, economies in transition, and developed economies.

3.2. *Variables*

The dependent variable in our statistical analysis is the international students' study field, coded as a nominal variable with three categories: 1 = STEM (science, technology, engineering, and mathematics), 2 = AHSS (arts, humanities, and social sciences), and 3 = medicine. The independent variables were gender, the gender equality index, and the country's development level. Data on students' gender was collected by the University of Oradea's Office for Budget and Student Records in a binary format, and thus, we recoded it into a variable with two categories: women (0) and men (1). The gender equality index is a variable that initially varied from 0 (no equality) to 1 (full equality). For analytical purposes, we standardized this variable (mean = 0, standard deviation = 1, Min = −2.56, Max = 1.50), the higher values showing more gender equality in the home country. The home country's development level represents a nominal variable with three categories: 1 = developing economies, 2 = economies in transition, and 3 = developed economies.

3.3. *Data Analysis*

The data analysis was carried out in two stages. First, we use descriptive statistics to summarize the sample characteristics and contingency tables to examine the association between variables. Second, we run a multinomial logistic regression model to assess the hypotheses and estimate the specific effects of students' gender, home country development level, and gender equality index on their field of study choice, controlling the effects of all other variables from the statistical model. We set the reference category for the dependent variable as STEM, which means that the B coefficients represent the log odds of choosing AHSS or medicine over STEM for a one-unit increase in the independent variable, maintaining the other variables constant. The exponential coefficients Exp(B) represent the odds ratio, namely how much the odds of choosing AHSS or medicine over STEM have changed for a one-unit increase of the independent variable, maintaining the other variables constant. The statistical significance level is at 0.05. We carried out the statistical analysis using SPSS version 20. Thus, since our outcome variable, namely, field of study, has three categories (STEM, AHSS, and medicine) the mathematical model applied in this article is the multinomial logistic regression. We set STEM as the reference category, and the mathematical formulas used to predict students' field of study are the following:

1. For AHSS versus STEM:

$$\log\left(\frac{\mathrm{P(Field\ of\ Study_i{=}AHSS)}}{\mathrm{P(Field\ of\ Study_i{=}STEM)}}\right) = \beta_0 + \beta_1 \times \text{Gender equality index}_i + \beta_2 \times \text{Gender}_i + \beta_3 \times \text{Developing economies}_i + \beta_4 \times \text{Economies in transition}_i$$

2. For Medicine versus STEM:

$$\log\left(\frac{\mathrm{P(Field\ of\ Study_i{=}Medicine)}}{\mathrm{P(Field\ of\ Study_i{=}STEM)}}\right) = \beta_0 + \beta_1 \times \text{Gender equality index}_i + \beta_2 \times \text{Gender}_i + \beta_3 \times \text{Developing economies}_i + \beta_4 \times \text{Economies in transition}_i$$

where:

- β_0 is the intercept.

- β_1 is the coefficient of the gender equality index.
- β_2 is the coefficient of gender.
- β_3 is the coefficient of developing economies.
- β_4 is the coefficient of economies in transition.
- i is the index for the *i*th observation in our dataset.
- The variables Gender, Developing economies, and Economies in transition, are dummy variables, which take the value 1 if the condition is true and 0 otherwise, for the *i*th observation. For example, $gender_i$ is 1 if the *i*th student is female and 0 if the *i*th student is male.

4. Results

4.1. Descriptive Statistics

Table 1 shows the descriptive statistics of the variables included in this study. We used a dataset comprising 984 international students from 57 countries, of which 530 (53.9%) are men and 454 (46.1%) are women. The most frequent field study is medicine (68.1%), followed by AHSS (24.2%) and STEM (7.7%). The country of origin gender equality index is a standardized variable (mean = 0, standard deviation = 1, Min = −2.56, Max = 1.50), with higher values indicating more gender equality. This variable was recodified from the gender gap original index supplied by WEF (2022). The most frequent country of origin development level was economies in transition (40.4%), followed by developing economies (36%) and developed economies (23.6%).

Table 1. Descriptive statistics.

Variables	N	%
Dependent variable–Field of Study		
STEM	76	7.7
AHSS	238	24.2
Medicine	670	68.1
Independent variables		
Gender		
Female	452	46.1
Male	530	53.9
Country of origin development level		
Developing economy	354	36.0
Economy in transition	398	40.4
Developed economy	232	23.6
Country of origin gender equality index (standardized)	N = 920 *, Mean = 0, SD = 1, Min = −2.56, Max = 1.50	

Note: * 64 cases are missing for this variable as there were no data available for the gender equality index for students coming from Iraq, Palestine, Russia, Syria, Sudan, and Yemen. Source: Authors' elaboration.

Table 2 shows the cross-tabulation of the students' gender and their study field, using a chi-square test of independence to analyze the data for the 984 students. The results show that there is a significant statistical association between students' field of study and their gender ($\chi^2(2) = 48.308$, $p < 0.001$, Cramer's V = 0.222). The percentage of female students studying STEM fields (4.4%) is significantly lower than that of the male students enrolled in STEM (10.6%). On the other hand, the proportion of female students pursuing degrees in AHSS (33.7%) is significantly higher than that of the male students in AHSS (16%). The female students' ratio in medicine (61.9%) is lower than that of the male students (73.4%), but in relative terms, the difference is not as big as in the case of other fields. These findings suggest significant gender gaps in students' field of study choice. More precisely, female students are more likely to choose AHSS and less likely to prefer STEM versus male students.

Table 2. Association between students' gender and field of study.

Field of Study	Gender		Total
	Female	Male	
STEM	20 (4.4%)	56 (10.6%)	76 (7.7%)
AHSS	153 (33.7%)	85 (16.0%)	238 (24.2%
Medicine	281 (61.9%)	389 (73.4%)	670 (68.1%)
Total	454 (100%)	530 (100%)	984 (100%)

Notes: ($\chi^2(2) = 48.308$, $p < 0.001$, Cramer's V = 0.222, N = 984). STEM = science, technology, engineering, and mathematics; AHSS = arts, humanities, and social sciences. Source: Authors' elaboration.

Table 3 underlines that there is a statistically significant relationship between students' field of study and their country of origin development level ($\chi^2(4) = 432.402$, $p < 0.001$, Cramer's V = 0.469). Most students from developing (94.4%) and developed (92.2%) economies have chosen medicine as their field of study, while only 30.7% of the students from economies in transition have chosen it. On the other hand, more than half of the students from the economies in transition (53.5%) have chosen AHSS as their field of study compared to 3.7% of the students from developing economies and 5.2% from developed economies. STEM is the least popular field of study among all students but has a significantly higher ratio among students from economies in transition (15.8%) than among students from developing (2%) and developed (2.6%) economies. These findings show that students' preferences for a field of study vary according to their home country's economic development levels.

Table 3. Students' field of study based on the level of development of their country of origin.

Field of Study	Country of Origin Development Level			Total
	Developing Economies	Economies in Transition	Developed Economies	
STEM	7 (2.0%)	63 (15.8%)	6 (2.6%)	76 (7.7%)
AHSS	13 (3.7%)	213 (53.5%)	12 (5.2%)	238 (24.2%)
Medicine	334 (94.4%)	122 (30.7%)	214 (92.2%)	670 (68.1%)
TOTAL	354 (100%)	398 (100%)	232 (100%)	984 (100%)

Notes: ($\chi^2(2) = 432.402$, $p < 0.001$, Cramer's V = 0.469, N = 984). STEM = science, technology, engineering, and mathematics; AHSS = arts, humanities, and social sciences. Source: Authors' elaboration.

4.2. Multinomial Logistic Regression

A multinomial regression model was used to examine the effects of the student gender, the country of origin development level, and the gender equality index on the choice of the field of study. The dependent variable is international students' choice for a field of study, which had three categories: STEM, AHSS, and medicine. The reference category is STEM. The independent variables are students' gender (0 = female, 1 = male), country of origin development level (1 = developing economies, 2 = economies in transition, 3 = developed economies), and gender equality index.

We have evaluated the model fit using AIC, BIC, −2 log-likelihood, and chi-square statistics. The final model has a significantly better match to data than the intercept-only model, as the chi-square model shows ($\chi^2(8) = 481.927$, $p < 0.001$). The pseudo-R-squared values also suggest that the model explains a moderate to a relatively large degree of the variance of the dependent variable (Cox and Snell = 0.408, Nagelkerke = 0.504, McFadden = 0.317). The classification table highlights that per whole, the model predicts correctly 75.9% of the cases. The model has a high accuracy for predicting AHSS and medicine but a low accuracy for predicting STEM.

The parameter estimates presented in Table 4 show the effects of each independent variable on the log odds of choosing AHSS or medicine over STEM, controlling the effects of other variables included in the model. The results indicate that the gender equality index has a positive and significant effect on the odds of choosing medicine over STEM (B = 1.027,

SE = 0.375, p = 0.006) but not on the likelihood of choosing AHSS over STEM (B = 0.528, SE = 0.418, p = 0.207). In other words, higher gender equality levels in students' countries of origin increase their probability of choosing medicine over STEM. However, it does not significantly influence their likelihood of choosing AHSS over STEM.

Table 4. The multinomial regression model predicting students' field of study.

Field of Study [a]	Predictors	B	Std. Error	Wald	df	Sig.	Exp(B)	95% Confidence Interval for Exp(B)	
								Lower Bound	Upper Bound
AHSS versus STEM	Intercept	0.172	0.517	0.110	1	0.740			
	Gender equality index	0.528	0.418	1.594	1	0.207	1.695	0.747	3.844
	Gender (female)	1.687	0.303	31.068	1	0.000	5.406	2.986	9.784
	Gender (male)	0 [b]			0				
	Developing economies	0.989	0.951	1.083	1	0.298	2.690	0.417	17.341
	Economies in transition	−0.010	0.567	0.000	1	0.986	0.990	0.326	3.011
	Developed economies	0 [b]			0				
Medicine versus STEM	Intercept	2.754	0.426	41.802	1	0.000			
	Gender equality index	1.027	0.375	7.505	1	0.006	2.792	1.339	5.822
	Gender (female)	1.795	0.316	32.216	1	0.000	6.021	3.239	11.191
	Gender (male)	0 [b]			0				
	Developing economies	2.135	0.807	6.999	1	0.008	8.455	1.739	41.111
	Economies in transition	−3.478	0.504	47.647	1	0.000	0.031	0.011	0.083
	Developed economies	0 [b]			0				

Note: [a] The reference category is STEM. [b] This parameter is set to zero because it is redundant. Source: Authors' elaboration.

On the other hand, students' gender has a positive and significant effect on the odds of choosing AHSS over STEM (B = 1.687, SE = 0.303, $p < 0.001$) and of choosing medicine over STEM (B = 1.795, SE = 0.316, $p < 0.001$). In other words, female students are over five times more likely to choose AHSS over STEM than male students, keeping all other variables constant (Exp(B) = 5.406). Similarly, female students are over six times more likely to choose medicine over STEM than male students (Exp(B) = 6.021).

The economic situation of the country of origin has mixed effects on the odds of choosing AHSS or medicine over STEM. Compared to the students coming from developed economies, the ones from developing countries have higher odds of choosing medicine over STEM (B = 2.135, SE = 0.807, p = 0.008) but not a significantly different probability of choosing AHSS over STEM (B = 0.989, SE = 0.951, p = 0.298). On the other hand, students from transition economies are less likely to choose medicine over STEM (B = −3.478, SE = 0.504, $p < 0.001$). However, they do not have a significantly different probability of choosing AHSS over STEM (B = −0.010, SE = 0.567, p = 0.986). In other words, students from developing economies are 8.455 times more likely to choose medicine over STEM as a field of study than students from developed countries (Exp(B) = 8.455). On the other hand, students from economies in transition are 32 times less likely to choose medicine over STEM than their counterparts from developed economies. These findings highlight the powerful effect of students' home country's economic development level on their choice between studying rather than STEM.

5. Discussion

The goal of this research paper was to examine the extent to which gender, country of origin development level, and gender equality affect the field of study choice in the case of the international students enrolled at the University of Oradea, Romania. More precisely, we explored how these micro and macro level factors predict international students' preference for one of the three major fields of study: science, technology, engineering, and mathematics (STEM); arts, humanities, and social sciences (AHSS); and medicine. The results support our argument that international students' academic choices are context-dependent, reflecting the complex interplay between gender, culture, and economic development. In the literature review section, we have operationalized this wide-ranging expectation into eight more specific hypotheses. We then evaluated these hypotheses with a multinomial regression model using the international students' dataset from the University of Oradea, complemented with contextual data regarding students' country of origin. As shown in Table 5, the results of our statistical analyses have confirmed five of the eight hypotheses, indicating the fact that gender, the country-of-origin development level, and gender equality index generally relate to the field of study choice in the case of the international students enrolled at the University of Oradea, Romania. Nevertheless, the statistical significance, the effect of size, and the direction of these relationships varied according to the level and category of each independent variable.

Table 5. Summary of hypotheses and results of statistical analyses.

Hypothesis	Hypothesis Status
H1A: Female international students are more likely to choose AHSS over STEM as their field of study than male international students.	Supported
H1B: Female international students are more likely to choose medicine over STEM as their field of study than male international students.	Supported
H2A: International students from countries with higher levels of gender equality are more likely to choose AHSS over STEM as their field of study than international students from countries with lower gender equality.	Not supported
H2B: International students from countries with higher levels of gender equality are more likely to choose Medicine over STEM as their field of study than international students from countries with lower gender equality.	Supported
H3A: International students from developing economies are more likely to choose AHSS over STEM as their field of study than international students from developed economies.	Not supported
H3B: International students from developing economies are more likely to choose medicine over STEM as their field of study than international students from developed economies.	Supported
H3C: International students from economies in transition are less likely to choose AHSS over STEM as their field of study than international students from developed economies.	Not supported
H3D: International students from economies in transition are less likely to choose medicine over STEM as their field of study than international students from developed economies.	Supported

Source: Authors' elaboration.

Our results showed that gender is a significant predictor of the field of study choice among international students. In support of hypotheses H1A and H1B, we found that female students are likelier to choose AHSS or medicine over STEM as fields of study than male students. These findings suggest that gender stereotypes and expectations play an enduring role in shaping students' options and decisions concerning their academic path. As we highlight in Figure 1, medicine is the preferred field of study for international female and male students. However, female students are relatively more likely to pursue fields perceived as more socially oriented, such as AHSS. In contrast, male students tend to be relatively more attracted by fields considered more technically oriented, such as STEM. These findings align with previous studies that found similar gender segregation patterns across fields of study among domestic students [6,10,13].

Further, our findings show that the gender equality index is a significant predictor, but only for choosing specific fields of study among international students. Supporting our hypothesis H2B, we found that international students from countries with higher gender equality levels are more likely to choose medicine as their study field over STEM than those from countries with lower gender equality levels. However, we did not find substantial support for hypothesis H2A since international students from countries with higher gender equality levels were not significantly more likely to choose AHSS over STEM as a study field than international students from countries with lower gender equality. These mixed findings suggest that fields of study are evaluated differently in various cultural contexts. For example, our findings suggest that students from countries with higher gender equality might perceive medicine as a more valuable and significant study field than science and engineering than those from countries with lower gender equality. However, international students from countries with gender equalitarian contexts do not seem to significantly assign more value to studying arts, humanities, and social sciences over science and engineering compared to students from countries with less gender equality.

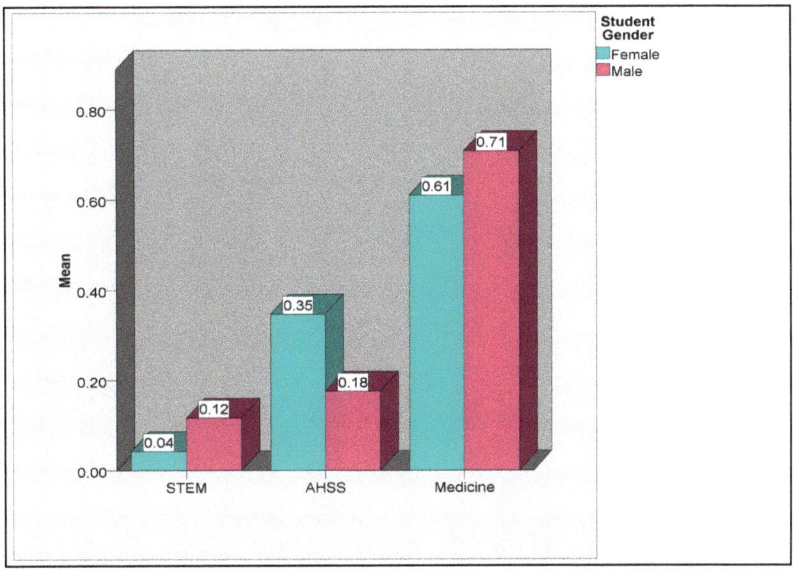

Figure 1. Mean of estimated probabilities for choosing STEM, AHSS, or medicine by gender.

Nevertheless, our findings support socialization theories, suggesting that the gender norms and contexts of students' home countries affect the field in which they choose to study abroad. More generally, our results also support the social value theory of education [76], according to which people choose their studies based on society's expectations. However, the social values and norms associated with different study fields may vary according to the gender equality levels in a specific country. In societies with higher gender equality levels, male and female students may have more diversity, flexibility, and freedom in their educational choices. On the other hand, in societies with high gender inequality, the socially prevalent gender stereotypes and expectations prescribe people's roles and shape their aspirations. There is more rigidity and conformity in male and female educational trajectories in these societies. Consequently, international students' preference for a field of study seems to depend on how gender equality in their home country is related to the demand and appreciation for skills in different sectors or occupations.

Our results also showed that the country of origin's development level significantly predicts the field of study choice among international students (see Figure 2). The human capital theory assumes that individuals are rational actors making educational choices based on the expected costs and benefits of different fields of study [65,66]. Our findings support this perspective as we found that students from developing economies are more likely to choose medicine over STEM than students from developed economies (hypothesis H3B) and that students from economies in transition are less likely to choose medicine over

STEM as a field of study, compared to students from developed economies (hypothesis H3D). However, our findings do not support hypothesis H3A, as students from developing economies were not significantly more likely to choose AHSS over STEM as a field of study than international students from developed economies. Moreover, we did not find support for hypothesis H3C since international students from economies in transition were not significantly less likely to choose AHSS over STEM as a field of study than international students from developed economies. Our results imply that labor market opportunities and rewards significantly shape students' academic choices. Thus, students from different economic contexts have different perceptions and preferences about the profitability and risks of different fields of study [45,67,68].

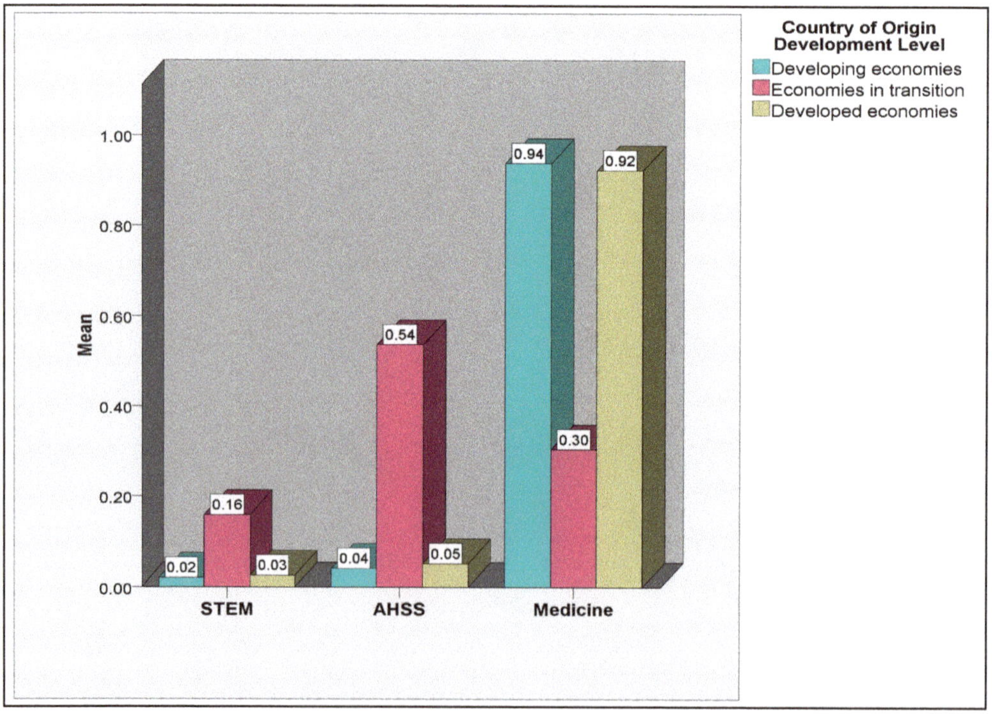

Figure 2. Estimated probabilities for choosing STEM, AHSS, or medicine by students' country of origin development level.

These findings also bring to light several interesting trends worth investigating and exploring more. For example, we described that medicine is the most preferred study field among international college students (over 68% are enrolled in medicine), regardless of gender, home country development level, or gender equality index. This pinpoints that international students consider medicine a prestigious field of study full of rewards. In addition, this pattern can display the more altruistic or humanitarian motivations of medical students. Furthermore, we ascertained that STEM is the least preferred field of study amongst international female students and those coming from gender-equalitarian societies. This suggests that international students perceive STEM as a difficult and uncertain field of study. This conclusion raises a few questions and has significant connotations for the University of Oradea and other higher education institutions that aim to draw and withhold international students and promote gender equality and diversity in STEM fields. The University's student body internationalization and diverseness efforts should focus on the prestige advancement and availability of AHSS and STEM field study programs in foreign languages. This could attract more international students interested in these fields and unknowledgeable of the Romanian language. Likewise, the University of Oradea should promote more intensely female role models who graduated and pursued successful careers in STEM so that it inspires more female students, both domestic and international, to choose STEM fields in which women are still belittled.

More generally, our study contributes to the academic literature concerning gender equality in higher education and allows a broader outlook which takes into consideration the way in which gender, culture, and economic development converge with the complex and dynamic process of the field of study choice among international students. In this context, we highlight the home country's role as a key moderator of gender gaps in major study fields such as STEM, AHSS, and medicine. Our findings advocate that gender segregation in higher education is not universal and inevitable but rather context-reliant, which reveals the interaction of individual factors and macrolevel settings. Consequently, we argue that policies and interventions to bridge gender gaps in different academic fields should be adjusted to the specific needs and features of various groups of students coming from different countries.

This study also has ramifications for academic institutions and decision-making factors that promote more inclusive and equitable higher education systems and job markets for domestic and international students coming from diverse backdrops. In this scope, we set forth the following four actions. First, academic institutions should provide more information and counseling to international students concerning various study fields, such as their specific requirements, benefits, and challenges. This would allow international students to make more documented and confident choices regarding their academic fields and professional careers. Secondly, academic institutions should also cater more support and resources to international students who pursue untraditional or underrepresented study fields for their gender or home country, such as mentoring, tutoring, counseling, or networking to enable international students to overthrow potential barriers or difficulties in the chosen fields and reinforce their academic success and satisfaction. Thirdly, the political decision-makers should intermediate more collaboration and exchanges among academic institutions and employers from different sectors and countries to increase skill and competency demand and supply in different fields of study. This would generate more opportunities and incentives both for domestic and international students to elect fields that match their skills and interests that are relevant and valuable for their countries or regions' social and economic development. Fourth, political decision-makers should also uphold more initiatives and programs aimed at fostering diverseness-related awareness and appreciation of different study fields and their contribution to the different countries and regions' sustainable economic and social development. This could diminish the existing stereotypes related to gender and chosen study field and would advocate for more inclusive and equitable higher education.

6. Conclusions

In the current paper, we probed how gender, home country economic development, and gender equality indexes affect the field of study choice of international students from the University of Oradea, Romania. We compared students' preferences for three academic fields of study: science, technology, engineering, and mathematics (STEM); arts, humanities, and social sciences (AHSS); and medicine. Generally, our results indicate that students' choices rely on the complex synergy between their gender and the cultural context and economic development level of their home country. More precisely, we discovered that gender is a significant predictor of field study choice among international students. Female students are more prone to choose AHSS or medicine over STEM compared to their male counterparts. This signifies that gender stereotypes and expectations impact students' choices. Furthermore, we found that the home country's gender equality index is a significant predictor of the international students' field study choice. However, we found mixed results for our hypotheses concerning the socialization effects of different gender contexts at a country level. Higher gender equality does not necessarily lead to more options for AHSS over STEM, but it is connected to more options for medicine over STEM. This implies that gender equality affects the flexibility and diversity of options for both genders in different ways, according to the field study.

We also found that the development level of the home country is a significant predictor of the chosen study field of international students. Nevertheless, we found mixed results regarding our hypotheses about the economic development effect. A lower level of development does not necessarily lead to more AHSS choices over STEM choices but is linked to more options for the field of medicine over STEM. On the other hand, coming from an economy in transition does not entice/drive/lead to less AHSS over STEM choices but is associated with a lesser preference for medicine over STEM. In other words, economic development levels affect students' preferences and motivations originating from different countries and regions, most probably influencing their perception of risks and rewards associated with educational outcomes of different study fields. In summary, our findings show that gender, country of origin development level, and gender equality index are significant predictors of the field of study choice among international students enrolled at the University of Oradea, Romania. However, our findings also reveal complex and sometimes contradictory gender segregation patterns per field of study among international students from different countries and regions. These patterns reflect the manifold interplay between gender, culture, and economic development in shaping students' preferences and decisions regarding their academic paths.

Recognizing the limitations of our current study and aiming at advancing knowledge about the gendered academic choice patterns of international students, we highlight the following specific directions for future research. First, to better understand the causal relations between the dependent and independent variables, we intend to adopt a longitudinal design which tracks the University of Oradea's international students' academic choices in time and in different contexts. For example, a panel survey could follow the same cohort of international students from admission to graduation and could measure their academic preferences, accomplishments, and results across different stages of their educational routes. Secondly, we aim to broaden the scope of this study by designing and administering a questionnaire for a representative sample of domestic and international students to compare their demographic characteristics, academic backdrops, motivations, expectations, experiences, and perceptions of their study fields. Third, to augment the external validity of results and their generalization scope, we shall search for a researchers' network to design and carry out a cross-national comparative study to probe the variations and resemblances of academic choice gendered patterns among international students of higher education institutions of different countries. Therefore, a cross-national comparison between international students of Romania and other countries could be achieved by attempting to delve into the institutional, educational, and social factors that influence students' academic choices. Last but not least, to complement the quantitative analysis methods used in the current study, a mixed-method approach could also be used to investigate the qualitative aspects and nuances of the international students' academic choice, such as their motivations, experiences, and perceptions. For example, semi-structured interviews and focus groups with a subsample of international students who chose different study fields could reveal the stories and meanings behind their academic choices.

Supplementary Materials: The following supporting information can be downloaded at: https://www.mdpi.com/article/10.3390/su152215820/s1, Dataset regarding the international students enrolled at the University of Oradea (2022–2023) supplemented with the country of origin macrolevel data (.sav format).

Author Contributions: Conceptualization, C.F.T., M.I.T., J.P. and G.W.W.; methodology, C.F.T., M.I.T., J.P. and G.W.W.; software, M.I.T.; validation, C.F.T., M.I.T., J.P. and G.W.W.; formal analysis, C.F.T., M.I.T. and J.P.; investigation, C.F.T., M.I.T., J.P. and G.W.W.; resources, C.F.T., M.I.T., J.P. and G.W.W.; data curation, M.I.T.; writing—original draft preparation, C.F.T., M.I.T., J.P. and

G.W.W.; writing—review and editing, C.F.T., M.I.T., J.P. and G.W.W.; visualization, C.F.T. and M.I.T.; supervision, C.F.T. and M.I.T.; project administration, C.F.T. and M.I.T.; funding acquisition, C.F.T. All authors have read and agreed to the published version of the manuscript.

Funding: This research received no external funding.

Institutional Review Board Statement: Not applicable (the study did not require the Institutional Review Board's Statement nor approval because it uses only publicly available and anonymized data).

Informed Consent Statement: Informed consent was obtained from all subjects involved in the study.

Data Availability Statement: Data supporting reported results can be found at: https://www3.wefrum.org/docs/WEF_GGGR_2022.pdf, accessed on 10 June 2023 and https://www.un.org/development/desa/dpad/wp-content/uploads/sites/45/WESP2022_ANNEX.pdf, accessed on 10 June 2023.

Acknowledgments: We are grateful to Krisztina Bronz from the University of Oradea's Office for Budget and Student Records for providing us with anonymized individual-level information regarding the international students enrolled at the University of Oradea during the 2022–2023 academic year.

Conflicts of Interest: The authors declare no conflict of interest.

References

1. Marginson, S. Limitations of human capital theory. *Stud. High. Educ.* **2019**, *44*, 287–301. [CrossRef]
2. Boldureanu, G.; Ionescu, A.M.; Bercu, A.-M.; Bedrule-Grigoruță, M.V.; Boldureanu, D. Entrepreneurship education through successful entrepreneurial models in higher education institutions. *Sustainability* **2020**, *12*, 1267. [CrossRef]
3. Mulang, H. The Effect of Competencies, Work Motivation, Learning Environment on Human Resource Performance. *Gold. Ratio Hum. Resour. Manag.* **2021**, *1*, 84–93. [CrossRef]
4. Wu, Y.-J.; Xu, X.; He, J. Gender, Educational Attainment, and Job Quality in Germany, Sweden, and the UK: Evidence from the 2015 European Working Conditions Survey. *Sustainability* **2021**, *13*, 13139. [CrossRef]
5. Aboramadan, M.; Albashiti, B.; Alharazin, H.; Dahleez, K.A. Human resources management practices and organizational commitment in higher education: The mediating role of work engagement. *Int. J. Educ. Manag.* **2020**, *34*, 154–174. [CrossRef]
6. Porter, C.; Serra, D. Gender differences in the choice of major: The importance of female role models. *Am. Econ. J. Appl. Econ.* **2020**, *12*, 226–254. [CrossRef]
7. Institute of International Education. *Open Doors 2019: Report on International Educational Exchange*; Institute of International Education: New York, NY, USA, 2019.
8. Rosa, R.; Clavero, S. Gender Equality in Higher Education and Research. *J. Gend. Stud.* **2022**, *31*, 1–7. [CrossRef]
9. Alonso, M.T.; Barba-Sánchez, V.; López Bonal, M.T.; Macià, H. Two Perspectives on the Gender Gap in Computer Engineering: From Secondary School to Higher Education. *Sustainability* **2021**, *13*, 10445. [CrossRef]
10. Makarova, E.; Aeschlimann, B.; Herzog, W. The Gender Gap in STEM Fields: The Impact of the Gender Stereotype of Math and Science on Secondary Students' Career Aspirations. *Front. Educ.* **2019**, *4*, 60. [CrossRef]
11. Kube, D.; Weidlich, J.; Jivet, I.; Kreijns, K.; Drachsler, H. "Gendered Differences versus Doing Gender": A Systematic Review on the Role of Gender in CSCL. *Unterrichtswissenschaft* **2022**, *50*, 661–688. [CrossRef]
12. Barone, C. Some Things Never Change: Gender Segregation in Higher Education across Eight Nations and Three Decades. *Sociol. Educ.* **2011**, *84*, 157–176. [CrossRef]
13. Charles, M.; Bradley, K. Equal but Separate? A Cross-National Study of Sex Segregation in Higher Education. *Am. Sociol. Rev.* **2002**, *67*, 573–599. [CrossRef]
14. Jacobs, J.A. Gender Inequality and Higher Education. *Rev. Sociol.* **1996**, *22*, 153–185. [CrossRef]
15. OECD. *The ABC of Gender Equality in Education*; OECD: Paris, France, 2015.
16. Smyth, E.; Steinmetz, S. Field of Study and Gender Segregation in European Labour Markets. *Int. J. Comp. Sociol.* **2008**, *49*, 257–281. [CrossRef]
17. Xie, Y.; Fang, M.; Shauman, K. STEM Education. *Annu. Rev. Sociol.* **2015**, *41*, 331–357. [CrossRef] [PubMed]
18. Barone, C.; Ortiz, L. Overeducation among European University Graduates: A Comparative Analysis of Its Incidence and the Importance of Higher Education Differentiation. *High. Educ.* **2011**, *61*, 325–337. [CrossRef]
19. Bobbitt-Zeher, D. The Gender Income Gap and the Role of Education. *Sociol. Educ.* **2007**, *80*, 1–22. [CrossRef]
20. Gerber, T.P.; Schaefer, D.R. Horizontal Stratification of Higher Education in Russia: Trends, Gender Differences, and Labor Market Outcomes. *Sociol. Educ.* **2004**, *77*, 32–59. [CrossRef]
21. Núñez, I.; Livanos, I. Higher Education and Unemployment in Europe: An Analysis of the Academic Subject and National Effects. *High. Educ.* **2010**, *59*, 475–487. [CrossRef]
22. Verdugo-Castro, S.; García-Holgado, A.; Sánchez-Gómez, M.C.; García-Peñalvo, F.J. Multimedia Analysis of Spanish Female Role Models in Science, Technology, Engineering and Mathematics. *Sustainability* **2021**, *13*, 12612. [CrossRef]

23. Casad, B.J.; Garasky, C.E.; Jancetic, T.R.; Brown, A.K.; Franks, J.E.; Bach, C.R. U.S. Women Faculty in the Social Sciences Also Face Gender Inequalities. *Front. Psychol.* **2022**, *13*, 792756. [CrossRef]
24. Van Veelen, R.; Derks, B. Equal Representation Does Not Mean Equal Opportunity: Women Academics Perceive a Thicker Glass Ceiling in Social and Behavioral Fields Than in the Natural Sciences and Economics. *Front. Psychol.* **2022**, *13*, 790211. [CrossRef]
25. Kerras, H.; Bautista, S.; Piñeros Perea, D.S.; de-Miguel Gómez, M.D. Closing the Digital Gender Gap among Foreign University Students: The Challenges Ahead. *Sustainability* **2022**, *14*, 12230. [CrossRef]
26. Ucar, I.; Torre, M.; Elías, A. Mind the Gender Gap: COVID-19 Lockdown Effects on Gender Differences in Preprint Submissions. *PLoS ONE* **2022**, *17*, e0264265. [CrossRef] [PubMed]
27. O'Connor, P.; Carvalho, T.; Vabø, A.; Cardoso, S. Gender in Higher Education: A Critical Review. In *The Palgrave International Handbook of Higher Education Policy and Governance*; Huisman, J., de Boer, H., Dill, D.D., Souto-Otero, M., Eds.; Palgrave Macmillan: London, UK, 2015; pp. 569–584.
28. Priyashantha, K.G.; De Alwis, A.C.; Welmilla, I. Gender Stereotypes Change Outcomes: A Systematic Literature Review. *J. Humanit. Appl. Soc. Sci.* **2023**, *5*, 450–466. [CrossRef]
29. Jonsson, J.O. Explaining Sex Differences in Educational Choice: An Empirical Assessment of a Rational Choice Model. *Eur. Sociol. Rev.* **1999**, *15*, 391–404. [CrossRef]
30. Torres-Ramos, S.; Fajardo-Robledo, N.S.; Pérez-Carrillo, L.A.; Castillo-Cruz, C.; Retamoza-Vega, P.d.R.; Rodríguez-Betancourtt, V.M.; Neri-Cortés, C. Mentors as Female Role Models in STEM Disciplines and Their Benefits. *Sustainability* **2021**, *13*, 12938. [CrossRef]
31. Wang, M.-T.; Degol, J.L. Gender Gap in Science, Technology, Engineering, and Mathematics (STEM): Current Knowledge, Implications for Practice, Policy, and Future Directions. *Educ. Psychol. Rev.* **2017**, *29*, 119–140. [CrossRef]
32. Hyde, J.S.; Lindberg, S.M.; Linn, M.C.; Ellis, A.B.; Williams, C.C. Gender Similarities Characterize Math Performance. *Science* **2008**, *321*, 494–495. [CrossRef]
33. Schiebinger, L.; Klinge, I.; Sánchez de Madariaga, I.; Schraudner, M.; Stefanick, M. Gendered Innovations in Science, Health, Medicine, Engineering, and Environment. In Proceedings of the JST Diversity Seminar, Tokyo, Japan, 16 March 2016.
34. Van De Werfhorst, H.G.; Sullivan, A.; Cheung, S.Y. Social Class, Ability and Choice of Subject in Secondary and Tertiary Education in Britain. *Br. Educ. Res. J.* **2003**, *29*, 41–62. [CrossRef]
35. Buchmann, C.; DiPrete, T.A.; McDaniel, A. Gender Inequalities in Education. *Annu. Rev. Sociol.* **2008**, *34*, 319–337. [CrossRef]
36. Altbach, P.G.; Knight, J. The Internationalization of Higher Education: Motivations and Realities. *J. Stud. Int. Educ.* **2007**, *11*, 290–305. [CrossRef]
37. Brooks, R.; Waters, J. *Student Mobilities, Migration and the Internationalization of Higher Education*; Palgrave Macmillan: London, UK, 2011.
38. UNESCO-IESALC. *Gender Equality: How Global Universities are Performing*; UNESCO International Institute for Higher Education in Latin America and the Caribbean and Times Higher Education: Paris, France, 2022; p. 45.
39. Mott, H. *Gender Equality in Higher Education: Maximising Impacts*; British Council: London, UK, 2022; p. 127.
40. Păunescu, M.; Vlăsceanu, L.; Miroiu, A. *Calitatea Învăţământului Superior din România: O Analiză Instituţională a Tendinţelor Actuale;* Polirom: Iaşi, Romania, 2011.
41. Hatos, A. *Sociologia Educaţiei;* Polirom: Iași, Romania, 2006.
42. Gavriluță, N. The Labor Market, Employability and Entrepreneurship in the Romanian Public Sector. *Transylv. Rev. Adm. Sci.* **2020**, *16*, 46–69. [CrossRef]
43. Eurostat. *Tertiary Education Statistics—Statistics Explained*; Eurostat: Luxembourg, 2023.
44. Altbach, P.G.; Reisberg, L.; Rumbley, L.E. *Trends in Global Higher Education: Tracking an Academic Revolution*; Brill: Leiden, The Netherlands, 2019.
45. Riegle-Crumb, C.; King, B.; Grodsky, E.; Muller, C. The More Things Change, the More They Stay the Same? Prior Achievement Fails to Explain Gender Inequality in Entry Into STEM College Majors Over Time. *Am. Educ. Res. J.* **2012**, *49*, 1048–1073. [CrossRef] [PubMed]
46. Xu, C.; Xiang, F.; Duan, R.; Miralles-Cardona, C.; Huo, X.; Xu, J. An Analysis of Factors Influencing Chinese University Students' Major Choice from the Perspective of Gender Differences. *Sustainability* **2023**, *15*, 14037. [CrossRef]
47. Liu, F.; Holme, P.; Chiesa, M.; AlShebli, B.; Rahwan, T. Gender Inequality and Self-Publication Are Common among Academic Editors. *Nat. Hum. Behav.* **2023**, *7*, 353–364. [CrossRef]
48. De Welde, K.; Stepnick, A. *Disrupting the Culture of Silence: Confronting Gender Inequality and Making Change in Higher Education;* Routledge: New York, NY, USA, 2023.
49. Cui, R.; Ding, H.; Zhu, F. Gender Inequality in Research Productivity during the COVID-19 Pandemic. *Manuf. Serv. Oper. Manag.* **2022**, *24*, 707–726. [CrossRef]
50. Kim, E.; Patterson, S. The Pandemic and Gender Inequality in Academia. *PS Political Sci. Politics* **2022**, *55*, 109–116. [CrossRef]
51. Kim, L.; Smith, D.S.; Hofstra, B.; McFarland, D.A. Gendered Knowledge in Fields and Academic Careers. *Res. Policy* **2022**, *51*, 104411. [CrossRef]
52. Butler, J. *Gender Trouble—Feminism and the Subversion of Identity;* Routledge: New York, NY, USA, 1990.
53. West, C.; Zimmerman, D.H. Doing Gender. *Gend. Soc.* **1987**, *1*, 125–151. [CrossRef]
54. Hill Collins, P. *Black Feminist Thought: Knowledge, Consciousness, and the Politics of Empowerment*, 2nd ed.; Routledge: New York, NY, USA, 2000.

55. Crenshaw, K.W. Mapping the Margins: Intersectionality, Identity Politics, and Violence against Women of Color. *Stanf. Law Rev.* **1991**, *43*, 1241–1299. [CrossRef]
56. Eccles, J.S.; Adler, T.F.; Futterman, R.; Goff, S.B.; Kaczala, C.M.; Meece, J.L.; Midgley, C. Expectancies, Values, and Academic Behaviors. In *Achievement and Achievement Motivation*; Spence, J.T., Ed.; W.H. Freeman: San Francisco, CA, USA, 1983; pp. 75–146.
57. Francis, B.; Archer, L.; Hodgen, J.; Pepper, D.; Taylor, B.; Travers, M.-C. Exploring the Relative Lack of Impact of Research on 'Ability Grouping' in England: A Discourse Analytic Account. *Camb. J. Educ.* **2017**, *47*, 1–17. [CrossRef]
58. Verdugo-Castro, S.; Sánchez-Gómez, M.C.; García-Holgado, A. University Students' Views regarding Gender in STEM Studies: Design and Validation of an Instrument. *Educ. Inf. Technol.* **2022**, *27*, 12301–12336. [CrossRef] [PubMed]
59. Eagly, A.H.; Wood, W. Social Role Theory of Sex Differences. In *The Wiley Blackwell Encyclopedia of Gender and Sexuality Studies*; Wiley Blackwell: Hoboken, NJ, USA, 2016; pp. 1–3.
60. Steele, C.M.; Aronson, J. Stereotype Threat and the Intellectual Test Performance of African Americans. *J. Personal. Soc. Psychol.* **1995**, *69*, 797–811. [CrossRef]
61. Bourdieu, P. *Distinction: A Social Critique of the Judgment of Taste*; Harward University Press: Cambridge, MA, USA, 1984.
62. Swidler, A. Culture in Action: Symbols and Strategies. *Am. Sociol. Rev.* **1986**, *51*, 273–286. [CrossRef]
63. Charles, M.; Bradley, K. Indulging Our Gendered Selves? Sex Segregation by Field of Study in 44 Countries. *Am. J. Sociol.* **2009**, *114*, 924–976. [CrossRef] [PubMed]
64. Inglehart, R.; Norris, P. *Rising Tide: Gender Equality and Cultural Change Around the World*; Cambridge University Press: Cambridge, UK, 2003.
65. Becker, G.S. *Human Capital: A Theoretical and Empirical Analysis with Special Reference to Education*; University of Chicago Press: Chicago, IL, USA, 1964.
66. Mincer, J. *Schooling Experience and Earnings*; National Bureau of Economic Research: New York, NY, USA, 1974.
67. Boudarbat, B.; Lemieux, T.; Riddell, W.C. The Evolution of the Returns to Human Capital in Canada, 1980-2005. *Can. Public Policy* **2010**, *36*, 63–89. [CrossRef]
68. Montmarquette, C.; Cannings, K.; Mahseredjian, S. How Do Young People Choose College Majors? *Econ. Educ. Rev.* **2002**, *21*, 543–556. [CrossRef]
69. Mirmirani, S.; Li, H.; Ilacqua, J.A. Health Care Efficiency in Transition Economies: An Application of Data Envelopment Analysis. *Int. Bus. Econ. Res. J.* **2008**, *7*, 47–56. [CrossRef]
70. Cornia, G.A.; Paniccià, R. *The Mortality Crisis in Transitional Economies*; Oxford University Press: Oxford, UK, 2000.
71. Gupta, S.; Verhoeven, M.; Tiongson, E.R. The Effectiveness of Government Spending on Education and Health Care in Developing and Transition Economies. *Eur. J. Political Econ.* **2002**, *18*, 717–737. [CrossRef]
72. Bell, D. *The Coming of the Post-Industrial Society: Venture in Social Forecasting*; Penguin Books: London, UK, 1976.
73. University of Oradea, U.O. International Students Enrolled at the University of Oradea in the 2022-23 Academic Year 2022.
74. WEF. *Global Gender Gap Report, World Economic Forum*; WEF: Geneva, Switzerland, 2022.
75. UN. *World Economic Situation and Prospects (WESP) 2022—Statistical Annex*; UN: New York City, NY, USA, 2022.
76. Boudon, R. *Education, Opportunity, and Social Inequality: Changing Prospects in Western Society*; Wiley-Interscience: New York, NY, USA, 1974.

sustainability MDPI

Article

Expert-Based Assessment of the Potential of Agroforestry Systems in Plain Regions across Bihor County, Western Romania

Ruben Budău [1], Andrei Apăfăian [2], Mihail Caradaică [3,4], Iulian A. Bratu [5], Claudia S. C. Timofte [6] and Cristian M. Enescu [7,*]

1 Department of Silviculture and Forestry Engineering, University of Oradea, 26 General Magheru Boulevard, 410048 Oradea, Romania; rbudau@uoradea.ro
2 Department of Ecology, "Marin Drăcea" National Institute for Research and Development in Forestry, 13 Cloșca Street, 50004 Brașov, Romania; apafaian.andrei@gmail.com
3 Department of International Relations and European Integration, National University of Political Studies and Public Administration, 30A Expoziției Boulevard, 1st District, 012104 Bucharest, Romania; mihai.caradaica@dri.snspa.ro
4 Tomorrow's Forest Foundation, 5 Mendeleev Street, 1st Floor, 2nd Office, 1st District, 010361 Bucharest, Romania
5 Department of Agricultural Sciences and Food Engineering, "Lucian Blaga" University of Sibiu, 7-9 Dr. Ion Ratiu Street, 550024 Sibiu, Romania; iulian.bratu@ulbsibiu.ro
6 Department of Law and Administrative Sciences, University of Oradea, 26 General Magheru Boulevard, 410048 Oradea, Romania; claudia.timofte@uoradea.ro
7 Department of Soil Sciences, University of Agronomic Sciences and Veterinary Medicine of Bucharest, 59 Mărăști Boulevard, 011464 Bucharest, Romania
* Correspondence: mihaienescu21@gmail.com

Abstract: Agroforestry systems are gaining increasing attention worldwide due to their several benefits both for landowners and also for the environment. Even though Romania has a great potential for adopting these systems on a large scale, only a few examples exist. The aim of this research was to highlight the main agroforestry systems that could be introduced in plain regions across Bihor County, Western Romania. A selection of the most suitable woody species and cereals was carried, based on available data and information. In order to select the most suitable combination, a set of eight criteria was considered and an Analytical Hierarchy Process Analysis was performed, with the aid of the Expert Choice Desktop (v. 11.5.1683) software package. The combinations that had the black locust as the main tree species scored better in comparison with the ones that had pedunculate oak as a main species. This research should be regarded as a first and important step in the analysis of several combinations of agroforestry systems that could be implemented across plain regions of Bihor County. Lastly, this proposed model could be replicated in similar studies aimed at selecting the most suitable agroforestry systems for certain sites. Future research should also consider criteria that account for various aspects, including the functional relationships of these future green spaces with nearby areas.

Keywords: Analytical Hierarchy Process; black locust; green spaces; land management; maize; pedunculate oak; sunflower

Citation: Budău, R.; Apăfăian, A.; Caradaică, M.; Bratu, I.A.; Timofte, C.S.C.; Enescu, C.M. Expert-Based Assessment of the Potential of Agroforestry Systems in Plain Regions across Bihor County, Western Romania. *Sustainability* **2023**, *15*, 15724. https://doi.org/10.3390/su152215724

Academic Editor: Ali Ayoub

Received: 4 October 2023
Revised: 25 October 2023
Accepted: 6 November 2023
Published: 8 November 2023

1. Introduction

Agroforestry (AF) systems include several land-use practices in which agricultural crops or pasture are integrated with woody plants (both tree and shrub species), which are regarded as an important component of agricultural land development [1,2].

There are several categories of common agroforestry practices worldwide and also in Europe [3–5] such as wood pastures, riparian buffer strips, hedgerows, windbreaks, grazed forests, intercropped and grazed orchards, forest farming and more novel silvoarable and silvopastoral practices and systems such as alley coppice, alley cropping and woodland chicken [6]. All these practices have the combination of trees and shrubs with crops and/or livestock in common, offering an integrated land-use system to landowners [7,8].

As this study is focused on agroforestry systems in Bihor County (Western Romania), an overview of the current practices across Europe is required to emphasis its relevance. The development of AF systems in Europe did not take off until now because of the different regulatory systems of the member states and a lack of knowledge on this issue that could enforce more EU regulation. In the European Union, agroforestry is mainly supported through the Common Agriculture Policy (CAP) [9]. The practices of the CAP in the 20th century discouraged agroforestry practices as CAP payments for crops or pastures were often reduced for parcels with scattered trees [10]. But as time passed, land-users and policymakers needed to adapt the CAP to the EU's climate agenda and to the growing enthusiasm in agroforestry around the world, including North America. Therefore, there has been a remarkable turn-around in the perception of the role and importance of trees on farmlands during the past two decades [10].

According to Burgess and Rosati [11], there are two CAP mechanisms for farm support in the EU: direct payments in Pillar I, which is completely financed by the European Commission, and payments that support rural development in Pillar II, whose measures have to be co-financed by the member states. However, different national legal constrains are slowing down the fund uptake for agroforestry systems. For examples, German farmers claiming Pillar I payments need to subdivide agroforestry areas into area containing trees and areas which do not have trees [12]. The possibility of registering the land as agroforestry would reduce bureaucracy and would allow farmers to operate more efficiently. There are also issues regarding the support for the maintenance of lone trees and hedgerows within rural development programs. The difficulties in monitoring the extent and quality of lone trees and hedgerows make payments in Pillar I difficult [13]. Within Pillar II there is one specific agroforestry measure in which the funds are dedicated for this kind of projects, and other 27 measures that may support agroforestry systems to some extent. Therefore, the recognition of agroforestry would be increased if the measures were collated together in one place.

Recently, due to its several benefits as many scientific studies have revealed, agroforestry has gained an increasing interest worldwide, including in Europe [14–17]. For example, within the recently adopted Common Agricultural Policy 2023-27, which entered into force on the 1 January 2023, agroforestry will be supported directly by four countries, namely Portugal, Germany, Greece and the Czech Republic, and indirectly through the support of landscape features from seventeen European countries, including Romania [18]. The higher importance of agroforestry in the CAP 2023-27 can also be understood as a mechanism to reach the objectives of the European Green Deal as agroforestry could increase carbon sequestration alongside cover crops. Therefore, the recent developments within the EU Green Deal show that the European Commission is realizing the potential of agroforestry and tries to take advantage of it.

Considering the fact that AF systems are a sustainable land management option that delivers market and non-market goods and services and the increasing support from the CAP programs, governments need to develop policies and actions that foster agroforestry within an EU policy framework [19]. In addition, a European AF strategy is needed to provide a proper framework that is recognized by the member states, although, the implementation of such strategy would require knowledge of these types of practices at the European level [19].

Besides the Common Agricultural Policy, agroforestry is a key element in other EU policy areas as it is reflected in the following strategies: the Farm to Fork Strategy, 2030, and the Biodiversity Strategy and EU Forest Strategy for 2030. The Farm to Fork Strategy (F2F) addresses the challenges of sustainable food systems, setting ambitions such as reducing the use and risk of chemical pesticides by 50%, decreasing the use of fertilizers by 20% and decreasing nutrient losses by 50%, having at least 25% of the EU's agricultural land under organic farming, or reducing EU sales of antimicrobials for farmed animals and in aquaculture by 50%, all of which aim to be achieved by 2030 [20]. All these objectives will need nature-based solutions and, along with storing carbon in soil, emphasize agroforestry as one of the most important tools.

The 2030 Biodiversity Strategy aims towards the protection and restauration of nature and putting the biodiversity on the path to recovery by 2030. The objectives of this strategy are to halt and reverse the decline of farmland birds and insects, halt soil degradation and to plant 3 billion trees by 2030 [21]. Thus, the 2030 Biodiversity Strategy strongly focus on making agriculture more nature-inclusive and nature-friendly, and mentions agroforestry as being a key tool [21]. The EU Forest Strategy for 2030 is more action-oriented and tries to accelerate the roll out of carbon farming practices through eco-schemes on agroforestry or rural development interventions and to boost research and innovation on agroforestry systems and other trees outside the forests [22].

Considering the new European developments on agroforestry, such as the CAP 2023-27 or the Forest and Biodiversity strategies, Romania has the potential to become one of the most important actors on the continent. Forest shelterbelts, grasslands with sparse trees, alley cropping, orchard meadows for both fodder and fruit production or even home-gardening are among the most common agroforestry systems worldwide and also in Romania [23–28].

These combinations of agricultural crops and forest plantations generate a broad range of benefits, such as: the development of rural communities and farms due to a sustainable production and livelihood improvement [29–33], increased biodiversity and biological control of the pests [34,35], improved and healthier soils [36,37], food production [38,39], carbon sequestration [1,40] and a veritable strategy to fight against changing climatic conditions. It was recently reported [41] that most of these benefits could be also provided by promoting agroecology techniques. In all these cases, special attention should also be given to water resource management which is critical for sustainable development of any type of social–ecological system [42].

In Romania, with the exception of field-protective forest shelterbelts that had their highest expansion during the middle of the last century, agroforestry systems represent a new concept, that is not perceived, for example, as an independent science [43]. In regards to the forest sector, the focus is on the natural regeneration of the stands and the maintenance of the composition of the stands as close as possible to the composition of natural forest types; by limiting the introduction of allochthonous tree or shrub species, for example [44]. In this context, in the last two decades, the areas afforested at national level decreased significantly [45]. But, by taking also into consideration the circular-bioeconomy transition which interferes with the forest sector [46], and the uneven distribution of forest lands across Romania, with the lowest presence in plain regions, of about 6.5% [47], we consider that agroforestry systems will play a significant role in rural development in several counties across Romania, especially in plain regions.

The aim of this study was to highlight one of the most important agroforestry systems that could be implemented in plain regions across Bihor County.

2. Materials and Methods

2.1. Study Design and Case Description

Bihor County (Figure 1) is situated in the north-western part of Romania, it has all the types of relief units, with an altitude ranging from 90 m a.s.l. (Ateaș-Cefa area) to 1849 m a.s.l. in the Bihorului Mountains. In this relief context, in regard to the forest land, Luvisoils and Cambisols are the representative soil classes. Luvisols (25%), eutric cambisols (22%), dystric cambisols (20%), preluvisols (17%) and entic podzols (5%) represent the most common forest soil types in Bihor County [48]. Across the county, 205,800 hectares are occupied by forests (around 27% of the total area of Bihor), of which 85% are occupied by deciduous forests. The National Forest Administration ROMSILVA, through the Bihor Forestry Department, manages an area of 115,260 hectares, of which 61,170 hectares are the property of the Romanian state [49].

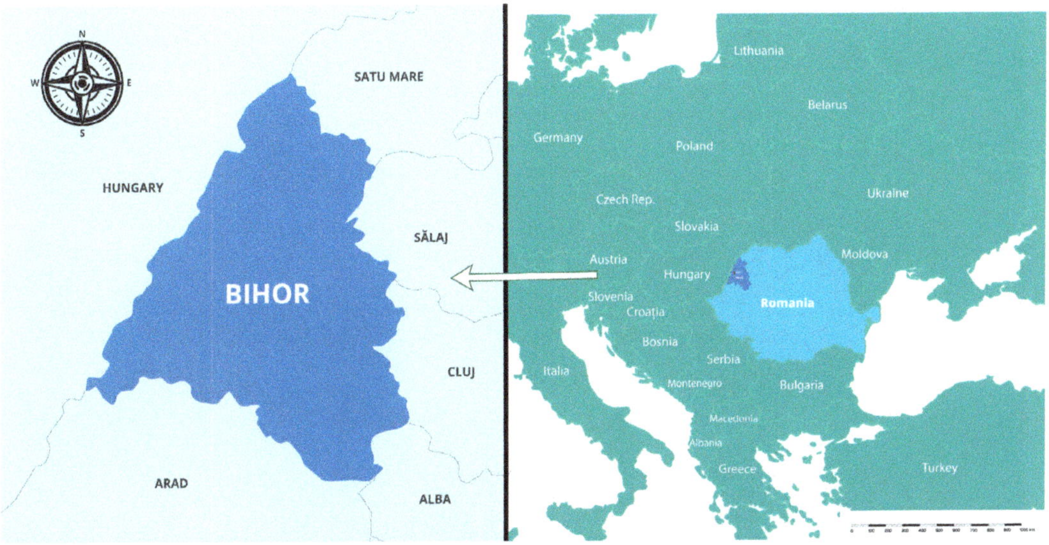

Figure 1. Location of Bihor County.

Bihor County belongs to the historical region of the Crișurilor Plain, which represents almost one quarter of the West Plain [50]. As the second largest plain by surface in Romania, the West Plain is bordered on the east by the West Hills and Occidental Carpathians, on the west by Hungary and Republic of Serbia, on the north by the Tur River, and on the south by the Republic of Serbia [51,52]. The altitude is generally low, predominantly under 100 m, and the groundwater level generally ranges between 0.5 and 3 m [53]. The region has a moderate, temperate continental climate, with strong oceanic influences generated by the dominant westerly winds. According to the meteorological data from the last five decades, the climate is characterized by average rainfall values of 620.0 mm, with a varying between 411.0 mm and 889.8 mm. The average air temperature values is 10.7 °C, with a minimum of 8.9 °C and a maximum of 12.45 °C, respectively [54]. The monthly average values ranged from −1.4 °C in January at Chișineu Criș meteorological station to 21.5 °C in July at Salonta meteorological station. The maximum and the minimum absolute temperature values recorded at Oradea meteorological station were 40.4 °C (in July 2007) and −22.5 °C (in January 1987), respectively [55]. All three meteorological stations are located along the studied area. Pedunculate oak (*Quercus robur* L.), Turkey oak (*Q. cerris* L.), Hungarian oak (*Q. frainetto* Ten.), European ash (*Fraxinus excelsior* L.), black locust (*Robinia pseudoacacia* L.), hornbeam (*Carpinus betulus* L.) and sessile oak [*Q. petraea* (Matt.) Liebl] represent the most common tree species across the Western Plain [51,52,56,57].

2.2. Case Study: Implementation of the Expert Model

To select the most suitable agroforestry system for plain regions across Bihor County, an Analytical Hierarchy Process (AHP) was performed.

AHP represents a multi-criteria decision analysis which is based on a theory of measurements focused on pairwise comparisons. Its aim is to decompose complex decision problems (i.e., the aim of this study: choosing the most suitable agroforestry system for the plain regions across Bihor County) into a hierarchy of sub-problems (i.e., the considered set of criteria), which can be deeply and independently analyzed. Thus, the alternatives (i.e., the proposed agroforestry systems) are compared one to each other and a scoring system is used [58–60].

Being simple to use, flexible and cost effective [61], AHP was widely used in several research fields in the last five decades. For example, in Romania, AHP was used to choose different tree and shrub species for the establishment of field shelterbelts [62] or to propose the most suitable solution for afforestation of sandy soils in Oltenia Plain and Carei Plain [47].

In this study, eight agroforestry systems were proposed, each of them being composed of a main tree species, a cereal, a secondary tree species and a shrub species. A 50 × 200 m rectangle, divided into 4 equal squares, was considered (Figure 2). We proposed this standardized model in order to be easy to assesses the concrete values regarding the yield, wood production, etc.

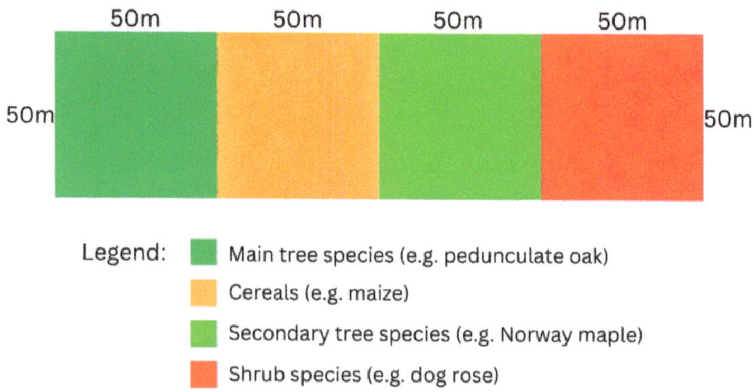

Figure 2. Design of the experimental plots.

Based on experts' opinions, eight alternatives of agroforestry systems were proposed (Table 1). The considered species were the following ones: pedunculate oak (*Quercus robur* L.)—Q.r., common sunflower (*Helianthus annuus* L.)—H.a., European ash (*Fraxinus excelsior* L.)—F.e., saskatoon berry (*Amelanchier alnifolia* Nutt.)—A.a., maize (*Zea mays* L.)—Z.m., Norway maple (*Acer platanoides* L.)—A.p., dog rose (*Rosa canina* L.)—R.c., common wheat (*Triticum aestivum* L.)—T.a., blackthorn (*Prunus spinosa* L.)—P.s., black locust (*Robinia pseudoacacia* L.)—R.p. and honey locust (*Gleditsia triacanthos* L.)—G.t., respectively. The selected woody species are also among the most often used in the afforestation of several categories of degraded terrains across Romania [63].

Table 1. The eight considered alternatives of agroforestry systems.

Alternative	Species			
1	Q.r.	H.a.	F.e.	A.a.
2	Q.r.	Z.m.	A.p.	R.c.
3	Q.r.	T.a.	F.e.	P.s.
4	Q.r.	H.a.	A.p.	A.a.
5	R.p.	Z.m.	G.t.	P.s.
6	R.p.	T.a.	G.t.	R.c.
7	R.p.	H.a.	F.e.	A.a.
8	R.p.	P.s.	F.e.	R.c.

In the case of tree and shrub species seedlings, according to the technical norms, a distance of 2 m between the rows and a distance of 1 m between the seedlings on the same row were adopted, which means that 1250 seedlings were considered to be planted in each square.

For cereal crops the average sowing density was 55,000–80,000 plants/ha for maize, 50,000–65,000 plants/ha for sunflower, using the 20 × 70 cm sowing scheme in both cases, and 450–550 plants/m^2 (180–280 kg/ha) in the case of wheat that is sown at an average distance of 12.5 cm between rows.

Moreover, in the case of the three selected cereals, concrete data regarding the yields/hectare and the costs/kilogram are available from the Romanian Institute of Statistics for the timeframe 2013–2022 [64] (Table 2).

Table 2. Yields and costs for common wheat, maize and common sunflower crops for the timeframe 2012–2022.

Species	Year									
	2013	2014	2015	2016	2017	2018	2019	2020	2021	2022
	Yield/hectare [kg/ha]									
Triticum aestivum	3985	3763	3783	2935	4110	4176	4361	4072	4173	3833
Zea mays	4781	4276	3106	4217	6117	7503	7519	6582	6300	2934
Helianthus annuus	1822	2003	1567	1978	2505	2069	2887	2369	2132	1620
	Cost/kilogram [Euro/Kg]									
Triticum aestivum	0.16	0.14	0.15	0.12	0.12	0.13	0.14	0.15	0.20	0.30
Zea mays	0.15	0.11	0.11	0.12	0.11	0.12	0.12	0.13	0.20	0.28
Helianthus annuus	0.23	0.21	0.29	0.26	0.25	0.24	0.24	0.31	0.52	0.56
	Cost/hectare [Euro/ha]									
Triticum aestivum	654	542	552	340	493	551	628	627	826	1150
Zea mays	698	479	342	506	697	900	887	882	1247	810
Helianthus annuus	419	421	454	522	636	492	699	739	1104	901

In order to select the most suitable agroforestry system for plain regions across Bihor County, a set of 8 criteria was proposed as follows:

1. The cost for planting/sowing (1—the highest cost ... 8—the lowest cost). This criterion considers the price of seedlings, the number of seedlings per hectare and the needed operations for planting manually and/or mechanized. According to the National Recovery and Resilience Plan, in a plain region, the cost for planting 1 ha of oak-dominated culture is 6379 EUR, while in the case of a black locust plantation, the cost is 5060 EUR, respectively [65]. In the case of cereals, there are minor differences, with the cost being 1100 EUR per hectare for wheat and 1200 EUR per hectare for maize and sunflower, respectively. These costs include plowing, land preparation, herbicide, seed cost, treatments, weeding in the case of maize and sunflower, fertilization and harvesting.
2. Yearly maintenance costs for the first 3 years (1—the highest cost ... 8—the lowest cost). This criterion includes the cost of grass cutting between the rows with trees/shrubs, soil mobilization around the tree/shrub seedlings and applying fertilizers/pesticides, where needed. In the case of an oak-dominated culture, the cost for the first year accounts for 2025 EUR/ha, which is similar to that of a black locust plantation. Differences between the two plantations appear in the second and third years: with 4089 EUR/ha and 2854 EUR/ha, in the case of an oak-dominated culture, and 2310 EUR/ha and 1226 EUR/ha for black locust, respectively.
3. Woody biomass production after 5 years (1—the lowest quantity ... 8—the highest quantity). This criterion is also correlated with the speed of growing of the woody species considered within this research.
4. Fruit and cereal production after 5 years (1—the lowest price ... 8—the highest price). In order to create the hierarchy, the number of fruits produced by the considered shrub species and the yield of the three cereal crops were taken into account.
5. Honey production (1—the lowest quantity ... 8—the highest quantity). In regard to this criterion, concrete data about honey production exist in the Romanian literature for the woody species, respectively: between 10–20 kg/ha for dog rose, 20 kg/ha in

the case of European ash and pedunculate oak, between 25–40 kg/ha for blackthorn, up to 250 kg/ha in the case of honey locust and around 1000 kg/ha for black locust plantations [66,67].

6. End product diversity (1—the lowest diversity ... 8—the highest diversity). This criterion takes into account the number and the diversity of derived products which may be obtained from certain morphological parts of the plants (e.g., leaf extracts, juice, etc.).
7. Resistance to abiotic/biotic threats (1—the lowest resistance ... 8—the highest resistance). The resistance to a broad spectrum of abiotic and biotic threats was assessed (e.g., drought, frosts, bugs, fungi, etc.).
8. Level of biodiversity (1—the lowest level ... 8—the highest level). This criterion takes into consideration the number of flora and fauna species that could appear and live in the environment generated by the proposed combinations of agroforestry systems.

2.3. Modeled Scenarios

In this study, like in the one used to assess the potential of certain non-wood forest products in six European Regions [68], the Expert Choice Desktop (v. 11.5.1683) software package was used and three scenarios were proposed.

In the first scenario, all 8 criteria received equal shares (i.e., 12.5%), meaning that they have an equal contribution in selecting the most suitable agroforestry system for plain regions across Bihor County.

In the second scenario, criterion "the cost for planting/sowing" and criterion "yearly costs for maintenance in the first 3 years" received a share of 20% each, while the remaining 6 criteria only 10% each, respectively.

Within the third scenario, criterion "fruits and cereal production" and criterion "level of biodiversity" received a share of 25% each, while the remaining 6 criteria received a share of 8.33% each, respectively.

One of the shortcomings of this method consists of the fact that, instead of dictating the 'right' choice, AHP assists decision-makers in identifying the option that aligns most effectively with their objectives and their perception of the issue. Therefore, AHP offers a logical structure for organizing a decision challenge, and the outcomes are associated with the expertise of the individuals who devised the hierarchy.

3. Results and Discussion

By summarizing the information from specialized manuals and studies, a brief description of the considered species was performed in accordance with the eight considered criteria. The following information is crucial for creating the hierarchy, by taking into consideration the considered criteria in different scenarios.

Pedunculate oak is one of the main hardwood species across Bihor County, covering one fifth of the forests managed by the Bihor Forestry Department. In the last few decades, it has been intensively studied across Europe and also Romania thanks to its great economic, ecological and social importance [69–71], being one of the most valued tree species in temperate forests [72]. It is a light-demanding species and it can tolerate a broad spectrum of site conditions, being resistant to dry winds and droughts [73,74]. In particular, its seedlings can grow in a wide range of shade levels, from heavy shade to direct light [75]. They also have a very good drought stress adaptive mechanism, by increasing their root system or by restricting their growth [76]. Pedunculate oak is also very appreciated in forest farming, being a key element for the production of truffles. Last but not least, it has been demonstrated that the content of soil organic matter was higher in a chernozem from an oak plantation in comparison with the same soil from a black locust plantation [77].

European ash is widespread across Central Europe, being a species which occurs in various types of broadleaf mixed forests [78]. In Bihor County it appears in mixed hardwood forests dominated mainly by oak species. It is a fast-growing species, which is able to grow in various sites with mean year temperatures ranging from 6.4 to 10.7 °C or annual amount of precipitation ranging from 400 to 760 mm, for example [79]. In regard to soil conditions, it prefers soils with pH values from 5 to 7.5, rich in moisture and well drained [80]. It was recently reported that, in mesic sites with acidic soils and lower content of soil organic matter, the crown defoliation was lower [81]. In several European countries, in the last few decades, crown defoliation was caused by the invasive pathogen *Hymenoscyphus fraxineus* [81]. Ash is sensitive to late spring frosts, severe winters and long-term drought events [82]. It was successfully used for the afforestation of several categories of degraded lands [83–86].

Norway maple is a widespread tree species across Central and Northern Europe [87–89]. It was introduced in North America where it become invasive, being a serious threat to native forests [87,90]. It has a vigorous juvenile growth rate, it is a shade- and drought-tolerant species and it is able to grow across a broad range of soil conditions, including soils rich in carbonates [87,89–93]. Across its natural distribution range, Norway maple is a good companion for pedunculate oak, where it creates a continuous secondary layer [73]. In the current context of climate change, which generates large-scale vitality loses and dieback in some of the main timber species, the importance of this species is expected to increase [94].

Black locust is native to North America, being the most planted allochthonous species in Romania, especially in sandy soils across south-western part of the country, but also across Bihor County [95]. It is regarded worldwide as a multipurpose tree, mainly due to its great adaptability to face different kinds of environmental stresses [96]. It has a very fast juvenile growth rate, a high annual production of fast-decomposing leaves which generate a high quantity of organic matter and a very good vegetative propagation system. It is a very shade intolerant and a thermophilous tree species [95,97–100]. Moreover, black locust plantations offer several ecosystem services, such as landscape rehabilitation, fuel wood and carbon sequestration [101].

Honey locust is also native to North America. In Romania, it is usually planted in association with black locust, as is the case in the plantations installed in sandy soils in southern Romania, for example, or in the composition of field protective forest shelterbelts. It can grow in differently degraded lands, including salt-affected terrains. It prefers direct exposure to sunlight and a mild climate [102,103], but it requires deep soils, with a moderate humus content [104]. In comparison to black locust, the honey locust also provided good results in carbonated soils [85].

Saskatoon berry is also a species originating from North America, which recently received an increasing amount of attention for cultivation in Romania, especially across western parts of the country [105]. This species can tolerate a broad range of site conditions, being able to grow in different soil types, with a pH ranging from 5.6 to 8, for example [106]. Moreover, it can easily be propagated both in vegetative and generative ways [107]. An adult shrub is able to produce between 4500 and 10,000 berries [108], which represent a good source of vitamins, nutrients, bioactive components and other micro- and macro-elements [105].

Blackthorn is one of the most common shrub species across Romania, including Bihor County. It is a light-demanding species, and it has a slow growing rate, usually reaching a height of 2–3 m. It can reproduce in both generative and vegetative ways, and its roots are deeply developed in depth, with numerous lateral branches [109]. It can grow in several types of soils [109], including acid soils [110]. Its fruits are very appreciated, due to their rich content of vitamins, sugars, minerals, organic acids, polyphenols, tannins, etc. [111].

Dog rose is a light-demanding and a drought-tolerant shrub species, with low requirements for soil conditions [62,112]. It is also very appreciated thanks to its reddish fruits, which have high content of vitamin C, carotenoids and polyunsaturated fatty acids [113]. In Romania, its berries are mainly used to produce juice, jelly, jam, wine and tea [114].

Maize is one of the most widely cultivated cereals worldwide, being famous for its economic value and superior nutritional properties [115,116]. It was introduced to Europe from Central America at the end of 15th century [117]. Particularly, in Romania, maize is of great interest, being a strategic cereal crop both for internal and foreign markets [118,119].

For example, more than 96,000 hectares were cultivated in 2022 across Bihor County [64]. Even if its cultivation is dependent on fertilizers [120], maize has several agrotechnical and agrobiological properties, such as: high yield, high resistance to drought and also to some pests and diseases, high economic benefits and total mechanization of agrotechnical and harvesting works [121,122].

Common wheat plays an important role in human nutrition worldwide [123,124], being the second or the third largest crop in the last two decades [125]. Thanks to its geographical location, climate and soils, Romania is cultivating wheat on approximately one quarter of its arable land [126], being one of the most important wheat producers in the European Union [127] after countries with higher areas (i.e., France, Germany and Poland) [128]. In Bihor County, the area cultivated with common wheat accounted for more than 83,000 hectares in 2022 [64]. In the current context of increasing temperatures, water deficit represents a major challenge to wheat productivity [129]. In this context, in order to protect the crops against drought, several techniques have recently been experimented, with arbuscular mycorriza—wheat association or applying biochar amendments—being among them [130,131]. However, nitrogen fertilization is mandatory to increase wheat yield [132].

Common sunflower crop is the third largest agricultural crop, after maize and wheat in Romania [133], being also one of the main crops across the West Plain [57]. A total area of 31,000 hectares was cultivated in 2022 across Bihor County [64]. In recent years, among the European Member States, Romania ranked first in terms of both production and cultivated area [134,135], and particularly in regards to the organic production area [136]. It is very appreciated as a honey plant and also due to its several industrial and food uses, such as the production of edible oil, bio-fuel and fodder [136,137].

An alternative AHP ranking for the eight criteria in the case of the eight alternatives (agroforestry systems), based on the information available in specialized manuals, scientific papers and authors' expertise, is given in Table 3.

Table 3. AHP alternative ranking.

Criterion No.	Alternative							
	Q.r.H.a. F.e.A.a.	Q.r.Z.m. A.p.R.c.	Q.r.T.a. F.e P.s.	Q.r.H.a. A.p.A.a.	R.p.Z.m. G.t.P.s.	R.p.T.a. G.t.R.c.	R.p.H.a. F.e.A.a.	R.p.P.s. F.e.R.c.
1	2	3	4	1	7	8	5	6
2	2	3	4	1	6	5	7	8
3	4	1	3	2	6	5	8	7
4	6	2	7	4	3	8	5	1
5	4	1	2	3	8	7	6	5
6	2	6	8	3	7	5	1	4
7	1	4	2	3	8	7	5	6
8	6	8	7	5	1	2	3	4

According to the AHP results, when all eight criteria received an equal share, the combination R.p.Z.m.G.t.P.s. (*Robinia pseudoacacia* + *Zea mays* + *Gleditsia triacanthos* + *Prunus spinosa*) proved to be the best alternative in terms of agroforestry systems that could be adopted in plain regions across Bihor County, accounting for 18.9% of the total global priority with respect to the goal (Figure 3, right part of the graph).

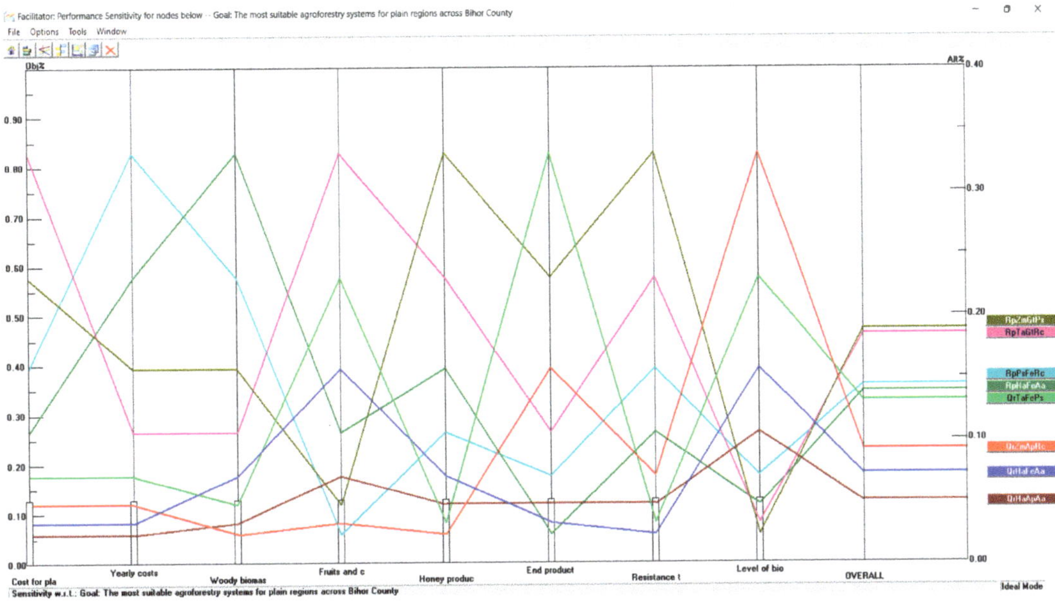

Figure 3. The ranking of the eight proposed agroforestry systems in the first scenario.

The second option was R.p.T.a.G.t.R.c. (*Robinia pseudoacacia* + *Triticum aestivum* + *Gleditsia triacanthos* + *Rosa canina*), which accounted for 18.4% of the total global priority with respect to the goal. Moreover, the four combinations with black locust as the main species scored in the first positions, while the agroforestry systems having pedunculate oak as main tree species situated in the last positions.

Within the second scenario, a switch in the first two proposals from scenario 1 was recorded, namely the combination R.p.T.a.G.t.R.c. (*Robinia pseudoacacia* + *Triticum aestivum* + *Gleditsia triacanthos* + *Rosa canina*) ranked first (Figure 4, right part of the graph), accounting for 19.1% of the total global priority with respect to the goal. However, the four combinations with black locust as main tree species were ranked in the first positions, meaning that the costs for planting and maintenance of these agroforestry systems are cheaper in comparison with the combinations based on pedunculate oak as main tree species. The combination Q.r.H.a.A.p.A.a. (*Quercus robur* + *Helianthus annuus* + *Acer platanoides* + *Amelanchier alnifolia*) was ranked in the last position.

In the third scenario, with two criteria accounting for 50% of the decision of choosing the most suitable agroforestry system, R.p.T.a.G.t.R.c. (*Robinia pseudoacacia* + *Triticum aestivum* + *Gleditsia triacanthos* + *Rosa canina*) was ranked first, accounting for 18.4% of the total global priority with respect to the goal, being followed by Q.r.T.a.F.e.P.s. (*Quercus robur* + *Triticum aestivum* + *Fraxinus excelsior* + *Prunus spinosa*) and R.p.Z.m.G.t.P.s. (*Robinia pseudoacacia* + *Zea mays* + *Gleditsia triacanthos* + *Prunus spinosa*), respectively (Figure 5).

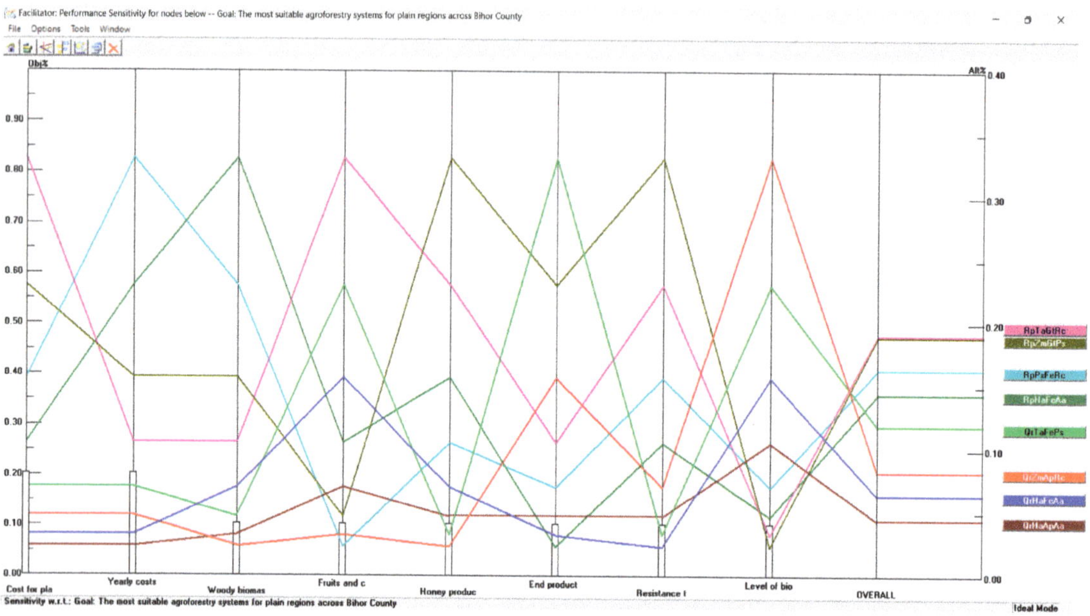

Figure 4. The ranking of the eight proposed agroforestry systems in the second scenario.

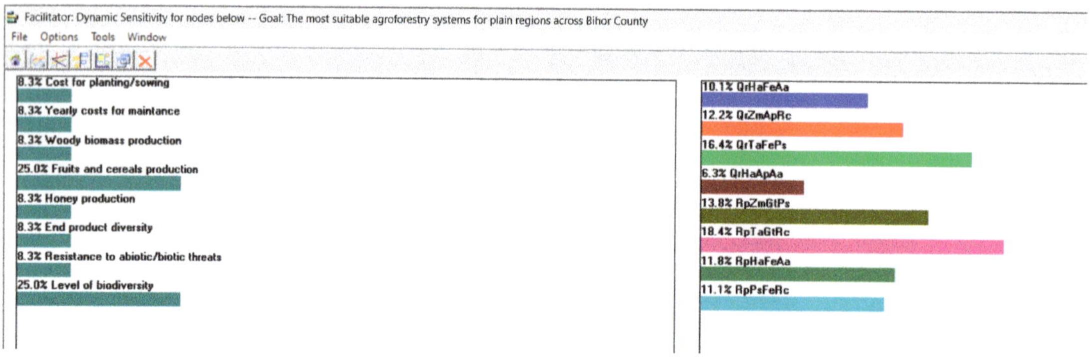

Figure 5. The ranking of the eight proposed agroforestry systems in the third scenario.

4. Conclusions

According to the literature review presented above, corroborated with the results of this study, we conclude that the plain region across Bihor County has a great potential for agroforestry systems. This potential could be better capitalized upon, for example, if more landowners will be aware of the benefits of these systems, on one hand, and if a favorable normative framework existed, on the other hand.

With respect to establishing agroforestry systems across the plain regions of Bihor County, combined with the goal of diversification, and addressing the effects of climate change, as well as addressing food crises or simply enhancing green spaces near localities, there is a pressing need for flexibility in choosing the species to be planted. However, this selection should be science based. Future research should also consider criteria that account for various aspects, including the functional relationships of these future green spaces with nearby areas. This holistic approach is crucial for integrated and sustainable development, with a primary focus on the development of rural communities.

Finally, it can be concluded that the combination of the Analytic Hierarchy Process and the Expert Choice Desktop (v. 11.5.1683) software package has proven to be a viable solution for selecting an agroforestry system that meets all the necessary criteria.

Author Contributions: Conceptualization, R.B., A.A., I.A.B. and C.M.E.; methodology, R.B. and C.M.E.; software, C.M.E.; validation, C.S.C.T., R.B., A.A. and I.A.B.; formal analysis, R.B., A.A. and C.M.E.; investigation, C.M.E., M.C. and C.S.C.T.; writing—original draft preparation, R.B., A.A., M.C. and C.S.C.T.; writing—review and editing, C.M.E., R.B., A.A., I.A.B. and M.C. All authors have read and agreed to the published version of the manuscript.

Funding: Publication of this research was funded by the University of Oradea.

Institutional Review Board Statement: Not applicable.

Informed Consent Statement: Not applicable.

Data Availability Statement: Data are contained within the article.

Conflicts of Interest: The authors declare no conflict of interest.

References

1. Cialdella, N.; Jacobson, M.; Penot, E. Economics of Agroforestry: Links between Nature and Society. *Agrofor. Syst.* **2023**, *97*, 273–277. [CrossRef]
2. Tubalov, A.A. Spatial Principles of Territories Selection for Priority Development of Agroforestry Complexes. *Forests* **2023**, *14*, 1225. [CrossRef]
3. Mosquera-Losada, M.R.; McAdam, J.H.; Romero-Franco, R.; Santiago-Freijanes, J.J.; Rigueiro-Rodróguez, A. Definitions and Components of Agroforestry Practices in Europe. In *Agroforestry in Europe: Current Status and Future Prospects*; Rigueiro-Rodróguez, A., McAdam, J., Mosquera-Losada, M.R., Eds.; Advances in Agroforestry; Springer: Dordrecht, The Netherlands, 2009; pp. 3–19, ISBN 978-1-4020-8272-6.
4. Mosquera-Losada, M.R.; Santiago-Freijanes, J.; Lawson, G.; Balaguer, F.; Vaets, N.; Burgess, P.; Rigueiro-Rodríguez, A. Agroforestry as Tool to Mitigate and Adapt to Climate under Lulucf Accounting. In Proceedings of the Agroforestry and Climate Change, Montpellier, France, 23 May 2016.
5. den Herder, M.; Burgess, P.; Mosquera-Losada, M.R.; Herzog, F.; Hartel, T.; Upson, M.; Viholainen, I.; Rosati, A. *Preliminary Stratification and Quantification of Agroforestry in Europe*; European Commission: Brussels, Belgium, 2015; pp. 1–53. Available online: https://www.agforward.eu/documents/M1_Stratification%20of%20agroforestry.pdf (accessed on 29 September 2023).
6. Herder, M.; Moreno, G.; Mosquera-Losada, M.R.; Palma, J.; Sidiropoulou, A.; Santiago-Freijanes, J.; Crous-Duran, J.; Paulo, J.; Tomé, M.; Pantera, A.; et al. Current Extent and Stratification of Agroforestry in the European Union. *Agric. Ecosyst. Environ.* **2017**, *241*, 121–132. [CrossRef]
7. Lundgren, B.O.; Raintree, J.B. *Sustained Agroforestry*; ICRAF Reprint Series; ICRAF: Nairobi, Kenya, 1983.
8. Leakey, R. Definition of Agroforestry Revisited. *Agrofor. Today* **1996**, *8*, 5–7.
9. Augère-Granier, M.-L. Agroforestry in the European Union. 2020. Available online: https://www.europarl.europa.eu/RegData/etudes/BRIE/2020/651982/EPRS_BRI(2020)651982_EN.pdf (accessed on 29 September 2023).
10. Nair, P.; Mohan Kumar, B.; Nair, V. *An Introduction to Agroforestry: Four Decades of Scientific Developments*; Springer: Cham, Switzerland, 2021.
11. Burgess, P.J.; Rosati, A. Advances in European Agroforestry: Results from the AGFORWARD Project. *Agrofor. Syst.* **2018**, *92*, 801–810. [CrossRef]
12. Tsonkova, P.; Mirck, J.; Böhm, C.; Fütz, B. Addressing Farmer-Perceptions and Legal Constraints to Promote Agroforestry in Germany. *Agrofor. Syst.* **2018**, *92*, 1091–1103. [CrossRef]
13. Santiago-Freijanes, J.; Rigueiro-Rodríguez, A.; Vazquez, J.A.; Moreno, G.; Herder, M.; Burgess, P.; Mosquera-Losada, M.R. Understanding Agroforestry Practices in Europe through Landscape Features Policy Promotion. *Agrofor. Syst.* **2018**, *92*, 1105–1115. [CrossRef]
14. Mușat, M.; Ciceoi, R.; Dolocan, C.; Argatu, G.; Bogdan, I.; Petcu, M. The Suitability of Southeastern Areas of Romania for the Establishment of Shelterbelts. *Agronomy* **2022**, *65*, 111–117.
15. Smith, L.G.; Westaway, S.; Mullender, S.; Ghaley, B.B.; Xu, Y.; Lehmann, L.M.; Pisanelli, A.; Russo, G.; Borek, R.; Wawer, R.; et al. Assessing the Multidimensional Elements of Sustainability in European Agroforestry Systems. *Agric. Syst.* **2022**, *197*, 103357. [CrossRef]
16. Dou, Y.; Li, Y.; Li, M.; Chen, X.; Zhao, X. The Role of Agroforestry in Poverty Alleviation: A Case Study from Nujiang Prefecture, Southwestern China. *Sustainability* **2023**, *15*, 12090. [CrossRef]
17. Philipp, S.M.; Zander, K. Orchard Meadows: Consumer Perception and Communication of a Traditional Agroforestry System in Germany. *Agrofor. Syst.* **2023**, *97*, 939–951. [CrossRef] [PubMed]

18. Buratti-Donham, J.; Venn, R.; Schmutz, U.; Migliorini, P. Transforming Food Systems towards Agroecology—A Critical Analysis of Agroforestry and Mixed Farming Policy in 19 European Countries. *Agroecol. Sustain. Food Syst.* **2023**, *47*, 1023–1051. [CrossRef]
19. Mosquera-Losada, M.; Santos, M.G.S.; Gonçalves, B.; Ferreiro-Domínguez, N.; Castro, M.; Rigueiro-Rodríguez, A.; González-Hernández, M.P.; Fernández-Lorenzo, J.L.; Romero-Franco, R.; Aldrey-Vázquez, J.A.; et al. Policy Challenges for Agroforestry Implementation in Europe. *Front. For. Glob. Chang* **2023**, *6*, 1127601. [CrossRef]
20. European Commission Communication from the Commission to the European Parliament, the Council, the European Economic and Social Committee and the Committee of the Regions. A Farm to Fork Strategy for a Fair, Healthy and Environmentally-Friendly Food System. Available online: https://eur-lex.europa.eu/legal-content/EN/TXT/?uri=CELEX:52020DC0381 (accessed on 29 September 2023).
21. European Commission Communication from the Commission to the European Parliament, the Council, the European Economic and Social Committee and the Committee of the Regions EU. Biodiversity Strategy for 2030 Bringing Nature Back into Our Lives. Available online: https://eur-lex.europa.eu/legal-content/EN/TXT/?qid=1590574123338&uri=CELEX:52020DC0380 (accessed on 29 September 2023).
22. European Commission Communication from the Commission to the European Parliament, the Council, the European Economic and Social Committee and the Committee of the Regions. New EU Forest Strategy for 2030. Available online: https://eur-lex.europa.eu/legal-content/EN/TXT/?uri=CELEX%3A52021DC0572 (accessed on 29 September 2023).
23. Elena, M.; Bitca, M.; Adrian, T.; Popovici, L. Alley Cropping, as a Model of System Adapted to Climate Change. *Ann. ORADEA Univ. Fascicle Manag. Technol. Eng.* **2022**, *37*, 213–220.
24. Dolocan, C.; Mușat, M.; Ciceoi, R.; Argatu, G.; Mușat, I.B.; Petcu, M. Aspects Regarding the Shelterbelts Establishment in Bărăgan Plain. *Sci. Pap.* **2022**, *65*, 54–60.
25. Mihăilă, E.; Tăulescu, E.; Tudora, A.; Bitca, M. Reasons for Maintaining and/or Introducing Trees on Grasslands. *Sci. Pap.* **2022**, *11*, 218–227.
26. Mihăilă, E.; Drăgan, D.; Marcu, C.; Costăchescu, C.; Dănescu, F.; Cojoacă, F.D. Elaboration the Substantiating Studies for the Necessity of Forest Shelterbelts to Protect the Field, Premise for Obtaining Funds for Their Realization. *Land Reclam. Earth Obs. Surv. Environ. Eng.* **2022**, *11*, 240–246.
27. Nishizawa, T.; Kay, S.; Schuler, J.; Klein, N.; Herzog, F.; Aurbacher, J.; Zander, P. Ecological–Economic Modelling of Traditional Agroforestry to Promote Farmland Biodiversity with Cost-Effective Payments. *Sustainability* **2022**, *14*, 5615. [CrossRef]
28. Sharma, R.; Mina, U.; Kumar, B.M. Homegarden Agroforestry Systems in Achievement of Sustainable Development Goals. A Review. *Agron. Sustain. Dev.* **2022**, *42*, 44. [CrossRef]
29. Gosling, E.; Reith, E.; Knoke, T.; Paul, C. A Goal Programming Approach to Evaluate Agroforestry Systems in Eastern Panama. *J. Environ. Manag.* **2020**, *261*, 110248. [CrossRef]
30. Marușca, T.; Taulescu, E.; Zevedei, P.M.; Andreoiu, A.C.; Comșia, C.C. Study on the Agroforestry System with Oak Trees (*Quercus robur* L.) in the Context of Changing Climate. *Agric. Silvic. Vet. Med. Sci.* **2020**, *9*, 47–54.
31. Copăcean, L.; Cojocariu, L.; Simon, M. Remote Sensing and Gis for Change Detection in the Agro- Forestry-Pastoral Space. *Res. J. Agric. Sci.* **2021**, *53*, 18–26.
32. Iqbal, W.; Siddiqui, T.; Ahmad, I.; Farooq, M. Effect of Allelochemicals Present in Leaf Litter of *Bombax ceiba* L. and *Populus deltoides* L. Tree Species on Wheat in Agroforestry System. *Appl. Ecol. Environ. Res.* **2022**, *20*, 4193–4209. [CrossRef]
33. Ntawuruhunga, D.; Ngowi, E.E.; Mangi, H.O.; Salanga, R.J.; Shikuku, K.M. Climate-Smart Agroforestry Systems and Practices: A Systematic Review of What Works, What Doesn't Work, and Why. *For. Policy Econ.* **2023**, *150*, 102937. [CrossRef]
34. Malschi, D.; Tărău, A.; Kadar, R.; Tritean, N.; Chețan, C. Climate Warming in Relation to Wheat Pest Dynamics and Their Integrated Control in Transylvanian Crop Management Systems with No Tillage and with Agroforestry Belts. *Romanian Agric. Res.* **2015**, *32*, 279–289.
35. Santos, M.; Cajaiba, R.L.; Bastos, R.; Gonzalez, D.; Petrescu Bakış, A.-L.; Ferreira, D.; Leote, P.; Barreto da Silva, W.; Cabral, J.A.; Gonçalves, B.; et al. Why Do Agroforestry Systems Enhance Biodiversity? Evidence From Habitat Amount Hypothesis Predictions. *Front. Ecol. Evol.* **2022**, *9*, 630151. [CrossRef]
36. Marușca, T.; Taulescu, E.; Memedemin, D. Preliminary Study of Agrosilvopastoral Systems from Romania. *Romanian J. Grassl. Forage Crops* **2020**, *22*, 25–32.
37. Soman, D.; Chavan, R.L. Soil Chemical Properties under Bamboo-Soybean Agroforestry System in Northern Transitional Zone, Karnataka. *Pharma Innov. J.* **2023**, *12*, 122–124.
38. Lehmann, L.M.; Smith, J.; Westaway, S.; Pisanelli, A.; Russo, G.; Borek, R.; Sandor, M.; Gliga, A.; Smith, L.; Ghaley, B.B. Productivity and Economic Evaluation of Agroforestry Systems for Sustainable Production of Food and Non-Food Products. *Sustainability* **2020**, *12*, 5429. [CrossRef]
39. Rolo, V.; Hartel, T.; Aviron, S.; Berg, S.; Crous-Duran, J.; Franca, A.; Mirck, J.; Palma, J.H.N.; Pantera, A.; Paulo, J.A.; et al. Challenges and Innovations for Improving the Sustainability of European Agroforestry Systems of High Nature and Cultural Value: Stakeholder Perspectives. *Sustain. Sci.* **2020**, *15*, 1301–1315. [CrossRef]
40. Kay, S.; Rega, C.; Moreno, G.; den Herder, M.; Palma, J.H.N.; Borek, R.; Crous-Duran, J.; Freese, D.; Giannitsopoulos, M.; Graves, A.; et al. Agroforestry Creates Carbon Sinks Whilst Enhancing the Environment in Agricultural Landscapes in Europe. *Land Use Policy* **2019**, *83*, 581–593. [CrossRef]

41. Goetz, A.; Hussein, H.; Thiel, A. Polycentric Governance and Agroecological Practices in the MENA Region: Insights from Lebanon, Morocco and Tunisia. *Int. J. Water Resour. Dev.* **2023**, *39*, 1–16. [CrossRef]
42. Hamidov, A.; Daedlow, K.; Webber, H.; Hussein, H.; Abdurahmanov, I.; Dolidudko, A.; Seerat, A.Y.; Solieva, U.; Woldeyohanes, T.; Helming, K. Operationalizing Water-Energy-Food Nexus Research for Sustainable Development in Social-Ecological Systems: An Interdisciplinary Learning Case in Central Asia. *Ecol. Soc.* **2022**, *27*, 12. [CrossRef]
43. Popovici, L.; Mihăilă, E.; Costăchescu, C.; Constandache, C. Can Agroforestry Systems Be Ordinary Practices in Romania? *Can Agrofor. Syst. Be Ordinary Pract. Rom.* **2018**, *61*, 263–268.
44. Tudoran, G.-M.; Cicșa, A.; Boroeanu, M.; Dobre, A.-C.; Pascu, I.-S. Forest Dynamics after Five Decades of Management in the Romanian Carpathians. *Forests* **2021**, *12*, 783. [CrossRef]
45. Enescu, C.M. Which Woody Species Should Be Used for Afforestation of Household Dumps Consisting of Demolition Materials Mixed with Organic Materials? *Agronomy* **2022**, *65*, 375–380.
46. Giurca, A.; Nichiforel, L.; Stăncioiu, P.T.; Drăgoi, M.; Dima, D.-P. Unlocking Romania's Forest-Based Bioeconomy Potential: Knowledge-Action-Gaps and the Way Forward. *Land* **2022**, *11*, 2001. [CrossRef]
47. Enescu, C.M. Sandy Soils from Oltenia and Carei Plains: A Problem or an Opportunity to Increase the Forest Fund in Romania? *Manag. Econ. Eng. Agric. Rural Dev.* **2019**, *19*, 203–206.
48. Lucian, D.; Aurelia, O.; Raluca, E.; Emilia, P.; Tamara, R.; Timiș-Gânsac, V. Chemical Properties of Forest Soils from Bihor County. *Nat. Resour. Sustain. Dev.* **2017**, *7*, 35–42.
49. Timiș-Gânsac, V.; Enescu, C.M.; Dinca, L.; Onet, A. The Management of Non-Wood Forest Products in Bihor County. *Nat. Resour. Sustain. Dev.* **2018**, *8*, 27–34. [CrossRef]
50. Marcu, C.; Dinca, L. The West Plain—A Physio-Geographical Characterisation Based on Data from Forest Management Plans. *Sci. Stud. Res.* **2021**, *30*, 59–62.
51. Dincă, L.; Cântar, I. Oak Forest from Romania's West Plain under Conservation Regime. *Res. J. Agric. Sci.* **2020**, *52*, 94–101.
52. Cântar, I.; Dinca, L. The Contribution of Forests from Counties Located in Romania's West Plain to The Area's Long Lasting Development. *Sustain. Dev. Res.* **2021**, *3*, 7–13. [CrossRef]
53. Dincă, L.; Chisalita, I.; Cantar, I.-C. Chemical Properties of Forest Soils from Romania West Plain. *Rev. Chim.* **2019**, *70*, 2371–2374. [CrossRef]
54. Cărbunar, M.; Mintaș, O.; Sabău, N.C.; Borza, I.; Stanciu, A.; Pereș, A.; Venig, A.; Curilă, M.; Cărbunar, M.L.; Vidican, T.; et al. Effectiveness of Measures to Reduce the Influence of Global Climate Change on Tomato Cultivation in Solariums—Case Study: Crișurilor Plain, Bihor, Romania. *Agriculture* **2022**, *12*, 634. [CrossRef]
55. Pereş, A.C.; Nandor, K. The Thermic Regime of the Crișurilor Plain. *Nat. Resour. Sustain. Dev.* **2013**, *3*, 399–404.
56. Oneț, A.; Teușdea, A.; Boja, N.; Domuța, C.; Oneț, C. Effects of Common Oak (*Quercus robur* L.) Defolition on the Soil Properties of an Oak Forest in Western Plain of Romania. *Ann. For. Res.* **2016**, *59*, 33–47. [CrossRef]
57. Dincă, L.; Timiș-Gânsac, V. Forests from the West Plain Forest Steppe—An Alternative for Agriculture? *Lucr. Ştiinţ.* **2021**, *64*, 95–100.
58. Enescu, C. Which Are the Most Important Non-Wood Forest Products in the Case of Ialomița County? *AgroLife Sci. J.* **2017**, *6*, 98–103.
59. Patel, N.; Blumberga, D. Insights of Bioeconomy: Biopolymer Evaluation Based on Sustainability Criteria. *CONECT Int. Sci. Conf. Environ. Clim. Technol.* **2023**, *27*, 323–338. [CrossRef]
60. Wang, S.; Zhang, Y.; Fan, J.; Zhang, H.; Fang, H. Comprehensive Sustainability Indicator for Land Resource-Carrying Capacity in a Farming-Pastoral Region. *Remote Sens.* **2023**, *15*, 3726. [CrossRef]
61. Debebe, Y.; Otterpohl, R.; Islam, Z. Remote Sensing and Multi-Criterion Analysis for Identifying Suitable Rainwater Harvesting Areas. *Acta Geophys.* **2023**, *71*, 855–872. [CrossRef]
62. Enescu, C. Which Shrub Species Should Be Used for the Establishment of Field Shelterbelts in Romania? *Agronomy* **2018**, *61*, 464–469.
63. Enescu, C. Shrub and Tree Species Used for Improvement by Afforestation of Degraded Lands in Romania. *For. IDEAS* **2015**, *21*, 3–15.
64. National Institute of Statistics. Vegetable Agricultural Production for the Main Crops, by Forms of Ownership, Macro-Regions, Development Regions and Counties 2023. Available online: http://statistici.insse.ro:8077/tempo-online/#/pages/tables/insse-table (accessed on 29 September 2023).
65. PNRR Planul Național de Redresare și Reziliență. Available online: https://pnrr.mmap.ro/wp-content/uploads/2022/11/Ghid-specific-PNRR_2022_C2-I.1.A_22.11.2022_APROBAT_MO.pdf (accessed on 29 September 2023).
66. Beldeanu, E. *Specii de Interes Sanogen Din Fondul Forestier. Editura Universităţii Transilvania Din Braşov*; Editura Universitatii Transilvania: Brașov, Romania, 2004; ISBN 978-973-635-380-2.
67. Ion, N. *Arbori Și Arbuști Meliferi*; Alex-Alex & Leti Press: Bucuresti, Romania, 2006.
68. Huber, P.; Kurttila, M.; Hujala, T.; Wolfslehner, B.; Sanchez-Gonzalez, M.; Pasalodos-Tato, M.; de-Miguel, S.; Bonet, J.A.; Marques, M.; Borges, J.G.; et al. Expert-Based Assessment of the Potential of Non-Wood Forest Products to Diversify Forest Bioeconomy in Six European Regions. *Forests* **2023**, *14*, 420. [CrossRef]
69. Curtu, A.; Șofletea, N.; Toader, A.; Enescu, C. Leaf Morphological and Genetic Differentiation between *Quercus robur* L. and Its Closest Relative, the Drought-Tolerant Quercus Pedunculiflora K. Koch. *Ann. For. Sci.* **2011**, *68*, 1163–1172. [CrossRef]

70. Molder, A.; Meyer, P.; Nagel, R.-V. Integrative Management to Sustain Biodiversity and Ecological Continuity in Central European Temperate Oak (*Quercus robur, Q. petraea*) Forests: An Overview. *For. Ecol. Manag.* **2019**, *437*, 324–339. [CrossRef]
71. Woziwoda, B.; Dyderski, M.K.; Kobus, S.; Parzych, A.; Jagodziński, A.M. Natural Regeneration and Recruitment of Native *Quercus robur* and Introduced *Q. rubra* in European Oak-Pine Mixed Forests. *For. Ecol. Manag.* **2019**, *449*, 117473. [CrossRef]
72. Vastag, E.; Cocozza, C.; Orlović, S.; Kesić, L.; Kresoja, M.; Stojnić, S. Half-Sib Lines of Pedunculate Oak (*Quercus robur* L.) Respond Differently to Drought through Biometrical, Anatomical and Physiological Traits. *Forests* **2020**, *11*, 153. [CrossRef]
73. Kruzhilin, S.N.; Taran, S.S.; Semenyutina, A.V.; Matvienko, E.Y. Growth Peculiarities and Age Dynamics of *Quercus robur* L. Formation in Steppe Region Conditions. *Kuwait J. Sci.* **2018**, *45*, 52–58.
74. Nechita, C.; Chiriloaei, F. Interpreting the Effect of Regional Climate Fluctuations on *Quercus robur* L. Trees under a Temperate Continental Climate (Southern Romania). *Dendrobiology* **2017**, *79*, 77–89. [CrossRef]
75. Sevillano, I.; Short, I.; Grant, J.; O'Reilly, C. Effects of Light Availability on Morphology, Growth and Biomass Allocation of *Fagus sylvatica* and *Quercus robur* Seedlings. *For. Ecol. Manag.* **2016**, *374*, 11–19. [CrossRef]
76. Deligoz, A.; Bayar, E. Drought Stress Responses of Seedlings of Two Oak Species (*Quercus cerris* and *Quercus robur*). *Turk. J. Agric. For.* **2018**, *42*, 114–123. [CrossRef]
77. Gorban, V.; Huslystyi, A. Changes in Selected Properties of Calcic Chernozem Due to Cultivation of *Robinia pseudoacacia* and *Quercus robur*. *Folia Oecologica* **2023**, *50*, 196–203. [CrossRef]
78. Fuchs, S.; Hertel, D.; Schuldt, B.; Leuschner, C. Effects of Summer Drought on the Fine Root System of Five Broadleaf Tree Species along a Precipitation Gradient. *Forests* **2020**, *11*, 289. [CrossRef]
79. Buksha, I.; Pyvovar, T.; Buksha, M.; Pasternak, V. Impact of Drought on the Forest Vegetation in North-Eastern Ukraine: The Long-Term Prognoses and Adaptation Measures. *Silva Balc.* **2019**, *20*, 27–38.
80. Constandache, C.; Tudor, C.; Popovici, L.; Dincă, L. Ecological Reconstruction of the Stands Affected by Drought from Meadows of Inland Rivers. *Sci. Pap.* **2022**, *11*, 76–84.
81. Turczański, K.; Rutkowski, P.; Dyderski, M.K.; Wrońska-Pilarek, D.; Nowiński, M. Soil pH and Organic Matter Content Affects European Ash (*Fraxinus excelsior* L.) Crown Defoliation and Its Impact on Understory Vegetation. *Forests* **2020**, *11*, 22. [CrossRef]
82. Roibu, C.-C.; Sfeclă, V.; Mursa, A.; Ionita, M.; Nagavciuc, V.; Chiriloaei, F.; Leșan, I.; Popa, I. The Climatic Response of Tree Ring Width Components of Ash (*Fraxinus excelsior* L.) and Common Oak (*Quercus robur* L.) from Eastern Europe. *Forests* **2020**, *11*, 600. [CrossRef]
83. Mănescu, M. Research on Evolution of Stands Installed on Degraded Lands on Dobrogea. *Analele ICAS* **2002**, *45*, 165–170.
84. Vlasin, H.D.; Budiu, V.; Stan, R. The Effect of Species and Antierosional Land Preparation of Eroded Lands on the Maintenance of Forestry Seedlings. *Bull. Univ. Agric. Sci. Vet. Med. Cluj-Napoca Hortic.* **2013**, *70*, 230–236.
85. Constandache, C.; Nistor, S. Preventing and Control of Soil Erosion on Agricultural Lands by Antierosional Shelterbelts. *Sci. Pap.* **2014**, *3*, 29–36.
86. Hoble, A.; Luca, E.; Dirja, M.; Marginas, P. The Variability of Biometric Measurments Used to Determine the Stand Adaptation of *Fraxinus excelsior* Trees after Afforestation of Degraded Land by Erosion. *Agricultura* **2015**, *93–94*, 109–117.
87. Webster, C.R.; Nelson, K.; Wangen, S.R. Stand Dynamics of an Insular Population of an Invasive Tree, *Acer Platanoides*. *For. Ecol. Manag.* **2005**, *208*, 85–99. [CrossRef]
88. Tomov, V.; Iliev, N.; Iliev, I. Analysis of the Forest Seed Production Base of *Acer platanoides* L. in Bulgaria. *For. Ideas* **2014**, *20*, 67–76.
89. Lazic, D.; George, J.-P.; Rusanen, M.; Ballian, D.; Pfattner, S.; Konrad, H. Population Differentiation in *Acer platanoides* L. at the Regional Scale—Laying the Basis for Effective Conservation of Its Genetic Resources in Austria. *Forests* **2022**, *13*, 552. [CrossRef]
90. Lapointe, M.; Brisson, J. A Comparison of Invasive *Acer platanoides* and Native *A. saccharum* First-Year Seedlings: Growth, Biomass Distribution and the Influence of Ecological Factors in a Forest Understory. *Forests* **2012**, *3*, 190–206. [CrossRef]
91. Martin, P.H.; Marks, P.L. Intact Forests Provide Only Weak Resistance to a Shade-Tolerant Invasive Norway Maple (*Acer platanoides* L.). *J. Ecol.* **2006**, *94*, 1070–1079. [CrossRef]
92. Banks, J.M.; Percival, G.C.; Rose, G. Variations in Seasonal Drought Tolerance Rankings. *Trees* **2019**, *33*, 1063–1072. [CrossRef]
93. Lovynska, V.; Holoborodko, K.; Ivanko, I.; Sytnyk, S.; Zhukov, O.; Loza, I.; Wiche, O.; Heilmeier, H. Heavy Metal Accumulation by *Acer platanoides* and *Robinia pseudoacacia* in an Industrial City (Northern Steppe of Ukraine). *Biosyst. Divers.* **2023**, *31*, 246–253. [CrossRef]
94. Fuchs, S.; Leuschner, C.; Mathias Link, R.; Schuldt, B. Hydraulic Variability of Three Temperate Broadleaf Tree Species along a Water Availability Gradient in Central Europe. *New Phytol.* **2021**, *231*, 1387–1400. [CrossRef] [PubMed]
95. Budău, R. *Cultura Salcâmului*; Editura Universității din Oradea: Oradea, Romania, 2023; ISBN 978-606-10-2255-7.
96. Kellezi, M.K.; Kortoci, Y. Comparison of Growth Rate of Black Locust (*Robinia pseudoacacia* L.) on Productive and Marginal Cultivated Lands for Sustainable Agroforestry Systems. *Ecol. Eng. Environ. Technol.* **2022**, *23*, 206–212. [CrossRef]
97. Rahmonov, O. The Chemical Composition of Plant Litter of Black Locust (*Robinia pseudoacacia* L.) and Its Ecological Role in Sandy Ecosystems. *Acta Ecol. Sin.* **2009**, *29*, 237–243. [CrossRef]
98. Rédei, K.; Csiha, I.; Keserű, Z.; Végh, Á.K.; Győri, J. The Silviculture of Black Locust (*Robinia pseudoacacia* L.) in Hungary: A Review. *South-East Eur. For.* **2011**, *2*, 101–107. [CrossRef]
99. Enescu, C. Black Locust (*Robinia pseudoacacia* L.)—An Invasive Neophyte in the Conventional Land Reclamation Flora in Romania. *Bulletion Transilv. Univ. Brasov Ser. II* **2013**, *6*, 23–30.

100. Nicolescu, V.-N.; Hernea, C.; Bakti, B.; Keserű, Z.; Antal, B.; Rédei, K. Black Locust (*Robinia pseudoacacia* L.) as a Multi-Purpose Tree Species in Hungary and Romania: A Review. *J. For. Res.* **2018**, *29*, 1449–1463. [CrossRef]
101. Ciuvăț, A.L.; Abrudan, I.V.; Ciuvăț, C.G.; Marcu, C.; Lorenț, A.; Dincă, L.; Szilard, B. Black Locust (*Robinia pseudoacacia* L.) in Romanian Forestry. *Diversity* **2022**, *14*, 780. [CrossRef]
102. Vilches, C.; Torremorell, A.; Rodriguez Castro, M.C.; Giorgi, A. Effects of the Invasion of Honey Locust (*Gleditsia triacanthos* L.) on Macrophytes and Algae of Pampean Streams (Argentina). *Wetlands* **2019**, *40*, 321–331. [CrossRef]
103. Enescu, C.M. Allochthonous Tree Species Used for Afforestation of Salt-Affected Soils in Romania. *Agronomy* **2020**, *63*, 74–79.
104. Constandache, C.; Nistor, S.; Ivan, V. Afforestation of Degraded Lands Inefficient for Agriculture from Southeast of Country. *Analele ICAS* **2006**, *49*, 187–204.
105. Budău, R.; Enescu, C.M. The Yield of a 11 Years Old Saskatoon Berry (*Amelanchier alnifolia* Nutt.) Culture from Arad County, Western Romania. *Sci. Pap.* **2022**, *22*, 87–92.
106. Sun, X.; Zhang, W.; Vassov, R.; Sherr, I.; Du, N.; Zwiazek, J.J. Effects of Elemental Sulfur on Soil pH and Growth of Saskatoon Berry (*Amelanchier alnifolia*) and Beaked Hazelnut (*Corylus cornuta*) Seedlings. *Soil Syst.* **2022**, *6*, 31. [CrossRef]
107. Hunková, J.; Libiaková, G.; Fejér, J.; Gajdošová, A. Improved *Amelanchier alnifolia* (Nutt.) Nutt. Ex M. Roem. Shoot Proliferation by Manipulating Iron Source. *Propag. Ornam. Plants* **2017**, *17*, 103–107.
108. Kuklina, A.G. Naturalization of Amelanchier Species from North America in a Secondary Habitat. *Russ. J. Biol. Invasions* **2011**, *2*, 103–107. [CrossRef]
109. Constandache, C.; Nistor, S.; Untaru, E. Research on the Behavior of Some Species of Trees and Shrubs Used in the Composition of the Protective Forest Shelterbelts in Southeastern Romania. *Rev. Silvic. Și Cineg.* **2012**, *30*, 35–47.
110. Rezzan, A.; Yildiz, O.; Sahin, H.; Eyupoğlu, O.; Mehtap, O.; Alpay Karaoğlu, Ş. Phenolic Components and Antioxidant Activity of *Prunus spinosa* from Gumushane, Turkey. *Chem. Nat. Compd.* **2015**, *51*, 346–349. [CrossRef]
111. Balta, I.; Sevastre, B.; Mireşan, V.; Taulescu, M.; Raducu, C.; Longodor, A.L.; Marchiş, Z.; Mariş, C.S.; Coroian, A. Protective Effect of Blackthorn Fruits (*Prunus spinosa*) against Tartrazine Toxicity Development in Albino Wistar Rats. *BMC Chem.* **2019**, *13*, 104. [CrossRef] [PubMed]
112. Soare, R.; Bonea, D.; Iancu, P.; Niculescu, M. Biochemical and Technological Properties of *Rosa canina* L. Fruits from Spontaneous Flora of Oltenia, Romania. *Bull. Univ. Agric. Sci. Vet. Med. Cluj-Napoca Hortic.* **2015**, *72*, 182–186. [CrossRef]
113. Tumbas Šaponjac, V.; Čanadanović-Brunet, J.; Četojević Simin, D.; Ćetković, G.; Đilas, S.; Gille, L. Effect of Rosehip (*Rosa canina* L.) Phytochemicals on Stable Free Radicals and Human Cancer Cells. *J. Sci. Food Agric.* **2012**, *92*, 1273–1281. [CrossRef]
114. Crișan, V.; Dincă, L.; Decă, S. The Most Important Forest Fruits from Vaslui County. *Ann. West Univ. Timisoara Ser. Biol.* **2021**, *24*, 3–10.
115. Suciu, A.L.; Chiș, A.C.; Barșon, G.; Tărău, A.; Morea, A.; Barb, A.; Crișan, I. Seed Pathogens Incidence in Eight Maize Hybrids (*Zea mays* L.) Cultivated in Transylvanian Plain. *LIFE Sci. Sustain. Dev.* **2021**, *2*, 104–109. [CrossRef]
116. Topan, C.; Nicolescu, M.; Simedru, D.; Becze, A. Complex Evaluation of Storage Impact on Maize (*Zea mays* L.) Quality Using Chromatographic Methods. *Separations* **2023**, *10*, 412. [CrossRef]
117. Revilla, P.; Alves, M.L.; Anđelković, V.; Balconi, C.; Dinis, I.; Mendes-Moreira, P.; Redaelli, R.; Ruiz de Galarreta, J.I.; Vaz Patto, M.C.; Žilić, S.; et al. Traditional Foods From Maize (*Zea mays* L.) in Europe. *Front. Nutr.* **2022**, *8*, 1–17. [CrossRef] [PubMed]
118. Bonea, D. Phenology, Yield and Protein Content of Maize (*Zea mays* L.) Hybrids as Affected by Different Sowing Dates. *Sci. Pap.* **2020**, *20*, 145–150.
119. Barșon, G.; Șopterean, L.; Suciu, L.A.; Crișan, I.; Duda, M.M. Evaluation of Agronomic Performance of Maize (*Zea mays* L.) under a Fertilization Gradient in Transylvanian Plain. *Agriculture* **2021**, *11*, 896. [CrossRef]
120. Adhikary, B.; Baral, B.; Shrestha, J. Productivity of Winter Maize as Affected by Varieties and Fertilizer Levels. *Int. J. Appl. Biol.* **2020**, *4*, 85–93. [CrossRef]
121. Suba, D.; Suba, T.; NEGRUȚ, G. Analysis of Some Perspective Maize Hybrids in Western Romania. *Life Sci. Sustain. Dev.* **2020**, *1*, 31–36. [CrossRef]
122. Dragomir, V.; Brumă, I.; Butu, A.; Tanasa, L.; Petcu, V.; Horhocea, D. An Overview of Global Maize Market Compared to Romanian Production. *Romanian Agric. Res.* **2022**, *39*, 535–544. [CrossRef]
123. Guță, B.-A.; Marin, D. Grain Yield and Yield Components on an Assortment of Winter Wheat (*Triticum aestivum* L.) Genotypes Cultivated under Conditions of A.R.D.S. Teleorman. *AAB Bioflux* **2020**, *12*, 45–51.
124. Kuneva, V.; Stoyanova, A.; Cojocaru, J.; Sturzu, R.; Meluca, C. Productive Capabilities of Promising Varieties of Wheat (*Triticum aestivum* L.). *Romanian Agric. Res.* **2023**, *40*, 209–214. [CrossRef]
125. Asseng, S.; Foster, I.; Turner, N.C. The Impact of Temperature Variability on Wheat Yields. *Glob. Change Biol.* **2011**, *17*, 997–1012.
126. Boiu-Sicuia, O.A.; Constantinescu, F.; Ursan, M.; Cornea, C.P. Microbial Inoculants Applied as Seed Treatments and Their Effect on Common Wheat *Triticum aestivum* L. *Ann. Univ. Craiova* **2019**, *49*, 38–43.
127. Golea, C.M.; Galan, P.-M.; Leti, L.-I.; Codină, G.G. Genetic Diversity and Physicochemical Characteristics of Different Wheat Species (*Triticum aestivum* L., *Triticum monococcum* L., *Triticum spelta* L.) Cultivated in Romania. *Appl. Sci.* **2023**, *13*, 4992. [CrossRef]
128. Cappelli, A.; Cini, E. Challenges and Opportunities in Wheat Flour, Pasta, Bread, and Bakery Product Production Chains: A Systematic Review of Innovations and Improvement Strategies to Increase Sustainability, Productivity, and Product Quality. *Sustainability* **2021**, *13*, 2608. [CrossRef]

129. Ahmad, A.; Aslam, Z.; Javed, T.; Hussain, S.; Raza, A.; Shabbir, R.; Mora-Poblete, F.; Saeed, T.; Zulfiqar, F.; Ali, M.M.; et al. Screening of Wheat (*Triticum aestivum* L.) Genotypes for Drought Tolerance through Agronomic and Physiological Response. *Agronomy* **2022**, *12*, 287. [CrossRef]
130. Abdi, N.; van Biljon, A.; Steyn, C.; Labuschagne, M.T. Bread Wheat (*Triticum aestivum*) Responses to *Arbuscular mycorrhizae* Inoculation under Drought Stress Conditions. *Plants* **2021**, *10*, 1756. [CrossRef] [PubMed]
131. Zaheer, M.S.; Ali, H.H.; Soufan, W.; Iqbal, R.; Habib-ur-Rahman, M.; Iqbal, J.; Israr, M.; El Sabagh, A. Potential Effects of Biochar Application for Improving Wheat (*Triticum aestivum* L.) Growth and Soil Biochemical Properties under Drought Stress Conditions. *Land* **2021**, *10*, 1125. [CrossRef]
132. Kadar, R.; Muntean, L.; Racz, I.; Ona, A.; Ceclan, A.; Hirişcău, D. The Effect of Genotype, Climatic Conditions and Nitrogen Fertilization on Yield and Grain Protein Content of Spring Wheat (*Triticum aestivum* L.). *Not. Bot. Horti Agrobot. Cluj-Napoca* **2019**, *47*, 515–521. [CrossRef]
133. Manole, D.; Jinga, V.; Marga, G.; Radu, I.; Ștefan, I.; Soare, S. New Edition on Sunflower Crop -Romanian Technology under Climate Change Conditions in Dobrogea. *Sci. Pap.* **2019**, *62*, 348–354.
134. Samoggia, A.; Perazzolo, C.; Kocsis, P.; Del Prete, M. Community Supported Agriculture Farmers' Perceptions of Management Benefits and Drawbacks. *Sustainability* **2019**, *11*, 3262. [CrossRef]
135. Stoicea, P.; Chiurciu, I.A.; Soare, E.; Iorga, A.M.; Dinu, T.A.; Tudor, V.C.; Gîdea, M.; David, L. Impact of Reducing Fertilizers and Pesticides on Sunflower Production in Romania versus EU Countries. *Sustainability* **2022**, *14*, 8334. [CrossRef]
136. Brumă, I.; Rodino, S.; Petcu, V.; Micu, M. An Overview of Organic Sunflower Production in Romania. *Romanian Agric. Res.* **2021**, *38*, 495–504. [CrossRef]
137. Popescu, A. Oilseeds Crops: Sunflower, Rape and Soybean Cultivated Surface and Production in Romania in the Period 2010–2019 and Forecast for 2020–2024 Horizon. *Sci. Pap.* **2020**, *20*, 467–478.

sustainability

MDPI

Article

Hydrochemical and Microbiological Investigations and the Therapeutic Potential of Some Mineral Waters from Bihor County, Romania

Ribana Linc [1,*,†], Emilia Pantea [2,†], Eugenia Şerban [2,†], Anca-Paula Ciurba (Pastor) [1,3,†] and Georgeta Serban [4,†]

1 Department of Geography, Tourism and Territorial Planning, Faculty of Geography, Tourism and Sports, University of Oradea, 1 University St., 410087 Oradea, Romania; anca-paula.ciurba@univ-lorraine.fr
2 Faculty of Environmental Protection, University of Oradea, 26 Gen. Magheru St., 410057 Oradea, Romania; emilia.pantea@uoradea.ro (E.P.); eugeniaserban@yahoo.com (E.Ş.)
3 Laboratory LOTERR-EA 7304, Department of Geography, University of Lorraine, 57045 Metz, France
4 Faculty of Medicine and Pharmacy, University of Oradea, 29 Nicolae Jiga, 410028 Oradea, Romania; getaserban_2000@yahoo.com
* Correspondence: ribanalinc@yahoo.com
† All authors contributed equally to this work.

Abstract: Water quality plays an important role for every sustainable social and economic system, as well as for maintaining human health. This study, carried out during 2022–2023, has as its main objective the physical–chemical and microbiological analysis of some underground water resources (two boreholes and a spring) with a natural mineral load from three areas in Bihor County (Romania), and the impact of their consumption on people's health. Therefore, six microbiological parameters and 17 physical–chemical indicators in three localities (Tămășeu, Sîntimreu, Pădurea Neagră) were analyzed. The results of the microbiological analysis indicate a type of water that respects the limits imposed by the legislation on natural mineral waters and potability. The physical–chemical indicators show that the hydrochemical type of the studied waters is predominantly bicarbonate, in association with calcium, magnesium, and sodium cations. The residents' perception on water quality and the effect on people's health was assessed through a questionnaire (23 items) addressed to the population of the three villages and neighboring localities. The results showed that the mineral waters from Sîntimreu and Pădurea Neagră are used frequently. Many respondents consider the local mineral waters as without quality-related problems and with beneficial effects regarding acute or chronic gastrointestinal conditions, such as gastritis, gastric ulcers, flatulence, or liver diseases.

Keywords: natural mineral water; hydrochemistry; microbiological analysis; spring; borehole; water quality; therapeutic effect; diseases

Citation: Linc, R.; Pantea, E.; Şerban, E.; Ciurba, A.-P.; Serban, G. Hydrochemical and Microbiological Investigations and the Therapeutic Potential of Some Mineral Waters from Bihor County, Romania. *Sustainability* **2023**, *15*, 15640. https://doi.org/10.3390/su152115640

Academic Editor: Agostina Chiavola

Received: 17 August 2023
Revised: 16 October 2023
Accepted: 24 October 2023
Published: 6 November 2023

1. Introduction

Water is one of the main sources of our existence on Earth. Alongside population growth and rising living standards, water consumption has also increased, which has led to a decrease in the resources available to ensure the daily requirement of fresh water. As a result, the identification of new water resources that can be the source of drinking water but can also bring benefits to human health is an important objective of the policies applied to this sector. Natural mineral waters might be valuable resources for the population of any country. Some authors [1,2] found a sudden increase in mineral water consumption worldwide, and that is supported by the very beneficial properties of these waters for human health [3]. According to [4], Romania has a diversified and significant range of natural mineral water resources for food consumption, and there is potential for their exploitation. The hydromineral sources are concentrated on the territory of 16 counties (Figure 1): Arad, Arges, Bihor, Brașov, Caraș-Severin, Covasna, Harghita, Hunedoara, Iași, Maramureș, Mureș, Neamț, Prahova, Satu-Mare, Suceava, and Timiș [5].

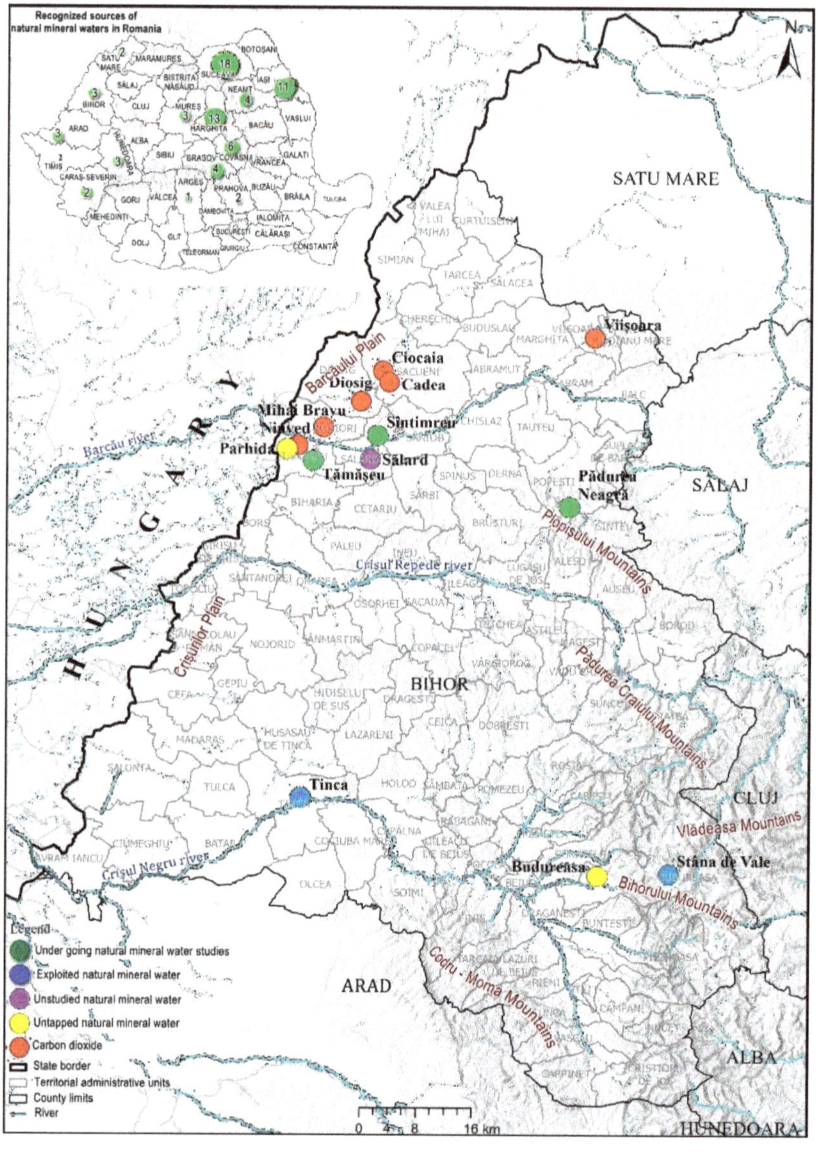

Figure 1. Bihor County: localities that have natural mineral water and carbon dioxide resources. In the inset—Romania and the counties that have natural mineral water resources and the number of sources recognized by the National Agency for Mineral Resources from Romania.

According to Feru [6], over 45% of natural mineral water deposits in Romania are located in carbonate rocks (limestone, calcareous conglomerates, etc.), about 25% are located in igneous rocks (pyroclastites and andesites), 25% in detritus sedimentary deposits, and 5% in sandstones and crystalline shales. Before the 1990s, the condition for mineral water to be bottled in Romania was that it had to contain a quantity of dissolved salts higher than 1000 mg/L, a carbon dioxide (CO_2) content higher than 500 mg/L, and have the therapeutic effects mentioned on the label. Nowadays, from Government Decision HG 1020/2005 [7] implementing Directive 80/777/EEC [8] (with subsequent additions), a distinction is made between the terms "natural mineral waters" and "medicinal waters", meaning that the medicinal waters must have scientifically recognized therapeutic actions, and can exceed the maximum allowed concentrations even for some undesirable and potentially toxic elements which have been proven to be responsible for the therapeutic effects (e.g., iron, manganese, arsenic, hydrogen sulphide). Teodoreanu and Gaceu [9] assume that 37% of Europe's mineral waters are owned by Romania, being spread over the three major relief steps (plains, hills, mountains) as springs and mineral borehole waters with different flow, temperatures, and chemical content.

Water has been used to promote health since ancient times. It was only in the 19th century that detailed chemical analyses brought information about mineral content, and thus revealed the medicinal potency and healing benefits of mineral waters in various pathologies [10,11]. Mineral waters have a beneficial effect on human health both through internal administration (crenotherapy) in the form of water ingestion or inhalation of aerosols, and through external administration (balneotherapy) by immersing the body in water [10,11]. Depending on their chemical content, mineral waters have been used for decades as curative agents in the treatment of various gastrointestinal conditions such as heartburn, dyspepsia, gastritis, ulcers, constipation, irritable bowel syndrome, and even gallbladder or liver diseases, as an adjuvant therapy to complement the effects of the drugs [11,12]. The reduced volume of urine or an increased urinary concentration of salts favors the formation of kidney stones. Crenotherapy is effective in the prevention and therapy of urolithiasis because a proper consumption of water washes the urinary tract, increases the urinary volume, increases the dilution of dissolved salts, and reduces microbial content and changes in urinary pH [13–15].

Crenotherapy has beneficial results in the therapy of respiratory pathologies. Steam or aerosol inhalation, as well as irrigation with high mineral content waters due to the several effects of minerals (antiseptic effect—sodium chloride, metabolic effects—iodine and iodides, analgesic and sedative effects—bromide and calcium) is an accessible method with long-term beneficial results for respiratory conditions such as allergies, general inflammation of the airways (chronic obstructive pulmonary disease—COPD), recurrent viral and bacterial infections, abnormalities in bronchial secretions, and impairment of lung and nasal functions [16]. Although the mode of action is still unclear, the anti-inflammatory, immunomodulatory, and mucolytic activities of mineral waters might make a substantial contribution to therapeutic effects, with the advantage of the lack of possible side effects of drugs usually used in the treatment of respiratory diseases (e.g., corticosteroids, antibiotics, bronchodilators). Inhalations with mineral water relieve the symptoms of lower and upper respiratory tract diseases, such as rhinitis, laryngitis, pharyngitis, rhinosinusitis, asthma, bronchitis, and pneumonia, and also improve patients' quality of life [12,16–18].

Sulphurous mineral waters in particular have multiple therapeutic effects. In addition to gastrointestinal and respiratory effects, since hydrogen sulfide is a small-sized gas molecule, it can be easily absorbed through the skin and mucosa, and gives systemic effects [19]. Thus, sulphurous waters are effective in conditions such as hypertension, heart disease, ischemia, or lower and upper urinary tract disorders [19–21]. During hydropinotherapy with sulphurous mineral waters, there is an increase in the secretion of insulin, leading to the burning of glucose [22]. In addition, a decrease in the concentration of reactive oxygen species was observed. The antioxidant properties are accompanied by anti-aging effects, as these mineral waters are able to prevent structural changes in DNA [22,23].

Bihor County is known in Romania as having significant resources of geothermal mineral waters that have been intensively studied [24–28], but there are few studies on the analysis of *cold natural mineral waters* [29–31]. Considering this fact and in correlation with the current trends regarding the sustainable management of water resources, the main purpose of this work is to provide the physical–chemical (pH, dry residue, chemical composition) and microbiological characteristics of some cold natural mineral waters from three sources in Bihor County (Tămășeu, Sîntimreu, Pădurea Neagră). Taking into account the consumption of these waters by the population of the respective areas, the first and second hypotheses of this study are as follows:

Hypothesis 1. *The physical characteristics and chemical composition of the water is a condition for its inclusion in the category of mineral water.*

Hypothesis 2. *The microbiological purity of mineral water is a condition for being admitted to human consumption.*

Since old documents cited by [32] suggest that lithium was the reason for the use and commercialization of mineral water from the Tămășeu for about 60 years, the third hypothesis is deduced as follows:

Hypothesis 3. *The lithium content is a condition for the mineral water from the Tămășeu borehole being classified as lithiniferous water.*

In addition, in order to highlight the potential impact of the consumption of these mineral waters on people's health, we aimed to find out the population's perception towards the use of water from the three sources. Therefore, a questionnaire addressed to the resident population of the three locations, as well as the neighboring villages and towns (totaling 12 localities), was studied and applied. A fourth hypothesis is deduced as follows:

Hypothesis 4. *Population awareness of the quality and possible therapeutic effects of the mineral waters from the three sources.*

To our knowledge, there are very few and old studies analyzing the water from the three sources [33–36]; therefore, our study brings current knowledge about the physical–chemical and microbiological characteristics of these waters, as well as their possible further exploitation for nutritional and therapeutic purposes. This information is useful to both the scientific and local communities.

2. Materials and Methods

2.1. Study Area

Located in northwestern Romania, Bihor County has various underground resources of recognized economic importance, such as coal and crude oil deposits concentrated especially in the Barcău river basin, as well as important geothermal water deposits spread throughout the county. Also concentrated in the Barcău basin, there are *several springs of cold natural mineral waters and resources of carbon dioxide.*

Due to CO_2 coming from deep in the fault lines of the foundation, in the *Western Plain of Bihor County,* there are several locations with bicarbonate water springs north of the Oradea city (at Sălard, Sîntimreu, Tămășeu), as well as in the Tinca resort located in the south of the county (here being exploited for therapeutic purposes). In the *Bihorului Mountains,* there are still natural mineral waters in the Budureasa commune (Izvorul Minunilor from the Stâna de Vale resort and Izvorul Cuciului/Hera), found on the annual list of recognized mineral waters in Romania [37]. There is also a spring in Pădurea Neagră holiday village, located in the *Plopisului Mountains,* from where people frequently supplied themselves with water for individual consumption.

In this study, we focused on three easily accessible sources of *cold natural mineral water*: two medium-depth boreholes in the plain area at ***Tămășeu*** and ***Sîntimreu,*** and a spring in the mountainous area of the ***Pădurea Neagră*** holiday village (Figure 1).

These sources have been known for a long time, and those in Tămășeu and Sîntimreu were used by the locals for their therapeutic properties especially in the late 19th–early 20th century. The boreholes were less used after the Second World War, so the locals now know little about the properties and qualities of these waters. However, until Romania joined the EU in 2007, these waters were used for individual consumption, but the opening of many hypermarkets that sell various types of water at relatively low prices, as well as the fact that these waters acquire a reddish color after a while, have considerably reduced the supply flow. Currently, the locals or people passing through the area continue to supply themselves even sporadically with natural mineral water from the boreholes in Tămășeu and Sîntimreu and the spring in Pădurea Neagră. Only on very hot summer days, a large number of people can be seen stocking up on water for consumption.

The benchmark of the study is the specifications of Directive 54/2009 [38], according to which natural mineral water can be differentiated from ordinary drinking water: *(a)* by its nature, characterized by its mineral components, oligoelements or other constituents and, possibly, by certain therapeutic effects; *(b)* by its original purity, both characteristics being kept unchanged due to the underground origin of the mineral water, which has been protected against all risks of pollution. The basic national legislation that constituted the point of reference was Government Decision HG 1020/2005 [7], supplemented and amended by HG 532/2010 [39], the main normative act that legislates the exploitation and commercialization of natural mineral waters in Romania.

2.2. Determination of Physical-Chemical Characteristics

The characteristics of mineral waters are the result of the mineralogical composition of the rocks crossed by the water, the crossing time, the type of waters encountered, the mixing rate, as well as the duration of this process [40].

2.2.1. Sampling of Natural Mineral Waters

In order to achieve the objectives of this study, two sampling sessions were carried out from the three locations in April 2022 and May 2023, respectively. Polyethylene terephthalate (PET) bottles of two liters were used to determine the physical–chemical indicators. The samples were analyzed within 24 h from sampling. To determine the microbiological characteristics, sterilized glass flasks of 0.5 L with glass stoppers were used. Sampling for microbiological analyses was carried out in accordance with Romanian Standard SR EN ISO 19458/2007—Water quality [41]. Figure 2 shows the sampling points related to each water source analyzed in this study.

Tămășeu Sîntimreu Pădurea Neagră

Figure 2. The natural mineral water boreholes from Tămășeu and Sîntimreu, and the spring from Pădurea Neagră holiday village. Water sampling.

2.2.2. The Methodology for Determining the Main Physical–Chemical Indicators

A variety of working techniques—specific to each parameter analyzed for the three mineral water sources—were used in this study. The qualitative evaluation was carried out by comparing the data obtained with the limits established by the national/international legislation. The quality of chemical analyses was ensured by the use of high purity reagents (Merck, Bucharest, Romania), experimental methods approved by standardization bodies, and laboratory equipment certified for high-precision chemical analyses. The main physical–chemical indicators analyzed are presented in Table 1.

Table 1. The main physical–chemical indicators.

Indicators/U.M.	Methods of Analysis	Reference Documents
Determination of pH	Water quality. Determination of pH (Hanna Instruments HI2020-02 multiparameter, Hanna Instruments, Nusfalau, Romania)	[42]
Conductivity (μS/cm)	Water quality. Determination of electrical conductivity (Conductometer HI6321, Hanna Instruments, Nusfalău, Romania)	[43]
Calcium, Magnesium (mg/L)	Water quality. Determination of calcium and magnesium content by flame atomic absorption spectrometry (Perkin Elmer Analyst 800, Perkin Elmer LLC, Rodgau, Germany)	[44]
Potassium, Sodium, (mg/L)	Water quality. Determination of sodium and potassium. Part 3. Determination of sodium and potassium by flame emission spectrometry (Perkin Elmer Analyst 800, Perkin Elmer LLC, Rodgau, Germany)	[45]
Lithium (mg/L)	Water quality. Determination of selected elements by inductively coupled plasma optical emission spectrometry (ICPE-9800, Shimadzu, Shimadzu Europa GmbH, Duisburg, Germany)	[46]
Iron (mg/L)	Water quality. Determination of iron content by flame atomic absorption spectrometry (Perkin Elmer Analyst 800, Perkin Elmer LLC, Rodgau, Germany)	[47]
Manganese (mg/L)	Water quality. Determination of manganese content by flame atomic absorption spectrometry (Perkin Elmer Analyst 800, Perkin Elmer LLC, Rodgau, Germany)	[48]
Chloride (mg/L)	Water quality. Determination of chloride content. Volumetric (Mohr Method)	[49]
Sulfate (mg/L)	Water quality. Turbidimetric determination of sulphates	[50]
Nitrates (mg/L)	Water quality. Determination of nitrate. Part 3. Spectrometric method using sulfosalicylic acid (HACH DR 3900 spectrophotometer—Hach, Romania)	[51]
Nitrites (mg/L)	Water quality. Determination of nitrites. Molecular spectrometric absorption method (HACH DR 3900 spectrophotometer—Hach, Romania)	[52]
HCO_3^- *(mg/L)*	Water quality. Determination of alkalinity. Part 2. Determination of carbonate alkalinity (volumetric method)	[53]
CO_2 *(mg/L)*	Water quality. Determination of carbon dioxide (Volumetric method)	[54]
H_2S *(mg/L)*	Water quality. Determination of sulphide content (Iodometric method)	[55]
Dry residue at 180 °C (mg/L)	Water, groundwater and wastewater. Determination of residue. Gravimetric method (Biobase BOV-T30C oven, Biobase, China)	[56]

Tools such as Piper and Stiff diagrams were used to process the analytical data obtained through the chemical analyses and to highlight the hydrochemical footprint of the samples [57].

2.3. The Methodology for Determining the Microbiological Load

Microbiological analysis of natural mineral water at the source can identify the presence of germs that can form colonies, on suitable culture media. This is particularly important for consumers, because after bottling, the number of germs can increase rapidly, reaching 104–105 CFU/mL (CFU = Colony Forming Unit) in 3–7 days, according to some authors [58]. Taking this into account, and correlated with the microbiological requirements provided by Government Decision HG 1020/2005 [7], supplemented and amended by HG 532/2010 [39], the microbiological study aimed to determine the microbiological parameters presented in Table 2.

Table 2. The main microbiological indicators.

Type of Analysis	Methods of Analysis	Reference Documents/ Work Technique
Determination of the total colony count (CFU) at 22 °C	Water quality. Enumeration of culture microorganisms. (Colony counting by seeding in agar-agar culture medium at a temperature of 22 °C, incubated for 72 h).	[59]
Determination of the total colony count (CFU) at 37 °C	Water quality. Enumeration of culture microorganisms. (Colony counting by seeding in agar-agar culture medium at a temperature at 37 °C incubated for 24 h).	[59]
Determination of Coliform bacteria and isolation of Escherichia coli species	Water quality. Enumeration of *Escherichia coli* and Coliform bacteria. Part 1: Membrane filtration method.	[60]
Determination of intestinal Enterococci	Water quality. Detection and enumeration of intestinal enterococci. Part 2: Membrane filtration method. (The determination was made after incubating the membrane for 48 h at 36 °C, on Slanetz–Bartley medium).	[44]
Pseudomonas aeruginosa	Water quality. Detection and enumeration of *Pseudomonas aeruginosa*. Membrane filtration method. (Pseudomonas agar/CN—agar medium membrane incubated at 36 °C for 49 h).	[61]
Clostridium perfringens	Water quality. Enumeration of *Clostridium perfringens*. Membrane filtration method.	[62]

2.4. Questionnaire Content

The objective of the questionnaire was to assess the residents' perception on the quality and effect on health of the mineral waters from the three studied water sources (Tămășeu, Sîntimreu, and Pădurea Neagră). Therefore, a questionnaire addressed to the population residing in the three villages and neighboring localities (12 localities in total) was applied.

The questionnaire was divided into two parts. The first part included 18 items related to mineral waters, through which it was sought to obtain two types of data: one on the quality of mineral waters and the other on their possible effect on human health. The second part of the questionnaire included five items related to the socio-demographic situation of the respondents (residence, age, education, occupation etc.).

The 18 items had as main elements: identification of the main source of drinking water of the residents; identification of problems affecting the quality of mineral waters; consumption of mineral water by residents (amount, frequency, duration); identification of acute or chronic conditions of the population improved by mineral water consumption; identification of the possible beneficial effects on health reported after drinking mineral water; the use of mineral water for other purposes (e.g., in the kitchen).

The questionnaire included multiple-choice questions, as well as open-ended questions. It contained a standard, concise introduction stating the confidentiality of answers and their exclusive purpose for scientific research. Each respondent was asked for his or her willingness to participate in the study. All persons over 15 years old of age who agreed to answer the questionnaire were included in this study.

The questionnaire was prepared and pre-tested in a selected community. Thus, 15 questionnaires were distributed to the respondents face-to-face, to find out if the items were clearly expressed. After reviewing some items, the questionnaire was distributed to the population between June and July 2023. It was applied online, using the Google Forms application (online forms and surveys creator), where a link was created that was distributed to the respondents via Internet. Since many inhabitants of the villages of Tămășeu, Parhida, Sîntimreu and Sălard are of Hungarian ethnicity, the questionnaire was distributed both in Romanian and Hungarian for better understanding by the respondents. The questionnaire was also applied face-to-face, although the number of respondents was smaller. The face-to-face investigation was done by the authors of this study and made them better understand the problems and the real perception of the population on the quality and effect of mineral waters on human health.

110 questionnaires were returned from respondents, of which 102 were validated (92.7%). Respondents of different ages, sexes, and education levels, who were randomly chosen in the field, took part in the survey. The data collected from the 23 items were statistically processed with the Google Forms application that generates automatic summaries based on percentage calculation.

Expected impact: through the present survey, we aimed to raise the awareness of the population about the physical–chemical characteristics and therapeutic potential of the mineral waters in the studied area.

The method of using the questionnaire to highlight the population's perception on a phenomenon is often used in specialized literature [63–67].

3. Results and Discussion

3.1. Research Background

The spring from Pădurea Neagră, in the Plopișului Mountains, is found in the outskirts of Popești commune, in the east of the holiday village, in a mixed hornbeam and beech forest, and flows towards the Bistra river. Mihăilă [68] explains the formation of the mineral spring. Aniței et al. [36] mentioned this spring as having oligomineral waters and Nicula [69] notes the high concentration of iron due to the host rocks and the presence of CO_2, which increases the water's ability to dissolve minerals.

Throughout time, the cold natural mineral waters of the bihorean plain have been used for both consumption (Tămășeu, Sîntimreu, Tinca) and for balneary treatment and crenotherapy (Tinca, Tămășeu, Sălard). Following the disappearance of some springs due to the flooding of the Barcău River (Sălard), the frequent change of owners (Tămășeu) [32], the decrease in flow rates (Tămășeu), and the tightening of legislative conditions for bottling and marketing natural mineral waters, bottling was abandoned.

In the *Tămășeu—Sîntimreu—Sălard sector*, the hydromineral deposit contains bicarbonated, magnesian waters with a CO_2 content of over 1.5 g/L. The mineralized aquifer layer from the polygenic sands was opened at a depth of 120 m, and flows freely with low flow rates (below 0.2 L/s). In the 1970s, this aquifer was exploited by the bottling station in the Tămășeu railway station [33].

According to Szabó [32], who cites many Hungarian documents about the natural mineral water from Tămășeu, the Hungarian press wrote in 1899 about the drilling of the well in Tămășeu and the eruption of a "highly carbonated sour water", as well as about the chemical analyses carried out at the Institute of Chemistry in the Budapest II University of Sciences. The way to promote Tămășeu mineral water—called "*Lithium healing water from Tămășeu*"—through an intense popularization campaign, was also detailed. In addition, the medicinal qualities of this water were highlighted, it being recommended for *kidney, bladder, stomach, and intestine diseases*, as well as for *bronchial asthma, gout, or diabetes*. It is mentioned that the water is also good for daily use as drinking water. The ownership and marketing rights have been held by various companies or individuals, and the name of the water has always been centred on the chemical element *lithium*, and has been named Lythia, Lythinia, and Lithymus. In the 1960s, Berlescu et al. [34] and Binștoc [35] mentioned Tămășeu both for the hydromineral treatment of the excretory system and for the cure of digestive diseases with alkaline carbonated waters, which reduce uric acid in the blood. Also, Pricăjan [33] briefly discussed the chemical composition of Bihor's natural mineral waters, namely those from Tămășeu and Pădurea Neagră. Aniței et al. [36] consider Tămășeu a place with natural healing factors, and an exciting lowland bioclimate.

There are currently two boreholes at Tămășeu at a short distance from each other, located on the edge of the village near the ruins of the railway station, in the Barcău river meadow at about 107 m altitude. The borehole the water samples were taken from has a depth of 110 m, and the local administration replaced a segment of the column at a depth of 40–50 m and installed a new wellhead in 2003 [32]. This borehole is protected above ground by a closed concrete tube fitted with a metal pipe, through which water flows at low flow rates. The second borehole reaches up to 220 m depth, and water supply work to the Parhida railway station is currently underway here. A methane gas pipeline and a storage station are located near the boreholes, and the surrounding farmland is cultivated with grain.

In the local development strategies of the Sălard commune, it is written that "before the world war the water from the borehole in Sîntimreu was bottled and it was useful for digestive diseases" [70] and that "living water has a therapeutic effect in the internal treatment of ailments of the digestive and hepatobiliary system" [71]. Aniței et al. only [36] mentioned Sîntimreu with mixed ferruginous, carbonated waters.

The current Sîntimreu borehole is located in the village near the school, at the base of the second terrace of the Barcău river at about 115 m in altitude. There were oak forests here in the past, but now these terraces are cultivated with vineyards. Due to modernization work, the borehole is now protected by a cover, and the water flows at high speed through two pipes with four outlets.

Currently, the natural mineral water from the Tămășeu—Sîntimreu—Sălard aquifer is unutilized and is discharged freely through the Tămășeu and Sîntimreu wells.

3.2. *Physical-Chemical Indicators*

3.2.1. Hydrochemical Characterization

The measured *pH values* for the analyzed waters fell within the limits specified by the legislative norms applied to both mineral waters and drinking water [5,6]. The analysis of the parameters reveals common aspects that give some particular characteristics to these waters. The pH values varied between a maximum value specific to the mineral water from Tămăseu (pH 7.5) to a minimum value found for the water from the Pădurea Neagră spring (slightly acidic, pH 6.5). The natural mineral water from Sântimreu has a pH value of 6.84 pH units. The lowest pH value determined in the water from the Pădurea Neagră spring is the result of a significant footprint of anions. Long-term consumption of water with a low pH can influence the solubility and bioavailability of some nutrients [2].

Total mineralization is a defining parameter that provides information about the potential use of the water, as well as the technical aspects that must be considered in exploitation. Related to this characteristic, Stoicescu [72] and Vernescu [73] mention that in Bihor County, there are a variety of acratic and hypertonic waters.

The quality of mineral waters depends on the concentration of dissolved mineral salts. Referring to this indicator, it was found that the highest mineralization (7128 mg/L) was obtained for the water of the borehole of Tămășeu in Barcăului Plain, and the lowest value was obtained for the water of the spring from Pădurea Neagră in the Plopișului Mountains (3086 mg/L) (Figure 3). According to Annex 3 to HG 1020/2005 [7] (with subsequent amendments and additions), and taking into account the content of mineral salts found, the natural mineral waters from the three localities fall into the category of highly mineralized natural mineral waters (residue content > 1500 mg/L). Highly mineralized waters are recommended to athletes and people performing physical activities at high temperatures, and are very useful to replenish the electrolytes lost through sweating.

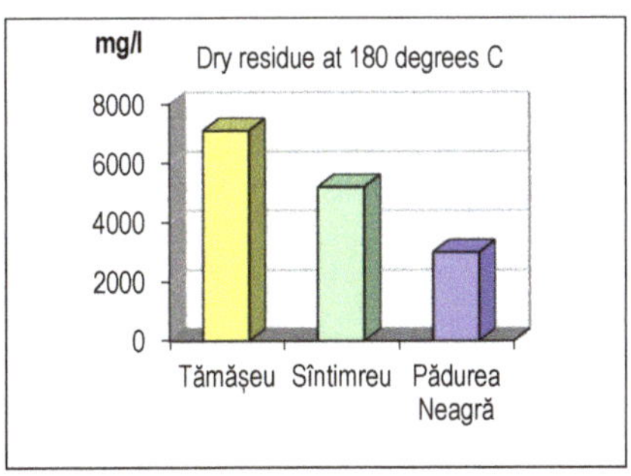

Figure 3. Dry residue at 180 °C.

According to Annex 3 of HG 1020/2005 [7] with subsequent amendments and additions, for a mineral water to be classified as acidulated, it is necessary to have a *CO_2 content* > 250 mg/L. Analyzing the results obtained for each mineral water, it was found that the mineral waters from Tămăşeu, Sîntimreu, and Pădurea Neagră are acidulated, with the water from Pădurea Neagră spring having the highest CO_2 value (2004.2 mg/L) (Figure 4). In addition, since the acidulated mineral waters from Pădurea Neagră and Sîntimreu spontaneously and visibly released carbon dioxide, they can be classified as effervescent waters.

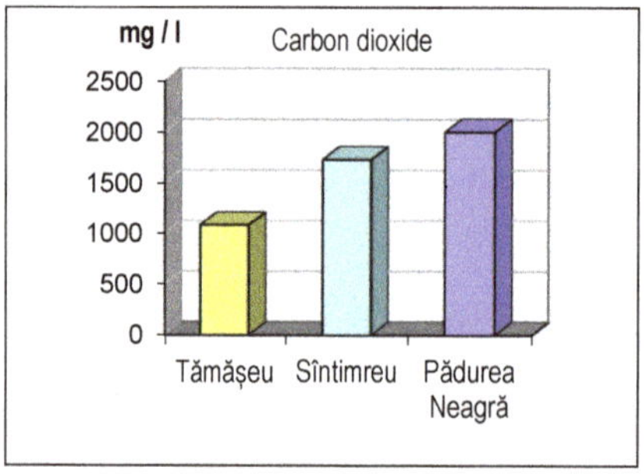

Figure 4. Carbon dioxide content.

3.2.2. Chemical Composition of Natural Mineral Waters

The chemical composition of mineral water, given by the specific content of dissolved salts, represents one of the main characteristics that can give it beneficial effects for health. The greatest non-uniformity in ion distribution is found in the cationic zone, where there are variations between the predominant characters of the mineral waters from the three locations. The sum of cations and anions dissolved in water, expressed in milliequivalents/liter, should be equal: (∑cations) = (∑anions). An imbalance of this equality can be attributed to errors in analytical determinations or electrolyte dissolution as a result of anthropogenic influences. The charts shown in Figure 5 give us clues about these aspects. The data presented in Figure 5 show that, among cations, calcium (Ca^{2+}), magnesium (Mg^{2+}), sodium (Na^{+}), and potassium (K^{+}) are predominant.

Calcium is predominantly found in the analyzed natural mineral waters, its origin being related to calcareous/dolomitic rocks. *Magnesium* is a cation present in natural mineral waters, usually together with calcium. It usually comes from dolomitic rocks. *Calcium* is an element with multiple functions in the human body, essential not only for the mineralization and health of the skeleton, but also for the proper functioning of the muscular system (regulates muscle contraction), the vascular system (mediates vasoconstriction and vasodilatation), and the nervous system (transmission of nerve impulses, ion exchange across cell membranes) [74,75]. Since adequate calcium intake is important for the body's health, there are many calcium-enriched food products and calcium supplements on the market. However, some supplements may have low absorption or be a source of gastric discomfort; therefore, mineral waters rich in calcium might be a potential source of calcium. The bioavailability of calcium from calcium-rich mineral waters is comparable or slightly higher than that of calcium from milk, with mineral waters providing over 40% of the recommended daily intake of calcium [75–77]. The consumption of mineral waters, especially alkaline mineral waters rich in bicarbonates and calcium, has a positive impact on bone mass and bone density, as well as in the prevention of osteoporosis. Therefore, calcium-rich mineral waters can be recommended as calcium supplements [74,75,77]. *Magnesium* is involved in various physiological processes, such as mitochondrial integrity, adenosine-triphosphate function, DNA synthesis, bone mineralization, or muscle activity [78]. Magnesium has a cardioprotective effect, and drinking hard mineral waters rich in magnesium and calcium reduces the risk of mortality from cardiovascular diseases [79].

These two cations have higher concentrations in the water from Sîntimreu and Pădurea Neagră (Figure 5).

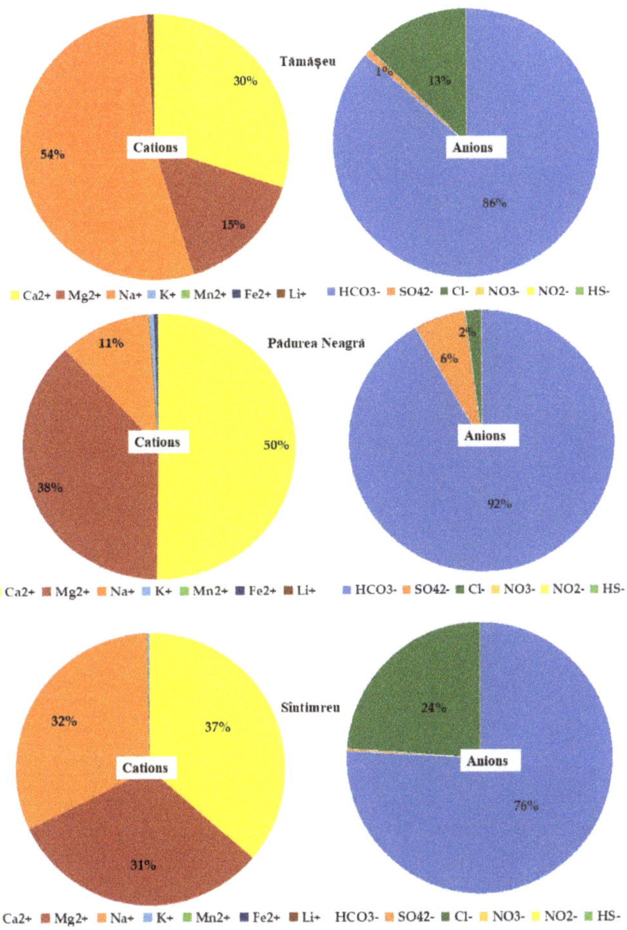

Figure 5. Ionic balances of waters from Tămășeu, Sîntimreu, and Pădurea Neagră.

Sodium, an essential element for human life, characterized by high reactivity, and can be found in most groundwater. Concentrations higher than 200 mg/L can influence the taste of water, which is not the case for the waters analyzed in this study, since the concentration of this cation was in the range of 47.8 mg/L (Tămășeu), 45.3 mg/L (Sîntimreu), and 9.2 mg/L (Pădurea Neagră). In general, sodium salts have no acute toxicity due to the ability of the healthy mature kidney to excrete sodium [6]. *Potassium* poisoning by ingestion is rare due to the kidney's ability to excrete it rapidly in the absence of pre-existing kidney damage, and large single doses usually induce vomiting [6]. Potassium was found in higher amounts in the water from Pădurea Neagră spring (1.35 mg/L).

Lithium in the mineral waters is related to the presence of clay muds where it was retained during its underground circulation. Lithiniferous waters are considered to be mineral waters with a lithium concentration of more than 3 mg/L [37]. In Romania, there are several mineral waters rich in lithium, such as the Cașin-Iacobeni spring (Perla Cașinului water, lithium concentration of 6.6 mg/L, the richest lithium water in Romania), the Harghita spring (6.2 mg/L lithium), the Maria spring from Malnaș-Băi (5.6 mg/L), and the Matilda spring from Bodoc resort (5.3 mg/L) [5,12,80]. Among the waters under study by us, the highest concentration of lithium (4.4 mg/L) was determined in the mineral water from the Tămășeu borehole, which can fit this water into the category of lithiniferous waters. Lithium salts (e.g., lithium carbonate) are implicated in the treatment of bipolar disorder, but these drugs have a low therapeutic index and multiple side effects [81]. The bioavailability of lithium in mineral waters is very high, and studies showed that drinking lithium-rich mineral waters can reduce the risk of suicide and prevent neurological disorders such as dementia or Alzheimer's disease [12,81].

Iron has an important role in the body. Ferruginous waters are administered as internal crenotherapy, freshly brought from the spring and during meals, because iron is inactivated in contact with air. Only bivalent iron is active and is absorbed in the presence of hydrochloric acid and vitamin C. Ferruginous waters are implicated in iron deficiency anemia, gastric achilia, gastric surgery, etc. [12]. Iron was found in significant amounts in the sample taken from Pădurea Neagră spring (0.57 mg/L), which is highlighted in Figure 5.

Regarding the anions, the preponderance of the bicarbonate anion (HCO_3^-) was found, followed by chloride (Cl^-) and sulphate (SO_4^{2-}) anions. The predominance of bicarbonate anions over those of strong acids (chlorides or sulphates) was found in all analyzed mineral waters. The bicarbonate ion is produced when water reacts with carbonate rocks such as limestone and dolomite. The mineral water from the Pădurea Neagră spring has the highest concentration of bicarbonate, followed by that from Tămășeu and Sîntimreu. An appreciable amount of sulphates were found in the water from the Pădurea Neagră spring (Figure 5).

Mineral waters rich in sodium or potassium bicarbonates can neutralize gastric acid, thus increasing the pH value of the stomach and reducing heartburn in adult patients. Considering the high tolerability, such mineral waters might be considered as alternative to drug treatments. In addition, depending on the time of administration (before or during the meal), carbonated alkaline waters can reduce or stimulate gastric secretions, influencing gastrointestinal motility and gallbladder activity [11,12,82,83]. Alkaline-earth waters contain the bicarbonate anion bound to calcium or magnesium cations, and their main applications are for digestive ailments (gastritis, colitis, gastric or duodenal ulcers, chronic enterocolitis, etc.). Since many of them have calcium as the main mineral, they are also effective in treating rickets and allergies [12]. Mineral waters rich in salts (e.g., bicarbonate, chlorine, sulphate, magnesium, and sodium) not only increased intestinal transit through laxative or purgative activity with beneficial effects for constipation, but also stimulated gallbladder contraction and bile acid elimination through stool, and decreased the level of serum total cholesterol and LDL-cholesterol. This therefore might be considered an alternative to hypocholesterolemic drugs [84–86]. In addition, bicarbonate-sulphate-alkaline-earth mineral waters have antitoxic and trophic effects on the liver [16].

A particular case is the water from Sîntimreu, with a strong smell of hydrogen sulphide. In addition to the effects determined by the existing cations and anions, hydrogen sulphide contributes to the properties of this water. Sulphurous waters stimulate gastric and bile secretion through choleretic and cholecystokinetic effects. Hydrogen sulphide protects the intestinal flora from various chemical attacks and therefore heals mucosal wounds and reduces local inflammation. Its cytoprotective activity, local vasodilation, and ability to promote new blood vessel growth also contributes to these effects. Sulphur is a constituent of some amino acids (e.g., cysteine), and through the thiol group of hydrogen sulphide, crenotherapy with sulphurous waters can heal lesions of the respiratory tract mucosa (bronchitis, chronic rhinitis, COPD) [11,19,87]. In addition, sulphurous waters have hypoglycemic and antioxidant properties. Both the antidiabetic and antioxidant effects are mainly due to hydrogen sulphide, and are beneficial in metabolic diseases such as diabetes [22,23].

In order to classify the natural mineral waters from the three localities from a hydrochemical point of view, the predominant ions (Na^+, K^+, Ca^{2+}, Mg^{2+}, HCO_3^-, Cl^-, SO_4^{2-})

were represented on the Piper diagram (Figure 6). The diagram shows the concentrations of the majority of cations and anions arranged on two separate trilinear diagrams, together with a central diagram (diamond) in which the points from the two trilinear diagrams are projected. In the case of the mineral water from the Pădurea Neagră spring, the location is found in the quadrant represented by the majority cation Ca^{2+} (calcium type). The natural mineral water sampled from Sîntimreu belongs to the specific area "no dominant type". The natural mineral water from Tămășeu is located in the quadrant with predominantly alkaline ions (sodium and potassium type). From an anionic point of view, all three types of natural mineral waters analyzed are in the area specific to the majority bicarbonate ion (bicarbonate type) (Figure 6).

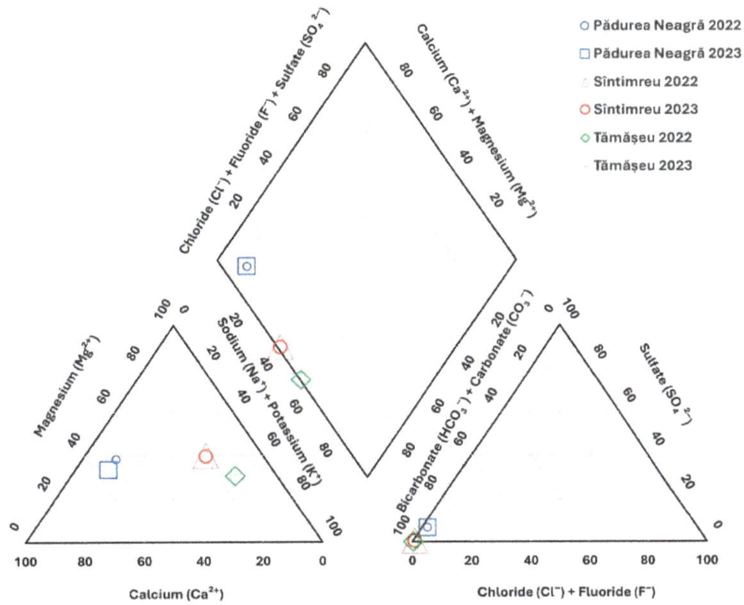

Figure 6. The Piper diagram of the studied natural mineral waters.

Mineral waters were classified according to the dominant constituents. In Figures 6 and 7 is the sum of cations $Ca^{2+} + Mg^{2+} > Na^{+} + K^{+}$, and the sum of anions $HCO_3^{-} + CO_3^{2-} > Cl^{-} + SO_4^{2-}$. The dominant cations are the alkaline-earth ones, and the dominant anion is bicarbonate.

As a general characteristic of the natural mineral waters analyzed in this study, according to Piper and Stiff representations (Figures 6 and 7), the hydrochemical fingerprint is as follows:

- The mineral water from Tămășeu is predominantly sodium bicarbonate (Na-HCO_3^{-}),
- The mineral water from Sîntimreu is mixed calcium-magnesian bicarbonate type (Ca-Mg-HCO_3^{-}),
- The mineral water from Pădurea Neagră is predominantly calcium bicarbonate (Ca-HCO_3^{-}).

The dominant hydrochemical type is specific to the aquifers in the three analyzed areas due to the fact that most of them are concentrated in reservoirs rich in carbonate. The mineralization of an underground water source can show variations within the specific limits of the lithology of the location area, and according to the literature, these variations can fall within the range of 15–25%.

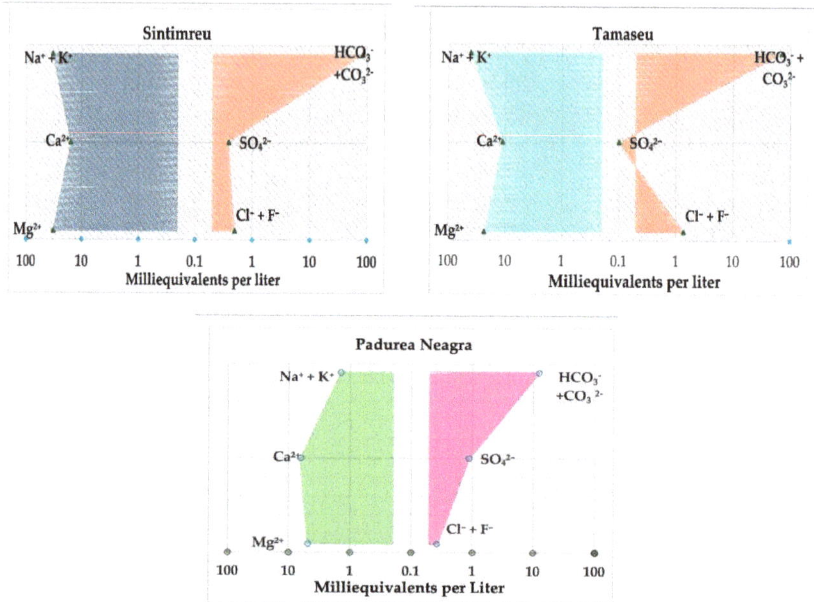

Figure 7. The Stiff diagram of the studied natural mineral waters.

3.3. Microbiological Characterization

In order to microbiologically characterize the natural mineral water samples from Tămășeu, Sîntimreu, and Pădurea Neagră, six indicators during the sampling campaign (2022–2023) were determined. As can be seen in Table 3, there are no pathogenic microorganisms that might represent a danger to human health in the studied samples. According to the national legislation for mineral waters, the reference values are 20 germs/mL at a temperature of 20–22 °C in 72 h and 5 germs/mL at a temperature of 37 °C in 24 h; therefore, the waters analyzed in this study comply with the quality conditions imposed by both drinking water and mineral-water-specific legislation.

Table 3. Microbiological indicators specific to mineral waters from Tămășeu, Sîntimreu, and Pădurea Neagră.

Microbiological Indicators	Unit	Results			HG 1020/2005 (Government Decision)	Law 458/2002
		Tămășeu	Sîntimreu	Pădurea Neagră		
The total colony count (CFU) at 22 °C	CFU/mL	No abnormal changes	No abnormal changes	No abnormal changes	100	100
The total colony count (CFU) at 37°C	CFU/mL	No abnormal changes	No abnormal changes	No abnormal changes	20	20
Total Coliform bacteria and Escherichia coli	CFU/250 mL	0	0	0	absent	0
Intestinal enterococci	CFU/250 mL	0	0	0	absent	0
Pseudomonas aeruginosa	CFU/250 mL	0	0	0	absent	0
Clostridium perfringens	CFU/50 mL	0	0	0	absent	0

3.4. Analysis of the Questionnaire

Below, we present the analysis of the questionnaire regarding *the residents' perception on the quality and effect on health of the natural mineral waters* from the local boreholes/springs under study (Tămășeu, Sîntimreu and Pădurea Neagră).

3.4.1. The Quality and the Effect of Mineral Waters on Health

Regarding the main source of drinking water of the inhabitants of the three localities and neighboring area who answered the questionnaire, statistical analysis shows that most of them (28%) use bottled water from the store, 20% of the respondents use the mineral water from the borehole in Sîntimreu, 18% use the mineral water spring from the holiday village Pădurea Neagră, and 18% use the water from the well dug in the yard of their house. Only 6% of the respondents use the mineral water from the borehole in Tămășeu village (Figure 8).

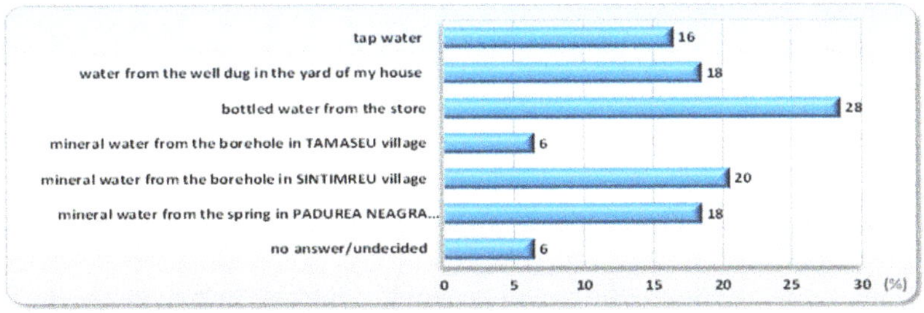

Figure 8. Distribution of respondents according to their main drinking water source.

The reason could lie in the fact that the borehole in Sîntimreu is easily accessible to the local population, being located near the center of the village and having a large water flow. On the other hand, the one in Tămășeu is located outside the village, about 500 m away, and the water flow is low. The use of water from the Tămășeu borehole in such a small percentage is also attributed, by some of those surveyed, to the gastrointestinal problems it causes (e.g., diarrhea). Many of the people questioned from the village of Tămășeu and the neighboring village of Parhida stated that they had not drunk mineral water from the borehole for several years. The current source of drinking water for those surveyed in Tămășeu village is either bottled water from the store or water from the local drinking water network (tap water). In Parhida, apart from the bottled water from the store, the respondents also use water from the borehole in the center of the village extracted from about 25 m deep (called "artesian" by the locals).

When asked what makes them choose a bottled water brand over local mineral water from spring/borehole when shopping, 38% of the respondents answered the taste, 24% the origin of the water source, 22% did not choose any answer, 16% the chemical composition of the water, and only 4% answered the advertisement. Some of them also specified the price as important in choosing the brand of bottled water.

Issues affecting the quality of mineral water from the local borehole/spring are viewed differently by the respondents. Most of them (46%) consider that the local mineral waters have no quality problems, 14% of the respondents believe that iron affects the water quality since it is at too high of a concentration, and 12% of them noted the taste or smell, respectively limestone (Figure 9). The fewest of the respondents consider salinity to be a problem (2%), solid particles in the water, or nitrates/nitrites (4%).

Regarding the frequency of mineral water consumption, most respondents (32%) drink mineral water daily from the local spring/borehole, 22% of them occasionally drink mineral water, 16% drank it once a month, 14% rarely drank it, and the fewest respondents drank it once every 2–3 days (2%) or once a week (6%) (Figure 10).

When asked how much mineral water they drink per day, most respondents (34%) did not give any answer, as many of them occasionally or rarely drink mineral water from the local borehole/spring. In total, 24% of the respondents consume more than 1 L/day, 16% about 1 L, and the rest of the respondents consume less than 1 L.

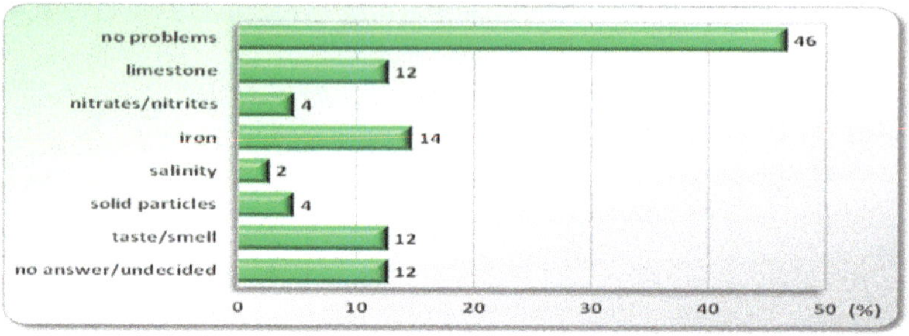

Figure 9. Respondents' perception on issues affecting the quality of mineral water.

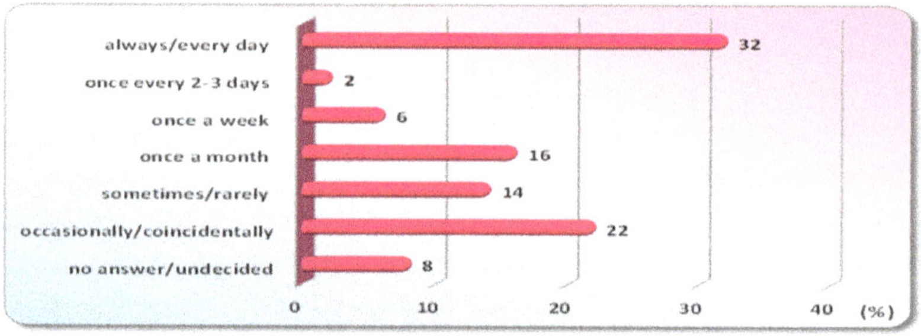

Figure 10. Distribution of respondents according to the frequency of mineral water consumption.

When asked how long they have been consuming mineral water from the local borehole/spring, most respondents (24%) said since childhood, 22% of respondents gave no answer, and 20% of them stated that they have been drinking mineral water for 10 years or less. Some respondents stated that they have been drinking the local mineral water for over 40, 50, or even 60 years.

When asked how long the mineral water stays clear, 38% of the respondents answered a few days, and 36% answered one day. Only 14% answered a few hours. After this period of time, deposits appear on the bottom of the container. Questioning the population, we found out that some locals from the village of Pădurea Neagră do not consume the mineral water from the spring in the holiday village precisely because of the deposits that occur on the bottom of the containers after a few days. As a result, they prefer to drink water from the village spring (spring water), which keeps its clarity and does not show deposits. The precipitation of some mineral salts is also present at the two boreholes (Tămășeu and Sîntimreu).

Asked if they add something to the mineral water in order not to change its color, taste or smell, the majority of respondents (70%) stated that they do not add anything, as they do not mind the organoleptic characteristics of the water. Additionally, 22% did not give any answer, and only 8% said they add lemon salt (or apple cider vinegar).

The majority of those surveyed (76%) believe that the mineral water from the local borehole/spring is much more hygienic/cleaner than the drinking water from the tap or the water from the well in the yard. Only 12% believe that it is not cleaner.

The majority of respondents (86%) stated that they do not suffer from any medical condition that requires them to drink more water than usual. Only 6% stated that they need to drink more water. For example, some of those surveyed consume more mineral water from the Sîmtimreu borehole when they have stomach aches.

The majority of those surveyed (70%) answered negatively when asked if they suffer/suffered from any acute condition for which the internal (drinking) or external (bathing) use of mineral water from the local borehole/spring made them feel an improvement in their health. Only 12% answered in the affirmative, mentioning that they suffer/suffered from conditions such as gastritis, gastric ulcers, flatulence, or liver conditions.

80% of the respondents answered negatively when asked if they suffer from any chronic condition for which the internal (drinking) or external (bathing) use of mineral water from the local borehole/spring makes them feel an improvement in their health. Only 8% answered affirmatively, declaring that the disease they suffer from is gastritis or gastric ulcers.

When asked how they feel after drinking mineral water from the local borehole/spring, 38% of respondents did not give any answer, 28% of them declared that mineral water gives them a feeling of well-being, good mood, and vitality (they do not feel melancholy, depressed, or angry), 20% of them answered that they felt more energetic, and 20% of them said that their digestive problems decreased (the activity of the digestive system improved). Smaller percentages of those surveyed show that their urinary tract problems (2%) and biliary disorders (2%) have decreased, or their cardiovascular system activity (2%) has improved (Figure 11). Some respondents stated that their liver conditions decreased. As shown above, some people questioned stated that they do not consume mineral water (especially the water from the Tămășeu borehole) because it causes them gastric disorders.

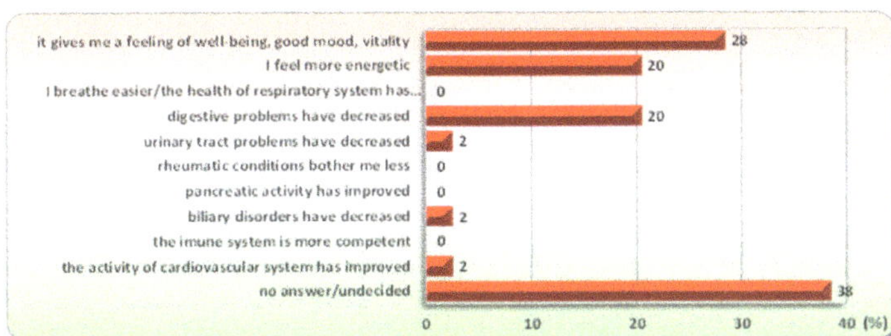

Figure 11. The effect on health perceived by the population after drinking mineral water.

Most of the surveyed people do not use the mineral water from the local borehole/spring in the kitchen for various reasons: 32% because the borehole/spring is too far and they only use the mineral water for drinking, 20% because the mineral water changes the taste of food/drinks, and 18% do not use it for other reasons (Figure 12). A total of 14% did not give any answer. Only 12% of respondents use the mineral water for food preparation, and a smaller percentage of them for the preparation of alcoholic (spritz, etc.) or non-alcoholic (juices, herbal infusions, etc.) drinks. The locals use the mineral water from the local borehole to make spritz, especially in Sîntimreu, where there are famous vineyards.

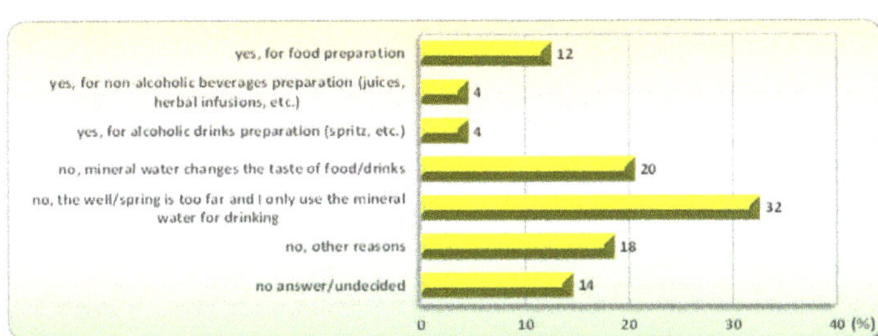

Figure 12. Distribution of respondents according to the use of mineral water in the kitchen.

Most of those surveyed (64%) drank mineral water from the local borehole/spring during the Covid pandemic (2020–2022), and 32% did not. Of the 64% respondents who affirmatively answered to the previous question, half (52.2%) stated that the mineral water did not help them not get infected with the SARS-CoV-2 virus or to recover faster, and only some of them (8.7%) stated that the water helped them. A total of 39.1% did not answer.

3.4.2. The Socio-Demographic Situation of the Respondents

Regarding *the socio-demographic situation* of the respondents, a fairly uniform distribution—in terms of their residence—between the three locations analyzed (Tămășeu, Sîntimreu, Pădurea Neagră) can be observed. Most respondents live in Pădurea Neagră and its surrounding localities (39%), followed by those in Sîntimreu and the neighboring villages (32%), then by those in Tămășeu and the neighboring localities (29%). The questionnaire was filled out by respondents from a total of 12 localities: two cities (Oradea and Marghita) and ten villages (Tămășeu, Parhida, Sălard, Sîntimreu, Mihai Bravu, Pădurea Neagră, Voivozi, Popești, Ciutelec, Mișca).

A fairly uniform gender distribution of those who answered the questionnaire can be observed (56% women and 44% men). Analyzing the age of the respondents (Figure 13), it can be seen that most fall into the age category 51–60 years (28%), followed by the 21–40 years category (24%), then the 41–50 years category (20%). Therefore, half of the respondents are aged between 41–60 years, and 20% of them are over 60 years old.

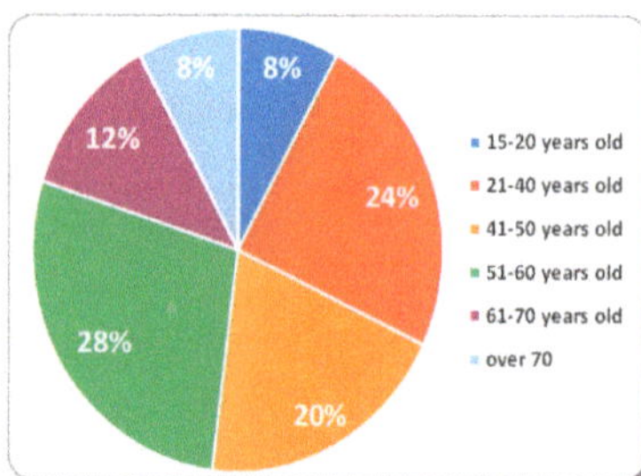

Figure 13. Distribution of respondents according to age.

Regarding the level of education of the respondents (Figure 14), 61.2% of them have a university education (22.4% master's studies, 20.4% bachelor studies, 18.4% doctoral studies), 28.6% have high school education, and 10.2% only have secondary school education. This proportion is explained by the fact that many respondents are educated young people or adults who filled out the questionnaire online. The elderly rural population (over 60 years old), with less education, does not use the Internet.

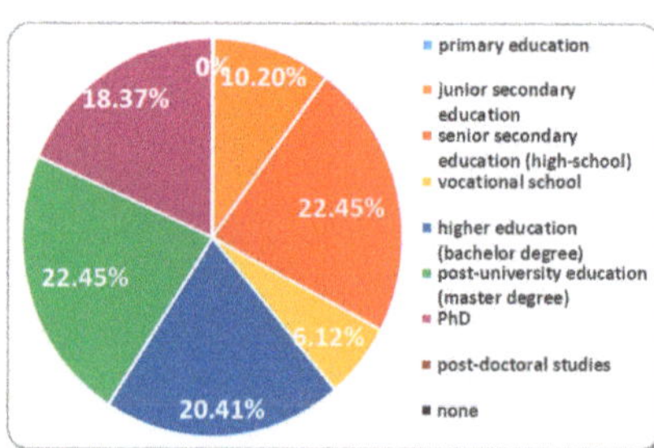

Figure 14. Distribution of respondents according to the level of education.

Most of those surveyed are permanent employees (69.4%), followed by retirees (14.3%). Temporary employees, pupils and students together represent 12.2% of those surveyed. Other social categories show very small percentages (administrators, farmers).

The face-to-face investigation helped us to report special cases, such as that of the mayor of Sălard commune, Mr. Nagy Miklós, who benefited greatly from the healing effect of the mineral water from the Sîntimreu borehole. He told us that in his youth, he underwent emergency surgery due to a perforated gastric ulcer, his medical condition being critical and having a minimal chance of survival. At the doctor's recommendation, he drank at least 2 L of mineral water/day from that borehole for about 3 months, and his health improved significantly. He stated that he was cured of gastric ulcers, and since then, he drinks that water daily.

When it is used as an internal cure (crenotherapy), the mineral water from the boreholes in Sîntimreu and Tămășeu initially causes gastrointestinal disorders (e.g., diarrhea) for about a week. After that, this effect disappears. It is possible that this effect is a result of the detoxification of the body, which the Lithinia Company also talks about on its website [88].

Several locals from Sîntimreu told us that there are many people over 90–95 years old (even over 100 years old—declared by the mayor of the Sălard commune) in their village. People from this area are long-lived, and one of the reasons might be drinking the local mineral water (another reason for their longevity could be drinking the local wine). The inhabitants of Sîntimreu village do not want the mineral water to be bottled. They do not want the local mineral water to be bottled and become the property of some company that would limit their access to water, meaning they would have to buy it from the store.

This study is not exhaustive and has some limitations, such as the sampling of a small number of samples, and the inability, at this time, to extend the study to the determination of other characteristics specific to natural mineral waters. Regarding the public opinion on the potential impact of these waters on health, we encountered some difficulties in communicating with the inhabitants of various ethnicities in the studied area. We also note the difficultness of speaking directly to all study participants. However, this study represents an important step for identifying new opportunities for the exploitation and management of natural mineral water sources in Bihor County, in the context of a sustainable management of this resource. In addition, through the questionnaire and discussions with the residents and local authorities, we think that we have taken a relevant step in the process of raising awareness in the population regarding the benefits that these natural mineral waters can bring for the therapy of various ailments.

The mineral waters from Tămășeu, Sîntimreu and Pădurea Neagră have microbiological and physical–chemical characteristics that comply with the current legislation specific to both drinking water and natural mineral waters, and our future research will aim to evaluate all the characteristics specified in HG 1020/2005 for including the water sources from Tămășeu, Sîntimreu and Pădurea Neagră in the category of natural mineral water sources recognized in Romania.

4. Conclusions

The microbiological and physical-chemical characteristics studied give us the opportunity to state that the waters from Tămășeu, Sîntimreu, and Pădurea Neagră represent a potential source of natural mineral water. The analysis of the questionnaire helped us to reveal people's opinion about these waters. Many respondents consider that the studied waters do not have quality problems and have beneficial effects, especially regarding gastrointestinal ailments, which could pave the way for an in-depth medical study on the therapeutic potential of these waters; this is one of the objectives of our future studies. Obtaining additional data for a detailed chemical characterization of these natural mineral water sources is another future goal. Taking into account the medical studies that show the high bioavailability of minerals from various mineral waters, we can consider these waters as accessible sources of calcium, magnesium, sodium, iron, bicarbonate, and sulfates with beneficial effects in the prevention or treatment of various diseases. A particular case is the mineral water from the Tămășeu borehole with a high lithium content, which might be recommended as complementary therapy in the treatment of psychiatric disorders. In addition, the water from Sîntimreu, due to its hydrogen sulphide content, could have beneficial effects for various pathologies. Correlated with historical data, our results indicate the hydraulic, nutritional, and therapeutic potential of these natural mineral waters, which is a benefit for the local population. The present study could increase the interest of the scientific community for the in-depth study of these waters and their inclusion on the list of natural mineral water resources in Romania. Located in an area with multiple types of hydromineral and hydrothermomineral resources with varied physical–chemical characteristics, Bihor County can offer new opportunities for their hydrological and therapeutic exploitation.

Author Contributions: Conceptualization, R.L., E.P., E.Ș., G.S. and A.-P.C.; Methodology, R.L., E.P., E.Ș., G.S. and A.-P.C.; Validation, R.L., E.P., E.Ș., G.S. and A.-P.C.; Investigation, E.P., E.Ș., G.S., A.-P.C. and R.L.; Writing—original draft preparation, R.L., E.P., E.Ș., G.S. and A.-P.C.; Writing—review and editing, R.L., E.P., E.Ș., G.S. and A.-P.C. All authors have read and agreed to the published version of the manuscript.

Funding: Publishing of this research was funded by University of Oradea.

Institutional Review Board Statement: Not applicable.

Informed Consent Statement: Not applicable.

Data Availability Statement: The data sets supporting the conclusions made are included in this article.

Conflicts of Interest: The authors declare no conflict of interest.

References

1. Rani, B.; Maheshwari, R.; Garg, A.; Prasad, M. Bottled Water—A Global Market Overview. *Bull. Environ. Pharmacol. Life Sci* **2012**, *1*, 1–4.
2. Bodor, K.; Bodor, Z.; Szép, A.; Szép, R. Classification and Hierarchical Cluster Analysis of Principal Romanian Bottled Mineral Waters. *J. Food Compos. Anal.* **2021**, *100*, 103903. [CrossRef]
3. Whelton, A.J.; Dietrich, A.M.; Burlingame, G.A.; Schechs, M.; Duncan, S.E. Minerals in Drinking Water: Impacts on Taste and Importance to Consumer Health. *Water Sci. Technol.* **2007**, *55*, 283–291. [CrossRef] [PubMed]
4. *Report on the Sectoral Investigation on the Exploitation Market of Natural Mineral Water Resources in Romania Initiated by the Order of the President of the Competition Council No. 177/29.02.2016*; European Commission: Brussels, Belgium, 2016.
5. Mititelu, M.; Morosan, E.; Udeanu, D. The Therapeutic Value of Mineral Waters. *Farmacist.ro* **2018**, *184*, 23–36. [CrossRef]
6. Feru, A. *Ghidul Apelor Minerale Naturale (Guide to Natural Mineral Waters)*; APEMIN Patronatul Apelor Minerale: Bucharest, Romania, 2012.
7. H.G. Nr. 1020/2005 Pentru Aprobarea Normelor Tehnice de Exploatare Și Comercializare a Apelor Minerale Naturale (Government Decision. No. 1020/2005 for the Approval of the Technical Norms for Exploitation and Commercialization of Natural Mineral Waters). Available online: http://www.namr.ro/wp-content/uploads/2013/02/hotarare1020.pdf (accessed on 13 June 2023).
8. Council Directive 80/777/EEC of 15 July 1980 on the Approximation of the Laws of the Member States Relating to the Exploitation and Marketing of Natural Mineral Waters. Available online: Http://Data.Europa.Eu/Eli/Dir/1980/777/Oj (accessed on 23 February 2023).
9. Teodoreanu, E.; Gaceu, O. *Turismul Balneoclimatic În România (Balneoclimatic Tourism in Romania)*; Editura Universității din Oradea: Oradea, Romania, 2013; ISBN 978-606-10-1083-7.
10. Moss, G.A. Water and Health: A Forgotten Connection? *Perspect. Public Health* **2010**, *130*, 227–232. [CrossRef] [PubMed]
11. Dumitrescu, M.; Iliescu, M.G.; Mazilu, L.; Micu, S.I.; Suceveanu, A.P.; Voinea, F.; Voinea, C.; Stoian, A.P.; Suceveanu, A.-I. Benefits of Crenotherapy in Digestive Tract Pathology. *Exp. Ther. Med.* **2022**, *23*, 122. [CrossRef] [PubMed]
12. Munteanu, C. *Ape Minerale Terapeutice (Therapeutic Mineral Waters)*; Editura Balneară: Bucharest, Romania, 2013.
13. Mennuni, G.; Serio, A.; Fontana, M.; Nocchi, S.; Costantino, C.; Tanzi, G.; Stornelli, G.; Fraioli, A. Prevention and Treatment of Nephrolithiasis: A Review on the Role of Spa Therapy. *Clin. Ter.* **2015**, *166*, e344–e356. [CrossRef]
14. Lo, J.A.; Kim, J.S.; Jo, M.J.; Cho, E.J.; Ahn, S.Y.; Ko, G.J.; Kwon, Y.J.; Kim, J.E. Impact of Water Consumption on Renal Function in the General Population: A Cross-Sectional Analysis of KNHANES Data (2008-2017). *Clin. Exp. Nephrol.* **2021**, *25*, 376–384. [CrossRef]
15. Siener, R.; Jahnen, A.; Hesse, A. Influence of a Mineral Water Rich in Calcium, Magnesium and Bicarbonate on Urine Composition and the Risk of Calcium Oxalate Crystallization. *Eur. J. Clin. Nutr.* **2004**, *58*, 270–276. [CrossRef]

16. Costantino, M.; Izzo, V.; Conti, V.; Manzo, V.; Guida, A.; Filippelli, A. Sulphate Mineral Waters: A Medical Resource in Several Disorders. *J. Tradit. Complement. Med.* **2020**, *10*, 320–326. [CrossRef]
17. Nimmagadda, S.V.; Schmale, I.L.; Man, L.-X. Crenotherapy as a Complementary and Integrative Treatment for Chronic Rhinosinusitis: A Systematic Review and Discussion of Current Evidence Limitations. *Am. J. Rhinol. Allergy* **2022**, *36*, 529–538. [CrossRef]
18. Cantone, E.; Maione, N.; Di Rubbo, V.; Esposito, F.; Iengo, M. Olfactory Performance after Crenotherapy in Chronic Rhinosinusitis in the Elderly. *Laryngoscope* **2015**, *125*, 1529–1534. [CrossRef]
19. Carbajo, J.M.; Maraver, F. Sulphurous Mineral Waters: New Applications for Health. *Evid.-Based Complement. Altern. Med.* **2017**, *2017*, 8034084. [CrossRef] [PubMed]
20. Holwerda, K.M.; Karumanchi, S.A.; Lely, A.T. Hydrogen Sulfide: Role in Vascular Physiology and Pathology. *Curr. Opin. Nephrol. Hypertens.* **2015**, *24*, 170–176. [CrossRef] [PubMed]
21. Feliers, D.; Lee, H.J.; Kasinath, B.S. Hydrogen Sulfide in Renal Physiology and Disease. *Antioxid. Redox Signal.* **2016**, *25*, 720–731. [CrossRef] [PubMed]
22. Costantino, M.; Conti, V.; Corbi, G.; Filippelli, A. Hydropinotherapy with Sulphurous Mineral Water as Complementary Treatment to Improve Glucose Metabolism, Oxidative Status, and Quality of Life. *Antioxidants* **2021**, *10*, 1773. [CrossRef]
23. Buret, A.G.; Allain, T.; Motta, J.-P.; Wallace, J.L. Effects of Hydrogen Sulfide on the Microbiome: From Toxicity to Therapy. *Antioxid. Redox Signal.* **2022**, *36*, 211–219. [CrossRef] [PubMed]
24. Ciurba, A.-P.; Haidu, I.; Ianc, D. Administrative Aspects Regarding the Valorisation of Geothermal Waters for Balneological Purposes in Bihor County, Romania. *Sustainability* **2023**, *15*, 10320. [CrossRef]
25. Tătar, F.C.; Dincă, I.; Linc, R.; Stupariu, M.; Bucur, L.; Stașac, M.; Nistor, S. Oradea Metropolitan Area as a Space of Interspecific Relations Triggered by Physical and Potential Tourist Actvities. *Sustainability* **2023**, *15*, 3136. [CrossRef]
26. Gavriliuc, R.; Rosca, M.; Cucueteanu, D. Geothermal Energy Use, Country Update for Romania. European Geothermal Congress. Available online: https://europeangeothermalcongress.eu/wp-content/uploads/2019/07/CUR-23-Romania.pdf (accessed on 15 November 2022).
27. Antal, C.; Setel, A. Exploitability of Geothermal Resources in Pannonia Depression. *Analele Univ. Din Oradea Fasc. Energetică* **2009**, *15*, 154–157.
28. Roba, C.; Olah, Ş. Încălzirea Serelor Din Livada (Județul Bihor) Cu Apă Geotermală (Heating the Greenhouses in Livada (Bihor County) with Geothermal Water). In *Environment and Progress*; Editura Efes: Cluj-Napoca, Romania, 2007; pp. 391–395.
29. Orășeanu, I. *Hidrogeologia Carstului Din Munții Apuseni (Karst Hydrogeology of the Apuseni Mountains)*; Editura Belvedere: Oradea, Romania, 2016.
30. Kahane, S.; Stoicescu, S.; Demayo, B.; Leonte, E. *Aspecte Experimentale Privind Acţiunea Diuretică a Apei Minerale de La Tinca, În Studii Şi Cercetări de Balneologie Şi Fizioterapie (Experimental Aspects Regarding the Diuretic Action of the Mineral Water from Tinca, in Balneology and Physiotherapy Studies and Research)*; Editura Medicală: Bucharest, Romania, 1965; Volume 7.
31. Orășeanu, I. *Consideraţii Privind Stabilitatea Chimică a Apei Unor Izvoare Din Zona Stâna de Vale (Masivul Vlădeasa) (Considerations Regarding the Chemical Stability of the Water of Some Springs in the Stâna de Vale Area (Vlădeasa Massif))*; 100 de ani de Hidrogeologie Modernă în România; edition II; Belvedere Publishing House: Oradea, Romania, 2000.
32. Szabó, J. *Ismeretlen Érmellék—Irasban És Kepben*; Varadinum Script: Oradea, Romania, 2018.
33. Pricăjan, A. *Apele Minerale Şi Termominerale Din România (Mineral and Thermal Mineral Waters from Romania)*; Editura Tehnică: Bucharest, Romania, 1972.
34. Berlescu, E.; Chiorean, T.; Cociașiu, E.; Constantinescu, D.; Dinculescu, T.; Opreanu, I.; Stoia, I.; Șdic, I.; Ștefănescu, S.; Suciu, D.; et al. *Indicații Și Contraindicații Pentru Tratamente În Stațiunile Balneo-Climaterice Din Republica Populară Romînă (Indications and Contraindications for Treatments in Balneo-Climate Resorts in the Romanian People's Republic)*; Institutul de Balneologie și Fizioterapie: Bucharest, Romania, 1960.
35. Binștoc, O. *Cura Cu Factori Naturali În Stațiunile Balneoclimaterice Și La Domiciliu (Cure with Natural Factors in Spas and at Home)*; Editura Medicală: Bucharest, Romania, 1962.
36. Aniței, L.; Băican, I.; Berlescu, E.; Chirvasie, L.; Conu, A.; Degeratu, C.; Dumitrescu, S.; Florian, M.; Samson Gheorghievici, S.; Gavrilescu, M.; et al. *Cura Balneoclimatică. Indicații Și Contraindicații (Balneoclimatic Cure. Indications and Contraindications)*; Editura Medicală: Bucharest, Romania, 1986.
37. Monitorul Oficial 103 Din 07.02.2023, Lista Apelor Minerale Naturale Recunoscute În România (Official Gazette 103 of 07.02.2023, List of Natural Mineral Waters Recognized in Romania). Available online: https://Lege5.Ro/Gratuit/Geztgmbwgu2te/Ordinul-Nr-116-2023-Privind-Aprobarea-Listei-Apelor-Minerale-Naturale-Recunoscute-in-Romania (accessed on 19 December 2022).
38. Directive 2009/54/EC of the European Parliament and of the Council of 18 June 2009 on the Exploitation and Marketing of Natural Mineral Waters. Available online: https://Eur-Lex.Europa.Eu/LexUriServ/LexUriServ.Do?Uri=OJ:L:2009:164:0045:0058:EN:PDF (accessed on 14 March 2023).
39. H.G. Nr. 532 Din 2 Iunie 2010 Privind Modificarea Si Completarea Normelor Tehnice de Exploatare Si Comercializare a Apelor Minerale Naturale, Aprobate Prin HG Nr. 1020/2005, Publicat În Monitorul Oficial Nr. 400 Din 16 Iunie 2010 (Government Decision No. 532 of June 2, 2010 Regarding the Modification and Completion of the Technical Norms for Exploitation and Commercialization of Natural Mineral Waters, Approved by HG No. 1020/2005, Published in the Official Gazette NO. 400 of June 16, 2010). Available online: https://Legislatie.Just.Ro/Public/DetaliiDocumentAfis/11958 (accessed on 20 November 2022).

40. Tapias, J.C.; Melián, R.; Sendrós, A.; Font, X.; Casas, A. Geochemical Characterisation and Health Concerns of Mineral Bottled Waters in Catalonia (North-Eastern Spain). *Water* **2022**, *14*, 3581. [CrossRef]
41. *SR EN ISO 19458/2007*; Water Quality. Sampling for Microbiological Analysis Asociatia de Standardizare Din România: București (Mohr Method). Romanian Standardization Association: Bucharest, Romania, 2007.
42. *SR ISO 10523: 2012*; Water Quality. Determination of PH. Asociatia de Standardizare Din România: Bucharest, Romania, 2012.
43. *SR EN 27888:1997*; Water Quality. Determination of Electrical Conductivity. Asociatia de Standardizare Din România (Romanian Standardization Association, Bucharest): Bucharest, Romania, 1997.
44. *SR EN ISO 7899-2:2002*; Water Quality. Detection and Enumeration of Intestinal Enterococci—Part 2: Membrane Fil-Tration Method. Asociatia de Standardizare Din România (Romanian Standardization Association): Bucharest, Romania, 2002.
45. *ISO 9964-3:1993*; Water Quality—Determination of Sodium and Potassium—Part 3: Determination of Sodium and Potassium by Flame Emission Spectrometry. ISO: Geneva, Switzerland.
46. *SR EN ISO 11885:2007*; Water Quality—Determination of Selected Elements by Inductively Coupled Plasma Optical Emission Spectrometry (ICP-OES). ISO: Geneva, Switzerland, 2007.
47. *SR 13315:1996/C91:2008*; Water quality. Determination of iron content by flame atomic absorption spectrometry. Asociația de Standardizare Din România (Romanian Standardization Association: Bucharest, Romania, 2008.
48. *SR 8662/2-1996*; Water Quality. Determination of Manganese Content. Atomic Absorption Spectrometric Method. Available online: https://Magazin.Asro.Ro/Ro/Standard/11710 (accessed on 4 November 2022).
49. *SR ISO 9297-2001*; Water Quality. Determination of Chloride Content. Titration with Silver Nitrate Using Chromate as Indicator (Mohr Method). Romanian Standardization Association: Bucharest, Romania, 2001.
50. *US.EPA 375.4-1978*; Methods for Chemical Analysis of Water and Wastes. National Institutes of Health: Bethesda, MD, USA. Available online: https://Www.Wbdg.Org/FFC/EPA/EPACRIT/Epa600_4_79_020.Pdf (accessed on 4 November 2022).
51. *SR ISO 7890-3/2000*; Calitatea Apei: Determinarea Conţinutului de Azotaţi: Standard Român: [Data Intrării În Vigoare] Martie 2000. Partea 3. Metoda Spectrometrică Cu Acid Sulfosalicilic (Water Quality: Determination of Nitrate Content: Romanian Standard: [Date of Entry into Force] March 2000. Part 3. Sulfosalicylic Acid Spectrometric Method). ISO: Geneva, Switzerland, 2000.
52. *SR EN 26777:2002/C91:2006*; Water Quality. Determination of Nitrite Content. The Method by Molecular Absorption Spectrometry. Organismul Național de Standardizare: Bucharest, Romania, 2002. Available online: https://Magazin.Asro.Ro/Ro/Standard/113232 (accessed on 28 October 2022).
53. *SR ISO 9963-2:1997/A99:2002*; Water Quality. Alkalinity Determination. Part 1: Determination of Total and Permanent Alkalinity. Romanian Standardization Association: Bucharest, Romania, 2002.
54. *SR 4450:1997*; Water Quality. Determination of Carbon Dioxide (Volumetric Method). Asociația de Standardizare Din România (Romanian Standardization Association): Bucharest, Romania, 1997.
55. *SR 7510-1997*; Water Quality. Determination of Sulphide Content (Iodometric Method). Asociația de Standardizare Din România (Romanian Standardization Association): Bucharest, Romania, 1997.
56. *STAS 9187-84*; Surface Water, Groundwater and Wastewater. Residue Determination. Asociatia de Standardizare Din România: (Romanian Standardization Association): Bucharest, Romania, 1984.
57. Piper and Stiff. Available online: https://Halfordhydrology.Com/Piper-and-Stiff/ (accessed on 23 April 2023).
58. Leclerc, H.; Moreau, A. Microbiological Safety of Natural Mineral Water. *FEMS Microbiol. Rev.* **2002**, *26*, 207–222. [CrossRef]
59. *SR EN ISO 6222:2004*; Water Quality. Enumeration of Culturable Micro-Organisms. Colony Count by Inoculation in a Nutrient Agar Culture Medium. Asociatia de Standardizare Din România (Romanian Standardization Association): Bucharest, Romania, 2004.
60. *SR EN ISO 9308:1:2015/A:2017*; Water Quality—Enumeration of Escherichia Coli and Coliform Bacteria—Part 1: Membrane Filtration Method for Waters with Low Bacterial Background Flora. Asociatia de Standardizare Din România (Romanian Standardization Association): Bucharest, Romania, 2017.
61. *SR EN ISO 16266/2008*; Water Quality. Detection and Enumeration of Pseudomonas Aeruginosa. Method by Mem-Brane Filtration. Asociatia de Standardizare Din România (Romanian Standardization Association): Bucharest, Romania, 2008.
62. *SR EN ISO 14189:2016*; Water Quality. Enumeration of Clostridium Perfringens. Method Using Membrane Filtration. Asociatia de Standardizare Din România (Romanian Standardization Association): Bucharest, Romania, 2016.
63. Madani, D.; Tarai, N.; Tir, C.; Benchikh, A. The State of Awareness about the Phenomenon of Desertification by the Inhabitants of the Region of Khoubana, M'sila, Algeria. *Nat. Resour. Sustain. Dev.* **2023**, *13*, 167–178. [CrossRef]
64. Wang, W.-C.; Lin, C.-H. A Model for Sustainable Tourism Development of Hot Spring Destinations Following Poverty Alleviation: Understanding the Tourists' Perspective. *Sustainability* **2021**, *13*, 9856. [CrossRef]
65. Cristescu, C.; Negreș, S.; Suciu, M.; Voicu, A.; Buda, V.; Suciu, L.; Proks, M.; Voicu, M. Study Regarding the Parents'use of Self-Medication among Children under 12 Years Old. *Farmacia* **2018**, *66*, 5. [CrossRef]
66. Roulet, L.; Asseray, N.; Foucher, N.; Potel, G.; Lapeyre-Mestre, M.; Ballereau, F. A Questionnaire to Document Self-Medication History in Adult Patients Visiting Emergency Departments. *Pharmacoepidemiol. Drug Saf.* **2013**, *22*, 151–159. [CrossRef]
67. Jain, P.; Sachan, A.; Singla, R.K.; Agrawal, P. Statistical Study on Self Medication Pattern in Haryana, India. *Indo Glob. J. Pharm. Sci.* **2012**, *2*, 21–35. [CrossRef]

68. Mihăilă, N. *Cercetări Hidrogeologice Și Hidrochimice În Cuprinsul Foilor Oradea—Aleșd (Hydrogeological and Hydrochemical Researches within the Oradea—Aleșd Sheets)*; Institutul Geologic: Bucharest, Romania, 1971.
69. Nicula, M.A. Caracteristici Geochimice Ale Apelor Termale Și Minerale Din Munții Apuseni (Geochemical Characteristics of Thermal and Mineral Waters from the Apuseni Mountains), Rezumatul Tezei de Doctorat. PhD Thesis, Universitatea Babes-Bolyai, Cluj-Napoca, Romania, 2021. Available online: https://www.frontiersin.org/articles/10.3389/feart.2021.648179/full (accessed on 14 November 2022).
70. Strategia de Dezvoltare Locală 2014–2020 a Unității Administrativ-Teritoriale Sălard (Local Development Strategy 2014–2020 of the Sălard Administrative-Territorial Unit). Available online: https://www.cjbihor.ro/wp-content/uploads/2021/11/Strategie_Bihor_2021-2027_final_pentru_consultare_publica.pdf (accessed on 4 November 2022).
71. Strategia de Dezvoltare Locală Pentru Perioada 2021–2027 a Unității Administrativ-Teritoriale Sălard (The Local Development Strategy for the Period 2021–2027 of the Sălard Administrative-Territorial Unit). Available online: https://www.google.com.hk/url?sa=t&rct=j&q=&esrc=s&source=web&cd=&cad=rja&uact=8&ved=2ahUKEwis{-}{-}-0k5OCAxVzywIHHXosD3oQFnoECBoQAQ&url=https%3A%2F%2Fwww.cjbihor.ro%2Fwp-content%2Fuploads%2F2021%2F02%2FStrategie_Bihor_2021-2027_aprobata_CJ_31_martie.pdf&usg=AOvVaw3zrtBoyRoaOXkAzCdNy4Fk&opi=89978449 (accessed on 4 November 2022).
72. Stoicescu, C. *Farmacodinamia Apelor Minerale de Cură Internă Din România (The Pharmacodynamics of Internal Cure Mineral Waters from Romania)*; Editura Academiei: Bucharest, Romania, 1982.
73. Vernesu, M. *Apele Minerale (Mineral Waters)*; Editura Tehnică: Bucharest, Romania, 1988.
74. Vannucci, L.; Fossi, C.; Quattrini, S.; Guasti, L.; Pampaloni, B.; Gronchi, G.; Giusti, F.; Romagnoli, C.; Cianferotti, L.; Marcucci, G.; et al. Calcium Intake in Bone Health: A Focus on Calcium-Rich Mineral Waters. *Nutrients* **2018**, *10*, 1930. [CrossRef]
75. Heaney, R.P. Absorbability and Utility of Calcium in Mineral Waters. *Am. J. Clin. Nutr.* **2006**, *84*, 371–374. [CrossRef]
76. Morr, S.; Cuartas, E.; Alwattar, B.; Lane, J.M. How Much Calcium Is in Your Drinking Water? A Survey of Calcium Concentrations in Bottled and Tap Water and Their Significance for Medical Treatment and Drug Administration. *HSS J.* **2006**, *2*, 130–135. [CrossRef]
77. Wynn, E.; Krieg, M.-A.; Aeschlimann, J.-M.; Burckhardt, P. Alkaline Mineral Water Lowers Bone Resorption Even in Calcium Sufficiency: Alkaline Mineral Water and Bone Metabolism. *Bone* **2009**, *44*, 120–124. [CrossRef]
78. Garzon, P.; Eisenberg, M.J. Variation in the Mineral Content of Commercially Available Bottled Waters: Implications for Health and Disease. *Am. J. Med.* **1998**, *105*, 125–130. [CrossRef]
79. Jiang, L.; He, P.; Chen, J.; Liu, Y.; Liu, D.; Qin, G.; Tan, N. Magnesium Levels in Drinking Water and Coronary Heart Disease Mortality Risk: A Meta-Analysis. *Nutrients* **2016**, *8*, 5. [CrossRef]
80. Cîmpean, C.D. O Nouă Metodă de Analiză Informațională a Apelor Minerale Medicinale (A New Method of Informational Analysis of Medicinal Mineral Waters), Rezumatul Tezei de Doctorat. PhD Thesis, UMF Carol Davila, Bucharest, Romania, 2020. Available online: https://Umfcd.Ro/Wpcontent/Uploads/2020/TEZA_DOCTORAT/CIMPEAN_I_CRISTINA_DANIELA/TEZA-Rezumat-Cimpean-Cristina.Pdf (accessed on 15 October 2022).
81. Seidel, U.; Baumhof, E.; Hägele, F.A.; Bosy-Westphal, A.; Birringer, M.; Rimbach, G. Lithium-Rich Mineral Water Is a Highly Bioavailable Lithium Source for Human Consumption. *Mol. Nutr. Food Res.* **2019**, *63*, 1900039. [CrossRef]
82. Linc, R.; Ilieș, D. Les Modifications Anthropiques Du Paysage Dans Le Bassin Hydrographique de Barcau (Nord-Ouest de La Roumanie). *Riv. Geogr. Ital.* **2000**, *107*, 3–16.
83. Labenz, J.; Anschütz, M.; Walstab, J.; Wedemeyer, R.-S.; Wolters, H.; Schug, B. Heartburn Relief with Bicarbonate-Rich Mineral Water: Results of the Randomised, Placebo-Controlled Phase-III Trial STOMACH STILL. *BMJ Open Gastroenterol.* **2023**, *10*, e001048. [CrossRef] [PubMed]
84. Capurso, A.; Solfrizzi, V.; Panza, F.; Mastroianni, F.; Torres, F.; Del Parigi, A.; Colacicco, A.M.; Capurso, C.; Nicoletti, G.; Veneziani, B.; et al. Increased Bile Acid Excretion and Reduction of Serum Cholesterol after Crenotherapy with Salt-Rich Mineral Water. *Aging* **1999**, *11*, 273–276. [CrossRef]
85. Crespo, P.-V.; Campos, F.; Leal, M.; Maraver, F. Effects of Sodium Chloride-Rich Mineral Water on Intestinal Epithelium. Experimental Study. *Int. J. Environ. Res. Public Health* **2021**, *18*, 3261. [CrossRef] [PubMed]
86. Mennuni, G.; Petraccia, L.; Fontana, M.; Nocchi, S.; Stortini, E.; Romoli, M.; Esposito, E.; Priori, F.; Grassi, M.; Geraci, A.; et al. The Therapeutic Activity of Sulphate-Bicarbonate-Calcium-Magnesiac Mineral Water in the Functional Disorders of the Biliary Tract. *Clin. Ter.* **2014**, *165*, e346–e352. [CrossRef]
87. Munteanu, C.; Munteanu, D.; Onose, G. Hydrogen Sulfide (H2S)-Therapeutic Relevance in Rehabilitation and Balneotherapy Systematic Literature Review and Meta-Analysis Based on the PRISMA Paradigm. *Balneo PRM Res. J.* **2021**, *12*, 176–195. [CrossRef]
88. Detoxification with Lithia Spring Water—Cleansing the Body. Available online: https://Lithinia.Ro/Detoxifiere-Cu-Apa-de-Izvor-Lithia-Curatarea-Organismului/ (accessed on 10 December 2022).

sustainability

MDPI

Article

Oradea's Cultural Event Management: The Impact of the 'Night of the Museums' on Tourist Perception and Destination Brand Identity

Said Mohamed Elhosiny [1], Thowayeb H. Hassan [1,2,*], Ioana Josan [3], Amany E. Salem [1,2], Mostafa A. Abdelmoaty [4], Grigore Vasile Herman [3,*], Jan Andrzej Wendt [3,5], Bekzot Janzakov [6], Hassan Marzok Elsayed Mahmoud [1,7] and Magdy Sayed Abuelnasr [1]

1 Social Studies Department, College of Arts, King Faisal University, Al Ahsa 31982, Saudi Arabia; safify@kfu.edu.sa (S.M.E.); asalem@kfu.edu.sa (A.E.S.); hmahmoud@kfu.edu.sa (H.M.E.M.); mabouelnasser@kfu.edu.sa (M.S.A.)

2 Tourism Studies Department, Faculty of Tourism and Hotel Management, Helwan University, Cairo 12612, Egypt

3 Department of Geography Tourism and Territorial Planning, University of Oradea, 410087 Oradea, Romania; ijosan@uoradea.ro (I.J.); jan.wendt@ug.edu.pl (J.A.W.)

4 StatisMed for Statistical Analysis Services, Giza 12573, Egypt; mostafa@statismed.com

5 Institute of Socio-Economic Geography and Spatial Management, University of Gdańsk, 4 Bażyńskiego Street, 80309 Gdańsk, Poland

6 Department of Food and Agricultural Economics, Faculty of Economics, Samarkand Branch of Tashkent State University of Economics, Samarkand 140103, Uzbekistan; becajaja@gmail.com

7 College of Arts, Minia University, Minia 61519, Egypt

* Correspondence: thassan@kfu.edu.sa (T.H.H.); gherman@uoradea.ro (G.V.H.); Tel.: +966-54-029-4550 (T.H.H.)

Citation: Elhosiny, S.M.; Hassan, T.H.; Josan, I.; Salem, A.E.; Abdelmoaty, M.A.; Herman, G.V.; Wendt, J.A.; Janzakov, B.; Mahmoud, H.M.E.; Abuelnasr, M.S. Oradea's Cultural Event Management: The Impact of the 'Night of the Museums' on Tourist Perception and Destination Brand Identity. *Sustainability* **2023**, *15*, 15330. https://doi.org/10.3390/su152115330

Academic Editor: Colin Michael Hall

Received: 4 September 2023
Revised: 17 October 2023
Accepted: 24 October 2023
Published: 26 October 2023

Abstract: Understanding how event quality influences destination identity and tourist satisfaction is crucial for creating memorable experiences and fostering positive perceptions of a destination. The 'Night of the Museums' event in Oradea, Romania, offers a unique cultural experience, but little research has been conducted to understand its impact on destination perception and tourist satisfaction. This study examines the relationship between event quality, destination identity, tourist satisfaction, and behavioral intentions within the context of Oradea's Night of the Museums event. Four domains of event quality (visit quality, interaction quality, outcome quality, and physical environment quality) are assessed for their impact on destination identity and tourist satisfaction. Results indicated that visit quality, interaction quality, and outcome quality significantly predicted destination identity, whereas interaction quality and the quality of the physical environment significantly influenced tourist satisfaction. Furthermore, positive event experiences increased the likelihood of tourists recommending the destination to others, which in turn significantly predicted the intention to revisit. These findings have implications for event organizers, destination management organizations, and policymakers seeking to enhance event quality, promote positive destination identity, and cultivate tourist satisfaction, ultimately leading to increased recommendations and revisitation.

Keywords: event quality; destination identity; tourist satisfaction; intention to recommend; intention to revisit

1. Introduction

Numerous tourists who wish to experience a destination's art, history, and culture visit museums, which are cultural institutions [1]. In recent years, numerous museums have created cultural events incorporating multimedia technologies, such as light and sound shows, to provide visitors with an engaging experience [2]. By attracting large crowds, especially younger generations, and families, these cultural events at museums can significantly impact a destination's tourism movement [1]. For instance, many people attend the Louvre Museum's annual 'Nuit Blanche' event, which features light installations, dance and musical performances, and interactive exhibits. The quality of the cultural events and experiences offered by the museums is crucial to attracting more tourists [2–4]. High levels of visitor satisfaction and enjoyment can result from highly interactive events, employ cutting-edge technologies such as stunning visual effects, and offer an exciting atmosphere [1,5]. For instance, the 'Sound and Light' show at the Pyramids of Giza uses laser projections, surround sound, and a dramatic retelling of ancient Egyptian history to captivate audiences [4]. Such an immersive experience is praised by visitors for its high levels of entertainment and emotional impact [5].

When tourists have a meaningful and memorable experience at a cultural event, they are more likely to return and recommend the museum to others [6]. There is an increase in repeat visitation and word-of-mouth referrals at museums that offer visitors positive emotions through innovative attractions and enrichment activities [7]. However, the success of such cultural events in promoting tourism is impossible without establishing a strong and alluring destination brand identity [2,8]. The brand identity of the destination influences the visitors' expectations and perceptions of the museum and events. If the brand identity can convey a sense of innovation, entertainment, and superior quality, it will be significantly more effective at attracting tourist interest and attendance [9]. For instance, through large-scale museum initiatives and annual cultural festivals, cities such as Singapore and Barcelona have garnered a reputation for being culturally innovative hotspots. Their powerful brand identities entice swarms of tourists in search of new and exciting cultural experiences.

Although previous research has examined the impact of cultural events and brand identity on tourism separately, few studies have investigated their combined effects. A well-known brand can amplify the result of a high-quality cultural event, whereas a weak brand can diminish the potential of even the most successful events [10]. To achieve maximum tourism promotion, aligning brand identity with event offerings is crucial to generate synergy. Therefore, this study investigates how the brand identity of a destination and the quality of cultural events, as measured by visit quality, interaction quality, outcome quality, and physical environment quality, influence tourists' perceptions and their intentions to revisit and recommend the museum. To examine the relationship between event quality, destination identity, tourist satisfaction, and behavioral intentions, this study employed a quantitative methodology. Specifically, data were collected through an online survey of visitors who attended the 'Night of the Museums' event in Oradea, Romania. The survey instrument measured respondents' perceptions of four domains of event quality (visit quality, interaction quality, outcome quality, and physical environment quality), as well as destination identity, satisfaction, intention to recommend, and intention to revisit. Structural equation modeling was then utilized to assess the conceptual framework and test the hypothesized relationships among these variables.

By leveraging its Night of the Museums cultural event, the city of Oradea has the chance to enhance its reputation as a cultural center in Eastern Europe. Examining how Oradea can optimize the alignment between its brand identity and event quality will generate actionable insights for increasing tourism impact.

Key strategies may include promoting the city's particular cultural heritage and experimental spirit through creative offerings at the Night of the Museums; implementing visual branding tactics that reinforce the brand identity throughout visitors' cultural event experiences [10]; and soliciting feedback to ensure that events continue to improve in terms of quality and relevance for tourists [1]. For Oradea to achieve its tourism and cultural development objectives, managing brand identity and event quality as complementary rather than separate elements will be essential. Based on the presented arguments, the following are two possible research questions:

- How does the alignment between the brand identity of a destination and the quality of its museum events affect tourist satisfaction and behavior intentions?

- How can the city of Oradea optimize the compatibility between its brand identity and the quality of the Night of the Museums?

2. Literature Review

2.1. Museum Event Quality

The quality of cultural events held in museums has a significant impact on tourists' experiences and satisfaction levels [11]. Events include arenas where interactions between show-makers and consumers occur in the context of particular physical environments [12]. Event characteristics such as interactivity, utilization of technology, excitement, and physical environment can work in tandem to produce engaging experiences that delight visitors, or ineffective ones that underwhelm them [13]. Experience quality can be assessed by the implication of functional and emotional aspects in equal contexts [14]. By strategically optimizing essential event qualities, museums can significantly enhance tourist satisfaction and achieve cultural promotion and revenue generation objectives [15]. Various types of characteristics may exist in museums. For example, quality of a visit refers to the overall experience and pleasure of attending an event. Essential aspects of a high-quality visit include an enjoyable experience, clear rules, an exciting mechanism, and a good balance [16]. When museum events are designed to provide visitors with an engaging yet manageable experience, visitor satisfaction increases [16]. For instance, an event that uses interactive multimedia displays to present artifacts engagingly without being overly complex will have a higher visit quality. Cronin and Taylor [17] empirically illustrated that service quality could be considered as a predictor of visitor satisfaction.

In Lithuania, service quality management in the museums was examined using the SERVQUAL model [18]. The study findings revealed that the most important service quality dimensions were reliability, tangibles, and empathy. Reliability of services was a key driver of visitor satisfaction. In relation to the present study on the Night of the Museums event, these results highlight the need to ensure reliable and consistent event execution. Tangibles like the physical environment and visual appeal can also influence perceptions of quality and satisfaction. Finally, empathy, i.e., caring customer service, is important, aligning with the role of interaction quality in the current framework. This indicates that focusing on service reliability, the venue aesthetics, and visitor interactions could enhance event quality and satisfaction. Furthermore, Hume and Mort [19] utilized focus groups and SERVQUAL to assess museum service priorities, revealing key factors like cleanliness, staff demeanor and engagement. Gil and Ritchie [20] explored museum visitors' quality expectations, finding that tangibles and reliability were ranked highly across various visitor segments. Respondents valued factors like parking facilities, interior comfort, and exhibit quality. Taken together, these studies indicate that consistent service quality can increase perceived value and satisfaction for museum visitors. This further supports the need to examine event quality constructs like interaction quality and physical environment quality in the Night of the Museums context [19,20].

Moreover, interaction quality, social engagement, and community building are opportunities [21]. Interaction quality will improve at events by facilitating meaningful interactions between attendees, encouraging networking and collaboration, and fostering an inclusive and welcoming environment [22]. Museums that host events that genuinely allow visitors to interact with one another will be perceived as more socially rewarding [21], thereby increasing visitor satisfaction. For instance, an event that incorporates group activities and discussions around thought-provoking exhibits will enhance interaction quality [23]. In addition, outcome quality refers to the advantages and consequences of attending an event [24]. Events with higher-quality outcomes will achieve their intended objectives, provide valuable learning and professional development experiences, resources, and information, and exceed expectations [24,25]. When museums create events that inspire visitors and equip them with new knowledge and insights, tourists will find the experience extremely valuable and impactful, thereby increasing their contentment [26,27].

For instance, an event that includes expert presentations, skill-building workshops, and access to research materials will enhance the quality of the outcome.

Additionally, physical environment quality relates to the convenience, organization, and services an event venue provides [28]. Physical environment quality includes a comfortable and inviting venue, a practical layout, high-quality audio–visual equipment, appropriate temperature and lighting, and food and beverage services [29]. Visitors will feel more at ease and cared for at museums that host events in well-designed spaces with the appropriate amenities and facilities, thereby increasing their satisfaction [27]. For example, an event with spacious rooms, advanced audio–visual equipment, comfortable seating, and refreshments will have a high-quality physical environment [28]. In conclusion, the quality of the visit, the quality of the interaction, the quality of the outcome, and the quality of the physical environment shape museum event experiences that appeal to tourists and meet their needs and expectations [29]. When these characteristics are high, visitors will be profoundly satisfied and perceive the event as having enduring meaning, purpose, and value [30]. This demonstrates that event quality significantly impacts visitor satisfaction [27]. Based on the arguments, we hypothesize that:

H1a. *Higher levels of Oradea's Night of the Museums event's visit quality will lead to greater tourist satisfaction.*

H1b. *Higher levels of Oradea's Night of the Museums event's interaction quality will lead to greater tourist satisfaction.*

H1c. *Higher levels of Oradea's Night of the Museums event's outcome quality will lead to greater tourist satisfaction.*

H1d. *Higher levels of Oradea's Night of the Museums event's physical environment quality will lead to greater tourist satisfaction.*

2.2. Brand Identity and Tourist Satisfaction

The brand identity of a destination is the distinct collection of associations that represent its essence and core values [31]. Brand identity is based upon organizational identity theory [32]. According to Kapferer, brand identity is divided into exterior and interior identities [33]. Interior identity is related to a consumer's self-image and the extent of emotional impact on the consumer, while the brand's exterior identity is whether brand has become part of the relationship or culture. According to Robinson and Clifford [34], citizen welfare and visitor satisfaction are strongly affected by city or local community image. Barnes et al. [35] claim that brand experience can be studied from four aspects: sensory, emotional, cognitive, and behavioral. Their study also reveals that the positive experience from a trip or any event in most cases fades away after six weeks [35]. Therefore, it is important to keep reminding the consumer about the brand.

Several studies have examined factors influencing tourist perceptions and satisfaction across major cities in Romania. Dumbrăveanu et al. [36] analyzed the Night of the Museums event held annually in Bucharest, Romania. Their study investigated how the event has developed over time to incorporate diverse cultural offerings and institutions beyond just museums. Surveys during the 2013 and 2014 events revealed that while the majority of attendees were local residents, an increasing number were domestic tourists interested in heritage and culture. The Night of the Museums provided a platform to showcase Bucharest's cultural assets and inject tourist activity into the city. However, the event has not yet reached the scale or international attendee levels of similar events in other European capitals. The study demonstrates the potential of museum events to catalyze cultural tourism and enhance destination image, which can inform Oradea's cultural programming aims. Muntean et al. [37] examined tourist satisfaction and loyalty for both foreign and domestic visitors in Bucharest. Their conceptual model assessed how expectations, motivations, perceived authenticity, infrastructure, safety, emotions, and desire to extend their stay impacted satisfaction and loyalty intentions. Key findings were that tourists' motivation to lengthen their stay strongly influenced expectations and motivations. Also, destination safety did not significantly predict satisfaction. The research uniquely identified tourists' positive emotions as a mediator between satisfaction and loyalty. Sidonia and Cristina [38] focused on cultural tourism motivations and perceptions of Romanian destinations. Through a survey of Bucharest residents, they found cultural heritage and attractions were central to travel decisions for regions like the Centre Region with rich cultural assets. Key cultural motivators were identified as historical sites, art galleries, religious sites, museums, ethnic traditions, and performances. However, the research highlighted that multiple motivations beyond just culture, like relaxation or family, can influence destination choice [38]. In another study, tourist perceptions and satisfaction with Cluj-Napoca, an important cultural and business destination in Romania, was analyzed [39]. Through interviews and data analysis, the study identified key tourist segments and examined how Cluj-Napoca's offerings align with target profiles. Cultural tourism was a major focus given the city's heritage attractions. The research provided insights into tourist motivations and the image of Cluj-Napoca as a leading Romanian destination [39].

Establishing a powerful brand identity is essential for museums to attract visitors by creating a memorable and compelling brand image in their minds [40]. Destinations cultivate their brand identity through marketing communications, logos, taglines, and visitor experiences [31]. A recent article by Matwiejczyk [41] examined place-branding strategies for Polish cities, finding that destination marketing and experiences influenced brand identity formation. Other studies have also revealed connections between branding and satisfaction. Kladou et al. [42] discussed brand communication tools like logos, advertising, and public relations that can shape destination image. They emphasized consistency across touchpoints. Oliveira and Panyik [43] highlighted the need for authenticity in place branding to generate emotional connections. When branding truthfully reflects local heritage and culture, tourist satisfaction increases. Together, these studies reinforce the potential impact of relevant, authentic branding on satisfaction via destination image formation [41–43]. An attractive brand identity will signal to tourists the benefits and qualities they can expect, such as an educational yet entertaining experience, interaction with rare artifacts, and a sophisticated environment [44]. The qualities of museum events, such as visit quality, interaction quality, outcome quality, and physical environment quality, collectively shape the brand identity of host destinations by facilitating profoundly satisfying experiences that embody the brand essence [40].

For instance, an event with high-tech interactive exhibits, opportunities to socialize and learn from other attendees [45], valuable insights from experts, and upscale facilities will represent the destination brand identity centered on a culturally enriching, innovative experience. When museum events deliver consistent brand-aligned experiences, it reinforces the destination brand image of host destinations in the minds of visitors and increases their satisfaction [46]. In other words, high-quality museum events shape a strong, differentiated destination brand identity [46], increasing overall museum satisfaction. It is also hypothesized that a strong brand identity of destinations directly and positively affects the satisfaction of museum visitors [47]. When visitors perceive the brand identity of a destination as appealing, it piques their interest in attending events and raises their expectations for a pleasant and rewarding experience. Actually, not all visitors of the event could be described as tourists, because a significant part of them lived nearby, but we decided to involve only tourists, namely, people who came from other provinces or countries and were planning to stay more than 24 h. Tourists were assumed not to have assessment bias. Consequently, following are three hypotheses based on the arguments:

H2a. *Higher levels of visit quality of Oradea's Night of the Museums event will positively influence the destination brand identity from tourists' perspectives.*

H2b. *Higher levels of interaction quality of Oradea's Night of the Museums event will positively influence the destination brand identity from tourists' perspectives.*

H2c. *Higher levels of outcome quality of Oradea's Night of the Museums event will positively influence the destination brand identity from tourists' perspectives.*

H2d. *Higher levels of physical environment quality of Oradea's Night of the Museums event will positively influence the destination brand identity from tourists' perspectives.*

H3. *A stronger destination brand identity of Oradea will positively influence tourist satisfaction.*

Satisfied visitors with an enjoyable and meaningful experience that meets their brand-shaped expectations are more likely to recommend the museum to others and return in the future [2,7]. Positive word of mouth and intention to return are crucial outcomes for museums, as they can attract new visitors and generate repeat business [7]. For museums in Oradea, increasing tourist satisfaction by enhancing event quality and reinforcing brand identity can result in favorable behavioral outcomes such as word of mouth and intention to return. Tourists who had an engaging and rewarding experience are more likely to spread positive word of mouth through their social networks and connections if they are pleased with their expertise [48]. Their desire to share a positive experience drives word of mouth [7,49]. When tourists hear positive recommendations from reputable sources, their perceptions of the museums become more favorable, and they desire to visit them in person [50]. This increases the likelihood that museum visitors will return in the future. This supports the following hypotheses:

H4. *Higher tourist satisfaction with Oradea's Night of the Museums event will increase their positive word-of-mouth recommendations.*

H5. *Increased positive word of mouth regarding Oradea's Night of the Museums event will increase tourists' intentions to return.*

Collectively, the above hypotheses can be summarized in Figure 1.

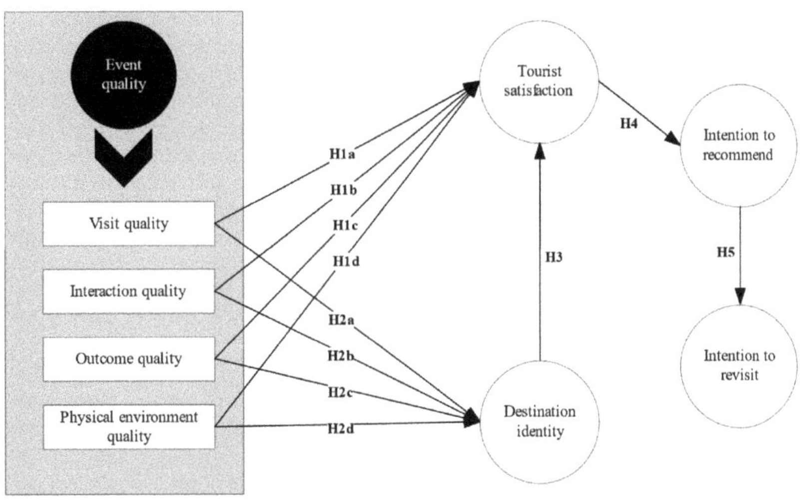

Figure 1. A framework of the study hypotheses.

3. Materials and Methods

3.1. Study Sample and Data Collection

The online survey was used as a means of data collection during the research. The advantage of this data collection method is that researchers can reach many people quickly and inexpensively. Designing and distributing surveys appropriately can be an effective tool for data analysis [51]. Data for the current study was collected using a convenient online survey of visitors who participated in the Night of the Museums event in Oradea, Romania, between June and July 2023. Museums in Oradea, a city located in Romania in a cross-border region inhabited by numerous minorities (Hungarian, Roma, German), are an important element of shaping identity for all residents of a multicultural city. Therefore, the case of "Night of the Museums" in Oradea is important for shaping attitudes of tolerance and supporting ethnic identity and cultural diversity. The organizers identified visitors who had registered in advance for the event as the target population for the study. By inviting only registered participants, the selection criteria ensured respondents had direct experience attending the event and could provide valuable insights into their perceptions and satisfaction.

While an online survey allows reaching a large sample, survey response rates are typically low. In this study, only a tiny percentage of registered visitors to the Night of the Museums completed the survey, limiting the generalizability of the results. To address this limitation and gain additional perspectives, semi-structured interviews were also conducted with a subset of 10 visitors who attended the event.

This case study of the Night of the Museums event in Oradea was chosen for several reasons. As a unique large-scale cultural attraction, examining visitors' experiences and perceptions of this event could provide insights into other destinations. Oradea was selected because the municipality invested significant resources into cultural programming, yet visitor satisfaction has not been comprehensively evaluated. Understanding the event's impact could help inform strategy to optimize cultural tourism outcomes. Therefore, examining visitor perceptions of an iconic event in this city serves a vital research purpose.

The survey link was electronically distributed by the organizers via email to around 150 registered participants both before and after the event. A total of 118 valid responses were received, indicating a response rate of about 78.6%. While this response rate is modest, it is not uncommon for online surveys and was considered adequate to analyze key constructs related to visitors' experiences. The online survey collection method had several advantages. By distributing the survey via email, the organizers were able to reach a large number of potential respondents with minimal cost and effort. The electronic format also allowed participants to complete the survey at a convenient time, which likely contributed to the satisfactory response rate. Additionally, utilizing a secure online survey platform maintained the anonymity of respondents' data and ensured the confidentiality of their responses.

The research also employs structural equation modeling to investigate relationship among variables. Structural equation modeling enables researchers to simultaneously model and assess the complex relationships among multiple dependent and independent variables. The method is widely used in survey analysis. It has several advantages including the possibility of studying complex patterns among the constructs, the analysis of structural relationships.

3.2. Construct Measures

Participants were invited to complete a self-administered survey that assessed their perceptions and experiences during the Night of the Museums event in Oradea. The survey included items measuring visit quality, interaction quality, outcome quality, physical environment quality, tourist satisfaction, destination identity, intention to recommend, and intention to revisit. The visit quality domain included items assessing the overall enjoyment, challenge, clarity of rules, excitement, and balance of the visit experience. Interaction quality covered perceptions of meaningful interactions, networking, collaboration, inclusivity, and welcoming atmosphere. Outcome quality measured the event's ability to achieve objectives, provide learning, impact development, offer useful resources, and exceed expectations. Physical environment quality examined the comfort, organization, audio–visual amenities, lighting, temperature, and food services of the venue. Destination identity items gauged feelings of community, connection to local culture, appreciation of attractions, belonging, and cultural learning. Tourist satisfaction items evaluated overall experience satisfaction, entertainment value, service quality, organization, and transportation. Intention to recommend was reflected through gauging the likelihood to recommend the event to others, on social media, and for vacations. Finally, intention to revisit incorporated interest in returning for future events, vacations, and to further explore attractions. Participants were asked to rate each item on a 5-point Likert scale, ranging from "strongly disagree" to "strongly agree." More details about the items and domains of the used survey are provided in Table 1. The survey also collected demographic information from participants, including age, gender, education, and employment status. The survey was administered online using a secure survey platform, and participants were assured of the confidentiality and anonymity of their responses.

Table 1. The survey used for data collection in the current study.

Scale	Item Code	Item
Visit quality	Vis_Q_1	The overall experience was enjoyable.
	Vis_Q_2	The visit was challenging but not frustrating.
	Vis_Q_3	The rules were clear and easy to understand.
	Vis_Q_4	The mechanism of the visit was exciting.
	Vis_Q_5	The visit was well balanced.
Interaction quality	Interact_Q_1	The event facilitated meaningful interactions with other attendees.
	Interact_Q_2	The event provided opportunities for networking.
	Interact_Q_3	The event encouraged collaboration and teamwork.
	Interact_Q_4	The event fostered a sense of community and inclusivity.
	Interact_Q_5	The event had a friendly and welcoming atmosphere.
Outcome quality	Outc_Q_1	The event achieved its intended objectives.
	Outc_Q_2	The event provided valuable learning experiences.
	Outc_Q_3	The event had a positive impact on my professional development.
	Outc_Q_4	The event provided useful resources and information.
	Outc_Q_5	The event exceeded my expectations.
Physical environment quality	Phys_Q_1	The event venue was comfortable and inviting.
	Phys_Q_2	The event was well-organized and easy to navigate.
	Phys_Q_3	The audio–visual equipment was of high quality.
	Phys_Q_4	The temperature and lighting were appropriate.
	Phys_Q_5	The event had effective food and beverage services.
Tourist satisfaction	Sat_1	Overall, I was satisfied with my experience visiting Oradea during the Night of the Museums event.
	Sat_2	The event met my expectations in terms of entertainment value.
	Sat_3	The quality of services (e.g., accommodation, dining) in Oradea during the event was satisfactory.
	Sat_4	The level of organization for the event was satisfactory.
	Sat_5	The transportation arrangements in Oradea during the event met my expectations.
Destination identity	Ident_1	I felt a strong sense of community spirit during my visit to Oradea for the Night of the Museums event.
	Ident_2	The event made me feel more connected to the people and culture of Oradea.
	Ident_3	My visit to Oradea for the Night of the Museums event enhanced my appreciation for the city's attractions and amenities.
	Ident_4	I felt like I was part of a special group of people while attending the Night of the Museums event in Oradea.
	Ident_5	My visit to Oradea for the Night of the Museums event helped me learn about the history and culture of the city.

Table 1. *Cont.*

Scale	Item Code	Item
Intention to recommend	Recom_1	I would recommend Oradea as a travel destination to my friends and family.
	Recom_2	I am likely to recommend Oradea specifically for other events.
	Recom_3	I would recommend Oradea for a vacation to anyone looking for a relaxing experience with cultural and natural attractions.
	Recom_4	I would recommend Oradea for similar or other events.
	Recom_5	I am willing to share my positive experience in Oradea with others on social media or travel forums.
Intention to revisit	Revis_1	I would consider revisiting Oradea as a travel destination in the future.
	Revis_2	I am likely to revisit Oradea specifically for other events.
	Revis_3	I would revisit Oradea for a vacation to experience more of what the city has to offer.
	Revis_4	I would revisit Oradea for similar or other events.
	Revis_5	I am interested in returning to Oradea in the future to explore more of the city's cultural and historical attractions.

3.3. Statistical Analysis

The data analysis was conducted utilizing RStudio (R version 4.3.0). Categorical data were summarized in terms of frequencies and percentages. To model the constructs employed in the study, a partial least squares structural equation modeling technique with bootstrapping was employed. Internal consistency reliability was assessed using Cronbach's alpha coefficients. To address the underlying assumptions of Cronbach's alpha, which assume equal indicator loadings, rhoC values were also used to express composite reliability [52], while rhoA was utilized as an additional conservative measure of internal consistency [53]. Convergent validity was evaluated by employing the average variance extracted (AVE), which assesses the extent to which each domain can converge to explain the variances of the indicators [54]. Discriminant validity was assessed by comparing the square roots of AVE to the correlations between different constructs and by utilizing the heterotrait–monotrait ratio (HTMT) of correlations [55]. The bootstrapped structural model employed a 1000-bootstrap method [56], and the results are reported as beta coefficients with their respective 95% confidence intervals (95% CIs). Statistical significance was indicated by a p-value of <0.05.

4. Results

4.1. Characteristics of the Respondents

We analyzed data of 118 participants in the current study. More than a half were females (60.2%) and employed (57.3%). Participants aged 18 to 30 years represented 42.4% of the sample, and 49.2% of respondents had a Bachelor's degree or higher (Table 2).

Table 2. Demographic characteristics of the participants.

Parameter	Category	N (%)
Gender	Male	47 (39.8%)
	Female	71 (60.2%)
Age	Below 18 years of age	9 (7.6%)
	18–30	50 (42.4%)
	31–40	24 (20.3%)
	41–50	23 (19.5%)
	51–60	7 (5.9%)
	61 and older	5 (4.2%)

Table 2. *Cont.*

Parameter	Category	N (%)
Education	Primary	11 (9.3%)
	Vocational	2 (1.7%)
	Secondary school	47 (39.8%)
	Higher	58 (49.2%)
Employment status	School student	15 (12.8%)
	Higher education student	27 (23.1%)
	Employed	67 (57.3%)
	Unemployed	2 (1.7%)
	Old-age pensioner/disabled pensioner	6 (5.1%)

4.2. Results of the Convergent Validity and Construct Reliability

In the bootstrapped model, five items were insignificantly loaded to their respective constructs. These included two items in the visit quality domain (Vis_Q_2 and Vis_Q_4), one item in the interaction quality domain (Interact_Q_3), one item in the outcome quality domain (Outc_Q_3), and one item in the physical environment quality domain (Phys_Q_5). The final bootstrapped model showed excellent reliability indicators (Figure 2 and Table 3). Mean bootstrap factor loadings were significant for all items (>0.50), and the rhoC and rhoA values were adequate (>0.70) [52,53]. The Cronbach's alpha coefficients varied from 0.711 to 0.878. Notably, the AVE values ranged from 0.517 to 0.673, indicating that the domains accounted for at least 51.7% of the variance observed in the indicators comprising each respective domain [54].

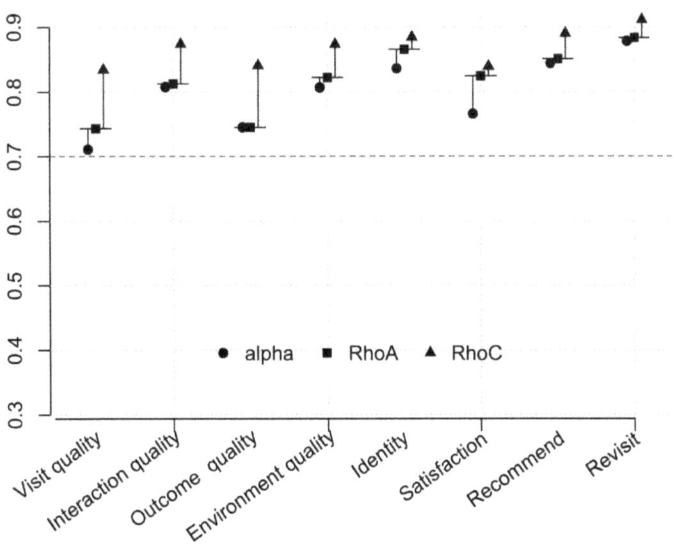

Figure 2. Results of the reliability indicators of the used constructs. The blue dashed line represents the 0.7 threshold above which reliability indicators are valid.

Table 3. Convergent validity and construct reliability.

Domains/Items	BFL	VIF	alpha	rhoC	rhoA	AVE
Visit quality			0.711	0.834	0.744	0.627
Vis_Q_1	0.795	1.436				
Vis_Q_3	0.709	1.409				
Vis_Q_5	0.838	1.325				

Table 3. *Cont.*

Domains/Items	BFL	VIF	alpha	rhoC	rhoA	AVE
Interaction quality			0.808	0.874	0.812	0.634
Interact_Q_1	0.848	2.254				
Interact_Q_2	0.782	2.210				
Interact_Q_4	0.807	1.838				
Interact_Q_5	0.726	1.461				
Outcome quality			0.745	0.841	0.745	0.570
Outc_Q_1	0.765	1.874				
Outc_Q_2	0.798	2.157				
Outc_Q_4	0.797	1.603				
Outc_Q_5	0.641	1.224				
Physical environment quality			0.807	0.873	0.822	0.634
Phys_Q_1	0.818	1.191				
Phys_Q_2	0.858	2.366				
Phys_Q_3	0.798	1.626				
Phys_Q_4	0.695	1.434				
Destination identity			0.836	0.884	0.865	0.609
Ident_1	0.779	1.712				
Ident_2	0.861	2.514				
Ident_3	0.796	1.854				
Ident_4	0.572	1.329				
Ident_5	0.851	2.197				
Tourist satisfaction			0.766	0.839	0.824	0.517
Sat_1	0.653	1.579				
Sat_2	0.854	2.118				
Sat_3	0.565	1.511				
Sat_4	0.837	1.874				
Sat_5	0.615	1.774				
Intention to recommend			0.844	0.890	0.851	0.620
Recom_1	0.752	1.732				
Recom_2	0.811	2.132				
Recom_3	0.840	2.248				
Recom_4	0.853	2.279				
Recom_5	0.666	1.444				
Intention to revisit			0.878	0.911	0.883	0.673
Revis_1	0.835	2.368				
Revis_2	0.851	2.758				
Revis_3	0.768	1.981				
Revis_4	0.855	2.349				
Revis_5	0.774	1.640				

VIF: variance inflation factor; Alpha: Cronbach's alpha; BFL: Bootstrapped factor loading; AVE: average variance extracted.

4.3. Outcomes of the Discriminant Validity

In terms of assessing discriminant validity, the square roots of the average variance extracted (AVE) were compared to the shared variance between constructs, as indicated by the inter-domain correlations. As depicted in Table 4, the square roots of AVE were found to be greater than the correlations between domains. Moreover, the bootstrapped heterotrait–monotrait ratio (HTMT) values, along with their corresponding 95% confidence intervals (CIs), did not surpass the threshold of 1, as reported in Table 5. This finding reinforces the confirmation of discriminant validity [55].

Table 4. Outcomes of the discriminant validity.

Parameter	1	2	3	4	5	6	7	8
1. Visit quality	0.792							
2. Interaction quality	0.438	0.796						
3. Outcome quality	0.454	0.604	0.755					
4. Environment quality	0.367	0.420	0.588	0.797				
5. Identity	0.496	0.590	0.619	0.500	0.780			
6. Satisfaction	0.450	0.579	0.572	0.578	0.566	0.719		
7. Intention to recommend	0.311	0.496	0.411	0.283	0.592	0.533	0.787	
8. Intention to revisit	0.248	0.421	0.343	0.287	0.610	0.467	0.652	0.820

The square roots of AVE are on the diagonal and inter-domain correlations are in the lower triangle.

Table 5. Outcomes of the bootstrapped HTMT.

Relationship	T Stat.	B-HTMT Values (95% CI)
Visit quality → Interaction quality	5.40	0.577 (0.375 to 0.779)
Visit quality → Outcome quality	4.64	0.645 (0.362 to 0.894)
Visit quality → Environment quality	3.55	0.493 (0.224 to 0.748)
Visit quality → Identity	4.58	0.618 (0.348 to 0.875)
Visit quality → Satisfaction	5.48	0.568 (0.377 to 0.768)
Visit quality → Recommend	2.80	0.403 (0.163 to 0.683)
Visit quality → Revisit	2.94	0.324 (0.153 to 0.530)
Interaction quality → Outcome quality	8.05	0.758 (0.564 to 0.932)
Interaction quality → Environment quality	3.88	0.517 (0.250 to 0.766)
Interaction quality → Identity	8.09	0.699 (0.508 to 0.847)
Interaction quality → Satisfaction	7.13	0.729 (0.500 to 0.910)
Interaction quality → Recommend	4.62	0.575 (0.295 to 0.783)
Interaction quality → Revisit	3.42	0.485 (0.202 to 0.742)
Outcome quality → Environment quality	6.41	0.747 (0.498 to 0.960)
Outcome quality → Identity	9.30	0.755 (0.577 to 0.903)
Outcome quality → Satisfaction	7.23	0.720 (0.513 to 0.911)
Outcome quality → Recommend	3.93	0.518 (0.280 to 0.773)
Outcome quality → Revisit	3.47	0.426 (0.212 to 0.663)
Environment quality → Identity	4.55	0.591 (0.338 to 0.849)
Environment quality → Satisfaction	7.11	0.715 (0.499 to 0.896)
Environment quality → Recommend	2.67	0.349 (0.132 to 0.610)
Environment quality → Revisit	2.88	0.337 (0.147 to 0.589)
Identity → Satisfaction	5.51	0.657 (0.415 to 0.883)
Identity → Recommend	5.87	0.683 (0.439 to 0.892)
Identity → Revisit	7.74	0.696 (0.510 to 0.870)
Satisfaction → Recommend	5.84	0.602 (0.388 to 0.787)
Satisfaction → Revisit	4.67	0.513 (0.306 to 0.719)
Recommend → Revisit	5.94	0.750 (0.484 to 0.972)

4.4. Structural Model

Results of the structural path indicated that the destination identity was significantly predicted by three domains of the event quality, including visit quality (β = 0.195, 95%

CI = 0.001 to 0.433, p = 0.044), interaction quality (β = 0.275, 95% CI = 0.076 to 0.430, p = 0.001), and outcome quality, as well as tourist satisfaction (β = 0.276, 95% CI = 0.066 to 0.462, p = 0.004). Only two domains of event quality significantly influenced tourist satisfaction, including the interaction quality (β = 0.258, 95% CI = 0.024 to 0.481, p = 0.023) and the quality of the physical environment (β = 0.295, 95% CI = 0.022 to 0.555, p = 0.023). Destination identity did not influence tourist satisfaction. However, satisfaction was an antecedent predictor of tourists' intention to recommend (β = 0.533, 95% CI = 0.291 to 0.703, p < 0.0001), which in turn predicted the intention to revisit the destination (β = 0.652, 95% CI = 0.458 to 0.839, p < 0.0001, Table 6).

Table 6. Results of the structural models.

Path	T Value	B (95% CI)	p	Hypothesis	Result
Visit quality → Satisfaction	1.037	0.109 (−0.095 to 0.340)	0.151	H1a	NS
Interaction quality → Satisfaction	2.021	0.258 (0.024 to 0.481)	0.023	H1b	Supp
Outcome quality → Satisfaction	0.764	0.101 (−0.121 to 0.393)	0.223	H1c	NS
Environment quality → Satisfaction	2.018	0.295 (0.022 to 0.555)	0.023	H1d	Supp
Visit quality → Identity	1.717	0.195 (0.001 to 0.433)	0.044	H2a	Supp
Interaction quality → Identity	3.063	0.275 (0.076 to 0.430)	0.001	H2b	Supp
Outcome quality → Identity	2.665	0.276 (0.066 to 0.462)	0.004	H2c	Supp
Environment quality → Identity	1.241	0.152 (−0.064 to 0.418)	0.109	H2d	NS
Identity → Satisfaction	0.802	0.149 (−0.224 to 0.493)	0.212	H3	NS
Satisfaction → Recommend	5.206	0.533 (0.291 to 0.703)	<0.0001	H4	Supp
Recommend → Revisit	6.377	0.652 (0.458 to 0.839)	<0.0001	H5	Supp

CI: confidence interval; Supp: supported; NS: Not supported.

5. Discussion and Limitations

Tourists' brand experiences affect their decision making process significantly [57]. Brand design, packaging, communication, and environment directly influence consumers on a subconscious level, triggering specific behavioral responses [58]. The relationship between quality dimensions and destination identity has long been established. The current study reported a relationship between three quality domains, namely visit quality (p = 0.044), interaction quality (p = 0.001), and outcome quality (p = 0.004), with destination identity (Table 4). These findings were consistent with the findings by multiple prior research papers [59–64]. First, Jin et al. [60] and Yamaguchi et al. [59], reported a positive association between event quality and destination image. Similarly, Moon et al. [61] found that destination image is positively influenced by perception of event quality, especially intangible factors. As for the interaction quality, Yang reported a positive association between interaction quality and image destination [62]. According to Yang [62], the quality of interactions is heavily influenced by sociable incidents, which shape tourists' cognitive image and, ultimately, influence the affective image of a destination. When it comes to outcome quality, Kim et al. [63] reported a significant association between the quality of information, which is part of outcome quality, with the destination image. In addition, Santana and Gosling emphasized that tourist experience is strongly influenced by the destination image [64]. Although some studies reported positive associations between physical environment qualities such as perceived atmosphere [65], quality of transport [66], infrastructure and facilities [67], destination safety and cleanliness [65,67], and the clean atmosphere with destination image [67], the current study did not report a significant association of the destination identity with the physical environment quality factors (Table 6). In their systematic review about measuring the service quality, Hartwig and Billert [66] argued that there are multiple models and tools utilized by researchers to measure service quality dimensions which can explain the variations in the study results that assessed the relation between service quality and other variables.

Competitive destinations must strive to achieve high tourist satisfaction since it has a significant impact on tourists' choice of holiday destination, and their future visits. The current study reported both the quality of interaction and the quality of environment to be associated with higher satisfaction among tourists, with $p = 0.001$ and $p = 0.023$, respectively (Table 6). In their study to assess the impact of multiple quality dimensions on customer satisfaction, Joon and Kim [67] reported interaction quality as a significant influencer on the customer satisfaction. Joon and Kim [67] explained that understanding the relationship between service quality and customer satisfaction is essential, as it affects customer experience and perception. Another study by Mustelier-Puig et al. [68] reported a significant influence of customer satisfaction on both service satisfaction and overall tourist satisfaction. They added that the overall satisfaction among customers is directly associated with the intention to revisit the destination [68]. As for the environmental quality, Zulvianti et al. reported a significant association between environmental quality, measured by perceived environmental value, and tourist satisfaction [69]. Nevertheless, they argued that environmental quality association with tourist satisfaction is mediated by sustainability tourism development [69]. In another paper, Naidoo et al. [70] found that destination attributes associated with physical environment quality, such as a peaceful environment, temperature, and weather, were more well perceived by tourists compared to other variables, and had an impact the on the overall perception of tourists on the destination and its image [70]. Zalejska-Jonsson and Wilhelmsson also explained that happiness and satisfaction are directly influenced by the quality of the environment as it affects individual health and wellbeing; however, they argued that the level of satisfaction can vary between people depending on individual and building characteristics [71].

Similar to destination identity and satisfaction, tourists' intention to recommend and to revisit a destination has been linked with their perception of a specific destination [72–74]. Intentions to revisit were positively linked with tourists' experience, satisfaction, and the destination identity in a recent study [75]. Considering the effects of these variables on individual perception, developing an understanding of the tourist perception represents a significant part of establishing the organizational strategy for future editions of any event [76–78]. According to Preko et al. [79], museums offer tourists both tangible and intangible experiences that can impact service quality and tourist satisfaction, and thus their behavior and intentions to revisit. The current study supports this finding, as we reported a significant relationship between tourists' satisfaction and recommendations and between recommendations and revisit intentions (Table 6). Moreover, Santana and Gosling explained that people who recommend a specific destination are more likely to revisit the destination [64]. This finding was consistent with the findings from our study, which reported a significant association between the destination recommendation variable and the intention to revisit ($p < 0.0001$). It is worth mentioning that Wang et al. found that the effects of destination-perceived quality are stronger for first-time visitors compared to later visits [80]. This underscores the need to conduct studies that assess the service quality impact on tourists upon their revisits rather than on newcomers.

When it comes to study limitations, the current study utilized a questionnaire as a data collection method, which increases the chances of response error. In addition, the lack of studies that assess the impact of quality dimensions and other variables on tourists' perception in a museum setting have slightly impacted the discussion part in the current study, and thus, we recommend conducting further studies that assess these associations in similar settings, and investigate the effect of multiple mediating variables that can mediate these relationships. Importantly, a limitation of the current study is the relatively modest sample size obtained through the survey methodology. While the 78.6% response rate for the online survey was satisfactory, the total number of valid responses analyzed was 118 participants. A larger overall sample would strengthen the conclusiveness and generalizability of the results. The limited sample size may have constrained the ability to detect small effects or completely accurately estimate the strengths of relationships between constructs. This is a common challenge for survey research restricted to a single event instance. For future research, deploying the survey to a larger target population, if possible across multiple event iterations, and adding an interview component would allow for collection of a larger dataset and mixed methods analysis.

Although the results of the study did not reflect how the different quality dimensions assessed relate to Oradea's brand identity, particularly in the context of the Night of the Museums event, prior research supports associations between event/service quality, satisfaction, and destination image/identity [56–63]. High-quality interactions, outcomes, and environment positively influenced visitors' perceptions of Oradea's identity as a cultural tourism destination (Table 4). Specifically, memorable social experiences and informative interactions could help shape cognitive and affective brand associations [61].

To better optimize identity compatibility, Oradea should focus on delivering consistently high-quality engagements and improvements across physical, staff, and program areas perceived by visitors as needing attention [64–70]. For example, addressing infrastructure, wayfinding, or crowding issues could enhance perceptions of environmental quality and satisfaction [66,69]. Strengthening educational components and sharing Oradea's history/culture more meaningfully may additionally cultivate cognitive and affective bonds with the city brand [62,63]. Monitoring and learning from visitor feedback on quality domains and branding impacts will be essential to refine strategic alignment over time. Doing so could reinforce Oradea's reputation management and cultural product competitiveness through a memorable Night of the Museums experiences [71,75,76,79,81].

6. Conclusions and Future Implications

6.1. Conclusions

Considering its role in a destination's survival and its impact on multiple levels, tourist perception is among the most investigated topics in tourism. Tourist perception and satisfaction at any event are influenced by several factors, and these factors can be influenced by a variety of factors as well. The current study reported a significant association between visit quality, interaction quality, and outcome quality domains on the destination identity. In addition, interaction quality and environment quality were directly associated with tourist satisfaction. Satisfied tourists were also more likely to recommend the destination to others, and with more intentions to revisit the same event in the future. Unlike previously reported associations, the relationship between satisfaction and destination identity with some of the quality domains were not significant.

It is therefore essential to assess the impact of specific factors and their domains, as well as the interrelation between these factors. In addition, developing effective marketing and operational strategies requires the collaboration of destination marketers, planners, and managers on all levels. Event organizers should focus on improving visit quality, interaction quality, and outcome quality. By ensuring that these aspects are well-executed, they can positively influence destination identity. This may involve investing in resources to enhance the visitor experience, providing engaging interactions, and delivering desirable outcomes. Furthermore, attention should be given to the quality of the physical environment where the event takes place. Creating an appealing and comfortable setting can significantly contribute to tourist satisfaction. Event organizers should consider factors such as aesthetics, cleanliness, accessibility, and comfort to create a positive atmosphere. Focusing on the destination identity, destination management organizations can work on promoting the unique aspects and cultural identity of the destination through events like the Night of the Museums. By highlighting the distinctive features, historical significance, and cultural heritage of Oradea, they can strengthen the destination's identity and appeal to tourists seeking authentic cultural experiences.

6.2. Theoretical Implications

This study makes several key contributions to theory on event quality, destination branding, and tourist behavior. By investigating the relationships between event quality domains, destination identity, satisfaction, and behavioral intentions, the research expanded theoretical knowledge on how memorable event experiences can shape brand image and drive outcomes. The findings empirically validated the ability of visit quality, interaction quality, and outcome quality to positively influence destination identity. This aligns with branding theories which propose that consistent, engaging brand experiences shape identity by creating meaningful associations in consumers' minds. Furthermore, the results supported satisfaction as a driver of vital post-purchase behaviors like intention to recommend and revisit. This builds on theoretical frameworks linking satisfaction to loyalty and relationships. Although destination identity did not directly predict satisfaction, this study provided an initial conceptualization of related constructs, which can guide future research to refine theoretical linkages. Overall, the research advanced understanding of strategies to align event quality with destination branding for maximum impact.

6.3. Managerial Implications

The results have several key practical implications for event organizers and destination managers in Oradea. Firstly, the findings suggest that event organizers should devote special attention to enhancing visit quality, interaction quality, and outcome quality when managing the Night of the Museums event. By focusing on the visitor experience, engagement opportunities, and learning value, they can shape perceptions of Oradea's cultural offerings. Secondly, the destination management organization should promote the distinctive features, heritage, and traditions of Oradea through branding strategies and partnerships with events like Night of the Museums. Strategic branding reinforced by memorable events can strengthen Oradea's identity as an authentic cultural destination in the minds of tourists. Thirdly, visitor satisfaction can be improved by ensuring high-quality interactions between staff, tourists, and local residents, as well as providing comfortable and visually appealing physical environments. Higher satisfaction can translate to positive word-of-mouth recommendations and repeat visitation. Finally, targeted marketing campaigns highlighting Oradea's cultural assets, incentives to share experiences, and personalized promotions based on visitor preferences may encourage future revisitation.

6.4. Place-Specific Implications

The current study's findings have place-specific implications that reflect Oradea's distinctive territorial context. Situated in western Romania, Oradea has a rich cultural heritage encompassing medieval architecture, artisanal crafts, diverse cuisines, and music. Applying the insights from this research within Oradea's local environment can enrich event experiences and strengthen destination branding. For instance, interactive Night of the Museums exhibits showcasing Oradea's history, landmarks, and multicultural communities could allow tourists to connect with place identity. Regional partnerships incorporating local businesses, artists, and cuisines would also authentically represent Oradea. Marketing initiatives tailored to domestic Romanian tourists represent another opportunity. Enhancing walkability and transit access could improve the visitor experience and perception of Oradea's offerings. Additionally, other cultural events in Oradea can apply learnings around interaction and outcome quality to better engage residents and visitors. Grounding the research implications locally will help Oradea solidify its brand identity as an emerging Romanian cultural hub.

Author Contributions: Conceptualization, T.H.H.; Methodology, S.M.E.; Software, I.J.; Validation, J.A.W.; Formal analysis, A.E.S.; Investigation, M.A.A. and B.J.; Resources, G.V.H.; Writing—original draft, H.M.E.M.; Writing—review & editing, M.S.A. All authors have read and agreed to the published version of the manuscript.

Funding: This work was supported by the Deanship of Scientific Research, Vice Presidency for Graduate Studies and Scientific Research, King Faisal University, Saudi Arabia [Grant No. 4640].

Institutional Review Board Statement: This study was conducted according to the guidelines of the Declaration of Helsinki and was approved by the Deanship of Scientific Research Ethical Committee, King Faisal University.

Informed Consent Statement: Not applicable.

Data Availability Statement: Data are available on request due to privacy/ethical restrictions.

Acknowledgments: We gratefully acknowledge the financial support provided by Grant No. 4640 from the Deanship of Scientific Research, Vice Presidency for Graduate Studies and Scientific Research, King Faisal University, Saudi Arabia.

Conflicts of Interest: The authors declare no conflict of interest.

References

1. Richards, G. Cultural tourism: A review of recent research and trends. *J. Hosp. Tour. Manag.* **2018**, *36*, 12–21. [CrossRef]
2. Liu, C.-R.; Liu, H.-K.; Lin, W.-R. Constructing Customer-based Museums Brand Equity Model: The Mediating Role of Brand Value. *Int. J. Tour. Res.* **2013**, *17*, 229–238. [CrossRef]
3. Dorrian, M. Museum atmospheres: Notes on aura, distance and affect. *J. Archit.* **2014**, *19*, 187–201. [CrossRef]
4. Goulding, C. The museum environment and the visitor experience. *Eur. J. Mark.* **2000**, *34*, 261–278. [CrossRef]
5. Easson, H.; Leask, A. After-hours events at the National Museum of Scotland: A product for attracting, engaging and retaining new museum audiences? *Curr. Issues Tour.* **2019**, *23*, 1343–1356. [CrossRef]
6. Kaya, S.M.; Afacan, Y. Effects of daylight design features on visitors' satisfaction of museums. *Indoor Built Environ.* **2017**, *27*, 1341–1356. [CrossRef]
7. de Rojas, C.; Camarero, C. Visitors' experience, mood and satisfaction in a heritage context: Evidence from an interpretation center. *Tour. Manag.* **2008**, *29*, 525–537. [CrossRef]
8. Konecnik, M.; Go, F. Tourism destination brand identity: The case of Slovenia. *J. Brand Manag.* **2007**, *15*, 177–189. [CrossRef]
9. Tsaur, S.-H.; Yen, C.-H.; Yan, Y.-T. Destination brand identity: Scale development and validation. *Asia Pac. J. Tour. Res.* **2016**, *21*, 1310–1323. [CrossRef]
10. Hanna, S.; Rowley, J.; Keegan, B. Place and Destination Branding: A Review and Conceptual Mapping of the Domain. *Eur. Manag. Rev.* **2020**, *18*, 105–117. [CrossRef]
11. Kirchberg, V.; Tröndle, M. Experiencing Exhibitions: A Review of Studies on Visitor Experiences in Museums. *Curator Mus. J.* **2012**, *55*, 435–452. [CrossRef]
12. Armbrecht, J. Event quality, perceived value, satisfaction and behavioural intentions in an event context. *Scand. J. Hosp. Tour.* **2021**, *21*, 169–191. [CrossRef]
13. Antchak, V.; Adams, E. Unusual venues for business events: Key quality attributes of museums and art galleries. *Int. J. Tour. Cities* **2020**, *6*, 847–862. [CrossRef]
14. Chen, M.-F.; Tung, P.-J. Developing an extended Theory of Planned Behavior model to predict consumers' intention to visit green hotels. *Int. J. Hosp. Manag.* **2014**, *36*, 221–230. [CrossRef]
15. Daskalaki, V.V.; Voutsa, M.C.; Boutsouki, C.; Hatzithomas, L. Service quality, visitor satisfaction and future behavior in the museum sector. *J. Tour. Herit. Serv. Mark. (JTHSM)* **2020**, *6*, 3–8.
16. Del Chiappa, G.; Andreu, L.; Gallarza, M.G. Emotions and visitors' satisfaction at a museum. *Int. J. Cult. Tour. Hosp. Res.* **2014**, *8*, 420–431. [CrossRef]
17. Cronin, J.J.; Taylor, S.A. Measuring Service Quality: A Reexamination and Extension. *J. Mark.* **1992**, *56*, 55–68. [CrossRef]
18. Burksiene, V.; Dvorak, J.; Valutiene, L. Quality Management in the Museums of Lithuania. *Cult. Manag. Sci. Educ.* **2022**, *6*, 9–20. [CrossRef]
19. Hume, M.; Sullivan Mort, G. The consequence of appraisal emotion, service quality, perceived value and customer satisfaction on repurchase intent in the performing arts. *J. Serv. Mark.* **2010**, *24*, 170–182.
20. Gil, S.; Ritchie, J.R.B. Understanding the museum image formation process: A comparison of residents and tourists. *J. Travel Res.* **2009**, *47*, 480–493.
21. Haywood, N.; Cairns, P. Engagement with an Interactive Museum Exhibit. In *People and Computers XIX—The Bigger Picture, Proceedings of HCI 2005, Edinburgh, UK, 1 July 2006*; Springer: London, UK, 2006; pp. 113–129.
22. Ciolfi, L.; Bannon, L. Designing Interactive Museum Exhibits: Enhancing visitor curiosity through augmented artefacts. In Proceedings of the ECCE11—Eleventh European Conference on Cognitive Ergonomics, Catania, Italy, 8–11 September 2002.
23. Pallud, J. Impact of interactive technologies on stimulating learning experiences in a museum. *Inf. Manag.* **2017**, *54*, 465–478. [CrossRef]
24. Wu, H.-C.; Li, T. An Empirical Study of the Effects of Service Quality, Visitor Satisfaction, and Emotions on Behavioral Intentions of Visitors to the Museums of Macau. *J. Qual. Assur. Hosp. Tour.* **2015**, *16*, 80–102. [CrossRef]
25. Anderson, D.; Storksdieck, M.; Spock, M. Understanding the long-term impacts of museum experiences. In *In Principle, In Practice: Museums as Learning Institutions*; AltaMira Press: Lanham, MD, USA, 2007; pp. 197–215.
26. Su, Y.; Teng, W. Contemplating museums' service failure: Extracting the service quality dimensions of museums from negative on-line reviews. *Tour. Manag.* **2018**, *69*, 214–222. [CrossRef]
27. Harrison, P.; Shaw, R. Consumer satisfaction and post-purchase intentions: An exploratory study of museum visitors. *Int. J. Arts Manag.* **2004**, *6*, 23–32.
28. Zhang, H.; Chang, P.-C.; Tsai, M.-F. How Physical Environment Impacts Visitors' Behavior in Learning-Based Tourism—The Example of Technology Museum. *Sustainability* **2018**, *10*, 3880. [CrossRef]

29. Jeong, J.-H.; Lee, K.-H. The physical environment in museums and its effects on visitors' satisfaction. *Build. Environ.* **2006**, *41*, 963–969. [CrossRef]
30. Kim, L.C.J. The Consumption of Museum Service Experiences: Benefits and Value of Museum Experiences. *J. Hosp. Mark. Manag.* **2009**, *18*, 173–196. [CrossRef]
31. Morgan, N.; Pritchard, A. Meeting the destination branding challenge. In *Destination Branding*; Routledge: New York, NY, USA, 2007; pp. 73–92.
32. Chiang, C.-T.; Chen, Y.-C. The Effect of Destination Brand Identity on Tourism Experience: The Case of the Pier-2 Art Center in Taiwan. *Sustainability* **2023**, *15*, 3254. [CrossRef]
33. Kapferer, J.-N. *Strategic Brand Management: New Approaches to Creating and Evaluating Brand Equity*; Simon and Schuster: New York, NY, USA, 1994.
34. Robinson, R.N.S.; Clifford, C. Authenticity and festival foodservice experiences. *Ann. Tour. Res.* **2012**, *39*, 571–600. [CrossRef]
35. Barnes, S.J.; Mattsson, J.; Sørensen, F. Remembered experiences and revisit intentions: A longitudinal study of safari park visitors. *Tour. Manag.* **2016**, *57*, 286–294. [CrossRef]
36. Dumbrăveanu, D.; Tudoricu, A.; Crăciun, A. The Night of Museums-a boost factor for the cultural dimension of tourism in Bucharest. *Hum. Geogr.—J. Stud. Res. Hum. Geogr.* **2014**, *8*, 55–63. [CrossRef]
37. Muntean, M.C.; Sorcaru, I.A.; Manea, L.D. Empirical evidence of tourist satisfaction and loyalty in the case of Bucharest—A partial least squares structural equation modelling approach. *J. Consum. Behav.* **2023**, *22*, 985–999. [CrossRef]
38. Sidonia, R.A.; Cristina, M. What motivates cultural tourists? An analysis of Bucharest inhabitants' motivation to visit the centre region. *Ann. Fac. Econ.* **2013**, *1*, 825–834.
39. Cosma, S.; Pop, C.; Negrusa, A. Tourists' perception regarding Cluj-Napoca as a cultural tourist destination. In Proceedings of the WSEAS International Conference on Cultural Heritage and Tourism (CUHT'08), Heraklion, Greece, 22–24 July 2008.
40. Ferreiro-Rosende, É.; Fuentes-Moraleda, L.; Morere-Molinero, N. Artists brands and museums: Understanding brand identity. *Mus. Manag. Curatorship* **2021**, *38*, 157–174. [CrossRef]
41. Matwiejczyk, A. Place Branding Approach within Functional Urban Areas—Evidence from Poland. *Sustainability* **2023**, *15*, 11872. [CrossRef]
42. Kladou, S.; Kavaratzis, M.; Rigopoulou, I.; Salonika, E. The role of brand elements in destination branding. *J. Destin. Mark. Manag.* **2017**, *6*, 426–435. [CrossRef]
43. Oliveira, E.; Panyik, E. Content, context and co-creation: Digital challenges in destination branding with references to Portugal as a tourist destination. *J. Vacat. Mark.* **2015**, *21*, 53–74. [CrossRef]
44. Lascu, D.-N.; Amesquita, E.R.S. Branding the controversial: Challenges in destination branding. *Innov. Mark.* **2011**, *7*, 20–28.
45. Caldwell, N.G. The emergence of museum brands. *Int. J. Arts Manag.* **2000**, *2*, 28–34.
46. Hede, A.-M. Branding museums in the global marketplace. In *Museum Marketing*; Routledge: New York, NY, USA, 2009; pp. 151–159.
47. Rodrigues, C. Chapter 4 The Destination Branding through Multisensory Experiences: Case Studies from Sweden. In *The Branding of Tourist Destinations: Theoretical and Empirical Insights*; Emerald Publishing Limited: Bingley, UK, 2018; pp. 61–82.
48. Hwang, J.; Lee, J. Relationships among Senior Tourists' Perceptions of Tour Guides' Professional Competencies, Rapport, Satisfaction with the Guide Service, Tour Satisfaction, and Word of Mouth. *J. Travel Res.* **2018**, *58*, 1331–1346. [CrossRef]
49. Saleh, M.I. Attribution Theory Revisited: Probing the Link Among Locus of Causality Theory, Destination Social Responsibility, Tourism Experience Types, and Tourist Behavior. *J. Travel Res.* **2022**, *62*, 1309–1327. [CrossRef]
50. Meng, B.; Han, H. Working-holiday tourism attributes and satisfaction in forming word-of-mouth and revisit intentions: Impact of quantity and quality of intergroup contact. *J. Destin. Mark. Manag.* **2018**, *9*, 347–357. [CrossRef]
51. Evans, J.R.; Mathur, A. The value of online surveys. *Internet Res.* **2005**, *15*, 195–219. [CrossRef]
52. Jöreskog, K.G. Simultaneous factor analysis in several populations. *Psychometrika* **1971**, *36*, 409–426. [CrossRef]
53. Dijkstra, T.K.; Henseler, J. Consistent partial least squares path modeling. *MIS Q.* **2015**, *39*, 297–316. [CrossRef]
54. Hair, J.F., Jr.; Hult, G.T.M.; Ringle, C.M.; Sarstedt, M. Assessing PLS-SEM Results—Part I: Evaluation of the Reflective Measeurement Models. In *A Primer on Partial Least Squares Structural Equation Modeling (PLS-SEM)*, 3rd ed.; Hair, J.F., Jr., Hult, G.T.M., Ringle, C.M., Sarstedt, M., Eds.; Sage Publications: Thousand Oaks, CA, USA, 2021.
55. Henseler, J.; Ringle, C.M.; Sarstedt, M. A new criterion for assessing discriminant validity in variance-based structural equation modeling. *J. Acad. Mark. Sci.* **2015**, *43*, 115–135. [CrossRef]
56. Streukens, S.; Leroi-Werelds, S. Bootstrapping and PLS-SEM: A step-by-step guide to get more out of your bootstrap results. *Eur. Manag. J.* **2016**, *34*, 618–632. [CrossRef]
57. Liu, L.; Xiang, Z.; Liu, Y.; Zach, F.J.; McGehee, N. Factors Influencing Exhibitor Satisfaction and Loyalty: A Meta-Analysis on the Chinese Exhibition Market. *Sustainability* **2020**, *12*, 8390. [CrossRef]
58. Baker, B. Consultation builds stronger brands. In *Tourism Branding: Communities in Action*; Emerald Group Publishing Limited: Bingley, UK, 2009; Volume 1, pp. 191–205.
59. Yamaguchi, S.; Akiyoshi, R.; Yamaguchi, Y.; Nogawa, H. Assessing the Effects of Service Quality, Past Experience, and Destination Image on Behavioral Intentions in the Spring Training Camp of a Japanese Professional Baseball Team. *J. Conv. Event Tour.* **2015**, *16*, 228–252. [CrossRef]

60. Jin, N.; Lee, H.; Lee, S. Event Quality, Perceived Value, Destination Image, and Behavioral Intention of Sports Events: The Case of the IAAF World Championship, Daegu, 2011. *Asia Pac. J. Tour. Res.* **2013**, *18*, 849–864. [CrossRef]
61. Moon, S.K.; Kim, M.; Jae Ko, Y.; Connaughton, D.P.; Hak Lee, J. The influence of consumer's event quality perception on destination image. *Manag. Serv. Qual. Int. J.* **2011**, *21*, 287–303. [CrossRef]
62. Yang, F.X. Tourist Co-Created Destination Image. *J. Travel Tour. Mark.* **2015**, *33*, 425–439. [CrossRef]
63. Kim, S.-E.; Lee, K.Y.; Shin, S.I.; Yang, S.-B. Effects of tourism information quality in social media on destination image formation: The case of Sina Weibo. *Inf. Manag.* **2017**, *54*, 687–702. [CrossRef]
64. Santana, L.D.; Sevilha Gosling, M.D. Dimensions of Image: A Model of Destination Image Formation. *Tour. Anal.* **2018**, *23*, 303–322. [CrossRef]
65. Loi, L.T.I.; So, A.S.I.; Lo, I.S.; Fong, L.H.N. Does the quality of tourist shuttles influence revisit intention through destination image and satisfaction? The case of Macao. *J. Hosp. Tour. Manag.* **2017**, *32*, 115–123. [CrossRef]
66. Hartwig, K.; Billert, M.S. Measuring service quality: A systematic literature review. In Proceedings of the Twenty-Sixth European Conference on Information Systems (ECIS 2018), Portsmouth, UK, 23–28 June 2018.
67. Joon, C.B.; Kim, H. The impact of outcome quality, interaction quality, and peer-to-peer quality on customer satisfaction with a hospital service. *Manag. Serv. Qual. Int. J.* **2013**, *23*, 188–204. [CrossRef]
68. Mustelier-Puig, L.C.; Anjum, A.; Ming, X. Interaction quality and satisfaction: An empirical study of international tourists when buying Shanghai tourist attraction services. *Cogent Bus. Manag.* **2018**, *5*, 1470890. [CrossRef]
69. Zulvianti, N.; Aimon, H.; Abror, A. The Influence of Environmental and Non-Environmental Factors on Tourist Satisfaction in Halal Tourism Destinations in West Sumatra, Indonesia. *Sustainability* **2022**, *14*, 9185. [CrossRef]
70. Naidoo, P.; Ramseook Munhurrun, P.; Ladsawut, J. Tourist satisfaction with Mauritius as a holiday destination. *Glob. J. Bus. Res.* **2010**, *4*, 113–123.
71. Zalejska-Jonsson, A.; Wilhelmsson, M. Impact of perceived indoor environment quality on overall satisfaction in Swedish dwellings. *Build. Environ.* **2013**, *63*, 134–144. [CrossRef]
72. Herman, G.V.; Banto, N.; Herman, L.M.; Ungureanu, M.; Kostilníková, K.; Josan, I. Perception, Reality and Intent in Bihorean Tourism, Romania. *Folia Geogr.* **2022**, *64*, 86–103.
73. Herman, G.V.; Banto, N.; Caciora, T.; Grama, V.; Trojan, J.; Ungureanu, M.; Furdui, S.; Garai, D.; Buhaş, R.; Buhaş, S. Considerations regarding the profile of tourists from Băile Felix–Băile 1 Mai Spa Destination, Romania. *Balt. J. Health Phys. Act.* **2021**, *13*, 15. [CrossRef]
74. Herman, G.V.; Banto, N.; Caciora, T.; Ungureanu, M.; Furdui, S.; Garai, L.D.; Grama, V. The Perception Of Bihor Mountain Tourist Destination; Romania. *Geogr. Pol.* **2021**, *94*, 573–587. [CrossRef]
75. Mishra, S.; Malhotra, G.; Arora, V.; Tiwari, S.R. Do Destination Identity, Memorable Experiences and Perceived Risks Matter to Revisit a Destination in the Age of Global Pandemic? *Paradigm* **2022**, *26*, 119–137. [CrossRef]
76. Herman, G.V.; Caciora, T.; Grama, V.; Baias, S.; BeatÓN, M.O.R.; Green, I.; Hassan, T.H.; Bulz, G.C.; Andriamampianina, H.S.; Gozner, M. Tourist Perception of the „Night of the Museums" Event. Case Study in Oradea Municipality, Romania. *Geoj. Tour. Geosites* **2023**, *47*, 486–492. [CrossRef]
77. Herman, G.V.; Grama, V.; Ilieș, A.; Safarov, B.; Ilieș, D.C.; Josan, I.; Buzrukova, M.; Janzakov, B.; Privitera, D.; Dehoorne, O. The Relationship between Motivation and the Role of the Night of the Museums Event: Case Study in Oradea Municipality, Romania. *Sustainability* **2023**, *15*, 1738. [CrossRef]
78. Marcu, F.; Ilieș, D.C.; Wendt, I.A.; Indrie, L.; Ilieș, A.; Burta, L.; Caciora, T.; Herman, G.V.; Todoran, A.; Baias, S. Investigations regarding the biodegradation of the cultural heritage. Case study of traditional embroidered peasant shirt (Maramures, Romania). *Rom. Biotechnol. Lett.* **2020**, *25*, 1362–1368. [CrossRef]
79. Preko, A.; Gyepi-Garbrah, T.F.; Arkorful, H.; Akolaa, A.A.; Quansah, F. Museum experience and satisfaction: Moderating role of visiting frequency. *Int. Hosp. Rev.* **2020**, *34*, 203–220. [CrossRef]
80. Wang, T.-L.; Tran, P.T.K.; Tran, V.T. Destination perceived quality, tourist satisfaction and word-of-mouth. *Tour. Rev.* **2017**, *72*, 392–410. [CrossRef]
81. Albu, A.V.; Sopota, D.; Caciora, T.; Herman, G.V.; Berdenov, Z.; Ilies, A.; Ilies, D.C.; Kecse, G.; Sturzu, B.; Ghergheles, C.G. Digitalization of garment in the context of circular economy. *Ind. Textila* **2021**, *30*, 31.

sustainability

MDPI

Article

The Sustainability Study Done for a Consolidation Work on a Historical Building

Mircea Raul Tudorica [1,2,*], Mihaela Teodora Toadere [1] and Corneliu Ioan Bob [3]

1 Department of Civil Engineering, Faculty of Constructions, Cadaster and Architecture, University of Oradea, 410058 Oradea, Romania; mihaelatoadere@yahoo.com
2 Doctoral School of Engineering Sciences, "Polytechnic" University of Timisoara, 300006 Timisoara, Romania
3 Department of Civil Engineering and Installations, Faculty of Civil Engineering, "Polytechnic" University of Timisoara, 300233 Timisoara, Romania; corneliu.bob@gmail.com
* Correspondence: tudorica.mircea@gmail.com

Abstract: A very important problem encountered all over the world and increasingly widespread is represented by sustainability. The construction field is responsible for a high environmental impact, for the entire duration of a building's operation, from the construction stage until its demolition. This paper presents a sustainability study, performed on an old historical building located in Romania—Arad County, which implied the consolidation of its resistance structure as a result of visible degradation. The study was performed using the Bob–Dencsak Calculation Model, which involved research into several specific parameters for each dimension separately (ecological, economic and social). Besides establishing the sustainability class for the consolidated building, an analysis was done on the impact that metal has as compared to reinforced concrete, thus resulting in the finding that metal is less sustainable than reinforced concrete, achieving growths of up to 42% for embodied energy and 28.50% of CO_2 emissions in the atmosphere. Finally, the paper offers recommendations for future sustainability assessment research with the aim of increasing the quality of life and minimizing the negative impact on the environment with minimal costs.

Keywords: sustainability study; embodied energy; GHG gas emissions; metal frame; reinforced concrete; environmental; economic dimension; social criteria; buildings and construction

Citation: Tudorica, M.R.; Toadere, M.T.; Bob, C.I. The Sustainability Study Done for a Consolidation Work on a Historical Building. *Sustainability* **2023**, *15*, 15285. https://doi.org/10.3390/su152115285

Academic Editor: Syed Minhaj Saleem Kazmi

Received: 15 September 2023
Revised: 20 October 2023
Accepted: 24 October 2023
Published: 25 October 2023

1. Introduction

Global warming due to greenhouse gas emissions is one of the most important problems the world is facing. In order to find a solution for this matter, we must keep in view the steps that must be taken in order to shift to a sustainable state, or to attain sustainable development. Over time, the meaning of the term sustainability has become more and more complex, but in the broadest sense sustainability means the capacity to continuously support a process over time [1]. The most widely agreed definition in the specialty literature is: "Sustainable development is development that meets the needs of the present without compromising the ability of future generations to meet their own needs" [2]. The Circular Economy has gained popularity as a viable tool for long-term development [3]. It facilitates collaboration between sectors where residues can be applied to a different process to create economic value, lessen waste, and reduce material consumption [4]. However, a crucial part of sustainable development is played by the global process of shelter provision. Up to 50% of the world's carbon emissions are reportedly caused by buildings when taking into account the entire building value chain, from production of raw materials through construction, use of buildings, and demolition [5]. In the European Union, buildings contribute to a sizable portion of final energy consumption and CO_2 emissions. By 2050, the EU Energy Performance of Buildings Directive hopes to have a stock of buildings that is both highly energy-efficient and carbon-free. This directive mandates that member states implement sustainable policies and practices to guarantee affordable solutions for reducing energy consumption and carbon emissions from both new and existing structures [6].

Buildings continue to be behind schedule for achieving carbon neutrality by 2050 despite the anticipated rebound in emission levels in 2021 being moderated by continued decarbonization of the power sector. To meet this goal, all new buildings and 20% of existing building stock must be completely carbon-free ready by 2030 [7]. Every existing building will need to undergo energy upgrades that combine increased operational energy efficiency, a switch to district heating systems that are powered by renewable energy sources without adding any carbon to the atmosphere, and the production and/or purchase of carbon-free renewable energy in order to fully decarbonize the building sector [8]. Factors like environmental pollution along with energy conservation have become crucial in the modern world due to the use of various energy sources and the worry that some of them will run out [9]. One approach in this sense is the use of biofuels and hydrogen fuel when they are produced with renewable processes. While biomass has a limited supply and needs to be used properly, electro-fuels have garnered a lot of attention. Although some fuels are constrained by carbon sources, electro-fuels are not constrained in the same way [10,11]. Thus, with the aim of reducing and estimating the energy consumption of a building [12,13], Ostergaard et al. focused on the status of renewable technologies, the role that renewable energy sources play in achieving sustainable development, the state of research into the long-term viability of renewable energy sources, and finally on incorporating renewable energy sources into low-carbon energy systems [14].

Several researchers have conducted extensive research on the energy and carbon design for zero energy communities [15,16] and have come to the conclusion that maximizing building sustainability through passive energy efficiency measures is a promising strategy and guarantees cost-effective solutions that reduce the operational energy use of buildings and associated CO_2 emissions [17,18]. For instance Bungau et al. propose measures by which the design, execution and operation of spaces, correctly and optimally, solves and possibly even eliminates the unsanitary and unhygienic conditions inside buildings in order to enhance standards for protecting the built environment [19]. Prada-Hanga et al. present how to optimize the energy consumption of an old building located on the campus of the University of Oradea, which is engaged in the "wave of renovations" [20]. Furthermore, strategies and solutions for 'green' and 'healthy' university campuses based on the application of green building principles (e.g., retrofitting old buildings, information management, environmental protection, and circular economy principles) have been evaluated in the literature [21].

Jie et al. looked into how the thickness of building envelope components affected the calculation of energy use and greenhouse gas emissions [22]. According to Nizam et al., embodied energy, transportation, and innovation linked to design processes are critical for ensuring the sustainability of building construction [23]. Boloni et al. explored a case study oriented in particular to recycled concrete, in the manner of which Eco-design, as a creative strategy in the construction and building sectors, Life Cycle Thinking and Life Cycle Assessment as fundamental sustainable development tools, and Construction and Demolition Waste recycling processes can encourage the circular economy in construction and building development [24]. However, recent studies highlight the significance of using a life cycle approach and taking the building's entire life cycle into account when aiming to adopt measures to enhance the sustainability performance of buildings [25,26].

Despite the variations in the standard definitions of "green buildings" and "sustainability," both ideas are frequently used interchangeably and in close proximity to one another. While "Green" is a particular term that frequently emphasizes products, people, and the impact on the environment, "sustainable" has a wider definition that includes the environmental, social, and economic pillars of sustainable development. These are the most used pillars in addressing sustainability issues [27–29]. The concept of the green building has witnessed substantial growth as a subject of research in recent times due to the increasing tensions arising from the extensive growth of construction and the deterioration of the natural environment [30]. Another close link is the historical interconnectedness between the concept of heritage and sustainable development [31]. Aspects related to politics, technology, and culture are also regarded as sub-domains of sustainable development [32,33]. A new systematic domain model with four dimensions (economic, ecological, political, and cultural) was recently proposed, which is in line with Agenda 21, UNESCO, and the United Nations, particularly the fourth dimension of sustainable development—culture [34]. The developing trend of sustainable cities and communities is greatly aided by innovations in the building and construction sector. Environmental impacts are divided into six levels by the construction and demolition (C&D) management hierarchy, a shift from low to high as follows: reduce, reuse, recycle, compost, incinerate, and landfill [35].

In the specialized literature, there are not many sustainability studies regarding the consolidation of old damaged buildings. In this sense, three buildings in the historic center of Viseu, Portugal, were chosen and used as case studies by Alemida et al. They used a Simplified Method for the Sustainability Assessment, starting from the MARS Calculation Model adopted in 2009, Portugal, which analyzes 19 coefficients from five coverage areas: Water—SA, Energy—SE, Materials—SM, Emissions—SAE and Cultural, Economic and Social Environments—CES [36]. Honarvar et al. investigated the architectural growth of two buildings in Iran's arid environment. According to the results, the evolution of architecture has resulted in a 78% reduction in energy consumption compared to previous architecture [37].

Finally, ensuring sustainability is a noble task which civil engineers should pursue, both in the design stage as well as during the consolidation process. The aim pursued in this paper is the calculation of the sustainability index of an old consolidated historical building, made of brick, using the Bob–Dencsak calculation model. Using the same calculation model in the article [38], a sustainability index BSI = 87.75 was obtained for an investment with the functionality of an agro-tourism guesthouse, placing the construction in the very good category of buildings from the sustainability point of view. This is due to the fact that a modern construction resulted, in terms of materials and efficient equipment, aspects that are completely missing from the studied building in this article. This model is a relatively new one, introduced in the specialized literature in 2010, being a complex and complete model that follows the analysis of 45 coefficients specific to the fundamental dimensions of sustainability. Due to the large number of coefficients, in the absence of accurate data, a common problem in the case of old buildings, they can be expressed based on experimental research, the result expressing a good classification, very close to the real situation of the construction from the point of view of sustainability. The article also presents an analysis of the impact that metal has compared to reinforced concrete, demonstrating that the use of metal is less durable than that of reinforced concrete, obtaining increases of up to 42% for embodied energy and 28.50% of CO_2 emissions in the atmosphere. The content of this article can be useful in the case of all buildings, but especially old and damaged ones, where there is a gap in the universal literature, leading to the best decisions regarding technical consolidation solutions, in order to meet global objectives in terms of sustainability.

2. Sustainability Models

2.1. Certificates and Models for Assessing the Sustainability of a Building

Worldwide, in the specialized literature, several complex models are presented for calculating the sustainability of a construction. They contain various parameters of several dimensions that influence the study of sustainability. The most frequently used sustainability models are presented in Figure 1 [39].

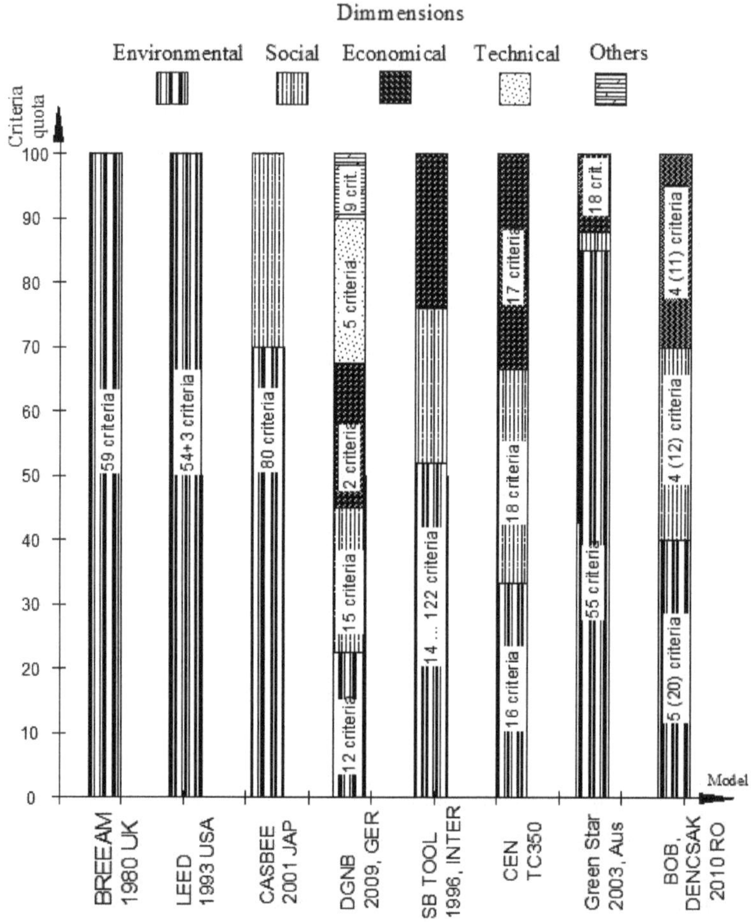

Figure 1. Sustainability Assessment Methods.

The Building Research Establishment Environmental Assessment Method (BREEAM) and Leadership in Energy and Environmental Design (LEED) are two of the most well-known sustainability models that are constantly being improved [40]. This expanding and re-scoping of what constitutes building sustainability is reflected in the ongoing emergence and development of numerous performance rating systems [41,42].

The average values of sustainability assessment methods for the main dimensions are shown in Table 1.

Table 1. Average values of sustainability assessment methods.

Environmental	Social	Cost/Economic
$e_{med} = 0.64$	$s_{med} = 0.18$	$c_{med} = 0.15$
$*e^1_{med} = 0.46$	$*s^1_{med} = 0.28$	$*c^1_{med} = 0.29$

* without including BREEAM and LEED criteria.

In Table 2 are presented the number of parameters for each dimension, their weight in the final result and the classification of construction from the perspective of sustainability [39,40].

Table 2. Classification of constructions from the sustainability point of view.

Sustainability Model	Ecological Criteria	Economic Criteria	Social Criteria	Construction Classification
BREEAM UK 1990 (59)	59 (100%)	-	-	Insufficient < 30 points Good enough 30–85 points Very Good > 85 points
LEED USA 1993 (57)	57 (100%)	-	-	Bronze Medal 40–49 points Silver Medal 50–59 points Gold Medal 60–79 points Platinum Medal > 85 points
CASBEE Japan 2001 (80)	56 (70%)	-	24 (30%)	C Class—grades < 0.5 B− Class -grades 0.5–1 B+ Class—grades 1–1.5 A Class–grades 1.5–3 S Class—grades > 3
SBTool Model International 1996 (14–122)	48%	24%	24%	Acceptable—score < 1 Good—score 1–3 Excellent—score > 3
CEN TC350 (51)	16 (33.3%)	17 (33.3%)	18 (33.3%)	Maximum score is 100 points. The classification being done on the score obtained.
Bob-Dencsak Romania 2010 (45)	21 (40%)	11 (30%)	13 (30%)	Very Good > 80 points (>4) Good 60–80 points (3–4) Acceptable 40–60 points (2–3) Insufficient < 40 points (<2)

2.2. The Bob–Dencsak Calculation Model, 2010 Romania

The Bob–Dencsak calculation model covers all 3 dimensions of the sustainability of a construction. The ecological dimension is divided into 5 groups of parameters: Energy—E_n, Gas emissions—G, Materials and resources—MR, Construction site works—CS and Land use and water consumption—LW. This dimension has the largest weight, i.e., 40% of the value of the sustainability index.

With a ratio of 30%, the economic dimension is characterized by the 4 groups of parameters: Cost—C, Construction Process—CP, Project Management—PM and Efficiency (duration of service)—Ef. Also representing 30% of the value of the sustainability index, the social dimension includes 4 parameters that take into account the Comfort—Cf, Air Quality—IAQ, Safety—Sa, and Accessibility/Adaptability—AA, respectively.

Thus, the Bob–Dencsak calculation model is a complete and complex model, which follows all 3 dimensions of sustainability both in the construction phase and in its exploitation phase.

The calculation formulas for each parameter mentioned above are not the subject of this article, and are presented in detail in [38]. The score obtained for each parameter was achieved by interpolating the value between the minimum and the optimum benchmark. The sustainability index BSI is the sum $p_i \times w_i$ for all three criteria:

$$\mathrm{BSI} = e \times 0.4 + c \times 0.3 + s \times 0.3 \tag{1}$$

$$e = \frac{\sum p_i^e \times w_i^e}{0.4} \tag{2}$$

$$c = \frac{\sum p_i^c \times w_i^c}{0.3} \tag{3}$$

$$s = \frac{\sum p_i^s \times w_i^s}{0.3} \tag{4}$$

where e represents the environmental criteria, c represents the economic criteria and s the social criteria.

3. Case Study

3.1. The Current Situation of the Consolidated Building

The subject of the sustainability study is an old structurally consolidated building, made of masonry, located on Revolutiei Avenue No. 55 in Arad County, Romania. It is classified as a monument building in the state heritage dating from 1870 according to the Inspection Certificate of Regional Construction Department West—Arad, presented in Figure 2.

Figure 2. Monument building located on Revolutiei Avenue No. 55.

The subject characteristics of the analyzed structure are as follows:

- Construction type: S+P+2E (basement, ground floor and two levels);
- The resistance structure of the building is made of brick masonry, with the following wall thicknesses: 30–90 cm at the basement of the building, 40–50 cm on the ground floor and 30–50 cm, both on the first and second floor of the building;
- The floors are made of metal beams with brick buttresses (prefabricated tiles);
- The roof structure is covered with ceramic tiles;
- The foundations are continuous under the walls, made of poor-quality concrete;

Due to the fact that there were no complete plans for regarding the damaged structure, measurements were made, in order to establish some plans on the parts with obvious and dangerous faults regarding strength and stability, as seen in Figure 3.

As shown in Figure 3, the building has suffered significant damage as a result of the soil failure that occurred between axes B 1-3 and E 1-3, with a higher tendency to collapse at the blind wall E 1-3, as well as the resistance structure's weakness due to the time it was completed. The presentation of the main causes of degradation or the seismic structural design are not the subject of this paper, but are detailed in the [43].

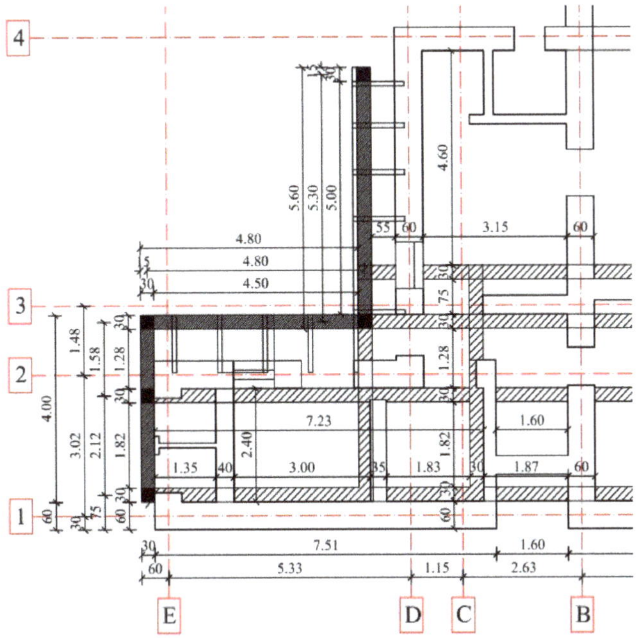

Figure 3. The section of the building analyzed in the study.

The consolidation work for the entire building was executed by MANYKON HOUSE Company from Arad—Romania, in two parts, due to the economic and social criteria. The first part started in August 2016 with the rehabilitation of the infrastructure and the second part continued in September 2017 with the rehabilitation of the building superstructure.

3.2. Materials Used in the Consolidation Process—Technical Data

Following the examples of technical expertise carried out in 2012, which lay behind the technical rehabilitation project No. 138/2014 drawn up by PERBO Company from Timisoara—Romania, the following phases were provided in order to strengthen the structure, in chronological order:

1. Taking down and eliminating the two trees that are close to the building;
2. 144 points of a cement–bentonite suspension will be injected beneath the current foundations;
3. Construction of independent foundations beneath the new reinforced concrete frame columns. Flat girdles measuring 30 × 5 cm (30 cm wide and 5 cm thick) are positioned at each level to connect the reinforced concrete RC frames to the existing structure.
4. Strengthening the cracked walls, by placing gibs 6 to 8 mm in diameter on the fissure channel and injecting epoxy resins SIKA REPAIR into them.

The economic part of the work for this consolidation was carried out by Civil Project Design Company from Oradea—Romania, which resulted in a total cost of 46,760 EURO.

The quantities of materials used for the consolidation, as well as technical data for the coefficients regarding the embodied energy and gas emissions for these materials, are given in Table 3 [43–45].

Table 3. Quantities of materials, technical data and coefficients used in the sustainability study.

Building Material	Quantity Volume (m^3) Weight (kg)	Embodied Energy Coefficient EE (MJ/kg)	GHG Emissions Coefficient EC-CO_2 (kgCO_2/kg)
Injection Cement CEM II/A-M (S-V) 42.5	28.3 m^3 52,638 kg	5	0.80
Concrete C16/20	46.3 m^3 111,120 kg	0.81	0.115
Timber, Softwood, air dried, roughswan	13.2 m^3 8448 kg	7.4	0.59
Steel section virgin U200 profile	0.05 m^3 392.5 kg	38	2.82
Mortar (Cement:sand mix 1:3)	3.18m^3 5247 kg	1.33	0.221
Steel Bar	0.6 m^3 4710 kg	29.20	2.59
Epoxy resin	0.006 m^3 11 kg	137	5.70

With the exception of the U200 metal profiles that were brought from a warehouse located 20 km from the site, all materials were brought from the nearest local building materials warehouse, located 10 km away. They were transported in trucks with a capacity of 3.5–20 t, which have the following characteristics: Embodied Energy coefficient EE = 4.60 MJ/tkm and GHG Emissions coefficient EC-CO_2 = 0.28 kg CO_2/tkm.

Embodied energies from non-renewable sources in building materials used for upkeep, renovation and replacement operations (En_3), are calculated with Formula (5).

$$En_3 = \frac{EE \times m}{A \times t} \quad (5)$$

Emissions of greenhouse gases (GHG) from building materials used for upkeep, renovation and replacement operations (G_3) are calculated with Formula (6).

$$G_3 = \frac{EC - CO_2 \times m}{A \times t} \quad (6)$$

The volume of the rehabilitated part of the building is 1834.62 cubic meters and the useful surface on all four levels is A = 632.56 square meters. The building was designed for a life cycle of 75 years (t = 75 years).

3.3. Reinforced Concrete Frame vs. Metal Frame

In the design stage of the intervention work, it was proposed to replace the monolithic reinforced concrete frame with a metal structure. Compared to the concrete frame, the metal version requires a much faster execution time as well as higher mechanical resistance, but it also has a series of disadvantages, such as special handling equipment that would have been very difficult to introduce in the inner courtyard of the building.

The differences between concrete and metal frames were studied, and in this sense Peyroteo et al. state that the reduction of steel is an advantage concerning the environmental impacts of the building, and that the reinforced concrete (RC) framed buildings have far less embodied energy and CO_2 emissions [46]. The same opinion is supported by Ranjbar et al., who concluded that RC-framed buildings use 5% less CO_2 for production and they are more effective than SS frames in terms of their operational electricity and gas consumption [47]. A completely different opinion was held by Zhong and Wu, who examined the effects of structural frames in Singapore on the environment and economic performance, where it is known that the gas emissions in reinforced concrete structures are 24–50% more than in steel structure frames [48].

In this research, the embodied energy consumption and the CO_2 gas emissions of the materials used in the building consolidation process in the existing reinforced concrete version and the metal frame version are highlighted, due to the fact that the production stage can account for up to 75% of the total energy consumption and carbon impact during the life cycle of a building [49–51].

Two solutions were proposed for making the metal frame, the first with a section of 200 × 200 mm, made up of corner profiles with equal sides of 60 × 60 × 6 mm, and the second with a section of 250 × 250 mm, also made up of corner profiles with equal sides 100 × 100 × 10 mm, both stiffened on each side from meter to meter with metal plates.

The mechanical characteristics of the two sections, such as M_{cap}—capable moment, T_{cap}—shear force capacity and section stiffness K, are presented in Table 4.

Table 4. Mechanical characteristics of the analyzed metal section proposed for the study.

Frame Section (mm)	M_{cap} (kNm)	T_{cap} (kN)	K (kNm)
Metal frame with 60 × 60 × 6 profile 200 × 200	58	464	1500
Metal frame with 100 × 100 × 10 profile 250 × 250	156	1290	4165

Due to the fact that the stiffness of the 200 × 200 mm section is three times lower than the reinforced concrete section, 30 × 30 cm having stiffness K = 4550 kNm, the study continued with the 250 × 250 mm section, the stiffness of which is comparable to the reinforced concrete section. The section of the metal frame analyzed in the study is presented in Figure 4.

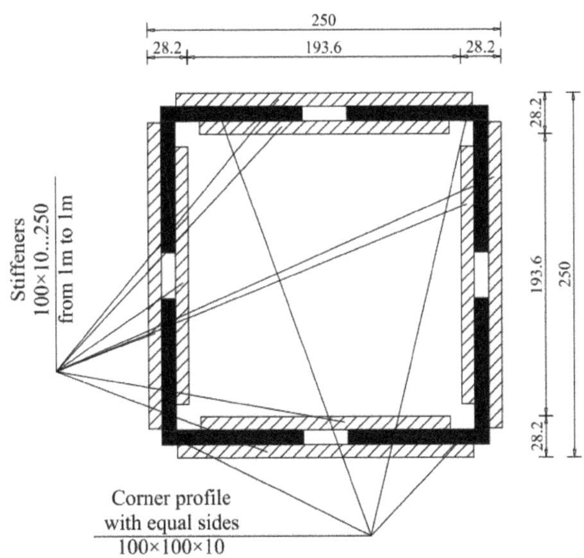

Figure 4. Section of the metal frame analyzed in the study—250 × 250 mm.

The materials used in the rehabilitation of the building, 2350 kg of reinforcement and 10.30 m^3 of concrete, were used to make the reinforced concrete frame. These materials will be removed from the calculation of embodied energy and CO_2 released into the atmosphere.

The metal frame proposed in the sustainability study consists of five pillars of 250 × 250 mm with a total height of 11 m, each being made of four corner profiles with equal sides of 100 × 100 × 10 mm. These profiles are arranged on the corners of the pillars, stiffened meter by meter on each side with metal plates of 250 × 100 × 10 mm.

Horizontally, the frame is made up of four beams arranged between the pillars of the frame (two beams of 1.85 m, one of 4.50 m, respectively, one with an opening of 5 m), with a total length of 13.20 m per level. They are arranged on all three levels of the building, having the same section as the frame pillars.

In cross-section, the effective area of the corner profile with equal sides is A_{ef} = 19.2 cm^2, resulting in 1.755 m^3 of steel required to make the frame in the metal version. The characteristics and the quantities of the rehabilitation work for the metallic version are presented in Table 5.

Table 5. Quantities of work required to make the metal frame, technical data and coefficients.

Building Material	Quantity Volume (m^3) Weight (kg)	Embodied Energy Coefficient EE (MJ/kg)	GHG Emissions Coefficient EC-CO_2 (kgCO_2/kg)
Injection Cement CEM II/A-M (S-V) 42.5	28.3 m^3 52,638 kg	5	0.80
Concrete C16/20	36 m^3 86,400 kg	0.81	0.115
Timber, Softwood, air dried, roughswan	13.2 m^3 8448 kg	7.4	0.59
Steel section virgin U200 profile	0.05 m^3 392.5 kg	38	2.82
Mortar (Cement:sand mix 1:3)	3.18 m^3 5247 kg	1.3 3	0.221
Steel Bar	0.30 m^3 2233 kg	29.20	2.59
Metal Frame 250 × 250 mm	1.755 m^3 13,777 kg	38	2.82
Epoxy resin	0.006 m^3 11 kg	137	5.70

Using Formulas (5) and (6) from the previous chapter, the following values are obtained for the embodied energy, En_3 = 21.45 MJ/m^2/year, and the content of CO_2 gases released into the atmosphere during the rehabilitation process, G_3 = 2.211 kgCO_2/m^2/year.

4. Results and Discussion

The result of the sustainability index was obtained using the Bob–Dencsak model, thus calculating all the parameters in question. The score obtained for each parameter was achieved by interpolating the value between the minimum and the optimum benchmark, as can be seen in Table 6.

Table 6. Values of sustainability parameters.

Parameter Name	Benchmark		Calculated or Estimated Value	Point Score w_i	Weight Factor p_i%	$p_i \times w_i$ Points
	w_i Min 20 Points	w_i Opt 100 Points				
En_1. (MJ/sqm/y) Initial embodied non-renewable energy in original construction materials	180.00	60.00	121.21	59.19	2.50	1.48
En_2. (MJ/sqm/y) Embodied non-renewable energy in all building operations facilities (HVAC)	1100.00	450.00	-	50	6.50	3.25
En_3. (MJ/sqm/y) Embodied energy from non-renewable sources in building materials used for upkeep, renovation, and replacement operations	40.00	15.00	12.34	100	2.00	2.00
En_4. (MJ/sqm/y) After-life non-renewable energy embedded in building materials	35.00	10.00	2.21	100	1.00	1.00
En_5. (%) Use of renewable energy sources	0.00	25.00	0.00	0	2.00	0.00
G_1. (kg CO_{2eq}/sqm/y) Initial GHG emissions	20.00	6.00	11.20	70.29	2.00	1.41
G_2. (kg CO_{2eq}/sqm/y) GHG emissions from all building operations facilities (HVAC)	93.00	10.00	-	50	4.00	2.00
G_3. (kg CO_2) GHG emissions from building materials used for upkeep, renovation, and replacement operations	3.00	1.00	1.581	76.76	1.00	0.77
G_4. (kg CO_2) End of life GHG emissions	1.90	0.60	0.22	100	1.00	1.00
G_5. (%) Impact of the roof's heat island	29.00	95.00	0.00	0	1.00	0.00
MR_1. (%) Reusing current materials, products and structural components, when it is possible	0.00	50.00	0.00	20	1.00	0.20
MR_2. (kg/m^3) Material efficiency	2000.00	900.00	679.13	100	2.00	2.00
MR_3. (%) The usage of recycled-content materials	0.00	30.00	0.92	22.45	2.00	0.45
MR_4. (km) Use of local resources	60.00	5.00	20	78.18	1.00	0.78
CS_1. (%) Waste on the site, generated by the building and demolition process	5.00	50.00	-	20	2.00	0.40
CS_2. (%) Dust created during construction	20.00	100.00	-	20	1.00	0.20
CS_3. (%) Construction-related noise production	105.00	70.00	-	50	1.00	0.50
LW_1. Land contamination	Yes	No	No	50	2.00	1.00
LW_2. (%) Land occupation ratio	>30	30.00	42	20	2.00	0.40
LW_3. (l/p/d) The amount of potable water used by building occupants	180.00	90.00	-	70	2.00	1.40
LW_4. (%) The ratio of grey or rain water use	0.00	30.00	-	20	1.00	0.20
	Total environmental criteria—e					20.44
C_1. (euro/sqm) Initial cost	650.00	300.00	683.00	20	5.00	1.00
C_2. (euro/sqm/y) Operational cost	40.00	5.00	22.50	60	5.00	3.00
C_3. (euro/sqm/y) Maintenance and Repair Cost	25.00	5.00	2.42	100	3.00	3.00
CP_1. (man × h/sqm) Total time for the construction of the building	120.00	55.00	52.94	100	2.50	2.50
CP_2. (euro/h) Production rate	6.00	15.00	13.74	88.80	2.50	2.22
CP_3.—Ca Construction Schedules	0.40	0.90	0.83	88.80	1.00	0.89
PM_1. (no. of documents) Initial documents	3.00	10.00	6.00	54.29	2.00	1.09
PM_2. (no. of documents) Documents of maintenance and operation	0.00	Yes	Yes	80	2.00	1.60
PM_3. Monitoring of performances	0.00	Yes	Yes	80	2.00	1.60
Ef_1. y Long service life	25.00	75.00	75.00	100	3.00	3.00
Ef_2. (%) Area efficiency	70.00	95.00	83.00	61.60	2.00	1.23
	Total economic criteria—c					21.13
Cf_1. PPD, PMV Thermal Comfort	<15	<6	8.40	78.70	4.00	3.15
Cf_2. Noise and acoustic Comfort	35.00 70.00	47.00 58.00	47.08 83.46	36.53	1.50	0.55
Cf_3. (%) Visual Comfort	0.50	3.00	1.57	54.24	1.50	0.81
IAQ_1. (%) VOC concentration in indoor air	0.30	0.80	0.55	60	1.00	0.60
IAQ_2. CO concentration in indoor air	Yes	No	No	80	2.00	1.60
IAQ_3. Ventilation efficiency in spaces with mechanical or natural ventilation	0.30	0.80	0.55	60	1.00	0.60
Sa_1. Protection against earthquake	RsI	RsIV	RsIII	73.30	7.00	5.13
Sa_2. (mm) Protection against flood	1000.00	6000.00	6000.00	100	4.00	4.00
Sa_3. Protection against fire	5.00	1.00	1.00	100	3.00	3.00

Table 6. *Cont.*

Parameter Name	Benchmark		Calculated or Estimated Value	Point Score w_i	Weight Factor p_i%	$p_i \times w_i$ Points
	w_i Min 20 Points	w_i Opt 100 Points				
AA_1. (Min) Public transportation availability and close proximity to user-specific amenities	30/50	5/10	5/10	100	1.50	1.50
AA_2. Lifetime homes	30.00	5.00	-	50	1.50	0.75
AA_3. Adaptability constraints imposed by structure	No	Yes	Yes	50	1.00	0.50
AA_4. Ability to change the type of energy supply in the future	No	Yes	Yes	75	1.00	0.75
Total social criteria—s						22.94

The sustainability index is represented by the sum of all three criteria. According to the importance they have in the model, the following values were obtained: e = 51.10 for the environmental dimension, c = 70.43 for the economic dimension and s = 76.47 for the social dimension. Using Formula (1) presented at the beginning of this article, a sustainability index BSI = 64.51 resulted, which classifies it in the category of good buildings from a sustainability point of view.

The graphical result of the parameters e, c and s are presented in Figure 5, where the hatched triangle represents the sustainability index.

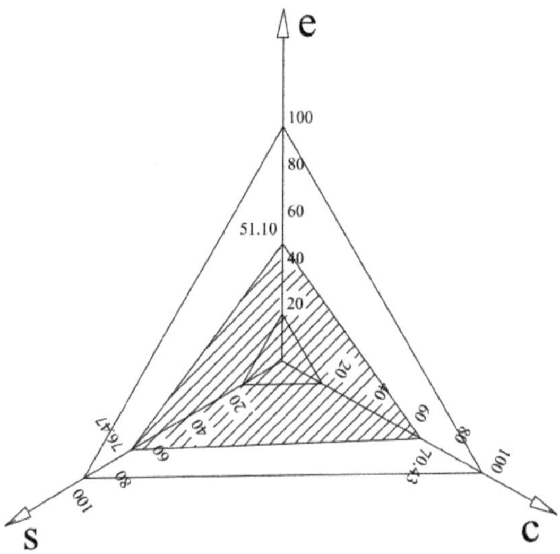

Figure 5. Graphic interpretation of the sustainability index BSI.

We can interpret this result more as "acceptable but good", because the value obtained is very close to the lower limit of the category that would decrease the performance level to "acceptable".

The result of the sustainability index would increase considerably by proposing and carrying out a thermal rehabilitation study regarding the building envelope, which would involve the use of energy-efficient equipment, such as a solar panels system, thus producing energy, together with an economic heating source, such as geothermal water; another relevant example is the use of heat pumps. This resolves both heating in the cold period of the year and cooling during the warm months through fan coil units. Besides the fact that they are efficient from an energy point of view, they also ensure the ventilation of the indoor air in all months of the year, leading to an increase in the results obtained in the social criterion through parameters that take into account the air quality IAQ and those of comfort Cf.

Even if the rehabilitation of the structure in the existing version with reinforced concrete is more sustainable than the analyzed metallic version, an important aspect is represented by the resistance of the metal frame, where the capable shear strength has the value T_{cap} = 1290 kN, which contributes to improving the parameters of the social criterion. However, the analyzed structure falls in both situations into the class of good buildings from the sustainability point of view.

This study could be a starting point to reach this objective, because worldwide there are many buildings of this kind. Considering this, special attention must be paid to all buildings, to new ones but especially to old ones, possibly monumental buildings, with sustainability as the main focus. Thus, we can notice the weak points as well as the strong points of the building, which through a separate study, depending on the data for the works that are going to be executed, could lead to a maximum sustainability index.

5. Conclusions

The result obtained after calculating the sustainability index BSI = 64.51 classifies the consolidated structure in the category of good buildings from a sustainability point of view. It is observed that the environmental dimension scored the lowest points e = 51.10, followed by the economic dimension with a result of c = 70.43, respectively, while the social dimension had a satisfactory score for the studied building of s = 76.47.

This is highlighted by the fact that, first of all, it is a very old building that was reconsolidated due to the structural degradation of the resistance structure, but also due to the disadvantages that the building presents from the perspective of the standards and updated requirements from the construction field, such as the fairly high levels and very thick walls made of underperforming materials. The results of the sustainability study clearly show the weak points of the building, such as energy consumption En and gas emissions released into the atmosphere G.

The parameter with the highest percentage, more precisely 7% of the sustainability index value, is represented by the factor which takes into account the seismic risk class Sa1. For structures included in seismic risk classes RsI and RsII, it is not possible to analyze their sustainability. Since it is necessary to strengthen their resistance structure in emergency mode, these buildings cannot be used under normal conditions. In conclusion, it would be imperative that, together with technical expertise and the technical project that includes structural consolidation solutions, the sustainability study should also be carried out to optimize all directly affected parameters to the maximum, so as to result in an efficient building from all three points of view, ecologically, economically and socially.

Another conclusion that can be supported following the study is represented by the fact that the rehabilitation of the structure in the reinforced concrete frame version is more sustainable from the point of view of the embodied energy and the carbon dioxide emissions released into the atmosphere, compared to the analyzed metallic version. In this way, percentages of 42% for embodied energy and 28.5% for CO_2 emissions were obtained in favor of the existing situation with a reinforced concrete frame.

Author Contributions: Conceptualization, M.R.T.; Formal analysis, M.R.T.; Investigation, M.R.T.; Methodology, C.I.B.; Resources, M.R.T., M.T.T. and C.I.B.; Supervision, C.I.B.; Validation, M.R.T. and M.T.T.; Visualization, M.T.T. and C.I.B.; Writing—original draft, M.R.T.; Writing—review & editing, M.R.T. All authors have read and agreed to the published version of the manuscript.

Funding: This research received no external funding.

Institutional Review Board Statement: Not applicable.

Informed Consent Statement: Not applicable.

Data Availability Statement: Data are available from the first and last author. The literature data are in accordance with the references inserted.

Conflicts of Interest: The authors declare no conflict of interest.

References

1. What Is Sustainability? Available online: https://www.investopedia.com/terms/s/sustainability.asp (accessed on 10 June 2023).
2. United Nations. *Report of the World Commission on Environment and Development: Our Common Future*; United Nations: New York, NY, USA, 1987.
3. Ghisellini, P.; Cialani, C.; Ulgiati, S. A review on circular economy: The expected transition to a balanced interplay of environmental and economic systems. *J. Clean. Prod.* **2016**, *114*, 11–32. [CrossRef]
4. Korhonen, J.; Honkasalo, A.; Seppälä, J. Circular Economy: The Concept and its Limitations. *Ecol. Econ.* **2018**, *143*, 37–46. [CrossRef]
5. Espuny, M.; Nunhes, T.V.; de Oliveira, O.J. How Is Building Sustainability Understood?—A Study of Research Papers and Sustainability Reports. *Sustainability* **2022**, *14*, 12430.
6. EU Directive. Directive (EU) 2018/844 of the European Parliament and of the Council of 30 May 2018 Amending Directive 2010/31/EU on the Energy Performance of Buildings and Directive 2012/27/EU on Energy Efficiency. 2018. Available online: https://www.legislation.gov.uk/eudr/2018/844 (accessed on 6 July 2023).
7. Kuloves, K.; Oberthür, S. Assessing the EU's 2030 Climate and Energy Policy Framework: Incremental change toward radical transformation? *Rev. Eur. Comp. Int. Environ. Law* **2020**, *29*, 151–166. [CrossRef]
8. Architecture 2030. Actions for a Zero Carbon Built Environment. 2023. Available online: https://architecture2030.org/existing-building-actions/ (accessed on 9 August 2023).
9. Seraji, M.A.N.; Ranjbar, Z.; Keshavarzzadeh, M.; Mousavi, S.; Zadehi, R. Biomass and solar energy in buildings, energetic dark greenhouse, an economic approach analysis. *Therm. Sci. Eng.* **2023**, *6*, 55–63. [CrossRef]
10. Ridjan, I.; Mathiesen, B.V.; Connolly, D. Terminology used for renewable liquid and gaseous fuels based on the conversion of electricity: A review. *J. Clean. Prod.* **2016**, *112*, 3709–3720. [CrossRef]
11. Kwon, P.S.; Ostergaard, P.A. Priority order in using biomass resources—Energy systems analyses of future scenarios for Denmark. *Energy* **2013**, *63*, 86–94. [CrossRef]
12. Zadehi, R.; Aslani, A.; Gitifar, A.; Farahani, O.N.; Yousefi, H. Application of Artificial Neural Network in predicting building's energy consumption. In Proceedings of the 8th International Conference on Technology and Energy Management (ICTEM), Mazandaran, Babol, Iran, 8–9 February 2023; Volume 10084336, pp. 1–5.
13. Prada, M.; Prada, I.F.; Cristea, M.; Popescu, D.E.; Bungău, C.; Aleya, L.; Bungău, C.C. New Solutions to Reduce Greenhouse Gas Emissions through Energy Efficiency of Buildings of Special Importance—Hospitals. *Sci. Total Environ.* **2020**, *718*, 137446. [CrossRef]
14. Ostergaard, P.A.; Duic, N.; Noorollahi, Y.; Kalogirou, S. Renewable energy for sustainable development. *Renew. Energy* **2022**, *199*, 1145–1152. [CrossRef]
15. Fouad, M.M.; Iskander, J.; Shihata, L.A. Energy, carbon and cost analysis for an innovative zero energy community design. *Sol. Energy* **2020**, *206*, 245–255. [CrossRef]
16. Balali, A.; Hakimelahi, A.; Valipour, A. Identification and prioritization of passive energy consumption optimization measures in the building industry: An Iranian case study. *J. Build. Eng.* **2020**, *30*, 101239. [CrossRef]
17. Gil-Baez, M.; Padura, Á.B.; Huelva, M.M. Passive actions in the building envelope to enhance sustainability of schools in a Mediterranean climate. *Energy* **2019**, *167*, 144–158. [CrossRef]
18. Friess, W.A.; Rakhshan, K. A review of passive envelope measures for improved building energy efficiency in the UAE. *Renew. Sustain. Energy Rev.* **2017**, *72*, 485–496. [CrossRef]
19. Bungau, C.C.; Prada, I.F.; Prada, M.; Bungau, C. Design and operation of construction: A healthy living environment—Parametric studies and new solutions. *Sustainability* **2019**, *11*, 6824. [CrossRef]
20. Prada-Hanga, I.F.; Bungau, C.C.; Scurt, A.A.; Cristea, M.; Prada, M.F. The Building Certification System—A Tool of Sustainable Development of University Campuses. *J. Appl. Eng. Sci.* **2023**, *13*, 105–112. [CrossRef]
21. Bungau, C.C.; Bungau, C.; Toadere, M.T.; Prada-Hanga, I.F.; Bungau, T.; Popescu, D.E.; Prada, M.F. Solutions for an Ecological and Healthy Retrofitting of Buildings on the Campus of the University of Oradea, Romania, Built Starting from 1911 to 1913. *Sustainability* **2023**, *15*, 6541. [CrossRef]
22. Jie, P.; Zhang, F.; Fang, Z.; Wang, H.; Zhao, Y. Optimizing the insulation thickness of walls and roofs of existing buildings based on primary energy consumption, global cost and pollutant emissions. *Energy* **2018**, *159*, 1132–1147. [CrossRef]
23. Nizam, R.S.; Zhang, C.; Tian, L. A BIM based tool for assessing embodied energy for buildings. *Energy Build.* **2018**, *170*, 1–14. [CrossRef]
24. Bonoli, A.; Zanni, S.; Serrano-Bernardo, F. Sustainability in Building and Construction within the Framework of Circular Cities and European New Green Deal. The Contribution of Concrete Recycling. *Sustainability* **2021**, *13*, 2139. [CrossRef]
25. Rivera, M.L.; MacLean, H.L.; McCabe, B. Implications of passive energy efficiency measures on life cycle greenhouse gas emissions of high-rise residential building envelopes. *Energy Build.* **2021**, *249*, 111202. [CrossRef]
26. Shadram, F.; Mukkavaara, J. Exploring the effects of several energy efficiency measures on the embodied/operational energy trade-off: A case study of swedish residential buildings. *Energy Build.* **2019**, *183*, 283–296. [CrossRef]
27. Khoshnava, M.; Rostami, R.; Valipour, A.; Ismail, M.; Rahmat, A.R. Rank of green building material criteria based on the three pillars of sustainability using the hybrid multi criteria decision making method. *J. Clean. Prod.* **2018**, *173*, 82–99. [CrossRef]

28. Awadh, O. Sustainability and green building rating systems: LEED, BREEAM, GSAS and Estidama critical analysis. *J. Build. Eng.* **2017**, *11*, 25–29. [CrossRef]
29. Capra, F.; Luisi, P.L. *The Systems View of Life: A Unifying Vision*; Cambridge University Press: Cambridge, UK, 2014.
30. Bungau, C.C.; Bungau, T.; Prada, I.F.; Prada, M.F. Green Buildings as a Necessity for Sustainable Environment Development: Dilemmas and Challenges. *Sustainability* **2022**, *14*, 13121. [CrossRef]
31. Bogdan, A.; Chambre, D.; Copolovici, D.M.; Bungau, T.; Bungau, C.C.; Copolovici, L. Heritage Building Preservation in the Process of Sustainable Urban Development: The Case of Brasov Medieval City, Romania. *Sustainability* **2022**, *14*, 6959. [CrossRef]
32. James, P. *Urban Sustainability in Theory and Practice: Circles of Sustainability*; Routledge: London, UK, 2014.
33. Magee, L.; Scerri, A.; James, P.; Thom, J.A.; Padgham, L.; Hickmott, S.; Deng, H.; Cahill, F. Reframing social sustainability reporting: Towards an engaged approach. *Environ. Dev. Sustain.* **2013**, *15*, 225–243. [CrossRef]
34. The Johannesburg World Summit on Sustainable Development. Recognized Cultural Diversity as the Fourth Pillar of Sustainable Development, Alongside the Economic, Social and Environment Pillars. 2002. Available online: https://www.un.org/en/conferences/environment/johannesburg2002 (accessed on 10 August 2023).
35. Peng, C.-L.; Scorpio, D.E.; Kibert, C.J. Strategies for successful construction and demolition waste recycling operations. *Constr. Manag. Econ.* **1997**, *15*, 49–58. [CrossRef]
36. Almeida, C.P.; Ramos, A.F.; Silva, J.M. Sustainability assessment of building rehabilitation actions in old urban centres. *Sustain. Cities Soc.* **2018**, *275*, 378–385. [CrossRef]
37. Honarvar, S.M.H.; Golabchi, M.; Ledari, M.B. Building circularity as a measure of sustainability in the old and modern architecture: A case study of architecture development in the hot and dry climate. *Energy Build.* **2022**, *275*, 112469. [CrossRef]
38. Tudorică, M.; Ghemiș, M.-T.; Bob, C. The Sustainability of a Building Made by using of Recycling Materials. *WSEAS Trans. Environ. Dev.* **2022**, *18*, 1532–1539.
39. Bob, C.; Dencsak, T. *Building Sustainability. Civil Engineering Approach*; Politehnica: Timisoara, Romania, 2010.
40. Díaz-López, C.; Carpio, M.; Martín-Morales, M.; Zamorano, M. Analysis of the scientific evolution of sustainable building assessment methods. *Sustain. Cities Soc.* **2019**, *49*, 101610. [CrossRef]
41. Doan, D.T.; Ghaffarianhoseini, A.; Naismith, N.; Zhang, T.; Ghaffarianhoseini, A.; Tookey, J. A critical comparison of green building rating systems. *Build. Environ.* **2017**, *123*, 243–260. [CrossRef]
42. Shan, M.; Hwang, B. Green building rating systems: Global reviews of practices and research efforts. *Sustain. Cities Soc.* **2018**, *39*, 172–180. [CrossRef]
43. Tudorica, M.; Bob, C. The influence of foundation soils cocerning the behavior of buildings. *J. Appl. Eng. Sci. Univ. Oradea Publ. House* **2015**, *5*, 87–93.
44. Heires, M. The international organization for standardization (ISO). *New Political Econ.* **2008**, *13*, 357–367. [CrossRef]
45. Mabdeh, S.; Ali, H.; Al-Momani, M. Life Cycle Assessment of Energy Retrofit Measures in Existing Healthcare Facility Buildings: The case of Developing Countries. *Int. J. Energy Econ. Policy* **2022**, *12*, 418–443. [CrossRef]
46. Peyroteo, A.; Silva, M.; Jalali, S. Life Cycle Assessment of Steel and Reinforced Concrete Structures: A New Analysis Tool. Available online: https://www.irbnet.de/daten/iconda/CIB11672.pdf (accessed on 14 August 2023).
47. Ranjbar, N.; Balali, A.; Valipour, A.; Yunusa-Kaltungo, A.; Edwards, R.; Pignatta, G.; Moehler, R.; Shen, W. Investigating the environmental impact of reinforced-concrete and structural-steel frames on sustainability criteria in green buildings. *J. Build. Eng.* **2021**, *43*, 103184. [CrossRef]
48. Zhong, Y.; Wu, P. Economic sustainability, environmental sustainability and constructability indicators related to concrete- and steel-projects. *J. Clean. Prod.* **2015**, *108*, 748–756. [CrossRef]
49. Shadram, F.; Mukkavaara, J.; Schade, J.; Sandberg, M.; Olofsson, T. Trade-off optimization of embodied versus operational carbon impact for insulation and window to wall ratio design choices: A case study. In Proceedings of the 10th International Conference on Sustainability in Energy and Buildings (SEB), Gold Coast, Australia, 24–26 June 2018; Volume 131, pp. 12–20.
50. Chau, C.K.; Leung, T.M.; Ng, W.Y. A review on Life Cycle Assessment, Life Cycle Energy Assessment and Life Cycle Carbon Emissions Assessment on buildings. *Appl. Energy* **2015**, *143*, 395–413. [CrossRef]
51. Giesekam, J.; Barrett, J.; Taylor, P.; Owen, A. The greenhouse gas emissions and mitigation options for materials used in UK construction. *Energy Build.* **2014**, *78*, 202–214. [CrossRef]

sustainability

MDPI

Article

In Vitro Propagation of Several Valuable Selections of *Robinia pseudoacacia* L. as a Fast and Sustainable Source for Wood Production

Ruben Budău †, Mariana Bei †, Cristian Onet *, Eliza Agud, Olimpia Smaranda Mintas, Adrian Ioan Timofte, Cristina Adriana Rosan , Vasile Laslo * and Simona Ioana Vicas

Faculty of Environmental Protection, University of Oradea, 26 Gen. Magheru Street, 410048 Oradea, Romania; rbudau@uoradea.ro (R.B.); mbei@uoradea.ro (M.B.); eliza.agud@uoradea.ro (E.A.); olimpia.mintas@uoradea.ro (O.S.M.); atimofte@uoradea.ro (A.I.T.); crosan@uoradea.ro (C.A.R.); svicas@uoradea.ro (S.I.V.)

* Correspondence: cristian.onet@uoradea.ro (C.O.); vasile.laslo@uoradea.ro (V.L.)

† These authors contributed equally to this work.

Abstract: There is an increasing trend in forest production towards planting rapid-growing trees as attractive, environmentally friendly energy sources. This study aimed to establish an alternative to the traditional propagation of a number of selections of *Robinia pseudoacacia* L. by developing an in vitro culture protocol. This study's topic is of great importance, and it reflects an ongoing concern at the University of Oradea's Faculty of Environmental Protection's sustainable research program. The explants from four forms (called S1, S2, S3, and S4), selected for their phenotypic characteristics, were inoculated on four culture media (Murashige–Skoog (MS), Anderson, Chée–Pool, and Driver and Kuniyuki Woody (DKW)) with the same phytohormonal balance. DKW medium proved to be the better support of morphogenic activity, and it was further tested under different phytohormonal balances. Different results were observed depending on the hormone content in the DKW environment. In the presence of 0.5 mg/L benzylaminopurine (BAP) and 0.04 mg/L aminoisobutyric acid (AIB), 91.5% of the explants developed an average of 4.45 ± 0.18 shoots, whereas the average upper shoot height (3.82 cm) was recorded on DKW medium with 0.5 mg/L BAP and 0.04 mg/L α-naphthaleneacetic acid (NAA). Auxin, 0.05 mg/L AIB, promoted root production (5.27 ± 0.15 roots/explant), while 0.1 mg/L NAA promoted root length. In conclusion, the S4 selection produced the greatest outcomes of all environmental variables in terms of both the number of shoots and their heights.

Keywords: black locust; *Robinia pseudoacacia* L.; in vitro propagation; tissue cultures; phytohormones

Citation: Budău, R.; Bei, M.; Onet, C.; Agud, E.; Mintas, O.S.; Timofte, A.I.; Rosan, C.A.; Laslo, V.; Vicas, S.I. In Vitro Propagation of Several Valuable Selections of *Robinia pseudoacacia* L. as a Fast and Sustainable Source for Wood Production. *Sustainability* **2023**, *15*, 15243. https://doi.org/10.3390/su152115243

Academic Editors: Alina Badulescu, Constantin Bungău, Delia Mirela Tit, Cosmin Mihai Vesa and Dorina Camelia Ilieș

Received: 3 October 2023
Revised: 21 October 2023
Accepted: 22 October 2023
Published: 25 October 2023

1. Introduction

Woody plants are usually resistant to in vitro propagation. However, studies to establish in vitro propagation protocols are needed to obtain high-quality micropropagated plants in a short amount of time. Micropropagation techniques are also important when the preservation of genetic resources is desired, when somaclonal variation is exploited, or when genetic engineering techniques are used.

Robinia pseudoacacia L. is a fast-growing, multi-purpose nitrogen-fixing tree that is used for timber and firewood as an important melliferous species and with very good results in the conservation and fight against soil erosion [1–3]. This species is mainly propagated via seeds. This method of propagation has the disadvantage that the plants obtained in this way are significantly uneven. Additionally, some Robinia plants produce few seeds. With a low yield for propagation, root or stem cuttings can also be used, but these methods are difficult and require tremendous efforts and a long time to grow, seeing as, in the case of root cuttings, the radicular biological material is limited, as it is collected in small quantities during the tree's rest period to avoid injury. The propagation of valuable biotypes in culture involves their identification and selection followed by in vitro micropropagation that ensures a very good uniformity of propagules. Several papers have been published that addressed the problem of in vitro multiplication of this species, with emphasis on the type of explants and the type and concentration of plant growth regulators [4–6]. Plants that are regenerated in vitro form roots more easily than cuttings taken directly from mature trees and can also be produced in large numbers [7,8]. The results that have been reported so far have described a series of difficulties related to the in vitro micropropagation of some Robinia selections [5,9,10]. As approaches to the in vitro regeneration of this species, the literature has outlined propagation from nodal explants and axillary buds [11–13], the formation of adventitious shoots from the callus, leaves, and hypocotyl fragments, cotyledons [14,15], and embryogenesis [16,17].

Optimizing mineral nutrients in the culture medium represents an important solution for enriching the micro-propagation protocol of plants, as these have a substantial influence on morphogenesis and organogenesis [18,19]. The ingredients which comprise culture mediums (organic and non-organic nutrients, such as carbohydrates, vitamins, and plant growth regulators) are determining factors for the quality of the final product obtained in any culturing protocol of plant cells [20]. Multiple studies have reported physiological disturbances and/or toxicity due to their lacking or excessive presence in the un-organic composition of the culture medium [21,22]. The most widely used culture medium for plant tissues is the MS medium [23,24]. Its composition makes it apt for elaborating new medium compositions, although it is often inadequate as it generates physiological disturbances, such as the necrosis of shoot tips and/or vitrification [25–27]. The first study relating to this topic, which focused on the in vitro propagation of acacia, was initially conducted in 1985 by Enescu and Jucan [10] in Romania. In the same year, Balla and Vértesy (1985) [9] reported the successful micropropagation of acacia in Hungary. In subsequent years, these studies have reported that the efficiency of regeneration is influenced both by the genotype and the composition and concentration of micro- and macro-elements, growth regulators, and other components of the culture medium. Initially, the MS culture medium was used, with other compositions being subsequently tested, with the predominant goal of obtaining well-rooted cuttings [28,29] In Romania, research was not focused on multiplication for the purposes of production of certain cultivars or selected acacia clones, with studies focusing foremost on establishing micropropagation protocols [2].

This study aimed to perfect the in vitro culture techniques used to multiply several valuable forms of acacia, emphasizing, as a novelty element, the influence that the composition of the culture medium has on the morphogenesis processes.

2. Materials and Methods

2.1. Acacia Selections

The four acacia selections were chosen in the O.S. Săcueni of the Bihor Forestry Directorate (latitude: 47°32′38″ north; longitude: 22°10′23″ east; altitude 140 m). Each selection was chosen by comparing it to at least four other trees with particular qualities and situations in its immediate area (on or within 25–30 m), with the condition that the age difference between them does not exceed ten years.

The main aim of biotype selection is productivity [30]; however, this is directly related to a complex of characteristics, such as tree height, diameter, branching tendency, adaptability to winter frosts, and resistance to diseases and pests.

The trunk/spindle shape in the biotypes investigated was defined via remarkable dimensions, the productivity of woody mass, straightness, and a height of at least 10 m from the ground level.

The phenotypic differences among the acacia selections (labeled as S1, S2, S3, and S4) are shown in Table 1.

Table 1. Phenotypic characterization of the acacia selections.

Acacia Selection	Height (m)	Diameter at a Height of 1.3 m	Volume (m^3)	Height at Which the First Lateral Branch Appears (m)
S1	24	52	2.103	17
S2	22	50	1.780	15
S3	19	46	1.310	13
S4	20	48	1.494	14

2.2. Culture Conditions

The biological material consisted of nodal explants obtained from the selected specimens (labeled as S1, S2, S3, and S4). After the shoots were collected from the selected specimens, they were placed in the growth chamber with the bases in water until the commencement of active growth. We collected the tips of the growing shoots from which the explants were shaped for inoculation.

Nodal explants were sterilized by washing using a tap water jet for 15 min, following which they were disinfected with 70% ethyl alcohol for 30 s and with 0.1% mercuric chloride for 5 min. Treatments were followed by 5 rinses with sterile distilled water, each lasting 5 min. The inoculation was carried out in a sterile hood (Microflow, Tamil Nadu, India) in 30 mL tubes with 8 mL of medium. Following inoculation, the tubes were placed in the Sanyo growth chamber, which was set to 20 °C at night and 25 °C during the day, with a photoperiod of 8 h in the dark and 16 h of light (3000 lx). We made 14 repetitions for each variant. The following aspects were evaluated after 28 days: shoot regeneration, callus formation, number of adventitious roots, and number of shoots formed per explant. The experimental design from which the stages that were completed in this investigation emerge is presented in Figure 1.

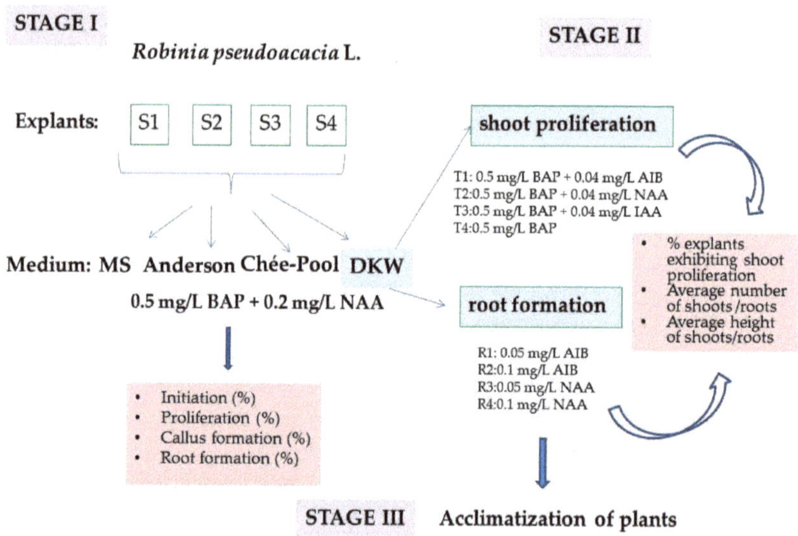

Figure 1. Experimental design. S1, S2, S3, and S4 acacia selections described in Table 1. MS—Murashige-Skoog medium, DKW—Driver and Kuniyuki Woody medium, BAP-6—benzylaminopurine, NAA-α—naphthaleneacetic acid, AIB-α—aminoisobutyric acid, and IAA—indole-3-acetic acid.

2.3. Shoot Proliferation and Root Formation

The influence of the composition in macro- and micro-elements of some culture media in supporting morphogenesis processes was analyzed using the culture media developed by Murashige Skoog (1962), Anderson (1975), Chée and Pool (1987), and DKW (Driver and Kuniyuki, 1984) (Table 2). All culture media used in this study were purchased from Duchefa Biochemie (Haarlem, the Netherlands).

Table 2. The chemical composition of the culture media used in the micro-multiplication stage of the species *Robinia pseudoacacia* L.

	Culture Media Composition			
Micro-Elements (mg/L)	**Murashige–Skoog**	**Chée–Pool**	**Anderson**	**DKW**
$CoCl_2 \cdot 6H_2O$	0.025	0.025	0.025	
$CuSO_4 \cdot 5H_2O$	0.025	0.025	0.025	0.25
FeNaEDTA	36.70	36.70	73.40	44.63
H_3BO_3	6.20	6.20	6.20	-
KI	0.83	-	0.30	4.80
$MnSO_4 \cdot H_2O$	16.90	0.85	16.90	33.80
$Na_2MoO_4 \cdot 2H_2O$	0.25	0.25	0.25	0.39
$ZnSO_4 \cdot 7H_2O$	8.60	8.60	8.60	17.0
Macro-Elements (mg/L)				
$Ca(NO_3)_2$	-	492.30	-	-
KH_2PO_4	170.00	170.00	-	265.0
KNO_3	1900.00	1900.00	480.0	-
$MgSO_4$	180.54	180.54	180.54	361.49
NH_4NO_3	1650.00	1650.00	400.0	1416.0
$CaCl_2$	332.02	-	332.02	112.50
NaH_2PO_4	-	-	330.60	-
$Ca(NO_3)_2 \cdot 2H_2O$	-	-	-	1664.64
K_2SO_4	-	-	-	1559.0

Culture media were brought to pH 5.6 before autoclaving at a pressure of 1.2 kgf/cm^2 for 20 min. After sterilization, 0.5 mg/L BAP and 0.2 mg/L NAA were added to each medium recipe through a 0.22 µm sterile filter (ISOLAB, Laborgeräte GmbH, Eschau, Germany).

2.4. *Acclimatization*

In order to acclimatize, the tubes with rooted plants were opened gradually, over several days, with the duration of opening increasing in order to reduce the relative humidity. This acclimatization stage lasted 14 days, after which the plants were removed from the culture medium and the agar on the roots was removed under running tap water. The culture substrate, consisting of a mixture of sand, peat, and perlite (1:1:1), was previously disinfected by spraying with 0.3% $KMnO_4$ [2]. The moisture provided to the substrate is a critical factor for plant survival. High humidity results in low aeration; the roots can be infected by mold and other pathogens; a lower substrate humidity dehydrates the leaves, and the seedlings die due to drought stress. During this stage, the relative humidity was maintained in the range of 70–80%.

2.5. *Statistical Analysis*

The results represent the means and standard deviation (SD). For each acacia selection, 14 repetitions were made in triplicate. Statistical significance between groups was determined with the one-way ANOVA test followed by Tukey's multiple comparison test, using GraphPad Prism. A value of $p < 0.05$ was considered statistically significant. Different letters for each sample indicate statistically significant differences. Data obtained on the proliferation and rhizogenesis of acacia selections were subjected to principal component analysis (PCA) using the statistical analysis software PAST, version 4.09, in order to establish the optimal in vitro culture medium for the development of acacia selections.

3. Results and Discussion

The type of inoculum, the culture conditions, and the ingredients of the culture media (inorganic and organic nutrients, vitamins, and plant growth regulators) are determining factors in any plant cell culture protocol [20].

Murashige–Skoog (MS) medium is the culture medium widely used in tissue culture laboratories [18], although relatively recent studies have revealed that it generates physiological disturbances, such as necrosis of the tip of the shoot and/or hyperhydricity [25,31]. It was hypothesized that the deaths of explants of some species inoculated on MS medium is due to the high concentrations of mineral salts, especially NH_4NO_3. For many plants, the high content of ammonium nitrogen ($NH4^+$) in the MS medium can increase the stress level in explant tissues, with a toxic effect on tissue proliferation. As a result, vitrified micro-shoots appear, and the regeneration potential of the culture is reduced [18]. Decreasing the concentration of NH_4NO_3 in the MS medium by 2–3 times lowers hyperhydricity in *Prunus avium* [32], *Phoenix dactylifera* [33], and *Aloe polyphylla* [34].

Compared to the Murashige–Skoog medium, in the composition of the Anderson medium, we have a reduction to about 1/4 of the concentrations of NH_4NO_3 and KNO_3. In the medium defined by Chée and Pool, the concentrations of chlorine and manganese are lower, and calcium chloride is replaced with calcium nitrate. Shoot multiplication was improved by excluding iodine and decreasing the manganese concentration. Compared to the Murashige–Skoog medium, the DKW medium lacks the micro-elements $CoCl_2 \cdot 6H_2O$ and KI, and $Ca(NO_3)_2 \cdot 2H_2O$ is added to the macro-elements as a source of calcium and nitrogen, while K_2SO_4 is replaced with KNO_3. When compared to MS basal salts, DKW media contains a similar ammonium/nitrate ratio but less total nitrogen [35].

Recent studies have revealed the potential of the DKW medium and that it may represent an alternative to MS in terms of multiple shoot induction from axillary buds and callus-mediated morphogenesis [36,37].

The composition of the culture medium is one of the recognized factors, with a significant influence on the evolution of the morphogenesis processes in the in vitro cultures of plants [35,38]. Under the conditions of the same phytohormonal balance, the morphogenesis processes evolved in different directions depending on the genotype and the chemical composition of the tested medium. The initiation of organogenesis was triggered in all environmental variants and in all selected genotypes, with values between 78–92% and with insignificant differences between variants (Table 3).

On DKW medium, 85–92% of explants produced between 4 and 5.6 shoots, while 50–64% of explants produced between 3.2 and 4.4 shoots on Chée–Pool medium. As a result, the culture medium x genotype interaction had a significant effect on the number of shoots produced. We did not record significant differences between the number of proliferated shoots from the explants of the selections inoculated on MS and Anderson media ($p = 0.89935$) (Figure 2a). The differences, determined with Tukey's multiple comparison test, were significant in terms of the average number of proliferated shoots/explant between MS and Chée–Pool ($p = 0.00151$), MS and DKW ($p = 0.00000$), Anderson and Chée–Pool ($p = 0.00024$), Anderson and DKW ($p = 0.00000$), and Chée–Pool and DKW ($p = 0.00896$).

The Chée–Pool medium, with an average of 3.9 shoots/explant, and DKW, with 4.5 shoots/explant, are the most advantageous to support the multiplication process (Figure 2b). The Tukey's multiple comparison test revealed statistically significant differences between MS and Chée–Pool ($p = 0.00248$), MS and DKW ($p = 0.00001$), Anderson and Chée–Pool ($p = 0.02901$), and Anderson and DKW ($p = 0.00014$).

The biotype had significant effects on the rate of initiation. There were significant differences between selections S3 and S1 ($p = 0.00433$), S4 and S1 ($p = 0.0004$), and S4 and S2 ($p = 0.0433$). Significant changes in proliferation capacity were found between S2 and S1 ($p = 0.0016$), S3 and S1 ($p = 0.0007$), and S4 and S1 ($p = 0.0002$). Several investigations have demonstrated that there are significant differences in the ability to initiate morphogenesis processes both between the different formulations of culture media and between the biotypes studied [39–42]. The biotype had a significant impact on the rate of initiation. Similarly, Juncker and Favre [43] showed that the genotype had a significant effect on micropropagation capabilities in a study on *Q. robur*, as some individuals perished during the initial culture phase, while others showed a gradual decline in vitality, and most of the youths displayed rapid growth over time.

Table 3. The effect of different types of culture media, under the conditions of the same phytohormonal balance, on the percentage of initiation in organogenesis, proliferation, callus formation, and rhizogenesis.

Medium	Acacia Selection	Initiation (%)	Proliferation (%)	Callus Formation (%)	Root Formation (%)
Murashige–Skoog	S1	85 ± 1.84 [a]	28 ± 2.73 [bc]	71 ± 3.26 [a]	-
	S2	85 ± 2.34 [a]	21 ± 1.23 [b]	71 ± 2.24 [a]	7 ± 0.23 [bc]
	S3	85 ± 3.42 [a]	21 ± 1.24 [b]	64 ± 1.82 [a]	14 ± 1.12 [ac]
	S4	92 ± 2.24 [a]	42 ± 2.36 [b]	42 ± 2.24 [a]	7 ± 0.14 [bc]
Anderson	S1	78 ± 2.11 [a]	28 ± 3.09 [b]	50 ± 2.38 [a]	-
	S2	85 ± 3.12 [a]	28 ± 2.12 [b]	64 ± 2.12 [a]	-
	S3	92 ± 2.16 [a]	21 ± 2.02 [b]	71 ± 3.24 [a]	-
	S4	92 ± 4.14 [a]	42 ± 3.46 [b]	64 ± 2.46 [a]	7 ± 0.48 [bc]
Chée–Pool	S1	78 ± 2.53 [a]	50 ± 2.97 [b]	35 ± 2.77 [b]	21 ± 2.12 [ac]
	S2	85 ± 3.18 [a]	57 ± 1.46 [b]	28 ± 1.86 [b]	21 ± 1.60 [ac]
	S3	85 ± 1.26 [a]	42 ± 3.21 [b]	35 ± 2.44 [bc]	28 ± 1.60 [ac]
	S4	92 ± 4.42 [a]	64 ± 2.48 [a]	42 ± 3.12 [ac]	21 ± 1.22 [ac]
DKW	S1	85 ± 3.45 [a]	78 ± 2.11 [a]	42 ± 3.45 [b]	21 ± 2.08 [ac]
	S2	85 ± 2.62 [a]	85 ± 2.08 [a]	21 ± 3.42 [b]	28 ± 0.68 [ac]
	S3	92 ± 4.20 [a]	78 ± 1.68 [a]	64 ± 4.12 [a]	35 ± 1.68 [ac]
	S4	92 ± 3.26 [a]	92 ± 2.82 [a]	21 ± 2.42 [b]	42 ± 2.24 [a]

Results are expressed as the mean ± SD (n = 14). Different lowercase letters indicate significant differences within the same column ($p < 0.05$).

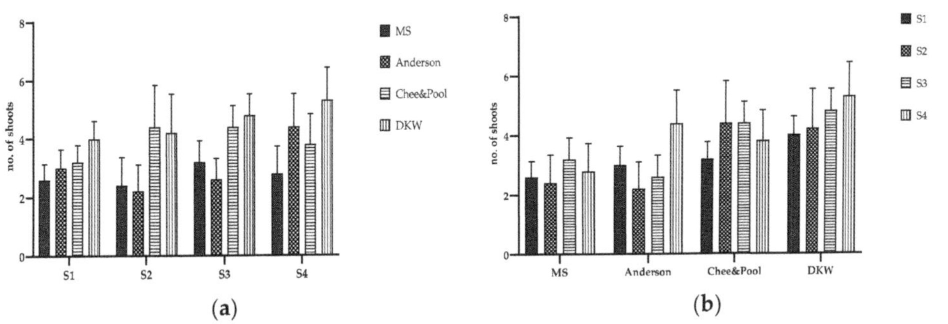

Figure 2. The average number of shoots formed on the tested media after 21 days after inoculation (**a**) genotype influence; (**b**) the influence of the culture environment.

The best callus initiation, with an average of 62% on the four inoculated genotypes, was recorded on the MS and Anderson media. $MgSO_4$, $CaCl_2$, and $MnSO_4$ are the essential nutrient macro-elements for explant growth in tissue culture and for callus formation [44,45]. Calcium chloride is the form of calcium commonly used in the composition of in vitro culture media, with roles in cellular pH, carbohydrate translocation, and callus induction [46]. The concentration of $CaCl_2$ is 332 mg/L in both MS and Anderson culture media. A significant reduction in the induction of calluses formed when Mg^{2+} is missing from the medium has been observed [47,48]. On the DKW medium, 34% of the inocula formed an organogenic callus (after 21 days of inoculation) at the level of the node immersed in the medium. After 5 weeks, calli were formed on this callus from which the shoots were later generated. The presence in the composition of the DKW medium of larger amounts of $MnSO_4$ (33.80 mg/L) and $MgSO_4$ (361.49 mg/L) than in the other tested environments favored the formation of this morphogenic callus.

With the exception of the S1 selection, between 7% and 14% of the inocula on the MS formed roots. On the Anderson medium, only 7% of the S4 explants initiated the rhizogenesis process, and in the other selections this process did not commence. On the Chée–Pool and DKW media, all selections generated roots. The average percentage of rooting in all four selections on the Chée–Pool medium was 22.75%, and on the DKW medium the average on the selections was 31.5%. The following selections with a rooting percentage above the average were observed: S3 on Chée–Pool (28%) and S4 on DKW (42%). Regarding the number of roots formed (Figure 3a), no significant differences between the number of roots formed on the explants inoculated between MS and Anderson ($p = 0.11923$) and between Chée–Pool and DKW ($p = 0.45856$) were recorded, as well as significant differences between the other media (MS and Chée–Pool, MS and DKW, Anderson and Chée–Pool, and Anderson and DKW). The average number of roots on explants inoculated on MS and Anderson was 1.4 ± 0.19, on Chée–Pool media 3.0 ± 0.54, and on DKW media 3.4 ± 0.47.

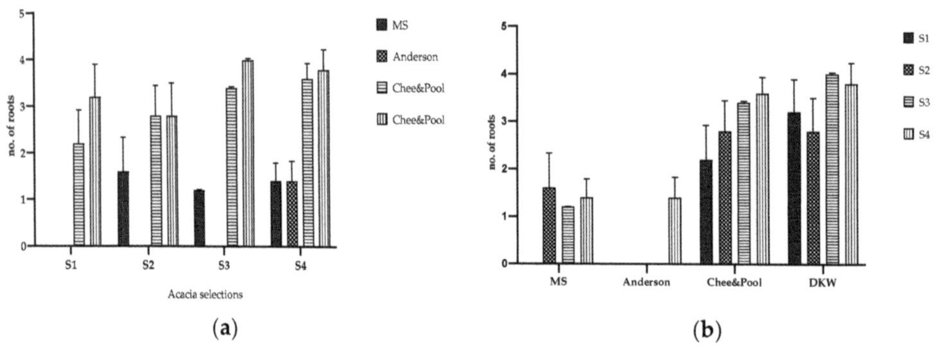

Figure 3. Average number of roots formed on the media tested after 21 days after inoculation (**a**) genotype influence; (**b**) the influence of the culture environment.

In S1 inoculums, the rhizogenesis process was only initiated on the Chée–Pool and DKW mediums, with an average of 2.7 ± 0.69 and 2.4 ± 0.64 roots/explant, respectively (Figure 3b). S2 and S3 did not form roots on Anderson medium. DKW medium is the only medium tested that stimulates the initiation of rhizogenesis in all four selections, ensuring an average number of 3.8 ± 0.56 roots/explant. Suggestive aspects regarding organogenesis on the culture media taken in the study are presented in Figure 4.

Based on the results obtained at this stage, we continued to pursue the possibility of modulating the morphogenesis processes using only the DKW medium with different phytohormonal balances. We carried out several experimental variants with the additions of BAP, NAA, AIB, and IAA. (Table 4). The effect of these combinations on shoot proliferation and root formation is presented in Tables 5 and 6.

Figure 4. Morphogenesis on the tested media. (**a**) Multiplication on DKW medium; (**b**) regeneration of shoots with basal callus on Chée–Pool medium; (**c**) formation of adventitious shoots with basal callus on Anderson medium; (**d**) the formation of shoots with friable, green basal callus on MS medium.

Table 4. The phytohormonal balances of the species *Robinia pseudoacacia* L. investigated in vitro on DKW medium.

Variant DKW		AP(mg/L) [1]	AIB (mg/L) [2]	NAA(mg/L) [3]	IAA (mg/L) [4]
Multiplication	T1	0.5	0.04	-	-
	T2	0.5	-	0.04	-
	T3	0.5	-	-	0.04
	T4	0.5	-	-	-
Rooting	R1	-	0.05	-	-
	R2	-	0.1	-	-
	R3	-	-	0.05	-
	R4	-	-	0.1	-

[1] BAP-6—benzylaminopurine; [2] AIB—indole-3-butyric acid; [3] NAA—1-naphthaleneacetic acid; [4] IAA—indole acetic acid.

The explants consisted of nodal segments of shoots, obtained from the previous experiment. This type of explant exhibits juvenile characteristics, are poor in endogenous contaminants, and give the possibility of ensuring a high homogeneity of the explants. The inoculation was performed in 30 mL tubes with 8 mL of medium. The tubes that were inoculated were incubated at a temperature of 24 ± 1 °C, with a photoperiod of 16/8 h using a white, cold, fluorescent light of 3000 lx. Eighteen replicas of each variant were made. After 5 weeks, the culture was evaluated. The results are presented in Table 5.

The highest percentages of regeneration were obtained in the T1 and T3 environmental variants, with statistically assured differences for $p < 0.05$ compared to the T2 and T4 variants. The largest number of shoots was generated by explants placed on DKW medium with the combination of the phytohormones BAP (0.5 mg/L) and AIB (0.04 mg/L). The most intense shoot growth was ensured by the combination of the phytohormones BAP (0.5 mg/L) and NAA (0.04 mg/L), with statistically ensured differences compared to the other environmental combinations. Our results confirm the observations of Barghchi, M. (1987) and Salem et al. (2022) [1,49] regarding the positive role of NAA in the elongation of shoots grown in vitro in *Robinia pseudoacacia* [1]. The best results, in all environmental variants, in terms of the number of stems and their heights, were recorded in the S4 selection. On average, this selection produced between 10–35% more shoots and 9–14% more roots than the other selections (Table 5). The better morphogenic responses of S4 is consistent with the studies that have stated that the genotype has a large impact on micropropagation ability [43]. The organogenic responses of inoculums to in vitro cultures are determined via the interaction between endogenous phytohormones and those added to the culture medium [50]. Exogenous cytokinins are essential for shoot formation during in vitro culture, but the use of synthetic cytokinins, such as 6-benzylaminopurine (BAP), can have a long-term residual effect, interfering with subsequent subcultures [51]. Auxin has a decisive role in regulating the spatial and temporal aspects of plant growth and development, being involved in the mechanism of the orientation of the cell division plane, before lateral root initiation, and in the formation of meristems [52,53].

Table 5. Shoot proliferation from stem nodal explants under the influence of different concentrations of phytohormones in DKW medium.

Phytohormonal Variant (mg/L)	Selection	% of Explants Exhibiting Shoot Proliferation	Average Number of Shoots	Average Height of Shoots (cm)
T1 BAP (0.5 mg/mL) + AIB (0.04 mg/mL)	S1	94 ± 3.30	3.8 ± 0.12	2.6 ± 0.8
	S2	89 ± 2.42	4.7± 0.28	2.8 ± 0.2
	S3	92 ± 2.02	4.1 ± 0.13	2.4± 0.2
	S4	91 ± 3.32	5.2 ± 0.18	3.5 ±0.12
Mean T1		91.5 ± 2.24 [a]	4.45 ± 0.18 [a]	2.82 ± 0.42 [b]
T2 BAP (0.5 mg/mL) + NAA (0.04 mg/mL)	S1	80 ± 2.40	2.8 ± 0.34	3.8 ± 0.08
	S2	86 ± 2.37	3.4 ± 0.32	3.8 ± 0.3
	S3	88 ± 3.82	3.2 ± 0.22	3.5 ± 0.22
	S4	90 ± 2.02	3.6 ± 0.21	4.2 ± 0.13
Mean T2		86 ± 1.86 [b]	3.25 ± 0.06 [b]	3.82 ± 0.58 [a]
T3 BAP (0.5 mg/mL) + IAA (0.04 mg/mL)	S1	92 ± 2.32	1.5 ± 0.08	2.7 ± 0.08
	S2	90 ± 3.12	2.2 ±0.22	2.4 ± 0.16
	S3	91 ± 2.32	1.8 ± 0.16	2.6 ± 0.2
	S4	90 ± 1.39	2.8 ± 0.14	3.2 ± 0.03
Mean T3		90.75 ± 1.92 [a]	2.07 ± 0.12 [c]	2.72 ± 0.08 [bc]
T4 BAP (0.5 mg/mL)	S1	78 ± 2.22	1.5 ± 0.15	1.8 ± 0.2
	S2	74 ± 2.80	1.2 ± 0.16	2.1 ± 0.12
	S3	80 ± 2.12	1.4 ± 0.14	2.4 ±0.23
	S4	82 ± 2.79	1.6 ± 0.04	2.4 ± 0.14
Mean T4		78.5 ± 2.02 [c]	1.42 ± 0.02 [d]	2.17 ± 0.12 [d]

Results are represented as means ± SD (n = 14); different lowercase letters indicate significant differences within the same column ($p < 0.05$).

The shoots obtained were shaped to lengths of 2.5–3.0 cm and placed on the variants of the culture medium for rooting. After 4 weeks, the culture was evaluated. The results are presented in Table 6.

For the significance threshold $p < 0.05$, the Tukey's HSD test did not reveal significant differences between the variants in terms of the percentage of explants that generated roots (Table 6).

The medium variants T1 and T3, with 0.05 mg/L AIB and NAA, ensured a higher number of roots per explant (Figure 5), statistically assured, than the variants T2 and T4 with 0.1 mg/L. It is worth noting that the variant T1, with 0.05 mg/L AIB, provided 35.1% more roots/explant than T2 (statistically assured $p = 0.0001$), and 17.8% more roots than T4 ($p = 0.02715$). The lengths of the roots were stimulated by their presence in the culture medium of NAA (0.05–0.1 mg/L). The selection S4 on the medium with 0.05 mg/L AIB provided the highest number of roots (5.8 ± 0.11) and the longest length of roots on the medium with 0.1 mg/L NAA.

Table 6. Plant rooting in vitro under the influence of different combinations of auxins in DKW medium.

Phytohormonal Variant (mg/L)	Selection	% of Explants Exhibiting Shoot Proliferation	Average Number of Roots/Explant	Average Length of Roots (cm)
R1 AIB (0.05 mg/L)	S1	70 ± 2.02	4.5 ± 0.16	3.8 ± 0.16
	S2	68 ± 3.12	5.0 ± 0.15	4.6 ± 0.14
	S3	82 ± 2.22	5.8 ± 0.18	3.8 ± 0.05
	S4	90 ± 2.80	5.8 ± 0.11	4.2 ± 0.25
Mean R1		77.5 ± 2.54 ^{a}	5.27 ± 0.15 ^{a}	4.1 ± 0.15 ^{b}
R2 AIB (0.1 mg/mL)	S1	67 ± 2.02	4.6 ± 0.16	2.2 ± 0.06
	S2	60 ± 3.32	3.8 ± 0.15	2.8 ± 0.17
	S3	86 ± 1.32	3.4 ± 0.28	2.5 ± 0.14
	S4	72 ± 3.62	3.8 ± 0.14	2.9 ± 0.25
Mean R2		71.25 ± 2.57 ^{a}	3.9 ± 0.18 ^{b}	2.6 ± 0.15 ^{c}
R3 NAA (0.05 mg/mL)	S1	68 ± 2.38	4.7 ± 0.26	4.5 ± 0.05
	S2	71 ± 2.52	4.6 ± 0.05	4.7 ± 0.15
	S3	67 ± 2.82	5.8 ± 0.19	4.2 ± 0.2
	S4	76 ± 1.62	5.6 ± 0.09	4.8 ± 0.24
Mean R3		70.5 ± 2.33 ^{a}	5.17 ± 0.14 ^{a}	4.55 ± 0.12 ^{a}
R4 NAA 0.1 (mg/mL)	S1	68 ± 2.02	4.4 ± 0.18	4.5 ± 0.14
	S2	66 ± 3.12	3.8 ± 0.04	4.9 ± 0.25
	S3	69 ± 1.02	4.6 ± 0.12	4.4 ± 0.05
	S4	68 ± 1.82	5.1 ± 0.13	5.0 ± 0.15
Mean R4		67.75 ± 2.32 ^{a}	4.47 ± 0.11 ^{b}	4.7 ± 0.14 ^{a}

Results are represented as means ± SD (n = 14); different lowercase letters indicate significant differences within the same column ($p < 0.05$).

(a) (b) (c) (d) (e)

Figure 5. Micromultiplication and rhizogenesis (**a**,**b**) and rooting (**c**–**e**) of *Robinia pseudoacacia* L. S4 inoculums on DKW medium.

3.1. Acclimatization of Plants Obtained In Vitro

An in vitro micropropagation system can only be considered successful after establishing the conditions of transfer and acclimatization to field conditions. For plants obtained in vitro, under rigorously controlled conditions, transfer to the greenhouse or field environment represents a great challenge due to a lower relative humidity level, higher light level, and septic environment [54]. Plants grown in vitro often show a low rate of photosynthesis and incomplete autotrophy, and these may be the reasons for the low survival rates of plants during the acclimation stage [55]. In addition, the nutrients in the culture media cause deviations in development, and repress or modulate several metabolic pathways differently than soil conditions [56]. As a result, plants grown in vitro develop small juvenile leaves, with a weak cuticular layer, defective stomata, and reduced photosynthetic capacity, while the roots have no or very few absorbent hairs [57]. The acclimatization of *Robinia pseudoacacia* plants obtained in vitro to an ex vitro environment by gradually exposing them to the relative humidity of the environment and to different light levels better facilitate the survival of young and physiologically sensitive plants when transferred to the soil. The conditions that are ensured at this stage help the plants to develop a fully functional root system and to better control their stomatal and cuticular transpiration [56].

It has been demonstrated by the authors of [57–62] that this approach improves stomatal physiology and increases the production of epicuticular wax. The same authors recommended that the level of relative humidity should not drop below 80% in order to permit good aeration of the culture. *Robinia pseudoacacia* L. adapted to ex vitro conditions is shown in Figure 6. In our study, the average percentage of survival after 4 weeks was 68%.

Figure 6. Plants of *Robinia pseudoacacia* L. adapted to ex vitro conditions.

3.2. Principal Component Analysis

Principal component analysis is a useful statistical technique for determining the correlations between variables. The PCA plot was used to analyze the following in vitro culture medium variables: MS, Anderson, Chée–Pool, and DKW. A 2D plot PCA (Figure 7) revealed the differences in the acacia selections based on the proliferation and rhizogenesis processes. The covariance matrix eigenvalues revealed that the set of the two principal components (PC1 and PC2) accounted for 88.919% of the total variance in the dataset in terms of the organogenesis process (Table 7). PC1 accounted 77.867% of the variance, with PC2 explaining the remainder of the variance, 11.052%. PC1 had an eigenvalue of 3.115, while PC2 had an eigenvalue of 0.442.

Table 7. The percentage of variance explained by each successive principal component.

PC	Eigenvalue	Cumulative Eigenvalues Explained by Each PC Percentage of Variance	Percentage of Variance Explained by Each PC (%)	Cumulative Percentage of Variance (%)
1	3.115	3.115	77.867	77.867
2	0.442	3.557	11.052	88.919
3	0.332	3.889	8.2959	97.215
4	0.111	4.000	2.7848	100.00

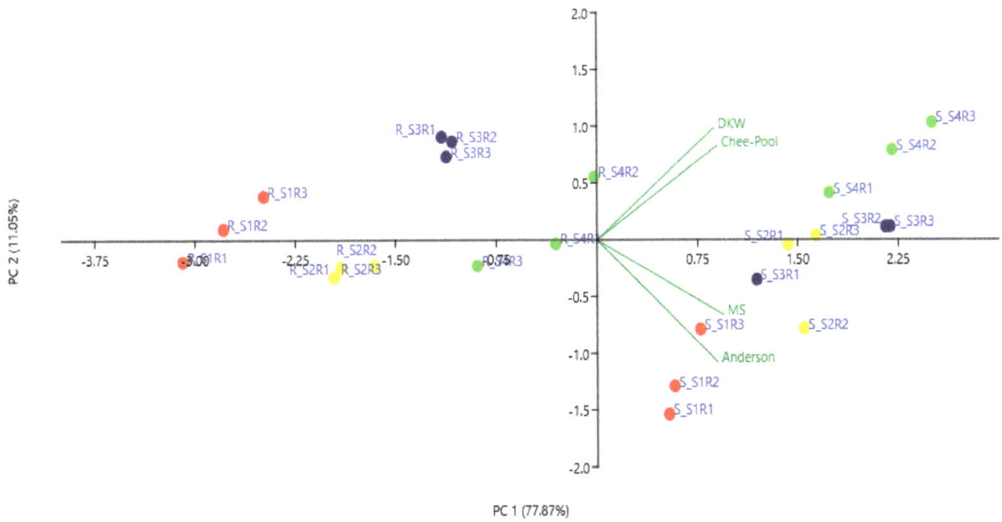

Figure 7. Principal component analysis (PCA) of the organogenesis process of four acacia selections. The red dots represent the percentage of S1 acacia selections' root production (samples with R as the first letter) and proliferation (samples with S from shoots), while the yellow dots represent S2, the blue dots represent S3, and the green dots represent S4.

4. Conclusions

The goal of this research was to test media, with specifications for woody plants, appropriate for large-scale cloning of acacia genotypes reluctant to traditional methods of vegetative multiplication. The vegetative response of the plants varied between the four selections, with a preference for the DKW basal salts. This tendency was manifested both in terms of callogenesis and rhizogenesis and led us to the conclusion that the DKW medium is more suitable for the in vitro multiplication of the *Robinia pseudoacacia* species L. than the other tested media. Considering the current climate context, the *Robinia pseudoacacia* L. species may become more significant due to its strong adaptation to poor and eroded soils, fluctuating temperatures, and positive economic impacts. In addition to providing a starting point for future research on the selection and evolution of this species, the current study can yield important insights for forestry practice and science.

Author Contributions: Conceptualization, R.B. and V.L.; methodology, V.L.; validation, A.I.T., O.S.M. and C.O.; formal analysis, V.L.; investigation, E.A., C.A.R. and M.B.; resources, M.B., R.B. and A.I.T.; writing—original draft preparation, V.L.; writing—review and editing, V.L., S.I.V.; supervision, S.I.V.; project administration, R.B. and V.L.; funding acquisition, R.B. and O.S.M. All authors have read and agreed to the published version of the manuscript.

Funding: The APC was funded by the University of Oradea.

Institutional Review Board Statement: Not applicable.

Informed Consent Statement: Not applicable.

Data Availability Statement: Not applicable.

Acknowledgments: The authors acknowledge the support provided by the University of Oradea through the grant "Excellence scientific research related to priority fields with capitalization through technology transfer: INO-TRANSFER-UO-2nd edition", project no. 260/11.11.2022.

Conflicts of Interest: The authors declare no conflict of interest.

References

1. Barghchi, M. Mass clonal propagation in vitro of *Robinia pseudoacacia* L. (Black locust) cv. 'Jaszkiseri'. *Plant Sci.* **1987**, *53*, 183–189. [CrossRef]
2. Nicolescu, V.-N.; Hernea, C.; Bakti, B.; Keserű, Z.; Antal, B.; Rédei, K. Black locust (*Robinia pseudoacacia* L.) as a multi-purpose tree species in Hungary and Romania: A review. *J. For. Res.* **2018**, *29*, 1449–1463. [CrossRef]
3. Du, B.; Pang, J.; Hu, B.; Allen, D.E.; Bell, T.L.; Pfautsch, S.; Netzer, F.; Dannenmann, M.; Zhang, S.; Rennenberg, H. N2-fixing black locust intercropping improves ecosystem nutrition at the vulnerable semi-arid Loess Plateau region, China. *Sci. Total Environ.* **2019**, *688*, 333–345. [CrossRef] [PubMed]
4. Zhang, J.; Liu, Y.; Wang, H. Micropropagation of Black Locust (*Robinia pseudoacacia* L.). In *Protocols for Micropropagation of Woody Trees and Fruits*; Jain, S.M., Häggman, H., Eds.; Springer: Dordrecht, The Netherlands, 2007; pp. 193–199. ISBN 978-1-4020-6351-0.
5. Szyp-Borowska, I.; Banha, C.; Wojda, T.; Szczygieł, K. Micropropagation of black locust (*Robinia pseudoacacia* L.) and genetic stability of long term cultivated plants. *Folia For. Pol.* **2016**, *58*, 13–19. [CrossRef]
6. Rédei, K.; Keserű, Z.; Rásó, J. Early evaluation of micropropagated black locust (*Robinia pseudoacacia* L.) clones in Hungary. *For. Sci. Pract.* **2013**, *15*, 81–84. [CrossRef]
7. Lyrene, P.M. Juvenility and Production of Fast-rooting Cuttings from Blueberry Shoot Cultures1. *J. Am. Soc. Hortic. Sci.* **1981**, *106*, 396–398. [CrossRef]
8. James, D.J. Adventitious root formation 'in vitro' in apple rootstocks (*Malus pumila*) I. Factors affecting the length of the auxin-sensitive phase in M.9. *Physiol. Plant.* **1983**, *57*, 149–153. [CrossRef]
9. Balla, I.; Vértesy, J. Experiences and problems related to the micropropagation of black locust. *Acta Hortic.* **1987**, *212*, 552. [CrossRef]
10. Enescu, V.; Jucan, A. *Problems of the "In Vitro" Micropropagation of Black Locust (Robinia pseudoacacia L.)*; Book of Abstracts I; International Society for Horticultural Science: Gembloux, Belgium, 1985.
11. Davis, J.M.; Keathley, D.E. Differential responses to in vitro bud culture in mature *Robinia pseudoacacia* L. (black locust). *Plant Cell Rep.* **1987**, *6*, 431–434. [CrossRef]
12. Davis, J.M. In vitro propagation of a black locust tree with an unusual phenotype. *Nitrogen Fixing Tree Res. Rep.* **1988**, *6*, 65–67.
13. Gatti, E.; Sgarbi, E. Micropropagation of *Quercus robur*: Explant sources and cultural conditions affect in vitro responses differently. *Acta Hortic.* **2015**, *1083*, 303–310. [CrossRef]
14. Arrillaga, I.; Merkle, S.A. Regenerating Plants from in Vitro Culture of Black Locust Cotyledon and Leaf Explants. *HortScience* **1993**, *28*, 942–945. [CrossRef]
15. Han, H.K.; Keathley, D.E. Regeneration of whole plants from seedling-derived callus of black locust (*Robinia pseudoacacia* L.). *Nitrogen Fixing Tree Res. Rep.* **1989**, *7*, 129–131.
16. Merkle, S.A.; Wiecko, A.T. Regeneration of Robiniapseudoacacia via somatic embryogenesis. *Can. J. For. Res.* **1989**, *19*, 285–288. [CrossRef]
17. In Park Chungbuk, J. Plant biotechnology research at forest fields in South Korea. *BMC Proc.* **2011**, *5*, 184. [CrossRef]
18. George, E.F.; Hall, M.A.; Klerk, G.-J.D. (Eds.) *Plant Propagation by Tissue Culture: Volume 1. The Background*, 3rd ed.; Springer: Dordrecht, The Netherlands, 2007; ISBN 978-1-4020-5004-6.
19. Reed, B.M.; Wada, S.; DeNoma, J.; Niedz, R.P. Mineral nutrition influences physiological responses of pear in vitro. *In Vitro Cell. Dev. Biol.-Plant* **2013**, *49*, 699–709. [CrossRef]
20. Hameg, R.; Arteta, T.A.; Landin, M.; Gallego, P.P.; Barreal, M.E. Modeling and Optimizing Culture Medium Mineral Composition for in vitro Propagation of Actinidia arguta. *Front. Plant Sci.* **2020**, *11*, 554905. [CrossRef]
21. Bresinsky, A.; Körner, C.; Kadereit, J.W.; Neuhaus, G.; Sonnewald, U. *Strasburger's Plant Sciences: Including Prokaryotes and Fungi*; Springer: Berlin, Heidelberg, 2013; ISBN 978-3-642-15517-8.
22. Nezami-Alanagh, E.; Garoosi, G.-A.; Landín, M.; Gallego, P.P. Computer-based tools provide new insight into the key factors that cause physiological disorders of pistachio rootstocks cultured in vitro. *Sci. Rep.* **2019**, *9*, 9740. [CrossRef]
23. Murashige, T.; Skoog, F. A Revised Medium for Rapid Growth and Bio Assays with Tobacco Tissue Cultures. *Physiol. Plant.* **1962**, *15*, 473–497. [CrossRef]
24. Bhojwani, S.S.; Dantu, P.K. Micropropagation. In *Plant Tissue Culture: An Introductory Text*; Springer: Uttar Pradesh, India, 2013; pp. 245–274. ISBN 978-81-322-1025-2.
25. Nezami-Alanagh, E.; Garoosi, G.-A.; Landín, M.; Gallego, P.P. Combining DOE With Neurofuzzy Logic for Healthy Mineral Nutrition of Pistachio Rootstocks in vitro Culture. *Front. Plant Sci.* **2018**, *9*, 1474. [CrossRef]
26. Nikam, T.D.; Mulye, K.V.; Chambhare, M.R.; Nikule, H.A.; Ahire, M.L. Reduction in hyperhydricity and improvement in in vitro propagation of commercial hard fibre and medicinal glycoside yielding *Agave sisalana* Perr. ex Engelm by NaCl and polyethylene glycol. *Plant Cell Tissue Organ Cult.* **2019**, *138*, 67–78. [CrossRef]
27. Gao, L.; Wang, L.; Sun, Z.; Li, H.; Wang, Q.; Yi, C.; Wang, X. Morusin shows potent antitumor activity for human hepatocellular carcinoma in vitro and in vivo through apoptosis induction and angiogenesis inhibition. *Drug Des. Dev. Ther.* **2017**, *11*, 1789–1802. [CrossRef] [PubMed]
28. Benková, E.; Bielach, A. Lateral root organogenesis—From cell to organ. *Curr. Opin. Plant Biol.* **2010**, *13*, 677–683. [CrossRef] [PubMed]

29. Rédei, K.; Csiha, I.; Keseru, Z.; Rásó, J.; Kamandiné Végh, Á.; Antal, B. Growth and Yield of Black Locust (*Robinia pseudoacacia* L.) Stands in Nyírség Growing Region (North-East Hungary). *Southeast Eur. For. SEEFOR* **2014**, *5*, 13–22. [CrossRef]
30. Károly, R.; Beatrix, B.; Kiss, T.; Takács, M.; Keserű, Z. Yield and crown structure characteristics in a black locust (*Robinia pseudoacacia* L.) stand: A case study—Short Communication. *J. For. Sci.* **2018**, *64*, 96–100. [CrossRef]
31. Mereti, M.; Grigoriadou, K.; Nanos, G.D. Micropropagation of the strawberry tree, *Arbutus unedo* L. *Sci. Hortic.* **2002**, *93*, 143–148. [CrossRef]
32. Daguin, F.; Letouze, R. Ammonium-induced vitrification in cultured tissues. *Physiol. Plant* **1986**, *66*, 94–98. [CrossRef]
33. El-Dawayati, M.M.; Zayed, Z.E. Controlling Hyperhydricity in Date Palm In Vitro Culture by Reduced Concentration of Nitrate Nutrients. In *Date Palm Biotechnology Protocols Volume I*; Al-Khayri, J.M., Jain, S.M., Johnson, D.V., Eds.; Methods in Molecular Biology; Springer: New York, NY, USA, 2017; Volume 1637, pp. 175–183; ISBN 978-1-4939-7155-8.
34. Ivanova, M.; Van Staden, J. Effect of ammonium ions and cytokinins on hyperhydricity and multiplication rate of in vitro regenerated shoots of Aloe polyphylla. *Plant Cell Tissue Organ Cult.* **2008**, *92*, 227–231. [CrossRef]
35. Phillips, G.C.; Garda, M. Plant tissue culture media and practices: An overview. *In Vitro Cell. Dev. Biol.-Plant* **2019**, *55*, 242–257. [CrossRef]
36. Driver, J.A.; Kuniyuki, A.H. In Vitro Propagation of Paradox Walnut Rootstock. *HortScience* **1984**, *19*, 507–509. [CrossRef]
37. Rahman, S.S. Establishment of In vitro micropropagation and callus formation protocols for large scale production of *Clitoria ternatea* L. *J. Med. Spice Plants* **2017**, *22*, 136–144.
38. Polivanova, O.B.; Bedarev, V.A. Hyperhydricity in Plant Tissue Culture. *Plants* **2022**, *11*, 3313. [CrossRef] [PubMed]
39. Puddephat, I.J.; Alderson, P.G.; Wright, N.A. Influence of explant source, plant growth regulators and culture environment on culture initiation and establishment of *Quercus robur* L. in vitro. *J. Exp. Bot.* **1997**, *48*, 951–962. [CrossRef]
40. Vieitez, A.M.; Corredoira, E.; Ballester, A.; Muñoz, F.; Durán, J.; Ibarra, M. In vitro regeneration of the important North American oak species Quercus alba, Quercus bicolor and Quercus rubra. *Plant Cell Tissue Organ Cult.* **2009**, *98*, 135–145. [CrossRef]
41. Kumar, M.; Sirohi, U.; Malik, S.; Kumar, S.; Ahirwar, G.K.; Chaudhary, V.; Yadav, M.K.; Singh, J.; Kumar, A.; Pal, V.; et al. Methods and Factors Influencing In Vitro Propagation Efficiency of Ornamental Tuberose (*Polianthes* Species): A Systematic Review of Recent Developments and Future Prospects. *Horticulturae* **2022**, *8*, 998. [CrossRef]
42. Abdelghaffar, A.M.; Soliman, S.S.; Ismail, T.A.; Alzohairy, A.M.; Latef, A.A.H.A.; Alharbi, K.; Al-Khayri, J.M.; Aljuwayzi, N.I.M.; El-Moneim, D.A.; Hassanin, A.A. In Vitro Propagation of Three Date Palm (*Phoenix dactylifera* L.) Varieties Using Immature Female Inflorescences. *Plants* **2023**, *12*, 644. [CrossRef] [PubMed]
43. Juncker, B.; Favre, J.M. Clonal effects in propagating oak trees via in vitro culture. *Plant Cell Tissue Organ Cult.* **1989**, *19*, 267–276. [CrossRef]
44. Alanagh, E.N.; Garoosi, G.; Haddad, R.; Maleki, S.; Landín, M.; Gallego, P.P. Design of tissue culture media for efficient Prunus rootstock micropropagation using artificial intelligence models. *Plant Cell Tissue Organ Cult.* **2014**, *117*, 349–359. [CrossRef]
45. Fallah Ziarani, M.; Tohidfar, M.; Navvabi, M. Modeling and optimizing in vitro percentage and speed callus induction of carrot via Multilayer Perceptron-Single point discrete GA and radial basis function. *BMC Biotechnol.* **2022**, *22*, 34. [CrossRef]
46. Amin, N.; Khattak, M.; Ahmad, I.; Ara, N.; Alam, A.; ALI, M. Corm and cormel size of gladiolus greatly influenced growth and development of subsequent corm production. *Pak. J. Bot.* **2013**, *2013*, 1407–1409.
47. Walker, K.A.; Sato, S.J. Morphogenesis in callus tissue of *Medicago sativa*: The role of ammonium ion in somatic embryogenesis. *Plant Cell Tissue Organ Cult.* **1981**, *1*, 109–121. [CrossRef]
48. Sarropoulou, E.; Sundaram, A.Y.M.; Kaitetzidou, E.; Kotoulas, G.; Gilfillan, G.D.; Papandroulakis, N.; Mylonas, C.C.; Magoulas, A. Full genome survey and dynamics of gene expression in the greater amberjack *Seriola dumerili*. *GigaScience* **2017**, *6*, gix108. [CrossRef] [PubMed]
49. Salem, J.; Hassanein, A.; El-Wakil, D.A.; Loutfy, N. Interaction between Growth Regulators Controls In Vitro Shoot Multiplication in Paulownia and Selection of NaCl-Tolerant Variants. *Plants* **2022**, *11*, 498. [CrossRef] [PubMed]
50. Chmielarz, P.; Kotlarski, S.; Kalemba, E.M.; Martins, J.P.R.; Michalak, M. Successful In Vitro Shoot Multiplication of *Quercus robur* L. Trees Aged up to 800 Years. *Plants* **2023**, *12*, 2230. [CrossRef] [PubMed]
51. Martins, J.P.R.; Santos, E.R.; Rodrigues, L.C.A.; Gontijo, A.B.P.L.; Falqueto, A.R. Effects of 6-benzylaminopurine on photosystem II functionality and leaf anatomy of in vitro cultivated *Aechmea blanchetiana*. *Biol. Plant.* **2018**, *62*, 793–800. [CrossRef]
52. Blakesley, D.; Weston, G.D.; Hall, J.F. The role of endogenous auxin in root initiation: Part I: Evidence from studies on auxin application, and analysis of endogenous levels. *Plant Growth Regul.* **1991**, *10*, 341–353. [CrossRef]
53. Marhavý, P.; Montesinos, J.C.; Abuzeineh, A.; Van Damme, D.; Vermeer, J.E.M.; Duclercq, J.; Rakusová, H.; Nováková, P.; Friml, J.; Geldner, N.; et al. Targeted cell elimination reveals an auxin-guided biphasic mode of lateral root initiation. *Genes Dev.* **2016**, *30*, 471–483. [CrossRef]
54. Hazarika, B.N. Acclimatization of tissue-cultured plants. *Curr. Sci.* **2003**, *85*, 1704–1712.
55. Faria, R.T.D.; Rodrigues, F.N.; Oliveira, L.D.V.R.; Müller, C. In vitro Dendrobium nobile plant growth and rooting in different sucrose concentrations. *Hortic. Bras.* **2004**, *22*, 780–783. [CrossRef]
56. Teixeira Da Silva, J.A.; Hossain, M.M.; Sharma, M.; Dobránszki, J.; Cardoso, J.C.; Zeng, S. Acclimatization of in Vitro-derived Dendrobium. *Hortic. Plant J.* **2017**, *3*, 110–124. [CrossRef]
57. Ziv, M. In vitro hardening and acclimatization of tissue culture plants. In *Plant Tissue Culture and its Agricultural Applications*; Withers, L.A., Alderson, P.G., Eds.; Butterworths: London, UK, 1986; pp. 187–203.

58. Miller, D. Weaning and growing-on of micropropagated plants. *Proc. Intern. Plant Prop. Soc.* **1983**, *33*, 253–256.
59. Pocock, S. Procedures and problems associated with the transfer of tissue-cultured plants. *Proc. Intern. Plant Prop. Soc.* **1983**, *33*, 316–320.
60. Short, K.C.; Wardle, K.; Grout, B.W.W.; Simpkins, I. In vitro physiology and acclimatization of aseptically cultured plantlets. In *Plant Tissue and Cell Culture Application to Crop Improvement*; Novák, F.J., Havel, L., Doležel, J., Eds.; Institute of Experimental Botany of the Czech Academy of Sciences: Prague, Czech Republic, 1984; pp. 475–486.
61. Wardle, K.; Dobbs, E.B.; Short, K.C. In Vitro Acclimatization of Aseptically Cultured Plantlets to Humidity. *J. Am. Soc. Hort. Sci.* **1983**, *108*, 386–389. [CrossRef]
62. Iliev, I. Techniques for Acclimatization of In Vitro Propagated Plants. Bogdanov, B., Denkova, S., Alexandrov, P., Zhelev, P., Eds.; Propagation of Ornamental Plants; Ministry of Education and Science Publishing House: Sofia, Bulgaria, 1994.

sustainability

MDPI

Article

Do Size and Ownership Determine the Willingness for Sustainable Innovations in Spa and Health Tourism? A Case Study on Baile Felix Spa Resort, Romania

Diana-Teodora Trip, Ramona Simut and Daniel Badulescu *

Department of Economics and Business, Faculty of Economic Sciences, University of Oradea, 410087 Oradea, Romania; diana.trip@csud.uoradea.ro (D.-T.T.); rsimut@uoradea.ro (R.S.)
* Correspondence: dbadulescu@uoradea.ro

Abstract: There is a considerable debate in the literature regarding the real impact of tourism on the environment and human communities, specifically with respect to tourism's openness to sustainable innovation. The way in which the different forms of tourism and entrepreneurial and managerial initiatives respond to the new economic, technological, social and environmental challenges acquire considerable importance for academics, practitioners, decision makers and consumers. The purpose of this research is to assess the potential for sustainable innovation in spa and health tourism in a relevant area in Romania through a survey-based study of both entrepreneurs who own a tourist accommodation unit, as well as top and middle managers from large tourism enterprises. The results of our research show that the development of spa and health tourism offers multiple examples of good practices, and that stimulating innovative entrepreneurial initiatives can lead to a sustainable reinvention of the tourism and local economy, in line with the international health tourism market. In the case of small businesses, even though sustainable initiatives are present, they are at an earlier stage and are less diversified compared to hotels, the scope of innovation is smaller and their motivation weaker. In other words, a large part of the objectives of modernization and sustainability in spa and health tourism (equipment, services, qualification and hiring of employees, sustainable and environmentally friendly orientation) are the responsibility of large hotel resorts. This research provides valuable insights into the ongoing debate surrounding the impact of tourism on the environment and local communities, with a specific focus on the potential for sustainable innovation in spa and health tourism within a significant region of Romania.

Keywords: sustainable innovation; ownership; business size; spa and health tourism; Baile Felix

Citation: Trip, D.-T.; Simut, R.; Badulescu, D. Do Size and Ownership Determine the Willingness for Sustainable Innovations in Spa and Health Tourism? A Case Study on Baile Felix Spa Resort, Romania. *Sustainability* **2023**, *15*, 14501. https://doi.org/10.3390/su151914501

Academic Editor: Harry Coccossis

Received: 12 September 2023
Revised: 27 September 2023
Accepted: 3 October 2023
Published: 5 October 2023

1. Introduction

Due to globalization, the emergence of a competitive environment has increased the importance of entrepreneurship in various fields [1], having an increasingly important role in analyzing and solving both macro and micro problems [2]. Entrepreneurship is also vital in the tourism sector [3], as tourism survival and development are closely linked to entrepreneurial activities and sustainability [4].

Tourism is influenced by changing consumer preferences and the emergence of new technologies [5]. Consumers are becoming less satisfied with the traditional concept of tourism services consisting of accommodation, catering and transportation, and instead demand more experience-based products [6], that requires an entrepreneurial behavior to innovate and create added value. Therefore, to be able to improve product quality by adding new features and tourist services and to penetrate new market segments, tourism depends on new, innovative companies. Unlike other industries, tourism has the relative advantage of not having to involve production processes and significant investments. For this reason, tourism is also considered an economic development strategy in less developed countries or areas [7].

The evolution of international tourism in recent decades, although impressive, can be characterized by the alternation of sustained growth with the shocks of global crises. Thus, in 2019, experts' forecasts for the coming decades predicted that tourism and travel would be one of the fastest growing sectors in the world in the early 21st century [8], and a continuous growth for the 2020–2030 period [9]. Globally, in the last pre-pandemic year (2019), tourism performed beyond expectations, with a record of nearly 1.5 billion international tourist arrivals. In developed countries, the growth rate exceeded the 2014–2019 world average, accounting for 57% of global arrivals and around 61% of international travel revenues [9]. According to World Tourism Organization [10] and World Trade Organization (WTO) [8] reports, international tourism export revenues reached USD 1700 billion in 2019. In 2019, tourism consolidated its position as the third largest global industry, with exports of around USD 1742 billion, after fuels and chemicals, but ahead of automobiles and food.

Export earnings from international tourism are an important source of foreign currency earnings for many destinations around the world. The contributions of tourism to employment and, subsequently, to solving, or at least alleviating, some social problems are widely recognized. The tourism sector generates millions of direct and indirect jobs and is a favorable field of action for private initiative. The majority of tourism enterprises (about 80%) are small- and medium-sized enterprises (SMEs) that employ a large proportion of women and young people. For example, in 2019, women represented 54% of the tourism workforce, compared to around 39% in the global economy [10,11].

Despite this sustained growth, 2020 brought "an unprecedented shock, challenges and an existential threat to the tourism sector" due to the COVID-19 pandemic [12], with travel and tourism being one of the most affected sectors of the global economy. According to UNWTO, international tourism registered a huge decline in 2020: international tourist arrivals (overnight stays) decreased by 74% compared to 2019, and in 2021 by around 72%.

The latest data released by UNWTO in early 2023 indicate that, looking ahead, international tourism is set to recover and improve its situation. Over 72% of UNWTO experts expect better performance in 2023, but almost two thirds of them (65%) believe that international tourism will not return to 2019 levels before 2024, or even later [13].

Critical challenges for the industry include managing changing consumer preferences, ensuring sustainability and adapting to crises. This article aims to assess the potential for sustainable innovation in spa and health tourism in one of the most developed spa regions in Romania, which could serve as a model for the wider tourism industry in Romania, or for other spa businesses internationally. The main goal is to meet these challenges and contribute to scientific knowledge.

Through this article, we aim to assess the sustainable innovation potential in the field of spa and health tourism in a relevant spa tourism area in Romania. The structure of this paper is as follows: after this introduction, we continue with a review of the main contributions on the relationship between tourism, innovation and sustainable development. Next, we will present the context of the research (the health and spa tourism subsector), the research methodology, followed by the main results and discussion. We will end with conclusions and highlight the contribution of this research to scientific knowledge on this topic and the main limitations of the research.

2. Related Works on Sustainable Innovation in Tourism and Firms' Characteristics

2.1. *Tourism, Innovation and Sustainable Development*

The last decades have seen increased concerns regarding the impact of tourism on the environment and society and, subsequently, the promotion of innovation and sustainable behaviors in domestic and international tourism.

Many studies have warned that the impact of tourism (direct or indirect) on a destination is often greater than anticipated. The positive aspects—economic development, jobs, income for businesses and the local community, cultural openness and the revival of some local traditions and customs, and the conservation and restoration of natural and heritage sites [14,15]—are counterbalanced by the existence of some negative effects, such as the leakage of income to developed countries, economic dependence, excessive consumption of resources, corruption and delinquency, environmental pollution, damage to social cohesion in the receiving communities, etc. [14,16,17]. For some researchers and governmental or non-governmental organizations, tourism has an ambivalent relationship with sustainability and sometimes it is seen as an "old-fashioned" practice, which can be indifferent or even aggressive towards the environment [18], being responsible, directly or indirectly, for about 8% of global greenhouse gas emissions, of which about half is due to the contribution of tourist transportation [19].

However, there are also optimistic perspectives: on one hand, technological innovations that offer governments and economic ventures opportunities to incorporate sustainable practices in tourism activities and, on the other hand, the definition and implementation of measures and policies to reduce the carbon footprint of tourism: the general decrease in tourist transportations, the development of local niche markets to distribute the pressure of long-distance transport, education, or public–private partnerships.

The innovation–sustainability–tourism relationship, although intensively invoked in the last decades, is far from being fully clarified and implemented in local, national and global practices and policies. Moreover, according to Hjalager [20], understanding innovation in tourism can be convergent—that is, tourism can learn from the main trajectory of innovation, or divergent—tourism is a separate sector, different from production and even other services, and the investigation of innovation in tourism must take into account other research tools and other perspectives, often multidisciplinary, in which the vision of sustainability (environmental, social) acquires increased importance. The particular characteristics of tourist services (intangibility, simultaneity of production and consumption, the fact that the tourist experience is made up and defined by multiple encounters with providers, the decisive role of information and expectations, the need for intensive labor and the quality of the labor force that influences the tourist experience) also gives a specific character to innovation in this sector [5].

Starting from the fundamental principle of sustainable development—meeting the needs of the present generation without jeopardizing the interests and needs of future generations [21]—tourism develops sustainably as long as it pays attention to local communities and the management of the natural environment [22], reconciling the quality of the tourist experience with the economic, social and cultural life of the residents and the natural environment in which it takes place [23–26]. The technological revolution, communications and the impact of the digital world, economic and social transformations are an undeniable reality of our time and, implicitly, the sustainable development of tourism cannot ignore these realities [27].

Innovation and entrepreneurship stimulate competitivity of any type of business; therefore, innovative entrepreneurship needs to integrate new business models and innovative applications so as to ensure the long-term sustainable development of tourism [17,28].

According to Teodorescu, Stancioiu, Ravar and Botos [29], the innovation and development of new products or services are very important elements for the differentiation and competitivity of a tourist destination, and even if many studies suggest a low permeability of the tourism sector to the innovation and dissemination of technological advances [30–32]; there are, however, many voices claiming that innovative practices in tourism exist and that they integrate technical advances, digital approaches and socio-cultural characteristics with sustainable development.

In tourism, sustainably oriented innovations have a considerable human and social component and often require new managerial and organizational structures, broad support from the local community, education and awareness of providers and tourists and internalization of practices oriented towards sustainable development [31].

The literature argues that the lower significance of innovation in tourism is only on the surface, when, in fact, it is less "visible", being more frequent in process and organizational innovation [33,34] and less in product innovation (which is usually more spectacular and perceptible). Likewise, it follows somewhat different patterns than those in production [35], and the dominance of mass tourism, which demands less sophisticated and relatively similar products, does not stimulate technological innovation but rather changes in behavior and processes [35]. At the same time, tourism innovations involving a relatively unsophisticated level of technology are relatively easy to imitate [20,35] and are accessible and open to competition [5].

Moreover, the vast majority of tourism ventures are small- and medium-sized companies with low capital that are risk-conscious and without the financial and human resources to support research and innovation [36]. In the case of a tourism company, the trend towards innovation is, therefore, positively linked to the size of the business and limited to incremental innovations and the acquisition of models and equipment necessary to increase efficiency, rather than necessarily promoting innovation [33]. Bramwell and Lane [31] also discuss the quality and qualification of human and managerial resources with lower levels of education and training or with various professional experiences, but insufficiently related to tourism and hospitality and moderate openness to technology. Reduced availability to cooperation, partnerships or alliances also diminishes the appetite for innovation and sustainability in the tourism industry [31,37,38]. This can be a serious disadvantage for the tourism industry that needs to be rapidly mitigated, with studies by Hjalager [20,39] finding that relationships and cooperation between tourism firms are essential for the transfer of knowledge, good practices and collective learning, and which could ultimately facilitate innovative ideas. Hjalager argues that regional sustainable innovation systems are capable of boosting institutional learning, building social capital and preparing companies and communities for global challenges and changes. This implies flexibility, openness to the new, while making the most of local connections and focusing on sustainability [39] for tourism.

2.2. The Relationship between Size, Ownership and Innovation and Sustainable Actions in Tourism

The relationship between the company size and innovation and sustainability is intensely debated in literature, with the vast majority of researchers considering that there is a positive relationship between the size of the company and innovation [40,41], which stems from market power, experience, economies of scale, access and control over financial and technical resources [42]. According to Aguilar-Fernández and Otegi-Olaso [43] (p. 5), "the literature reveals a trend towards the notion of a positive relationship between the size of a company and its innovative activities towards sustainability". In the tourism industry, large companies, based on their financial strength and expertise and knowledge of legislation and tax facilities, explore, test and implement innovation to a much greater extent than small ones. They are able to manage several innovative projects simultaneously, spread the risk and absorb the considerable associated costs with innovation through higher volumes of sales, employees and stakeholders with varied knowledge, skills and experience [44].

On the other hand, the realities and opportunities of the contemporary economy also support different views, according to which small businesses now seem to be more innovative than large businesses due to flexibility [45,46], reduced bureaucracy [47], using a business-to-business configuration [48], or alliances with other SMEs to complement and improve their capabilities [49,50]. This trend is also identified in the tourism sector [51], highlighting hotels as the most active and innovative sub-segment of this industry [6]. Finally, other studies consider that no obvious long-term relationship can be established between company size and innovation within enterprises in the tourism industry [44].

Many of the identified studies argue that small businesses and family businesses in tourism are more socially and community-responsible [52–54], and that they calculate and project the future of their business by considering the future of the community and the quality of the destination where they operate [55]. Chen [56] and Dwyer [55] argue that place satisfaction increases the proactive participation of residents (people and businesses) in regional tourism development, and small family-owned businesses adjust their voluntary sustainability practices according to legitimacy, competitiveness and environmental responsibility [57]. This appears to be even more visible in tourism sub-sectors and destinations where ecological and social considerations have a major influence on their sustainable engagement, such as rural tourism [58] or spa and healthcare tourism [59,60]. Qasem, Mohammed, Battisti and Ferraris [61] found a significant positive association between firm sustainable investments and institutional investor ownership in the tourism sector. Companies with foreign shareholders seem to be more interested in sustainable actions than the local ones. Moreover, there is a significant, positive association between managerial ownership and firm sustainable investments, confirming, at least for the contextual case studied, that firm ownership has an influence on firm sustainable involvement in the tourism industry.

2.3. *The Role and the Importance of Spa and Health Tourism*

Health tourism is a relatively new concept and has been in use for around two decades [62]. However, travelling to another country for health care is not a new phenomenon. People have travelled for treatment and revitalization throughout history—the Romans travelled to thermal baths, pilgrims visited the Dead Sea for the therapeutic benefits of the water, and Asians went to thermal springs for relaxation and socialization [63]. Health tourism is one of the fastest growing segments on the global tourism market [64], with the growing demand for healthcare services creating opportunities for entrepreneurs. According to Global Healthcare Resources, approximately 11 million people travel to different parts of the world for medical care each year [65].

The landscapes, climate and natural resources in Romania have contributed to the wide recognition of spa treatment and health destinations of this country [66]. After being among the favorite destinations of foreign tourists in Romania between 1970 and 1990, spa tourism continues to be a major segment of the health tourism market in Romania [66]. Over the last two decades, several factors (economic, political, changing trends) have led to the decline in spa resorts and to a decrease in the number of foreign tourists. However, tradition, international certifications, employee qualifications, healing natural resources, modernization of infrastructure [67], accommodation and treatment facilities, the implementation of modern spa services and a lower level of tariffs can be a competitive advantage in revitalizing this sector in Romania's economy. There is a growing interest in changing the way people take care of their health, the development of health and wellness tourism in Romania and in Europe in general, driving to the re-evaluation of the positioning and promotion strategies of spa and health resorts, by supplementing the spa offer with innovative [68] and, at the same time, sustainable services [69].

In this context, we would like to investigate the following general questions: on one hand, whether the goals of starting and developing sustainable businesses in spa tourism differs according to the characteristics of the firms; on the other hand, how do managers/owners in this field perceive the factors that support or hinder their attempts to implement sustainable innovations.

Starting from the above questions and taking into account the main contributions from the literature, especially those that substantiate the links between the firm's main characteristics and the sustainable innovations in the tourism sector, we defined four hypotheses:

Hypothesis 1. *The perception of the factors and opportunities underlying a sustainable business in the spa tourism differs according to the size and ownership type of the tourism ventures;*

Hypothesis 2. *The objectives set when launching a tourism business vary according to the size and ownership type of the tourism ventures;*

Hypothesis 3. *The perception of the factors determining the market launch of a new sustainable service/product differs according to the size and ownership type of the tourism ventures;*

Hypothesis 4. *The perception of the factors with a major negative impact in supporting sustainable innovations within the spa tourism businesses differs according to the size and ownership type of the tourism ventures.*

3. Methodology

Our research was carried out among the tourism companies in Baile Felix Spa Resort, the largest Romanian spa resort, located in the northwest of Romania [66,70]. After a period of stagnation, the last decade witnessed a significant increase in the number of entrepreneurial initiatives and jobs in this area through the expansion and modernization of tourist accommodation structures, food, leisure and treatment services, contributing to the prosperity of the local economy and increasing the attractivity of the area.

In order to investigate the sustainable innovative potential in the field of spa and health tourism in Baile Felix area, we conducted field research between 1–31 October 2022 based on a semi-structured interview and an opinion questionnaire. The questionnaire comprised 12 questions, covering open and multiple-choice questions. A 5-level ordinal scale (Likert scale) was used to construct the closed questions, while Cronbach's Alpha was used to measure the reliability of the items underlying the questionnaire. In the case of the instrument, the Cronbach's Alpha was 0.631, higher than 0.600 [71]. We targeted a representative sample made up of entrepreneurs who own a tourist accommodation unit in Baile Felix area (authorized individuals and SMEs) and, respectively, top and middle managers in large tourism enterprises (over 250 employees). In the reports provided by the Ministry of Entrepreneurship and Tourism (Romania) [70], 203 units were found, which differ based on the type of unit, star rating, number of beds, the type of economic venture, as well as the services and facilities offered.

Of these 203 units, we were able to contact 137, obtaining 91 answers (for a response rate of 66.4%), of which 29 belong to top and middle managers in the hotel sector, representing 15 hotels, and the other 62 people are entrepreneurs and owners of guesthouses, villas and rooms for rent. As the present research focuses on the respondents' attitudes, willingness and experiences related to innovation and sustainability in spa and health tourism, we consider only the responses of entrepreneurs and managers who have introduced new sustainable services/facilities/products in the last 2 years, i.e., 60 respondents out of a total of 91 valid responses.

In Figure 1, we present a research methodology diagram that contains the data source, the procedure, the statistical tests and methods that will be applied to the groups, identifying the inputs and possible outputs.

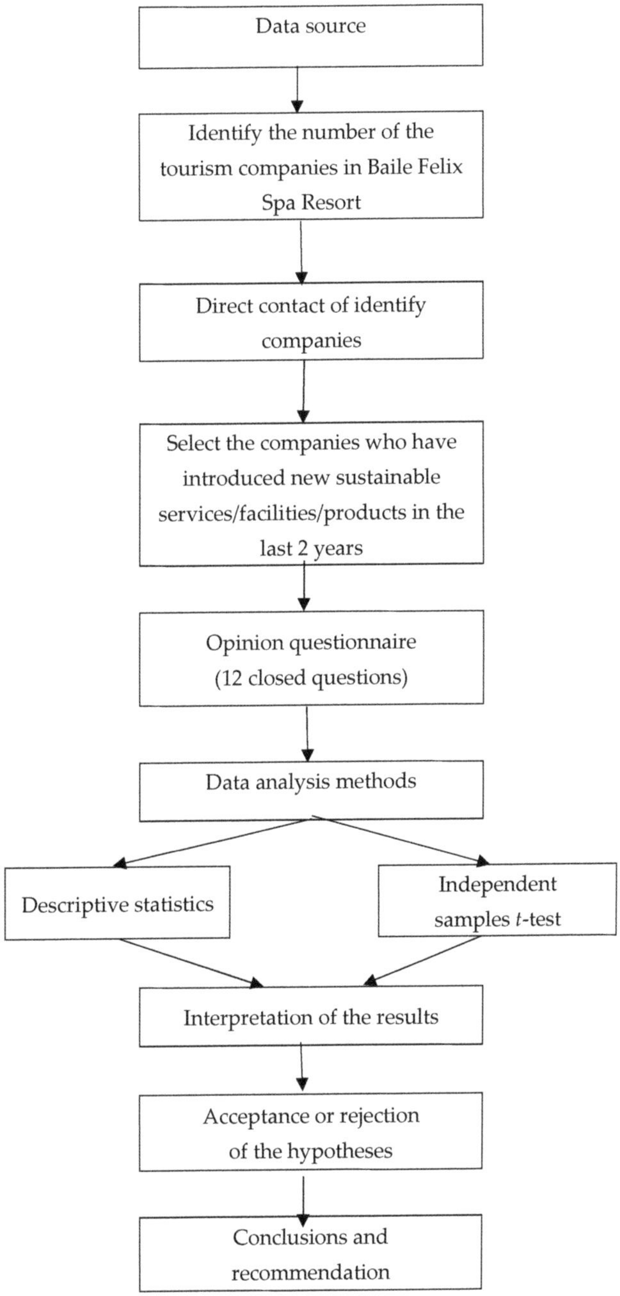

Figure 1. Research Methodology Diagram.

4. Results and Discussions

We have carried out a statistical analysis of the two independent samples (*group 1—Hotels*, and *group 2—Guesthouses, villas, rooms for rent*), starting from the assumption that the type of organization, the ownership and the size of the company impact the willingness for sustainable innovations in this sector.

In order to test the four hypotheses to determine whether there are significant differences between the entrepreneurial perspectives underlying a particular type of accommodation unit, we applied the independent samples *t*-test.

Hypothesis 1. *The perception of the factors and opportunities underlying a sustainable business in spa tourism differs according to the size and ownership type of the tourism ventures.*

According to the results (Table 1), we can see that entrepreneurs and small business owners are more interested in exploiting current opportunities that are essential for strengthening their market position, and less interested in maintaining continuity with the past features of the sector. The top three options in the case of guesthouses were *financial independence, the desire to succeed, to do things better and the expansion of sustainable business opportunities in the area*, and in the case of large hotels' managers, the options were *the desire to succeed, to do things better, to provide innovative services in a field that was showing signs of aging/capping, and to offer a service already existing on the market, but in a sustainable and innovative way.*

Table 1. Descriptive statistics for the perception of the factors and opportunities underlying a sustainable business in spa tourism, according to the type of venture.

	Type of Unit	N	Mean	Std. Dev.	Std. Error Mean
The desire to succeed, to do things better	**Hotel**	21	**3.857**	1.852	0.404
	Guesthouse	39	2.846	2.020	0.323
Expansion of sustainable business opportunities in Baile Felix area	Hotel	21	2.714	2.028	0.443
	Guesthouse	39	3.051	2.025	0.324
Financial independence	Hotel	21	2.143	1.852	0.404
	Guesthouse	39	**3.564**	1.944	0.311
Flexible schedule	Hotel	21	1.190	0.873	0.190
	Guesthouse	39	2.026	1.769	0.283
The decision to be self-employed	Hotel	21	1.571	1.434	0.313
	Guesthouse	39	2.744	2.009	0.322
Offering a service already existing on the market, but in a sustainable and innovative way	Hotel	21	3.095	2.047	0.447
	Guesthouse	39	2.333	1.910	0.306
Providing innovative services in a field that was showing signs of aging/capping	Hotel	21	3.667	1.932	0.422
	Guesthouse	39	1.718	1.555	0.249
Cost reduction by implementing innovations and sustainable technologies	Hotel	21	1.571	1.434	0.313
	Guesthouse	39	1.718	1.555	0.249

Note: The highest mean for each category was highlighted in bold; Source: authors contribution.

Regarding the less relevant factors and opportunities, we can mention *the decision to be self-employed*, in the case of hotels (explained by the fact that many of these respondents are managers, and not entrepreneurs or hotel owners), while the provision of new services in a decaying domain is not a concern in the case of guesthouses (with no affirmative answers), suggesting that small, privately owned and managed businesses are mainly focused on the fast capitalization of current opportunities. In comparison, this item (*providing new services in a field that was showing signs of aging/capping*) is among the top three answers among hotel managers.

According to the results of the independent samples *t*-test (Table 2), we can state that there are significant differences between *the factors and opportunities underlying a sustainable business in spa tourism* from the perspective of hotel managers and of small entrepreneurs managing a guesthouse or villa. In the case of the factor *providing new innovative services in a field that was showing signs of aging/capping,* the value of the t-statistic equals 3.980 ($p < 0.01$), a statistically significant result that shows that the two-population means are not equal. Also, considering the positive sign of the t-statistic, we can state that the mean number of subjects in group 1 (hotels) is higher than the mean number of the subjects in group 2 (guesthouses), we can conclude that this is an important factor in developing *a sustainable business in spa tourism* for hotels. Significant differences were also identified in the case of the factors: *the desire to succeed, to do things better* ($t = 1.953$, $p = 0.057$) *financial independence* ($t = -2.786$, $p = 0.008$); *flexible schedule* ($t = -2.446$, $p = 0.018$); and *the decision to be self-employed* ($t = -2.611$, $p = 0.012$). In the case of the last three factors the mean among the guesthouses is higher than the mean among hotels. Therefore, we can state that guesthouse owners place a significantly greater emphasis on *financial independence, flexible schedule* and also on *the decision to be self-employed* when running their own business, while concerns related to sustainability fall into the background. The results confirm the first hypothesis of this study.

Table 2. *t*-test results: the perception of the factors and opportunities underlying a sustainable business in spa tourism, according to the type of venture.

		Levene Test		*t*-Test				
		F	Sig.	t	Df	Sig. (2-tailed)	Mean Diff.	Std. Error Diff.
The desire to succeed, to do things better	**Hotel**	**7.405**	**0.009**	**1.902**	**58**	**0.062**	**1.010**	**0.531**
	Guest house			**1.953**	**44.279**	**0.057**	**1.010**	**0.517**
Expansion of sustainable business opportunities in Baile Felix area	Hotel	0.693	0.408	−0.614	58	0.541	−0.337	0.548
	Guest house			−0.614	41.019	0.543	−0.337	0.548
Financial independence	**Hotel**	**1.429**	**0.237**	**−2.745**	**58**	**0.008**	**−1.421**	**0.517**
	Guest house			**−2.786**	**42.841**	**0.008**	**−1.421**	**0.510**
Flexible schedule	**Hotel**	**26.280**	**0.000**	**−2.028**	**58**	**0.047**	**−0.835**	**0.411**
	Guest house			**−2.446**	**57.712**	**0.018**	**−0.835**	**0.341**
The decision to be self-employed	**Hotel**	**32.836**	**0.000**	**−2.364**	**58**	**0.021**	**−1.172**	**0.495**
	Guest house			**−2.611**	**53.288**	**0.012**	**−1.172**	**0.448**
Offering a service already existing on the market, but in a sustainable and innovative way	Hotel	2.405	0.126	1.437	58	0.156	0.761	0.530
	Guest house			1.407	38.678	0.167	0.761	0.541
Providing innovative services in a field that was showing signs of aging/capping	**Hotel**	**6.180**	**0.016**	**4.249**	**58**	**0.000**	**1.948**	**0.458**
	Guest house			**3.980**	**34.196**	**0.000**	**1.948**	**0.489**
Cost reduction by implementing innovations and sustainable technologies	Hotel	0.531	0.469	−0.357	58	0.722	−0.146	0.409
	Guest house			−0.366	44.045	0.716	−0.146	0.399

Note: Statistically significant results were highlighted in bold; Source: authors contribution.

Hypothesis 2. *The objectives set when launching a tourism business vary according to related to the size and ownership type of the tourism ventures.*

The results show that, although the order of priorities in the case of guesthouses is different from that of hotels, the outline of options is similar, related to a valid entrepreneurial vision oriented towards performance—motivation, identification of needs and expectations, satisfying these needs under market conditions and profit-making (Table 3).

Table 3. Descriptive statistics for the objectives set when launching a sustainable tourism business, according to the type of venture.

	Type of Unit	N	Mean	Std. Dev.	Std. Error Mean
Personal fulfillment	Hotel	21	2.143	1.852	0.404
	Guesthouse	39	3.872	1.824	0.292
The desire to provide added value through innovative and sustainable elements in spa and health tourism	**Hotel**	**21**	**4.048**	**1.746**	**0.381**
	Guesthouse	39	2.744	2.009	0.322
Making profit	Hotel	21	3.857	1.852	0.404
	Guesthouse	**39**	**4.487**	**1.355**	**0.217**
Restoring the economic dynamism of tourism in the area through innovative and sustainable businesses	Hotel	21	2.905	2.047	0.447
	Guesthouse	39	2.538	1.971	0.316
Identification and correction of sources of dissatisfaction related the services provided	Hotel	21	3.857	1.852	0.404
	Guesthouse	39	2.744	2.009	0.322

Note: The highest mean for each category was highlighted in bold; Source: authors contribution.

The first three options in the case of guesthouses were making a profit (mean = 4.48), personal fulfillment (mean = 3.872) and the desire to provide added value through innovative and sustainable elements in spa tourism and the identification and correction of sources of dissatisfaction related the services provided in spa resorts, both with an equal average of 0.744. In the case of hotels, the order was the desire to provide added value through innovative and sustainable elements in spa tourism (mean = 4.048); making a profit (mean = 3.857); respectively, the identification and correction of sources of dissatisfaction related to the services provided in spa resorts (mean = 3.857). We find that motivations are influenced by the nature of the activity: modernization and performance in the case of hotel managers, and respectively personal fulfillment and performance (profit) for entrepreneurs and small business owners. Moreover, a more general motivation, which goes beyond the career or personal business objectives, namely restoring the economic dynamism of the area (through innovative and sustainable businesses), does not seem to be a priority for either category, which is not surprising as this objective of revival could rather be perceived as an objective of the public regional or national authorities, rather than an objective of the business sector.

The *t*-test results (Table 4) show that there were significant differences between *the objectives underlying the launch of a business* from the point of view of hotel managers, compared to guesthouse owners, in the case of *personal fulfilment* (t statistic = −3.484, p = 0.001); specifically, this factor is more important in the case of guesthouse owners. Another significant difference was found in the case of *the desire to provide added value through innovative and sustainable elements in spa tourism*, which is more important for hotel managers (t-statistic = 2.615, p = 0.012). The results partially confirm the second hypothesis of this study.

The same order of priorities was found for the factors influencing the market launch of a new service/facility/product: *market needs; attracting new customer segments* and *new trends in consumption*, as well as relatively similar scores, between guesthouse owners and hotel managers (Table 5). No significant differences were found for the variable with the lowest score (i.e., *how much are potential clients willing to pay for my services),* reflecting that managers and entrepreneurs alike are interested in solvable demand for future (innovative) products they will put on the market. The results are in line with our expectations of a market orientation defined by a certain prudence, which is otherwise justified by recent events (i.e., crises) and the high variability of demand in this field.

Hypothesis 3. *The perception of the factors determining the market launch of a new sustainable service/product differs according to the type of tourism ventures.*

Table 4. *t*-test results: the objectives set when launching a sustainable tourism business according to the type of venture.

		Levene Test		*t*-Test				
		F	Sig.	T	df	Sig. (2-tailed)	Mean Diff.	Std. Error Diff.
Personal fulfillment	**Hotel**	**0.003**	**0.953**	**−3.484**	**58**	**0.001**	**−1.729**	**0.496**
	Guest house			**−3.468**	**40.527**	**0.001**	**−1.729**	**0.498**
The desire to provide added value through innovative and sustainable elements in spa tourism	**Hotel**	**10.94**	**0.002**	**2.506**	**58**	**0.015**	**1.304**	**0.520**
	Guest house			**2.615**	**46.310**	**0.012**	**1.304**	**0.498**
Making profit	Hotel	8.440	0.005	−1.507	58	0.137	−0.630	0.418
	Guest house			−1.374	31.801	0.179	−0.630	0.458
Restoring the economic dynamism of tourism businesses in the area through innovative and sustainable businesses	Hotel	1.022	0.316	0.677	58	0.501	0.366	0.541
	Guest house			0.670	39.747	0.507	0.366	0.546
Identification and correction of sources of dissatisfaction related to the services provided	Hotel	5.859	0.019	1.134	58	0.261	0.601	0.530
	Guest house			1.163	44.079	0.251	0.601	0.516

Note: Statistically significant results were highlighted in bold; Source: authors contribution.

Table 5. Descriptive statistics for the perception of the factors determining the launch of a new sustainable service/product according to the type of tourism venture.

	Type of Unit	N	Mean	Std. Dev.	Std. Error Mean
Market needs	**Hotel**	**21**	**4.429**	**1.434**	**0.313**
	Guesthouse	**39**	**4.897**	**0.641**	**0.103**
Market areas unexplored by the competitors	Hotel	21	2.333	1.932	0.422
	Guesthouse	39	2.436	1.944	0.311
Attracting new customer segments	Hotel	21	4.238	1.609	0.351
	Guesthouse	39	4.077	1.707	0.273
New trends in consumption	Hotel	21	3.476	1.990	0.434
	Guesthouse	39	3.051	2.025	0.324
How much are potential clients willing to pay for my services	Hotel	21	2.143	1.852	0.404
	Guesthouse	39	2.538	1.971	0.316

Note: The highest mean for each category was highlighted in bold; Source: authors contribution.

The results of the independent samples *t*-test (Table 6) indicate that both groups carefully planned the context and the timing for the launch of new services/facilities/products. This is an interesting and encouraging finding because it shows the clear and realistic positioning of the two categories towards the market and consumers, even when the resources and business models are not similar. This is also a positive signal for new innovative businesses in the tourism sector in Baile Felix area, as it contradicts previous assumptions that the spa tourism sector has a moderate permeability to innovation. However, the willingness to innovate must also be interpreted through the lens of actual resources, i.e., finding a balance between intentions and possibilities. The large budgets required for investments in equipment, in specialized constructions and the training of qualified personnel are not accessible to all tourism ventures, regardless of their desires, plans and objectives. The third hypothesis of the study is not confirmed.

Table 6. *t*-test results: the perception of the factors determining the market launch of a new sustainable service/product according to the type of tourism venture.

		Levene Test		*t*-Test				
		F	**Sig.**	**T**	**df**	**Sig. (2-tailed)**	**Mean Diff.**	**Std. Error Diff.**
Market need	Hotel	13.743	0.000	−1.752	58	0.085	−0.468	0.267
	Guest house			−1.424	24.378	0.167	−0.468	0.329
Market areas unexplored by the competitors	Hotel	0.160	0.691	−0.195	58	0.846	−0.102	0.525
	Guest house			−0.196	41.286	0.846	−0.102	0.524
Attracting new customer segments	Hotel	0.530	0.469	0.356	58	0.723	0.161	0.453
	Guest house			0.362	43.227	0.719	0.161	0.445
New trends in consumption	Hotel	2.164	0.147	0.780	58	0.439	0.424	0.544
	Guest house			0.784	41.698	0.438	0.424	0.542
How much potential clients are willing to pay for my services	Hotel	2.633	0.110	−0.757	58	0.452	−0.395	0.522
	Guest house			−0.772	43.364	0.445	−0.395	0.512

Source: authors contribution.

Hypothesis 4. *The perception of the factors with a major negative impact in supporting sustainable innovations within the spa tourism businesses differs according to the size and ownership type of the tourism ventures.*

According to the results presented in Table 7, we can state that in the case of hotels, the factors with the greatest negative impact in supporting sustainable innovations are: *Insufficient fiscal support for innovative and sustainable initiatives in tourism* (mean = 4.619) and *Lacking or insufficient funds supporting innovation and promoting new sustainable products/services* (mean = 4.429). In the case of guesthouses, we note that the factors with the greatest negative impact are: *Lacking or insufficient qualified personnel in sustainable tourism activities* (mean = 4.795), *Lacking or insufficient funds supporting innovation and promoting sustainable products/services* (mean = 4.692) and, finally, *Insufficient fiscal support for innovative and sustainable initiatives in tourism* (mean = 4.385).

Table 7. Descriptive statistics for the factors with a major negative impact in supporting sustainable innovations within the spa tourism businesses according to the type of venture.

	Type of Unit	N	Mean	Std. Dev.	Std. Error Mean
Lacking or insufficient funds supporting innovation and promoting sustainable products/services	**Hotel**	21	**4.429**	1.434	0.313
	Guesthouse	39	**4.692**	1.080	0.173
Lacking or insufficient qualified personnel in specific sustainable tourism activities	Hotel	21	3.476	1.990	0.434
	Guesthouse	**39**	**4.795**	**0.894**	**0.143**
Customers wary of new, sustainable products	Hotel	21	2.714	2.028	0.443
	Guesthouse	39	1.615	1.462	0.234
Personnel with no interest in promoting new sustainable products	Hotel	21	1.381	1.203	0.263
	Guesthouse	39	1.308	1.080	0.173
Insufficient fiscal support for innovative and sustainable initiatives in tourism	**Hotel**	**21**	**4.619**	**1.203**	**0.263**
	Guesthouse	**39**	**4.385**	**1.462**	**0.234**

Note: The highest mean for each category was highlighted in bold; Source: authors contribution.

According to the results of the *t*-test (Table 8), we can state that there are significant differences between hotel and guesthouse managers regarding several items pertaining to the factors with a major negative impact on sustainable innovation in spa tourism. In the case of the factor *Lacking or insufficient qualified personnel in sustainable tourism ac-*

tivities, the mean score is lower in the hotel group compared to the guesthouse group (t-statistic = −2.883, p = 0.008); suggesting that this factor is more important in the case of guesthouses seeking sustainable innovations.

Table 8. *t*-test: The factors with major negative impact in supporting sustainable innovation within the spa tourism businesses according to the size and ownership type of venture.

		Levene Test		*t*-Test				
		F	Sig.	T	df	Sig. (2-tailed)	Mean Diff.	Std. Error Diff.
Lacking or insufficient funds supporting innovation and promoting sustainable products/services	Hotel	2.544	0.116	−0.803	58	0.425	−0.263	0.328
	Guest house				−0.738	32.479	0.466	0.357
Lacking or insufficient qualified personnel in sustainable tourism activities	**Hotel**	**61.334**	**0.000**	**−3.544**	**58**	**0.001**	**−1.318**	**0.372**
	Guest house			**−2.883**	**24.427**	**0.008**	**−1.318**	**0.457**
Customers wary of new, sustainable products	**Hotel**	**16.420**	**0.000**	**2.418**	**58**	**0.019**	**1.098**	**0.454**
	Guest house			**2.195**	**31.460**	**0.036**	**1.098**	**0.500**
Personnel with no interest in promoting new sustainable products	Hotel	0.230	0.634	0.241	58	0.811	0.073	0.304
	Guest house			0.233	37.407	0.817	0.073	0.314
Insufficient fiscal support for innovative and sustainable initiatives in tourism	Hotel	1.695	0.198	0.628	58	0.532	0.234	0.373
	Guest house			0.666	48.358	0.508	0.351	0.355

Note: Statistically significant results were highlighted in bold; Source: authors contribution.

Another significant difference was found for the variable *the customers are wary of new, sustainable products* (t = 2.195, p = 0.036). In this case, the mean of subjects in group 1 (hotels) is higher, which shows that this factor is more relevant for hotels. Therefore, we can state that the fourth hypothesis of the research is partially confirmed.

5. Conclusions

The results of the study suggest that entrepreneurs and small business owners prioritize exploiting current opportunities to strengthen their market position rather than focusing on the sector's past. In the case of small tourist accommodations such as villas and guesthouses, the top three options regarding the factors and opportunities that motivate owners' aspirations for a sustainable business were financial independence, the desire for success, and the expansion of sustainable business opportunities in the area. For hotels, the options were a desire to succeed, to innovate in an aging or saturated field, and to deliver existing services in a sustainable and innovative manner. Likewise, the goals for a sustainable tourism business tend to differ when comparing hotel managers and guesthouse owners. More specifically, the motivations of hotel managers are primarily modernization and performance, while small business owners and entrepreneurs prioritize personal fulfillment and profit. The goal of restoring economic dynamism to the area through innovative and sustainable businesses does not appear to be a priority for any category, as it is commonly perceived as an objective for regional or national public authorities rather than the business sector. Additionally, we found that both hotel managers and entrepreneurs prioritize market needs, attracting new customer segments and responding to new consumer trends in their decision when to launch new services, facilities or products. In the case of the factors with the greatest negative impact on supporting sustainable innovations, hotel managers complained foremost about the lack of fiscal support for innovative and sustainable initiatives, as well as insufficient funds for supporting innovation and promoting sustainable products and services. These results can be a signal for decision makers at local, regional or national level in designing legislative, fiscal and organizational measures to support innovative efforts in tourism. For guesthouses, the main obstacles are insufficient staff involved in sustainable tourism activities, the insufficient funds to support innovation and promote sustainable products/services and, finally, insufficient fiscal support for innovative and sustainable initiatives in tourism. The only factor that differs significantly between hotel managers and guesthouse owners is the lack of qualified staff in sustainable tourism activities, with the latter group considering it a more pressing difficulty. This finding highlights the importance of qualified staff in supporting sustainable innovations in spa tourism businesses, especially for guesthouses.

The results of our research largely confirm the finding of several scholars as Sundo et al. [6], Aguilar-Fernández and Otegi-Olaso [43] or Chipunza [44], according to which large companies are more oriented towards the development and implementation of innovations and sustainable practices, in the tourism sector as well as in general. At the same time, our findings partially confirm the contributions that claim that small and family businesses in tourism are more social and community-responsible [52–54], more involved in regional and local tourism development and more environmentally responsible [56,57], quickly adapting sustainable strategies and practices to the specificities of tourism subsectors [58–60].

We believe that this study contributes to existing scientific knowledge by exploring and comparing the motivations of guesthouse owners and hotel managers in the context of spa tourism. It highlights the different priorities between these two groups and how they align with their respective roles. The study also highlights the association between motivations and the nature of the activity, revealing the importance of modernization and performance for hotel managers, as well as personal fulfillment and profit for entrepreneurs and small business owners. This research underscores the importance of addressing market needs, money segments and responding to emerging consumer trends when planning the launch of new offerings. In addition, the study shows the need to find a balance between innovation aspirations and practical considerations such as financial constraints. Considering both market orientation and resource constraints, the studies provide valuable information for entrepreneurs, managers and decision makers in the tourism sector.

Our study has several limitations. The first limitation refers to the sample size. Another limitation would be related to its regional perspective: the study focused on guesthouses and hotels from a single geographical region, the Baile Felix area (Romania). Thus, we can conclude that it does not fully represent the perspectives and priorities of entrepreneurs in other regions or different types of accommodation businesses. Also, given that the study was based on self-reported data from entrepreneurs and small business owners, we could consider that these responses could be subject to subjective interpretations.

Future research could explore a comparative analysis of different regions in Romania that are known for spa and health tourism. As future goals, we propose to expand the analysis to the level of several regions so that we can also determine regional disparities, unique challenges and success factors at the level of each region. This broader perspective would provide a more comprehensive picture of the country's spa tourism landscape.

Author Contributions: Conceptualization, D.-T.T. and D.B.; Formal analysis, D.-T.T. and R.S.; Methodology, D.-T.T., R.S. and D.B.; Supervision, D.B.; Writing—original draft, D.-T.T., R.S. and D.B.; Writing—review and editing, D.-T.T., R.S. and D.B. All authors have read and agreed to the published version of the manuscript.

Funding: This research received no external funding.

Data Availability Statement: Data available by request.

Conflicts of Interest: The authors declare no conflict of interest.

References

1. Low, M.; MacMillan, I. Entrepreneurship: Past research and future challenges. *J. Manag.* **1988**, *14*, 139–161. [CrossRef]
2. Baumol, W. Entrepreneurship in economic theory. In *The American Economic Review, The Eightieth Annual Meeting of The American Economic Association*; American Economic Association: Nashville, TN, USA, 1968; Volume 58, pp. 64–71.
3. Solvoll, S.; Alsos, G.; Bulanova, O. Tourism Entrepreneurship—Review and Future Directions, Scandinavian Journal of Hospitality and Tourism. *Scand. J. Hosp. Tour.* **2015**, *15* (Suppl. 1), 120–137. [CrossRef]
4. Crnogaj, K.; Rebernik, M.; Bradac Hojnik, B.; Omerzel Gomezelj, D. Building a model of researching the sustainable entrepreneurship in the tourism sector. *Kybernetes* **2014**, *43*, 377–393. [CrossRef]
5. Hall, C.; Williams, A. *Tourism and Innovation*, 2nd ed.; Routledge: London, UK, 2020.

6. Sundbo, J.; Orfila-Sintes, F.; Sorensen, F. The innovative behaviour of tourism firms—Comparative studies of Denmark and Spain. *Res. Policy* **2007**, *36*, 88–106. [CrossRef]
7. Johannesson, G.; Huijbens, E. Tourism in times of crisis: Exploring the discourse of tourism development in Iceland. *Curr. Issues Tour.* **2010**, *13*, 419–434. [CrossRef]
8. World Trade Organization (WTO). *World Trade Statistical Review*; World Trade Organization: Geneva, Switzerland, 2020.
9. OECD. *Tourism Trends and Policies 2020*; OECD Publishing: Paris, France, 2020.
10. World Tourism Organization. *International Tourism Highlights*, 2020th ed.; UNWTO: Madrid, Spain, 2021.
11. World Travel & Tourism Council (WTTC). Economic Impact Reports. Available online: https://wttc.org/Research/Economic-Impact (accessed on 4 January 2022).
12. UNWTO. World Tourism Barometer. Available online: https://www.e-unwto.org/doi/pdf/10.18111/wtobarometereng.2020.18.1.3 (accessed on 12 August 2021).
13. UNWTO. Tourism Recovery Tracker. Available online: https://www.unwto.org/tourism-data/unwto-tourism-recovery-tracker (accessed on 31 March 2023).
14. Dwyer, L.; Forsyth, P.; Spurr, R. Estimating the impacts of special events on an economy. *J. Travel Res.* **2005**, *43*, 351–359. [CrossRef]
15. Kim, W.; Jun, H.; Walker, M.; Drane, D. Evaluating the perceived social impacts of hosting large-scale sport tourism events: Scale development and validation. *Tour. Manag.* **2015**, *48*, 21–32. [CrossRef]
16. León-Gómez, A.; Ruiz-Palomo, D.; Fernández-Gámez, M.; García-Revilla, M. Sustainable Tourism Development and Economic Growth: Bibliometric Review and Analysis. *Sustainability* **2021**, *13*, 2270. [CrossRef]
17. Tsekouropoulos, G.; Gkouna, O.; Theocharis, D.; Gounas, A. Innovative Sustainable Tourism Development and Entrepreneurship through Sports Events. *Sustainability* **2022**, *14*, 4379. [CrossRef]
18. Badulescu, D.; Simut, R.; Simut, C.; Badulescu, A. Tourism at the Crossroads between Well-Being, Public Health and the Environment: Panel Data Evidence from the European Union. *Int. J. Environ. Res. Public Health* **2022**, *19*, 12066. [CrossRef]
19. Lenzen, M.; Sun, Y.Y.; Faturay, F.; Ting, Y.P.; Geschke, A.; Malik, A. The carbon footprint of global tourism. *Nat. Clim. Chang.* **2018**, *8*, 522–528. [CrossRef]
20. Hjalager, A. A review of innovation research in tourism. *Tour. Manag.* **2010**, *31*, 1–12. [CrossRef]
21. Brundtland, G. *Report of the World Commission on Environment and Development: Our Common Future*; United Nations General Assembly document A/42/427: Geneva, Switzerland, 1987.
22. Ruhanen, L. Local government: Facilitator or inhibitor of sustainable tourism development? *J. Sustain. Tour.* **2013**, *21*, 80–98. [CrossRef]
23. Bachleitner, R.; Zins, A. Cultural Tourism in Rural Communities: The Residents' Perspective. *J. Bus. Res.* **1999**, *44*, 199–209. [CrossRef]
24. Pan, S.Y.; Gao, M.; Kim, H.; Shah, K.J.; Pei, S.L.; Chiang, P.C. Advances and challenges in sustainable tourism toward a green economy. *Sci. Total Environ.* **2018**, *635*, 452–469. [CrossRef] [PubMed]
25. Doxey, G.V. A causation theory of visitor-resident irritants: Methodology and research inferences. In Proceedings of the Travel and Tourism Research Association Sixth Annual Conference Proceedings, Salt Lake City, UT, USA, 8–11 September 1975; pp. 195–198.
26. Rivera, M.; Croes, R.; Lee, S. Tourism development and happiness: A residents' perspective. *J. Destin. Mark. Manag.* **2016**, *5*, 5–15. [CrossRef]
27. Higgins-Desbiolles, F. Sustainable tourism: Sustaining tourism or something more? *Tour. Manag. Perspect.* **2018**, *25*, 157–160. [CrossRef]
28. Streimikiene, D.; Svagzdiene, B.; Jasinskas, E.; Simanavicius, A. Sustainable tourism development and competitiveness: The systematic literature review. *Sustain. Dev.* **2021**, *29*, 259–271. [CrossRef]
29. Teodorescu, N.; Stancioiu, A.; Ravar, A.; Botos, A. Creativity and innovation—Sources of competitive advantage in the value chain of tourism enterprises. *Theor. Appl. Econ.* **2015**, *22*, 35–48.
30. Bramwell, B. Governance, the state and sustainable tourism: A political economy approach. *J. Sustain. Tour.* **2011**, *19*, 459–477. [CrossRef]
31. Bramwell, B.; Lane, B. Towards innovation in sustainable tourism research? *J. Sustain. Tour.* **2012**, *20*, 1–7. [CrossRef]
32. Guttentag, D. Airbnb: Disruptive innovation and the rise of an informal tourism accommodation sector. *Curr. Issues Tour.* **2015**, *18*, 1192–1217. [CrossRef]
33. Nordli, A. Information use and working methods as drivers of innovation in tourism companies. *Scand. J. Hosp. Tour.* **2018**, *18*, 199–213. [CrossRef]
34. Camisón, C.; Monfort-Mir, V. Measuring innovation in tourism from the Schumpeterian and the dynamic-capabilities perspectives. *Tour. Manag.* **2012**, *33*, 776–789. [CrossRef]
35. Sundbo, J. Management of innovation in services. *Serv. Ind. J.* **1997**, *17*, 432–455. [CrossRef]
36. Teixeira, S.; Ferreira, J. A bibliometric study of regional competitiveness and tourism innovation. *Int. J. Tour. Policy* **2018**, *8*, 214–243. [CrossRef]
37. Nordli, A. Measuring innovation in tourism with Community Innovation Survey: A first step towards a more valid innovation instruments. *Scand. J. Hosp. Tour.* **2017**, *17*, 423–440. [CrossRef]

38. Dabbous, A.; Tarhini, A. Does sharing economy promote sustainable economic development and energy efficiency? Evidence from OECD countries. *J. Innov. Knowl.* **2021**, *6*, 58–68. [CrossRef]
39. Hjalager, A. Stages in the economic globalization of tourism. *Ann. Tour. Res.* **2007**, *34*, 437–457. [CrossRef]
40. Schumpeter, J. *Capitalism, Socialism, and Democracy*, 3rd ed.; Harper Perennial Modern Thought: New York, NY, USA, 1942.
41. Audretsch, D.; Thurik, R. *Linking Entrepreneurship to Growth*; OECD Science, Technology and Industry Working Papers; OECD Publishing: Paris, France, 2001.
42. Acs, Z.; Audretsch, D. Innovation, Market Structure and Firm Size. *Rev. Econ. Stat.* **1987**, *69*, 567–575. [CrossRef]
43. Aguilar-Fernández, M.; Otegi-Olaso, J. Firm Size and the Business Model for Sustainable Innovation. *Sustainability* **2018**, *10*, 4785. [CrossRef]
44. Chipunza, L. Does firm size matter in innovation in small accommodation businesses in developing economies? *Afr. J. Hosp. Tour. Leis.* **2019**, *8*, 1–12.
45. Laforet, S. Effects of size, market and strategic orientation on innovation in non high-tech manufacturing SMEs. *Eur. J. Mark.* **2009**, *43*, 188–212. [CrossRef]
46. Orser, B.; Hogarth-Scott, S.; Riding, A. Performance, Firm Size and Management Problem Solving. *J. Small Bus. Manag.* **2020**, *38*, 22–29.
47. Cohen, W.; Klepper, S. Firm Size and the Nature of Innovation within Industries: The Case of Process and Product R&D. *Rev. Econ. Stat.* **1996**, *78*, 232–243.
48. Lee, K.; Go, D.; Park, I.; Yoon, B. Exploring Suitable Technology for Small and Medium-Sized Enterprises (SMEs) Based on a Hidden Markov Model Using Patent Information and Value Chain Analysis. *Sustainability* **2017**, *9*, 1100. [CrossRef]
49. Klewitz, J.; Hansen, E. Sustainability-oriented innovation of SMEs: A systematic review. *J. Clean. Prod.* **2014**, *65*, 57–75. [CrossRef]
50. Yoon, B.; Shin, J.; Lee, S. Open innovation projects in SMEs as an engine for sustainable growth. *Sustainability* **2016**, *8*, 146. [CrossRef]
51. Ateljevic, I.; Doorne, S. Staying with the Fence: Lifestyle Entrepreneurship in Tourism. *J. Sustain. Tour.* **2000**, *8*, 378–392. [CrossRef]
52. Godfrey, P. The relationship between corporate philanthropy and shareholder wealth: A risk management perspective. *Acad. Manag. Rev.* **2005**, *30*, 777–798. [CrossRef]
53. Peake, W.; Cooper, D.; Fitzgerald, M.; Muske, G. Family business participation in community social responsibility: The moderating effect of gender. *J. Bus. Ethics* **2017**, *142*, 325–343. [CrossRef]
54. Glowka, G.; Zehrer, A. Tourism Family-Business Owners' Risk Perception: Its Impact on Destination Development. *Sustainability* **2019**, *11*, 6992. [CrossRef]
55. Dwyer, L.; Edwards, D.; Mistilis, N.; Roman, C.; Scott, N. Destination and enterprise management for a tourism future. *Tour. Manag.* **2009**, *30*, 63–74. [CrossRef]
56. Chen, N.L. Residents' place satisfaction and place attachment on destination brand-building behaviors: Conceptual and empirical differentiation. *J. Travel Res.* **2018**, *57*, 1026–1041. [CrossRef]
57. Bansal, P.; Roth, K. Why companies go green: A model of ecological responsiveness. *Acad. Manag. J.* **2000**, *43*, 717–736. [CrossRef]
58. Kallmuenzer, A.; Nikolakis, W.; Peters, M.; Zanon, J. Trade-offs between dimensions of sustainability: Exploratory evidence from family firms in rural tourism regions. *J. Sustain. Tour.* **2018**, *26*, 1204–1221. [CrossRef]
59. Badulescu, D.; Badulescu, A. Medical Tourism: Between Entrepreneurship Opportunities and Bioethics Boundaries. *Iran. J. Public Health* **2014**, *43*, 406–415.
60. Marian, M.; Darabaneanu, D.; Chirodea, F.; Toca, C. Analysis of Social Support as an Argument for the Sustainable Construction of the European Community Space. *Sustainability* **2022**, *14*, 7448. [CrossRef]
61. Qasem, A.; Mohammed, A.; Battisti, E.; Ferraris, A. Ownership structure and firm sustainable investments: Evidence from emerging markets. *Int. J. Emerg. Mark.* **2023**. *ahead-of-print*. [CrossRef]
62. Garcia-Altés, A. The development of Health Tourism Services. *Ann. Tour. Res.* **2005**, *32*, 262–266. [CrossRef]
63. Global Wellness Institute. Global Wellness Tourism Economy. 2018. Available online: https://globalwellnessinstitute.org/wp-content/uploads/2018/11/GWI_GlobalWellnessTourismEconomyReport.pdf (accessed on 1 March 2021).
64. Romanova, G.; Vetitnev, A.; Dimanche, F. Health and Wellness Tourism (Chapter 7). In *Tourism in Russia: A Management Handbook*; Dimanche, F., Andrades, L., Eds.; Emerald Publising Limited: Bingley, UK, 2015; pp. 231–287.
65. Fortune Business Insights. Medical Tourism Market Size, Share and Global Trend By Type, By Treatment, and Geography Forecast till 2026. Available online: https://www.globenewswire.com/news-release/2020/06/03/2043057/0/en/Medical-Tourism-Market-to-Gain-Impetus-from-Availability-of-Medical-Treatment-at-Minimal-Cost-Fortune-Business-Insights.html (accessed on 21 February 2021).
66. Teleki, N.; Munteanu, L. *Spa Tourism in Romania Balneo-Turistică/Spa Tourism in Romania*; Royal Company Publishing: Bucharest, Romania, 2012.
67. Lopes, A.P.; Rodríguez-López, N. Application of a Decision-Making Tool for Ranking Wellness Tourism Destinations. *Sustainability* **2022**, *14*, 15498. [CrossRef]
68. Szromek, A.R.; Polok, G. A Business Model for Spa Tourism Enterprises: Transformation in a Period of Sustainable Change and Humanitarian Crisis. *J. Open Innov. Technol. Mark. Complex.* **2022**, *8*, 72. [CrossRef]
69. Könnyid, L.; Váradi, Z.; Nagy, Z.; Ilyés, N.; Horváth, O.H. The Changes in the Demographic Characteristics and Spatial Structure of Tourism Demand in the West Balaton Region's Spa Cities. *Sustainability* **2022**, *14*, 10531. [CrossRef]

70. Ministry of Entrepreneurship and Tourism (Romania) | Classified Tourist Reception Structures with Accommodation Function in Băile Felix area. Available online: http://turism.gov.ro/web/autorizare-turism (accessed on 12 August 2022).
71. Cronbach, L.; Shavelson, R. My Current Thoughts on Coefficient Alpha and Successor Procedures. *Educ. Psychol. Meas.* **2004**, *64*, 391–418. [CrossRef]

sustainability

MDPI

Article

The Relationship between Physical Activity Level and Sociodemographic Factors in Romanian Adults in the Post-COVID-19 Pandemic Period

Elena Ioana Iconaru [1,†], Emilian Tarcau [2,*,†], Manuela Mihaela Ciucurel [3,†], Liviu Draghici [4,†] and Constantin Ciucurel [1,†]

1 Department of Medical Assistance and Physical Therapy, University Center of Pitesti, National University for Science and Technology Politehnica Bucuresti, 110040 Pitești, Romania; ioana.iconaru@upit.ro (E.I.I.); constantin.ciucurel@upit.ro (C.C.)

2 Department of Physical Education, Sport and Physical Therapy, University of Oradea, 410087 Oradea, Romania

3 Department of Psychology, Communication Sciences and Social Assistance, University Center of Pitesti, National University for Science and Technology Politehnica Bucuresti, 110040 Pitești, Romania; manuela.ciucurel@upit.ro

4 Department of General Surgery, Sf. Ioan Emergency Clinical Hospital, Carol Davila University of Medicine and Pharmacy, 050474 București, Romania; liviudraghici@hotmail.com

* Correspondence: emilian.tarcau@yahoo.com; Tel.: +40-770104471

† The authors contributed equally to the work.

Abstract: Background: This study examines how socio-demographic factors relate to post-pandemic physical activity patterns among Romanian adults. Methods: A cross-sectional study explores post-COVID-19 physical activity levels (PAL) and their correlation with socio-demographic factors in Romanian adults ($n = 237$, average age 28.23 ± 9.91 years). An online questionnaire covering constitutional, socio-demographic, and physical activity-related variables was administered for data collection. Data analysis involves descriptive and inferential statistics, including Kendall's tau correlation, along with multinomial regression analyses. Results: Noteworthy correlations emerged, including a robust association ($r = 0.79$, $p < 0.001$) between testing and history of clinical signs of COVID-19; a significant moderate correlation between health status and PAL compared to the period before the pandemic ($\tau = 0.56$, $p < 0.001$); and significant moderate correlation between health status and current PAL ($\tau = -0.51$, $p < 0.001$). Multinomial regression underscores an intricate relationship; testing for COVID-19 relates to clinical sign severity, health status changes influence post-pandemic PAL, and self-perceived health associates with current PAL ($p < 0.001$). Conclusions: Revealing significant links between PAL and socio-demographic factors among adults in Romania's post-pandemic landscape, this study emphasizes the interaction between health changes and activity involvement. It also highlights the potential to guide interventions for rehabilitation and healthier living.

Keywords: activity patterns; lifestyle; post-pandemic; sociodemographic characteristics; Romanian population

Citation: Iconaru, E.I.; Tarcau, E.; Ciucurel, M.M.; Draghici, L.; Ciucurel, C. The Relationship between Physical Activity Level and Sociodemographic Factors in Romanian Adults in the Post-COVID-19 Pandemic Period. *Sustainability* **2023**, *15*, 13488. https://doi.org/10.3390/su151813488

Academic Editor: Gianpiero Greco

Received: 11 August 2023
Revised: 3 September 2023
Accepted: 7 September 2023
Published: 8 September 2023

1. Introduction

The COVID-19 pandemic, declared by the World Health Organization on 2020, has had profound and widespread impacts on health, the environment, psychology, education, and the global economy [1]. This pandemic has affected humanity significantly, particularly vulnerable individuals. It resulted in reduced physical condition, quality of life, physical activity levels (PAL), healthcare access, health service quality, safe food availability, sleep quality, life satisfaction, vitamin D production, and healthy lifestyle. Additionally, there has been a rise in falls among the elderly, depression, loneliness, social injustice and inequality, bodily pain, coronaphobia, and the risk of secondary diseases [2].

Different authors reported that individuals who maintained or increased their PAL during the lockdown showed improved quality of life, highlighting the importance of promoting physical activity, especially in challenging situations like the pandemic [3]. The findings offer valuable guidance for shaping policies and interventions to promote physical activity engagement in challenging circumstances, like the pandemic, by underscoring its positive influence on quality of life [4]. This calls for the development of specific strategies, such as eHealth (electronic health)/mHealth (mobile health) approaches, to encourage individuals to stay physically active during such periods. Subsequent research should focus on identifying factors that predict changes in health behaviors compared to maintaining them following stressful life events, such as the pandemic and its associated restrictions [3].

One compelling argument for conducting our study is the potential to identify crucial insights that can inform public health policies and interventions. Understanding how socio-demographic factors influence PAL in the aftermath of the COVID-19 pandemic can help develop targeted strategies to improve overall population health and well-being [5]. Moreover, as the pandemic has introduced unique challenges and disruptions to daily life, investigating the impact on physical activity can provide valuable knowledge to guide individuals, communities, and policymakers in fostering healthier behaviors and resilience in times of crisis. By addressing this research question, we can contribute to a deeper understanding of the complex interplay between socio-demographic factors and physical activity, ultimately leading to evidence-based measures that promote healthier lifestyles and long-term post-pandemic rehabilitation.

Existing evidence indicates a general decline in physical activity due to the COVID-19 pandemic, but this impact may vary among different sub-populations. However, there is a lack of comprehensive understanding regarding the specific social and physical factors that can encourage physical activity while minimizing safety risks during pandemics [6]. In the context of the Romanian population, there is a need to explore these aspects further to bridge the existing knowledge gaps and better inform public health interventions.

As we shift our focus to the post-pandemic era, it is imperative to recognize the paramount importance of acknowledging the potential enduring implications stemming from diminished physical activity during lockdown periods. As restrictions ease, there may be a change in physical activity patterns, and it becomes essential to assess how the population's activity levels evolve. Understanding the factors influencing physical activity in the aftermath of the pandemic is vital for designing effective interventions to promote healthier lifestyles.

Moreover, considering the socio-demographic characteristics unique to the Romanian population, tailored strategies should be developed to address the specific challenges and opportunities for enhancing PAL. This could involve leveraging technology, such as eHealth/mHealth solutions, to reach and engage diverse age groups and socio-economic backgrounds effectively. Furthermore, considering the influence of pandemic-related psychological stressors, initiatives centered on mental health support and coping mechanisms might assume a pivotal role in fostering physical activity engagement.

Collaboration between public health authorities, researchers, and community organizations is essential in implementing sustainable initiatives aimed at revitalizing physical activity behaviors in the post-pandemic era. By utilizing data-driven approaches and considering the specific needs of the Romanian population, we can create a healthier and more resilient society with improved well-being and quality of life.

This study aimed to conduct a comprehensive cross-sectional, descriptive, and epidemiological investigation by administering an online survey to assess the PAL and explore the underlying socio-demographic factors among the adult population in Romania. The research seeks to contribute to the creation of valuable information resources essential for managing the uncertainties associated with a novel disease, such as the COVID-19 pandemic. Understanding the link between physical activity and socio-demographic characteristics in the post-pandemic period can provide valuable insights for public health interventions and policy-making to promote a healthier lifestyle and well-being in the Romanian population.

As a research hypothesis, this study postulates that socio-demographic factors may be predictive of variations in individual physical activity levels. Additionally, the research hypothesizes that the experience of the COVID-19 pandemic might have led to changes in physical activity patterns, with potential implications for health outcomes.

2. Materials and Methods

2.1. Participants

This research was conducted on a cohort of adult participants (n = 237), with an average age of 28.23 $\pm$ 9.91 years, comprising 82 men and 155 women, falling within an age range of 18 to 49 years. Prior to data collection, we ensured that each participant provided online written informed consent, adhering to the ethical guidelines governing research involving human subjects. The study was granted approval by the Research Ethics Committee of the Research Center for Promoting Excellence in Professional Training, University of Pitesti, under reference number 168/6 February 2023.

The study was conducted on a sample of the Romanian adult population, with an initial recruitment of 310 volunteers who completed an online questionnaire. Among them, 237 individuals (a response rate of 90.9%) met the criteria for inclusion in the study, which required residing in Romania and providing complete responses on the questionnaire. Furthermore, as exclusion criteria, we considered the condition that subjects did not experience acute illnesses in the last week, given that our survey assesses physical activity over the past 7 days.

To ensure standardized testing, a group of students from University of Pitesti received training and were responsible for administering the test. Subsequently, each student selected 3-5 acquaintances (such as family members or friends) who also underwent the online survey procedure. The research aimed to focus exclusively on the adult population, including individuals aged between 18 and 49 years. By concentrating on this adult stage, the study pursued to gain a relevant perspective on the behavior, preferences, and specific needs of this age group, considering they are in a significant phase of life with responsibilities and meaningful experiences.

2.2. Data Acquisition

For online data collection, the PsyToolkit software (https://www.psytoolkit.org/) (accessed on 4 March 2023) was used [7,8]. The survey can be completed on a PC, laptop, tablet, or smartphone (with Android or iOS operating system) in approximately 10 min and includes three mandatory parts:

I. Introduction of the study title and purpose, along with obtention of online informed consent from the participants.

II. Socio-demographic data that include 18 variables. These variables cover the following aspects: age, weight (W), height (H), sex, level of education, professional status, type of residence, health status, level of anxiety related to one's health, history of chronic illnesses, occurrence of acute illness in the past week, testing for COVID-19, history of clinical signs of COVID-19, COVID-19 vaccination, smoking status, dietary habits, health status compared to the period before the pandemic, and level of physical activity compared to the period before the pandemic.

Within this section of the survey, there are 3 numerical continuous variables (age, weight, and height) and 15 nominal variables that were converted to numerical dummy values. The nominal variables in the study were obtained through close-ended questions, where respondents were asked to choose only one answer option for each question. To calculate the body mass index (BMI), we utilized the data from W (in kg) and H (in m) according to the following formula:

$$\mathrm{BMI} = \frac{\mathrm{W}}{\mathrm{H}^2}. \tag{1}$$

III. Physical Activity Questionnaire—The International Physical Activity Questionnaire—Short Form (IPAQ). The IPAQ short form is an instrument with reasonable to moderate validity, primarily designed for population surveillance of PAL among adults aged 15 to 69 years [9]. The IPAQ evaluates physical activity over the past seven days, covering four main domains: leisure, domestic and gardening (yard), work, and transport [10]. The IPAQ includes three specific types of physical activity: walking, moderate-intensity, and vigorous-intensity activities [11].

From the IPAQ, we gathered both categorical and continuous indicators of physical activity. The IPAQ scoring protocol allows us the calculation of IPAQ scores expressed in MET minutes/week as a continuous variable. This was achieved by assigning weights to each type of activity based on their energy requirements [12]. To compute the IPAQ scores, we summed up the duration (in minutes) and frequency (days) of the mentioned activities. The data collected using IPAQ as a self-report measure enabled us to classify individuals into three categories of PAL: 1 = low PAL, 2 = moderate PAL, and 3 = high PAL [13].

Completing the questionnaire allows for the assessment of an individual's PAL, offering an opportunity for self-reflection on health status, health behaviors, and the influence of the COVID-19 pandemic on these variables. After categorizing the participant into one of the three mentioned PAL categories, personalized feedback is provided regarding their status. Consequently, if the subject achieves a high level of physical activity, they are encouraged to sustain this positive behavior. Conversely, if the observed level of physical activity is low or moderate, the patient receives recommendations to enhance their engagement in physical activities.

2.3. *Statistical Analysis of Data*

The responses gathered from the survey were subjected to scientific analysis using the IBM SPSS 26.0 software (IBM Corp., Armonk, NY, USA) [14]. The data underwent descriptive statistical procedures, including calculations of mean, standard deviation, and frequency distribution. Additionally, inferential statistical methods, such as Kendall's tau correlation analysis, and multinomial regression analysis, were employed to derive further insights from the data.

3. Results and Discussion

The primary aim of our research was to identify potential associations and patterns between physical activity participation and socio-demographic variables, shedding light on the possible impact of the pandemic on activity behaviors. To achieve this, we conducted a comprehensive analysis of data collected from 237 participants. Before delving into the complex relationship between physical activity and socio-demographic factors, we provide an overview of the baseline descriptive statistical indicators for the study participants.

Table 1 presents a comprehensive overview of the essential characteristics pertaining to the study participants, focusing specifically on continuous numerical variables. The respondents' average age was 28.23 ± 9.91 years, and the age range encompassed individuals aged 19 to 49 years. The sample comprised 237 participants, consisting of 82 men and 155 women, resulting in a sex ratio of 0.53.

Table 1. Summary of the characteristics of study participants (n = 237), continuous variables.

Variable	Age (Years)	W (kg)	H (cm)	BMI (kg/m^2)	MET Minutes/Week
mean	28.23	69.79	170.42	23.99	7547.81
SD	9.91	13.78	9.07	4.18	7930.51

Note—W: weight; H: height; BMI: body mass index; MET: metabolic equivalents; SD: standard deviation; n: group size.

Based on the participants' BMI values, the study sample exhibited a diverse nutritional status. The majority (67.1%) fell within the normal weight range, indicating a significant proportion of individuals with a healthy weight. However, it is important to note that

32.9% of the participants were categorized as underweight (3.8%), pre-obese (17.3%), or displaying different degrees of obesity (obesity class I: 10.5% and obesity class II: 1.3%). These findings suggest the presence of various nutritional challenges within the sample, underscoring the need for further attention and potential interventions to address the distinct health requirements of these individuals.

The study participants' physical activity was quantified and expressed in MET minutes/week score. This metric serves as a robust measure to assess their weekly PAL. Upon analyzing the data, we found that the mean MET minutes/week score for the participants was 7547.81 ± 7930.51. This indicates a considerable variability in PAL among the study participants.

From Table 2, which summarizes the nominal variables collected through the questionnaire applied to the study participants, it can be observed that the majority of them (73.4%) were university graduates or students pursuing a master's or a PhD degree. Regarding professional status, 45.6% were employed or self-employed, 51.9% were unemployed students or householders, and a small proportion (1.3%) were retired. Most participants resided in urban areas (70%) compared to rural areas (30%). In terms of health status, 28.7% rated their health as excellent, while 15.7% reported fair or poor health. Approximately 53.6% experienced moderate health-related anxiety. The data show that 11.8% of participants reported a history of chronic illnesses, while none reported an occurrence of acute illness in the past week. The majority had not tested positive for COVID-19 (58.6%) and had not been diagnosed with the disease (53.2%). Vaccination rates were 19% for one dose, 38.8% for two doses, and 12.7% for three doses. The study also explored lifestyle factors, such as smoking status and dietary habits, as well as self-reported changes in health and PAL compared to the pre-pandemic period.

Table 2. Summary of the characteristics of study participants (n = 237), nominal variables.

Code	Variable	Answers	Frequency (n)	Percent (%)
Q1	sex	male	82	34.6
		female	155	65.4
Q2	level of education	secondary education graduate	24	10.1
		high school or vocational school student	39	16.5
		university graduate or student/master's/PhD candidate	174	73.4
Q3	professional status	employed/self-employed	108	45.6
		unemployed/householder	3	1.3
		unemployed student	123	51.9
		retired	3	1.3
Q4	type of residence	urban	166	70
		rural	71	30
Q5	health status	excellent	68	28.7
		very good	65	27.4
		good	67	28.3
		fair	30	12.7
		poor	7	3
Q6	level of anxiety related to one's health	not anxious at all (restless)	96	40.5
		moderately anxious (restless)	127	53.6
		very anxious (restless)	14	5.9

Table 2. *Cont.*

Code	Variable	Answers	Frequency (n)	Percent (%)
Q7	history of chronic illnesses	yes	28	11.8
		no	209	88.2
Q8	occurrence of acute illness in the past week	yes	0	0
		no	237	100
Q9	testing for COVID-19	I have tested positive multiple times, with an interval of more than one month	14	5.9
		I have tested positive only once	84	35.4
		I have never tested positive	139	58.6
Q10	history of clinical signs of COVID-19	I have had the illness at least once with severe clinical signs and/or hospitalization	14	5.9
		I have had the illness with moderate/mild clinical signs, once or multiple times	82	34.6
		I have tested positive, but I was asymptomatic	15	6.3
		I have not been diagnosed with COVID-19	126	53.2
Q11	COVID-19 vaccination	yes, with one dose of vaccine	45	19
		yes, with two doses of vaccine	92	38.8
		yes, with three doses of vaccine	30	12.7
		no	70	29.5
Q12	smoking status	yes	77	32.5
		no	160	67.5
Q13	dietary habits	omnivore (mixed diet, including meat and vegetarian options)	231	97.5
		lacto/ovo vegetarian (no meat, but includes dairy and/or eggs)	5	2.1
		vegan (strictly plant-based)	1	0.4
Q14	health status compared to the period before the pandemic	weaker, as I have experienced health issues following COVID-19	19	8
		weaker, but unrelated to COVID-19 illness	42	17.7
		the same	149	62.9
		better	27	11.4
Q15	PAL compared to the period before the pandemic	lower, as I have experienced health issues following COVID-19	6	2.5
		lower, but unrelated to COVID-19 illness	53	22.4
		the same	129	54.4
		higher	49	20.7
Q16	PAL	low	45	19
		moderate	47	19.8
		high	145	61.2

Note—PAL: physical activity level; n: group size; Low PAL: individuals not meeting criteria for higher categories; Moderate PAL: engaging in 3 or more days of vigorous-intensity activity for at least 20 min each day, or participating in 5 or more days of moderate-intensity activity and/or walking for at least 30 min each day, or combining these activities to achieve a minimum of 600 MET-minutes/week; High PAL: involves either engaging in vigorous-intensity activity on 3 or more days, totaling at least 1500 MET-minutes/week, or participating in a combination of walking, moderate-intensity, or vigorous-intensity activities for at least 3000 MET-minutes/week [9,11–13].

The data provided in Table 2 indicate also the distribution of PAL among the study participants. Out of the total respondents, 19% had a low PAL, 19.8% had a moderate PAL, and the majority, 61.2%, exhibited a high PAL. This distribution indicates that a substantial number of participants regularly engaged in physical activity, with most falling into the high PAL category. Understanding these fitness and behavioral patterns could offer valuable insights into the participants' overall health and well-being. Further analysis exploring the relationship between PAL and other variables could provide deeper understanding of the impact of physical activity on various aspects of health. These insights pave the way for robust inferential statistical analyses to uncover potential connections and patterns within the dataset.

Next, we conducted a correlational analysis to explore the interrelationships among the nominal variables in our dataset. To assess these associations, we utilized Kendall's tau correlation coefficient, a non-parametric measure well-suited for our data. Additionally, we calculated the statistical significance of the correlations to determine their meaningfulness. To aid in the identification of potential patterns and associations, we constructed a correlation matrix, which visually represents the relationships between key pairs of variables (see Table 3).

Table 3. Key correlations (Kendall's tau correlation coefficient, *r*) between recorded variables and level of statistical significance, *p* ($n = 237$).

Variable	Q1	Q5	Q6	Q7	Q9	Q10	Q11	Q14	Q15	
Q1	1									
Q5	0.21 *	1								
Q6	0.21 *	0.21 *	1							
Q7	−0.18 *	−0.23 *	−0.17 *	1						
Q9	−0.07	−0.18 *	−0.06	0.02	1					
Q10	−0.15 *	−0.20 *	−0.05	0.06	0.79 *	1				
Q11	0.04	−0.05	−0.04	0.07	0.07	0.07	1			
Q14	−0.18 *	−0.30 *	−0.23 *	0.16 *	0.23 *	0.20 *	0.21 *	1		
Q15	−0.10	−0.31 *	−0.23 *	0.18 *	0.15 *	0.13 *	0.11	0.56 *	1	
Q16	−0.10	−0.51 *	−0.15 *	0.11	0.29 *	0.25 *	0.02	0.16 *	0.24 *	1

Note: Q1: sex; Q5: health status; Q6: level of anxiety related to one's health; Q7: history of chronic illnesses; Q9: testing for COVID-19; Q10: history of clinical signs of COVID-19; Q11: COVID-19 vaccination; Q14: health status compared to the period before the pandemic; Q15: PAL compared to the period before the pandemic: Q16: PAL; * $p < 0.05$ was considered statistically significant (2-tailed); *n*: group size.

Moving forward with our analysis, we delve into the important correlations obtained from our dataset, with a particular focus on the most significant ones. This sheds light on how certain factors may influence others within the dataset. One highly significant correlation, with a coefficient τ of 0.79, was observed between the variables Q9 (testing for COVID-19) and Q10 (history of clinical signs of COVID-19). This result indicates a robust association between the frequency of positive COVID-19 tests and the severity of clinical signs experienced by the participants. Specifically, participants who reported multiple positive COVID-19 tests with intervals of more than one month tended to have a history of the illness with severe clinical signs and/or hospitalization. Conversely, those who tested positive only once or never tested positive generally had a history of mild or moderate clinical signs of the illness or were asymptomatic.

It is important to emphasize that this significant correlation does not imply causation, and other factors may also influence the relationship between the frequency of positive COVID-19 tests and the severity of clinical signs. Further in-depth analysis and investigations are warranted to gain a more comprehensive understanding of this association.

Moreover, our analysis revealed another notable finding—a significant correlation of 0.56 between Q14 (health status compared to the period before the pandemic) and Q15 (PAL compared to the period before the pandemic). This value suggests a moderate positive association between the participants' reported changes in health status and their PAL after the pandemic. Specifically, participants who experienced a decline in health status following COVID-19 or due to other unrelated factors also tended to exhibit lower PAL levels at present, after the pandemic. This implies that changes in health status may be linked to changes in PAL in the post-pandemic period.

Lastly, we observed a moderate negative correlation (−0.51) between health status (Q5) and the current level of physical activity (PAL) (Q16). This indicates that participants with better health (higher values in Q5, e.g., "excellent" or "very good") tended to exhibit higher levels of physical activity (higher values in Q16, e.g., "high"). In essence, a healthier state is associated with increased levels of physical activity.

Apart from these significant correlations, other obtained Kendall nonparametric correlations were found to be weak. These results suggest a weak or negligible association between the respective variables, and no significant linear trends were observed in the analyzed data.

In summary, our analysis has offered valuable insights into the interconnectedness among different variables within our dataset. Certain correlations have exhibited robust associations, while others have displayed less substantial or insignificant links. These findings contribute to our understanding of the factors influencing different aspects of the participants' health and behavior, emphasizing the importance of further investigations to unveil the complex dynamics within the dataset.

Continuing with the statistical analysis, we performed a multinomial logistic regression to examine the associations between the nominal variables that showed significant Kendall correlations. The purpose of this regression was to assess the effects of these variables on the different categories of the dependent variable. By employing this regression technique, we aimed to gain valuable insights into the factors influencing the outcomes observed in our study. The results of the multinomial logistic regression are presented and discussed in the following sections.

The first multinomial logistic regression analysis was conducted to explore the relationship between variables Q9 (testing for COVID-19) and Q10 (history of clinical signs of COVID-19). In this analysis, the dependent variable was Q10, which represents the history of clinical signs of COVID-19, categorized into four levels. The independent variable in this analysis was Q9, representing the frequency of COVID-19 testing, categorized into three levels. The multinomial logistic regression allowed us the examination of how the frequency of positive COVID-19 tests (Q9) influenced the history of clinical signs of COVID-19 (Q10) experienced by the participants. The results of the regression analysis provided information on the odds ratios and significance levels for each level of the dependent variable (history of clinical signs of COVID-19) compared to the reference category "I have not been diagnosed with COVID-19".

By using the multinomial logistic regression, we aimed to understand whether there was a significant association between the frequency of positive COVID-19 tests and the severity of clinical signs experienced by the participants. This analysis helped us identify whether participants who tested positive more frequently were more likely to experience severe clinical signs and/or hospitalization compared to those who tested positive only once or never tested positive.

As a result, the likelihood ratio tests for the multinomial regression analysis indicated a statistically significant relationship between the predictor variable "COVID-19 testing" and the dependent variable "history of clinical signs of COVID-19" ($\chi^2 = 227.195$, df = 3, $p < 0.001$). These tests were performed to evaluate the overall model fit and demonstrated that the inclusion of the "COVID-19 testing" variable significantly improved the model's ability to explain the variability in the clinical signs of COVID-19 categories.

The results from the multinomial logistic regression analysis revealed a significant relationship between the frequency of COVID-19 testing (Q9) and the history of clinical signs of COVID-19 (Q10) categories (Table 4). As indicated by the negative coefficients for the "testing for COVID-19" variable in all three clinical sign categories, higher frequencies of positive COVID-19 tests were associated with an increased likelihood of experiencing more severe clinical signs, both moderate/mild and severe, compared to the reference category "I have not been diagnosed with COVID-19". The odds ratios and corresponding 95% confidence intervals further confirmed these associations, highlighting the elevated risk of developing severe clinical signs and/or hospitalization among participants with more frequent positive test results.

Table 4. Results of multinomial logistic regression analysis for the relationship between the independent variable Q9 (testing for COVID-19) and the dependent variable Q10 (history of clinical signs of COVID-19)—parameter estimation (n = 237).

History of Clinical Signs of COVID-19 [a]		B	Std. Error	Wald	df	Sig.	Exp (B)	95% Confidence Interval for Exp (B)	
								Lower Bound	Upper Bound
I have had the illness with moderate/mild clinical signs, once or multiple times	Intercept	17.579	3.085	32.481	1	0.001			
	Testing for COVID-19	−6.729	1.055	40.668	1	0.001	0.001	0.001	0.009
I have had the illness at least once with severe clinical signs and/or hospitalization	Intercept	15.616	3.240	23.225	1	0.001			
	Testing for COVID-19	−6.631	1,160	32.696	1	0.001	0.001	0.001	0.013
I have tested positive, but I was asymptomatic	Intercept	14.587	3.222	20.490	1	0.001			
	Testing for COVID-19	−6.099	1.134	28.914	1	0.001	0.002	0.001	0.021

[a] The reference category is: I have not been diagnosed with COVID-19.

Our findings prompt a broader contextual analysis informed by recent research. A significant study explored the interrelation between COVID-19 testing patterns and the gravity of clinical presentations, aligning with our own results. This investigation delved into the correlation between testing approaches and the severity of clinical indications, underscoring the importance of comprehensive testing strategies [15].

Other researchers have similarly examined the intricate interplay between testing methodologies and their practical implications. For instance, in a related study, it was observed that testing patterns hold valuable insights into the dynamics of COVID-19 diagnosis within an academic medical center [16]. The ongoing nature of the pandemic underscores the need for continuous assessment, as the researchers emphasized the importance of further analysis utilizing up-to-date data to corroborate existing findings and provide a robust basis for future testing guidelines. Notably, a significant proportion of subjects in their study underwent multiple rounds of testing, which resonates with our own findings and highlights the potential nuances and complexities that arise from repeated testing. This collective body of research contributes to a comprehensive understanding of COVID-19 testing strategies, emphasizing the importance of informed decision-making in healthcare settings. As the scientific community continues to navigate the evolving landscape of the pandemic and beyond, these insights remain instrumental in shaping effective testing protocols and enhancing diagnostic practices.

The intricate relationship revealed between COVID-19 testing frequency and clinical signs in our study aligns with recent findings underscoring the significance of comprehensive diagnostic approaches for SARS-CoV-2 [17]. Notably, the use of markers indicating viral load emerges as essential, offering substantial insights into disease prognosis and informing healthcare choices [18].

In the broader context of global testing efforts, it is worth noting that the lack of systematic testing across the world has posed challenges to the accuracy of epidemiological data, particularly during the initial stages of the outbreak [19]. Within this framework, the observation that individuals who received positive test results were more prone to indicate greater severity of symptoms compared to those who yielded negative outcomes [20] further accentuates the significance of comprehensive testing approaches. Additionally, the relationship between test positivity and the vaccination status of subjects should also be taken into account due to potential immunological interferences that may occur [21,22].

Reflecting on the concluded pandemic, our study's insights cast light on the intricacies of COVID-19 testing patterns. The correlation unveiled between testing frequency and clinical outcomes offers a retrospective perspective that guides our understanding of the past. Moving forward, our results hold relevant significance in shaping future healthcare strategies. The observed prevalence of repeated testing underscores the need for adaptable testing frameworks, which will likely continue to play a pivotal role in managing potential health challenges [23]. This forward-thinking strategy aligns with a broader perspective on healthcare preparedness, highlighting the significance of versatile strategies that can be effectively employed across diverse medical scenarios, thus reimagining pandemic readiness to bolster global health and ensure our shared trajectory towards lasting well-being [24,25]. As we transition beyond the pandemic era, our findings contribute to evidence-based decision-making, fostering resilient testing protocols that are well-equipped to address forthcoming health dynamics.

The next multinomial regression analysis was between variables Q14 (health status compared to the period before the pandemic) and Q15 (PAL compared to the period before the pandemic). In this analysis, Q14 served as the independent variable, representing the perceived change in health status due to COVID-19, categorized into four levels. Q15 was the dependent variable, representing the change in PAL compared to the period before the pandemic, also categorized into four levels. The analysis aimed to investigate how changes in health status (Q14) were associated with variations in PAL (Q15), providing insights into how perceived health alterations might relate to shifts in physical activity patterns during the pandemic.

The outcomes of the likelihood ratio tests in the context of the multinomial regression analysis unveiled a notable statistical linkage between the independent variable "health status compared to the period before the pandemic" and the dependent variable "PAL compared to the pre-pandemic period" (χ^2 = 98.919, df = 3, p < 0.001). These tests were executed with the intent to evaluate the holistic model suitability. It was evident from the results that the inclusion of the "health status compared to the period before the pandemic" variable substantially augmented the model's efficacy in comprehending the fluctuations in PAL relative to the period preceding the pandemic.

The results presented in Table 5 underscore a significant relationship between perceived changes in health status and variations in levels of physical activity (PAL) after the pandemic. More specifically, it is observed that individuals who report a lower health status due to COVID-19-related issues or unrelated causes are associated with a decreased likelihood of maintaining the same or higher levels of physical activity compared to the pre-pandemic period.

This finding suggests that changes in health status have a substantial impact on individual physical activity behavior in the post-pandemic period. Individuals who have experienced a decline in health status, either due to COVID-19 or other reasons, appear to be less likely to sustain previous or higher levels of physical activity compared to that in the pre-pandemic period. This association can be explained by the fact that changes in health status may influence the individuals' capacity and motivation to actively engage in physical activities, regardless of whether these changes are directly related to COVID-19 or not. Thus, this study highlights the importance of assessing health status in the context of its influence on physical activity and underscores the need for an integrated approach in promoting health and physical activity during periods of significant change, such as those generated by a pandemic.

Our study corresponds with the findings of numerous investigations that elucidate the pervasive adverse influence of COVID-19 on physical activity. In the context of COVID-19, physical inactivity can pose greater risks than the viral infection itself, with alarming health implications [26]. Notably, this is particularly relevant considering the established role of physical activity in both the mitigation and management of COVID-19 [27]. Together, these findings emphasize the necessity of comprehensive approaches that consider health status, social factors, and physical environments in promoting physical activity amidst post-pandemic challenges [28,29].

Table 5. Results of multinomial logistic regression analysis for the relationship between the independent variable Q14 (health status compared to the period before the pandemic) and the dependent variable Q15 (PAL compared to the period before the pandemic)—parameter estimation (n = 237).

PAL Compared to the Period before the Pandemic [a]		B	Std. Error	Wald	df	Sig.	Exp (B)	95% Confidence Interval for Exp (B)	
								Lower Bound	Upper Bound
Lower, as I have experienced health issues following COVID-19	Intercept	10.075	1.871	29.011	1	0.001			
	Health status compared to the period before the pandemic	−4.345	0.721	36.338	1	0.001	0.013	0.003	0.053
Lower, but unrelated to COVID-19 illness	Intercept	11.160	1.595	48.972	1	0.001			
	Health status compared to the period before the pandemic	−3.801	0.523	52.824	1	0.001	0.022	0.008	0.062
The same	Intercept	8.652	1.461	35.057	1	0.001			
	Health status compared to the period before the pandemic	−2.467	0.459	28.936	1	0.001	0.085	0.035	0.208

[a] The reference category is higher.

Given the well-documented positive effects of physical activity on both physical and mental health [30], our findings, in alignment with Wilson et al. [31], emphasize the importance of promoting and facilitating safe and regulated physical activity opportunities. Additionally, the transition of sports offerings to the digital realm, as explored by Parker et al. [32], offers a creative avenue to maintain engagement with physical activity, even in times of restrictions. Overall, the collective body of research underscores the necessity for adaptable strategies that encourage physical activity while navigating the evolving landscape of COVID-19 pandemic-related challenges. This is further accentuated by recognizing the concurrent pandemic of cardiovascular diseases, underscoring the urgency for holistic health approaches and acknowledging the detrimental effects of sedentary behavior [33,34].

In contrast to the transient decrease in exercise engagement during COVID-19 restrictions and subsequent return to pre-pandemic activity levels, as demonstrated in some studies [35,36], our investigation delves more deeply into the enduring repercussions of health status changes on post-pandemic physical activity behaviors. Moreover, some researchers have observed an alteration in physical activity patterns during periods of relaxation, with the decline in physical activity primarily arising from mobility constraints, while activities of moderate-to-vigorous intensity remain unaffected [37]. Our findings elucidate a more intricate interplay between health status and the continuity of physical activity. Specifically, our study unveils that individuals experiencing declines in health status, regardless of COVID-19, are less likely to maintain their previous or heightened levels of physical activity after the pandemic. This nuanced perspective extends the discourse regarding the repercussions of the pandemic on physical activity by highlighting the enduring impact of health-related factors, encompassing immune system dysfunction and deterioration in mental well-being [38]. In conclusion, our study contributes to the evolving understanding of post-COVID-19 physical activity patterns, underscoring the enduring role of health status transitions in shaping activity levels in the aftermath of the pandemic.

The final multinomial regression analysis examines the relationship between the variables "health status" (Q5) and the "current PAL" (Q16). In this analysis, the independent variable is "health status" (Q5), which represents individual self-perceived health status categorized into five levels. The dependent variable is the "current PAL" (Q16), which categorizes individual PAL into three categories: low, moderate, and high. The analysis aims to investigate how variations in individuals' self-reported health status (Q5) are associated with different PALs (Q16), providing insights into how perceived health status might relate to current physical activity patterns.

The outcomes of the likelihood ratio tests in the context of the multinomial regression analysis revealed a significant connection between the independent variable "health status" (Q6) and the dependent variable "PAL" (Q16) (χ^2 = 87.565, df = 2, $p < 0.001$). These tests were conducted with the aim of assessing the overall appropriateness of the model. The results indicated that the inclusion of the "health status" variable (Q6) substantially improved the model's effectiveness in understanding the variations in PAl (Q16). This statistical linkage suggests that changes in health status have a noteworthy influence on the shifts in individual physical activity patterns.

From Table 6, the coefficients associated with the variable "health status" (Q6) provide important insights into the relationship between individuals' self-perceived health status and their current PAL. The coefficients indicate the direction and strength of the effect that changes in health status have on the odds of being in different PAL categories.

Table 6. Results of multinomial logistic regression analysis for the relationship between the independent variable Q6 (health status) and the dependent variable Q16 (PAL)—parameter estimation (n = 237).

PAL [a]		B	Std. Error	Wald	df	Sig.	Exp (B)	95% Confidence Interval for Exp (B)	
								Lower Bound	Upper Bound
Low	Intercept	−5.834	0.766	58.015	1	0.001			
	Health status	1.751	0.249	49.296	1	0.001	5.761	3.533	9.393
Moderate	Intercept	−3.204	0.498	41.333	1	0.001			
	Health status	0.915	0.191	23.066	1	0.001	2.497	1.719	3.628

[a] The reference category is: High.

The coefficient for "Health status" in the "Low" PAL category is 1.751 ($p < 0.001$), while in the "Moderate" PAL category, it is 0.915 ($p < 0.001$), both significant at a high level of statistical significance. These coefficients suggest that as individuals' self-perceived health status decreases, the odds of being in the "Low" or "Moderate" PAL categories increase compared to the reference category, which is the "High" PAL category. In other words, individuals reporting lower health status are more likely to have lower or moderate levels of physical activity compared to those with higher perceived health status. The lower coefficients for the "Moderate" PAL category also indicate that the impact of health status on activity levels is slightly weaker in this category compared to the "Low" PAL category.

Overall, the statistical linkage between health status and PAL underscores the importance of considering health-related factors when analyzing variations in PAL. These findings offer valuable insights into the complex interplay between health and physical activity behaviors, highlighting the need for tailored interventions and strategies to promote and sustain healthy levels of physical activity, particularly among individuals with lower health status.

It is intriguing to examine our findings in light of other research. Accordingly, conclusions from another study [39] underscore the intricate relationship between physical activity, physical fitness, and health status. Their findings highlight that equivalent levels of physical activity can yield varying effects on physical fitness or health status due to contextual factors, content, and the purpose of the activity. Their study emphasizes that mere movement is not the sole determinant of fitness or health benefits. Consequently, these collective insights emphasize the critical need not only to grasp the connection between health status and PAL [40], as our study accomplishes, but also to recognize the multi-faceted nuances of physical activity's influence on overall fitness and health. Ultimately, this intricate relationship manifests at the level of individuals' well-being, where those maintaining high activity levels are poised to experience elevated states of well-being [41].

By contextualizing our findings, we acknowledge the intricate interplay encompassing health status, physical activity patterns, fitness outcomes, and overall well-being [42]. Studies link insufficient physical activity to noncommunicable diseases, cognitive impairments, and reduced quality of life [43,44]. The reciprocal health–physical activity connection is clear: better health drives physical activity, and physical activity enhances biological states. The benefits of physical activity extend beyond physiology, encompassing quality of life, self-esteem, and systemic effects like endorphin release [45]. These insights support holistic, tailored approaches for promoting healthy activity, especially for those with compromised health [46]. Modern strategies to combat sedentary behavior are based on motivating patients to engage in regular activity, creating conducive environments, tracking disease incidence among active and inactive patients, and disseminating exemplary approaches [47]. These interventions hold particular relevance in the context of an aging population [48]. Furthermore, their importance is accentuated within the current global landscape as humanity navigates the post-COVID-19 pandemic transition [49].

A brief final mention should be made regarding certain correlations in Table 3, which are statistically significant but of weak intensity. Thus, our analysis revealed statistically significant but weak correlations (Kendall's tau of -0.23) between "the level of anxiety related to one's health" (Q6) and "health status compared to the period before the pandemic" (Q14) and "physical activity level compared to the period before the pandemic" (Q15). These results suggest that while there is a connection between anxiety and changes in health and physical activity, this link is not particularly strong. This implies that while anxiety could contribute somewhat to the pandemic's negative impact, it is only one factor among many others. Socio-economic status, coping strategies, social support, and personal resilience likely interact with anxiety to collectively shape participant experiences [50]. In essence, our findings underscore the complexity of these interactions. While neuroticism-related traits could play a role, a comprehensive understanding of the overall quality of life during and after the pandemic requires considering multiple contributing factors [51]. Further research could delve into these mechanisms for deeper insights.

While we did not specifically focus on athletes, our findings can be contextualized within the framework of research that examines the impact of the COVID-19 outbreak on the mental health of individuals engaged in various sports. This broader context aligns with the approach taken by some authors who emphasize the significant mental health challenges faced by athletes during the pandemic, driven by factors such as isolation and disrupted training [52]. Furthermore, it appears that psychological distress related to the COVID-19 pandemic, including concerns about both sports and careers, is more pronounced in team sports compared to individual sports [53]. These observations resonate with the focal points of our study, underscoring a potential avenue for future exploration.

In summary, our research emphasizes the intricate interplay between health status, patterns of physical activity, and sociodemographic variables, underlining the importance of tailored strategies to encourage healthier activity levels, particularly among individuals at risk due to sedentary behaviors. By exploring the complex connections among health-related variables, socio-demographic factors, and post-pandemic physical activity behaviors, this study provides new insights into the complex interdependencies within this multifaceted context. Through thorough analysis and interpretation of the collected data, this research expands the current understanding of the factors influencing physical activity behaviors, underscoring their significance in advancing post-pandemic public health strategies. The study's innovative perspectives not only contribute to scholarly knowledge but also have practical implications for designing effective interventions to enhance physical activity engagement in the aftermath of the pandemic.

As with any study, certain limitations must be acknowledged. Despite the meticulous design and implementation of our research methodology, several constraints are worth noting. First, the study was conducted exclusively within the Romanian adult population, limiting the generalizability of the findings to other cultural or geographical contexts. Second, the data collection relied on self-report measures, which could introduce recall bias or social desirability effects, potentially affecting the accuracy of the reported physical activity behaviors. Third, the cross-sectional nature of the study design prevents us from establishing causal relationships between the examined variables. Furthermore, while efforts were made to ensure a diverse participant pool, there may still exist some degree of selection bias, as participants were recruited through acquaintances.

4. Conclusions

Our study delved into the intricate interplay between physical activity engagement and socio-demographic factors in the post-pandemic context among Romanian adults. The main findings underscore the essential role of health-related factors in shaping post-pandemic physical activity behaviors. The significant impact of health status changes, whether prompted by COVID-19 or other factors, on individual activity levels showcases the enduring interrelation between health transitions and participation in physical activity. These insights are further underscored by the multinomial regression outcomes, revealing the intricate interplay of COVID-19 testing frequencies and clinical sign severity, the influence of health status shifts on post-pandemic PAL, and the noteworthy link between self-perceived health status and current engagement in physical activity ($p < 0.001$). These important findings highlight the necessity for holistic strategies that integrate health considerations, aiding the formulation of evidence-based approaches to foster and maintain healthy engagement in physical activity.

To summarize, it is important to acknowledge the limitations of our study. These comprise a singular focus on the Romanian adult population, utilization of self-report measures, adoption of a cross-sectional design, and the potential for selection bias. Certain avenues for future research stemming from these limitations include expanding the study's scope to encompass diverse cultural and geographical contexts, employing more objective measures in addition to self-report data, adopting longitudinal designs to explore causal relationships, and employing randomized sampling methods to mitigate potential selection bias.

Author Contributions: Conceptualization, E.I.I., E.T., M.M.C., L.D. and C.C.; methodology, E.I.I., E.T., M.M.C., L.D. and C.C.; validation, E.I.I., E.T., M.M.C., L.D. and C.C.; formal analysis, E.I.I., E.T., M.M.C., L.D. and C.C.; resources, E.I.I., E.T., M.M.C., L.D. and C.C.; data curation, E.I.I., E.T., M.M.C., L.D. and C.C.; writing—original draft preparation, E.I.I., E.T., M.M.C., L.D. and C.C.; writing—review and editing, E.I.I., E.T., M.M.C., L.D. and C.C.; visualization, E.I.I., E.T., M.M.C., L.D. and C.C.; supervision, E.I.I., E.T., M.M.C., L.D. and C.C.; project administration, E.I.I., E.T., M.M.C., L.D. and C.C. All authors have read and agreed to the published version of the manuscript.

Funding: This research received no external funding.

Institutional Review Board Statement: The study was conducted in accordance with the Declaration of Helsinki, and approved by the Ethics Committee of the Research Centre for Promoting Excellency in Professional Training, University of Pitesti (reference number 168/6 February 2023).

Informed Consent Statement: Informed consent was obtained from all subjects involved in the study.

Data Availability Statement: The data are available on request from the corresponding author. All data relevant to the study are included in the article.

Conflicts of Interest: The authors declare no conflict of interest.

References

1. Miyah, Y.; Benjelloun, M.; Lairini, S.; Lahrichi, A. COVID-19 Impact on Public Health, Environment, Human Psychology, Global Socioeconomy, and Education. *Sci. World J.* **2022**, *2022*, 5578284. [CrossRef]
2. Şevgin, O.; Alptekin, H.K.; Şevgin, K. The Weak Ring of the COVID-19 Pandemic: The Elderly (A Literature Review). *J. Geriatr. Med. Gerontol.* **2021**, *7*, 111. [CrossRef]
3. Niermann, C.; Bollenbach, L.; Kanning, M. Perceiving reduced physical activity during COVID-19 lockdown is related to lower quality of life: A cross-sectional study with young adults. *Ger. J. Exerc. Sport Res.* **2022**, *52*, 472–476. [CrossRef]
4. Symons, M.; Meira Cunha, C.; Poels, K.; Vandebosch, H.; Dens, N.; Alida Cutello, C. Physical Activity during the First Lockdown of the COVID-19 Pandemic: Investigating the Reliance on Digital Technologies, Perceived Benefits, Barriers and the Impact of Affect. *Int. J. Environ. Res. Public Health.* **2021**, *18*, 5555. [CrossRef] [PubMed]
5. Ciucurel, C.; Iconaru, E.I. An Epidemiological Study on the Prevalence of the Clinical Features of SARS-CoV-2 Infection in Romanian People. *Int. J. Environ. Res. Public Health* **2020**, *17*, 5082. [CrossRef] [PubMed]
6. Park, A.H.; Zhong, S.; Yang, H.; Jeong, J.; Lee, C. Impact of COVID-19 on physical activity: A rapid review. *J. Glob. Health* **2022**, *12*, 05003. [CrossRef] [PubMed]
7. Stoet, G. PsyToolkit—A software package for programming psychological experiments using Linux. *Behav. Res. Methods* **2010**, *42*, 1096–1104. [CrossRef]
8. Stoet, G. PsyToolkit: A novel web-based method for running online questionnaires and reaction-time experiments. *Teach. Psychol.* **2017**, *44*, 24–31. [CrossRef]
9. Craig, C.L.; Marshall, A.L.; Sjöström, M.; Bauman, A.E.; Booth, M.L.; Ainsworth, B.E.; Pratt, M.; Ekelund, U.; Yngve, A.; Sallis, J.F.; et al. International physical activity questionnaire: 12-country reliability and validity. *Med. Sci. Sports Exerc.* **2003**, *35*, 1381–1395. [CrossRef]
10. Ciucurel, C.; Iconaru, E.I. The Relationship between the Frontal QRS-T Angle on ECG and Physical Activity Level in Young Adults. *Int. J. Environ. Res. Public Health* **2023**, *20*, 2411. [CrossRef]
11. Manta, A.; Cojocaru, E.; Leon-Constantin, M.M.; Mastaleru, A.; Roca, M.; Rusu, C.; Cojocariu, S.A.; Mitu, F. IPAQ-L and CPET usefulness in a North-Eastern Romanian population undergoing cardiac rehabilitation. *Appl. Sci.* **2021**, *11*, 5483. [CrossRef]
12. Ainsworth, B.E.; Haskell, W.L.; Herrmann, S.D.; Meckes, N.; Bassett, D.R., Jr.; Tudor-Locke, C.; Greer, J.L.; Vezina, J.; Whitt-Glover, M.C.; Leon, A.S. 2011 Compendium of Physical Activities: A second update of codes and MET values. *Med. Sci. Sports Exerc.* **2011**, *43*, 1575–1581. [CrossRef]
13. International Physical Activity Questionnaire. Guidelines for Data Processing and Analysis of the International Physical Activity Questionnaire (IPAQ)—Short and Long Forms. 2005. Available online: https://sites.google.com/view/ipaq/home (accessed on 7 February 2023).
14. IBM Corp. *IBM SPSS Statistics for Windows*; Version 26.0, Released 2019; IBM Corp.: Armonk, NY, USA, 2019.
15. Wohl, D.A.; Barzin, A.H.; Napravnik, S.; Davy-Mendez, T.; Smedberg, J.R.; Thompson, C.M.; Ruegsegger, L.; Gilleskie, M.; Weber, D.J.; Whinna, H.C.; et al. COVID-19 symptoms at time of testing and association with positivity among outpatients tested for SARS-CoV-2. *PLoS ONE* **2021**, *16*, e0260879. [CrossRef] [PubMed]
16. Salerno, S.; Zhao, Z.; Prabhu Sankar, S.; Salvatore, M.; Gu, T.; Fritsche, L.G.; Lee, S.; Lisabeth, L.D.; Valley, T.S.; Mukherjee, B. Patterns of repeated diagnostic testing for COVID-19 in relation to patient characteristics and outcomes. *J. Intern. Med.* **2021**, *289*, 726–737. [CrossRef]
17. Shafie, M.H.; Antony Dass, M.; Ahmad Shaberi, H.S.; Zafarina, Z. Screening and confirmation tests for SARS-CoV-2: Benefits and drawbacks. *Beni. Suef. Univ. J. Basic Appl. Sci.* **2023**, *12*, 6. [CrossRef]
18. Rabaan, A.A.; Tirupathi, R.; Sule, A.A.; Aldali, J.; Mutair, A.A.; Alhumaid, S.; Muzaheed; Gupta, N.; Koritala, T.; Adhikari, R.; et al. Viral Dynamics and Real-Time RT-PCR Ct Values Correlation with Disease Severity in COVID-19. *Diagnostics* **2021**, *11*, 1091. [CrossRef] [PubMed]
19. Šuljagić, V.; Đurić-Petković, D.; Lazić, S.; Mladenović, J.; Rakonjac, B.; Opačić, D.; Ljubenović, N.; Milojković, B.; Radojević, K.; Nenezić, I.; et al. Epidemiological Predictors of Positive SARS-CoV-2 Polymerase Chain Reaction Test in Three Cohorts: Hospitalized Patients, Healthcare Workers, and Military Population, Serbia, 2020. *Int. J. Environ. Res. Public Health* **2023**, *20*, 3601. [CrossRef]
20. Morlock, R.; Morlock, A.; Downen, M.; Shah, S.N. COVID-19 prevalence and predictors in United States adults during peak stay-at-home orders. *PLoS ONE* **2021**, *16*, e0245586. [CrossRef] [PubMed]
21. Zhao, Z.; Salerno, S.; Shi, X.; Lee, S.; Mukherjee, B.; Fritsche, L.G. Understanding the Patterns of Serological Testing for COVID-19 Pre- and Post-Vaccination Rollout in Michigan. *J. Clin. Med.* **2021**, *10*, 4341. [CrossRef] [PubMed]
22. Song, S.; Zang, S.; Gong, L.; Xu, C.; Lin, L.; Francis, M.R.; Hou, Z. Willingness and uptake of the COVID-19 testing and vaccination in urban China during the low-risk period: A cross-sectional study. *BMC Public Health* **2022**, *22*, 556. [CrossRef] [PubMed]
23. Goss, C.W.; Maricque, B.B.; Anwuri, V.V.; Cohen, R.E.; Donaldson, K.; Johnson, K.J.; Powderly, W.G.; Schechtman, K.B.; Schmidt, S.; Thompson, J.J.; et al. SARS-CoV-2 active infection prevalence and seroprevalence in the adult population of St. Louis County. *Ann. Epidemiol.* **2022**, *71*, 31–37. [CrossRef] [PubMed]
24. Lal, A.; Abdalla, S.M.; Chattu, V.K.; Erondu, N.A.; Lee, T.L.; Singh, S.; Abou-Taleb, H.; Vega Morales, J.; Phelan, A. Pandemic preparedness and response: Exploring the role of universal health coverage within the global health security architecture. *Lancet Glob. Health* **2022**, *10*, e1675–e1683. [CrossRef]

25. Coccia, M. Sources, diffusion and prediction in COVID-19 pandemic: Lessons learned to face next health emergency. *AIMS Public Health.* **2023**, *10*, 145–168. [CrossRef] [PubMed]
26. Segura-Ortí, E.; Martínez-Olmos, F.J.; Ródenas-Pascual, Á.; Guillem-Giménez, E.; Vercher-Narbona, V.; Piñón-Ruiz, M.J.; Garcia-Testal, A. Impact of COVID-19 Pandemic on Health-Related Quality of Life and Physical Activity of Patients in Hemodialysis. *Healthcare* **2022**, *10*, 2217. [CrossRef]
27. Yang, J.; Li, X.; He, T.; Ju, F.; Qiu, Y.; Tian, Z. Impact of Physical Activity on COVID-19. *Int. J. Environ. Res. Public Health* **2022**, *19*, 14108. [CrossRef] [PubMed]
28. Yang, X.; Lo, K. Environmental health research and the COVID-19 pandemic: A turning point towards sustainability. *Environ Res.* **2021**, *197*, 111157. [CrossRef]
29. Wunsch, K.; Kienberger, K.; Niessner, C. Changes in Physical Activity Patterns Due to the COVID-19 Pandemic: A Systematic Review and Meta-Analysis. *Int. J. Environ. Res. Public Health* **2022**, *19*, 2250. [CrossRef]
30. Ciucurel, C.; Iconaru, E.I. The importance of sedentarism in the development of depression in elderly people. *Procedia Soc. Behav. Sci.* **2012**, *33*, 722–3726. [CrossRef]
31. Wilson, D.; Driller, M.; Johnston, B.; Gill, N. The effectiveness of a 17-week lifestyle intervention on health behaviors among airline pilots during COVID-19. *J. Sport Health Sci.* **2020**, *10*, 333–340. [CrossRef]
32. Parker, K.; Uddin, R.; Ridgers, N.D.; Brown, H.; Veitch, J.; Salmon, J.; Timperio, A.; Sahlqvist, S.; Cassar, S.; Toffoletti, K.; et al. The use of digital platforms for adults' and adolescents' physical activity during the COVID-19 pandemic (our life at home): Survey study. *J. Med. Internet Res.* **2021**, *23*, e23389. [CrossRef]
33. Vinciguerra, M.; Romiti, S.; Sangiorgi, G.M.; Rose, D.; Miraldi, F.; Greco, E. SARS-CoV-2 and Atherosclerosis: Should COVID-19 Be Recognized as a New Predisposing Cardiovascular Risk Factor? *J. Cardiovasc. Dev. Dis.* **2021**, *8*, 130. [CrossRef]
34. Wang, C.; Wang, J. Ethnolinguistic diversity and the spread of communicable diseases: A cross-country study on the COVID-19 pandemic. *Health Promot. Int.* **2022**, *37*, daac082. [CrossRef]
35. Schöttl, S.E.; Schnitzer, M.; Savoia, L.; Kopp, M. Physical Activity Behavior During and After COVID-19 Stay-at-Home Orders—A Longitudinal Study in the Austrian, German, and Italian Alps. *Front. Public Health* **2022**, *10*, 901763. [CrossRef]
36. Jago, R.; Salway, R.; House, D.; Walker, R.; Emm-Collison, L.; Sansum, K.; Breheny, K.; Reid, T.; Churchward, S.; Williams, J.G.; et al. Short and medium-term effects of the COVID-19 lockdowns on child and parent accelerometer-measured physical activity and sedentary time: A natural experiment. *Int. J. Behav. Nutr. Phys. Act.* **2023**, *20*, 42. [CrossRef]
37. Mason, M.R.; Hudgins, J.H.; Campbell, M.S.; Biddle, M.J.; Ickes, M.J.; Dugan, A.; Bollinger, L.M. Changes in physical activity during the initial stages of the COVID-19 pandemic. *J. Sports Sci.* **2022**, *40*, 116–124. [CrossRef] [PubMed]
38. Wafi, A.M.; Mosleh, A.A.; Mutaen, A.A.; Hakami, D.F.; Baeshen, R.A.; Rajhi, S.A.; Alamri, J.M.; Alqassim, A.Y. The Impact of COVID-19 on Maintaining Physical Activity and Its Associated Factors among Medical Students. *Int. J. Environ. Res. Public Health.* **2022**, *19*, 15752. [CrossRef]
39. Schmidt, S.C.E.; Tittlbach, S.; Bös, K.; Woll, A. Different Types of Physical Activity and Fitness and Health in Adults: An 18-Year Longitudinal Study. *Biomed. Res. Int.* **2017**, *2017*, 1785217. [CrossRef]
40. Kapoor, G.; Chauhan, P.; Singh, G.; Malhotra, N.; Chahal, A. Physical Activity for Health and Fitness: Past, Present and Future. *J. Lifestyle Med.* **2022**, *12*, 9–14. [CrossRef]
41. Rampisheh, Z.; Ramezani, M.; Khalili, N.; Massahikhaleghi, P.; Hoveidamanesh, S.; Darroudi, S.; SoleimanvandiAzar, N.; Tayefi, B. Physical Activity and Well-being Status among Employees of University of Medical Sciences. *Med. J. Islam. Repub Iran.* **2022**, *36*, 97. [CrossRef] [PubMed]
42. Roychowdhury, D. Using Physical Activity to Enhance Health Outcomes Across the Life Span. *J. Funct. Morphol. Kinesiol.* **2020**, *5*, 2. [CrossRef] [PubMed]
43. Park, J.H.; Moon, J.H.; Kim, H.J.; Kong, M.H.; Oh, Y.H. Sedentary Lifestyle: Overview of Updated Evidence of Potential Health Risks. *Korean J. Fam. Med.* **2020**, *41*, 365–373. [CrossRef] [PubMed]
44. Alnofaiey, Y.H.; Atallah, H.M.; Alrawqi, M.K.; Alghamdi, H.; Almalki, M.G.; Almaleky, J.S.; Almalki, K.F. Correlation of Physical Activity to Mental Health State and Grade Point Average Among Medical Students in Saudi Arabia: A Cross-Sectional Study. *Cureus* **2023**, *15*, e40253. [CrossRef]
45. Anokye, N.K.; Trueman, P.; Green, C.; Pavey, T.G.; Taylor, R.S. Physical activity and health related quality of life. *BMC Public Health* **2012**, *12*, 624. [CrossRef]
46. Ball, K.; Carver, A.; Downing, K.; Jackson, M.; O'Rourke, K. Addressing the social determinants of inequities in physical activity and sedentary behaviours. *Health Promot Int.* **2015**, *30* (Suppl. S2), ii8–ii19. [CrossRef] [PubMed]
47. Tuso, P. Strategies to Increase Physical Activity. *Perm. J.* **2015**, *19*, 84–88. [CrossRef] [PubMed]
48. Ramalho, A.; Petrica, J. The Quiet Epidemic: An Overview of Emerging Qualitative Research Trends on Sedentary Behavior in Aging Populations. *Healthcare* **2023**, *11*, 2215. [CrossRef]
49. Hughes, D.C.; Orchard, J.W.; Partridge, E.M.; La Gerche, A.; Broderick, C. Return to exercise post-COVID-19 infection: A pragmatic approach in mid-2022. *J. Sci. Med. Sport* **2022**, *25*, 544–547. [CrossRef] [PubMed]
50. Kindred, R.; Bates, G.W. The Influence of the COVID-19 Pandemic on Social Anxiety: A Systematic Review. *Int. J. Environ. Res. Public Health* **2023**, *20*, 2362. [CrossRef]
51. Kroencke, L.; Geukes, K.; Utesch, T.; Kuper, N.; Back, M.D. Neuroticism and emotional risk during the COVID-19 pandemic. *J. Res. Pers.* **2020**, *89*, 104038. [CrossRef] [PubMed]

52. Shukla, A.; Dogra, D.K.; Bhattacharya, D.; Gulia, S.; Sharma, R. Impact of COVID-19 outbreak on the mental health in sports: A review. *Sport Sci. Health* **2023**, *20*, 1–15. [CrossRef]
53. Håkansson, A.; Moesch, K.; Kenttä, G. COVID-19-related impact on mental health and career uncertainty in student-athletes-Data from a cohort of 7025 athletes in an elite sport high school system in Sweden. *Front. Sports Act. Living* **2022**, *4*, 943402. [CrossRef] [PubMed]

sustainability

Article

Multicriteria Quantification of the Compatibility of the Targets from Romania's Relevant Strategies with the European Green Deal

Adrian Ioan Felea [1], Ioan Felea [2,*] and Calin Radu Hoble [2]

[1] Energy Engineering Department, University of Oradea, 410087 Oradea, Romania; adi_f11@yahoo.com
[2] Doctoral School of Engineering Sciences, University of Oradea, 410087 Oradea, Romania; calin.hoble@gmail.com
* Correspondence: ifelea@uoradea.ro

Abstract: This study deepens and further concretizes an analysis conducted in a prior study highlighting Romania's goals and objectives in comparison with the sustainable development principles established through the adoption of the European Green Deal (EGD) at the European Union (EU) level. The second section of this paper presents this study's methodology, aiming to highlight the quantification of the compatibility of the targets of Romania's pertinent strategies with the EGD after evoking the principles of sustainable development (SD) and—in this context—the importance of achieving the objectives set by the EGD. The third section of this paper presents our findings, and the final section offers conclusions drawn from our analysis.

Keywords: strategy; sustainability; target; compatibility; quantification; convergence

Citation: Felea, A.I.; Felea, I.; Hoble, C.R. Multicriteria Quantification of the Compatibility of the Targets from Romania's Relevant Strategies with the European Green Deal. *Sustainability* **2023**, *15*, 13386. https://doi.org/10.3390/su151813386

Academic Editor: Grigorios L. Kyriakopoulos

Received: 14 June 2023
Revised: 28 July 2023
Accepted: 22 August 2023
Published: 7 September 2023

1. Introduction

This paper expands upon the authors' previous concerns regarding the analysis of the relevant development strategies of Romania (RDSR) to indicate the degree to which they include the principles of SD and their compatibility with the strategies of the EU in this domain, that is, mainly with respect to the EGD.

In this context, we consider it useful to reiterate the statement made by Felea and Felea [1], according to which the full integration of Romania into the EU implies much more prompt reactions in terms of the compatibility of national policies and strategies with those of the community. The most important direction of compatibility concerns the way in which society develops in general and with respect to the economy. Sustainable development is a well-known concept that has been implemented at the EU level since the 1990s [2–7].

The publication of Brundtland's "Our Common Future" report in 1987 (under the auspices of the United Nations [3]) yielded the first definition of sustainability: "sustainable development is development that aims to satisfy the needs of the present without jeopardizing future generations' ability to satisfy their own needs." We cannot discuss sustainable development without harmonizing the three pillars that underpin it. As a result, sustainable development arises where the three principles intersect (Figure 1).

Each of the three components of sustainable development presented in Figure 1 includes a series of objectives (Figure 2). Next, we will summarize the objectives of the economic, social, and ecological systems as they are described by the economist M. Munasinghe in his work published in 1993 [8].

The way the concept of sustainable development is structured—the three pillars—but also the objectives generated by each pillar make this concept very complex, generating many opinions about how it can be implemented considering the constraints arising from both the social and environmental domains with respect to the economic area. This concept is detailed in Felea's work [9], where a set of related indicators can also be found.

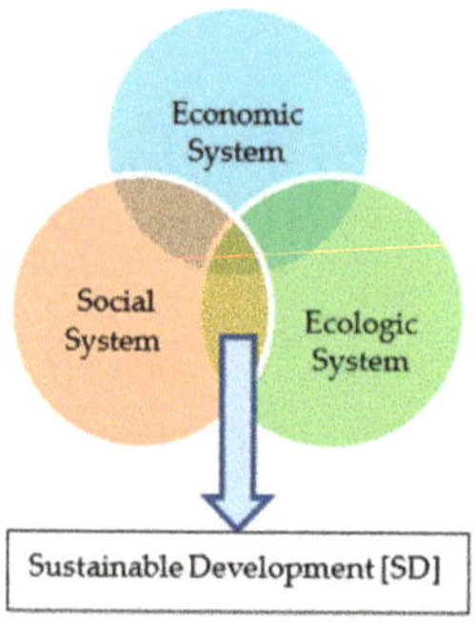

Figure 1. Interdependence between the three pillars of sustainable or lasting development (source: [7]).

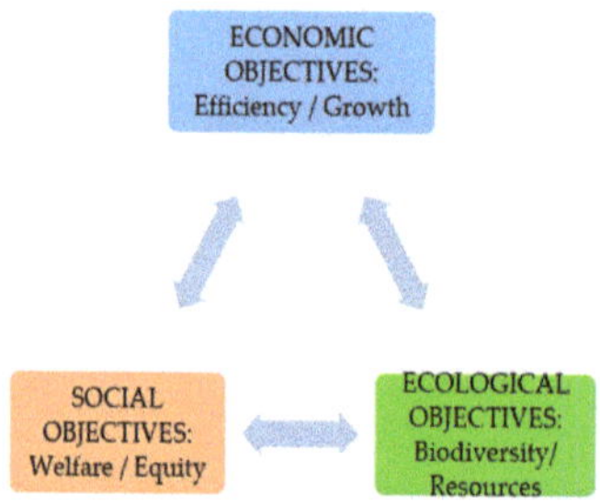

Figure 2. The three components of sustainable development (source: [8]).

Establishing a set of SD indicators is currently a major concern at the EU and member-state levels, as evidenced by many documents developed at the EU level, such as the document published by Bubbico and Djkstra [10] on the social aspects of SD. As a result, the European Commission created a set of SD indicators [11] that have been broken down into three tiers and concentrate on ten actual key societal issues.

Consequently, based on the European model, a three-level national system of SD indicators was created. It is in accordance with the EU's system of indicators, which is managed by the National Institute of Statistics. Based on this set of indicators, the National Institute of Statistics performs multilevel periodic evaluations of Romania, often via comparison with the EU. Notably, the analysis we carry out in this paper cannot be found in the National Institute of Statistics or Eurostat reports; thus, this paper provides greater depth with regard to the way in which the concept of SD is structured. Effectiveness and sustainable development have multiple links and mutual conceptual and operational interactions, and these links and interactions relate to current concepts as well. When analyzing the relationship between the two concepts, i.e., sustainable development and competitiveness, we must start with how the concept of sustainability is structured, namely, the fact that it is supported by three pillars: economic, social, and environmental pillars. One of the most relevant questions that can be asked is related to the order of priority of the three pillars. It is almost obvious that the economic pillar is the main element of the concept, considering its effects on human life. After careful consideration, we can conclude that the effects of the economic pillar, which lead to an increase in standards of living and economic development, are incompatible with the ability to achieve the desired goals of the other two pillars: social inclusion and environmental protection. The goal of shaping socioeconomic processes based on the values promoted by the EU finds its expression in the social model: "People represent the most important asset of Europe and must be the central point of the policies of the Union"; this objective has been in place since 2000 in the Lisbon Strategy [12]. Consequently, the interdependent relationship between economic competitiveness, a pillar, and sustainable development, a comprehensive and broad concept, is essential. The necessary analysis for establishing the hierarchy in which the three pillars of sustainable development are positioned must start with this relationship. If this order is not achieved by starting with the environmental and social pillars, the concept of sustainability risks remaining in the idea phase. When focusing on the economic aspect and considering it as the primordial aspect, it is impossible to comprehensively account for the other two pillars, which are only mentioned at a theoretical level in future strategies out of a desire to keep concerns related to appearance at the level perceived through the concept of SD, as also revealed in the article published by Felea in 2015 [13].

The most relevant EU development strategies [12,14–18] from the last 30 years positioned SD in the foreground, constantly adapting to the development stage and the objective restrictions identified at community and international levels. Humanity has become aware, especially in the last 10 years, of the primary importance of the impact of economic processes on climate change, for which the sustainable development strategies (SDSs) have introduced appropriate restrictions. The effects of climate change are well known [19,20]: extreme weather conditions, melting of glaciers, rising sea levels, ocean acidification, and depletion of biodiversity. In order to limit global warming to an acceptable threshold from the point of view of the Intergovernmental Panel on Climate Change, i.e., to 1.5 °C, from this group of specialists' viewpoints, it is essential that humanity reaches CO_2 emission neutrality by 2050.

The Paris Agreement on Climate Change [21], signed by 195 countries and the EU, establishes this vital objective for humanity. Concerns are focused on the greenhouse gas (GHG) with the largest share (81% of total GHG), namely CO_2. According to the Sixth Assessment Report of the Intergovernmental Panel on Climate Change (IPCC) [22], other GHGs that matter in terms of environmental impact and SD and that are emitted as a result of human activities are methane (11%), nitrous oxide (5%), and hydrocarbons (2%).

2. Literature Review

The specialized literature is increasingly rich in theses, reports, and scientific works that deal with the important and broad theme of SD, in various specific directions. For example, the prestigious MDPI database currently comprises 421 journals, out of which at least 16 dedicate many works to SD, the most relevant being *Sustainability*, *Energies*, *Climate*, *Economies*, *Ecologies*, *Clean Technologies*, *Environments*, *Hydrogen*, *Smart Cities*, *Waste*, and *Water*. In terms of the subject treated in this paper, in this framework, we evoke some works with particular relevance for the subject assumed by us.

Thus, in [23], it is emphasized that, since there are no regulations on climate change at the macroregional level, currently, each region in the EU adopts its own climate strategies, avoiding macroregional and global effects. The author campaigns for the integration of regional strategies in European SD strategies. It is concluded that the regulations and mechanisms applied at the regional and monosectoral levels must be assumed and integrated into the European sustainability strategies. At present, it is admitted that for the operationalization of SD principles, the key factor is the transformation of energy processes to become much cleaner compared with those currently in use. In this respect, hydrogen is considered an energy agent with a great perspective. Thus, political factors and specialists see hydrogen as a carrier of energy to allow the decarbonization of sectors that are difficult to reduce.

In 2003, Dillman and Heinonen [24] approached the issue of hydrogen-synthesizing technologies on a less optimistic note, based on currently accessible energy development models, concluding that a "safe" value for GHG emissions by 2050 would be equivalent to the cumulative consumption of 8–12% of the remaining carbon budget. According to the authors of this paper, the excessive use of green hydrogen involves large amounts of energy from renewable sources, which can have a negative impact on the consumption of material and energy resources.

Van der Spek et al. [25] discussed the anticipated function of hydrogen in the energy transition and pointed out that a unique dynamic has been observed over time in this context. As a result, whereas previously hydrogen and/or electricity production (EIE) were primarily thought of as clean fuels for automobiles, the main interest now stems from hydrogen's versatility in easing the transition to CO_2 neutrality, where addressing emissions from intricate industrial processes is of the utmost importance. The authors argue for a robust strategy to enable the hydrogen economy.

The key to decarbonization and the operationalization of the crucial goal of "net zero emissions" is the production of the energy types that mankind now consumes, namely electrical, thermal, and mechanical, from green resources. The potential for green energy production and pollution reduction in rural regions is discussed by Borowski and Barwici [26]. The findings of research on the viability of employing environmentally friendly energy for production while also utilizing toxic waste produced on rural farms are provided. When it comes to environmental protection, the generation of biogas can be extremely important, particularly when there is an excess of animal waste and products from slaughterhouses. The work's writers underline the numerous benefits, including monetary, energetic, environmental, and agronomic ones.

Aiming to establish useful correlations for the whole economy, Nuno, Carlos Leitão [27] address the relationship between fundamental economic indicators and GHG emissions for a 46-year period, investigating the impact of this relationship on the Portuguese economy. The authors' conclusion is that the empirical study demonstrates that the increase in the intensity of trade contributes to effects on the environment and energy consumption, with a negative impact on CO_2 emissions.

According to Nurrohman et al. [28], a comprehensive approach to urban planning is necessary to increase urban resilience in the face of climate change. The paper's authors advise applying the research findings with reference to two cities in Indonesia because they believe that municipal officials' motivation and initiative, along with the proper restrictions on environmental impact, are the most powerful factors of influence in the process of developing and implementing urban development strategies.

According to Javier et al. [29], the design and execution of urban regeneration strategies require the use of holistic solutions, which are necessary for the transformation of cities into smarter ones. The authors created an urban regeneration model through the EU-funded REMOURBAN project, which was carried out in Valladolid (Spain). This model integrates the main influencing factors to identify the impact on the main SD characteristics of sustainable and smart cities: sustainable neighborhoods, sustainable urban mobility, and smart infrastructures and processes. The authors propose, at the city level, a holistic, integrated evaluation regarding SD.

In accordance with Guido C. Guerrero-Liquet et al. [30], for the implementation of SD, renewable sources are key pillars. Since the process of increasing the share of renewable energy sources requires appreciable technical and financial efforts, the authors of this paper address the issues of risk management within the projects of renewable sources based on solar energy, elaborating a specific guide applicable in similar situations.

Peri et al. [31] aim at more comprehensive concerns specific to sustainability, referring to the three very important aspects—water, energy, and food—and to their interconnections. They use a multiobjective model, applicable in such complex processes that are of great importance for SD.

Similarly, Marttunen and Mika et al. [32] address the sustainable evolution of human civilization, stating that water security demands guaranteeing economic, social, and environmental sustainability. The authors propose models for the assessment of water supply security, establishing specific indicators to take into account the interdependencies of the three essential resources: water, energy, and food. The model also aims to simulate the applicability framework in Finland.

We consider particularly relevant the conclusions of Sajid, Ali, and Jang Choon-Man [33], which reaffirm the well-known and recognized truth according to which the large-scale adoption of renewable energy technologies leads, in addition to reducing GHG, to increasing the standard of living in isolated areas where classic electrical systems are not accessible. The authors of this paper elaborate on a project model, including technical evaluation and economic analysis, for a wind-solar hybrid system dedicated to supplying electric energy to isolated areas [33]. In addition to the specific equipment of the wind-solar system, this system is also equipped with energy storage installations in accumulator batteries and pumps.

The work of Gregory N. et al. [34], which deals with the projection of sustainability through the prism of intersectoral legislation at the government level, is noteworthy since it relates to our own issues. Interest in the water-energy nexus, a theory that describes how these two resources are connected, has grown along with worries about energy security and water scarcity. This paper allows us to understand, through a comparative analysis of two case studies, how the management policies of the two key sectors—water and energy—can influence the SD of some regions and states. The authors propose a methodology for studying the integration of climate change policies, considering the connection between these two very important resources.

Asumadu et al. [35] deal with the same topical issue—global warming—which they put in the context of the inevitable limitations of economic performance. The authors of this paper analyzed the relationship between GHG emissions and energy development based on renewable resources in OECD countries, considering industrial evolution. The circular economy is a key factor in reducing GHG and implementing EGD.

According to Romero-Perdomo, Felipe et al. [36], the circular economy is essential for reducing GHG and mitigating and then stopping climate change. The authors highlight—through the analysis of 789 specialized works—four essential directions in scientific research [36]: EC practices, bioeconomy, climate and energy, and sustainability and natural resources.

In the view of Matak and Nikola et al. [37], to be successful in the implementation of SD strategies, first of all, local and regional initiatives and adhesions are essential. The authors of this paper draw attention to the complexity of the mathematical model of the specific processes and propose solving the problem through nonlinear regression algorithms.

In [38], Ciot analyzed the administrative premises related to the processes that will be involved in the implementation of the EGD in Romania. To achieve this objective, the author proposed an analysis model that takes into account three levels and dimensions: strategic, administrative, and operational, a model which—in the author's opinion—is applicable at the European and Romanian level.

The same difficulties are also addressed by Doskocil [39], who suggests a novel and suitable strategy for the SD of the EU member states, respecting the EGD objectives. The report, which proposed a multicriteria evaluation approach with four criteria and indicators for tracking progress toward green growth, used the OECD public database. The author claims that this strategy gives management authorities a tool to assess the Green Deal's level of development.

The present work is in line with the authors' concerns regarding the analysis of Romania's strategies compared with the SDSs of the EU, with the aim of identifying solutions that lead to the compatibility of national strategies with those of the community and to increasing the compatibility of the national economy [40]. In order to meet the current requirements—including the objectives of the Paris Agreement—at the EU level, the EGD was adopted, through which the EU aims to make the community climate neutral by 2050. With all the damages caused to the foundations of SD by Russia through the consequences of the barbaric war it waged against Ukraine, the European Civilized Community is determined to apply and implement the objectives and targets of the Paris Agreement and the GEP Environmental team. The European Civilized Community's member states must work together for this to happen, and critical to this is the analysis undertaken at the national level on these issues. This work's goal is also to achieve that. Through the comparative analysis of the Significant National Strategies under the aspect of the impact on the climate, and RDSR in relation to the EGD, the author of [1] indicated the need to update the RDSR, starting from the objectives and targets identified in the EGD. To achieve the purpose of this work, that of quantification, we will take only the identified targets, as these are quantitative amounts, the targets being formulated only as desired with linguistic value.

We specify that the National Recovery and Resilience Plan (NRRP) and the Significant National Strategies for Climate Impacts (SNSC) are included in the RDSR. We also specify that in this paper, we use both the terms Sustainable Development and Sustainability, which we define as having a broader meaning than SD and encompassing not only growth but also stagnation or decrease, with the objective being the effective sustainability of the processes.

3. Materials and Methods

From [1], we took the targets with a direct impact on the fulfillment of the vital objective "net zero emissions", identified in the related EGD legislation and from other EU documents congruent with it, and compared them with the targets intended to achieve the same objectives, which are written in the RDSR. The steps of the work methodology are highlighted in Figure 3.

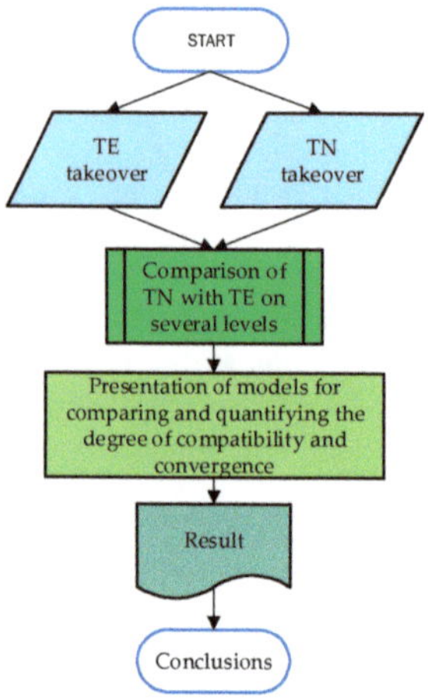

Figure 3. The steps of the work methodology.

The EU's goal of becoming climate neutral by 2050, set by GEP Environmental, is crucial because it calls for an economy with zero net GHG emissions, which implies that those emissions will be offset by CO_2 sequestration. This is especially true given that the EU is currently the third-largest GHG emitter in the world [41]. Energy conversion processes are responsible for approximately 80% of GHG emissions, which justifies focusing on these processes, mainly through a substantial increase in the share of renewable energy resources (RER) and the significant improvement of energy efficiency (EE). Table 1 lists the targets identified in the EGD's applied legislation and other documents congruent with it, drawn up at the EU level.

Table 1. Targets from the EGD application legislation and other relevant EU documents.

No.	Field	Targets Assumed at EU Level (TE)
1.	Environment [EN]	• TE1: The GHG-neutral European space = reduction by 2050 of net GHG emissions by 100% compared with 1990. • TE2: Net reduction by 2030 of GHG emissions by 55% compared with 1990. • TE3: 13% reduction in GHG intensity by 2030. • TE4: 90% reduction in GHG emissions in transport by 2050.
2.	Energy [E]	• TE5: Installed power in wind and photovoltaic EE sources to be close to 10 TW. • TE6: By 2050, ElE, H_2, and synthetic fuels will represent 50% of the energy mix. • TE7: By 2050, ElE produced from RRE will represent 80–90% of total EE. • TE8: By 2050, clean H_2 will account for 12% of final energy consumption. • TE9: By 2050, ElE consumption will double, accounting for 50% of final energy consumption with an increase of 1%/year. • TE10: RRE will represent 40% of the EU's energy mix in 2030. • TE11: Increase by 1.1%/year of the share of RRE in industry. • TE12: Increase by 2.1%/year of the share of RRE in the centralized heating and cooling processes. • TE13: Installed power in offshore wind turbines: 60 GW in 2030 and 300 GW in 2050. • TE14: Reduction in primary and final energy consumption by 39% and 36%, respectively, by 2030. • TE15: During 2024–2030, EU member states will reduce final energy consumption by 1.5%/year.
3.	Renewable energy industry [I]	• TE16: Low-carbon hydrogen production reaches 150 million tons. • TE17: Electrolyzers for the production of H_2 from renewable energy resources: minimum 6 GW by 2024 and minimum 40 GW by 2030.
4.	Transport [T]	• TE18: Electric cars will exceed 60% of classic car sales.
5.	Buildings [B]	• TE19: 49% of the energy consumed in buildings comes from RRE by 2030. • TE20: Reducing the energy consumption of buildings by 1.7%/year.

Romania has signed all international and community documents regarding the environment and climate, including the Paris Agreement. There are many areas included in the GEP environmental-related legislation in which Romania has substantial negative gaps compared with the EU average. In summary, at the level of 2020 [1]:

- In terms of GHG emissions, Romania is well below the EU average, with an emission intensity of 537.6 g equiv. CO_2/1 EUR, which can be compared with the EU average of 266.5 g equiv. CO_2/1 EUR.
- Under the energy efficiency aspect, energy productivity had a value of 5.2 EUR/kg in Romania, which can be compared with the EU average of 8.57 EUR/kg.
- Under the aspect of the valorization of renewable energy resources, the gross final consumption index of RER had values of 24.48% in Romania and 22.09% in the EU.
- The municipal waste recycling rate was 13.7% in Romania and 47.20% on average in the EU. In order to obtain the overall vision, the synthesis presented in [1] was completed with specific sectoral targets taken from the NRRP [42], thus obtaining the national targets listed in Table 2.

Comparing and quantifying the compatibility of TEs with TNs is a meticulous operation that carries the risk of some errors occurring. The comparisons and evaluations are suitable and were created only for the targets that are found in both categories: in the EU according to the EGD and in Romania according to the RDSR. For this reason, it is not possible to make an explicit analysis regarding the national targets (TNs) that concern waste (TN6, TN7). We specify the fact that, even if the recovery of waste cannot be identified as an explicit target in the EGD, this concern is intense in many EU states, and is materialized through the reduction in GHG and the increase in energy produced from RRE.

Table 2. Targets registered in the RDSR.

No.	Field	Romania's Adopted Targets (TN)
1.	Environment [EN]	• TN1. Reducing GHG emissions from industry by 30% by 2035 through increasing energy efficiency. • TN2. Reduction in GHG emissions by 40% (compared with 1990) by 2030. • TN3. Reduction in GHG by 43.9% compared with 2005. • TN4. GHG reduction of 80% by 2050 compared with 1990. • TN5. Reduction in GHG emissions by 40% by 2030 and 60% by 2050 (compared with 1990). • TN6. Recycling of 40% of municipal waste by 2027 and 55% by 2037. • TN7. Recycling of 65% of packaging waste by 2025 and 70% by 2030.
2.	Energy [E]	• TN8. RER will have a share of 27% of gross energy consumption in 2030. • TN9. The share of RER energy in the final gross energy consumption will be 30.7%. • TN10. The share of RER in total primary energy will be 37.9% in 2030. • TN11. The production of ElE from RER will be 37.5% by 2030 and 37.8% by 2050. • TN12. ElE will represent 19.5% of gross energy consumption by 2030 and 23.6% by 2050. • TN13. The share of ElE in the final energy consumption will be 19% by 2030 and 25% by 2050. • TN14. Increasing energy efficiency by at least 27% compared with the status quo scenario by 2030. • TN15. Reduction in primary energy consumption by 45.1% and final consumption by 40.4% (compared with the 2007 PRIMES projection). • TN16: The power installed in RER (wind and solar) will be 0.95 GW by 2024 and 3GW by 2026.
3.	Renewable energy industry [I]	• TN17: Electrolyzers for the production of H_2 from RER: minimum 0.1 GW, with a production of 10 thousand tons/year by 2026. • TN18: Realization of 1870 km hydrogen distribution network by 2026. • TN19: Construction of a high-efficiency cogeneration plant with a power of 300 MW by 2026. • TN20: Construction of a production and assembly line for storage batteries with a productivity of 2 GW/year by 2025. • TN21: Realization of production and assembly line of cells and photovoltaic panels with the productivity of 200 GW/year by 2025. • TN22: Realization of an electricity storage capacity with the power of 240 MW by 2025.
4.	Transport [T]	• TN23: Realization of 30 thousand charging stations for electric vehicles by 2026. • TN24: Construction of 1339 km of modernized roads by 2026. • TN25: Construction of 2851 km of new/modernized railway infrastructure and 262 new electric rolling stock by 2026. • TN26: Construction of 12.7 km of new subway lines by 2026. • TN27: Scrapping of at least 250 thousand highly polluting vehicles (including Euro 3) and the purchase of at least 29.5 thousand vehicles with zero emissions by 2026.
5.	Buildings [B]	• TN28: Energetic renovation of at least 4.36 million square meters of residential buildings by 2026. • TN29: Energetic renovation of at least 2.31 million square meters of public buildings by 2026.

At first glance, we find that the tables of the two sets of targets:

- Are structured on the same five domains, which we rank according to their importance on OV, as follows: environment (EN), energy (E), SRE industry (I), transport (T), and buildings (B).
- Are numerically unequal (19 TE and 29 TN), with the following distribution by domains:

$$
\begin{array}{c c} & \begin{array}{ccccc} EN & E & I & T & B \end{array} \\ \begin{array}{c} TE \\ TN \end{array} & \begin{bmatrix} 4 & 11 & 2 & 1 & 2 \\ 7 & 9 & 6 & 5 & 2 \end{bmatrix} \end{array} \quad (1)
$$

- Contain some targets that are unclear, irrelevant, or superfluous, being included, practically, in others that are more comprehensive. For this reason, for the comparative numerical evaluations, we will work with a set of clear, relevant, comprehensive targets, such as this: TE—all, TN—those registered in matrix (2).

$$
TN \begin{bmatrix} EN \to \{2,4\} \\ E \to \{10,11,13,15,16\} \\ I \to \{17\} \\ T \to \{23,27\} \\ B \to \{28,29\} \end{bmatrix} \quad (2)
$$

- Are unequal in terms of importance regarding the impact on vital objectives. Here, we are not referring to the importance viewed in terms of the scope of the geographical and economic area to which the target refers (EU—very large, RO—small) but to the importance in terms of the effect on GHG reduction, that is, to the scale to which the target is applied. From this point of view, we consider that the targets can be grouped into three categories, in which—for the purpose of quantification—we will give them different importance factors (IFs), as follows:

$$
IF = \begin{cases} 1\text{—for those in the E field and in the EN field with direct reference to GHG;} \\ 0.7\text{—for those from fields I, T, and B;} \\ 0.5\text{—for those in the EN field without direct reference to GHG;} \end{cases} \quad (3)
$$

To compare targets that are measured in absolute units [tons, GW et al.], we define and evaluate the absolute values of the economic relevance factor (RF) and, respectively, the relative values (RRFs) for each TN as follows:

$$
RF = \begin{cases} TEV/GDP_E - \text{for EU;} \\ TNV/GDP_N - \text{for RO;} \end{cases} \quad (4)
$$

$$
RRF = RF_N/RF_E \quad (5)
$$

where:

(TEV, TNV)—the assumed values for TE and TN, respectively;

(GDP_E, GDP_N)—the value of the gross domestic product at the level of the EU and Romania respectively [EUR];

(RF_E, RF_N)—the factor of economic relevance at the level of the EU and Romania.

We evaluate the characterization indicators of the compatibility and convergence of TNs with TEs. Obviously, for this purpose, the reference is TE, and TNs that do not have a correspondent in TE have only an intrinsic value regarding the national level of concern that we characterize through a score, evaluated by domains (SD) and at the national level (SN), as follows:

$$
\begin{cases} SD_i = \sum_{j \in \{i\}} IF_{ij} \\ SN = \sum_{i=1}^{5} SD_i \end{cases} \quad (6)
$$

IF_{ij}—importance factor of target "j" in domain "i".

In order to highlight the degree of compatibility of the set of TNs with the reference set of TEs, for each domain, we resort to the vector expression method:

$$
DC = (a, b, \ldots) \quad (7)
$$

where (a, b, c, . . .)—numbers that reflect the coverage of successive TEs by TN, without highlighting the degree of coverage, with the following meaning:

zero (0)—TE not covered by TN;

one (1)—TE covered by TN.

For the five domains, we gather the related vectors in a matrix.

The evaluation of the degree of convergence involves the use of the numerical values of the targets (TE, TN) and of the characterization factors (RF, RRF). The set of values for the degree of convergence of the targets for each domain (DCT) is highlighted as follows:

$$\mathrm{DCT} = \begin{cases} \left\{ \frac{\mathrm{TNV}}{\mathrm{TEV}} \right\} \text{—for assumed targets in } [\%] \\ \mathrm{RRF} \text{—for targets assumed with absolute values} \end{cases} \tag{8}$$

Based on the individual values, the values of the cumulative degrees of convergence are calculated by domains (DCD) and at the national level (DCN):

$$\begin{cases} \mathrm{DCD} = \frac{1}{\mathrm{NTED}} \sum_{j \in i}^{\mathrm{NTN}} \mathrm{DCT}_{ij} \times \mathrm{IF}_{ij} \\ \mathrm{DCN} = \frac{1}{5} \sum_{i=1}^{5} \mathrm{DCD}_i (8) \end{cases} \tag{9}$$

NTED—the number of TEs in a domain.

4. Results

Through the processing of the existing data by applying the presented model, the results presented below were obtained. The values obtained for the DC indicator are presented in Table 3.

Table 3. DC indicator values.

No.	Domain	DC-Vector
1.	Environment [EN]	(1, 1, 0, 0)
2.	Energy [E]	(1, 0, 1, 0, 1, 1, 0, 0, 0, 1, 0)
3.	Renewable energy industry [I]	(1, 1)
4.	Transport [T]	(1)
5.	Buildings [B]	(0, 1)

The evaluated score for TN, according to the presented model, has the following values:

$$\begin{matrix} \mathrm{D} \\ \mathrm{SD} \end{matrix} \begin{bmatrix} \mathrm{EN} & \mathrm{E} & \mathrm{I} & \mathrm{T} & \mathrm{B} \\ 6 & 9 & 4.2 & 3.5 & 1.4 \end{bmatrix} \tag{10}$$

$$\mathrm{SN} = 24.1 \tag{11}$$

The values obtained for DCT and DCD, determined in accordance with the model described and based on the values assumed by TN and TE, are summarized in Table 4, and in Figures 4 and 5, we show graphically, by targets and domain, the level of convergence of the RDSR with the EGD.

Table 4. DCT and DCD indicator values.

No.	Domain	DCT Values	DCD Values
1.	Environment [EN]	{0.8, 0.73, 0, 0}	0.38
2.	Energy [E]	{0.27, 0, 0.45, 0, 0.2, 0.95, 0, 0, 0, 1.14, 1.14}	0.38
3.	Renewable energy industry [I]	{0.1, 0.7}	0.28
4.	Transport [T]	{1}	0.7
5.	Buildings [B]	{0, 1}	0.25

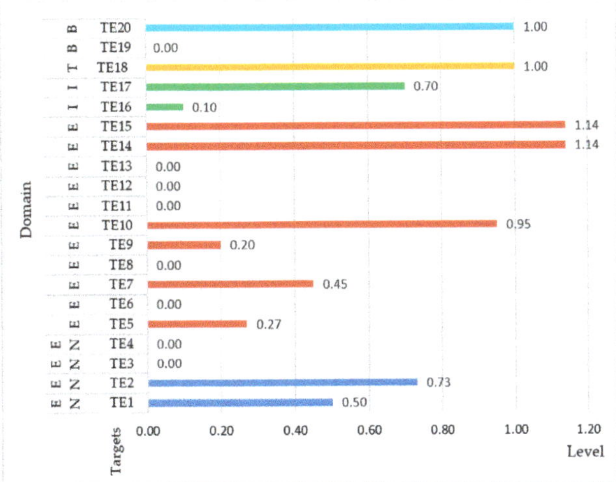

Figure 4. The level of convergence by domain.

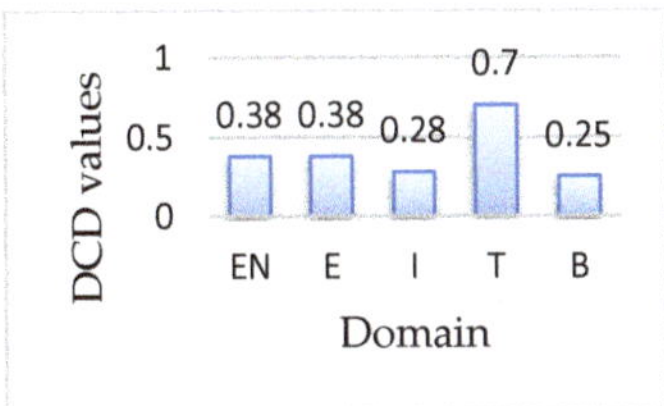

Figure 5. The level of convergence by domain of the RDSR with the EGD. The national value of the degree of convergence is DCN = 0.398.

5. Discussion

5.1. The Purpose of the Work

This paper is intended to illustrate the current state, the gaps that still need to be filled, and the steps Romania must take in order to completely achieve the goals and targets outlined in the EGD [41] through the research conducted and synthesized in this document. The future of Romania depends on maintaining the economy in a sustainable manner. The fundamental documents created by the EU for this goal are the EGD and related legislation. Due to the EGD's recent adoption at the EU level, the member states have not yet synchronized their national strategies with the EGD [41], as seen in the example of Romania. Due to the limited amount of highly specialized literature available on the topic addressed in this research—the analysis of the compatibility of national plans with the

EGD—it is impossible to compare the findings of this paper with those of other studies of a similar nature.

5.2. Romania's National Strategies

Romania has many national strategies, the most important from the perspective of EGD objectives being [43]: the national strategy for the sustainable development of Romania 2030, the development strategy of Romania in the next 20 years, the National Integrated Plan for Energy and Climate Change 2021–2030, and Romania's Energy Strategy 2019–2030 with the perspective of 2050. These strategies mainly target the 2020–2030 time frame, not even being connected from the temporal perspective of the EGD, which has the essential objectives and targets quantified for two important stages: 2030 and 2050.

The significant national strategies in terms of the impact on the climate, recently completed with the National Recovery and Resilience Plan [42], are comprehensive in terms of the areas addressed by the legislation implementing the EGD. As a result of the opportunity created by the EU, Romania developed and is implementing the National Recovery and Resilience Plan [42], which will implement a series of concrete targets regarding the modernization of the national economy, with a major impact on reducing GHG, utilizing clean energy, and harnessing renewable energy resources. The National Recovery and Resilience Plan is more adapted to the demands of the EGD than the national strategies in terms of its impact on the climate.

The significant national strategies in terms of the impact on the climate [43] and the National Recovery and the Resilience Plan [42] have objectives consistent with EGD objectives [41], but they have some superfluous targets and a significant number of targets that do not synchronize with those assumed at the EU level.

5.3. Level of Targets

Among the 29 targets with an impact on vital objectives identified in the significant national strategies in terms of the impact on the climate and the National Recovery and the Resilience Plan, only 12 are in correspondence with 11 similar targets from the EU legislation, structured as follows: 2 in the environment, 5 in energy, 1 in the renewable energy resources industry, 2 in transport, and 2 in buildings. It is obvious that the significant national strategies in terms of their impact on the climate drafting groups did not coordinate while taking into account the discrepancies between current national targets that have the same object (for example, GHG).

The targets (TNs) in The significant national strategies in terms of the impact on the climate and the National Recovery and the Resilience Plan are significantly more modest than the corresponding ones in the EU application legislation, in most directions: environmental impact through GHG and recyclable waste, energy efficiency, the share of RER in the energy mix, and through the share of electricity across all energy agents. There is one exception—the TN, which refers to the reduction in energy consumption (total and in buildings) and, in relation to this, increasing the use of electric cars. This work determines the directions in which Romania must act to close the gaps and evaluates the degree of convergence of the current discrepancies. The degree of convergence is insufficient for each of the following five domains: environment (38%), energy (38%), the industry for renewable energy resources (28%), transportation (70%), and buildings (25%). As a result, the overall degree of convergence is only close to 40%.

6. Conclusions

We believe it is imperative that Romania revises its national strategies and adopts a National Green Deal to bring the national targets (TNs) to the same level as those in the EU (TEs), with the fundamental objective (OF) of net zero emissions by 2050, in light of the current situation and the significance of the vital objective assumed by the EU (net zero emissions).

We think the working methodology described in this paper is universal in the sense that it may be used for analyses of a comparable nature in any EU member state, which is a good incentive for additional study, including comparisons of analyses among EU member states. The writers of this paper sought to advance the topic by examining changes at the national, EU, and state levels, as well as by enlarging the scope of analytical inquiry through the use of fuzzy information processing, among other techniques.

Author Contributions: Conceptualization, I.F.; methodology, A.I.F. and I.F.; validation, A.I.F. and I.F.; formal analysis, A.I.F.; investigation, I.F.; resources, C.R.H.; data curation, I.F.; writing—original draft preparation, I.F.; writing—review and editing, C.R.H.; visualization, A.I.F.; supervision, I.F. All authors have read and agreed to the published version of the manuscript.

Funding: This research received no external funding.

Institutional Review Board Statement: Not applicable.

Informed Consent Statement: Not applicable.

Data Availability Statement: The analyzed data can be found on the websites of the European Commission, the European Union, and the Romanian government. No link was created to present the results of the research.

Conflicts of Interest: The authors declare no conflict of interest.

References

1. Felea, I.; Felea, A.I. Analysis of the compatibility of Romania's relevant strategies with the European Green Deal, Oradea. *J. Sustain. Energy* **2022**, *13*, 89–93.
2. Artis, M.J.; Nixson, F. The Economics of the European Union. In *Policy and Analysis*, 4th ed.; Oxford University Press: Oxford, UK, 2007.
3. The Brundtland Report. Paper Presented at UN. 1987. Available online: http://www.un-documents.net/wced-ocf.htm (accessed on 3 April 2023).
4. European Commission. European Commission Documents: Competitiveness Advisory Group. Climate action. Enhanding European Competitiveness. 1995. Available online: http://ec.europa.eu/clima/politics/brief/eu/index_en.htm (accessed on 3 April 2023).
5. United Nations. Human Development Report. 1999. Available online: https://digitallibrary.un.org/record/286004#record-files-collapse-header (accessed on 3 April 2023).
6. European Commission. *A Final Report for The European Commission Directorate-General Regional Policy—A Study on the Factors of Regional Competitiveness*; European Commission Documents; Cambridge Econometrics, University of Cambridge: Cambridge, UK, 2003.
7. Popescu, I.; Bondrea, A. Constantinescu. In *Sustainable Development. A Romanian Perspective*; Economic Publishing House: Bucharest, Romania, 2005.
8. Munasinghe, M. *Environmental Economics and Sustainable Development*; World Bank Paper; The World Bank: Washington, DC, USA, 1993.
9. Felea, A.I. Increasing the Competitiveness of the Romanian Economy in the Context of Sustainable Development. Ph.D. Thesis, University of Oradea, Oradea, Romania, 2014.
10. Bubbico, R.L.; Dijkstra, L. The European regional Human Development and Human Poverty Indices. *Reg. Focus* **2011**, *2*, 1–10.
11. Eurostat. 2023. Available online: https://ec.europa.eu/eurostat (accessed on 6 April 2023).
12. Lisbon Strategy 2000. Available online: http://circa.europa.eu/irc/opoce/fact_sheets/info/data/policies/lisbon/article_7207_ro.htm (accessed on 3 April 2023).
13. Felea, A.I. *Competitiveness and Sustainable Development. An Integrated Strategic Approach at the Level of Romania and the EU*; Publishing House of the University of Oradea: Oradea, Romania, 2015; ISBN 978-606-10-1529-0.
14. Single European Act. 1987. Available online: https://eur-lex.europa.eu/legal-content/EN/TXT/?uri=celex%3A11986U%2FTXT (accessed on 4 April 2023).
15. Maastricht Treaty. 1993. Available online: https://www.europarl.europa.eu/about-parliament/en/in-the-past/the-parliament-and-the-treaties/maastricht-treaty (accessed on 4 April 2023).
16. Gothenburg Summit. 2001. Available online: https://www.consilium.europa.eu/media/20983/00200-r1en1.pdf (accessed on 6 April 2023).
17. European Commission. Sustainable Development Strategy. 2006. Available online: https://data.consilium.europa.eu/doc/document/ST-10917-2006-INIT/en/pdf (accessed on 6 April 2023).
18. Europe 2020 Strategy. 2010. Available online: https://eur-lex.europa.eu/LexUriServ/LexUriServ.do?uri=COM:2010:2020:FIN:en:PDF (accessed on 6 April 2023).

19. European Union. Climate change—European Climate Deal. 2019. Available online: https://climate-pact.europa.eu/index_en (accessed on 6 April 2023).
20. European Commission. Consequences of climate change. 2023. Available online: https://climate.ec.europa.eu/climate-change/consequences-climate-change_en (accessed on 6 April 2023).
21. European Council. The Paris Agreement on Climate Change—Consilium. Available online: https://www.consilium.europa.eu/en/policies/climate-change/paris-agreement/ (accessed on 3 April 2023).
22. Sixth Assessment Report of the Intergovernmental Panel on Climate Change. Available online: https://www.ipcc.ch/2021/08/09/ar6-wg1-20210809 (accessed on 6 April 2023).
23. Cattivelli, V. Macro-Regional Strategies, Climate Policies and Regional Climatic Governance in the Alps. *Climate* **2023**, *11*, 37. [CrossRef]
24. Dillman, K.; Heinonen, J. Towards a Safe Hydrogen Economy: An Absolute Climate Sustainability Assessment of Hydrogen Production. *Climate* **2023**, *11*, 25. [CrossRef]
25. van der Spek, M.; Banet, C.; Bauer, C.; Gabrielli, P.; Goldthorpe, W.; Mazzotti, M.; Munkejord, S.T.; Røkke, N.A.; Shah, N.; Sunny, N.; et al. Perspective on the hydrogen economy as a pathway to reach net-zero CO_2 emissions in Europe. *Energy Environ. Sci.* **2022**, *15*, 1034–1077. [CrossRef]
26. Borowski, P.; Barwici, J. Efficiency of Utilization of Wastes for Green Energy Production and Reduction of Pollution in Rural Areas. *Energies* **2023**, *16*, 13. [CrossRef]
27. Dillman, K.; Heinonen, J. Testing the Role of Trade on Carbon Dioxide Emissions in Portugal. *Economies* **2021**, *9*, 22.
28. Wijaya, N.; Nitivattananon, V.; Shrestha, R.P.; Kim, S.M. Drivers and Benefits of Integrating Climate Adaptation Measures into Urban Development: Experience from Coastal Cities of Indonesia. *Sustainability* **2020**, *12*, 750. [CrossRef]
29. Antolín, J.; de Torre, C.; García-Fuentes, M.Á.; Pérez, A.; Tomé, I.; Mirantes, M.L.; Hoyos, E. Development of an Evaluation Framework for Smartness and Sustainability in Cities. *Sustainability* **2020**, *12*, 5193. [CrossRef]
30. Guerrero-Liquet, G.C.; Sánchez-Lozano, J.M.; García-Cascales, M.S.; Lamata, M.T.; Verdegay, J.L. Decision-Making for Risk Management in Sustainable Renewable Energy Facilities: A Case Study in the Dominican Republic. *Sustainability* **2016**, *8*, 455. [CrossRef]
31. Peri, M.; Vandone, D.; Baldi, L. Volatility Spillover between Water, Energy and Food. *Sustainability* **2017**, *9*, 1071. [CrossRef]
32. Marttunen, M.; Mustajoki, J.; Sojamo, S.; Ahopelto, L.; Keskinen, M. A Framework for Assessing Water Security and the Water–Energy–Food Nexus—The Case of Finland. *Sustainability* **2019**, *11*, 2900. [CrossRef]
33. Ali, S.; Jang, C.-M. Optimum Design of Hybrid Renewable Energy System for Sustainable Energy Supply to a Remote Island. *Sustainability* **2020**, *12*, 1280. [CrossRef]
34. Sixt, G.N.; Strambo, C.; Zhang, J.; Chow, N.; Liu, J.; Han, G. Assessing the Level of Inter-Sectoral Policy Integration for Governance in the Water–Energy Nexus: A Comparative Study of Los Angeles and Beijing. *Sustainability* **2020**, *12*, 7220. [CrossRef]
35. Sarkodie, S.A.; Ajmi, A.N.; Adedoyin, F.F.; Owusu, P.A. Econometrics of Anthropogenic Emissions, Green Energy-Based Innovations, and Energy Intensity across OECD Countries. *Sustainability* **2021**, *13*, 4118. [CrossRef]
36. Romero-Perdomo, F.; Carvajalino-Umaña, J.D.; Moreno-Gallego, J.L.; Ardila, N.; González-Curbelo, M. Research Trends on Climate Change and Circular Economy from a Knowledge Mapping Perspective. *Sustainability* **2022**, *14*, 521. [CrossRef]
37. Matak, N.; Mimica, M.; Krajačić, G. Optimising the Cost of Reducing the CO_2 Emissions in Sustainable Energy and Climate Action Plans. *Sustainability* **2022**, *14*, 3462. [CrossRef]
38. Ciot, M.-G. On European Green Deal and Sustainable Development Policy (the Case of Romania). *Sustainability* **2021**, *13*, 12233. [CrossRef]
39. Doskočil, R. The Multicriteria Assessment of the Green Growth in the Context of the European Union's Green Deal. *Amfiteatru Econ.* **2022**, *24*, 739–757. [CrossRef]
40. Felea, I.; Adrian, I.F. The Multidimensional Characterization of Romania's Energetic Competitiveness, 12th Edition—FOREN. In *Tomorrow's Energy: From Vision to Reality*; World Energy Council Central & Eastern Europe Energy Forum: Bucharest, Romania, 2014.
41. The European Green Deal. 2019. Available online: https://commission.europa.eu/strategy-and-policy/priorities-2019-2024/european-green-deal_en (accessed on 7 April 2023).
42. National Recovery and Resilience Plan. Ministry of Investments and European Projects of Romania. 2021. Available online: https://mfe.gov.ro/pnrr/ (accessed on 7 April 2023).
43. The Romanian Government. 2023. Available online: https://www.gov.ro/ (accessed on 7 April 2023).

sustainability

MDPI

Article

Chemometric Analysis-Based Sustainable Use of Different Current Baking Wheat Lots from Romania and Hungary

Adrian Vasile Timar , Alin Cristian Teusdea *, Cornelia Purcarea , Adrian Nicolae Vuscan *, Adriana Ramona Memete and Simona Ioana Vicas

Department of Food Engineering, Faculty of Environmental Protection, University of Oradea, 410048 Oradea, Romania; atimar@uoradea.ro (A.V.T.); cpurcarea@uoradea.ro (C.P.); memeteadriana25@gmail.com (A.R.M.); svicas@uoradea.ro (S.I.V.)

* Correspondence: ateusdea@uoradea.ro (A.C.T.); avuscan@uoradea.ro (A.N.V.)

Abstract: Wheat is the most important raw material for bakery industries. Real-time grain quality assessment could increase bakery product quality and baking efficiency. The quality assessment of wheat grains can be conducted using modern and non-invasive techniques based on near-infrared spectrophotometry (NIRS) methods for the assessment of gluten content (WetGL), protein content, Zeleny index (ZelenyIdx), grain humidity (Ur), etc. The topic covered in the study is of current interest, is a part of sustainable research, and involves aspects of food quality, one of the concerns addressed by the University of Oradea's Department of Food Engineering. The present study was carried out in 2020 on eleven wheat lots from Romania and Hungary. Following the NIRS analyses, the results show varied quality for the Romanian and Hungarians wheat lots. The Romanian variety Crisana recorded the highest values for quality parameters, being similar to the Hungarian variety Bekes from Hajdu Bihar County. The statistical analysis was carried out using multivariate analysis (multivariate analysis of variance (MANOVA), canonical variate analysis (CVA) and hierarchical cluster analysis (HCA)), which highlighted which of the batches of wheat grains can be mixed to obtain a raw material of high quality for the bakery industry.

Keywords: wheat grains; near-infrared spectrophotometry; quality chemical analysis; multivariate analysis of variance

Citation: Timar, A.V.; Teusdea, A.C.; Purcarea, C.; Vuscan, A.N.; Memete, A.R.; Vicas, S.I. Chemometric Analysis-Based Sustainable Use of Different Current Baking Wheat Lots from Romania and Hungary. *Sustainability* **2023**, *15*, 12756. https://doi.org/10.3390/su151712756

Academic Editor: Ada Margarida Correia Nunes Da Rocha

Received: 22 May 2023
Revised: 29 July 2023
Accepted: 17 August 2023
Published: 23 August 2023

1. Introduction

Wheat cropping is a longtime activity of humankind. Baking wheat milling factories transform the grain into flour, which is then sold to bread-making facilities. The assessment of wheat grain quality started to be conducted using modern and non-invasive techniques in order to reduce the time and increase the precision of wheat evaluation. The application of modern non-destructive testing methods on grain quality testing had important implications for grain production, distribution and sale. Using modern non-destructive testing methods to assess the grain quality can reduce costs and improve the detection accuracy [1–4]. The new approach in wheat grain quality control is based on NIRS methods. Those methods are recommended because of their precision, reliability, repeatability, cost effectiveness, and superior data management and especially because they are not time consuming [5,6]. There are several studies that have demonstrated that NIRS is widely used in determining the quality and safety of agricultural products, such as the assessment of aflatoxin and fungal contamination or the identification of pest insect species found in various grains [2,5–8].

This assessment equipment is very reliable, can operate in severe environments like harvesting plots in the field, warehouses, mills and even cars or containers. That allows the assessment of the wheat grains from the beginning and provides good batch quality for delivery.

The application of modern non-destructive testing methods on grain quality testing could reduce costs and improve detection accuracy; furthermore, when combined with modern computer technology and data processing, it would play an important role in agriculture. Compared with traditional methods of chemical analysis, the main technical characteristics of near-infrared spectroscopy (NIRS) are fast analysis, simultaneous determination of multi-component samples without pretreatment, non-destructive analysis, distance measurement and real-time analysis, low analysis cost and simple operation. The prospective uses of NIRS are very promising, and it could be applied in the rapid detection and online detection of hazardous substances, other constituents of the grains (even pests) and nutrients combined with the new instrument and chemical analysis methods developed or under development. In this way, compared with classical methods, the new NIRS equipment is superior [2,5,6,8].

Another application of FT-NIR (Fourier transform near-infrared) and FT-MIR (mid-infrared) spectroscopy in combination with chemometric methods was developed by De Girolamo et al. (2018) to identify and classify a large volume of wheat bran contaminated with DON (Deoxynivalenol) [9].

Data resulting from the measurements are safely recorded on internal storage drive, memory stick or other storage devices and can be delivered via the internet to specific locations. The most important advantages of this method are the short measuring time, nondestructive analysis and the very good ratio cost/effectiveness. An operator with average training can conduct the measurement in less than two minutes [5,6,8].

Chemical analysis used as chemometric analysis is also a statistical method, but in contrast with classical statistics, it is focused on the extraction of functional information among variables [10,11]. In recent decades, chemometry techniques have been intensively applied to the simultaneous determination of compounds in a specified sample [10,12–15].

Process analytical technology instruments are used in the real-time inspection of a process line in all food sectors. They include measurement tools, process–product analyzers and sensors. As process analytical technology tools developed from univariate measuring methods (temperature, pH, color and pressure) to multivariate measuring systems, it was determined that this system might provide information regarding the physical, chemical and biological properties of the materials being investigated [16].

Also, European regulations were laid down considering these parameters as key parameters and there are specifically noticed levels of these parameters and actions related to deviations from these values, decreased prices, restrictions in public intervention schemes from an acquisitions point of view, export banning, etc. The values for the studied parameters laid down by Regulation (EU) no. 1272/2009 of the Commission from 11 December 2009 establishing the rules for implementing Regulation (EC) no. 1234/2007 of the Council regarding the purchase and sale of agricultural products within the public intervention scheme and Romanian Standard SR 13548/2013 are the following: Organoleptical properties—specifically for healthy products according to variety and lots description, hectolitre mass (Mh)—minim 75 kg [17,18], 1000 grains mass (M1000) between 40–60 g, MS between 1.19–1.28 g/cm^3 [19,20], protein content—10.5%, WetGL—22%, Ur—14.5%, ZelenyIdx—minim 22 mL [17,18].

Chemometrics is applied to solve both descriptive and predictive problems in experimental natural sciences, especially in chemistry [14,21]. In descriptive applications, properties of chemical systems are modeled with the intent of learning the underlying relationships and structure of the system. In predictive applications, properties of chemical systems are modeled with the intent of predicting new properties or behavior [10,21].

Developing tools for evaluating the quality of the raw material in order to generate high quality products is especially essential for the bakery industry, where tons of products are produced every day [16].

Our main goal in this work was to show that it is possible to predict and rank grain quality using physical and NIRS information (protein content, WetGL, Ur and Zeleny Idx) as well as modern statistical analysis. Another objective was to demonstrate how the results of the chemometric analysis might be used to determine which wheat lots should be combined to produce a high-quality product with excellent baking capabilities. This work is novel considering that it integrates non-invasive analytical techniques with multivariate analysis to acquire relevant information for the estimation of products with applications in the bakery sector.

2. Materials and Methods

2.1. The Biological Material

The present study was carried out in 2020, using both wheat lots imported to Romania from Hajdu Bihar County, Hungary, and wheat lots grown in Bihor County, Romania. Eleven wheat batches from Romania and Hungary were obtained for study (Table 1). The number of repetitions was 8 for each batch taken into the study.

Table 1. Wheat lots' sample coding.

No.	Sample Code	Sample Location
1.	DROPIA	Romania, Săcuieni
2.	ALEX_LOT1	Romania, Săcuieni
3.	ALEX_LOT2	Romania, Săcuieni
4.	ALEX_LOT3	Romania, Săcuieni
5.	CRIŞANA	Romania, Săcuieni
6.	BEKES	Hungary, Berettyóújfalu
7.	MV_KOLO	Hungary, Berettyóújfalu
8.	MV_VEKNI	Hungary, Berettyóújfalu
9.	MV_MARSHALL	Hungary, Berettyóújfalu
10.	MV_EMESE	Hungary, Berettyóújfalu
11.	MV_MAGDALENA	Hungary, Berettyóújfalu

For better data management, the samples were coded. The sample codes for each wheat lot are listed in Table 1.

2.2. Methods

Following the harvest, the grain lots were cleaned with a winnower and grader and then stored for a short time (5 days) before analysis. The samples were analyzed the same day after being received in the laboratory. Preparation of samples was conducted by removing foreign parts in accordance with European legislation [17,18,22].

The quality parameters evaluated in our study were Mh, M1000, MS, Protein content, WetGL, Ur and ZelenyIdx. The parameters were chosen because of their relevance for flour quality [7,23]. Wheat lots were collected and analyzed in eight replicates for each wheat lot ($n = 8$). The methods used for analysis are in accordance with Romanian standards and are cited from latest scientific studies [21,24]. Sampling was conducted with cylindrical probes from the surface to the bottom layers.

2.2.1. Organoleptic Analysis

Organoleptic analysis was conducted just in order to select the proper samples. The organoleptic test was carried out using 5 testers who are members of the University of Oradea's Food Engineering Department who are taking part in the "Sensory Analysis of Food Products" course and have competence in this field of study. For the evaluation of wheat lots, the panel approach was used in terms of smell, taste, color and appearance. The samples were graded, and acceptance was granted for samples that met the minimum criteria according to each lot (Romanian Standard SR 13548:2013, ISO 7970:2021—Wheat (*Triticum aestivum* L.)—Specification and Romanian National Standards for Wheat STAS 6253-80) [22,25,26]. The ISO 7970 standard specifies minimum requirements for wheat (*Triticum aestivum* L.) grains intended for human consumption and traded internationally. It is also applicable to the local wheat trade. Impurities, damaged wheat grains, broken grains, low-value wheat, milled grains, immature grain, black point grain, grain attacked by pests and other elements are detailed in the site https://www.iso.org/standard/75731.html (accessed on 13 December 2022) [22]. The scoring was conducted in panels with grades from 1 to 10. The lowest score was 1, and the highest was 10. The acceptance value for each parameter was 7.

The samples were classified as accepted, and the samples that met the minimum criteria according to each variety and the Romanian Standard SR 13548:2013, ISO 7970:2021—Wheat (*Triticum aestivum* L.)—Specifications and Romanian National Standards for Wheat STAS 6253-80 [17,18] were studied.

2.2.2. Chemical Analysis

For the chemical parameters, an NIRS AgriCheck spectrometer from Bruins InstrumentsTM (KPM Analytics Technology Drive, Westborough, MA 01581) was used. This instrument uses near-infrared light to evaluate the sample spectra and determine the composition of the sample. The wavelength range used was from 730 mn to 1100 nm with a wavelength data increment of 0.5 nm. The AgriCheck is an NIRS cost-effective solution for grain and seed analysis. This approach is similar to other recent studies that have studied applications of NIRS to the analysis of wheat, oilseed and rice components, pests and even contaminants [27]. The chemical and rheological parameters determination was conducted using whole grains as samples. The samples were cleaned, and all foreign parts and damaged grains were removed. Then, samples were placed in the feeder hopper, and the following measurements were made: ZelenyIdx (mL), WetGL (%), Protein content (%) and UR (%). For statistical analysis, data were exported from the database to the computer.

2.2.3. Physical Analysis

Granomat™ (PFEUFFER GMBH, Flugplatzstraße 70 97318 Kitzingen, Germany) and Acculab™ analytical scales (4802 Glenwood Rd. Brooklyn, NY 11234, USA) were used to determine the physical parameters, which also provide very good precision, reliability and repeatability. Data were processed with an Acer Aspire™ 5733 laptop (Acer's EMEA headquarters, Lugano, Switzerland). The physical parameters determination was conducted using whole grains as samples. The samples were cleaned, and all foreign parts and damaged grains were removed. After that, they were weighed (600 g) and put in the feeder hopper, and there were measured the temperature, weight and reflection [23,24,28].

The M1000 (1000 grains mass) was measured using the Acculab™ analytical scale; for M1000, the 1000 grains were counted in 5 repetitions and weighed.

The MS (specific mass) was determined using a pycnometer and 2 g of grains weighed with the Acculab™ analytical scale that were immersed in toluene.

The values were calculated using following Equation (1) [24]:

$$\gamma = \frac{m_1}{m_1 + m_2 + m_3}\rho \quad g/cm^3 \tag{1}$$

m_1—mass of the weighed sample, g;
m_2—mass of pycnometer with liquid, g;
m_3—mass of pycnometer with sample and liquid, g;
ρ—MS of the toluene used, g/cm^3.

2.3. Statistical Analysis

Each parameter/variable was subjected to one-way ANOVA (p = 0.05), and furthermore to post hoc multiple pairwise sample means comparisons using Tukey's test. This test generates comparison results as letters accompanying the mean values (Table 2). Different letters along the columns denote statistically significant differences (p = 0.05) between wheat lots. The multivariate analysis included principal component analysis (PCA), multivariate analysis of variance (MANOVA, p = 0.05), canonical variates analysis (CVA) and hierarchical cluster analysis (HCA) with complete linkage and Euclidean distance options performed with the P.A.S.T. version 3.05 statistical package. Canonical variates analysis (CVA) biplot and Canonical variates analysis (CVA) 3D graphical representationwere generated with MATLAB Software v2022b.3.

Table 2. One-way ANOVA (p = 0.05) results of analyzed wheat parameters.

Wheat Lots	Mh (kg/hL)	M1000 (g)	MS (g/cm^3)	WetGL (%)	Protein (%)	Ur (%)	ZelenyIdx (mL)
DROPIA	77.49 ^{a,b} ± 0.04	47.00 ^{c} ± 0.00	1.23 ^{c,d} ± 0.00	25.11 ^{f} ± 0.21	13.69 ^{a} ± 0.05	12.29 ^{e} ± 0.05	62.88 ^{f} ± 1.89
ALEX_LOT1	77.89 ^{a} ± 0.04	47.36 ^{a,b,c} ± 0.29	1.25 ^{b,c} ± 0.03	25.11 ^{f} ± 0.30	13.42 ^{b,c,d,e} ± 0.04	12.23 ^{e} ± 0.06	75.25 ^{a} ± 0.71
ALEX_LOT2	78.16 ^{a} ± 0.05	47.24 ^{b,c} ± 0.04	1.11 ^{h} ± 0.05	24.89 ^{f} ± 0.09	12.99 ^{g} ± 0.19	12.81 ^{b} ± 0.05	73.13 ^{a,b} ± 1.13
ALEX_LOT3	76.96 ^{b,c} ± 0.29	47.49 ^{a,b} ± 0.20	1.15 ^{f,g,h} ± 0.02	25.34 ^{d,e} ± 0.00	13.58 ^{a,b,c} ± 0.12	12.26 ^{e} ± 0.05	71.50 ^{b,c} ± 1.31
CRISANA	76.36 ^{c}± 1.21	47.60 ^{a,b} ± 0.00	1.19 ^{d,e,f} ± 0.02	26.01 ^{c} ± 0.13	13.23 ^{e,f} ± 0.02	12.49 ^{d} ± 0.11	68.38 ^{c,d,e} ± 1.85
BEKES	77.58 ^{a,b} ± 0.19	47.24 ^{b,c} ± 0.04	1.20 ^{d,e} ± 0.01	26.79 ^{b} ± 0.15	13.60 ^{a,b} ± 0.08	12.57 ^{c,d} ± 0.06	70.50 ^{b,c} ± 2.33
MV KOLO	78.09 ^{a} ± 0.14	47.76 ^{a} ± 0.07	1.17 ^{e,f,g} ± 0.01	26.07 ^{c} ± 0.03	13.41 ^{b,c,d,e} ± 0.07	12.33 ^{e} ± 0.10	68.88 ^{c,d} ± 2.10
MV VEKNI	77.73 ^{a} ± 0.15	47.26 ^{b,c} ± 0.04	1.30 ^{a} ± 0.04	25.36 ^{d} ± 0.09	13.52 ^{a,b,c,d} ± 0.19	12.22 ^{e} ± 0.05	62.00 ^{f} ± 1.77
MV MARSHALL	78.09 ^{a} ± 0.15	47.05 ^{c} ± 0.54	1.29 ^{a,b} ± 0.02	25.12 ^{e,f} ± 0.08	13.38 ^{c,d,e,f} ± 0.07	13.01 ^{a} ± 0.08	65.00 ^{e,f} ± 4.2
MV EMESE	77.48 ^{a,b} ± 0.46	47.62 ^{a,b} ± 0.46	1.23 ^{c,d} ± 0.01	27.15 ^{a} ± 0.07	13.36 ^{d,e,f} ± 0.12	12.77 ^{b} ± 0.14	65.50 ^{d,e,f} ± 3.02
MV MAGDALENA	78.05 ^{a} ± 0.15	47.55 ^{a,b} ± 0.22	1.14 ^{g,h} ± 0.01	26.12 ^{c} ± 0.09	13.19 ^{f,g} ± 0.20	12.63 ^{c} ± 0.06	62.13 ^{f} ± 0.83

Results are expressed as the mean ± SD (n = 8). Different letters presented in superscript denote statistically significant differences (p = 0.05) between wheat lots.

The proposed multivariate sequence was used to get the proper number of clusters. In this way, we know which clusters maximize which parameters' values, and so we can prescribe which wheat lot is suitable for different bakery and bread products. Also, these results can provide information about how certain wheat lots can be mixed from different clusters to provide a specific or desired flour quality. Furthermore, the MANOVA uses the canonical axes of the samples to calculate the multiple pairwise comparisons in a multivariate way. This is another the reason we considered the CVA in conjunction with MANOVA. The HCA was used because its results corelate very well with the CVA and MANOVA clustering results and provide the dissimilarity distance threshold that generates the CVA and MANOVA clusters. Additionally, the clustering evolution with respect to the dissimilarity distance can be followed on the HCA dendrogram. This information can also provide the "similarity" between the clusters, information that is needed when wheat lots are mixed.

3. Results and Discussion

The parameters taken into account in this study were the following: Organoleptical properties, Mh, M1000, MS, Protein content, WetGL, Ur and ZelenyIdx. Their significance is related to processing technological requirements [21,27–29], particularly in terms of flour component quality, and the latest studies showing that they have nutritional impact in human intake, even from a functional point of view [27,29–31]. Recent studies have confirmed the correlations between wheat parameters and the quality of bakery products [32,33].

3.1. Organoleptic Analysis

All samples submitted for analysis passed the organoleptic test. With this fast test, the operator may avoid examining and processing the damaged grains. All organoleptic parameters were in accordance with the requirements; the scores for all samples ranged from 9 to 10 for each of the parameters.

3.2. Physical, Chemical and Rheological Parameters

The results according to physical, chemical and rheological parameters of the studied wheat grains recorded during the study experiences are presented in Table 2.

Grain moisture content does not directly affect grain quality but can indirectly affect quality since grain will spoil at moisture contents above that recommended for storage (15%). The Ur was not considered a key parameter because it can be corrected easily but was taken into calculation because it affects the rest of the parameters [28,34,35]. The Ur value in wheat lots tested ranged from 12.22% to 13.01% (Table 2). The grinding is affected by some of the parameters that are not strictly related with quality and for this reason they are also kept in calculations.

There are key parameters like protein content and ZelenyIdx present in European regulations related with wheat trading. These include price ceiling, price reductions of lots with lower than standard parameters and rejecting the acquisitions. Using these parameters from the farmers' point of view can provide them with a useful tool for meeting the highest possible criteria in their production.

According to Table 2, the wetGL content in wheat lots from Romania ranged from 25.11% to 26.01%, whereas wheat types from Hungary ranged from 25.12% to 26.72%. The DROPIA wheat lot had the highest protein content, followed by the BEKES lot, at 13.69% and 13.60%, respectively. In the case of the examined lots, the ZelenyIdx value ranged from 62% (at MV VEKNI wheat lot) to 75.25% (at ALEX LOT 1 wheat lot).

On the other hand, in this study a different approach for storage facilities operators and actors in the bakery sector was presented. Table 3 shows the result of comparing parameters related to physical parameters (M1000, MS, Ur and Mh) and chemical parameters (WetGL, Protein and ZelenyIdx) important for storage operators from storage safety and grain values and only chemical parameters (WetGL, Protein and ZelenyIdx) important for actors in the bakery sector. As a result, all supply chain actors' needs from farm to fork can be addressed.

Table 3. Wheat lots ranking: by Lp-norm (p = 2.5) between protein content (%) and ZelenyIdx content (left part of the table), by Lp-norm (p = 2.5) between all analyzed parameters (middle part of the table) and by Lp-norm (p = 2.5) between protein, WetGL content (%) and ZelenyIdx content (right part of the table). Data columns implied in Lp-norm were subjected to min-max normalization.

Wheat Lots	Lp Norm (p = 2.5)	Rank	Wheat Lots	Lp Norm (p = 2.5)	Rank	Wheat Lots	Lp Norm (p = 2.5)	Rank
a *				b *		c *		
ALEX_LOT1	0.840	1	CRISANA	0.707	1	DROPIA	0.793	1
ALEX_LOT3	0.785	2	MV KOLO	0.697	2	ALEX_LOT1	0.715	2
BEKES	0.769	3	MV MARSHALL	0.676	3	MV KOLO	0.702	3
DROPIA	0.758	4	ALEX_LOT1	0.669	4	BEKES	0.672	4
ALEX_LOT2	0.636	5	DROPIA	0.651	5	MV VEKNI	0.645	5
MV VEKNI	0.578	6	ALEX_LOT3	0.628	6	MV MARSHALL	0.550	6
MV KOLO	0.564	7	MV MAGDALENA	0.617	7	ALEX_LOT3	0.541	7
MV MARSHALL	0.443	8	MV EMESE	0.576	8	MV MAGDALENA	0.499	8
MV EMESE	0.428	9	MV VEKNI	0.558	9	ALEX_LOT2	0.450	9
CRISANA	0.424	10	BEKES	0.543	10	CRISANA	0.378	10
MV MAGDALENA	0.215	11	ALEX_LOT2	0.483	11	MV EMESE	0.376	11

* The columns from the table represent the following: (a) by Lp-norm between protein and ZelenyIdx content, (b) by Lp-norm between all analyzed parameters and (c) by Lp-norm between protein, WetGL and ZelenyIdx content.

Table 3 shows that the Romanian wheat lots provide a good level of production. The results are shown better in Figure 1 that reveals the CVA distribution of the parameters. The wheat lots Alex and Dropia are very well situated and show the highest values. The Hungarian lot Vekni is the same group but considering the fertilization, its potential is lower. These results confirm the adaptability of the Romanian cultivars even when there is a lack of agro technique, according to the latest studies [36]. There are three groups that are emphasized in the Figure 1.

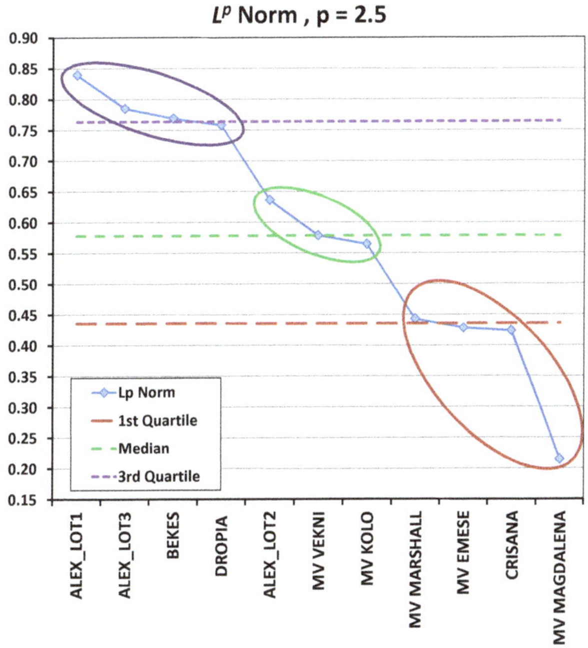

(a)

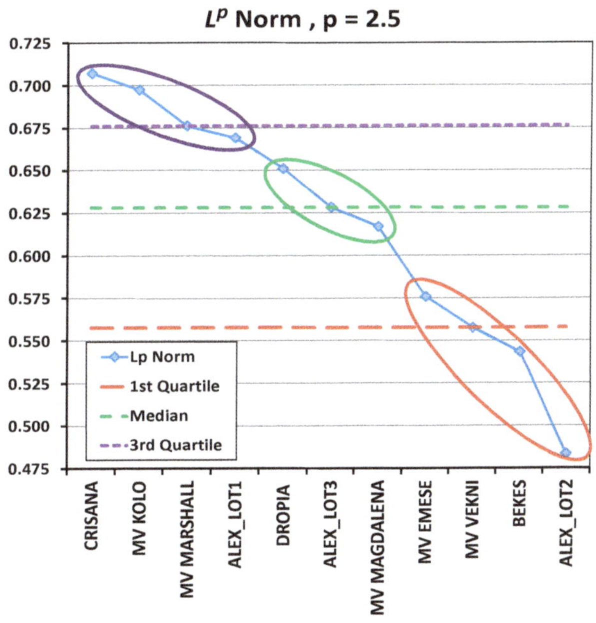

(b)

Figure 1. *Cont.*

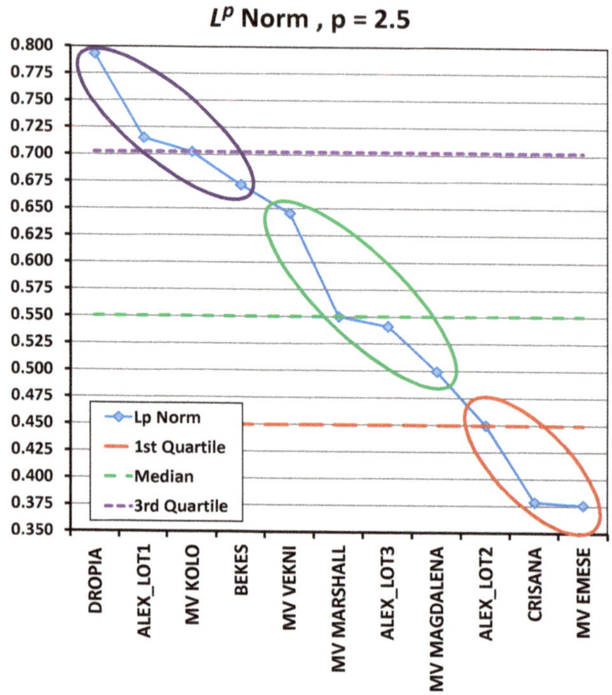

(c)

Figure 1. Graphical representation of wheat lots ranking: (**a**) by Lp-norm between protein and ZelenyIdx content, (**b**) by Lp-norm between all analyzed parameters and (**c**) by Lp-norm between protein, WetGL and ZelenyIdx content. Data columns implied in Lp-norm were subjected to min-max normalization. The ellipses gathers the lots that are corresponding to groups surrounding the 1st quartile, median and 3rd quartile (as mentioned in the legend).

3.3. Multivariate Analysis

Multivariate data analysis is the evaluation of data comprised of several variables collected from a variety of samples. The aim of multivariate data analysis is to identify all of the variations in the data matrix studied. As a result, chemometric tools attempt to discover correlations between samples and variables in a given data set and convert them to new latent variables [37].

The multivariate analysis was performed in order to get the clustering information for the wheat lot samples. Hierarchical cluster analysis (HCA) was performed. As HCA input, the scores of first two canonical axes of the canonical variates analysis (CVA) were considered [38–40].

The ideal composition of wheat for bakery products is based on diverse recipes, milling and production technologies and additives or even on traditions, sensory perceptions or just marketing and is just asserted theoretically.

The quality of durum wheat flour depends on the type and amount of gluten proteins and starch, while flour's nutritional value rests on metabolite contents such as polyphenols [41]. The nutrition composition of red and white wheat (from U.S. data) indicates that the intact grains are nearly identical (in case of similar cropping technologies), with only minor differences between micronutrients such as iron, zinc, phosphorus and potassium (slightly higher in white wheat), and magnesium, copper and niacin (slightly higher in red wheat) [42,43].

The generally accepted minimal requirements are provided by National and EU regulation [17,18] and due to there being no ideal proper composition for different kinds of bread or bakery products, the aim is to exceed the legally established parameters.

In some studies from the literature, multivariate analysis was performed by investigating the data with principal component analysis (PCA) [38,44]. FTIR spectroscopy combined with chemometric data processing in terms of PCA is an effective method in food-related research for the development of novel food quality indicators [45]. The initial data (i.e., measured parameters values) were standardized before PCA processing. The total explained variance of the first two components was around 50%. In this situation, the significance of the clustering process using the PCA scores would not have the appropriate statistical consistence. This was the reason for performing the canonical variates analysis (CVA) to obtain the clustering information.

The CVA, or discriminant function analysis (DFA), is based on Eigen-analysis of the multivariate data (several variables at the same time), as with the PCA, but the axes (i.e., discriminant functions) maximize the among-groups covariance matrix. Thus, this multivariate method's main advantage is that it performs the best discrimination between the groups of wheat lots. Also, the CVA is related to multivariate analysis of variance (MANOVA) [38,40]. From the MANOVA table (Table 4), the most important information is the p-values of the pairwise comparisons made in five dimensions (which equal the variables number). In Table 4 were highlighted the pairwise comparisons with $p > 0.05$, to emphasize the lot pairs that have no statistically significant differences.

Table 4. Multivariate analysis of variance (MANOVA) results among wheat samples groups with Bonferroni-corrected p-values.

MANOVA *p*-Values	DROPIA	ALEX_LOT1	ALEX_LOT2	CRISANA	BEKES	MV_KOLO	MV_VEKNI	MV_MARSHALL	ALEX_LOT3	MV_EMESE	MV_MAGDA
DROPIA		0.0164	0.0002	0.0004	0.0001	0.0011	0.6116	0.0012	0.0175	0.0000	0.0005
ALEX_LOT1	0.0164		0.0004	0.0009	0.0002	0.0053	0.0173	0.0002	0.1666	0.0000	0.0003
ALEX_LOT2	0.0002	0.0004		0.0004	0.0000	0.0002	0.0001	0.0011	0.0010	0.0000	0.0007
CRISANA	0.0004	0.0009	0.0004		0.0207	0.0781	0.0007	0.0002	0.0086	0.0024	0.1014
BEKES	0.0001	0.0002	0.0000	0.0207		0.0342	0.0002	0.0000	0.0005	0.0908	0.0081
MV_KOLO	0.0011	0.0053	0.0002	0.0781	0.0342		0.0025	0.0001	0.0379	0.0008	0.0454
MV_VEKNI	0.6116	0.0173	0.0001	0.0007	0.0002	0.0025		0.0007	0.0060	0.0000	0.0006
MV_MARSHALL	0.0012	0.0002	0.0011	0.0002	0.0000	0.0001	0.0007		0.0002	0.0000	0.0005
ALEX_LOT3	0.0175	0.1666	0.0010	0.0086	0.0005	0.0379	0.0060	0.0002		0.0000	0.0022
MV_EMESE	0.0000	0.0000	0.0000	0.0024	0.0908	0.0008	0.0000	0.0000	0.0000		0.0021
MV_MAGDA	0.0005	0.0003	0.0007	0.1014	0.0081	0.0454	0.0006	0.0005	0.0022	0.0021	

The highlighted values mark the pairwise comparisons with $p > 0.05$, to emphasize the lot pairs that have no statistically significant differences.

Figure 2 presents the CVA biplot with the first two axes (Axis 1 and Axis 2). The biplot combines the samples' CVA scores and the variables' loadings that generate the discriminant functions or axes. The total variance explained by first two axes is 76.41% and is much higher than the PCA; furthermore, the total variance explained by the first three canonical axes (Figure 3) is 91.88%. In this way, the possible clustering information generated with the scores from these two CVA axes has a high statistical significance level.

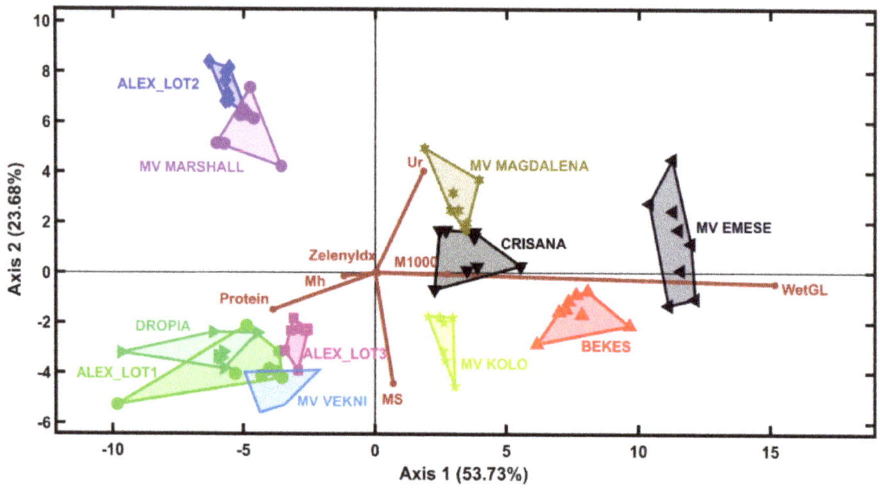

Figure 2. Canonical variates analysis (CVA) biplot. Sample points were represented as convex hull polygons with different color rendering. Figure was generated with MATLAB Software v2022b.3.

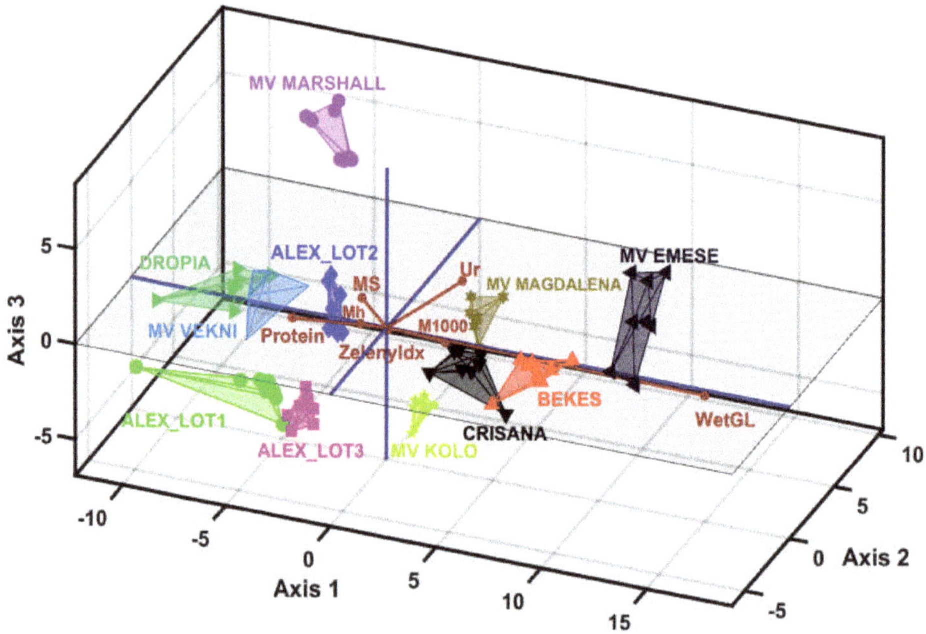

Figure 3. Canonical variates analysis (CVA) 3D graphical representation of samples group scores. Figure was generated with MATLAB Software v2022b.3.

The graphical representation of the variables' loadings was completed with vectors emerging from the coordinates' frame origin (Figure 2). Their end points denote the direction of the positive loading of the variable or the most abundant content direction described by the variables.

Based on that, the samples that are near a variable vector end point will display the most abundant content of that variable. On the other hand, the samples that are near the opposed direction of a variable vector end point will display the less abundant content of that variable. In other words, the variable vectors are performing the contrast of samples' content, and thus the samples' discrimination is conducted. The CVA is generating the multivariate "contrast" or discrimination of the samples. In our experimental design, the variables' vectors have a range distribution over the four quadrants of the coordinates' frame. This reveals qualitative and quantitative differences between the wheat lots with respect to the considered variables.

The wheat samples Dropia, Alex_Lot1, Alex_Lot3, MV_Vekni, MV_Kolo and BEKES are positively loaded with the protein content and MS vectors and thus present the most abundant content of protein and specific mass. The wheat samples Dropia, Alex_Lot1, Alex_Lot3, MV_Vekni, MV_Marshall and Alex_Lot2 are positively loaded with the ZelenyIdx vector and thus present the highest level of ZelenyIdx.

The wheat samples MV_Marshall and Alex_Lot2 are positively loaded with the Mh vector and thus present the highest level of Mh. The wheat samples MV_Magdalena, Crisana, MV_Emese, MV_Marshall and Alex_Lot2 are positively loaded with the Ur vector and thus present the highest level of moisture (Ur). The wheat samples MV_Magdalena, Crisana, MV_Emese, Bekes and MV_Kolo are positively loaded with the WetGL and M1000 vectors and thus present the highest level of wet gluten (WetGL) and thousand grains mass (M1000).

The wheat samples MV_Marshall and Alex_Lot2 are positively loaded with the Mh vector and thus present the highest level of Mh. The wheat samples MV_Magdalena, Crisana, MV_Emese, MV_Marshall and Alex_Lot2 are positively loaded with the Ur vector and thus present the highest level of moisture (Ur).

The wheat samples from the first group—MV_Magdalena, Crisana, MV_Emese, Bekes and MV_Kolo—are positively loaded with the WetGL and M1000 vectors and thus present the highest level of wet gluten (WetGL) and thousand grains mass (M1000).

There are also another three groups that are visible in Figure 1. The results are also confirmed by Figure 2. The second group is formed from the Alex_Lot 2 and Marshal lots and also shows good properties regarding protein content and ZelenyIdx. The third group that is formed from Crisana, Magdalena and Kolo is not so valuable because of its lower ZelenyIdx. The fourth group formed from Emese and Bekes, despite having high WetGL, is poor in total protein content and rheological properties as shown by ZelenyIdx being lower.

Figures 2 and 3 show the mixing possibility in order to correct the quality parameters of the studied batches. In this way, the optimal combination is Group 1 and Group 3, Group 1 and Group 4, Group 2 and Group 3 and Group 1 and Group 4. Mixing group Group 1 and Group 2 is not effective because of their similar properties.

In Figures 2 and 3, the biplots combine the samples' canonical coordinates (Axis 1, Axis 2 and Axis 3) with the variable vectors. Each vector points out the highest-level content of the variables. The samples that are neighbored with the vector end point performed at a high level of that parameter. In our DOE (Design of Experiment) there are several vectors that point out some wheat samples. In this way, the sample profiles define the wheat baking quality. Furthermore, if one needs a different predefined baking quality (described by our DOE), one can mix the wheat samples based on the parameter vectors.

One of the main goals of the multivariate analysis was to assess which wheat lots have the highest content of protein and the highest values of ZelenyIdx. From the CVA biplot, it can be noticed that the wheat samples group Dropia, Alex_Lot1, Alex_Lot3 and MV_Vekni are simultaneously positively loaded with these two variables' vectors. This partially validates the results of Lp-norm (p = 2.5) ranking (Table 3 and Figure 1) based on the protein and ZelenyIdx variables' values.

From a legal point of view, according to the abovementioned European regulation regarding official trading of grains and intervention policies, the results shown are the following. Storage operators are focused on many parameters from a storage safety point of view, up to quality parameters that are leading to market value. In this way M1000, MS, Ur, Mh, WetGL, Protein and ZelenyIdx are key parameters for these actors. The values and results after conducting the statistical analysis showed that CRISANA and ALEX_LOT1 from the Romanian lots and MV KOLO and MV MARSHALL from the Hungarian lots are leading due to M1000 and MS, with Ur as a regulator because of the storage costs that are strictly related to the volumes and surfaces of the storage infrastructure and the energy costs of drying for providing storage safety. Because the storage facilities of the last decades are very effective and because the processing of wheat has allowed the correction of chemical parameters, the actors from the wheat grain storage field are mostly cost oriented, and this is shown by their requirements referring to the low volume and surface demands and low water amount removed for the storage requirements. For the actors in the bakery sector, the correlation between WetGL, Protein and ZelenyIdx parameters is significant due to the technological requirements (fermentation, dough development, crumb development and nutritional values). In this way, the ranking showed that Dropia and ALEX_LOT1 from the Romanian lots and MV KOLO and BEKES are the best positioned. Data from Table 2 show that protein content is the main asset but is correlated with the gluten percentage. The ZelenyIdxat very high values are acting as a controller and can compensate for the previous slightly lower parameters' values.

In the first five places, above the Lp-norm median, the CVA results can be recognized. The single small discrepancy on MV_Vekni is due to the fact that CVA is performing the grouping for seven variables and the Lp-norm was calculated only for two variables. The HCA process also designated this wheat lot group as a valid cluster (Figure 4). The other HCA generated clusters generally agreed with the CVA grouping and MANOVA pairing. The discrepancies between the HCA and CVA and MANOVA results derive from the fact the HCA dendrogram and CVA biplot (Figure 2) are performed in two dimensions from the seven-dimensional analysis results. In contrast, the MANOVA exclusively generates results only from the seven-dimensional analyses, with no possibility of reducing the number of variables.

The hierarchical cluster analysis (HCA) dendrogram that reveals the mathematically based group structures is shown in Figure 4.

The Magdalena and Crisana lots are also a good option, especially when using drying systems that will increase the quality out of the concerns regarding contamination that can irreversibly affect the batch quality.

The CVA and HCA small discrepancies are due to the fact the CVA biplot is generated as a 2D projection on the Axis 1 and Axis 2 plane from seven dimensions of data, and HCA has the input of only the scores of Axis 1 and Axis 2 from the CVA. This can be seen in Figure 3 where the CVA scores are 3D graphically represented over Axis 1, Axis 2 and Axis 3. Here the wheat samples Alex_Lot2 and MV_Marshall are separated over Axis 3 (result validated by MANOVA, Table 4), but in HCA there are contained in one cluster.

Our results reveal that there are differences between Romanian and Hungarian wheat lots.

This confirms studies of other environmental conditions and allows quantification of the results, as the correlated statistical method used allows us to predict and emphasize the grains' quality. Also, real-time assessment of quality parameters through non-invasive and whole-grain methods based on NIRS and Granomat allow prediction of mixing batches in order to obtain the optimal grain quality in the new batches. This was based on the measurements and showed also the direct correlation between parameters.

Multivariate analysis was performed on wheat lot samples in order to assess the wheat lots that have the highest content of protein and highest level of ZelenyIdx at the same time. The canonical variates analysis (CVA) biplot emphasizes wheat lots Dropia, Alex_Lot1, Alex_Lot3 and MV_Vekni in response to this demand. Also, the hierarchical cluster analysis (HCA) and multivariate analysis of variance (MANOVA) results validate and offer consistent statistical significance accuracy for the HCA conclusion.

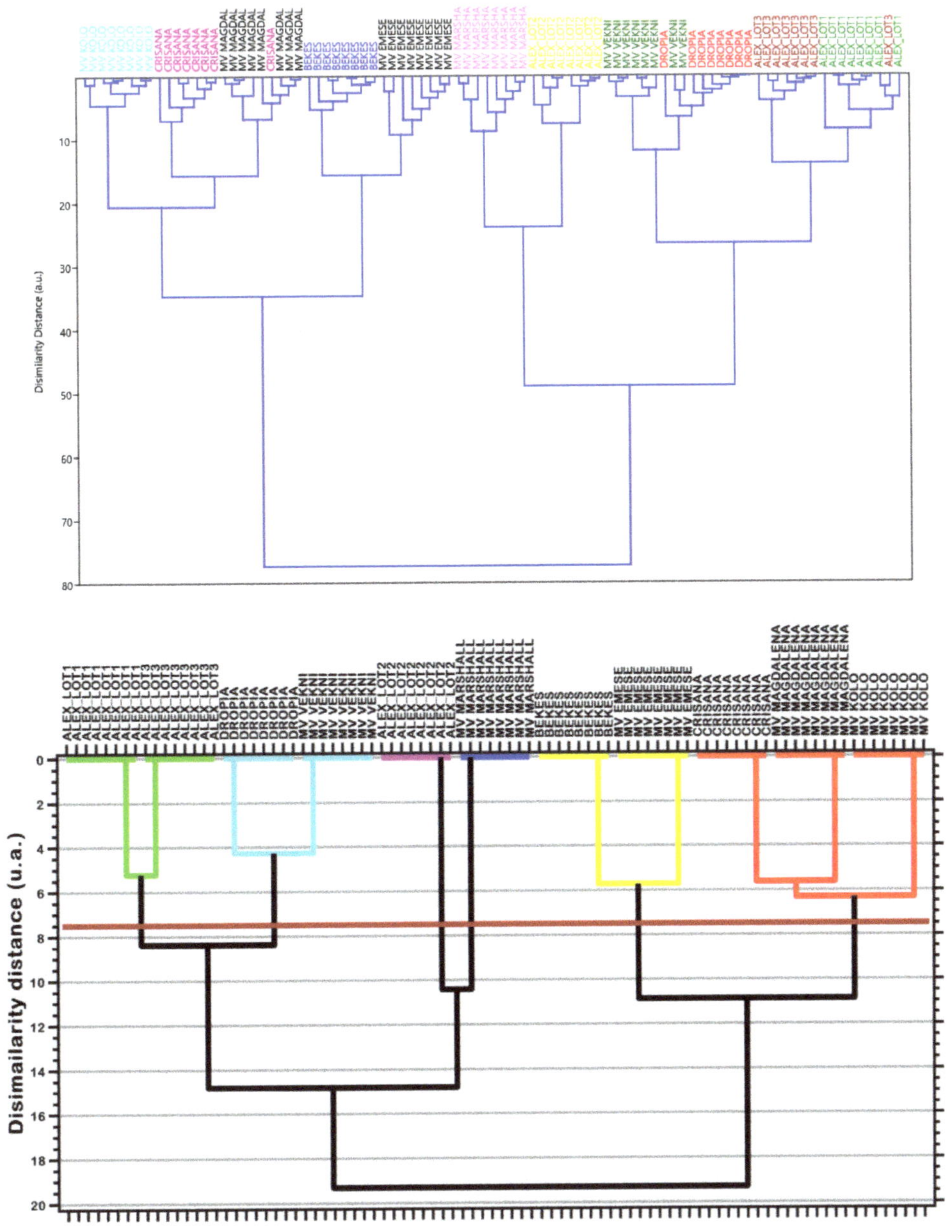

Figure 4. Hierarchical cluster analysis (HCA) dendrogram. The HCA method used the "complete-linkage" option to compute the results. The red colored horizontal line marks the dissimilarity distance threshold that generates the lots clusters in MANOVA ($p = 0.05$) results. Each lots cluster is marked, also, with different color.

4. Conclusions

Multivariate analysis was performed on all wheat samples in order to assess the wheat lots that have the highest content of protein and highest value of ZelenyIdx at the same time. The canonical variates analysis (CVA) biplot emphasizes that wheat lots

Dropia, Alex_Lot1, Alex_Lot3 and MV_Vekni were in the same cluster and are suitable for use in mixing batches as correctors. Also, the hierarchical cluster analysis (HCA) and multivariate analysis of variance (MANOVA) results validate and offer consistent statistical significance accuracy for the HCA conclusion. The statistical methods used allows us to predict and emphasize the quality of grain batches in real time. Also, those methods allow the predicting of batch mixing in order to obtain the optimal grain quality in new batches. This was based on the measurements and showed also the direct correlation between parameters. Finally, the obtained results demonstrate that the statistical methods used allow predicting and highlighting the quality of the grains. With this cross-validation approach, other relevant parameters specific to each batch can be introduced, such as pest presence or various contaminants.

Author Contributions: Conceptualization, A.V.T., S.I.V. and A.N.V.; methodology, A.C.T., C.P., A.N.V. and A.V.T.; formal analysis, A.C.T. and A.V.T.; investigation, A.C.T., C.P., A.N.V., A.R.M. and A.V.T.; writing—original draft preparation, A.C.T. and A.V.T.; writing—review and editing, A.V.T., A.R.M. and S.I.V.; visualization, A.V.T.; supervision, A.V.T. and S.I.V.; funding acquisition, A.N.V. and A.V.T. All authors have read and agreed to the published version of the manuscript.

Funding: Publishing of this research was funded by University of Oradea.

Institutional Review Board Statement: Not applicable.

Informed Consent Statement: Not applicable.

Data Availability Statement: Not applicable.

Acknowledgments: The research has been supported through the grant competition "Scientific Week of the University of Oradea", project number 121/06.25.2021.

Conflicts of Interest: The authors declare no conflict of interest.

References

1. Mahesh, B. Non-Destructive Testing. *Int. J. Res. Eng. Sci. Manag.* **2020**, *2*, 2581–2792.
2. Wang, F.; Xi, Z.; Yang, J.; Yang, X. Research Progress of Grain Quality Nondestructive Testing Methods. In *Computer and Computing Technologies in Agriculture VI: 6th IFIP WG 5.14 International Conference, CCTA 2012, Zhangjiajie, China, 19–21 October 2012*; Li, D., Chen, Y., Eds.; Springer: Berlin/Heidelberg, Germany, 2013; pp. 255–262.
3. Banach, J.K.; Majewska, K.; Grzywińska-Rąpca, M.; Żuk-Gołaszewska, K. The Use of a Non-Invasive Electrical Method to Assess the Chemical Composition, Hardness, and Color of Durum Wheat Grain Cultivated in an Integrated System. *Sustainability* **2022**, *14*, 3834. [CrossRef]
4. Hussain, N.; Sun, D.-W.; Pu, H. Classical and Emerging Non-Destructive Technologies for Safety and Quality Evaluation of Cereals: A Review of Recent Applications. *Trends Food Sci. Technol.* **2019**, *91*, 598–608. [CrossRef]
5. Johnson, J.B. An Overview of Near-Infrared Spectroscopy (NIRS) for the Detection of Insect Pests in Stored Grains. *J. Stored Prod. Res.* **2020**, *86*, 101558. [CrossRef]
6. Li, X.; Zhang, L.; Zhang, Y.; Wang, D.; Wang, X.; Yu, L.; Zhang, W.; Li, P. Review of NIR Spectroscopy Methods for Nondestructive Quality Analysis of Oilseeds and Edible Oils. *Trends Food Sci. Technol.* **2020**, *101*, 172–181. [CrossRef]
7. Zhu, F.; Yao, H.; Hruska, Z.; Kincaid, R.; Brown, R.; Bhatnagar, D.; Cleveland, T. Integration of Fluorescence and Reflectance Visible Near-Infrared (VNIR) Hyperspectral Images for Detection of Aflatoxins in Corn Kernels. *Trans. ASABE* **2016**, *59*, 785–794. [CrossRef]
8. Tao, F.; Yao, H.; Hruska, Z.; Burger, L.W.; Rajasekaran, K.; Bhatnagar, D. Recent Development of Optical Methods in Rapid and Non-Destructive Detection of Aflatoxin and Fungal Contamination in Agricultural Products. *TrAC Trends Anal. Chem.* **2018**, *100*, 65–81. [CrossRef]
9. De Girolamo, A.; Cervellieri, S.; Cortese, M.; Porricelli, A.C.R.; Pascale, M.; Longobardi, F.; von Holst, C.; Ciaccheri, L.; Lippolis, V. Fourier Transform Near-Infrared and Mid-Infrared Spectroscopy as Efficient Tools for Rapid Screening of Deoxynivalenol Contamination in Wheat Bran: Fourier Transform near-Infrared and Mid-Infrared Spectroscopy. *J. Sci. Food Agric.* **2019**, *99*, 1946–1953. [CrossRef]
10. Horvat, D.; Magdic, D.; Jurkovic, Z.; Simic, G.; Drezner, G.; Kurtanjek, Z. Chemometric and Image Analysis of Croatian Wheat Cultivars and Their Bread Making Quality. In *ITI 2002, Proceedings of the 24th International Conference on Information Technology Interfaces (IEEE Cat. No. 02EX534), Cavtat, Croatia, 24–27 June 2002*; IEEE: Manhattan, NY, USA, 2002.
11. Magdić, D.; Horvat, D.; Jurkovic, Z.; Rezica, S.; Kurtanjek, Z. Chemometric Analysis of High Molecular Mass Glutenin Fractions and Image Data of Bread Crumb Structure from Croatian Wheat Cultivars. *Food Technol. Biotechnol.* **2002**, *40*, 331–341.

12. Hucl, P.; Briggs, C.; Shirtliffe, S.; Beres, B.; Spaner, D.; Dyck, A.; Gerard, G. Increasing Grain Yield While Maintaining Baking Quality in Canada Western Red Spring Wheat. *Can. J. Plant Sci.* **2022**, *102*, 973–983. [CrossRef]
13. Zivancev, D. Characterization of Bread Wheat Cultivars (Triticum aestivum L.) by Glutenin Proteins. *Cereal Res. Commun.* **2012**, *41*, 133–140. [CrossRef]
14. Kurtanjek, Z.; Horvat, D.; Drezner, G.; Magdić, D. Prediction of Wheat Baking Quality Based on Gliadin Fractions and HMW-GS Data by Chemometric Analysis (PLS Modelling). *Acta Aliment.* **2013**, *42*, 609–617. [CrossRef]
15. Jańczak-Pieniążek, M.; Buczek, J.; Kaszuba, J.; Szpunar-Krok, E.; Bobrecka-Jamro, D.; Jaworska, G. A Comparative Assessment of the Baking Quality of Hybrid and Population Wheat Cultivars. *Appl. Sci.* **2020**, *10*, 7104. [CrossRef]
16. Jerome, R.E.; Singh, S.K.; Dwivedi, M. Process Analytical Technology for Bakery Industry: A Review. *J. Food Process Eng.* **2019**, *42*, e13143. [CrossRef]
17. Publications Office of the European Union CELEX1, Commission Regulation (EU) No 1272/2009 of 11 December 2009 Laying down Common Detailed Rules for the Implementation of Council Regulation (EC) No 1234/2007 as Regards Buying-in and Selling of Agricultural Products under Public Intervention. Available online: http://op.europa.eu/en/publication-detail/-/publication/275f9aed-b70c-4d31-9d79-20f103f83667/language-en (accessed on 16 January 2023).
18. Official Journal of the European Union Council Regulation (EC) No 1234/2007 of 22 October 2007 Establishing a Common Organisation of Agricultural Markets and on Specific Provisions for Certain Agricultural Products (Single CMO Regulation); 2007; Volume 299. Available online: https://eur-lex.europa.eu/legal-content/EN/TXT/?uri=celex:32007R1234 (accessed on 16 January 2023).
19. Martin, C.; Herrman, T.J.; Loughin, T.; Oentong, S. Micropycnometer Measurement of Single-Kernel Density of Healthy, Sprouted, and Scab-Damaged Wheats. *Cereal Chem.* **1998**, *75*, 177–180. [CrossRef]
20. Chang, C.S. Copyright since 1988 by the American Association of Grain Chemists, Inc. Measuring Density and Porosity of Grain Kernels Using a Gas Pycnometer. *Cereal Chem.* **1988**, *65*, 13–15.
21. Yanchev, I.; Ivanova, D.; Ivanov, V. Effect of the Meteorological Conditions in Two Ecological Regions on Common Wheat Development and Productivity. *J. Environ. Prot. Ecol.* **2014**, *15*, 965–972.
22. International Organization for Standardization ISO 7970:2021 Wheat (Triticum aestivum L.)—Specification. Available online: https://www.iso.org/standard/75731.html (accessed on 16 January 2023).
23. Jimenez-Pulido, I.J.; Daniel, R.; Perez, J.; Martínez-Villaluenga, C.; De Luis, D.; Martín Diana, A.B. Impact of Protein Content on the Antioxidants, Anti-Inflammatory Properties and Glycemic Index of Wheat and Wheat Bran. *Foods* **2022**, *11*, 2049. [CrossRef] [PubMed]
24. Timar Adrian. *Grains Quality Control—Practical Guide*; Oradea Publishing House: Oradea, Romania, 2008.
25. *ASRO SR 13548:2013*; Grâu Comun (Triticum aestivum). Asociatia de Standardizare din România: București, Romania, 2013.
26. *STAS 6253-80*; Seminţe pentru Consum. Determinarea Caracteristicilor Organoleptice. Organismul National de Standardizare: Bucharest, Romania, 1980.
27. Cecchini, C.; Antonucci, F.; Costa, C.; Marti, A.; Menesatti, P. Application of Near-Infrared Handheld Spectrometers to Predict Semolina Quality. *J. Sci. Food Agric.* **2021**, *101*, 151–157. [CrossRef]
28. Hassoon, W.H.; Dziki, D.; Miś, A.; Biernacka, B. Wheat Grinding Process with Low Moisture Content: A New Approach for Wholemeal Flour Production. *Processes* **2020**, *9*, 32. [CrossRef]
29. Guerrini, L.; Napoli, M.; Mancini, M.; Masella, P.; Cappelli, A.; Parenti, A.; Orlandini, S. Wheat Grain Composition, Dough Rheology and Bread Quality as Affected by Nitrogen and Sulfur Fertilization and Seeding Density. *Agronomy* **2020**, *10*, 233. [CrossRef]
30. Cecchini, C.; Bresciani, A.; Menesatti, P.; Pagani, M.A.; Marti, A. Assessing the Rheological Properties of Durum Wheat Semolina: A Review. *Foods* **2021**, *10*, 2947. [CrossRef]
31. Cappelli, A.; Cini, E.; Guerrini, L.; Masella, P.; Angeloni, G.; Parenti, A. Predictive Models of the Rheological Properties and Optimal Water Content in Doughs: An Application to Ancient Grain Flours with Different Degrees of Refining. *J. Cereal Sci.* **2018**, *83*, 229–235. [CrossRef]
32. Schopf, M.; Wehrli, M.C.; Becker, T.; Jekle, M.; Scherf, K.A. Fundamental Characterization of Wheat Gluten. *Eur. Food Res. Technol.* **2021**, *247*, 985–997. [CrossRef]
33. Radawiec, A.; Rutkowska, B.; Tidaback, J.A.; Gozdowski, D.; Knapowski, T.; Szulc, W. The Impact of Selenium Fertilization on the Quality Characteristics of Spring Wheat Grain. *Agronomy* **2021**, *11*, 2100. [CrossRef]
34. Xu, M.; Jin, Z.; Simsek, S.; Hall, C.; Rao, J.; Chen, B. Effect of Germination on the Chemical Composition, Thermal, Pasting, and Moisture Sorption Properties of Flours from Chickpea, Lentil, and Yellow Pea. *Food Chem.* **2019**, *295*, 579–587. [CrossRef]
35. De Bondt, Y.; Liberloo, I.; Roye, C.; Windhab, E.J.; Lamothe, L.; King, R.; Courtin, C.M. The Effect of Wet Milling and Cryogenic Milling on the Structure and Physicochemical Properties of Wheat Bran. *Foods* **2020**, *9*, 1755. [CrossRef] [PubMed]
36. Kirchev, H.; Muhova, A. Phenological Development of Triticale Varieties Depending on the Weather Conditions. In Proceedings of the XXIII International Symposium on Biotechnology, Rome, Italy, 10–11 December 2018; pp. 57–63.
37. Kumar, N.; Bansal, A.; Sarma, G.S.; Rawal, R.K. Chemometrics Tools Used in Analytical Chemistry: An Overview. *Talanta* **2014**, *123*, 186–199. [CrossRef]
38. Popa, M.; Morar, D.; Adrian, T.; Teuşdea, A.; Popa, D. Study Concerning the Pollution of the Marine Habitats with the Microplastic Fibres. *J. Environ. Prot. Ecol.* **2014**, *15*, 916–923.

39. Purcarea, C.; Laslo, V.; Memete, A.R.; Agud, E.; Miere (Groza), F.; Vicas, S.I. Antigenotoxic and Antimutagenic Potentials of Proline in Allium Cepa Exposed to the Toxicity of Cadmium. *Agriculture* **2022**, *12*, 1568. [CrossRef]
40. Ntumi, S. Reporting and Interpreting Multivariate Analysis of Variance (MANOVA): Adopting the Best Practices in Educational Research. *J. Res. Educ. Sci.* **2021**, *12*, 48–57. [CrossRef]
41. Graziano, S.; Marmiroli, N.; Visioli, G.; Gullì, M. Proteins and Metabolites as Indicators of Flours Quality and Nutritional Properties of Two Durum Wheat Varieties Grown in Different Italian Locations. *Foods* **2020**, *9*, 315. [CrossRef] [PubMed]
42. Grafenauer, S.; Miglioretto, C.; Solah, V.; Curtain, F. Review of the Sensory and Physico-Chemical Properties of Red and White Wheat: Which Makes the Best Whole Grain? *Foods* **2020**, *9*, 136. [CrossRef] [PubMed]
43. Doblado-Maldonado, A.F.; Pike, O.A.; Sweley, J.C.; Rose, D.J. Key Issues and Challenges in Whole Wheat Flour Milling and Storage. *J. Cereal Sci.* **2012**, *56*, 119–126. [CrossRef]
44. Memete, A.R.; Teusdea, A.C.; Timar, A.V.; Vuscan, A.N.; Mintaș, O.S.; Cavalu, S.; Vicas, S.I. Effects of Different Edible Coatings on the Shelf Life of Fresh Black Mulberry Fruits (Morus nigra L.). *Agriculture* **2022**, *12*, 1068. [CrossRef]
45. Durazzo, A.; Kiefer, J.; Lucarini, M.; Camilli, E.; Marconi, S.; Gabrielli, P.; Aguzzi, A.; Gambelli, L.; Lisciani, S.; Marletta, L. Qualitative Analysis of Traditional Italian Dishes: FTIR Approach. *Sustainability* **2018**, *10*, 4112. [CrossRef]

sustainability

Article

Development of an Algorithm for Textile Waste Arrangement

Liliana Indrie [1,*], Julieta Ilieva [2], Zlatin Zlatev [2], Simona Tripa [1] and Amalia Sturza [3]

1 Department of Textiles, Leather and Industrial Management, Faculty of Energy Engineering and Industrial Management, University of Oradea, 410058 Oradea, Romania; simona.tripa@didactic.uoradea.ro
2 Faculty of Technics and Technologies, Trakia University, 8602 Yambol, Bulgaria; zhulieta.ilieva@trakia-uni.bg (J.I.); zlatin.zlatev@trakia-uni.bg (Z.Z.)
3 Faculty of Construction, Cadastre and Architecture, University of Oradea, 410058 Oradea, Romania; asturza@uoradea.ro
* Correspondence: lindrie@uoradea.ro

Abstract: With the constant availability of new designs at extremely low prices, the production and disposal of clothing have increased significantly, leading to the need for the sustainable management of processes. The implementation of established craft practices in modern sustainable mass production requires the development and application of software and hardware computer tools as well as production machinery. Although the management of textile waste for interior design articles is addressed in the scientific literature by various techniques, there is still limited data and strategies based on the use of specific algorithms. Therefore, in this research, an algorithm is proposed, with the help of which textile waste resulting from upholstery production can be reused in the creation of interior decoration parts. The algorithm is implemented in the GNU Octave 6.4 programming environment, which makes it easily redistributable and accessible. The algorithm consists of a total of six stages, offering an option for arranging the textile elements and analyzing their color characteristics. The arrangement is performed with a Voronoi diagram, and the colors are represented by a four-color circle. Moreover, data on waste textile fabrics are presented, as well as their application in the conception of interior design elements. The proposed algorithm allows designers to focus on the visual design rather than compatibility checks and constraints. The present paper provides an algorithm for reusing textile wastes, which come in a variety of shapes and colors and are produced throughout the fabric cutting phase of upholstery manufacturing, in order to identify the most optimal combinations in matching irregular waste shapes and combinations of colors, create a suitable pattern for new interior design items, and contribute to improving the sustainable management of textile waste that is produced in considerable amounts.

Keywords: design tools; craft; fabrication; sustainability; circular economy; algorithm; textile waste

Citation: Indrie, L.; Ilieva, J.; Zlatev, Z.; Tripa, S.; Sturza, A. Development of an Algorithm for Textile Waste Arrangement. *Sustainability* **2023**, *15*, 11399. https://doi.org/10.3390/su151411399

Academic Editor: Maxim A. Dulebenets

Received: 10 June 2023
Revised: 19 July 2023
Accepted: 20 July 2023
Published: 22 July 2023

1. Introduction

Accelerated population growth, increasingly fast-paced and consumption-based lifestyles, exponential advances in technology, and the development of industry and agriculture to meet the growing needs of mankind have the unfortunate consequence, first and foremost, of generating waste in huge quantities. In the past few decades, global waste production has increased substantially, and currently there are no indications of a decline. Moreover, less than 20% of waste is recycled annually, with enormous amounts still improperly managed [1].

Among the most numerous and hazardous industrial wastes are chemicals from many industrial fields, with some of the most relevant being the medical sector [2,3] and textile industry [4]. The resulting toxic products of these industries [5–7] pollute the environment and reach the human body through food, breathing, direct contact, etc., generating or aggravating numerous diseases [8].

The textile industry, with its many stages of manufacturing and the large quantities of items produced, contributes to the overconsumption of natural resources. In 2020, the average EU textile consumption per individual required 400 square meters of land, 9 square meters of water, and 391 kg of raw materials, resulting in a carbon footprint of approximately 270 kg. It has been determined that the global apparel and textile sector worldwide consumed 79 billion cubic meters of water in 2015, while the EU's entire economy required 266 billion cubic meters in 2017 [4].

From agricultural and petrochemical production (i.e., fiber generation) to manufacturing, distribution, and the retail sector, the textile and fashion industries have a lengthy and intricate supply chain. Every single manufacturing process has a direct effect on the environment due to the use of water, materials, chemicals, and energy. Pre-consumer waste in the garment sector, also known as production waste, is generated throughout the production of textiles and apparel and consists of yarn, fiber, and fabric waste. Fabric waste is generated throughout the cutting stage of garment making and is affected by the conceptualization of the flat patterns and the overall garment design. Furthermore, garment assembly errors may result in substantial garment waste [9].

In the textile industry, the largest amount of waste is created in the cutting process [10]. Cut waste occurs because of gaps in the marking plan between pattern components [11]. These wastes are called raw material wastes and are determined by the characteristics of the product (i.e., shape and size of the patterns), the properties of the textile materials (i.e., width, design, dimensional changes, etc.) and the structure of the order (i.e., the number of pieces in the order). In recent years, the concept of circular economy and associated policies have widely addressed resource use, production, consumption, and waste [12].

Textile waste is used in the production of wadding for car doors and household accessories such as pillows, napkins, tablecloths, carpets, etc. [13]. The study of Sai et al. [14] aimed at designing and producing artifacts for interior decoration from waste fabrics. Governmental decision makers, entrepreneurs, and manufacturing industries are putting a greater emphasis on devising sustainable strategies to minimize textile waste generated during the production process [15,16].

Several strategies are available to address waste in order to reduce it. Textile reuse encompasses a variety of methods for extending the lifespan of textile products by shifting them to new owners, with or without adjustment. In contrast, textile recycling typically refers to the reuse of pre- or post-consumer textile waste in the production of novel textile or non-textile items [17]. Because preventing an issue is preferable to solving it, this assertion is congruent with waste management strategies. Considering the numerous advantages that waste generation prevention provides over the treatment of waste, waste source reduction and waste minimization can be viewed as a more suitable and sustainable waste-decreasing strategy. The concept involves preventing waste from being created before it is produced. This system-wide approach to waste management prevents waste rather than merely administering it. Integrated waste management is a relatively new concept in the realm of waste management that incorporates new dimensions into methods for dealing with waste and improves numerous traditional procedures for handling it. Integrated waste management incorporates technical as well as non-technical elements of waste administration [18].

From the review of available literature sources, it can be summarized that in recent years, from the point of view of sustainable development and the circular economy, more and more authors are directing our attention to the use of waste materials from clothing production in the design of modern fashion accessories, household elements, and works of art. There is also a trend towards the introduction of well-established craft practices in modern serial production. Botsman and Rogers [19] suggested that designers strike a balance between the needs of consumers and the common interests of society, prioritizing the ecological effects of the products.

Different algorithms [20,21] and procedures have been proposed to facilitate modern designs using craft techniques. These algorithms do not fully cover the possibilities of applying waste materials from clothing production to create new and original products.

For this reason, the aim of the present work is to propose such an algorithm by which waste textile fabrics can be used in the contemporary design of products for interior decoration. The research proposes such models by reusing the textile waste resulting from the process of cutting tapestries, leading to new solutions to finding the best combinations of textile waste, considering that many approaches do not involve algorithms. These are the result of matching irregular shapes of waste and color combinations to achieve a suitable pattern for new interior design items (i.e., tablecloths, pillow covers, chair covers). The novelty of this research lies in the development of an algorithm that enables the reuse of textile waste, provides visual modeling capabilities, integrates into designers' workflows, and allows a focus on visual design rather than compatibility checks and constraints. It addresses key aspects of sustainable design practices and provides valuable insights for researchers, practitioners, and industry professionals striving for more sustainable and efficient design processes in fashion and interior design.

2. Literature Methodology

The present paper's design was based on a literature search methodology that assessed the correlation between textile waste from cutting processes and the sustainable management of their reuse. The search methodology involved two approaches: an advanced search in three databases (ScienceDirect, SpringerLink, and Web of Science) with wide scientific coverage involving the use of the Boolean operator AND for the numerical evaluation of articles involved in the above-mentioned correlation (Figure 1), and a more detailed search in the SpringerLink, providing detailed information on research areas and subdisciplines (Figure 2) [22].

Even though the number of articles evaluating textile waste, textile waste from cutting processes, and the sustainable management of textile reuse per se is relatively large, the number of publications that have evaluated certain aspects resulting from their correlation is very small, opening research directions and facilitating areas where there are still unmet needs due to not having sufficient studies.

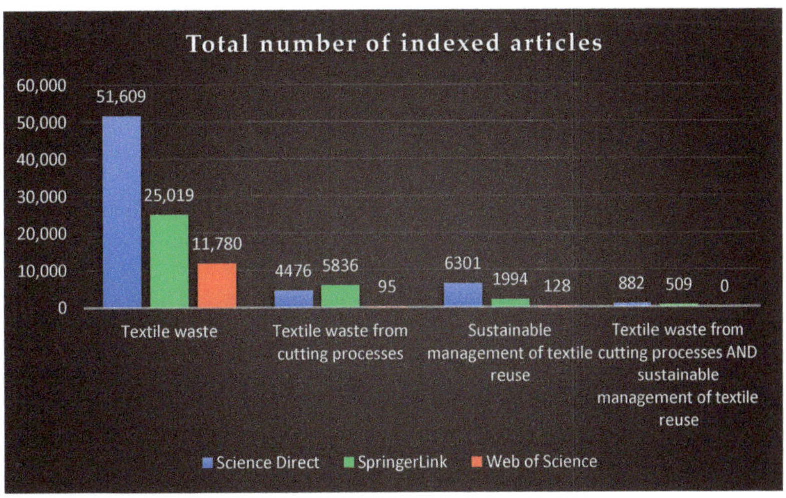

Figure 1. Total number of publications displayed for specific search terms in specific scientific databases.

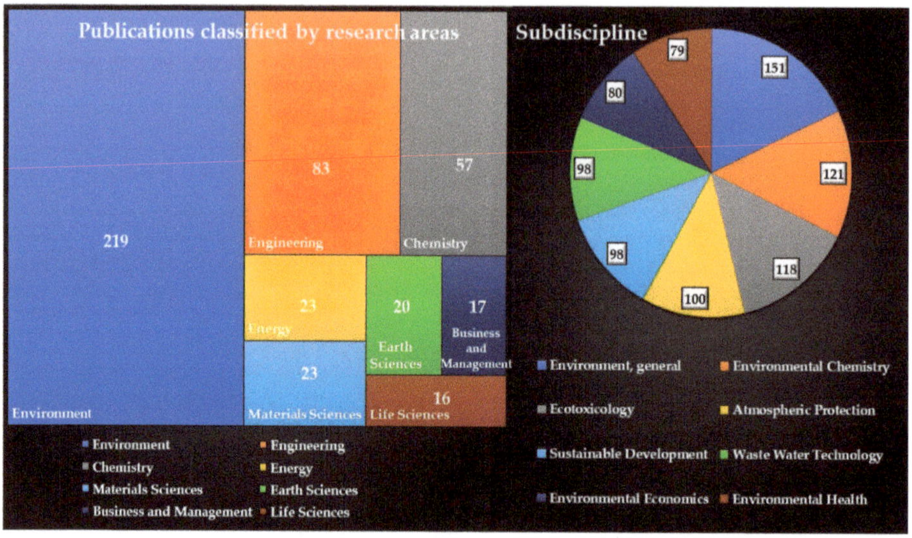

Figure 2. Publications in the field included in different research areas and subdisciplines provided by SpringerLink for the searching algorithm "textile waste from cutting processes AND sustainable management of textile reuse".

3. Material and Methods

The waste textile fabrics were procured from the Westbridge Manufacturing SRL company from Oradea, Romania, specializing in cutting and sewing fabric upholstery for sofas. The objective of this article was to analyze ways of reutilizing waste fabrics of various shapes and colors, which are presented in a general form in Figure 3.

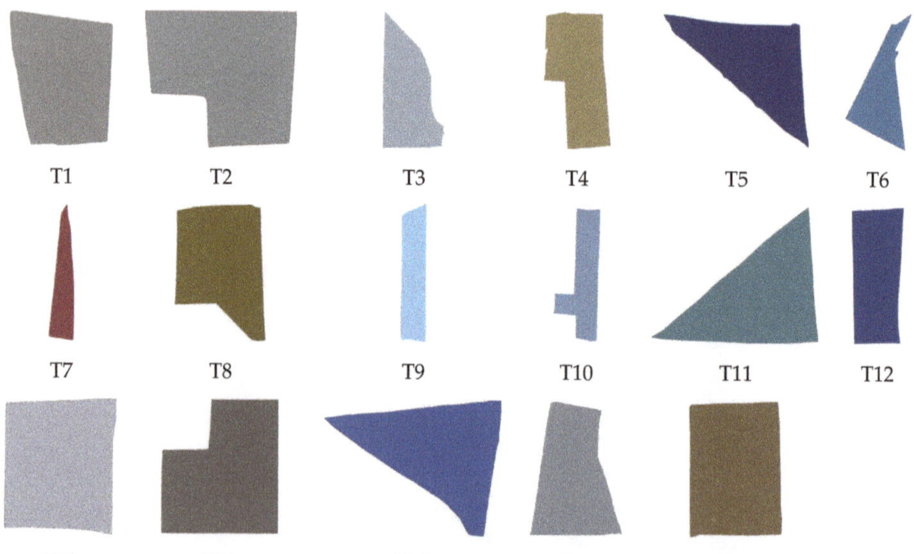

Figure 3. Textile waste of different shapes and colors, as the subject of this study, to be reused as interior design pieces.

The implementation of image processing algorithms and procedures was carried out in the GNU Octave 6.4 programming environment (GNU Octave, https://octave.org/, accessed on 10 February 2023). It is a free analog of the popular Matlab environment, version 2017b (The Mathworks Inc., Natick, MA, USA) and offers most of the features of Matlab, with some minor differences.

In the algorithm proposed in this article, the images of the textile fabrics are represented as a matrix, with the dimensions nxm. The number of rows and columns in this matrix is determined as follows:

$$n = \sqrt{N} \begin{cases} if\ n = int(n),\ m = n \\ if\ n \neq int(n),\ m = n + 1 \end{cases} \quad (1)$$

where N is the number of images; n is the number of rows; m is the number of columns of the image matrix. If n is an integer, then $n = m$ and the matrix is square.

A Voronoi diagram is used in the image processing algorithm. In general, obtaining this diagram can be described by the distance $d(x,A) = \inf\{d(x,a) \mid aEA\}$, which defines that between the variable x and the subsample variable A,

$$R_k = \{x \in X | d(x, P_k \leq d(x, P_j),\ for\ all\ j \neq k\} \quad (2)$$

where X is a metric space with distance function d; K is a set of indices; $(P_k)k_E K$ is a sequence of complete subsamples in the space X. R_k is the Voronoi domain; P_k is the set of all points in the given space; P_j are the remaining points; j is an index that is different from k.

The conversion from the RGB color model to the Lab color model is carried out with function "*rgb2lab*" in Octave. In this programming environment, the conversion between the two-color models is performed under a luminance level setting of D65 and observer 2 degree.

4. Results and Discussion

The result of the research was an algorithm that was developed for the image processing of waste textile fabrics coming from upholstery production. The algorithm proposes ways of reutilizing the waste fabrics (which come in various shapes and colors) generated during the fabric cutting stage of the upholstery production to find the best combinations in matching irregular shapes of waste and color combinations and to achieve a suitable pattern for new items of interior design. The algorithm consists of a total of six stages (Figure 4) and offers an option for arranging the textile elements and analyzing their color characteristics. The created pattern design was used to simulate several elements of interior design, such as a pillow and an armchair. The arrangement was carried out using a Voronoi diagram, and the colors are represented by a four-color circle.

Stage 1. Images of waste materials are located in a common folder, as files with the extension *.PNG. These are loaded by the main program and arranged in an nxn1 matrix, which can be square or rectangular, depending on the number of images. The "*subplot*" command from the Octave programming environment is used. The resulting figure is saved in the folder with the extension *.JPG. This image is again loaded by the algorithm for further processing.

Stage 2. The loaded image is converted to black and white, with an appropriate binarization threshold. Using the "*regionprops*" function, the centers of gravity for each element are obtained.

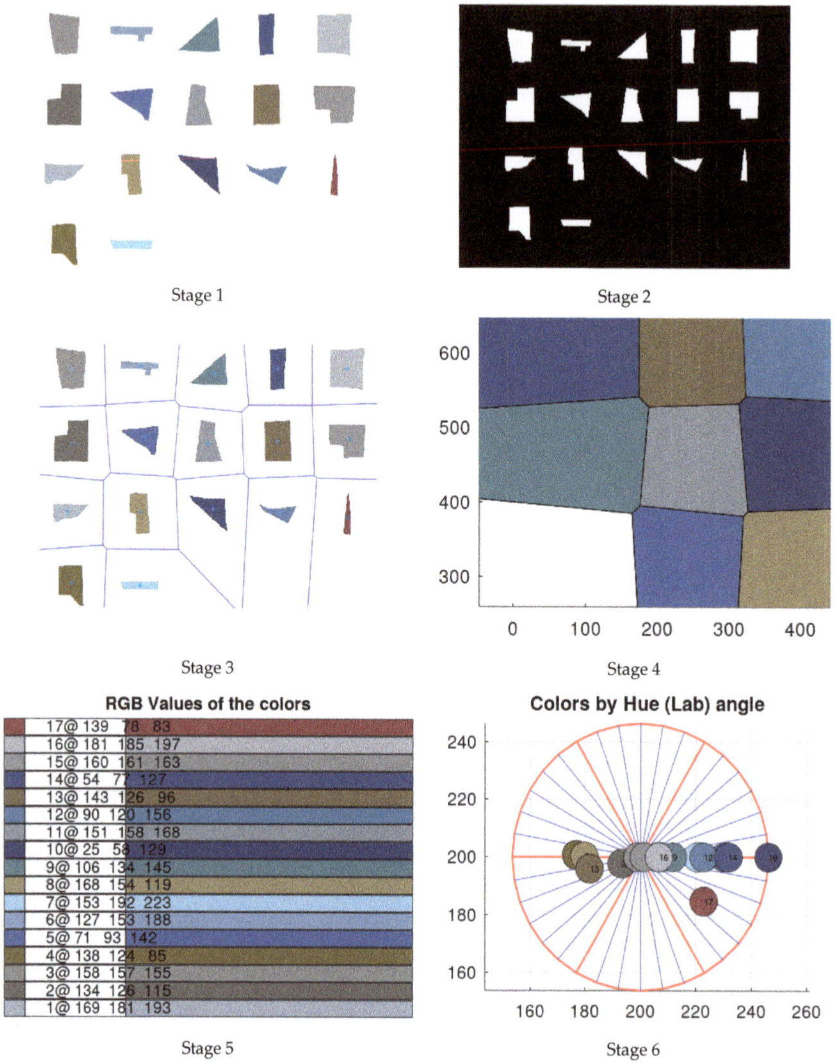

Figure 4. The algorithm for image processing of textile waste carried out in 6 specific steps.

Stage 3. The command *"[v1,c1] = voronoin ([x,y],{"Qbb"});"* is used to obtain a Voronoi diagram. This diagram is calculated according to the points representing the centers of gravity for each textile element. The items themselves are plotted in the diagram. This is done to guide the user on the appropriate placement of elements when creating designs with them.

Stage 4. Using the command *"patch(v1(c1{d},1),v1(c1{d},2),col1(d,:))"*, the Voronoi diagram is recreated, but with areas filled with the colors of the elements.

Stage 5. The colors of the elements with their RGB values are visualized. These values can be used in the analysis and visualization of the developed designs from the textile items.

Stage 6. The colors of the pieces are applied to the Lab four-color circle. The function of this visualization helps the designer to easily judge whether the colors are contrasting or related. Thus, s/he can judge how the combinations between them will look when creating designs with the textile articles.

The algorithm is presented as a GNU Octave code in Appendix A. The workability of the proposed algorithm in processing images of waste materials from textile fabrics used in upholstery production has been verified. Fabric waste comes in various shapes and colors; the resulting colors of textile waste products are related, related–contrasting, or completely contrasting.

Table 1 shows data on the color components of textile waste fabrics. The values of the color components vary considerably, as follows: from dark and light blue, light and dark red, gray, brown, and various shades of green.

Table 1. Values of RGB and Lab color components of waste textile fabrics.

CC TW	R	G	B	L	a	b	CC TW	R	G	B	L	a	b
T1	169	181	193	73.11	−1.73	−7.50	**T10**	25	58	129	26.21	15.76	−43.48
T2	134	126	115	53.20	1.03	7.08	**T11**	151	158	168	64.85	−0.46	−6.05
T3	158	157	155	64.76	−0.02	1.17	**T12**	90	120	156	49.50	−1.01	−22.79
T4	138	124	85	52.37	−0.67	23.27	**T13**	143	126	96	53.61	1.76	18.75
T5	71	93	142	39.78	6.09	−29.68	**T14**	54	77	127	33.19	7.15	−30.97
T6	127	153	188	62.44	−0.67	−21.08	**T15**	160	161	163	66.22	0.03	−1.16
T7	153	192	223	75.95	−5.75	−19.67	**T16**	181	185	197	75.17	0.94	−6.53
T8	168	154	119	63.97	−0.41	20.28	**T17**	139	78	83	40.51	26.21	8.34
T9	106	134	145	54.19	−7.27	−9.15	-	-	-	-	-	-	-

CC—color component; TW—textile waste with corresponding number; R, G, and B—color components from RGB color model; L, a, b—color components from Lab color model.

Examples of interior design applications of the sorted textile waste were offered. For this purpose, the online applications Art of Where (https://artofwhere.com, accessed on 23 February 2023) and Bags of Love (https://www.bagsoflove.co.uk, accessed on 27 February 2023) were used to simulate several interior design elements, which include a pillow and an occasional armchair (Figure 5), based on the pattern obtained from the algorithm design.

The pattern was based on a square shape, and the fabric's joints worked together to create a mixture of joints and patterns that mixed nicely with the fabric's many colors. The occasional armchair is suitable to decorate a living room. The proposed model is a new version of this type of armchair using pieces of fabric from textile waste. In this way, they can be used sustainably for many years. The various colors of the pillow can help improve the physical and emotional state of the user and should be chosen to provide the maximum comfort.

Via computational imaging and color analysis, the present article provides an investigation into the reuse of textile waste resulting from upholstery production. A six-stage algorithm was developed for the visual processing of textile waste. The algorithm uses image segmentation approaches, Voronoi diagrams, and color analysis to arrange and assess the color properties of textile elements. A pattern based on a square shape was used to simulate a few elements of interior design, such as pillows and armchairs.

The successful application of the algorithm was proven by analyzing the color components of waste textile fabrics. With the proposed algorithm, the recommendations of Zhang [23], regarding the implementation of craft production techniques in the modern sustainable design of interior design accessories, can be fulfilled. The authors presented a table with RGB and Lab color component values, demonstrating the vast array of colors present in the waste fabrics, such as various hues of blue, red, gray, brown, and green. By converting the RGB color model to the Lab color model, the algorithm enabled a more thorough analysis of the color characteristics, enabling designers to readily evaluate color combinations and contrasts.

(a) (b)

Figure 5. Home décor designs based on the pattern obtained from the algorithm. (**a**). Occasional armchair; (**b**). Pillow 18 × 18.

The results of the study demonstrate the viability of using waste textiles in sustainable design implementations. By repurposing these elements, designers can contribute to waste reduction and the preservation of the environment, while designing visually enticing products with a distinct identity. The algorithm's application to interior design elements such as pillows and armchairs demonstrates its applicability and relevance in actual design scenarios.

In addition, the authors highlight the potential advantages of using textile waste products to enhance consumers' physical and psychological health. While offering optimum comfort, the various hues of the patterns may improve the aesthetic appeal of interior spaces. This aspect is consistent with current studies on the psychological effects of colors [24,25], which demonstrate that color choices can affect mood, well-being, and the overall user experience.

Contemporary commercial reasoning in the textile industry is founded on expanding production and sales, rapid manufacturing, poor product quality, and a short lifespan of products, which all result in unsustainable consumption, rapid material turnover, considerable waste, and extensive ecological effects. Consequently, manufacturing processes and patterns of consumption need to be modified. In order to obtain certain improvements, the participation of all stakeholders is essential: textile manufacturers need to invest in clean technology, clothing companies must develop new business patterns, consumers must modify their purchasing habits, and lawmakers must adjust regulations and global business standards. The emphasis is on key strategies for establishing an entirely novel model for sustainable industries, such as limiting growth, reducing waste, and supporting a circular economy. Good management of post-consumer textile waste is mandatory for attaining a zero-waste target [26].

The implications of technology transfer may additionally play a role in the management of textile waste. By transferring novel algorithms and strategies and integrating software tools, creators are able to effectively repurpose textile waste and create novel and distinctive products. Moreover, the transfer of technology may enable creators to improve their design procedures, increase their efficiency, and concentrate on visual aesthetics, as opposed to being limited by compatibility checks and constraints. In addition, it supports sustainable manufacturing practices, assists the circular economy, and addresses the industry's critical textile waste challenge [27,28].

By reusing garment production waste in the creation of interior design elements, the algorithm helps reduce waste in the fashion industry, contributing to sustainability. The algorithm allows designers to utilize scraps effectively, maximizing the use of available resources and minimizing the need for additional materials. By repurposing textile waste, designers can potentially reduce the costs associated with purchasing new materials. This is particularly beneficial for designers working on a limited budget or for businesses aiming to optimize their production expenses.

By using textile waste resulting from the main production process to create interior design elements, the costs related to the purchase of raw material are completely eliminated. In addition, companies can considerably reduce the total amount of waste generated, which will be reflected in a reduction in costs related to the collection and storage of textile waste in landfills. For a company that only makes interior design elements from the waste from another company, the main advantage is that these materials will be attained for free; the company only bears the costs of managing and sorting them, which are much lower than having to purchase the necessary materials.

The algorithm is implemented in the GNU Octave programming environment, which is redistributable and easily accessible. Designers can utilize the algorithm without significant barriers, promoting its widespread adoption and potential for collaboration. The algorithm offers designers the option to arrange textile elements using the Voronoi diagram, a mathematical tool for partitioning space that enables creative freedom and flexibility in designing interior elements and clothing accessories. The algorithm includes the analysis of color characteristics, representing them with a four-color circle, providing designers with insights into color combinations and assisting them in creating visually appealing designs.

By automating compatibility checks and constraints, the algorithm allows designers to focus on the visual design aspect rather than spending time on manual checks. This streamlines the design process and improves overall efficiency. The proposed tool seamlessly integrates into designers' workflow, making it user-friendly and easily adaptable, facilitating the design of basic combinations of textile pieces, and enhancing the overall productivity and output of designers. This presents certain advantages and limitations (Figure 6).

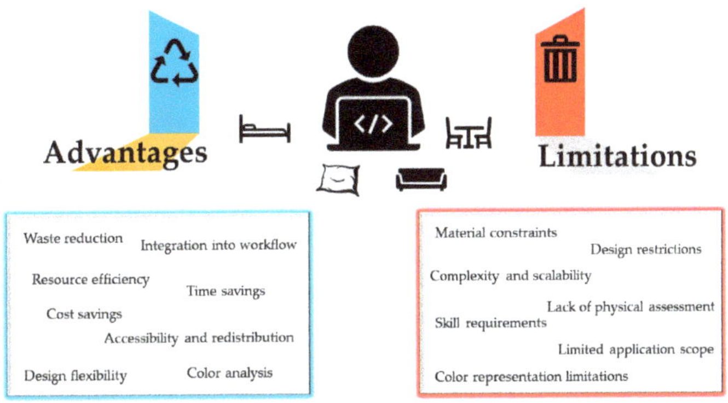

Figure 6. Strengths (blue) and limitations (red) of the algorithm.

The algorithm relies on the availability of waste textile fabrics from garment production. If there is a limited supply or if the quality of the waste fabric is insufficient, it may restrict the options for designers and limit the feasibility of the algorithm. The algorithm imposes restrictions on the geometry of the fabric elements used and the order in which they are arranged. This may limit the design possibilities and creative freedom for designers who prefer more unconventional or complex designs. The algorithm requires designers to have a certain level of technical skills and familiarity with the GNU Octave programming environment. Hence, designers who are not comfortable with programming or software tools may face challenges in utilizing the algorithm effectively.

The algorithm is specifically designed for the creation of interior design elements using garment production waste, so it may not be directly applicable to other design domains or industries, which limits its versatility and broader adoption. While the algorithm includes color analysis using a four-color circle representation, this may not capture the full complexity and nuances of color in interior design. Designers who require more precise color matching or intricate color schemes may find the algorithm's color representation inadequate. The algorithm focuses primarily on the visual design aspect, allowing designers to focus on aesthetics rather than compatibility checks and constraints. However, it does not account for physical factors such as fabric durability, texture, or performance, which may be important considerations for interior design and clothing accessories. The algorithm consists of six stages, which may introduce complexity and potentially slow down the design process as well as present limitations regarding the scalability of the algorithm in dealing with larger quantities of waste fabric or more intricate design requirements.

The main strength of this research is its novelty, considering that the new algorithm facilitates and allows textile waste recovery. As a kind of limitation, the proposed algorithm can be further improved in order to achieve a higher level of successful automation as part of the patchwork techniques. This will further refine the methods proposed by Yuan et al. [29] and Minda [30] as solutions for using waste textile fabrics.

The presented algorithm's application to the repurposing of waste textile fabrics indicates a promising strategy for sustainable design practices. This study offers valuable insights into image processing techniques, color analysis, and interior design applications, demonstrating possible applications for waste reduction and innovative textile waste utilization. Additional research and investigation in this field could lead to the development of innovative solutions that support a more sustainable and aesthetically appealing design industry.

The limitations of this work can be solved in further research by the following means:

- While this research focuses on upholstery production waste, efforts can be made to adapt the algorithm for other aspects of textile waste management. This could involve exploring additional waste streams, such as garment manufacturing, textile recycling, or consumer textile waste, and developing specific modules or adaptations to address the unique challenges in these areas.
- To enhance the algorithm's effectiveness and applicability, it can be integrated with emerging technologies. For example, artificial intelligence techniques can be leveraged to improve the accuracy and efficiency of waste classification or prediction. Virtual reality or simulation tools can be used to create immersive experiences or virtual environments for waste management planning and decision making. Collaborating with experts in these fields can help identify specific opportunities and facilitate the incorporation of elements that can enhance the algorithm's capabilities.
- The algorithm can be designed to be flexible and customizable, allowing users to adapt it to their specific contexts. This could involve developing a modular architecture that can be easily modified or extended to accommodate different waste management scenarios. Providing a user-friendly interface or API (Application Programming Interface) that allows users to configure parameters, select relevant features, or incorporate domain-specific knowledge could increase the algorithm's adaptability and usefulness across different applications.
- The algorithm can be regularly updated and improved based on feedback and real-world validation. Discussions with industry stakeholders, waste management professionals, and researchers could be useful to gather insights, validate the algorithm's performance, and incorporate lessons learned into future iterations. This iterative approach would ensure that the algorithm remains relevant and effective as new challenges and technologies emerge in the field of textile waste management.

The algorithm presents both advantages which give it the potential for increased applicability and limitations in the form of unmet needs that will have to be addressed in future research directions. Advanced algorithms for optimization, such as hybrid heuristics [31], metaheuristics [32], adaptive algorithms, self-adaptive algorithms, and island algorithms, may play a crucial role in resolving difficult decision-making issues in a variety of domains, including the textile industry. These algorithms provide successful strategies for solving complex real-world issues while offering optimal or near-optimal solutions [33,34].

In fields such as medicine [35,36], online education, scheduling, multi-objective optimization, transportation, and data classification, complex algorithms for optimization enhance the use and distribution of resources, adapt to unpredictable circumstances, minimize costs, improve decision-making processes, and increase overall efficiency [37]. Furthermore, the algorithms can simultaneously consider multiple objectives, limitations, and factors, enabling more thorough and effective decision making [38]. As a future scientific direction, the performance and scalability of our method could be evaluated by comparing the proposed method to the most advanced algorithms. This comparison would shed light on the efficacy of sophisticated optimization algorithms in resolving textile waste management issues and provide a direction for future developments in the field.

5. Conclusions

To address uncontrollable fabric waste issues, the upholstery sector and fashion industry need a new perspective that eliminates wasteful fabric practices. In the present work, an algorithm is proposed with which patterns and designs can be calculated using scraps of fabric produced from cutting textile fabrics from upholstery production. Through the proposed tool, a visual model of the used textile waste and a few combinations were offered. Fabric elements are arranged into a finished pattern of the final product. Thus, it is anticipated that this study will bridge the gap between academic research and efforts by the upholstery sector to reduce cut waste.

The presented tool fits into designers' workflow and greatly facilitates the design of basic combinations of textile pieces. The algorithm allows designers to focus on the visual design rather than compatibility checks and constraints.

The results of this research can be practically applied in interior design, fashion design, production planning, consumer decision making, design education, and sustainability initiatives within the fashion and interior design industries.

The algorithm's effectiveness and applicability could be further enhanced by incorporating innovative technologies such as artificial intelligence, virtual reality, or advanced simulation tools. Continuing this work through collaboration with industry stakeholders and integration with emerging technologies could advance the algorithm's effectiveness, broaden its applicability, and contribute to sustainable design practices in the fashion and interior design industries.

Author Contributions: Conceptualization, L.I. and Z.Z.; investigation, L.I., Z.Z., J.I., A.S. and S.T.; writing—original draft preparation, L.I., J.I. and Z.Z.; writing—review and editing, L.I. and Z.Z. All authors have read and agreed to the published version of the manuscript.

Funding: This research was funded by the University of Oradea within the framework of the grants competition "Scientific Research of Excellence Related to Priority Areas with Capitalization through Technology Transfer: INO-TRANSFER-UO-2nd Edition" projects no. 232/2022 and 236/2022.

Institutional Review Board Statement: Not applicable.

Informed Consent Statement: Not applicable.

Data Availability Statement: The data will be available upon request to the corresponding author.

Acknowledgments: The research undertaken was made possible by the equal scientific involvement of all the authors concerned.

Conflicts of Interest: The authors declare no conflict of interest.

Appendix A

List of the software in GNU Octave

1	2	3

tic
clc, clear all, close all
pkg load image
 	
d1=dir('*.png'); a=numel(d1)
 	
a1=(1:a).';
n=sqrt(size(a1,1))
if n==round(n)
n = round(sqrt(size(a1,1)))
n1=n
else
n=round(sqrt(size(a1,1)))
n1=n+1
endif
 	
figure
hold on
for b=1:a
subplot(n,n1,b)
imshow(d1(b).name)
end
print -djpg figure1.jpg
hold off
 	
i=imread('figure1.jpg')
i1=im2bw(i,0.95)
i2=imcomplement(i1)
i2=bwareaopen (i2, 100)
figure
imshow(i2)
 	
ss=regionprops(i2)
 	
for j=1:a
c(j,1:2)=ss(j).Centroid
end
 	
x=c(:,2); y=c(:,1)
 	
figure
hold on
imshow(i)
voronoi(y,x)
axis equal
hold off
 	
figure
hold on
x1=round(x);y1=round(y)
 	
col=impixel (i, y1, x1)
col1=(col/255)
 	
*v*1, *c*1
=voronoin ([x,y],{"Qbb"});
 	
for d = 1:a
patch(v1(c1{d},1),v1(c1{d},2), col1(d,:))
end
hold off
axis equal
	%Creating Lab Wheel
	
	rgb=col
	numcolors=a
	num_rects=numcolors
	
	rect_colors = rgb/255
	rgb1=rgb
	
	j=0
	for l=1:length(rgb(:,1))
	j=j+1
	mn=min(rgb(j,1),rgb(j,2))
	Iw(j)=min(mn,rgb(j,3))
	mn1=min(1-rgb(j,1),1-rgb(j,2))
	Ib(j)=min(mn1,1-rgb(j,3))
	end
	Iw=transpose(Iw);Ib=transpose(Ib)
	Iw=[Iw, Iw, Iw]
	Ib=[Ib, Ib, Ib]
	
	rgb2=rgb1; map1=rgb1/255
	lab=rgb2lab(map1)
	
	figure
	title('RGB Values of the colors')
	m=0
	for i=1:num_rects
	rectangle('Position', [0,i+m,10,num_rects], 'FaceColor', rect_colors(i,:));
	rectangle('Position', [.5,i+m,2.5,num_rects], 'FaceColor', 'w');
	ylim([0 num_rects^2+num_rects])
	axis off
	str1=num2str(rgb1(i,:))
	str2=num2str(i)
	text(1,i+m+num_rects/2,[str2 '@ ' str1])
	m=m+num_rects
	end
	
	figure
	title('Colors by Hue (Lab) angle')
	hold on
	
	t=360/pi
	nn=0
	for bn=1:length(map1(:,1))
	nn=nn+1
	if lab(nn,1)<0&&lab(nn,3)<0
	degr(nn)=atan(lab(nn,3)/lab(nn,2))*t
	end
	if lab(nn,1)==0
	degr(nn)=(pi/4)*t
	end
	if lab(nn,1)>0&&lab(nn,3)>0
	degr(nn)=(pi/2)-atan(lab(nn,3)/lab(nn,2))*t
	end
	end
		degr=transpose(degr)
		r=rgb2(:,1); g=rgb2(:,2);
		b=rgb2(:,3)
		rx=sqrt(lab(:,2).^2+lab(:,3).^2)
		
		ri=rgb2(:,1)
		gi=rgb2(:,2)
		bi=rgb2(:,3)
		delta=sqrt(ri.^2+gi.^2+bi.^2)
		
		a=200;b=200;r=max(rx);t=pi/180
		
		fi3=0:10:360
		x3=a+r*cos(fi3*t)
		y3=b+r*sin(fi3*t)
		plot(x3,y3,'r','linewidth',2)
		x3=transpose(x3)
		y3=transpose(y3)
		a1=repmat(a,length(x3),1)
		
		for i=1:length(a1)
		x4(i,:)=[a1(i),x3(i)]
		y4(i,:)=[a1(i),y3(i)]
		end
		plot(x4',y4','b')
		
		fi3=0:60:360
		x3=a+r*cos(fi3*t)
		y3=b+r*sin(fi3*t)
		x3=transpose(x3)
		y3=transpose(y3)
		a1=repmat(a,length(x3),1)
		for i=1:length(a1)
		x5(i,:)=[a1(i),x3(i)]
		y5(i,:)=[a1(i),y3(i)]
		end
		plot(x5',y5','r','linewidth',2)
		
		r=rx
		fi=transpose(degr)
		
		for i=1:num_rects
		
		x(i)=a+r(i)*cos(fi(i)*t)
		y(i)=b+r(i)*sin(fi(i)*t)
		x=transpose(x)
		y=transpose(y)
		
		fi4=0:360
		r1=5
		x6=x(i)+r1*cos(fi4*t)
		y6=y(i)+r1*sin(fi4*t)
		fill(x6,y6,map1(i,:))
		%circles(x(i),y(i),wh,'facecolor', map1(i,:))
		str2=num2str(i)
		text(x(i),y(i),str2)
		
		axis equal
		grid on
		end
		toc

References

1. Global Waste Generation—Statistics & Facts. Available online: https://www.statista.com/topics/4983/waste-generation-worldwide/#topicOverview (accessed on 19 June 2023).
2. Bungau, S.; Bungau, C.; Tit, D.M. Studies on the Last Stage of Product Lifecycle Management for a Pharmaceutical Product. *J. Environ. Prot. Ecol.* **2015**, *16*, 56–62.
3. Bungau, S.G.; Suciu, R.N.; Bumbu, A.G.; Cioca, G.; Tit, D.M. Study on Hospital Waste Management in Medical Rehabilitation Clinical Hospital, Baile Felix. *J. Environ. Prot. Ecol.* **2015**, *16*, 980–987.
4. The Impact of Textile Production and Waste on the Environment. Available online: https://www.europarl.europa.eu/news/en/headlines/society/20201208STO93327/the-impact-of-textile-production-and-waste-on-the-environment-infographics (accessed on 10 January 2023).
5. Tit, D.M.; Bungau, S.G.; Nistor-Cseppento, D.C.; Copolovici, D.M.; Buhas, C.L. Disposal of Unused Medicines Resulting from Home Treatment in Romania. *J. Environ. Prot. Ecol.* **2016**, *17*, 1425–1433.
6. Bungau, S.; Tit, D.M.; Fodor, K.; Cioca, G.; Agop, M.; Iovan, C.; Cseppento, D.C.; Bumbu, A.; Bustea, C. Aspects Regarding the Pharmaceutical Waste Management in Romania. *Sustainability* **2018**, *10*, 2788. [CrossRef]
7. Sarwar, T.; Khan, S. Textile Industry: Pollution Health Risks and Toxicity. In *Textile Wastewater Treatment*; Muthu, S.S., Khadir, A., Eds.; Springer Nature Singapore: Singapore, 2022; pp. 1–28, ISBN 978-981-19-2832-1.
8. Filipoiu, D.C.; Bungau, S.G.; Endres, L.; Negru, P.A.; Bungau, A.F.; Pasca, B.; Radu, A.F.; Tarce, A.G.; Bogdan, M.A.; Behl, T.; et al. Characterization of the Toxicological Impact of Heavy Metals on Human Health in Conjunction with Modern Analytical Methods. *Toxics* **2022**, *10*, 716. [CrossRef]
9. Niinimäki, K.; Peters, G.; Dahlbo, H.; Perry, P.; Rissanen, T.; Gwilt, A. The Environmental Price of Fast Fashion. *Nat. Rev. Earth Environ.* **2020**, *1*, 189–200. [CrossRef]
10. Aus, R.; Moora, H.; Vihma, M.; Unt, R.; Kiisa, M.; Kapur, S. Designing for Circular Fashion: Integrating Upcycling into Conventional Garment Manufacturing Processes. *Fash. Text.* **2021**, *8*, 34. [CrossRef]
11. Enes, E.; Kipöz, Ş. The Role of Fabric Usage for Minimization of Cut-and-Sew Waste within the Apparel Production Line: Case of a Summer Dress. *J. Clean. Prod.* **2020**, *248*, 119221. [CrossRef]
12. Jacometti, V. Circular Economy and Waste in the Fashion Industry. *Laws* **2019**, *8*, 27. [CrossRef]
13. Rahman, S.S.; Siddiqua, S.; Cherian, C. Sustainable Applications of Textile Waste Fiber in the Construction and Geotechnical Industries: A Retrospect. *Clean. Eng. Technol.* **2022**, *6*, 100420. [CrossRef]
14. Sai, A.; Acquaye, R. Recycling Fabric Waste into Functional Interior Decoration Pieces. *J. Art Des.* **2022**, *2*, 15–32. [CrossRef]
15. Cuc, S.; Bungau, C. Creating Competitive Advantage through Sustainable Value Chain: Insights on Automotive and Textile Industry. Management between Profit and Social Responsibility. In Proceedings of the Review of Management and Economic Engineering International Management Conference, Napoca, Romania, 18–20 September 2014; pp. 186–196.
16. Cuc, S.; Tripa, S.; Bungau, C. Strategies for Increasing Competitiveness of the Romanian Textile and Clothing Industry. In Proceedings of the 5th Review of Management and Economic Engineering International Management Conference, Cluj Napoca, Romania, 22–23 September 2016; p. 100.
17. Sandin, G.; Peters, G.M. Environmental Impact of Textile Reuse and Recycling—A Review. *J. Clean. Prod.* **2018**, *184*, 353–365. [CrossRef]
18. Hussain, C.M.; Paulraj, M.S.; Nuzhat, S. Source Reduction and Waste Minimization—Concept, Context, and Its Benefits. In *Source Reduction and Waste Minimization*; Elsevier: Amsterdam, The Netherlands, 2022; pp. 1–22, ISBN 978-0-12-824320-6.
19. Botsman, R.; Rogers, R. *What's Mine Is Yours: The Rise of Collaborative Consumption*; Harper Collins: New York, NY, USA, 2010.
20. Zandberga, A.; Kalnins, S.; Gusca, J. Decision-Making Algorithm for Waste Recovery Options. Review on Textile Waste Derived Products. *Environ. Clim. Technol.* **2023**, *27*, 137–149. [CrossRef]
21. Prasad, N.H.; Rajyalakshmi, G.; Reddy, A.S. A Typical Manufacturing Plant Layout Design Using CRAFT Algorithm. *Procedia Eng.* **2014**, *97*, 1808–1814. [CrossRef]
22. SpringerLink Database. Available online: https://link.springer.com/ (accessed on 17 May 2023).
23. Zhang, W. Design for the Sustainment of Traditional Making Practices: A Research Study in Central China. *Des. J.* **2021**, *24*, 137–147. [CrossRef]
24. Hussain, A. Colour Psychology in Art: How Colour Impacts Mood. *Art Des. Rev.* **2021**, *09*, 301–308. [CrossRef]
25. Elliot, A.J. Color and Psychological Functioning: A Review of Theoretical and Empirical Work. *Front. Psychol.* **2015**, *6*, 368. [CrossRef]
26. Garavaglia, J.; Markoski, M.M.; Oliveira, A.; Marcadenti, A. Grape Seed Oil Compounds: Biological and Chemical Actions for Health. *Nutr. Metab. Insights* **2016**, *9*, 59–64. [CrossRef]
27. Craiut, L.; Bungau, C.; Negru, P.A.; Bungau, T.; Radu, A.-F. Technology Transfer in the Context of Sustainable Development—A Bibliometric Analysis of Publications in the Field. *Sustainability* **2022**, *14*, 11973. [CrossRef]
28. Craiut, L.; Bungau, C.; Bungau, T.; Grava, C.; Otrisal, P.; Radu, A.F. Technology Transfer, Sustainability, and Development, Worldwide and in Romania. *Sustainability* **2022**, *14*, 15728. [CrossRef]
29. Jinlong, Y.; Yue, W.; Yanxiang, C.; Yuanyuan, Z.; Huiyi, W.; Yan, W. Research on Innovative Design of Patchwork Art Based on the Concept of Sustainable Development—Take the Creation of "HUI Yu Men Ting" as an Example. *J. Phys. Conf. Ser.* **2021**, *1790*, 12026. [CrossRef]

30. Minda, Q. The Application of Patchwork Art in the Design of Decorative Art. *Art Perform. Lett.* **2022**, *3*, 63–67.
31. Singh, E.; Pillay, N. A Study of Ant-Based Pheromone Spaces for Generation Constructive Hyper-Heuristics. *Swarm Evol. Comput.* **2022**, *72*, 101095. [CrossRef]
32. Pasha, J.; Nwodu, A.L.; Fathollahi-Fard, A.M.; Tian, G.; Li, Z.; Wang, H.; Dulebenets, M.A. Exact and Metaheuristic Algorithms for the Vehicle Routing Problem with a Factory-in-a-Box in Multi-Objective Settings. *Adv. Eng. Inform.* **2022**, *52*, 101623. [CrossRef]
33. Dulebenets, M.A. An Adaptive Polyploid Memetic Algorithm for Scheduling Trucks at a Cross-Docking Terminal. *Inf. Sci.* **2021**, *565*, 390–421. [CrossRef]
34. Dulebenets, M.A. A Diffused Memetic Optimizer for Reactive Berth Allocation and Scheduling at Marine Container Terminals in Response to Disruptions. *Swarm Evol. Comput.* **2023**, *80*, 101334. [CrossRef]
35. Ala, A.; Chen, F. Alternative Mathematical Formulation and Hybrid Meta-Heuristics for Patient Scheduling Problem in Health Care Clinics. *Neural Comput. Appl.* **2020**, *32*, 8993–9008. [CrossRef]
36. Whelehan, D.F.; Conlon, K.C.; Ridgway, P.F. Medicine and Heuristics: Cognitive Biases and Medical Decision-Making. *Ir. J. Med. Sci.* **2020**, *189*, 1477–1484. [CrossRef]
37. Chen, M.; Tan, Y. SF-FWA: A Self-Adaptive Fast Fireworks Algorithm for Effective Large-Scale Optimization. *Swarm Evol. Comput.* **2023**, *80*, 101314. [CrossRef]
38. Gholizadeh, H.; Fazlollahtabar, H.; Fathollahi-Fard, A.M.; Dulebenets, M.A. Preventive Maintenance for the Flexible Flowshop Scheduling under Uncertainty: A Waste-to-Energy System. *Environ. Sci. Pollut. Res.* **2021**, 1–20. [CrossRef]

sustainability

MDPI

Article

Organically Cultivated Vine Varieties—Distinctive Qualities of the Oils Obtained from Grape Seeds

Manuel Alexandru Gitea [1,†], Daniela Gitea [2,*], Delia Mirela Tit [2,3,*], Simona Gabriela Bungau [2,3,†], Mihaela Alexandra Bogdan [2,†], Andrei-Flavius Radu [3], Francisc Vasile Dulf [4] and Manuela Bianca Pasca [2]

1 Department of Agriculture and Horticulture, Faculty of Environmental Protection, University of Oradea, 410048 Oradea, Romania; giteamanuel@yahoo.com
2 Department of Pharmacy, Faculty of Medicine and Pharmacy, University of Oradea, 410028 Oradea, Romania; sbungau@uoradea.ro (S.G.B.); mihaela.alexandra.bogdan@gmail.com (M.A.B.); biancapasca28@yahoo.com (M.B.P.)
3 Doctoral School of Biomedical Sciences, University of Oradea, 410087 Oradea, Romania; andreiflavius.radu@uoradea.ro
4 Faculty of Agriculture, Department of Environmental and Plant Protection, University of Agricultural Sciences and Veterinary Medicine, 400372 Cluj-Napoca, Romania; francisc.dulf@usamvcluj.ro
* Correspondence: gitea_daniela@yahoo.co.uk (D.G.); dtit@uoradea.ro (D.M.T.)
† Equal contribution to the first author.

Abstract: Grape seeds, which have an increased concentration of high-quality compounds in their oil, are the byproduct of the grape processing industry. The purpose of this study is to evaluate the physico-chemical and bioactive profile of grape seed oil (GSO) obtained by extraction with n-hexane, using three different techniques and coming from two varieties of grapes. DPPH and ABTS radical scavenging ability assessments, and CUPRAC and FRAP assays, were used to determine the oil's antioxidant properties, whereas the total phenolic content (TPC) was determined by applying an adapted version of the Folin–Ciocalteu technique. Utilizing a coupling method of gas chromatography and mass spectrometry, 14 fatty acids have been identified by analyzing their methylated intermediates. GSOs were characterized by a high content of polyunsaturated acids (PUFAs) (69.25–80.32%), of which linoleic acid stands out (66.97 and 79.88%), followed by monounsaturated acids (MUFAs) (16.64–19.59%), with the representative being oleic acid (15.20–17.86%) and then saturated acids (SFAs) (9.26–15.53%), through the palmitic acid (6.29–9.82%). GSO from Merlot samples recovered by MW had the greatest ratio of fatty acids with hypo-/hypercholesterolemia (H/H) values (14.09). The atherogenicity index and thrombogenicity index ranges for red GSO were 0.278–0.393 and 0.242–0.268, respectively, and for white GSO, 0.401–0.440 and 0.256–0.268, respectively. The oil from the red grape variety has the highest quantity of total polyphenols regardless of the extraction method (1.263–2.035 mg GAE/g vs. 0.918–1.013 mg GAE/g). Through the DPPH and FRAP methods, the results were similar (8.443–14.035 μmol TE/g oil and 6.981–13.387 μmol TE/g oil, respectively). The best results were obtained by the CUPRAC method (8.125–19.799 μmol TE/g oil). The assessment of the grape varieties revealed that they are appropriate for making edible GSO, which was endorsed by our results.

Keywords: grape seed oil; ecological culture; bioactive profile; Fetească Regală; Merlot; CUPRAC method; Folin–Ciocalteu method

Citation: Gitea, M.A.; Gitea, D.; Mirela Tit, D.; Bungau, S.G.; Bogdan, M.A.; Radu, A.-F.; Dulf, F.V.; Pasca, M.B. Organically Cultivated Vine Varieties—Distinctive Qualities of the Oils Obtained from Grape Seeds. *Sustainability* **2023**, *15*, 11037. https://doi.org/10.3390/su151411037

Academic Editors: Marc A. Rosen and Vita Di Stefano

Received: 22 May 2023
Revised: 29 June 2023
Accepted: 12 July 2023
Published: 14 July 2023

1. Introduction

One of the oldest and most cultivated plant species is the grapevine (*Vitis vinifera* L.), which is extremely valuable both for its fruits and for the wines that are produced. Creating wine means using about a quarter of the entire world's grape harvest. In the case of Europe, viticulture plays a significant role in the economy of several countries, with France, Italy, and Spain leading the market in the field [1]. Currently, the viticulture field is in continuous development and modernization, and wine production is one of the most developed industries based on agriculture, experiencing a rapid evolution, especially after the country's accession to the European Union in 2007 [2]. The result was that Romania ranks among the top 15 wine-producing countries in the world [3].

The production of large amounts of byproducts like grape marc, grape seed, grape skin, grape stem, and grape leaf also occurs during the cultivation and processing of grapes. Of these byproducts, a large amount is rich in phenolic chemicals, which are essential to human physiology and used in the food industry, pharmaceutical industry, cosmetics, etc. [4,5]. A significant quantity (10–12%) of the solid byproducts of grape processing are grape seeds [6], where several beneficial substances can be found. However, for a long time, they were mainly burned and used to feed animals, being considered agricultural waste [7].

Grape seed oil (GSO), which is found in quantities between 7–20% in grape seeds, is frequently used, especially in cosmetic formulas, and is abundant in essential fatty acids. Palmitic, linoleic, oleic, and stearic acids make up much of the fatty acid composition of GSO. Also, GSO contains one of the most abundant naturally occurring forms of tocols, especially tocopherols, which are highly potent oil-soluble antioxidants. Due to its nutritional benefits and favorable sensory qualities, this oil is currently very common for consumption. The high concentration of important fatty acids, natural antioxidants, and phytochemicals in this oil makes it both a helpful dietary supplement and a cosmetic product [8]. Furthermore, antioxidants provide an essential defense mechanism against oxidative damage and play a key role in contributing to the improvement of the management of numerous diseases [9].

Phenols are organic compounds that have antioxidant properties and are found in grapes, particularly in the seeds and extracts of grapes. Anthocyanins, proathocyanidins, flavonols, and flavan-3-ols (that are part of the flavonoid family), as well as stilbenes and phenolic acids (which are not in the flavonoid family), are some of the most significant grape polyphenols. The various families can exist in either conjugated or free forms, and each one is distinct from the others in terms of the degree of hydroxylation, the way the hydroxy groups are substituted (glycosylation, methylation, or acylation), and even creating adducts between them (e.g., condensed tannins; anthocyanins with phenolic acids). These data clarify why grape polyphenols have such a wide chemical variety [10,11]. GSO's water-soluble phenolic content is relatively low; however, using the right oil extraction techniques can raise the phenol level of oils [12]. By using colorimetric techniques and standard curves generated after testing, known quantities of isolated polyphenol molecules, like gallic acid or catechin, and the total amount of polyphenols can be determined [13]. Total seed polyphenols are usually determined by colorimetric methods with Folin–Ciocalteu reagent [14,15]. The ability of an antioxidant to counteract oxidation is determined by the Folin–Ciocalteu reaction, a test that relies on electron transfer [16]. It is frequently used to determine the amount of total phenol/polyphenol contained in foods produced from plants and biological specimens [17].

The methods of extracting GSO that are the most frequently used are pressing and methods involving organic solutions (Bligh and Dyer or Soxhlet). Although solvent extraction produces a better yield, it has the disadvantages of a longer processing time, the presence of potentially harmful residues in the finished product, and reduced nutritional properties of the oil. Cold extraction of oils is frequently linked with decreased final production [18]. Pressurized liquid extraction, microwave (MW)-assisted extraction, ultrasound (US)-assisted extraction, and supercritical fluid extraction (SFE), which employs fluids and CO_2, are additional techniques for extracting oil. US extraction uses negative pressure after US treatment, while MW extraction employs nonionizing electromagnetic waves that are converted to thermal energy. Both methods target cell wall destruction to make extraction easier [19,20].

Through this research, we wanted to highlight and provide the distinctive qualities (physicochemical characteristics, fatty acid content, antioxidant potential, functional value, and total phenolic content) of GSO obtained from two varieties of vines, namely Fetească

Regală (a Romanian-specific white variety) and Merlot (an international red variety), cultivated in an ecological culture system. GSOs have been extracted with n-hexane using three different techniques (US-assisted extraction, cold extraction under stirring, and MW-assisted extraction). According to our knowledge, there are very few studies [21] that characterize the GSO obtained from Fetească Regală, and our research is unique considering the approach. The results offer new perspectives and research opportunities for the incorporation of these extracts into different functional foods, pharmaceuticals, or cosmetic products.

2. Results

2.1. Macroscopic and Microscopic Analyses

The macroscopic analysis shows that the two varieties of *Vitis vinifera* L. differ in the color of the fruits, size, and number of seeds. In the Fetească Regală variety (Figure 1), the seeds are pyriform in shape and dark brown in color. The surface is smooth with a ridge on the back, the tip is discoidal, the size is 4–8 mm long, and the taste is bitter. In most grapes, we find two seeds in one fruit. In the Merlot variety, the shape of the seeds is still pyriform, the color is dark brown, the surface is smooth, with a ridge on the back surface, the tip is discoidal, the size is 4–6 mm long, and the taste is bitter. Most fruits have three seeds.

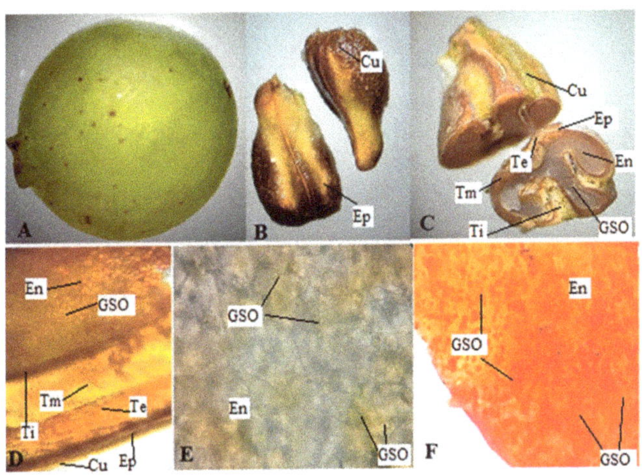

Figure 1. *Vitis vinifera* L.—Fetească Regală (**A**)Fruit; (**B**) Seeds (kernels); (**C**) Transverse section of seeds (ob.4×); (**D**) Transverse section of seed showing regions of cuticle, epidermis, outer integument, mid-integument, and inner integument (ob.10×); (**E**,**F**)Transverse section through the endosperm, highlighting the oil droplets (ob.40×); Cu, cuticle; Ep, epidermis; Te, external integument; Ti, intertegument; Tm, medium integument; En, endosperm; GSO, grape seed oil.

The microscopic analysis was carried out on cross-sections through the seeds of both varieties, highlighting five areas: cuticle and epidermis; outer integument or soft seed coat, composed of parenchyma tissue; middle integument or hard seed coat, composed of two layers of cells; inner skin; endosperm and embryo (Figure 2a–f). A thin cuticle that is not very developed can be observed, followed by the epidermis made up of rectangular cells. The outer skin is made up of parenchymal cells. The epidermis and outer integument form a soft layer of cells that covers the seed. Two layers of rectangular, thin-walled cells follow, which constitute the middle tissue or the middle integument. The inner integument lies between the mid-integument and the endosperm. The center of the seed consists of parenchymatous tissue, surrounded by integuments and containing the embryo sac. The presence of oil droplets in all the parenchymal tissues that make up the endosperm is highlighted by the staining technique [22].

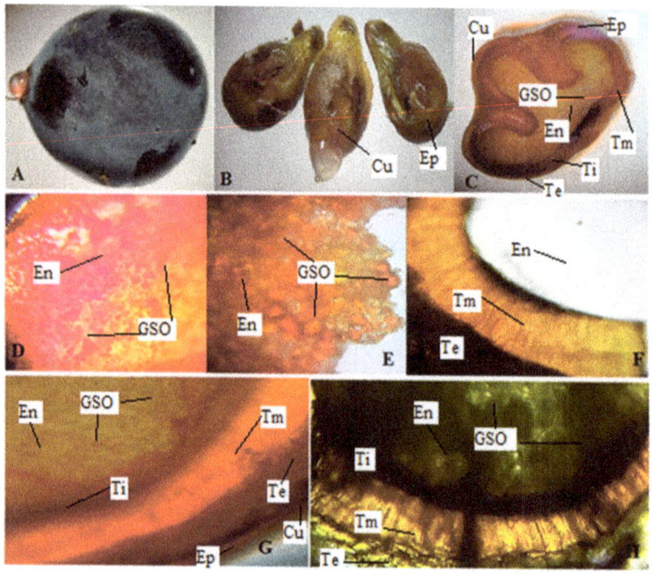

Figure 2. *Vitis vinifera* L.—*Merlot* (**A**) Fruit; (**B**) Seeds (kernels); (**C**) Transverse section of seeds (ob.4×); (**D**) Transverse section through the endosperm, highlighting the oil droplets (ob.40×); (**E–H**) Transverse section of seed showing regions of cuticle, epidermis, outer integument, mid-integument, and inner integument (ob.40×). Cu, cuticle; Ep, epidermis; Te, external integument; Ti, inter tegument; Tm, medium integument; En, endosperm; GSO, grape seed oil.

2.2. Extraction Yield and Physical Indices

An extraction yield has been determined for each GSO extraction based on the dried material in grams. The highest yield was obtained through solvent extraction method by stirring at room temperature (11.67–12.52%), followed by MW (10.52–10.79%) and then US extraction (9.02–10.73). The yield was significantly different ($p < 0.05$), depending on the extraction method used, for all samples (Table 1). GSOs have a medium density (0.8732–0.9147 g/mL).

Table 1. Physico-chemical characterization of GSOs and extraction yields.

GSO Sample	Amount of Seed/Solvent	Oil Quantity (g)	Oil Volume (mL)	Density (g/mL)	Yield (% *w/w*)
GSO_M_US	30 g/250 mL	2.707 ± 0.01^{a}	3 ± 0.10^{a}	0.9023 ± 0.02^{a}	9.02 ± 0.05^{a}
GSO_M_stirring		3.503 ± 0.12^{b}	3.7 ± 0.20^{b}	0.9467 ± 0.02^{a}	11.67 ± 0.38^{b}
GSO_M_MW	20 g/400 mL	2.157 ± 0.04^{c}	2.3 ± 0.10^{b}	0.937 ± 0.03^{a}	10.79 ± 0.20^{c}
GSO_FR_US	30 g/250 mL	3.220 ± 0.08^{a}	3.6 ± 0.10^{a}	0.8944 ± 0.04^{a}	10.73 ± 0.27^{a}
GSO_FR_stirring		3.755 ± 0.16^{b}	4.3 ± 0.10^{b}	0.8732 ± 0.04^{a}	12.52 ± 0.52^{b}
GSO_FR_MW	20 g/400 mL	2.104 ± 0.10^{c}	2.3 ± 0.10^{c}	0.9147 ± 0.03^{a}	10.52 ± 0.50^{c}

GSO, grape seed oil; M, Merlot; US, ultrasound; MW, microwave; FR, Fetească Regală. Data are reported as mean ± SD; all determinations were made in triplicate; ^{a,b,c,} significant difference between data for GSO obtained from the same variety by three different methods (US, stirring, MW), for one test, by applying Tukey's test for $p < 0.05$. Results from a wide range of sources show no statistically different means for superscripts with the same letter.

The oil extracted from the grape seeds has a yellowish, brown-yellow, or yellow greenish color (Figure 3).

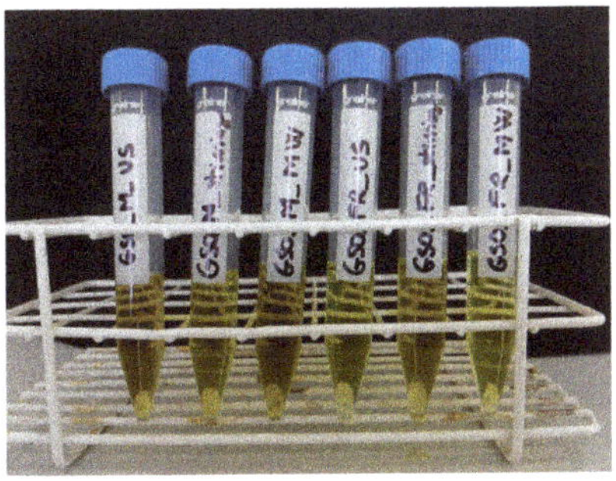

Figure 3. Grape seed oil samples.

2.3. Fatty Acid Composition and Functional Quality

Monounsaturated fatty acids (16.64–19.59%), saturated fatty acids (9.26–15.53%), and polyunsaturated fatty acids (67.37–80.32%) were all present in the samples, as shown by the findings presented in Table 2. The most common saturated fatty acid was palmitic acid (6.29–9.82%), which was followed by stearic acid (2.75–5.32%). The percentage of monounsaturated fatty acids ranged from 15.20% to 17.86% of total fatty acids. In the studied samples, linoleic acid was the most prevalent polyunsaturated fatty acid, making up between 66.97 and 79.88% of the total fatty acid content. No matter the grape variety, or the extraction method, a total of 14 fatty acids were found in all samples. For the oil obtained from the Merlot variety, ΣSFAs were significantly different ($p < 0.05$), depending on the extraction method, with the lowest quantity being obtained by MW (Table 2).

Table 2. Fatty acid concentrations (%) in white and red GSOs, according to various extraction methods, expressed as percentages.

Fatty Acids		Rt	(M+)	GSO_M_US	GSO_M_ Stirring	GSO_M_ MW	GSO_FR_ US	GSO_FR_ Stirring	GSO_FR_ MW
(6:0)	Hexanoic	7.804	130	0.07 ± 0.01	0.08 ± 0.01	0.07 ± 0.01	0.02 ± 0.01	0.07 ± 0.01	0.05 ± 0.00
(14:0)	Myristic	18.308	242	0.03 ± 0.00	0.06 ± 0.01	0.02 ± 0.00	0.07 ± 0.01	0.05 ± 0.00	0.07 ± 0.01
(16:0)	Palmitic	21.224	270	8.26 ± 0.71	9.82 ± 0.10	6.29 ± 0.70	8.78 ± 0.9	8.69 ± 0.90	9.48 ± 0.75
(18:0)	Stearic	24.46	298	4.66 ± 0.38	5.32 ± 0.50	2.75 ± 0.30	4.07 ± 0.40	4.15 ± 0.03	4.04 ± 0.30
(20:0)	Arachidic	28.852	326	0.23 ± 0.01	0.23 ± 0.02	0.13 ± 0.01	0.15 ± 0.01	0.15 ± 0.01	0.17 ± 0.01
(22:0)	Behenic	35.515	354	0.05 ± 0.00	0.03 ± 0.00	0.01 ± 0.00	0.05 ± 0.00	0.05 ± 0.00	0.04 ± 0.00
	ΣSFAs	-	-	13.30 ± 0.21 [a]	15.53 ± 0.37 [b]	9.26 ± 0.52 [c]	13.15 ± 0.71	13.17 ± 0.45	13.90 ± 1.03
16:1(n-9)	cis-7 hexadecenoic	21.579	268	0.05 ± 0.00	0.05 ± 0.00	0.05 ± 0.00	0.08 ± 0.00	0.05 ± 0.01	0.08 ± 0.01
16:1(n-7)	Palmitoleic	21.674	268	0.12 ± 0.01	0.12 ± 0.01	0.12 ± 0.01	0.15 ± 0.01	0.14 ± 0.01	0.20 ± 0.15
18:1(n-9)	Oleic	24.97	296	15.84 ± 1.04	17.86 ± 1.46	15.20 ± 0.90	15.79 ± 1.40	15.58 ± 1.35	17.37 ± 1.50
18:1(n-7)	Vaccenic	25.075	296	1.20 ± 0.95	1.29 ± 1.5	0.88 ± 0.05	0.81 ± 0.07	0.79 ± 0.15	0.89 ± 0.10
20:1(n-9)	11-eicosenoic	25.075		0.25 ± 0.01	0.27 ± 0.02	0.41 ± 0.03	0.24 ± 0.01	0.19 ± 0.01	0.20 ± 0.01
	ΣMUFAs	-	-	17.45 ± 0.32	19.59 ± 0.77	16.64 ± 0.45	17.08 ± 0.41	16.75 ± 0.53	18.73 ± 0.16
18:2(n-6)	Linoleic	26.051	294	68.74 ± 5.30	79.88 ± 8.14	73.72 ± 6.40	69.40 ± 6.01	69.76 ± 5.60	66.97 ± 5.03
20:2(n-6)	Eicosadienoic	31.088	322	0.06 ± 0.00	0.05 ± 0.00	0.03 ± 0.00	0.05 ± 0.00	0.05 ± 0.00	0.07 ± 0.00
18:3(n-3)	α-linolenic	27.446	292	0.45 ± 0.01	0.40 ± 0.01	0.34 ± 0.02	0.33 ± 0.02	0.28 ± 0.01	0.34 ± 0.02
-	ΣPUFAs	-	-	69.25 ± 0.46	80.32 ± 0.51	74.10 ± 0.35	69.77 ± 0.12	70.09 ± 0.42	67.37 ± 0.26
-	n-3	-	-	0.45 ± 0.01	0.40 ± 0.01	0.34 ± 0.02	0.33 ± 0.02	0.28 ± 0.01	0.34 ± 0.02
-	n-6	-	-	68.80 ± 0.57	79.93 ± 0.30	73.75 ± 0.18	69.45 ± 0.32	69.81 ± 0.22	67.03 ± 0.30
-	n-3/n-6	-	-	0.0065	0.005	0.0046	0.0047	0.004	0.005
-	PUFAs/SFAs	-	-	5.21	5.17	8.00	5.31	5.32	4.85

GSO, grape seed oil; SFA, saturated fatty acids; MUFA, monounsaturated fatty acids; PUFA, polyunsaturated fatty acids; Rt, retention time of the methyl esters of the fatty acids; M+, Molecular ion of the methyl esters of the fatty acids; [a,b,c,] significant difference between data, for one test, by applying Tukey's test for $p < 0.05$. Data are reported as mean ± SD; all determinations were made in triplicate; US, ultrasound; MW, microwave.

Figure 4 presents the GC-MS chromatogram for the fatty acid profile of grape seed oil obtained by stirring.

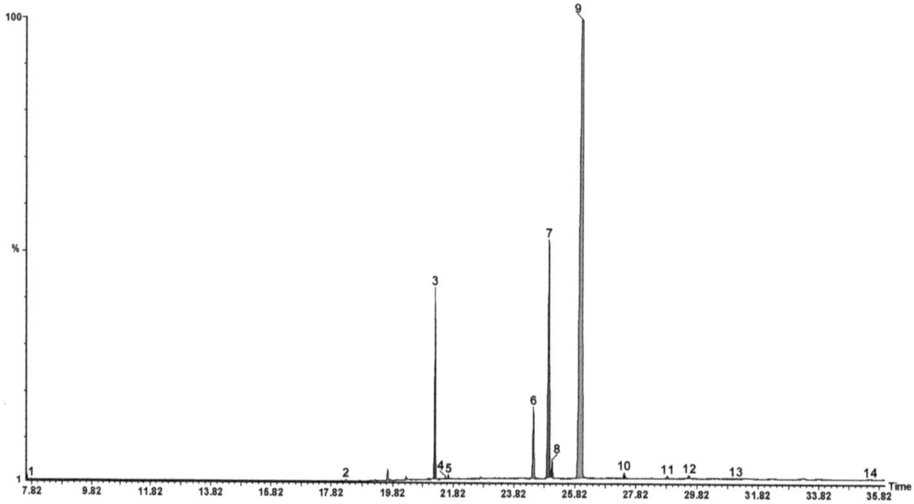

Figure 4. Chromatogram obtained by gas chromatography–mass spectrometry for the methylated derivatives of fatty acids of grape seed oil obtained by stirring from Fetească Regală. 1, (6:0); 2, (14:0); 3, (16:0); 4, 16:1(n-9); 5, 16:1(n-7); 6, (18:0); 7, 18:1(n-9); 8, 18:1(n-7); 9,1 8:2(n-6); 10, 18:3(n-3); 11, (20:0), 12, 20:1(n-9), 13, 20:2(n-6); 14, (22:0).

The analysis of GSO functional characteristics may benefit from taking into account the fatty acid chemical profile. As only three SFAs are hypercholesterolemic, the polyunsaturated/saturated fatty acids ratio (PUFA/SFA) is frequently used to assess indicators for common symptoms of cardiovascular impairments (thrombogenicity and atherogenicity) [23]. Therefore, we determined the atherogenicity index (AI), thrombogenicity index (TI), and hypo- and hypercholesterolemic fatty acid ratios (H/H). Red GSOs' H/H values varied from 9.89 to 14.09, showing significant differences ($p < 0.01$) between the three samples obtained by using the extraction methods. White GSOs' H/H values ranged from 8.83 to 9.76, with significant differences between the oil obtained by MW compared to those obtained by stirring and US. Atherogenicity index AI range for red GSO was 0.278–0.393, and for white GSO, it was 0.401–0.440. Red GSO thrombogenicity index (TI) rates ranged from 0.242–0.268, and white GSO from 0.256–0.268. AI and TI values did not differ significantly, regardless of the extraction method (Table 3).

Table 3. GSO functional quality indicators derived from various extraction methods.

Sample	H/H	AI	TI
GSO_M_US	10.20 ^{a}	0.381 ^{a}	0.291 ^{a}
GSO_M_stirring	9.89 ^{b}	0.393 ^{a}	0.298 ^{a}
GSO_M_MW	14.09 ^{c}	0.278 ^{a}	0.196 ^{a}
GSO_FR_US	9.63 ^{a}	0.404 ^{a}	0.219 ^{a}
GSO_FR_stirring	9.76 ^{a}	0.401 ^{a}	0.292 ^{a}
GSO_FR_MW	8.83 ^{b}	0.440 ^{a}	0.309 ^{a}

AI, atherogenicity index; TI, thrombogenicity index; H/H: ratio between hypocholesterolemic and hypercholesterolemic fatty acids; ^{a,b,c,} significant difference between data for GSO obtained from the same variety by three different methods (US, stirring, MW), for one test, by applying Tukey's test for $p < 0.05$.

2.4. Determination of Total Phenolic Content

The GSOs' overall polyphenol content ranged from 0.918 to 2.035 mg GAE/g, with significant variations depending on the extraction method only in the case of GSOs from Merlot. GSO_M_MW had the highest content of polyphenols, significantly different from GSO_M_stirring and GSO_M-US (2.035 vs. 1.533 mg and 1.263 mg GAE/g GSO_M_US, $p < 0.01$, respectively).

In GSOs from Fetescă Regală, the highest content of polyphenols was recorded in the oil obtained by stirring, which was insignificantly higher compared to the other samples (1.013 mg GAE/g vs. 0.938 mg GAE/g GSO_FR_US and 0.918 MW, $p > 0.05$, respectively). The Merlot variety polyphenol content was found to be significantly higher ($p < 0.01$), regardless of the extraction method (Figure 5).

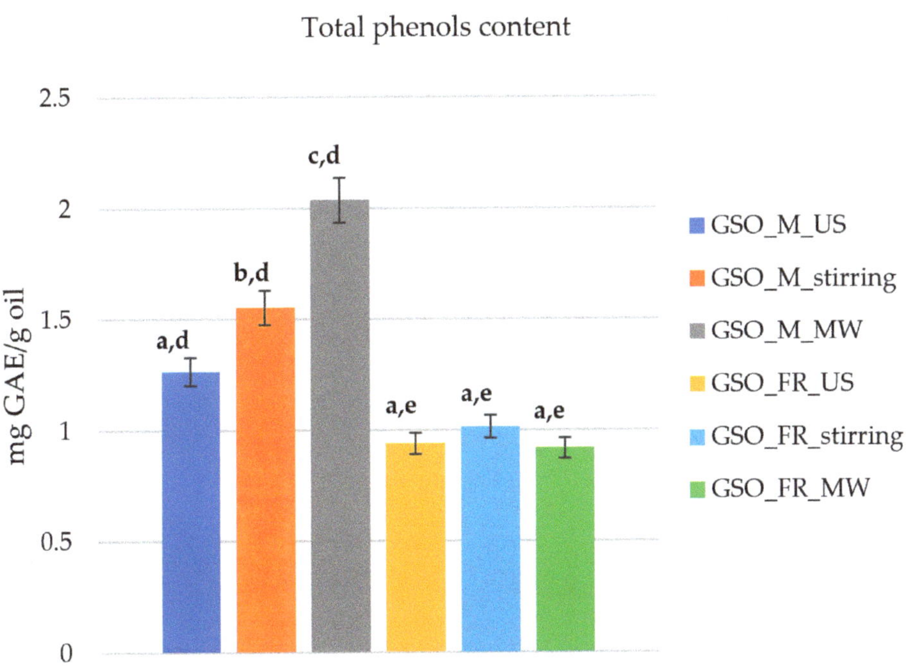

Figure 5. Determination of the total phenolic content (mg GAE/g oil) of grape seed oils obtained by Ultrasound-Assisted Extraction (US), Conventional Extraction (stirring), and Microwave-Assisted Extraction (MW), from the Merlot and Fetească Regală varieties. a, b, c, significant difference between data for GSO obtained from the same variety by three different methods (US, stirring, MW); d, e, significant difference between data for GSO obtained from each variety by the same methods (US, stirring, MW), for one test, by applying Tukey's test for $p < 0.05$.

2.5. Antioxidant Capacity Determination of Grape Seed Oils

The highest antioxidant capacity was found in GSO from Merlot, obtained by MW extraction. Through the DPPH and FRAP methods, the results were similar. Using DPPH, we obtained values between 8.443–14.035 µmol TE/g oil, and the FRAP method revealed values between 6.129–13.387 µmol TE/g oil. The best results were obtained by the CUPRAC method, with values between 5.993–19.799 µmol TE/g oil, while the worst results were obtained through ABTS (1.124–4.025 µmol TE/g oil) (Table 4).

The ABTS method, commonly used as an electron acceptor in various tests to measure antioxidant activity, can be used in the analysis of GSO but with lower results than other methods.

Table 4. Antioxidant capacity of grape seed oils.

GSO Sample	DPPH (μmol TE/g Oil)	FRAP (μmol TE/g Oil)	CUPRAC (μmol TE/g Oil)	ABTS (μmol TE/g Oil)
GSO_M_US	8.553 ± 0.076 [a]	9.677 ± 0.913 [a]	13.498 ± 2.849 [b]	4.025 ± 0.205 [a]
GSO_M_stirring	14.035 ± 0.554 [b]	6.921 ± 0.193 [b]	9.840 ± 2.799 [b]	1.595 ± 0.079 [b]
GSO_M_MW	10.307 ± 0.526 [c]	13.387 ± 0.374 [c]	19.799 ± 0.733 [c]	1.908 ± 0.134 [c]
GSO_FR_US	7.675 ± 0.080 [a]	1.040 ± 0.270 [a]	5.993 ± 0.441 [a]	2.768 ± 0.107 [a]
GSO_FR_stirring	8.443 ± 0.225 [b]	6.129 ± 0.170 [b]	14.636 ± 0.330 [b]	1.124 ± 0.056 [b]
GSO_FR_MW	7.905 ± 0.021 [a]	6.981 ± 0.120 [c]	8.125 ± 0.022 [c]	1.255 ± 0.025 [b]

GSO, grape seed oil; DPPH, 2,2′-diphenyl-1-picrylhydrazyl radical; FRAP, ferric reducing ability of plasma; CUPRAC, cupric ion reducing antioxidant capacity; ABTS, 2,2′-azino-bis-3-ethylbenzthiazoline-6-sulphonic acid. Data are reported as mean ± SD; all determinations were made in triplicate; [a,b,c,] significant difference between data for GSO obtained from the same variety by three different methods (US, stirring, MW), for one test, by applying Tukey's test for $p < 0.05$. Results from a wide range of sources show no statistically different means for superscripts with the same letter.

The Pearson correlation matrix was used to assess the connection between the antioxidant capability and total phenol content (TP) of GSO samples. TP was positively and significantly correlated with antioxidant capacity measured by FRAP, DPPH, and CUPRAC for most samples. No correlation was observed between TP and antioxidant capacity, regardless of the method, for oil obtained from the Fetească Regală variety by MW extraction. Negative non-significant correlations were obtained between TP and ABTS for oils obtained from the Merlot variety by US extraction and the Fetească Regală variety by US and MW (Table 5).

Table 5. Pearson correlation coefficients of TPC and antioxidant capacity of GSO samples.

Antioxidant Capacity	Pearson Correlation	GSO_M_US	GSO_M_ Stirring	GSO_M_ MW	GSO_FR_ US	GSO_FR_ Stirring	GSO_FR_ MW
FRAP	r	**1.000 ***	**1.000 ***	**1.000 ***	**1.000 ***	**1.000 ***	0.991
	p	**0.002**	**0.008**	**0.003**	**0.019**	**0.010**	0.085
DPPH	r	0.993	**1.000 ***	**1.000 ***	**1.000 ***	**1.000 ***	0.524
	p	0.076	**0.005**	**0.003**	**0.017**	**0.006**	0.649
CUPRAC	r	**1.000 ***	**1.000 ***	**1.000 ***	**1.000 ***	0.871	−0.989
	p	**0.003**	**0.005**	**0.002**	**0.019**	0.327	0.096
ABTS	r	−0.712	0.222	0.453	−0.978	0.196	−0.704
	p	0.495	0.857	0.700	0.134	0.874	0.503

* Correlation is significant at the 0.005 level (2-tailed); bolded values, significant values. TP, total polyphenols; r, Pearson coefficient; US, ultrasound; MW, microwave.

3. Discussion

Since a very large amount of seeds is produced worldwide, considering the beneficial effect of consuming GSO on health and the possible uses in other industries, the extraction of oil from grape seeds (*V. vinifera* L.) is presently the widest use of these seeds [24]. An important role in the safety of the use of products, which contain active principles from plants, is attributed to the source and quality of the plant material. Many factors can affect the quality and, consequently, the value of phyto-complexes, such as light exposure, temperature, water availability, nutrients, collection period and time, collection method, drying, packaging, storage, and transportation of raw materials. For this reason, it is of utmost importance to have raw materials with consistent and reproducible quality standards [25].

In this study, GSOs obtained from two varieties of grapes have been analyzed, a red one (Merlot, an international variety) and a white one (Fetească Regală, Romanian variety), obtained from an ecological vine culture, with all the grapes being harvested at maturity.

The seeds were analyzed from the point of view of their macro-/microscopic characteristics to identify the potential differences between the two varieties. As far as we know, this type of sections analysis was performed for the first time in this study for the varieties we are referring to. The microscopic analysis of the cross-sections through the grape seeds did not reveal any differences between the two varieties. The presence of oil drops in the endosperm was located/highlighted.

According to earlier research, the percentage of oil in grape seeds is between 6% and 20% [26–28]. The chemical composition is primarily influenced by the stage of seed maturation, various environmental cultivation conditions, and to a lesser extent, the seed extraction procedure [6,29].

In our study, the oil was extracted using novel methods like MW, US, and cold extraction, using n-hexane as a solvent. These techniques present the advantage of shorter extraction times and lower temperatures, which slow down the degradation of thermolabile oily components during the procedure [23]. The yield of GSO extraction varied slightly depending on the extraction method and grape varieties. The highest yield was obtained through the solvent extraction method by stirring at room temperature, followed by MW and then US extraction. This order is kept for both grape varieties.

Moreover, it must be mentioned that the yield of GSO extraction can vary depending on several factors, such as the type of grapes, the growing conditions, the extraction method used, and the quality of the grape seeds. However, on average, the yield of GSO extraction can vary greatly, and in some cases, it can be 10–15% or 25–30% [30,31]. The use of an organic solvent, which has been removed by evaporation, ensures a better yield for oil extraction than the use of a press, where the extraction temperature cannot be controlled, being quite high due to the pressing force [30,31]. Although no prior studies to support the MW technique's suitability for isolating oils from grape seeds were found in the literature, Dimić et al. compared three methods of obtaining GSO and found that the MW technique demonstrated excellent extraction yield [23]. Our findings thus support the utilization of MW for this objective.

An essential factor in determining a vegetable oil or fat's nutritional value and potential for industrial use is the quantification of each fatty acid. The varieties and proportions of fatty acids in vegetable oils have a significant impact on their physicochemical and nutritional properties [32]. It is important to observe that cultivation conditions and grape variety may have a significant impact on the fatty acid composition of GSOs [24].

In the present study, the profile of fatty acids is similar to that found in data from the literature [23]. The oil obtained from the two grape varieties, regardless of the extraction technique, is characterized by a high content of polyunsaturated acids (PUFAs) (69.25–80.32%), of which linoleic acid stands out (66.97 and 79.88%), followed by monounsaturated acids (MUFAs) (16.64–19.59%), with the representative being oleic acid (15.20–17.86) and then saturated acids (SFAs) (9.26–15.53%), through the palmitic acid (6.29–9.82%). According to earlier research, PUFAs predominated in the grape seed samples with concentrations of 69.27–74.88%, followed by MUFAs, varying between 13.53–18.62%, and SFAs with concentrations from 11.28 to 12.27%. Considering the fatty acid profiles, stearic acid, which varied from 3.79 to 4.37%, followed palmitic acid, which dominated the category of SFAs with 7.20–7.93%. Oleic acid, which made up 13.39–18.47% of the MUFA contents, was identified as the main acid from the samples [23].

Regardless of the variety or the extraction technique, linoleic acid was discovered to be the most prevalent fatty acid among those discovered, contributing 68.74% to 79.88% in the Merlot variety and 66.97 to 69.76% in the Fetească Regală variety. Viktória Kapcsándi et al. [12], using the Soxhlet extraction method with petroleum ether as a solvent for 3 h, obtained from the Merlot variety an oil with a linoleic acid content of 72.47%, a value close to the linoleic acid content of the oil obtained by us simply by shaking the powder from grape seeds of the Merlot variety with n-hexane (79.88%). Additionally, Dilsat Bozdogan Konuskan et al. examined total phenolic contents, the fatty acid profiles, and antioxidant activity of the oil obtained from grape seeds of the Merlot variety and other varieties using solvent and cold-pressed acquisition techniques. The GSOs obtained through solvent extraction had the greatest concentrations of linoleic acid, the most prevalent fatty acid, antioxidant activity, and total phenolic content [33]. GSO is an oil with high linoleic acid content. Due to its emollient properties, linoleic acid has benefits in moisturizing and strengthening the skin's protective barrier [24].

The data on functional quality indices showed that GSOs from Merlot samples recovered by MW had the greatest H/H values (14.09). Since this index shows how fatty acids affect cholesterol metabolism, a higher level is preferred in nutrition. For example, beneficial oils like sesame and olive oils have lower values than linseed, which has a similar H/H index to grape oils (13.24) [34]. The AI and TI ranges for red GSO were 0.278–0.393 and 0.242–0.268, respectively, and 0.401–0.440 and 0.256–0.268, respectively, for white GSO. Since they result in oil with an excellent functional and nutritional composition, AI and TI values near zero are preferred. Reduced AI and TI have an impact on preventing coronary diseases [35]. Dimić et al. found that white and red GSOs had an average AI of 0.085, which is less than our results report [23].

Although phenolic compounds are poorly soluble in oily phases, extraction does transfer a tiny quantity of them from the solid matrix to the oil [36]. In our study, we observed that the oil obtained from the red grape variety has the highest number of total polyphenols regardless of the extraction method (1.263–2.035 mg GAE/g vs. 0.918–1.013 mg GAE/g). According to earlier research by Kapcsándi et al., the overall polyphenol content of GSOs ranged from 0.24 to 1.13 mg GAE/g for Merlot being 0.97 mg GAE/g, lower than in this study [12].

DPPH and ABTS radical scavenging ability assessments, CUPRAC and FRAP assays, as well as other tests were used to determine the oil's antioxidant properties. These tests were all carried out three times to ensure their reproducibility and to obtain a more complete picture of the oil's antioxidant activity. DPPH and FRAP are chemical compounds used as standards in antioxidant tests to determine the antioxidant capacity of various substances, including GSO. Through the DPPH and FRAP methods, the results were similar (8.443–14.035 μmol TE/g oil and 6.981–13.387 μmol TE/g oil, respectively). The best results were obtained by the CUPRAC method (8.125–19.799 μmol TE/g oil). The ABTS method, commonly used as an electron acceptor in various tests to measure antioxidant activity, showed the lowest results (1.124–4.025 μmol TE/g oil). Mollica et al. came to a different conclusion, indicating that the oil produced from the Montepulciano variety exhibited minimal action in various antioxidant bioassays and no activity in the DPPH and ABTS assays [37]. But through the method used, we demonstrated the antioxidant action of GSO, with the oil obtained from the Merlot variety having the best antioxidant activity. The results could vary because each grape variety contains a different type of phenolic compound or because these compounds have been associated with other compounds, making them more complex or insoluble, like long chains of cutin and suberin, which prevent a direct reaction with the DPPH radical [38].

Antioxidant capacity and total polyphenol concentration showed a strong and favorable correlation, except for ABTS, for most samples, regardless of the method of extraction and variety.

It is undeniable that oil making constitutes an advantageous and sustainable utilization of grape seeds considering the wine sector's continuous waste growth, as well as all the GSO's benefits for health. According to the findings of the current research, GSO is an important source of constituents with antioxidant activity. Furthermore, this investigation highlights a characteristic of a Romanian indigenous grape variety that has not yet received enough attention. We must emphasize, as a strong point, the novelty of the study carried out on an organic native vine crop (Fetească Regală), with the limitation consisting in the lack of detailing of the phenolic profile and the fact that solvent residues were not analyzed.

4. Materials and Methods

4.1. Description of the Plant Material

The vineyard from which the plant material was obtained is part of the vineyard area of the Crişana and Maramureş Wine Regions, Bihor County, Romania, located at 47°16′10.1″ N 22°08′04.2″ E, and has as an activity the cultivation of vines in ecological system (Figure 6) (Ecological certificate 22/162368/1458484 dated 30 September 2022). The varieties grown on the farm are Fetească Regală (on an area of 3.00 ha) and Merlot (on an area of 1.45 ha).

(**a**) (**b**)

Figure 6. Map with the grapevine plantation; (**a**) satellite image [39], (**b**) terrestrial image (first author's personal archive).

Fetească Regală (Figure 7a) is a Romanian variety of white wine grapes belonging to the category of semi-aromatic varieties, like Chardonnay. Merlot (Figure 7b) is a grape variety used to produce red wines and is cultivated in most wine-growing regions of the world. The year 2022 was a favorable year for grapevine culture, thanks to the good weather conditions resulting in productions of 8 tons/ha for the Fetească Regală variety and 6 tons/ha for the Merlot variety. To obtain wine from the Fetească Regală variety, a traditional method was used, which involves reception of the grapes, crushing and de-stemming (evacuation of the bunches), sulfiting, mustering, separation of the wine by pressing the marc, and evacuation of the pomace. For the Merlot variety, the grapes are received, crushed and de-stemmed (clusters are removed), sulfiting, mustering, maceration, fermentation on the marc until the end of the lactic fermentation (approximatively one month), and pressing the pomace and removing the residues.

a b

Figure 7. *Vitis vinifera* L. varieties: (**a**) Fetească Regală, (**b**) Merlot (first author's personal archive).

The samples were comparatively analyzed from the point of view of the macro- and microscopic characteristics and the extraction yield. The oils obtained by three extraction methods were analyzed from the point of view of the fatty acids and total polyphenol content and of the antioxidant action.

4.2. Seed Samples

The white grapes of Fetească Regală and red grapes from the Merlot variety were harvested in the fall of 2022. After removing the bunches, the grapes were crushed, and then those from the Fetească Regală variety were left to macerate for a short time while those from the Merlot variety were left to macerate and ferment on the marc. Then, it was pressed, and the marc was removed. From the marc, the seeds were selected by sifting with sieves of different sizes and air dried at room temperature until the moisture content was below 10% [40]. After drying, the seeds were crushed into powder in a grinder for 30 s to a particle size of less than 0.5 mm. The 2 varieties of grape seeds are kept in the herbarium of the Pharmaceutical Botany Laboratory, Pharmacy Department, Faculty of Medicine and Pharmacy, University of Oradea, for further research.

4.3. Macroscopic and Microscopic Analyses

The macroscopic characters of the seeds were studied, considering the Romanian Pharmacopoeia, 10th edition [41], by observing the organoleptic characters: appearance (size, color), scent, and flavor. Microscopic control was carried out on the fresh vegetable product, included in the elderberry marrow, and sectioned with the help of a blade. Cross-sections were then cleared and stained. Staining was performed to highlight the types of tissues that make up the seed or to locate specific compounds (oil). The processing consisted of making cross-sections through the seeds, which were then pigmented with a hydroalcoholic solution of Genevez reagent (Congo red and chrysoidine). The sections were kept for 5 min, then washed several times with distilled water to remove excess dye. The Sudan III reagent was used for the histochemical localization of the fatty oil from the seeds [25]. An Optika microscope model C-B10+ (BG-Italy, 24010 Ponteranica, Italy) equipped with a OpticamB10 digital camera (BG-Italy, 24010 Ponteranica, Italy) was used to make observations and capture pictures.

4.4. Chemicals and Reagents

n-Hexane was purchased from Merck KGaA, Darmstadt, Germany. Trolox (6-hydroxy-2,5,7,8-tetramethylchroman-2-carboxylic acid), 1,1-Diphenyl-2-picrylhydrazyl-hydrate (DPPH), 2,4,6-Tris(2-pyridyl)-S-triazine (TPTZ), Neocuproine (2.9-dimethyl-1, and 10-phenantroline) were purchased from Sigma Aldrich, St. Louis, MO, in the United States. 2,2′-Azino-bis(3-ethylbenzothiazoline-6-sulfonic acid) diammonium salt 98% (ABTS) was purchased from Thermo Fisher Scientific, Waltham, MA, USA. Dimethyl sulfoxide pure was purchased from Chempur, Jana Lortza, Poland. Gallic acid 98%, Folin–Ciocalteu reagent, iron (III) chloride hexahydrate ($FeCl_3 \cdot 6H_2O$), copper (II) chloride ($CuCl_2$), and sodium carbonate (Na_2CO_3) were purchased from Carl Roth GmbH + Co. KG, Karlsruhe, Germany. Methanol was purchased from Chimreactiv SRL, Bucuresti, Romania. The lipid standards used from preparation of fatty acid methyl esters (FAMEs) were purchased from Sigma–Aldrich, St. Louis, MO, USA.

4.5. Extraction Techniques

4.5.1. Ultrasound-Assisted Extraction

For the US extraction, 30 g of grape seed powder from both varieties were used, over which 250 mL of hexane was added (Table 5). The samples were placed in an Elma Elmasonic S100H (Elma Schmidbauer GmbH, Singen, Germany) ultrasonic bath, with a power of 550 W, frequency of 37 kHz, and at a constant temperature of 30 °C, for 90 min [23]. Afterwards, the extracts were filtered under vacuum, and the solvent was removed with a rotary evaporator under vacuum at 40 °C. The obtained oil was placed in glass vials and kept at 4 °C until further determinations.

4.5.2. Cold Extraction under Stirring

In total, 30 g of grape seed powder from the 2 varieties were subjected to extraction, together with 250 mL of hexane for 90 min, at room temperature, by simply stirring on an IKAmag RCT magnetic stirrer (IKA®-Werke GmbH & Co. KG, Staufen, Germany). After filtering, the extraction solvent was evaporated under pressure, at a bath temperature of 40 °C and a rotational speed of 30 rpm, in a rotating evaporator made by Heidolph Instruments, Berlin, Germany. The obtained oil was placed in glass vials and kept at 4 °C until further determinations [36].

4.5.3. Microwave-Assisted Extraction

MW-assisted extraction was performed, according to the literature [42,43] with some modifications. The experiments were performed by using a Microwave Extraction Reactor (Betameg Invest SRL, Bucharest, Romania), with adjustable power between 0–900 W and frequency of 2450 MHz. An amount of 20 g of grape seed powder was introduced into the cartridge. An amount of 400 mL of hexane (Table 6) was introduced into the solvent vessel. The matrix-to-solvent ratios were modified in accordance with the constraints of the experimental design [19,36].

Table 6. Experimental factors for producing GSO under various handling conditions.

Sample	Extraction Technique	Process Condition
	Red grape seeds Merlot (M)	
GSO_M_US	Ultrasound-Assisted Extraction	Solvent: n-hexan, 37 kHz, 30 °C, 90 min
GSO_M_stirring	Conventional Extraction	Solvent: n-hexan, 90 min, room temperature
GSO_M_MW	Microwave-Assisted Extraction	Solvent: n-hexan, 300 W, 30 s ON, 8 min OFF, 40 min, 38 °C
	White grape seeds Fetească Regală (FR)	
GSO_FR_US	Ultrasound-Assisted Extraction	Solvent: n-hexan, 37 kHz, 30 °C, 90 min
GSO_FR_stirring	Conventional Extraction	Solvent: n-hexan, 90 min, room temperature
GSO_FR_MW	Microwave-Assisted Extraction	Solvent: n-hexan, 300 W, 30 s ON, 8 min OFF, 40 min, 38 °C

The extraction was performed at MW irradiation power of 300 W, and the irradiation time was 30 s, followed by a break of 8 min. The total extraction period was 40 min, with 6 successive irradiations being carried out every 8 min. The extractive solution was subjected to the process of removing the solvent in a rotary evaporator (Heidolph Instruments, Berlin, Germany) under vacuum at 40 °C. The obtained oil was placed in dark bottles and kept at 4 °C.

4.5.4. Extraction Yield and Physical Indices

After the extraction of the oil using the 3 methods, the physical characterization was carried out, determining the extraction yield (%), density as a ratio between mass (g) and volume (mL), refractive index, and color. Following each extraction, the oil yield was determined using Equation (1).

$$\text{Yield}\ (\%) = m_{oil}/m_{seed} \times 100 \quad (1)$$

The refractive index of GSO was determined at room temperature with the Abbe Refractometer, OPTIKA model 2WAJ (Optika SRL, Ponteranica, Italia) [36].

4.6. Chemical Characterization of Grape Seed Oil

4.6.1. Fatty Acid Determination from GSO

Total lipid extracts were trans esterified into fatty acid methyl esters (FAMEs) using the acid-catalyzed procedure and analyzed with a gas chromatograph (GC) coupled to a mass spectrometer (MS) (PerkinElmer Clarus 600 T GC-MS; PerkinElmer, Inc., Shelton, CT, USA) [44].

The system was equipped with a Supelcowax 10 capillary column (60 m × 0.25 mm i.d., 0.25 μm film thickness; Supelco Inc., Bellefonte, PA, USA), and the operating conditions were as follows: injector temperature 210 °C; helium carrier gas flow rate 0.8 mL/min; injection volume 1 μL; split ratio 1:24; oven temperature 140 °C (hold 2 min) to 220 °C at 7 °C/min (hold 23 min); electron impact ionization voltage 70 eV; trap current 100 μA; ion source temperature 150 °C; mass range 22–395 *m/z* (0.14 scans/s with an intermediate time of 0.02 s between the scans). The methylated fatty acid peak identification was based on comparison of both retention time and MS of the unknown peak to those of known standards (37 components FAME Mix, Supelco no. 47885 47885 U) and with data provided by MS database (NIST MS Search 2.0). The amount of each fatty acid was calculated as the individual peak area percentage from the total fatty acid content.

4.6.2. Functional Quality

Three metrics derived and computed from fatty acid (FA) profiles were used to assess the functional value of GSOs. Equation (2) was used to determine the ratio of FAs with hypo- and hypercholesterolemia (H/H) [45].

$$\frac{H}{H} = \frac{C18:1 + C18:2 + C18:3}{C14:0 + C16:0} \quad (2)$$

The thrombogenicity index (TI) and atherogenicity index (AI) were also determined using Equations (3) and (4) [46,47].

$$AI = \frac{C14:0 + 4(C16:0)}{\sum MUFA + \sum \omega - 3 + \sum \omega - 6} \quad (3)$$

$$TI = \frac{C14:0 + C16:0 + C18:0}{0.5(\sum MUFA) + 3\sum \omega - 3 + 0.5\sum \omega - 6 + \left(\frac{\sum \omega - 3}{\sum \omega - 6}\right)} \quad (4)$$

Linoleic acid is C18:2, α-linolenic acid is C18:3, myristic acid is C14:0, oleic acid is C18:1, palmitic acid is C16:0, and stearic acid is C18:0. ΣMUFA is the total of monounsaturated FAs, $\Sigma\omega - 6$ is the total of polyunsaturated fatty $\omega - 6$ acids, and $\Sigma\omega - 3$ is the total of polyunsaturated $\omega - 3$ FAs.

4.6.3. Determination of Total Phenolic Content

The total phenolic content of GSO was measured by using a modified Folin–Ciocalteu method. First, GSO was diluted 1:1 (*v/v*) in dimethyl sulfoxide (sample) [48]. In total, 200 μL of recently prepared Folin–Ciocalteu reagent (1:10 dilution (*v/v*)), 100 μL GSO, 1700 μL of distilled water, and 7.5% Na_2CO_3 solution were combined. The mix was left to sit at room temperature in the dark for two hours. The absorbance was recorded at 765 nm using the spectrophotometer (PG Instruments Ltd., Leicestershire, UK) and using gallic acid as a reference; the findings were reported in milligrams of gallic acid equivalent (GAE) per g of oil (y = 0.0236x + 0.008, R^2 = 0.9998) [49].

4.7. *Antioxidant Capacity Determination of Grape Seed Oil*

The antioxidant capacity of GSO was measured using four different methods. Before starting the determinations, GSO obtained by the three methods was diluted 1:1 (*v/v*) in dimethyl sulfoxide (sample) [50].

4.7.1. DPPH (2,2-Diphenyl-1-Picryl-Hydrazyl-Hydrate) Assay

The radical DPPH scavenging ability of GSO was calculated using a technique taken from the scientific literature [49]. A quantity of 100 μL of sample was blended with 2800 μL of recently made 80 μM DPPH methanol solution, and the mixture was then incubated at room temperature for precisely 30 min in the dark. The absorbance was assessed at 517 nm, and the radical scavenging capacity was determined using Equation (5), where A_0 represents the absorbance of blank, and A_1 represents the absorbance of the sample under investigation.

$$\% \text{ Radical Scavenging Activity} = [(A_0 - A_1)/A_0] \times 100 \quad (5)$$

A calibration curve was designed by plotting the amount of% DPPH inhibition that was scavenged against the concentration of a standard antioxidant (0.125–4 mM Trolox) (y = 5.2384x + 0.3069, R^2 = 0.9983). Outcomes were indicated in terms of µmol Trolox equivalent (TE)/g of oil.

4.7.2. FRAP (Ferric Reducing Antioxidant Power) Assay

In essence, 100 µL of sample was combined with 2000 µL of distilled water and 500 µL of the FRAP working solution, which was made up of recently prepared solutions of 20 mM $FeCl_3 \cdot 6H_2O$ solution, 10 mM 2,4,6-tripyridyl-S-triazine solution (TPTZ), and 300 mM acetate buffer, pH3.6, in ratios of 10:1:1 (*v*/*v*/*v*). This mixture was kept at room temperature and in the dark for an hour. The data were presented as µmol TE/g of oil after the absorbance was determined at 595 nm. Trolox (0.03125–0.5 mM) served as a reference solution [49], and the regression equation coefficient of the measurement for the calibration curve was R^2 = 0.9956 (y = 1.6549x + 0.3522).

4.7.3. ABTS (2,20-Azino-Bis [3-Ethylbenzothiazolin-6-Sulfonic Acid]) Assay

A technique optimized from the literature was used to assess the sample's capacity to scavenge ABTS radicals [49]. Basically, ABTS solution 7 mM and potassium persulphate solution 2.45 mM were combined, and the mixture was left in the dark for 12 h to create the $ABTS^{\bullet +}$ cation radical. After that, the obtained ABTS solution was diluted in a phosphate buffer with a pH of 6.7 so that the absorbance at 730 nm would be 0.70 ± 0.02. The absorbance was measured at 730 nm exactly 1 min after adding100 µL of sample to 2400 µL of diluted ABTS cation radical solution. The antioxidant capacity of GSO was expressed as µmol Trolox equivalent (TE)/g of oil. Using 0.6–40 µM Trolox as reference, and the calibration curve was generated (y = 1902.9x + 3.0018, R^2 = 0.9989).

4.7.4. CUPRAC (Cupric Reducing Antioxidant Capacity) Assay

The procedure involves mixing an antioxidant extract with a copper (II) chloride solution (1×10^{-2} M), an ammonium acetate aqueous buffer (pH 7), and a Neocuproine (2.9-dimethyl-1, 10-phenantroline) alcoholic solution (7.5×10^{-3} M), then determining the absorbance at 450 nm after 30 min [49]. Thus, the procedure was carried out by adding 100 µL of sample, 1 mL of Neocuproine solution, 1 mL of $CuCl_2$ solution, 1 mL of ammonium acetate buffer, and then 4.1 mL of water. The findings were given in µmol of TE per gram of oil by using Trolox as reference (0.0156–0.25 mM), and the calibration curve was generated (y = 3.826x + 0.008, R^2 = 0.9988).

4.8. Statistical Analysis

The statistical analysis was performed using one-way ANOVA and the Tuckey test via SPSS statistical package (version 25, Chicago, IL, USA). To evaluate relationships between TPC and GSO antioxidant potential, Pearson correlation coefficient (r) was used. Data are reported as mean ± SD; all determinations were made in triplicate.

5. Conclusions

According to the findings, independent of the technique of extraction, grape seeds from the two cultivars unde r study were a good source of high-quality GSOs, rich in linoleic acid, with high amounts of total phenolics and good antioxidant capacity. However, further research involving extraction without organic solvents and determining the chemical profile of compounds with antioxidant potential is needed to accurately determine the bioactive properties of the oils obtained from the varieties studied. The results offer new perspectives and research opportunities to incorporate these extracts into pharmaceuticals, different functional foods, or cosmetic products. The option of making use of the waste in seed oil extraction seems to be extremely profitable, given the abundance of grape pomace in Romania.

Author Contributions: Conceptualization, M.A.G. and M.B.P.; Data curation, M.A.G., M.A.B. and F.V.D.; Formal analysis, D.G. and D.M.T.; Investigation, M.A.G., D.G., M.A.B., F.V.D. and M.B.P.; Methodology, D.G. and D.M.T.; Software, A.-F.R.; Supervision, D.M.T. and S.G.B.; Validation, S.G.B.; Writing—original draft, D.G., D.M.T., S.G.B., A.-F.R. and M.B.P.; Writing—review and editing, D.M.T., S.G.B. and A.-F.R. All authors have read and agreed to the published version of the manuscript.

Funding: The research has been funded by the University of Oradea, within the Grants Competition "Scientific Research of Excellence Related to Priority Areas with Capitalization through Technology Transfer: INO—TRANSFER—UO", Project No. 308/2021.

Institutional Review Board Statement: Not applicable.

Informed Consent Statement: Not applicable.

Data Availability Statement: Data are available from the two first authors. The literature data are in accordance with the references inserted.

Acknowledgments: This work was supported by a grant from the Ministry of Research, Innovation and Digitization, CNCS-UEFISCDI, project number PN-III-P4-PCE-2021-G750, within PNCDI in. The authors would like to thank S.C. Gentis Prest SRL for the materials used for experiments, as well as Research grant—no. 127/25.06.2021 (using microwave apparatus obtained through this project).

Conflicts of Interest: The authors declare no conflict of interest.

References

1. Sargolzaei, M.; Rustioni, L.; Cola, G.; Ricciardi, V.; Bianco, P.A.; Maghradze, D.; Failla, O.; Quaglino, F.; Toffolatti, S.L.; De Lorenzis, G. Georgian Grapevine Cultivars: Ancient Biodiversity for Future Viticulture. *Front. Plant Sci.* **2021**, *12*, 630122. [CrossRef] [PubMed]
2. Antoce, A.O.; Calugaru, L.L. Evolution of grapevine surfaces in Romania after accession to European Union—period 2007–2016. *BIO Web Conf.* **2017**, *9*, 1–7. [CrossRef]
3. OIV: Organisation Internationale de la Vigne et du vin Global Economic Vitiviniculture Data. Available online: https://www.oiv.int/ (accessed on 21 March 2023).
4. Oprea, O.; Apostol, L.; Bungau, S.B.; Cioca, G.; Samuel, A.D.; Badea, M.; Gaceu, L. Researches on the chemical composition and the rheological properties of wheat and grape epicarp flour mixes. *Rev. Chim.* **2018**, *69*, 70–75. [CrossRef]
5. Badea, M.; di Modugno, F.; Floroian, L.; Tit, D.M.; Restani, P.; Bungau, S.; Iovan, C.; Badea, G.E.; Aleya, L. Electrochemical strategies for gallic acid detection: Potential for application in clinical, food or environmental analyses. *Sci. Total Environ.* **2019**, *672*, 129–140. [CrossRef]
6. Shinagawa, F.B.; de Santana, F.C.; Torres, L.R.O.; Mancini-Filho, J. Grape seed oil: A potential functional food? *Food Sci. Technol.* **2015**, *35*, 399–406. [CrossRef]
7. Lucarini, M.; Durazzo, A.; Romani, A.; Campo, M.; Lombardi-Boccia, G.; Cecchini, F. Bio-Based Compounds from Grape Seeds: A Biorefinery Approach. *Molecules* **2018**, *23*, 1888. [CrossRef]
8. Argon, Z.U.; Celenk, V.U.; Gumus, Z.P. Cold pressed grape (*Vitis vinifera*) seed oil. In *Cold Pressed Oils: Green Technology, Bioactive Compounds, Functionality, and Applications*; Academic Press: New York, NY, USA, 2020; pp. 39–52.
9. Abdel-Daim, M.M.; Abo-EL-Sooud, K.; Aleya, L.; Bungau, S.G.; Najda, A.; Saluja, R. Alleviation of Drugs and Chemicals Toxicity: Biomedical Value of Antioxidants. *Oxid. Med. Cell. Longev.* **2018**, *2018*, 6276438. [CrossRef]
10. Hornedo-Ortega, R.; González-Centeno, M.R.; Chira, K.; Jourdes, M.; Teissedre, P.-L.; Hornedo-Ortega, R.; González-Centeno, M.R.; Chira, K.; Jourdes, M.; Teissedre, P.-L. Phenolic Compounds of Grapes and Wines: Key Compounds and Implications in Sensory Perception. In *Chemistry and Biochemistry of Winemaking, Wine Stabilization and Aging*; IntechOpen: London, UK, 2020; pp. 1–26.
11. Gouot, J.C.; Smith, J.P.; Holzapfel, B.P.; Walker, A.R.; Barril, C. Grape berry flavonoids: A review of their biochemical responses to high and extreme high temperatures. *J. Exp. Bot.* **2019**, *70*, 397–423. [CrossRef]
12. Kapcsándi, V.; Hanczné Lakatos, E.; Sik, B.; Linka, L.Á.; Székelyhidi, R. Characterization of fatty acid, antioxidant, and polyphenol content of grape seed oil from different *Vitis vinifera* L. varieties. *OCL* **2021**, *28*, 1–6. [CrossRef]
13. Galili, S.; Hovav, R. Determination of Polyphenols, Flavonoids, and Antioxidant Capacity in Dry Seeds. In *Polyphenols in Plants: Isolation, Purification and Extract Preparation*; Academic Press: New York, NY, USA, 2014; pp. 305–323.
14. Folin, O.; Ciocalteu, V. On Tyrosine and Tryptophane Determinations in Proteins. *J. Biol. Chem.* **1927**, *73*, 627–650. [CrossRef]
15. Singleton, V.L.; Orthofer, R.; Lamuela-Raventós, R.M. Analysis of total phenols and other oxidation substrates and antioxidants by means of Folin-Ciocalteu reagent. *Methods Enzymol.* **1999**, *299*, 152–178.

16. Pallag, A.; Bungau, S.G.; Tit, D.M.; Jurca, T.; Sirbu, V.; Honiges, A.; Horhogea, C. Comparative Study of Polyphenols, Flavonoids and Chlorophylls in *Equisetum arvense* L. Populations. *Rev. Chim.* **2016**, *67*, 530–533.
17. Lamuela-Raventós, R.M. Folin–Ciocalteu method for the measurement of total phenolic content and antioxidant capacity. In *Measurement of Antioxidant Activity and Capacity: Recent Trends and Applications*; John Wiley & Sons, Ltd.: Hoboken, NJ, USA, 2017; pp. 107–115.
18. Duba, K.; Fiori, L. Supercritical CO_2 extraction of grape seeds oil: Scale-up and economic analysis. *Int. J. Food Sci. Technol.* **2019**, *54*, 1306–1312. [CrossRef]
19. Krishnaswamy, K.; Orsat, V.; Gariépy, Y.; Thangavel, K. Optimization of Microwave-Assisted Extraction of Phenolic Antioxidants from Grape Seeds (*Vitis vinifera*). *Food Bioprocess Technol.* **2013**, *6*, 441–455. [CrossRef]
20. Makarova, N.V.; Valiulina, D.F.; Eremeeva, N.B. Comparative studies of extraction methods of biologically-active substances with antioxidant properties from grape seed (*Vitis vinifera* L.). *Proc. Univ. Appl. Chem. Biotechnol.* **2020**, *10*, 140–148. [CrossRef]
21. Cristea, O.; Catana, L.; Bracacescu, C.; Baltatu, C.V.; Vladutoiu, L.; Grigore, I.; Ungureanu, N. Processing methods of wine by-products to obtain grape seed oil and grape seed flour as functional ingredients for food fortification. In Proceedings of the Actual Tasks on Agricultural Engineering, Zagreb, Croatia, 2–4 March 2021; pp. 445–453.
22. Aziz, R.; Popescu, M.L.; Mihele, D. Contributions to the Pharmacognostical Study on Grapes Hamburg Cultivar. *Farmacia* **2008**, *56*, 571–576.
23. Dimić, I.; Teslić, N.; Putnik, P.; Kovačević, D.B.; Zeković, Z.; Šojić, B.; Mrkonjić, Ž.; Čolović, D.; Montesano, D.; Pavlić, B. Innovative and Conventional Valorizations of Grape Seeds from Winery By-Products as Sustainable Source of Lipophilic Antioxidants. *Antioxidants* **2020**, *9*, 568. [CrossRef]
24. Gitea, M.A.; Bungau, S.G.; Gitea, D.; Pasca, B.M.; Purza, A.L.; Radu, A.-F. Evaluation of the Phytochemistry-Therapeutic Activity Relationship for Grape Seeds Oil. *Life* **2023**, *13*, 178. [CrossRef]
25. Nowshehri, J.A.; Bhat, Z.A.; Shah, M.Y. Pharmacognostic Standardisation and Phytochemical Evaluation on the Seeds of Two *Vitis Vinefera* L. Varieties Grown in Kashmir Valley, India. *Pharmacogn. J.* **2016**, *8*, 465–470. [CrossRef]
26. Özcan, M.M.; Al Juhaimi, F.; Gülcü, M.; Uslu, N. Geçgel Determination of Bioactive Compounds and Mineral Contents of Seedless Parts and Seeds of Grapes. *S. Afr. J. Enol. Vitic.* **2017**, *38*, 212–220.
27. Tangolar, S.G.; Özoğul, Y.; Tangolar, S.; Torun, A. Evaluation of fatty acid profiles and mineral content of grape seed oil of some grape genotypes. *Int. J. Food Sci. Nutr.* **2009**, *60*, 32–39. [CrossRef] [PubMed]
28. Al Juhaimi, F.; Geçgel; Gülcü, M.; Hamurcu, M.; Özcan, M.M. Bioactive Properties, Fatty Acid Composition and Mineral Contents of Grape Seed and Oils. *S. Afr. J. Enol. Vitic.* **2017**, *38*, 103–108. [CrossRef]
29. Garavaglia, J.; Markoski, M.M.; Oliveira, A.; Marcadenti, A. Grape Seed Oil Compounds: Biological and Chemical Actions for Health. *Nutr. Metab. Insights* **2016**, *9*, 59–64. [CrossRef] [PubMed]
30. Rombaut, N.; Savoire, R.; Thomasset, B.; Castello, J.; Van Hecke, E.; Lanoisellé, J.L. Optimization of oil yield and oil total phenolic content during grape seed cold screw pressing. *Ind. Crops. Prod.* **2015**, *63*, 26–33. [CrossRef]
31. Li, H.; Fu, X.; Deng, G.; David, A.; Huang, L. Extraction of oil from grape seeds (*Vitis vinifera* L.) using recyclable CO_2-expanded ethanol. *Chem. Eng. Process.-Process Intensif.* **2020**, *157*, 108147. [CrossRef]
32. Nehdi, I.A. Cupressus sempervirens var. horizentalis seed oil: Chemical composition, physicochemical characteristics, and utilizations. *Ind. Crops Prod.* **2013**, *41*, 381–385. [CrossRef]
33. Konuskan, D.B.; Kamiloglu, O.; Demirkeser, O. Fatty Acid Composition, Total Phenolic Content and Antioxidant Activity of Grape Seed Oils Obtained by Cold- Pressed and Solvent Extraction. *Indian J. Pharm. Educ. Res.* **2019**, *53*, 144–150. [CrossRef]
34. Hashempour-Baltork, F.; Torbati, M.; Azadmard-Damirchi, S.; Savage, G.P. Chemical, Rheological and Nutritional Characteristics of Sesame and Olive Oils Blended with Linseed Oil. *Adv. Pharm. Bull.* **2018**, *8*, 107–113. [CrossRef]
35. Pinto, R.H.H.; Menezes, E.G.O.; Freitas, L.C.; Andrade, E.H.d.A.; Ribeiro-Costa, R.M.; Silva, J.O.C., Jr.; Carvalho, R.N., Jr. Supercritical CO_2 extraction of uxi (Endopleura uchi) oil: Global yield isotherms, fatty acid profile, functional quality and thermal stability. *J. Supercrit. Fluids* **2020**, *165*, 104932. [CrossRef]
36. De Souza, R.d.C.; Souza Machado, B.A.; de Abreu Barreto, G.; Leal, I.L.; dos Anjos, J.P.; Umsza-Guez, M.A. Effect of Experimental Parameters on the Extraction of Grape Seed Oil Obtained by Low Pressure and Supercritical Fluid Extraction. *Molecules* **2020**, *25*, 1634. [CrossRef]
37. Mollica, A.; Scioli, G.; Valle, A.D.; Cichelli, A.; Novellino, E.; Bauer, M.; Kamysz, W.; Llorent-Martínez, E.J.; De Córdova, M.L.F.; Castillo-López, R.; et al. Phenolic Analysis and In Vitro Biological Activity of Red Wine, Pomace and Grape Seeds Oil Derived from *Vitis Vinifera* L. cv. Montepulciano d'Abruzzo. *Antioxidants* **2021**, *10*, 1704. [CrossRef]
38. Herrero, M.; Castro-Puyana, M.; Mendiola, J.A.; Ibañez, E. Compressed fluids for the extraction of bioactive compounds. *TrAC Trends Anal. Chem.* **2013**, *43*, 67–83. [CrossRef]
39. Google Map with the Grapevine Plantation. Available online: https://www.google.com/maps/place/47°16$'$10.6%22N+22°08$'$04.2%22E/@47.26962,22.1327663,578m/data=!3m2!1e3!4b1!4m13!1m8!3m7!1s0x4747ca07a65c2231:0x4fcaa48c15f119dc!2sSîniob+417192!3b1!8m2!3d47.2638987!4d22.128266!16s%2Fg%2F1q6j717xq!3m3!8m2!3d47.26962!4d22.1 (accessed on 8 April 2023).
40. Dabetic, N.M.; Todorovic, V.M.; Djuricic, I.D.; Stankovic, J.A.A.; Basic, Z.N.; Vujovic, D.S.; Sobajic, S.S. Grape Seed Oil Characterization: A Novel Approach for Oil Quality Assessment. *Eur. J. Lipid Sci. Technol.* **2020**, *122*, 1900447. [CrossRef]

41. Pharmacognostic determinations, Macroscopic and microscopic control of plant products. In *Romanian Pharmacopoeia*; Bucharest Medical Publishing House: Bucharest, Romania, 1993; pp. 1051–1053.
42. Hong, N.; Yaylayan, V.A.; Vijaya Raghavan, G.S.; Paré, J.R.J.; Bélanger, J.M.R. Microwave-assisted Extraction of Phenolic Compounds from Grape Seed. *Nat. Prod. Lett.* **2001**, *15*, 197–204. [CrossRef]
43. Bandici, L.; Teusdea, A.C.; Soproni, V.D.; Hathazi, F.I.; Arion, M.N.; Molnar, C.O.; Vicas, S.I. The Influence of Microwave Treatments on Bioactive Compounds and Antioxidant Capacity of *Mentha piperita* L. *Materials* **2022**, *15*, 7789. [CrossRef]
44. Dulf, F.V.; Vodnar, D.C.; Toşa, M.I.; Dulf, E.H. Simultaneous enrichment of grape pomace with γ-linolenic acid and carotenoids by solid-state fermentation with Zygomycetes fungi and antioxidant potential of the bioprocessed substrates. *Food Chem.* **2020**, *310*, 125927. [CrossRef]
45. Santos-Silva, J.; Bessa, R.J.B.; Santos-Silva, F. Effect of genotype, feeding system and slaughter weight on the quality of light lambs: II. Fatty acid composition of meat. *Livest. Prod. Sci.* **2002**, *77*, 187–194. [CrossRef]
46. Cunha, V.M.B.; da Silva, M.P.; de Sousa, S.H.B.; Bezerra, P.D.N.; Menezes, E.G.O.; da Silva, N.J.N.; Banna, D.A.D.d.S.; Araújo, M.E.; de Carvalho, R.N., Jr. Bacaba-de-leque (Oenocarpus distichus Mart.) oil extraction using supercritical CO_2 and bioactive compounds determination in the residual pulp. *J. Supercrit. Fluids* **2019**, *144*, 81–90. [CrossRef]
47. Ulbricht, T.L.V.; Southgate, D.A.T. Coronary heart disease: Seven dietary factors. *Lancet* **1991**, *338*, 985–992. [CrossRef]
48. Eroglu, E.; Girgin, S.N. A unique phenolic extraction method from olive oil macerate of *Hypericum perforatum* using DMSO: Assessment of in vitro anticancer activity, LC-MS/MS profile, total phenolic content and antioxidant capacity. *S. Afr. J. Bot.* **2021**, *139*, 6–11. [CrossRef]
49. Schlesier, K.; Harwat, M.; Böhm, V.; Bitsch, R. Assessment of antioxidant activity by using different in vitro methods. *Free Radic. Res.* **2002**, *36*, 177–187. [CrossRef] [PubMed]
50. Fratianni, F.; D'Acierno, A.; Ombra, M.N.; Amato, G.; De Feo, V.; Ayala-Zavala, J.F.; Coppola, R.; Nazzaro, F. Fatty Acid Composition, Antioxidant, and in vitro Anti-inflammatory Activity of Five Cold-Pressed *Prunus* Seed Oils, and Their Anti-biofilm Effect Against Pathogenic Bacteria. *Front. Nutr.* **2021**, *8*, 775751. [CrossRef] [PubMed]

sustainability

MDPI

Article

Solutions for an Ecological and Healthy Retrofitting of Buildings on the Campus of the University of Oradea, Romania, Built Starting from 1911 to 1913

Constantin C. Bungau [1,2], Constantin Bungau [3,4,*], Mihaela Teodora Toadere [5], Ioana Francesca Prada-Hanga [1,2], Tudor Bungau [6,*], Daniela Elena Popescu [7] and Marcela Florina Prada [5]

1 Department of Architecture, Faculty of Constructions, Cadaster and Architecture, University of Oradea, 410058 Oradea, Romania
2 Doctoral School, Technical University of Cluj-Napoca, 400114 Cluj-Napoca, Romania
3 Doctoral School of Engineering Sciences, University of Oradea, 410087 Oradea, Romania
4 Department of Engineering and Management, University of Oradea, 410087 Oradea, Romania
5 Department of Civil Engineering, Faculty of Constructions, Cadaster and Architecture, University of Oradea, 410058 Oradea, Romania
6 Civil, Industrial and Agricultural Constructions Program of Study, Faculty of Constructions, Cadastre and Architecture, University of Oradea, 410058 Oradea, Romania
7 Department of Computers and Information Technology, Faculty of Electrical Engineering and Information Technology, 410087 Oradea, Romania
* Correspondence: bungau@uoradea.ro (C.B.); tudor.bungau@gmail.com (T.B.)

Abstract: On university campuses, retrofitting studies have historically concentrated on individual buildings (or building components) instead of the entire campus. In the present paper, we examine how an incorporated strategic planning strategy might be used to investigate the socio-technical construction of a campus retrofit operation throughout multiple scale/sectors. The campus of the University of Oradea (CUO), Romania, with its beginnings in the 1910s was investigated using its new master plan. The developed strategies for a "green" and "healthy" campus depict a CUO redesign involving complex solutions for the green renovation of old buildings. In addition, the improvement effects of the modernization interventions already carried out were analyzed and quantified. Sixteen buildings (30% of the built area) were consolidated/rehabilitated/modernized, and/or equipped in the last decade, seven educational spaces being included in the circuit of the buildings fund (totaling 5491.59 sq m). For the renovated spaces, energy consumption was reduced by 20–88% and CO_2 emissions by 41.82–86%, depending on the specifics of each space. The reconfiguration, rehabilitation, and energy efficiency of the entire heating system of CUO (which uses geothermal water as a specific characteristic) significantly improved (20% decrease in energy and 21% decrease in geothermal water consumption). Our findings offer new directions and design solutions for the ecological modernization of other outdated university campuses, highlighting new perspectives in the green university campuses' management, as a way to implement sustainability in the higher education environment. Data presented give professionals in the field (architects, designers, engineers, planners, and decision makers) a clear picture of the benefits due to ecological renovation, also offering the necessary tools to implement new solutions for reducing the impact of urban areas on the environment.

Keywords: ecological renovation; green retrofitting; university campus; green building; sustainable development; renewable energy; thermal water; University of Oradea

Citation: Bungau, C.C.; Bungau, C.; Toadere, M.T.; Prada-Hanga, I.F.; Bungau, T.; Popescu, D.E.; Prada, M.F. Solutions for an Ecological and Healthy Retrofitting of Buildings on the Campus of the University of Oradea, Romania, Built Starting from 1911 to 1913. *Sustainability* **2023**, *15*, 6541. https://doi.org/10.3390/su15086541

Academic Editor: Firoz Alam

Received: 22 February 2023
Revised: 3 April 2023
Accepted: 10 April 2023
Published: 12 April 2023

1. Introduction

Rehabilitation of existing buildings, taking into account compliance with ecological construction requirements, leads to improvement of the environmental attributes of the buildings [1]. Therefore, this sector has great potential to reduce CO_2 emissions since traditional buildings increase environmental pollution by generating a large amount of waste during their life cycle [2,3]. Globally, the construction industry consumes 40% of total energy production, 12–16% of all water available, 32% of nonrenewable and renewable resources, 25% of all timber, 40% of all raw materials, produces 30–40% of all solid wastes, and emits 35–40% of CO_2, which has resulted in a rising global awareness of the importance of sustainability in the construction industry [4]. Therefore, the reduction of emissions in the field of construction is necessary to mitigate climate changes, which are already extremely noticeable [5–7].

The past decade has witnessed an increased interest in sustainable, healthy, or green building (GB) concepts and practices across the globe [8]. Generally, sustainable building design considers the impact of buildings on their habitat from multifaceted dimensions, including ecology, economy, and society. It involves formal as well as informal initiatives expanded by private industry, professional organizations, and governments. Such endeavors have resulted in the development of improved energy codes, low environmental impact materials, design guides, renewable energy and resources, and the concept of analyzing consequences of design choices over the building's entire life cycle [9,10].

Academic institutions dedicated to education, research, and community services, can have a particularly visible and important role in sustainable development, as well as in the management of measures taken to mitigate/stop climate change [11,12]. Moreover, universities have a fundamental role in helping society understand and get involved in the current environmental conditions and climate change. The challenges mentioned by the international frameworks (i.e., Green Deal [13], the framework for achieving climate neutrality [14]) are focused in the short/medium term on achieving climate neutrality. In the current period, numerous higher education institutions have clearly established objectives that are based on sustainability, emphasizing the role of universities on the path of transforming the environment in which people live and carry out their activities towards sustainability. Most of these initiatives are carried out through operational activities, to raise awareness among interested parties, but also through educational and research activities (e.g., greening/cleaning of campuses; curriculum redesign; strengthening of networks built in local/regional/continental/international areas, etc.) to positively influence human behavior and attitude towards the environment [15]. Mutual cooperation between different stakeholders can contribute to the improvement of sustainable development activities [16].

The green campus can thus be seen as a community with the unitary goal of increasing the efficiency of energy use, supporting, and saving natural resources, improving the quality of the environment, etc. It is obvious that these objectives can only be achieved through an education focused on sustainable development, as well as implementing projects based on a healthy, ecological lifestyle, and with care for everything that surrounds us [17].

Considering the above, the ecological modernization of old buildings must be seen as an effective solution, with a practical approach, in view of sustainable urban development. The retrofitting of old buildings is different from the modern design given to new ecological buildings, emphasizing the location of the site within a settlement (urban or rural), adaptability, functionality, technology, and taking into account the existing climate [18].

Due to its complex heritage, Romania possesses many remnants of its cultural history. The architectural heritage of the University of Oradea is a significant component of the local cultural heritage, though a part of it still needs to be acknowledged, restored, and preserved [19,20].

Sustainable retrofitting and green restoration of old buildings on university campuses began to receive considerable interest because of their promise to enhance energy performance and decrease emissions of greenhouse gases, but extensive research is required due to the limited number of publications on this subject [21,22].

Researching the scientific literature to identify publication trends and research gaps is essential in this promising field for a future based on sustainable development. In this regard, the relevance of the topic chosen for the study was assessed by applying search algorithms in three large and scientifically validated databases (i.e., SpringerLink,

ScienceDirect, and Web of Science). The results displayed in each database (Figure 1) underline the low number of publications that have addressed the concepts of green retrofitting, ecological renovation, and geothermal water as a heating agent; moreover, when using the Boolean operator AND, which filters and associates these concepts with university campuses, the results obtained are even more drastically reduced, reaching only one result displayed in a Web of Science search. Moreover, the whole working context applied to the situation in Romania shows a major deficit of publications evaluating university campuses in these integral parts of sustainable development, thus identifying certain knowledge gaps. In Romania, there is a significant gap in knowledge regarding the need for green retrofitting studies of old buildings on university campuses. The only resulting paper identified in the Web of Science for the application of geothermal water as a heating agent in university campuses in Romania belongs to a part of the authors' collective [23], the use of geothermal water in renewable energy is an essential and rare aspect for a university campus in general (no paper in the world provided studies related to retrofitting buildings in a university campus, built in different years during more than a century, some of them changing their initial function, and heated with geothermal water), and even more so for a "green" one.

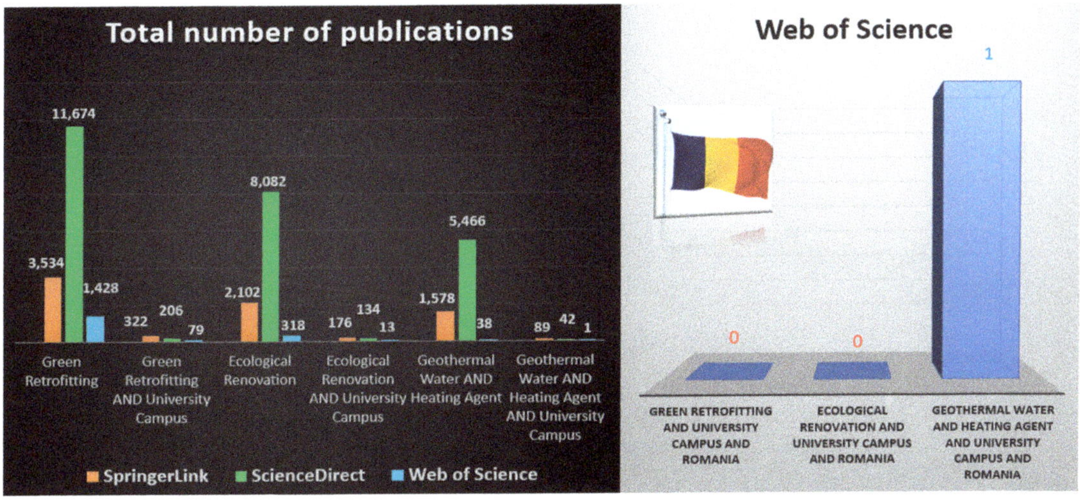

Figure 1. Linking the very low number of publications evaluating the green renovation/retrofitting of university campuses, and especially in the case of the deficit registered in Romania, with the identification of the research gap and the need for new research.

The idea of this study came naturally from the current context detailed above and from the daily needs of the campus of the University of Oradea (CUO). The main purpose of the research was to analyze the works in progress dedicated to the modernization of this 150-year-old campus, a campus developed in different stages and periods. Another goal was to summarize coherent and effective principles for approaching the conversion of an existing urban complex, with a certain functionality (such as a university campus) into a "green" built complex ("green" campus), in which daily activities are carried out in a healthy way. The authors aim to demonstrate that this is achieved at the micro level (for example through interventions in each building) as well as at the "macro" level, through a global and holistic approach, on various levels, through integrated management.

The present study proposes a distinct approach to address some of the research gaps identified in the literature search process by applying sustainable development principles to the design of a green university campus and highlighting the advantages offered in a distinct way, since as Figure 1 shows, it is the first of its kind in Romania (meeting all the key points we have approached in this research—e.g., green retrofitting of numerous buildings with multiple uses, ecological renovation of very old buildings whose prior utility has not been correlated with the university, as well as the application of renewable energy principles through the implementation of geothermal water as a heating agent, with application to an entire university campus). Hence, the suggested approach can serve as a useful guide for finding viable retrofitting solutions for both older and more contemporary university campuses, as well as other campus types that match this perspective.

2. Materials and Methods

2.1. Presentation of the University Campus

The CUO, located in northwest Romania, latitude: 47°02′26.40″ N, longitude: 21°55′7.19″ E [24] (Figure 2), is very close (17.1 km) to the border between Romania and Hungary, is almost 150 years old, was originally intended as a Gendarmerie School [25], and has developed and expanded over time.

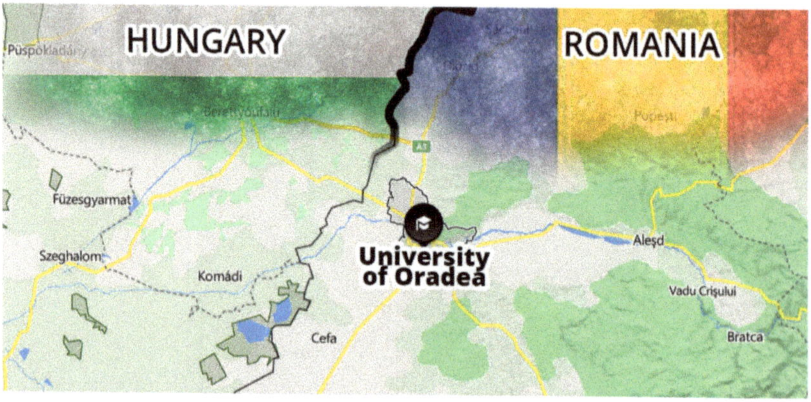

Figure 2. Satellite image of the location of University of Oradea, Romania.

The current campus is a complex urban architectural composition, which includes the historic campus and later buildings. The historic campus is composed of a unitary urban architectural ensemble, Category A—historical monument (architectural monument of national interest from a value point of view) [26], the reference work of avant-garde secession by the architect Vágó Jószef. Around it, in different stages, the Pedagogical Institute campus was developed (between the 1960s and 1970s), and later the buildings of today's UO were built (in the years 2000–2010).

In the last decade, the management of the UO has focused on renovation, rehabilitation, modernization, and retrofitting of existing buildings. All these activities are part of the extensive development actions of a green campus, remaining at the same time focused on the legislation/directives in force [27].

The study was carried out on the site of the Central CUO, which occupies a land area of 222,307.00 square meters (sq m). A satellite Google Maps image [28] of the site is presented in the Figure 3. The existing buildings on this site have a constructed area of 33,456.74 sq m, with various height regimes (one to five levels) and a developed area of 90,002.40 sq m. The buildings were constructed with different concepts, having different structures, depending on the stage in which they were built; consequently, for exploitation, they have different consumptions.

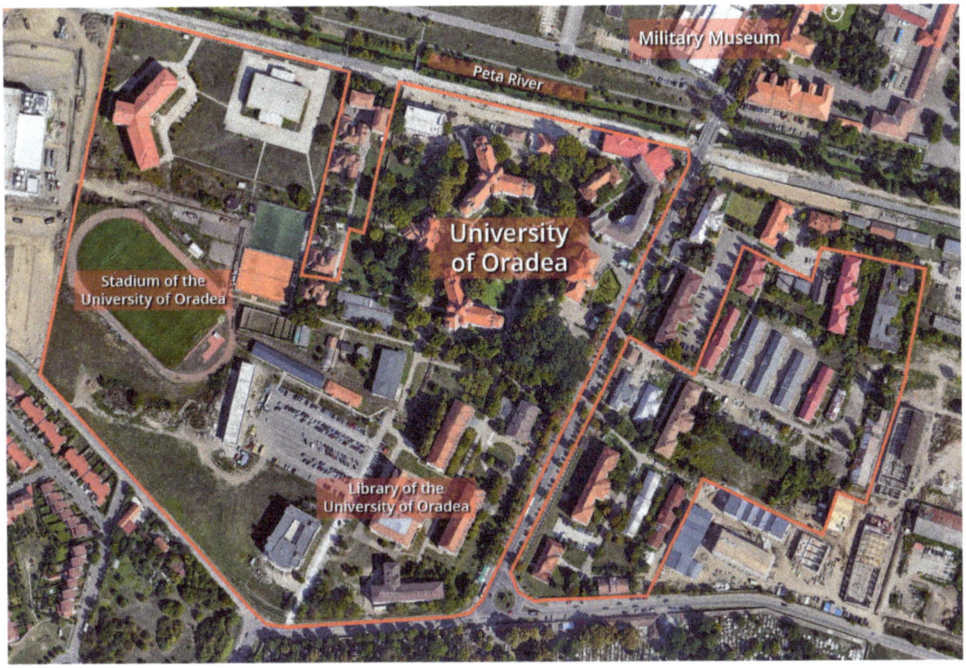

Figure 3. The site of the Central Campus of the University of Oradea.

The interventions on the infrastructure of the university campus, highlighted mainly in the present paper, refer to the retrofitting of some buildings with an age of 100 years, but also to the thermal and/or energetic rehabilitation of other buildings, all of which represent a quota of 30% of the area of existing buildings. This specific building was part of the "Smart Campus University of Oradea" project, financed both from European funds and from its own revenues.

2.2. Data Analysis

The method of the study consisted of several stages following the master plan of the campus. To begin with, data analysis was conducted by focusing on the existing documents in the state archives, university documents, European, and international directives in the context of the study [23]. This stage involved the identification and description of the strategies and solutions for a "green" and "healthy" campus, created and implemented by the UO management in various stages of the reconversion or rehabilitation of the buildings under study. Next, the results obtained from the retrofitting of very old buildings and the thermal and/or energy rehabilitation of other existing buildings were analyzed. To this end, existing documentation at the General Administrative Directorate/Sustainable Campus Directorate of the UO was consulted and studied, as follows: the "Smart Campus—University of Oradea—stage I" project, from whose Energy Audit [29–31] (including the Energy Performance Certificates from both the beginning and from the reception of the rehabilitation works of the E-F body were considered [32]), interpreting their results and observing the differences in the parameters. Later, the indicators provided by these certificates were used to calculate the impact of the interventions to be modernized on energy consumption and CO_2 emissions, comparing the values of initial parameters with those resulting after the rehabilitation. Thus, the results of energy efficiency were studied on a developed built surface of 7915.29 sq m.

From the "Smart campus—University of Oradea—stage II" project, information was taken from the historical study. The campus network rehabilitation project and how to implement it was studied. Data on consumption monitoring from existing reports within the University's Investments and Consolidations Office were analyzed, processed, and interpreted. Based on the records of consumption before and after the thermal rehabilitation, the impact of the thermal rehabilitation on the thermal energy consumption was calculated.

To provide solutions for the construction/rehabilitation of a building to become a GB, as well as for the solutions for the development of a green campus, the authors of the paper corroborated the information of the indicated references with those of the energy efficiency projects implemented in the UO.

3. Results

3.1. Strategies and Solutions for a "Green" and "Healthy" Campus—The Case of the University of Oradea

Starting from 2013, the management of the UO, at all its levels, was oriented towards creating a sustainable education by transforming the existing campus into a "green" and "healthy" one. At this stage, the investments were dedicated especially to the thermal efficiency of the buildings. In the following years, a new stage, with a better determined trajectory towards a "green" and "healthy" campus (with more complex works related to its infrastructure) took shape; therefore, through a holistic approach, Oradea City Hall, in collaboration with the UO's management team, developed in 2018 a sustainable development strategy for the university, respectively a regional urban plan called "Masterplan" [33].

The project was developed by an architect with European experience in the design of university campuses, and it states that, "The concept starts from the idea of a green spine with the role of a connector along the entire length of the campus. This spine unites the historical part of the campus with the proposed new areas, giving the whole complex a unitary aspect from a compositional point of view. The proposed new functions relate to the existing ones through an integrated system of public and green spaces". A figure from this Masterplan presents the approach to the project (Figure 4) [33].

Figure 4. Masterplan—Campus of the University of Oradea. Reprinted/adapted with permission from Ref. [33]. 2023, Năstase, R.P.

From the point of view of the infrastructure (buildings and their related utilities) of the university campus, the management team's actions focused on:

- Efficient use of the existing buildings fund and their "recycling" through:

- Structural rehabilitation;
- Introduction into the circuit of educational spaces of buildings abandoned over time and unused with their reconversion;
- Application of GB principles to all interventions on the existing buildings, through the actions described in Figure 5 [34].

- Imposing the "Green", "Smart," and nearly zero energy building "NZEB" criteria on new constructions proposed during the design, execution, and operation period.
- Efficient use of renewable energies: existing geothermal water on the CUO site, geothermal energy (use of ground-water heat pumps), outdoor air energy (air-to-air heat pumps), solar energy (photovoltaic panels).
- Instituting an integrated system to ensure heating of the buildings of the entire campus. This is possible through efficient use of the existing renewable energy source on site: geothermal water as the primary agent, residual geothermal water with the equipment of thermal plants with heat exchangers and heat pumps [8].

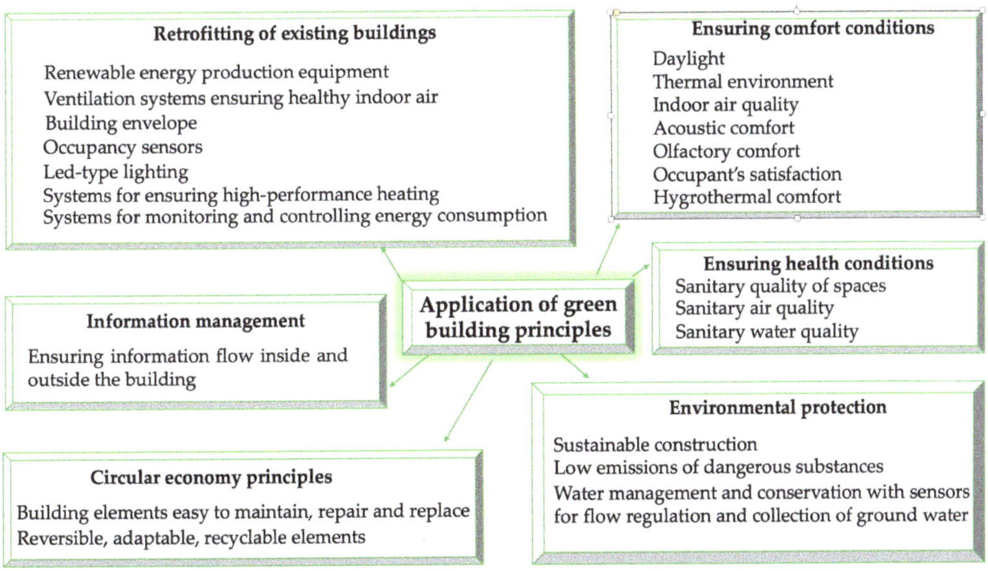

Figure 5. Application of GB principles.

At the same time, in the context of the circular economy and protection of the environment, re-injection of residual geothermal water into the custom drilled locations was ensured. Consequently, two thermal power plants were equipped with heat exchangers for use of geothermal water as the primary agent and high-performance heat pumps equipped with monitoring and control systems for the use of residual geothermal water. The principles are illustrated in Figure 6.

Another series of important measures for development in a sustainable context are:

- Metering of buildings (thermal energy, cold water, hot water, electricity).
- Rehabilitation of external heating networks and sanitary networks (cold water, hot water, and wastewater).
- Putting into operation a system of rainwater recovery to be used for sprinkling green spaces and outdoor sports fields.
- Implementation of an integrated and efficient IT system, considering the explosive growth of information, the emphasis on digitization, which involved numerous resources (facilities, hardware, infrastructure, energy, personnel). UO data centers have significant annual consumption, but there are studies showing that energy savings

can be achieved (about 46%) using integrated concepts, residual heat, and optimized operating temperatures [33].

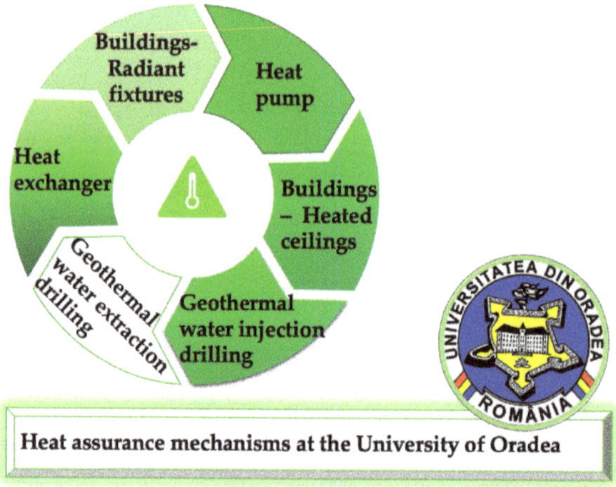

Figure 6. Principle scheme for ensuring heat for the University of Oradea Campus.

Data and communication are essential to the conduct of UO activities [35]. Hence, for efficient operation at the university level, an intranet and extranet communication solution is provided based on a wide range of latest generation technologies that ensure safe, high-speed access. The IT services offered in the context of UO (Figure 7) conditions offer both high quality and appropriate information/data security [36]. The solutions implemented by the IT department offer both a high level of security and confidentiality of information/data, constituting a true spine for the development of a real smart campus [34].

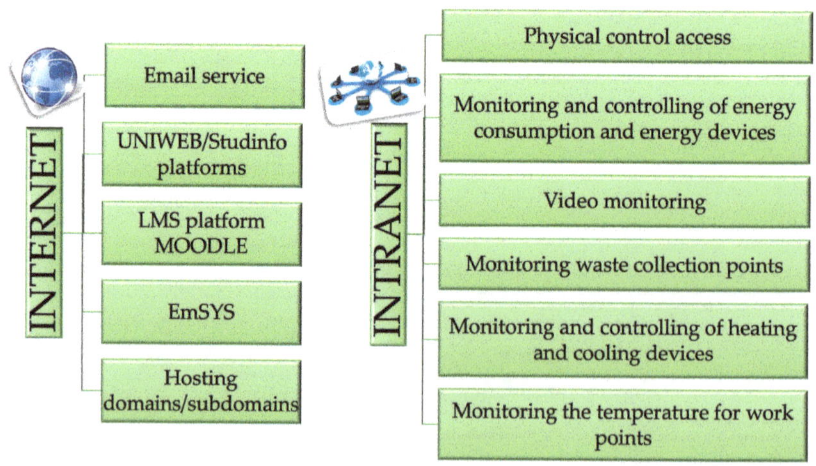

Figure 7. IT services for UO.

From the point of view of applying the principles of a circular economy, waste management is implemented (organic waste treatment, inorganic waste treatment, toxic waste treatment, and sewage disposal). The waste management programs within the UO are designed to reduce pollution, protect and be friendly to the environment, as well as to encourage recycling by the whole community. Current and future waste management policies at the level of any institution are intertwined with the waste management strategy of the locality of the institution. Considering both the constant decrease in the cost of new technologies and the numerous wireless software/technologies available to facilitate the intelligent management of waste, UO is considering substantial improvement of the current management system of all types of waste [37], since in the CUO, 15 faculties carry out their activity, and some of them (Faculty of Medicine and Pharmacy, Faculty of Sciences (including Chemistry, Biology), Engineering, etc.) produce specific waste (tissue samples, medical and pharmaceutical [38,39], chemical, electronical, etc.). This vision is also included in Figure 7. At the same time, the UO's sustainable development strategy includes a program to reduce the use of paper and plastic on campus.

When rehabilitating existing buildings and constructing new buildings, as well as during the phase of ensuring mobility within the campus, facilities for people with disabilities, special needs and/or maternity care are considered.

From the point of view of the distribution of urban green space, through the Masterplan (Figure 4) a "green rail" is proposed, which increases the existing generous green area, by abolishing morally and physically worn buildings. Their rehabilitation would be unsustainable. The considerable number of trees on campus contribute significantly to the air quality, comfort, and health of campus life. In the CUO, there are protected plant species, which are given special attention. A dedicated service ensures the maintenance of the university's green spaces.

In terms of ensuring mobility within the campus, special attention has been paid to the arrangement of pedestrian walkways, so that vehicles have access only to the parking lots arranged in the marginal areas. Bicycle parking racks are distributed around the campus and an electric scooter station is interconnected with the city network. The municipality extended the tram line from the city to the campus, thus ensuring easy and non-polluting access to the CUO. Through the same Masterplan (Figure 4) it is proposed to close the street that currently divides the campus into two areas, by transforming it into a pedestrian street.

Regarding study programs dedicated to sustainable and intelligent development, the following are offered at the UO:

- Degree program: "Engineering of renewable energy systems";
- Master's programs: "Smart and sustainable constructions", "Environmental management, evaluation, conservation, and protection", "Renewable energies", "Energy systems management", "Biodiversity and ecosystem monitoring", etc.

In other study programs, disciplines applied to sustainable development and climate change, regarding the circular economy, are supported.

The teaching philosophy encourages friendly explanation and understanding, as well as tools using some of the most advanced technologies. The university is equipped with smart boards, video projectors, 3D printers, and other appropriate equipment (Figure 8).

In addition, within the university, two educational platforms ("eLearning—Moodle" and Microsoft—Teams platform) are operable through which teaching activities can be carried out under the best conditions. The UO's IT service is available to assist both teaching staff and students in their operation. At the same time, as shown in Figure 7, a platform created by UO is operable, namely Studinfo/UNIWEB platform, for management of didactic activities with students (grades, fees, etc.).

For economic and administrative management, related staff are trained to use dedicated electronic/computer applications. Special interest is currently given to digitization, which is supported by European funds for which the UO applied.

Regarding research-innovation activity, through the Oradea University Library, students and teaching staff have access to the "Anelis Plus" database, which connects them to the latest publications. This activity is stimulated by the university by awarding authors of innovations, valuable articles, students participating in various competitions, or conferences.

In the geopolitical context of today, the University of Oradea has all the tools to carry out its activity both in a pandemic situation and in other extreme/special situations. In the CUO, conferences, scientific and cultural activities, and extracurricular activities are organized in which students and teaching staff are trained, so that life on the student campus is harmonious, balanced, and attractive.

(a) **(b)**

Figure 8. (**a**) "Aula Magna"; (**b**) "Smart Mat" laboratory.

The administrative management of the university has special involvement in maintaining the existing sports fields and in developing a sports base, ensuring to the students and teaching staff the possibility of accessing them both during sports classes and during the time dedicated to recreation.

The opening of the UO to internationalization provides another corridor for students and teaching staff towards an attitude anchored in European and world current affairs.

3.2. Results following Intervention on Existing Buildings

Table 1 highlights the consolidation/rehabilitation/retrofitting, modernization/endowment activities on 16 existing buildings, during the last decade. These buildings represent 30% of the developed built surfaces existing in the CUO. Example of Energy Performance Certificates (for the original building vs. the thermally + energetically rehabilitated building) is presented in the Figure 9.

Table 1. Interventions on the infrastructure of the University of Oradea Campus (buildings) from 2012 to present, by results.

Name of Building/Function/Developed Area (sq m)/Year(s) of Construction	Type of Intervention/Year	Results
	Energy efficient	
Body C1: Building L—Geothermal laboratory—Thermal power plant/1915.83/1970	Reconversion/ 2020–2022	Geothermal laboratory for practical work and research activities, as well as a thermal plant equipped with a heat exchanger and heat pumps for the central CUO
Body C3: Building J—Department of International Relations, Council Room/846.85/1911–1913	Consolidation/ Reconversion/ 2018–2020	Initially having the function of a canteen, later used as a university library, after a period of 20 years of abandonment, this building was returned to the university circuit, having the function of offices (for the Department of International Relations) and also having a Council Room for the UO; being a historical monument building, it did not lend itself to thermal rehabilitation. Heating is accomplished with geothermal water—a renewable source; the building is included in energy performance class "C" [32]
Body C15: Building B-N/706.26/1912–1914	Retrofitting/Equipment/ 2014–2015	In recent years, the building had been abandoned and fell into disrepair; after the retrofitting (through enveloping it and energy efficiency, being equipped with a heat pump), it was put into the circuit of the research activity

Table 1. *Cont.*

Name of Building/Function/Developed Area (sq m)/Year(s) of Construction	Type of Intervention/Year	Results
	Green and towards NZEB	
Body C4: Building E-F—Educational spaces/7915.29/1993–1995	Rehabilitation/ Modernization/ Equipment/2020–2022	Analyzing the Energy Performance Certificates for the original building vs. the thermally + energetically rehabilitated building, comparing the initial energy consumption vs. that resulting from the rehabilitation (e.g., Figure 9a,b), it is found that the energy consumption of the building has decreased by 77%, an 86% decrease in CO_2 emissions as well as its transition from energy class C to energy Class A [29,40]
Body C5: Building D—Rectorate Building/1182.52/ 1993–1995	Rehabilitation/ Modernization/ 2021–2022	Analyzing the Energy Performance Certificates for the original building vs. the thermally + energetically rehabilitated building, comparing the initial energy consumption vs. that resulting from the rehabilitation, it is found that the energy consumption of the building has decreased by 88%, a 76% decrease in CO_2 emissions as well as its transition from energy class C to energy Class A [30,41]
Body C20: Building P—Gym/ 869.04/1968	Rehabilitation/ Modernization/ Equipment/2018–2019	Analyzing the results of energy audit and comparing the initial energy consumption vs. those resulting from the rehabilitation, it is found that the energy consumption of the building has decreased by 63%, a 51% decrease in CO_2 emissions [31]
Body C: Building V2— "Smart-Mat" Research laboratory/673.00/1950–1960	Consolidation/Retrofitting/ Equipped/2013–2014	Analyzing the results of energy audit and comparing the initial energy consumption vs. that resulting from the rehabilitation, it is found that the energy consumption of the building has decreased by 42.43%, a 41.82% decrease in CO_2 emissions [42]
Body C4: Building V3—Research laboratory/1413.22/ 1950–1960	Expansion/Consolidation/ Retrofitting/Equipped 2020–2022	The building, being used for >40 years as a warehouse, did not benefit from utilities; according to the Energy Performance Certificate (EPC) from the construction' reception, the current estimated energy consumption is 66.34 kWh/m^2 year, and the equivalent CO_2 emission index is 19.08 kg/m^2 year, the values is very low, and the interior comfort is high
Body C8: Building B-H/ 368.00/1912–1914	Retrofitting/ Equipment/ 2020–2022	The buildings, being used for >40 years as warehouses, do not benefit from utilities. Following the retrofitting, buildings with low consumption and high comfort were obtained. According to the EPCs of the buildings, the current estimated energy consumption is 66.55 kWh/m^2 year, and the CO_2 equivalent emission index is 16.55 kg/m^2 year [43]
Body C11: Building B-K/ 808.61/1912–1914		The current estimated energy consumption is 49.51 kWh/m^2 year, and the CO_2 equivalent emission index is 15.03 kg/m^2 year [43]
Body C12: Building B-L/760.62/1912–1914		The current estimated energy consumption is 52.83 kWh/m^2 year, and the CO_2 equivalent emission index is 14.93 kg/m^2 year [43]
Body C13: Building B-M/761.87/1912–1914		The current estimated energy consumption is 51.44 kWh/m^2 year, and the CO_2 equivalent emission index is 13.99 kg/m^2 year [43]

Table 1. *Cont.*

Name of Building/Function/Developed Area (sq m)/Year(s) of Construction	Type of Intervention/Year	Results
	Thermally efficient	
Body C25: Building C1—Student dorm/4793.32/ 1970	Thermal rehabilitation/2016–2017	Analyzing and comparing the consumptions before and after the thermal rehabilitation, a reduction in energy consumption of 20% is appreciated, according to consumption registration documents—General Administrative Directorate—UO.
Body C21: Building C2—Student dorm/4793.32/ 1967	Thermal rehabilitation/2021	
Body C26: Building T—Educational spaces, research laboratories/3007.45/ 1966–1967	Thermal rehabilitation/2016	Analyzing the results of the energy audit comparing the initial energy consumption vs. that resulting from the rehabilitation, it is found that the energy consumption of the building has decreased by 42.43%, and a 41.82% decrease in CO_2 emissions [42]
Body C3: Building X—Educational spaces, offices/814.00/ 1912–1914	Rehabilitation/ Modernization/ Equipment/2018–2022	Analyzing and comparing the consumptions before and after the thermal rehabilitation, a reduction in energy consumption of 20% is appreciated, according to consumption registration documents—General Administrative Directorate—UO.

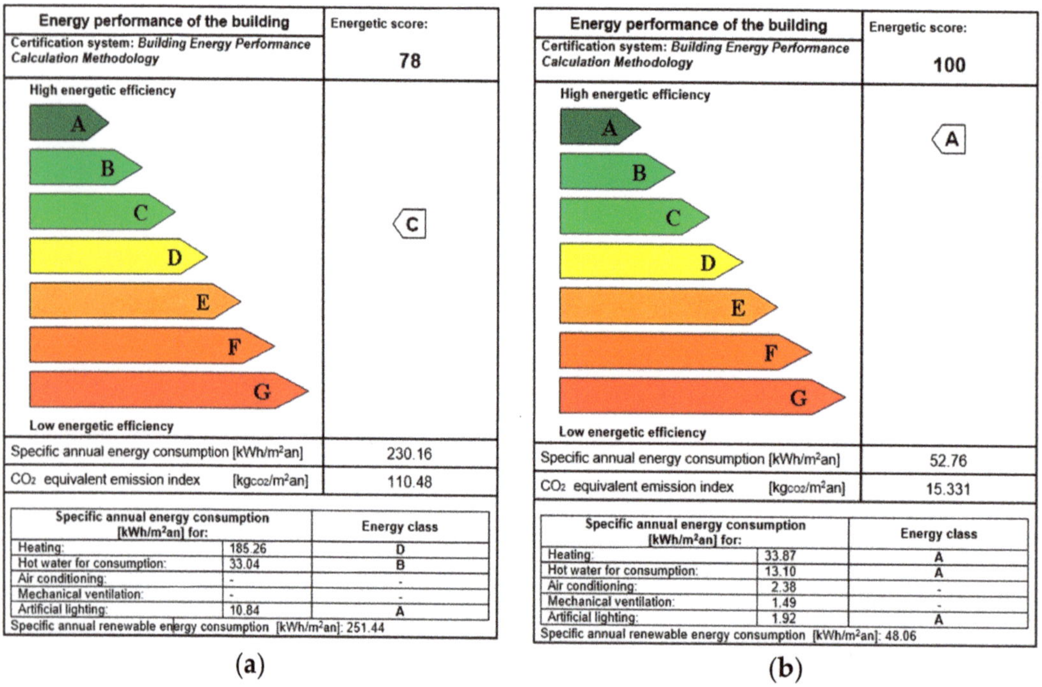

Figure 9. Energy performance certificate: (**a**) initial building; (**b**) thermally and energetically rehabilitated building.

3.3. *Results following the Retrofitting of Seven Buildings on the CUO*

In the last decade, seven educational spaces were added to the circuit of buildings fund, totaling a developed area of 5491.59 sq m. These buildings were abandoned or used as warehouses. Complying with the criteria of the circular economy, the buildings were "recycled" through "retrofitting" and were equipped. They are currently being used as spaces for education and research. Figure 10 shows two of these buildings, which were included in the Hungarian—Romanian "Smart Mat" (Figure 10(a1,b1)) and "Smart Campus—UO" (Figure 10(a2,b2)) projects, resulting in educational spaces and research laboratories (Figure 10(c1,c2)). On the roof of the "Smart Mat" building (Figure 10(b2)), the photovoltaic panels can be seen. These were arranged for use of renewable solar energy in addition to the geothermal energy.

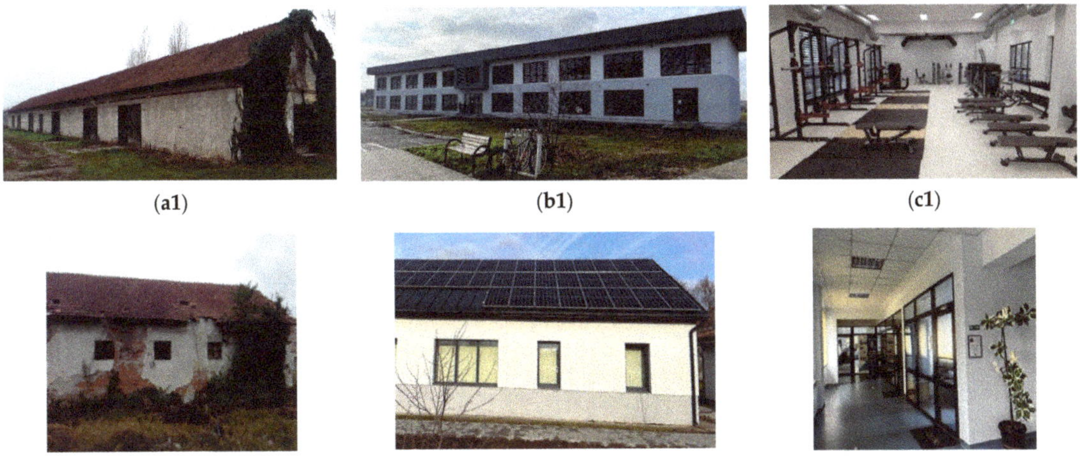

(**a1**) (**b1**) (**c1**)

(**a2**) (**b2**) (**c2**)

Figure 10. Two cases of: (**a1**,**a2**) initial buildings—deposits; (**b1**,**b2**) retrofitted buildings; (**c1**,**c2**) research laboratory inside the renovated buildings.

3.4. Results following the Intervention on the Existing Hot Water Networks and following the Reconfiguration of the Heating System

The entire heating system of CUO has been reconfigured, rehabilitated, and made more efficient, resulting in two modern thermal plants (Figure 11) equipped with high-performance equipment with monitoring and control systems, geothermal water (renewable energy) being used/reused. Respectively, after the residual geothermal water from heating a part of the buildings, the resulting geothermal water is re-injected into the injection well.

(**a**) (**b**)

Figure 11. Thermal power plants: (**a**) geothermal laboratory; (**b**) laboratory of renewable energies.

With the rehabilitation solutions described above, the reduction of geothermal water consumption m^3/m^2 year is 21%, given that only 30% of the used spaces have been rehabilitated (according to the descriptions in Table 1) and, at the same time, only 10% of areas have been introduced into the circuit of education and research spaces.

4. Discussion

Since 2013, the University of Oradea has gradually integrated sustainable development strategies and actions into teaching, research, infrastructure, and campus operations.

The CUO's management activity is focused on European decision makers and targets, with a focus on climate change policy. The actions and legislative packages issued by both the United Nations and the European Union form the framework for the proposed targets. In this legislative and geopolitical context presented, the UO's management team is concentrating efforts to provide a sustainable education on a "green", "healthy", and sustainable campus. The term "Smart Campus" has been coined to describe projects that are being developed to raise funds. Part of the funding for these projects comes from the European Union, with funds coming from projects such as the Regional Operational Program (POR) 2014–2020. Priority Axis 10. Improving the educational infrastructure, Investment Priority 10.1. Investments in education and training, including vocational training, for the acquisition of skills and lifelong learning through the development of education and training infrastructure; Specific Objective 10.3. Increasing the relevance of university tertiary education in relation to the labor market and competitive economic sectors. As part of the "Smart Campus—University of Oradea" [44] project carried out under this financing, existing buildings were "recycled" through consolidation and "retrofitting", the buildings were thermally and energetically rehabilitated, the investments committed to renovation of the buildings were from the existing construction stock (Figure 12).

Figure 12. "Smart Campus-UO", European funding, 2020–2023.

Regarding the construction of a GB or the transformation of an existing building into a sustainable building, the authors of this study propose the scheme of approach highlighted in Figure 13. The proposal starts from the concept of a GB in the context of climate change and the circular economy, taking into account the solutions and results highlighted in the present work.

The image suggests the ideal case, namely construction concept designed and arranged on multiple levels, starting with the base:

- The legal framework, as well as management at the organizational level (state, administrative territorial units).
- A favorable economic environment (including funding, construction activity creditors) will play a decisive role in the realization of a sustainable construction or complex, but only in the context of a healthy and extensive educational and research system that would provide additional support.
- The ground floor of the designed building: with an adequate design and a specialized execution, depending on the financial power of the investors, the building will be able to meet the requirements of a sustainable one.

- The body of the projected building: GB, NZEB, and smart construction that is subject to the circular economy. Within the structure of the building, elements of composition and building equipment, building functions, mode of operation, and use are arranged on different levels.
- The designers (architects and engineers) must have the capacity to select the geometric shape of the building (a compact volume structure is recommended), to which they must associate a suitable envelope. Depending on the abilities of the designer/architect and the financial resources of the investor, the building should ideally be projected in such a way that it can easily be adapted to other functions—with minimal financial implications.
- The architectural solutions and materials used will be sustainable and green, contributing to the creation of a healthy environment within the building. The building will be equipped to meet the performance and functionality requirements, with guaranteed consumption efficiency and renewable energy input taken into consideration. It is very important to prepare the building user for the administration and operation of a Smart building system. The quality of the comfort provided, the satisfaction of the occupants, the appearance and the identity imprinted on the building represent the parameters of the building's functions and aesthetics. The initial cost of the building and the cost of its maintenance and operation are the parameters that become the focus of the beneficiary.
- The roof of the projected building is represented by the results of building operation: the amount of energy consumed, pollution, low carbon emissions, garbage, and ultimately the idea of post-use of the building—all in the context of circular economy and sustainability.

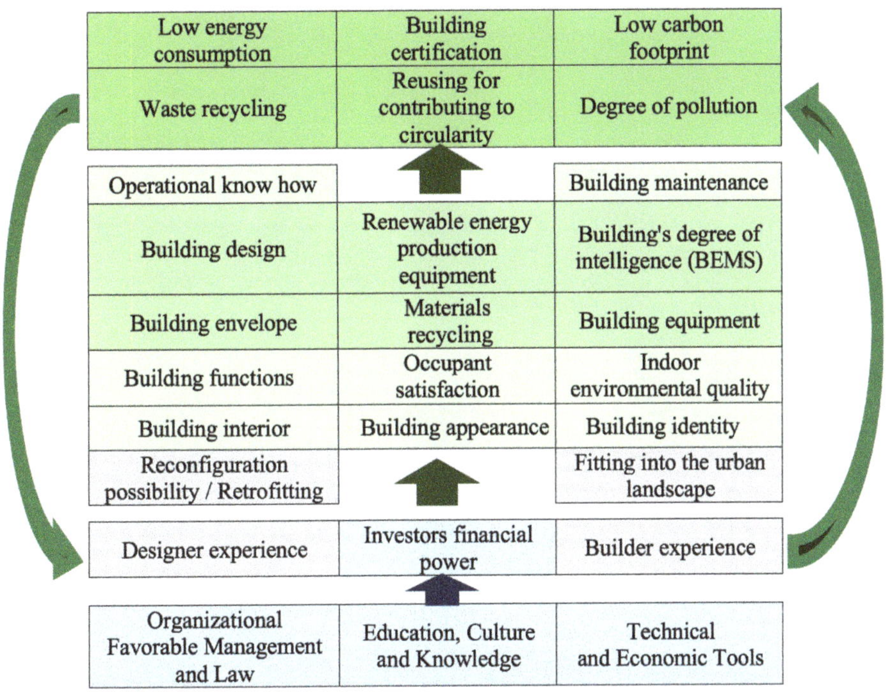

Figure 13. Factors helping in obtaining a green building in the circular economy.

All the aforementioned data are suggested in conceiving a summarizing hybrid model for the design or conversion of an existing building into a sustainable building, as it is presented in Figure 14.

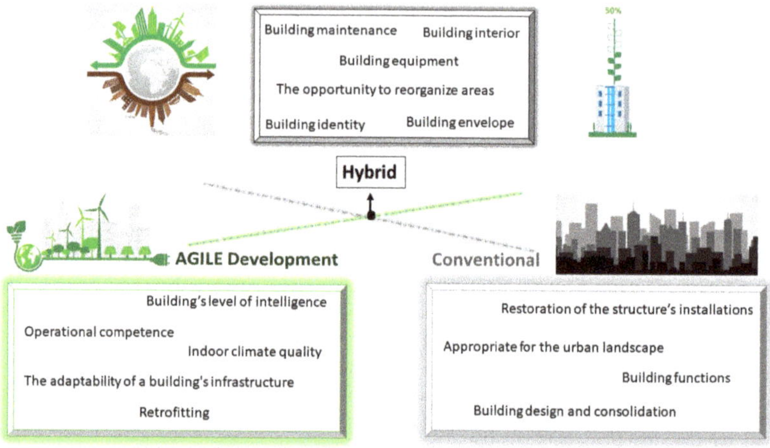

Figure 14. Hybrid model for design or conversion of an existing building into a sustainable building.

In this theoretical hybrid approach, a conventional design and planning stage can be followed by an agile development process. Moreover, this approach allows for the development and execution of a detailed plan while providing adaptability in the implementation phase, maximizing the benefits of both strategies [45].

Climate change mitigation targets are set at international, European, and national levels. These are taken over in the development strategies of the territorial administrative units, in our case—in the development strategy of the UO. The present paper emphasizes that the defining elements, respectively the criteria for the certification of GB, settlements and further—of green university campuses are applicable and constitute an incentive for both beneficiaries and investors. By applying for certification in the UI Green Metric system [46,47], UO understood and directed its development strategy to meet the criteria of a "green" campus (Figure 15), to provide students and teachers with a healthy environment, to have a sustainable development, and to help combat climate change.

Figure 15. UI Green Metric World University Rankings Certificate, from 2021.

To advance in this ranking, respectively to advance in the "Green" and "Heathy" development of its campus, UO will at least follow the criteria highlighted in this research, starting from the material/patrimonial components of the campus (buildings, land, equipment, endowments, utilities) reaching a holistic approach to education, student life on campus, educational and research activity and their results.

In this context, the UO has a very good collaboration with the local authorities. UO also initiated and responded to the request to enter clusters/consortia/alliances dedicated to sustainable/green development actions. The European Universities Alliance for Sustainability: responsible growth, inclusive education and environment (EU GREEN) can be mentioned here. The following nine universities are part of this alliance: University of Parma, Wroclaw University of Environmental and Life Sciences, University of Extremadura, Otto von Guericke University Magdeburg, University of Évora, University of Angers, University of Gävle, University of Oradea and Atlantic Technical University [48].

The actions for the sustainable development of an existing university campus and its orientation towards a "green" campus must be located on different levels highlighted in Figure 16. This scheme proposes a gradual holistic approach, starting from the simple material elements that define the campus and reaching its spirit and purpose. Thus, the challenges of a "green" campus address the site, the buildings, respectively the infrastructure, and the biodiversity by:

- Furnishing and equipping the site with a special respect for nature and the natural;
- The sense of safety created, the healthy ambience and the comfort in which the entire activity is carried out;
- The quality of life inside the campus;
- The services provided to students and teaching staff;
- The educational and relational system offered by the research and innovation conditions created through technology transfer approaches [49,50].

Infrastructure	Circular economy and resource conservation	Health, quality of the indoor environment safety	Monitoring, education & research	Management	Environmental Governance
Green elements. Green housing and buildings Energy and climate change Clean and efficient energy policies CO_2 reduction strategy Services Recreation areas Green transport promotions Green land use policies Intelligent management system – BMS	Retrofitting Recover Reuse Reduce Regenerate Rethink Waste reduction policies. Waste recycling. Policies of the renovation wave and use longer. Renewable energy.	Healthy Safe Culturally vibrant & happy Sanitation policy Clean air policies Comfort Quality of water Testing / monitoring indoor and outdoor comfort conditions Total area on campus covered in planted and forest vegetation	Education Research Technology and innovation Information, motivation, and guidance Educational platforms / IT email service / intranet services Embrace creativity Innovative programs in energy and climate change	Green policies Cyber security Digitization Responsibility of public finances Financing green development Partnerships, Alliances Resilient campus policies Green urban planning. Transparency and open data.	Harmonize with nature. Green action plan. Green management. Green growth Biodiversity sustainable Environmental monitoring. Environmental management Waste- water system treatment Water efficiency treatment policies Climate neutrality

Figure 16. New Green campus conceptual model including as promoting factor the clean infrastructure.

All these principles and criteria are subject to the desire to ensure a future for the following generations, obeying the principles of the circular economy, enhancing adaptation to climate change, recovery, and green growth, obeying international and European policies.

The CUO redesign initiated and led to a framework that integrates strategies and complex solutions that cover buildings with multiple destinations (for education—laboratories and lecture halls, student dormitories, warehouses, offices, etc.). It should also be emphasized that it is specific to CUO that it contains buildings built throughout the last century, in different stages: 1910–1920, 1960–1970, as well as buildings built relatively recently, 1990–2000. Moreover, the constructions addressed originally had various functions, most of them different from the ones they have today. Another very special feature of CUO is the existence of geothermal water (renewable energy) that the campus benefits from as a heating agent.

It is undeniable that other already published papers have also evaluated and quantified the impact of modernizing university campuses [51–53]. However, their different approaches, the numerous differences between the campuses at the geoclimatic, architectural level, the construction materials used, the age of the buildings, the heating agent used, etc. make it very difficult to compare the results obtained. However, as a leitmotif, the positive impact of modernization, especially related to the consumption of thermal and electrical energy, but also on the communities served, respectively, the need for sustainable development, is emphasized in this research.

5. Conclusions

The study deepens the case of UO which is the beneficiary of renewable energy (geothermal water), used/reused as a heating agent and domestic hot water. In addition, the authors summarize the CUO heating/cooling principle, proposing it as an effective solution of sustainable development, for this type of location. With the intervention solutions on some of the buildings and networks of the university campus presented in the paper (energy efficiency, thermal efficiency, rehabilitation, renovation, retrofitting, modernization) a 20% reduction in the specific energy consumption for heating alone was obtained, under the conditions of a relevant contribution (10%) of educational spaces obtained through retrofitting. Corroborating all the interventions on the existing infrastructure described, it is obvious that the "wave of renovation" will contribute significantly to reach the targets set regarding sustainable development.

Through this paper, not only have we framed the "green" evolution of CUO, but we came up with new aspects compared to those presented in the specialized literature, namely:

- The efficiency and modernization of some buildings, older than 100 years (heritage buildings), are described;
- Buildings with various destinations (for students, professors, staff, and materials) are addressed in the same work;
- Constructions changing their initial destination are presented;
- A summarizing scheme depicting the main aspects which help in the construction of a GB or in the transformation of an existing building into a sustainable building (Figure 13);
- A green campus conceptual model (Figure 16), including as a promoting factor the clean infrastructure (both for new and existing buildings, a coherent approach solution was provided, in the context of the circular economy, so that the result of the construction/intervention on the existing construction is that of a GB).

In addition, through the practical solutions presented, the study demonstrates that a "recipe"/solution for a "green" and "healthy" campus is the application of the principles enunciated in the policies of sustainable and "green" development on different levels. Regarding the case study brought to attention (CUO), as future directions, it would be the combination of thermal and energy efficiency solutions and the introduction of renewable energies from various sources, in order to reduce energy consumption and CO_2 emissions.

Finally, there is an obvious need to continue the management of this campus in a "green" vision, respectively to intensify the application of the principles and concepts synthesized for "green" development. It is noticeable that the solutions applied to CUO and presented in the paper can be taken over and adapted according to the specifics of each campus, considering the types of interventions on buildings, their age, original vs. final function, heating agent, etc.

Author Contributions: Conceptualization, C.C.B. and C.B.; Data curation, C.C.B. and I.F.P.-H.; Formal analysis, C.C.B. and M.T.T.; Investigation, C.C.B., I.F.P.-H. and T.B.; Methodology, C.C.B., C.B., T.B. and M.F.P.; Project administration, C.B. and M.F.P.; Resources, M.F.P.; Software, C.C.B. and D.E.P.; Supervision, C.B.; Validation, M.T.T. and M.F.P.; Visualization, C.C.B. and M.F.P.; Writing—original draft, C.C.B., I.F.P.-H. and T.B.; Writing—review & editing, C.C.B. and T.B. All authors have read and agreed to the published version of the manuscript.

Funding: This research received no external funding.

Institutional Review Board Statement: Not applicable.

Informed Consent Statement: Not applicable.

Data Availability Statement: The data used in the work are available within the General Administrative Directorate/Sustainable Campus Directorate//Investments and Consolidations Office and Heritage Office, of the University of Oradea, Romania, being included as appropriate: in the construction books, the technical documentation of construction interventions, the energy audits within them, the documents for the reception of the renovation works, the energy performance certificates drawn up at the initiation of the rehabilitation works and at their reception, the technical projects authorized and put into operation, as well as on the data basis of monitoring and controlling the utilities consumption within the University of Oradea.

Acknowledgments: The authors express their gratitude to the University of Oradea for providing the facilities required by this research.

Conflicts of Interest: The authors declare no conflict of interest.

References

1. Zakaria, R.; Foo, K.S.; Zin, R.M.; Yang, J.; Zolfagharian, S. Potential Retrofitting of Existing Campus Buildings to Green Buildings. *Appl. Mech. Mater.* **2012**, *178–181*, 42–45. [CrossRef]
2. Nguyen, H.D.; Macchion, L. Risk management in green building: A review of the current state of research and future directions. *Environ. Dev. Sustain.* **2022**, *25*, 2136–2172. [CrossRef]
3. Berardi, U. A cross-country comparison of the building energy consumptions and their trends. *Resour. Conserv. Recycl.* **2017**, *123*, 230–241. [CrossRef]
4. Darko, A.; Zhang, C.; Chan, A.P.C. Drivers for green building: A review of empirical studies. *Habitat Int.* **2017**, *60*, 34–49. [CrossRef]
5. Ahmed, N.; Abdel-Hamid, M.; Abd El-Razik, M.M.; El-Dash, K.M. Impact of sustainable design in the construction sector on climate change. *Ain Shams Eng. J.* **2021**, *12*, 1375–1383. [CrossRef]
6. Bogdan, A.; Chambre, D.; Copolovici, D.M.; Bungau, T.; Bungau, C.C.; Copolovici, L. Heritage Building Preservation in the Process of Sustainable Urban Development: The Case of Brasov Medieval City, Romania. *Sustainability* **2022**, *14*, 6959. [CrossRef]
7. Wang, X.; Stewart, M.G.; Nguyen, M. Impact of climate change on corrosion and damage to concrete infrastructure in Australia. *Clim. Change* **2012**, *110*, 941–957. [CrossRef]
8. Bungau, C.C.; Bungau, T.; Prada, I.F.; Prada, M.F. Green Buildings as a Necessity for Sustainable Environment Development: Dilemmas and Challenges. *Sustainability* **2022**, *14*, 13121. [CrossRef]
9. Ahn, Y.H.; Pearce, A.R.; Wang, Y.; Wang, G. Drivers and barriers of sustainable design and construction: The perception of green building experience. *Int. J. Sustain. Build. Technol. Urban Dev.* **2013**, *4*, 35–45. [CrossRef]
10. Zhai, Z.J.; McNeill, J.S. Roles of building simulation tools in sustainable building design. *Build. Simul.* **2013**, *7*, 107–109. [CrossRef]
11. Cordero, E.C.; Centeno, D.; Todd, A.M. The role of climate change education on individual lifetime carbon emissions. *PLoS ONE* **2020**, *15*, e0206266. [CrossRef] [PubMed]
12. Prada, M.; Popescu, D.; Bungau, C. Building Education, Source of Energy Saving in Romania. In Proceedings of the 15th National Technical-Scientific Conference on Modern Technologies for the 3rd Millennium, Oradea, Romania, 27–28 November 2015; pp. 157–162.

13. COM/2019/640. Communication from the Commission to the European Parliament, the European Council, the Council, the European Economic and Social Committee and the Committee of the Regions. *The European Green Deal*. Available online: https://eur-lex.europa.eu/legal-content/EN/TXT/HTML/?uri=CELEX:52019DC0640&from=EN (accessed on 2 February 2023).
14. COM/2020/80. Proposal for a Regulation of the European Parliament and of the Council Establishing the Framework for Achieving Climate Neutrality and Amending Regulation (EU) 2018/1999 (European Climate Law). Available online: https://eur-lex.europa.eu/legal-content/EN/TXT/HTML/?uri=CELEX:52020PC0080&from=EN (accessed on 2 February 2023).
15. Mahdee, J.; Abu Bakar, N.; Oh, V.; Seng, K. Green campus universities: Case studies on problems and prospects. *F1000Research* **2022**, *11*, 1200. [CrossRef]
16. Ilies, M.; Ilies, D.C.; Ilies, A.; Josan, I.; Ilies, G. The gateway of Maramureş land: Geostrategical implications in space and time. *Ann. Ser. Hist. Sociol.* **2010**, *20*, 469–480.
17. Noor, S.N.A.M.; Loong, L.K.; Naamandadin, N.A. An Insight of Challenges in Implementing Green Campus: ACase Study of Universiti Malaysia Perlis. *J. Adv. Res. Eng. Knowl.* **2019**, *7*, 34–40.
18. Li, B.; Zhao, X.; Luo, Y. Design Strategies for Green Retrofitting of a University Student Dormitory in the Hot-humid and Less-windy Climate. In *Proceedings of the 2022 International Conference on Green Building, Civil Engineering and Smart City*; Springer Nature: Singapore, 2022; Volume 211, pp. 22–31.
19. Herman, G.V.; Caciora, T.; Ilies, D.C.; Ilies, A.; Deac, A.; Sturza, A.; Sonko, S.M.; Suba, N.S.; Nistor, S. 3D Modeling of the Cultural Heritage: Between Opportunity and Necessity. *J. Appl. Eng. Sci.* **2020**, *10*, 27–30. [CrossRef]
20. Grama, V.; Ilies, G.; Safarov, B.; Ilies, A.; Caciora, T.; Hodor, N.; Ilies, D.C.; Kieti, D.; Berdenov, Z.; Josan, I.; et al. Digital Technologies Role in the Preservation of Jewish Cultural Heritage: Case Study Heyman House, Oradea, Romania. *Buildings* **2022**, *12*, 1617. [CrossRef]
21. Tsantili, A.; Koronaki, I.; Polydoros, V. Maximizing Energy Performance of University Campus Buildings through BIM Software and Multicriteria Optimization Methods. *Energies* **2023**, *16*, 2291. [CrossRef]
22. Carbonara, E.; Tiberi, M. Assessing Energy Performance and Economic Costs of Retrofitting Interventions in a University Building. In Proceedings of the EEEIC 2016—International Conference on Environment and Electrical Engineering, Florence, Italy, 7–10 June 2016; IEEE: Piscataway, NJ, USA, 2016; pp. 1–5.
23. Prada, M.; Popescu, D.E.; Bungau, C.; Pancu, R. Parametric Studies on European 20-20-20 Energy Policy Targets in University Environment. *J. Environ. Prot. Ecol.* **2017**, *18*, 1146–1157.
24. University of Oradea. Satellite Image. Available online: https://latitude.to/articles-by-country/ro/romania/71587/university-of-oradea (accessed on 2 February 2023).
25. University of Oradea—Former Gendarmerie School Vágó József. Available online: http://enciclopediavirtuala.ro/monument.php?id=186 (accessed on 2 February 2023).
26. National Heritage Institute. List of Historical Monuments in Romania. Available online: https://patrimoniu.ro/monumente-istorice/lista-monumentelor-istorice (accessed on 14 February 2023).
27. Infographic—Renovation Wave: Creating Green Buildings for the Future. Available online: https://www.consilium.europa.eu/en/infographics/renovation-wave/ (accessed on 27 December 2022).
28. University of Oradea. Google Maps. Available online: https://www.google.com/maps/place/Universitatea+din+Oradea/@47.0446787,21.9179081,17z/data=!4m6!3m5!1s0x4746479564d81847:0xda97ccf1bcea0869!8m2!3d47.0446751!4d21.9201021!16zL2 0vMDU4NWow (accessed on 15 January 2023).
29. Tivadar, D.; Cristea, M. *Energy Audit—Documentation for the Approval of Intervention Works on Building E–F*; University of Oradea: Oradea, Romania, 2018; p. 20.
30. Tivadar, D.; Cristea, M. *Energy Audit*; University of Oradea: Oradea, Romania, 2020; p. 20.
31. Csizmadia, Z. *Energy Audit—Technical Project-Rehabilitation of Pavilion P and Gym-Building C 20*; University of Oradea: Oradea, Romania, 2018; pp. 13–21.
32. Sabau, E. *Energy Performance Certificate*; University of Oradea: Oradea, Romania, 2022; p. 1.
33. Năstase, R.P. Masterplan for the University Campus of Oradea through Oradea City Hall. Available online: https://www.anuala.ro/en/projects/2019/196/ (accessed on 10 February 2023).
34. Bungău, C.C.; Prada, I.F.; Prada, M.; Bungău, C. Design and Operation of Constructions: A Healthy Living Environment-Parametric Studies and New Solutions. *Sustainability* **2019**, *11*, 6824. [CrossRef]
35. IT Infrastructure: Forging the Future Efficiently and Sustainably. IT Solutions for Companies. Available online: https://www.green.ch/en/blog/detail/it-infrastructure-forging-the-future-efficiently-and-sustainably (accessed on 24 January 2023).
36. Popescu, D.E.; Prada, M.F.; Dodescu, A.; Hemanth, D.J.; Bungau, C. A Secure Confident Cloud Computing Architecture Solution for a Smart Campus. In Proceedings of the 7th International Conference on Computers Communications and Control, ICCCC, Oradea, Romania, 8–12 May 2018; IEEE: Piscataway, NJ, USA, 2018; pp. 240–245.
37. Popescu, D.E.; Bungau, C.; Prada, M.; Domuta, C.; Bungau, S.; Tit, D.M. Waste management strategy at a public university in smart city context. *J. Environ. Prot. Ecol.* **2016**, *17*, 1011–1020.
38. Tit, D.M.; Bungau, S.G.; Nistor-Cseppento, D.C.; Copolovici, D.M.; Buhas, C.L. Disposal of unused medicines resulting from home treatment in Romania. *J. Environ. Prot. Ecol.* **2016**, *17*, 1425–1433.
39. Bungau, S.G.; Suciu, R.N.; Bumbu, A.G.; Cioca, G.; Tit, D.M. Study on Hospital Waste Management in Medical Rehabilitation Clinical Hospital, Baile Felix. *J. Environ. Prot. Ecol.* **2015**, *16*, 980–987.

40. Tivadar, D.; Cristea, M. *SMART Campus Project. Technical Project within the SMART Campus Project*; University of Oradea: Oradea, Romania, 2020; p. 1.
41. Tivadar, D.; Cristea, M. *Energy Efficiency of D Building*; Technical Project; University of Oradea: Oradea, Romania, 2020; p. 1.
42. Prada, M. *Energy Audit—Feasibility Study within the Project "Infrastructure for Research in Smart Spaces and Energy Efficient Processes"*; University of Oradea: Oradea, Romania, 2013; pp. 19–32.
43. Cristea, M. *Energy Performance Certificate at Building Reception*; SMART Campus Project; University of Oradea: Oradea, Romania, 2022; p. 1.
44. Regional Operational Programme 2014–2020, North-West Region of Romania, Northern Transylvania. Available online: https://www.nord-vest.ro/wp-content/uploads/2017/09/Brosura-POR.pdf (accessed on 15 January 2023).
45. Lalmi, A.; Fernandes, G.; Souad, S.B. A conceptual hybrid project management model for construction projects. *Procedia Comput. Sci.* **2021**, *181*, 921–930. [CrossRef]
46. UI GreenMetric. Overall Rankings 2022. Available online: https://greenmetric.ui.ac.id/rankings/overall-rankings-2022 (accessed on 15 January 2023).
47. UI GreenMetric World University Rankings. Available online: https://chancellery.utm.my/sustainable/ui-greenmetric/ (accessed on 8 January 2023).
48. European Universities Alliance for Sustainability: Responsible Growth, Inclusive Education and Environment (EU GREEN). Available online: https://www.eugreenalliance.eu/ (accessed on 15 January 2023).
49. Craiut, L.; Bungau, C.; Negru, P.A.; Bungau, T.; Radu, A.-F. Technology Transfer in the Context of Sustainable Development—A Bibliometric Analysis of Publications in the Field. *Sustainability* **2022**, *14*, 11973. [CrossRef]
50. Craiut, L.; Bungau, C.; Bungau, T.; Grava, C.; Otrisal, P.; Radu, A.F. Technology Transfer, Sustainability, and Development, Worldwide and in Romania. *Sustainability* **2022**, *14*, 15728. [CrossRef]
51. Chen, G.; Cheng, L.; Li, F. Integrating Sustainability and Users' Demands in the Retrofit of a University Campus in China. *Sustainability* **2022**, *14*, 10414. [CrossRef]
52. Eldin, S.S.; Elwan, M.M.; Bayoumi, A.A. Retrofitting Campus Outdoor Space Based on Thermal Performance: Case Study Seberbay University Campus, Tanta City. *Civ. Eng. Archit.* **2021**, *9*, 2505–2516. [CrossRef]
53. Aboulnaga, M.; Wanas, A.; Hammad, M.; Hussein, M. Sustainability of Higher Educational Buildings: Retrofitting Measures to Enhance Energy Performance—The Case of AASTMT Business Management School Building, Egypt. In *Mediterranean Green Buildings and Renewable Energy: Selected Papers from the World Renewable Energy Network's Med Green Forum*; Springer International Publishing: Cham, Switzerland, 2017; pp. 117–150.

sustainability

MDPI

Article

Indoor Air Quality Perception in Built Cultural Heritage in Times of Climate Change

Dorina Camelia Ilies [1], Grigore Vasile Herman [1], Bahodirhon Safarov [2], Alexandru Ilies [1], Lucian Blaga [1], Tudor Caciora [1,*], Ana Cornelia Peres [3], Vasile Grama [1], Sigit Widodo Bambang [4], Telesphore Brou [5], Francois Taglioni [5], Thowayeb H. Hassan [6,7] and Mallik Akram Hossain [8]

1 Department of Geography, Tourism and Territorial Planning, Faculty of Geography, Tourism and Sport, University of Oradea, 1 Universitatii Street, 410087 Oradea, Romania; dilies@uoradea.ro (D.C.I.); gherman@uoradea.ro (G.V.H.); ilies@uoradea.ro (A.I.); lblaga@uoradea.ro (L.B.); vgrama@uoradea.ro (V.G.)
2 Department of Digital Economy, Samarkand State University, Samarkand 140105, Uzbekistan; safarovb@rambler.ru
3 Department of Environmental Engineering, Faculty of Environmental Protection, University of Oradea, Magheru Street 26, 410087 Oradea, Romania; peresana@uoradea.ro
4 Department of Geography Education, Faculty of Social Sciences and Law, Universitas Negeri Surabaya, Surabaya 60213, Indonesia; bambangsigit@unesa.ac.id
5 Department of Geography, Planning, Environment and Development, University of Reunion Island, BP 7151, CEDEX 9, Reunion Island, 97715 Saint-Denis, France; telesphore.brou@univ-reunion.fr (T.B.); francois.taglioni@univ-reunion.fr (F.T.)
6 Department of Social Studies, College of Arts, King Faisal University, Al Ahsa 31982, Saudi Arabia; thassan@kfu.edu.sa
7 Tourism Studies Department, Faculty of Tourism and Hotel Management, Helwan University, Cairo 12612, Egypt
8 Department of Geography and Environment, Jagannath University, Dhaka 1100, Bangladesh; mallik.a@geography.jnu.ac.bd
* Correspondence: tudor.caciora@yahoo.com; Tel.: +40-740941144

Citation: Ilies, D.C.; Herman, G.V.; Safarov, B.; Ilies, A.; Blaga, L.; Caciora, T.; Peres, A.C.; Grama, V.; Bambang, S.W.; Brou, T.; et al. Indoor Air Quality Perception in Built Cultural Heritage in Times of Climate Change. *Sustainability* **2023**, *15*, 8284. https://doi.org/10.3390/su15108284

Academic Editor: Reza Daneshazarian

Received: 16 April 2023
Revised: 12 May 2023
Accepted: 15 May 2023
Published: 19 May 2023

Abstract: Low quality in a museum's internal microclimate can induce both the deterioration of the exhibit collections, as well as affecting the health of visitors, employees and restorers. Starting from this premise, the present study aims to study the perception of visitors and employees of Darvas-La Roche Museum House (Romania) in relation to the air quality in the exhibition spaces. Their opinions were analyzed based on a questionnaire comprising 11 items aimed at understanding the influence of the indoor environment on the health of individuals, the degree of disturbance induced by the indoor air, if they experienced symptoms of illness after visiting the museum, etc. The obtained data were analyzed statistically in the SPSS 28 program, using tests such as coefficient, analysis of variance (ANOVA) and model summary, in order to obtain correlations between the sets of variables. The results obtained indicate that the majority of respondents perceived the indoor air quality as good, but there were also exceptions (approximately 20% of the respondents), which indicated different symptoms induced by the indoor air. Most of those (%) affected stated that they had pre-existing conditions, wear contact lenses or are smokers. In their case, the statistical-mathematical analyses indicated strong correlations between the ailments they suffer from and the appearance of certain discomforts (caused by too low or too high temperature, dust or dry air, etc.) and disease symptoms (nasal congestion, eye and skin irritations, coughs, migraines, frequent colds, etc.).

Keywords: perception; indoor air quality; museums; cultural heritage; tourism; human health; microclimate change

1. Introduction

Cultural tourism is one of the emerging activities in urban environments, which has seen a constant development recently at urban tourist destinations [1–5]. One of the basic pillars on which cultural tourism is premised is represented by museums [6–8]. Museums diversify tourist activity, thus contributing to increasing the length of stay in a tourist destination, improving the image of the destination and, finally, increasing the satisfaction of tourists by improving the lived experiences [9].

In the Oradea Tourist Destination, there are 12 museums and museum collections, all operating in heritage buildings, some with historical monument status [10]. These buildings have existed without much attention from researchers regarding the internal microclimate [11]. Recently, there have been numerous studies that refer both to the conservation of heritage buildings and their interiors [12–16], and to indoor air quality [17–21]. Monitoring and analyzing the concentration of pollutants that decrease air quality is paramount to prevent the potential risk to human health caused by sick building syndrome [22–25]. In order to assess the danger to which the citizens are subjected, it is also necessary to determine the microbial load of the surfaces and the air inside the building [26–36].

The climate change—indoor air quality—public health nexus, known and argued over since the first decade of the 21st century [37–40], has today become an emerging issue, the understanding and solution of which largely depends on maintaining a balance between the use of heritage buildings and their preservation over time [41,42]. In this sense, EU-project FP7 "Climate for Culture" [43], which continues the research activity focused on climate change and heritage from the FP6 Noah's Ark project, based on two moderate scenarios regarding gas emissions. Scenarios A1B and RCP4.5 [44], carried out by the Intergovernmental Panel on Climate Change (IPCC), forecast indoor climates in historic buildings until the year 2100. The results of the project show that the temperature increases forecast for outdoor climates in Europe (between 1–3 °C for RCP4.5 and 2–4.5 °C for A1B) and extreme weather events will accelerate building degradation rates and raise air conditioning costs. Increases in the temperature inside buildings (indoor temperature) are expected in Sweden, Norway, Denmark, Holland, central Romania, the Alps, Italy, on the Adriatic coast and Greece. An increase in indoor temperature is also indicated by the study carried out for the southern part of England (Southern England) by Lankester and Brimblecombe [45], and the analysis carried out by González et al. [46], demonstrates that these temperature changes will affect the thermal comfort of visitors. Vardoulakis et al. [47] shows that the measures aimed at reducing greenhouse gas emissions have positive effects for public health, but there are also secondary effects, in the sense that the sealing of buildings to increase energy efficiency can lead to an increase in pollutant concentrations inside them. This means an increase in energy consumption for ventilation and air conditioning systems [46,48–50] and smart building technologies, with a possible indirect effect of increasing the degree of pollution outside [51].

In order to accredit and operate the heritage buildings as museum institutions, they have undergone some structural (recompartmentalization of the interiors) and functional (change of the original purpose for which they were built) modifications. Often, these changes affected the optimal functionality of the buildings, represented by microclimatic variations and deviations (temperature, relative humidity, carbon dioxide, suspended particles, brightness, formaldehyde concentration and volatile organic compounds) and biological (bacteria, molds and fungi) [26,52].

Indoor air quality has a decisive role for human health, for a healthy working environment and leisure time. The level of pollutant concentration, the indoor microclimate, as well as the exposure time are determinants both for human health and for the interiors of heritage buildings and their exhibits [53–56]. Air pollution inside heritage buildings, as well as an inadequate indoor microclimate, can negatively affect the health of visitors and employees [52,57–60] and can have consequences on the degree of preservation inside the building [61,62].

Too high temperatures can cause symptoms of physical and mental exhaustion, while low temperatures can lead to constriction of blood vessels and chills. Air humidity has significant effects on the quality of life over time. Too much humidity produces a favorable environment for the development of mold and considerably increases the risk of allergies and bronchial asthma, leading to fatigue, lack of concentration and headaches. In contrast, low humidity leads to drying of the nasal mucosa, skin irritations, allergies and drying of the skin. For the well-being of the human body, the relative humidity should be between 40–70% [63].

Indoor air pollution is a risk factor for people's health, the most exposed being workers and restaurateurs who spend longer time indoors. Among the manifested symptoms are respiratory problems, allergies and decreased work capacity [64,65]. Tourists, even if they spend a shorter time inside heritage buildings, can be affected by the microclimate and indoor air quality [66–68]. Among their manifested symptoms are dizziness, vomiting, headache, fatigue, eye irritation and skin rashes, coughs, etc. [69,70]. Indoor air pollution with suspended particles, especially with $PM_{2.5}$ and PM_{10}, has been associated with the intensification of bronchial asthma and cognitive disorders [71–73].

Indoor air quality also damages in time the interiors of heritage buildings by degrading walls and equipment. Temperature variations, as well as high or low humidity, contribute to damage to equipment and specific objects (religious, artistic, etc.). A high humidity of over 70% and a reduced ventilation is conducive for the growth of fungi [66].

The state of health is an interesting and important aspect that must be taken into account when the problem of the perception of indoor air quality is raised, in parallel with the knowledge of the consumption behavior of potential visitors [74,75]. The perception of indoor air quality refers to how each individual, thanks to the endosomatic instruments with which he/she is naturally endowed, perceives the air quality as good or otherwise [76–78]. In this context, the aim of the present study is to study the perception of indoor air quality in the Darvas-La Roche Museum in Oradea, Romania (Figures 1 and 2). This is a heritage building, built in the Art Nouveau style, between 1911 and 1912. Starting in 2020, after an extensive restoration aimed at restoring its former beauty and glory, Darvas-La Roche House reopened its doors in the form of an Art Nouveau museum. Within it, there are numerous important pieces of furniture, clothing, crockery, as well as other household items dating from La Belle Époque.

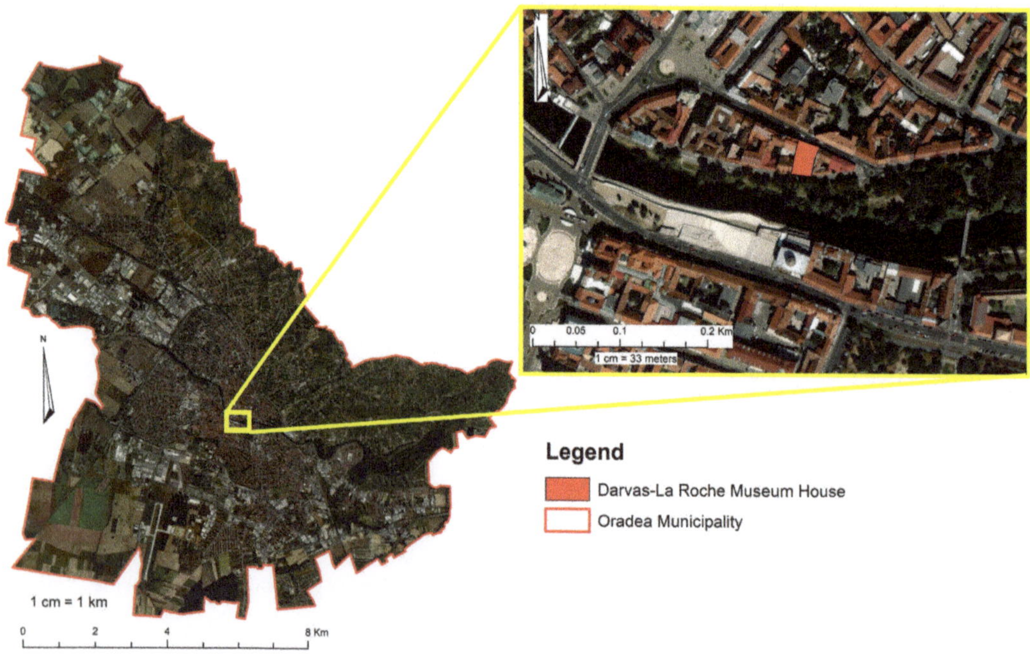

Figure 1. The location of Darvas-La Roche Museum House in the Oradea Municipality.

Figure 2. The main facade of Darvas-La Roche Museum House.

The working hypothesis is that air quality influences the perception of visitors and staff, thus good air quality will be reflected in good perception and vice versa. We note that the present study is a continuation of other studies aimed at understanding air quality [66] and the bacteriological microflora inside buildings [79]. Regarding indoor air quality, previous studies have revealed that the indoor environment can be quite unstable, mainly due to wide and very frequent fluctuations in temperature, relative humidity (RH), carbon dioxide (CO_2) and particles in suspension (PM). Exceedances of the international standards in force were also recorded for pollutants such as SO_2, O_3, NO_2, NO and H_2S, but these were sporadic and limited in terms of quantity. The pollutants VOC and HCHO have a high degree of risk for human health and the integrity of the exhibits, considering that during the measurement period, they exceeded the allowed limits by 28% (VOC) and 125% (HCHO).

At the same time, in the research conducted by Ilies et al. [79], the interior of the museum was monitored to determine the bacteriological microflora in the air and on the surfaces for establishing the degree of security for the health of museum employees, restorers and visitors. The results obtained emphasize the presence of some fungi and bacteria, among them *Alternaria* spp., *Aspergillus* spp., *Penicillium* spp., *Cladosporium* spp. and *Botrytis* spp., while the degree of contamination of the rooms was high to very high (between 524 and 3674 CFU/m^3). These results indicate a high degree of risk for human health, considering that some types of identified fungi can cause health problems for people with low sensitivity, while also being able to amplify already existing conditions.

Taking into account the issues identified in previous studies, and to extend existing information, the current study aims to understand the perception that museum employees and visitors have of the quality of the microclimate inside the Darvas-La Roche House. Thus, this series of studies not only aims to identify quantitatively and qualitatively the main indicators of the internal microclimate and bacteriological microflora, but also aims to obtain valuable information from those actively involved inside the museum. The data obtained are important as they include both employees and restaurateurs (those who spend up to 8 h a day/6 out of 7 days per week in this environment), as well as visitors (who spend between 1 and 4 h in this environment); starting from the premise that their experiences in relation to the inner environment are definitely different.

2. Materials and Methods

Considering the fact that poor quality air can negatively affect humans, the study also took into account the analysis of their perception of the air quality inside the exhibition spaces of Darvas-La Roche House. This study is mainly based on primary data from the questionnaire survey. Secondary sources of data were also used to supplement primary data sources. In the period September 2023–March 2023, 250 questionnaires were administered to study the perception of both employees and visitors of the museum in terms of air quality. The target group was chosen in such a way as to cover all age groups and genders, to include both people sensitive to impure air, under treatment, as well as healthy people, people who come into daily contact with the indoor environment and people who visit the museum only occasionally, etc.

The opinions of the visitors and staff were analyzed based on a questionnaire comprising 11 items, which aimed to determine the influence of the indoor environment on the health of individuals, the degree of disturbance induced by the indoor air, if they experienced symptoms of illness after visiting the museum, etc. At the same time, for an accurate interpretation, the analysis of the details that take into account the age of the respondents, their occupation, the number of visits they have made so far inside the Darvas-La Roche House and the average duration of a visit were taken into account.

Regarding the questions on the disturbing factors inside the exhibition spaces, they had a choice between multiple answer options, prepared following the identification of shortcomings at the site. Irregularities in terms of temperature and air humidity, unpleasant odors, dust in suspension and the presence of molds were thus taken into account. At the same time, the visitors were also questioned regarding the symptoms they felt after visiting the museum, such as: headaches, vomiting, repeated coughing and sneezing, eye and skin irritations, severe fatigue, etc.

In order to determine the effects of indoor air on human health, the data obtained from the respondents were entered into the SPSS 28 program, where various analyses and statistical calculations were carried out. Among the statistical calculations, coefficient, analysis of variance (ANOVA) and model summary are applied in this study.

3. Results and Discussion

3.1. The Results of the Questionnaire Implementation

Among the 250 respondents, 108 (43%) were men, and 142 (57%) were women. Most of them, 71 respondents, were aged between 19–25 years, followed by those between 14–18 years (62 respondents) and the 26–35 age group (52 respondents). The smallest group comprised those over 61, represented by only 12 respondents. About 11% declared that they were workers, 10% did not have a stable job (they were mainly represented by pupils and students), 8% were staff with higher education, and 7% were entrepreneurs and 7% were workers in the tourism and services industry (Figure 3). The majority of respondents (220–88%) visited Darvas-La Roche House less than five times, 16 respondents (6%) visited more than 10 times, while 12 (5%) between five and 10 times (Figure 4a). According to the respondents, visiting the museum mostly lasted (50% of the cases) between 1 and 2 h, 101 (40%) of them usually visited it in 2–4 h, while only 23 (9%) needed more than four hours for a visit (Figure 4b).

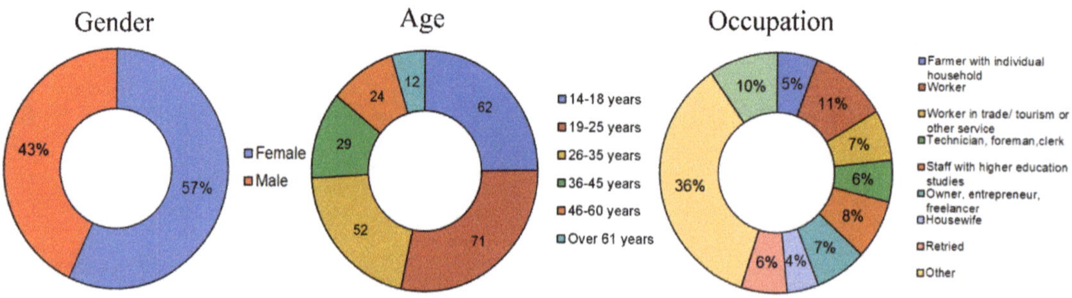

Figure 3. The characteristics of the 250 respondents in terms of gender, age group and occupation.

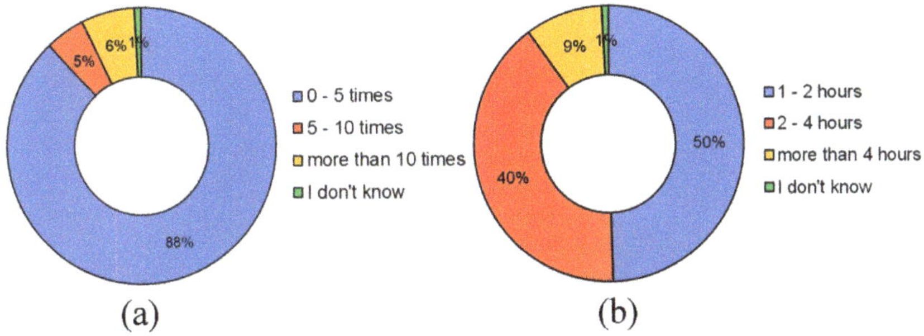

Figure 4. The number of visits and the visiting time of Darvas-La Roche House ((**a**)—How many times have you visited Casa Darvas La Roche? (**b**)—Approximately how many hours did it take to visit the Darvas La Roche House?).

Approximately 77% of the respondents were satisfied with the indoor air quality, 23% evaluating its quality as very good, while 54% recognized it as good. On the contrary, 13% of them claim that the air is poor, while 5% stated that it is very poor (Figure 5a). Among those who report problems regarding air quality, they claimed that they felt most acutely throughout the activities (36%), at the end of the activities (11%) but also at the beginning (9%) (Figure 5b).

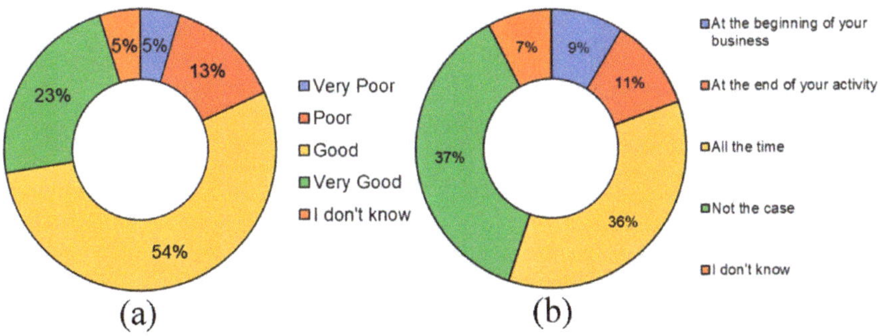

Figure 5. Evaluation of the indoor air quality by the 250 respondents ((**a**)—In general, how do you assess the air quality in the exhibition spaces? (**b**)—If you noticed problems with the air quality in the exhibition spaces, when do you think they are more pronounced?).

Based on the data presented in Figure 6a, it can be explained that, generally, the respondents never experienced disease symptoms when they visited the exhibition spaces. Of the 24 respondents out of 250, approximately 9.8% often felt dizziness/fainting. More than 54% never felt any disease symptoms during a visit to the exhibition room. The most common ailments that respondents felt were headache, dry throat, cough, fatigue and eye irritation; but these symptoms appeared only sometimes, without being based on a well-established pattern. Regarding the influence of the quality of the indoor microclimate on visitors and employees, most of them declared that they never encountered any inconveniences. Some claimed that sometimes they suffered due to high temperature differences, dry and unventilated air, dust in suspension and unpleasant smells. Other respondents (47) stated that they quite often encountered air that was too dry and dusty (Figure 6b).

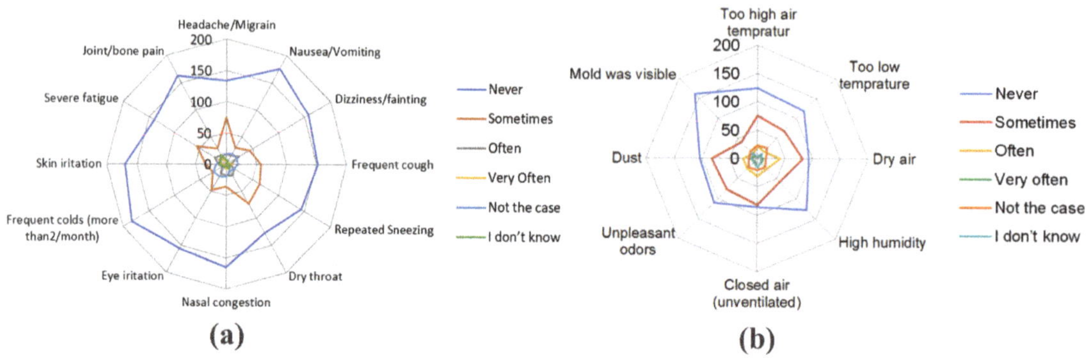

Figure 6. (**a**)—The symptoms induced to visitors and employees by the internal microclimate of Darvas-La Roche House; (**b**)—The influence of indoor air quality of Darvas-La Roche House, potentially harmful to human health.

3.2. *Implementation of Statistical Analyses*

Furthermore, to determine the effect of the respondent's health on the discomfort and disease symptoms when visiting the exhibition spaces, two statistical calculations were applied. The results show that there is a strong correlation between the pre-existing conditions/habits of visitors and employees and the symptoms they face after spending a certain amount of time inside the exhibition spaces.

3.2.1. The Influence between Health and Discomfort in the Exhibition Spaces

Based on the data presented in Table 1, it can be seen that smokers have an effect on discomfort in exhibition spaces seen from the significance value of very high temperature ($0.023 < 0.05$), closed and unventilated air ($0.000 < 0.05$) and dust ($0.013 < 0.05$). Wearing contact lenses influences discomfort in exhibition spaces, seen from the significance value of too low temperature ($0.048 < 0.05$), high humidity ($0.010 < 0.05$) and unventilated ($0.000 < 0.05$). Medical treatment to which the respondents were subjected contributed to the disturbance created in the exhibition spaces by the significance value of unpleasant odors ($0.015 < 0.05$) and visible mold ($0.000 < 0.05$); while health problems affected discomfort by significance value of too high temperature ($0.006 < 0.05$), dry air (0.018), high humidity (0.001), unventilated air ($0.000 < 0.05$), unpleasant odors ($0.002 < 0.05$), dust ($0.001 < 0.05$) and molds ($0.006 < 0.05$). All this indicates a positive correlation between the analyzed variables, so that the influence between health and discomfort is one of reciprocity and interdependence. We want to note that the significance value, the p-value, is the coefficient utilized in this argument. The p-value is used in this situation to establish whether there is a significant association between the variables being studied and their impact on discomfort in exhibition spaces, such as smoking, using contact lenses, receiving medical treatment, and having health issues. The null hypothesis, according to which there is no association between the variables under investigation, is tested using the p-value. A statistically significant association exists between the variables if the p-value is less than 0.05. According to this argument, there is a substantial correlation between smoking, using contact lenses, receiving medical attention, and health issues and their impact on discomfort in exhibition spaces. All of the p-values stated in this argument are less than 0.05.

Table 1. Health and bother in exhibition spaces based on coefficient.

	Smoker	Wearing Contact Lenses	Medical Treatment	Health Problem
Too high air temperature	0.023	0.183	0.091	0.006
Too low temperature	0.652	0.048	0.684	0.405
Dry air	0.505	0.863	0.188	0.018
High humidity	0.134	0.010	0.448	0.001
Closed air (unventilated)	0.0	0.0	0.485	0.0
Unpleasant odors	0.907	0.075	0.015	0.002
Dust	0.013	0.644	0.852	0.001
Visible mold	0.349	0.485	0.0	0.006

The statistical analysis's findings suggest a significant linear association between the number of severe smokers and the discomfort felt in the exhibition area. The null hypothesis (H0) was rejected, indicating a statistically significant correlation between the two variables. This study underscores the need for initiatives to lower smoking prevalence in such situations and has significant implications for understanding the factors influencing smoking behavior in public settings. The data also showed a substantial linear link between the discomfort felt in the exhibition space and additional elements such as contact lens use, medical care, and health issues. In this instance, the rejection of H0 shows that these characteristics are also significant predictors of discomfort felt in public areas. This finding emphasizes the significance of considering various variables when developing treatments to enhance public health outcomes. Overall, our results shed important light on the intricate interplay between personality traits and environmental variables affecting public health behaviors. This study can guide focused actions to lower smoking prevalence and enhance overall public health outcomes by identifying important determinants of bother experienced in exhibition settings. More research is required to fully understand these links and create successful ways to promote healthy behaviors in public environments (Table 2).

Table 2. Health and discomfort in exhibition spaces based on test.

	Smoker	Wearing Contact Lenses	Medical Treatment	Health Problem
Too high air temperature Too low temperature Dry air High humidity Closed air (unventilated) Unpleasant odors Dust Visible mold	0.000 < 0.05	0.002 < 0.05	0.000 < 0.05	0.000 < 0.05

The study's summary model sheds light on the degree of correlation between different variables, according to the Adjusted R-Square correlation coefficient, which is used to gauge the strength of the correlation between smoker variables and Y. Variable X influences Y to the extent of 10.9%, with other variables influencing the remaining percentage. The Adjusted R-Square correlation coefficient for medical treatment variables is 0.113, which shows a strong link between these and Y. Variable X controls Y by 11.3%. Wearing contacts, on the other hand, offers a weaker connection with Y, with an Adjusted R-square correlation coefficient of 0.065 and variable X having a 6.5% effect on Y. The study also shows that, as demonstrated by an Adjusted R-Square correlation coefficient of 0.24, health problem factors reveal a much higher association level with Y than any other variable evaluated. This implies that, while other factors affect the remaining percentage, variable X significantly impacts Y by 24.6%. It is critical to emphasize that these findings substantially influence healthcare practitioners and decision-makers in creating efficient interventions and strategies to deal with smoking-related health issues and medical care. Additionally, by being aware of the different degrees of connection between various variables, healthcare professionals can better adapt their treatment to fit each patient's specific needs and circumstances (Table 3).

Table 3. Health and bother in exhibition spaces based on model summary.

	Smoker	Wearing Contact Lenses	Edical Treatment	Health Problem
Too high air temperature Too low temperature	R = 370	R = 308	R = 0.376	R = 0.520
Dry air High humidity	R-square 0.137	R-square 0.095	R-square 0.142	R-square 0.271
Closed air (unventilated) Unpleasant odors	Adjusted R-square 0.109	Adjusted R-square 0.065	Adjusted R-square 0.113	Adjusted R-square 0.246
Dust Visible mold	DW 1.812	DW 2.094	DW 1.768	DW 1.139

A multiple linear regression test using the Enter method is to enter all predictor variables with sig t >. The value of the variable regression coefficient is as follows. The multiple linear regression equation model proves that:

- Smoker = 2.322 − 0.321 (too high temperature) + 0.033 (too low temperature) − 0.062 (dry air) + 0.150 (high humidity) + 0.211 (closed air unventilated) + 0.010 (Unpleasant odors) + 0.187 (dust) − 0.061 (visible mold);
- Wearing Contact Lenses = 1.286 − 0.131 (too high temperature) − 0.101 (too low temperature) + 0.011 (dry air) + 0.182 (high humidity) + 0.138 (closed air unventilated) − 0.106 (Unpleasant odors) + 0.024 (dust) + 0.032 (visible mold);
- Medical Treatment = 1.261 − 0.178 (too high temperature) + 0.022 (too low temperature) + 0.091 (dry air) + 0.056 (high humidity) + 0.029 (closed air unventilated) − 0.153 (Unpleasant odors) − 0.010 dust + 0.244 (visible mold);
- Health Problems = 0.648 − 0.269 (too high temperature) − 0.042 (too low temperature) + 0.151 (dry air) + 0.219 (high humidity) + 0.294 (closed air unventilated) − 0.185 (Unpleasant odors) − 0.178 (dust) + 0.124 (visible mold).

3.2.2. The Influence between Health and Disease Symptoms in the Exhibitions Space

Based on the data presented in Table 4, it could be stated that smoking has an effect on symptoms in exhibition spaces seen from the headache significance value (0.000 < 0.05), nausea (0.001 < 0.05), dizziness (0.022 < 0.05), dry throat (0.000 < 0.05) and skin irritation (0.007 < 0.05). Those who wear contact lenses declared that they feel affected inside the museum by dizziness (0.000 < 0.05), dry throat (0.000 < 0.05), nasal congestion (0.000 < 0.05) and joint/bone pain (0.029 < 0.05). Those under treatment indicated increased sensitivity to dizziness (0.041 < 0.05), repeated sneezing (0.000 < 0.05) and skin irritation (0.000 < 0.05). Finally, health problems affected symptoms in exhibition spaces seen from the significance value of dizziness (0.044 < 0.05), dry throat (0.000 < 0.05), skin irritation (0.000 < 0.05) and joint/bone pain (0.000 < 0.05).

The current study aimed to examine the connections between symptoms in an exhibition setting and significant smoking variables, as well as other potential influences such as contact lens use, medical care, and health issues. With a Sig value of 0.000, less than the preset alpha limit, the statistical analysis's findings showed that the linear model between smokers and symptoms in the exhibition space was significant. According to this result, H0—which contends that no conclusive link exists between smokers and symptoms in the exhibition space—can be ruled out. Additionally, the linear model between symptoms in the exhibition space and contact lens use, medical care and health issues yielded comparable results. Given that each component has a Sig value of 0.000, which is less than, H0 can also be rejected for each. These results indicate a strong correlation between these factors and the symptoms people experience in exhibition venues.

Table 4. Health and disease symptoms in exhibition spaces based on coefficient.

	Smoker	Wearing Contact Lenses	Medical Treatment	Health Problem
Headache/migraine	0.0	0.907	0.850	0.536
Nausea/vomiting	0.001	0.874	0.216	0.860
Dizziness/fainting	0.022	0.0	0.041	0.044
Frequent cough	0.543	0.900	0.583	0.801
Repeated sneezing	0.229	0.385	0.0	0.057
Dry throat	0.0	0.0	0.784	0.0
Nasal congestion	0.990	0.0	0.807	0.455
Eye irritation	0.633	0.255	0.668	0.074
Frequent colds	0.981	0.669	0.732	0.476
Skin irritation	0.007	0.050	0.0	0.0
Severe fatigue	0.656	0.625	0.919	0.200
Joint/bone pain	0.464	0.029	0.197	0.0

It is important to remember that smoking has been a significant risk factor for several health issues, including cancer and respiratory disorders (CDC, 2021). Smokers may develop more symptoms when exposed to environmental elements such as those in exhibition rooms. Similarly, wearing contact lenses or having specific medical conditions may worsen the sensations people experience in these situations. These findings significantly impact public health initiatives intended to lessen exposure to hazardous environmental elements in exhibition venues. Targeted interventions can be created to limit exposure and enhance overall health outcomes for those who frequent these places by identifying the precise risk variables linked to symptom exacerbation (Table 5).

Table 5. Health and disease symptoms in exhibition spaces based on AVONA test.

	Smoker	Wearing Contact Lenses	Medical Treatment	Health Problem
Headache/migraine Nausea/vomiting Dizziness/fainting Frequent cough Repeated sneezing Dry throat Nasal congestion Eye irritation Frequent colds Skin irritation Severe fatigue Joint/bone pain	0.00 < 0.05	0.00 < 0.05	0.00 < 0.05	0.00 < 0.05

The summary model provides an overview of the Adjusted R-Square correlation coefficient, which measures the level of relationship between medical treatment variables and their impact on Y. The coefficient for medical treatment variables is 0.145, indicating that variable X affects Y by 14.5%, while other variables influence the remaining percentage. Similarly, the summary model reveals that smoker variables have a correlation coefficient of 0.255, meaning that variable X affects Y by 25.5%. On the other hand, the summary model also shows that wearing contact lenses has a correlation coefficient of 0.290, indicating that variable X affects Y by 29%. This suggests a stronger relationship between wearing contact lenses and their impact on Y than medical treatment and smoker variables (Table 6).

Table 6. Health and disease symptoms in exhibition spaces based on model summary.

	Smoker	Wearing Contact Lenses	Medical Treatment	Health Problem
Migraine Nausea/vomiting Dizziness/fainting Frequent cough Repeated sneezing Dry throat Nasal congestion Eye iritation Frequent colds Skin iritation Severe fatigue Joint/bone pain	R = 0.539 R-Square 0.290 Adjusted R-Square 0.255 DW 1.679	R = 0.570 R-Square 0.324 Adjusted R-Square 0.290 DW 2.260	R = 0.431 R-Square 0.186 Adjusted R-Square 0.145 DW 1.622	R = 0.681 R-Square 0.464 Adjusted R-Square 0.437 DW 1.440

Additionally, the summary model highlights that health problem variables have a correlation coefficient of 0.437, meaning that variable X affects Y by 43.7%. This indicates a significant relationship between health problems and their impact on Y. Overall, these findings suggest that different variables have varying degrees of influence on Y and should be considered when analyzing data related to medical treatment, smoking habits, wearing contact lenses and health problems. Further research could explore these relationships in more detail to better understand how these factors affect outcomes in various contexts.

A multiple linear regression test using the enter method is to enter all predictor variables with sig $t > \alpha$. The value of the variable regression coefficient is as follows:

- Smoker = 3.151 + 0.829 (headache) − 0.446 (nausea) − 0.692 (dizziness) + 0.113 (frequent cough) + 0.128 (repeated sneezing) + 0.306 (dry throat) + 0.004 (nasal congestion) − 0.058 (eye irritation) + 0.002 (frequent colds) − 0.221 (skin irritation) − 0.044 (severe fatigue) + 0.045 (joint/bone pain);
- Wearing Contact Lenses = 0.999 + 0.012 (headache) + 0.014 (nausea) − 1.138 (dizziness) − 0.016 (frequent cough) + 0.061 (repeated sneezing) + 0.254 (dry throat) + 0.936 (nasal congestion) − 0.092 (eye irritation) + 0.028 (frequent colds) − 0.107 (skin irritation) − 0.032 (severe fatigue) + 0.090 (joint/bone pain);
- Medical Treatment = 1.976 + 0.023 (headache) + 0.135 (nausea) − 0.496 (dizziness) − 0.082 (frequent cough) + 0.544 (repeated sneezing) + 0.017 (dry throat) + 0.058 (nasal congestion) − 0.041 (eye irritation) + 0.027 (frequent colds) − 0.262 (skin irritation) − 0.008 (severe fatigue) + 0.064 (joint/bone pain);
- Health problems = 0.871 + 0.061 (headache) + 0.061 (nausea) − 0.395 (dizziness) + 0.030 (frequent cough) + 0.131 (repeated sneezing) + 0.522 (dry throat) − 0.144 (nasal congestion) − 0.140 (eye irritation) + 0.046 (frequent colds) − 0.252 (skin irritation) − 0.081 (severe fatigue) + 0.211 (joint/bone pain).

4. Conclusions

The findings of the study revealed that the perception of employees and visitors on the quality of the microclimate inside Darvas-La Roche House is mostly good; most respondents viewed the indoor air as good (54%) or very good (23%), only a small percentage perceiving it as poor (13%) or very poor (5%). Among those who categorized the indoor microclimate as inappropriate, 37% felt discomfort throughout the indoor activity, while 11% reported discomfort at the end of the tour and 9% at the beginning of the activity. Most of the respondents symptoms disappeared after leaving the museum, but in some cases, it persisted throughout the day.

The respondents associated most of the symptoms with too high or too low temperature, dry and unventilated air, as well as a large amount of dust in suspension. All these mainly led to the appearance of migraines, severe fatigue, dizziness, frequent coughing, repeated sneezing, dry throat and eye irritation among the visitors. Most of the respondents who declared that they experienced such symptoms after spending time in the museum also indicated that they have health problems, are under medical treatment, wear contact lenses or are smokers. Statistical analyses have indicated a strong correlation between pre-existing conditions and the variables related to disease symptoms (nasal congestion, eye and skin irritations, coughs, migraines, frequent colds, etc.) and/or discomfort sensations (dry air, humidity sea, unpleasant smells, etc.) induced by the internal microclimate.

In conclusion, for most of the respondents, the indoor microclimate in Darvas-La Roche House did not pose any potential harmful effect. Of the 20% who were affected, most reported pre-existing conditions and health problems. Thus, it can be assumed that the internal microclimate is not favorable for those who face health problems, having the potential to accentuate the manifestations of pre-existing conditions.

Regarding the limitations of the present study, it does not take into account the identification of the exhibition spaces where the discomfort is more pronounced, taking into account that the influence on human health is not constant over the entire museum. At the same time, it was impossible to monitor the medium and long-term effects that the indoor microclimate has on the visitors, in respect to what extent the symptoms experienced pass after leaving the museum or not.

Author Contributions: Conceptualization, D.C.I., T.C. and G.V.H.; Methodology, A.I. and S.W.B.; Software, S.W.B. and L.B.; Validation, T.B. and F.T.; Writing: original draft preparation, A.C.P., V.G., T.H.H., B.S. and M.A.H.; Writing: review and editing, T.C., A.I. and G.V.H.; Visualization, D.C.I.; Supervision, D.C.I. and M.A.H. All authors have read and agreed to the published version of the manuscript.

Funding: This work was supported by the Deanship of Scientific Research, Vice Presidency for Graduate Studies and Scientific Research, King Faisal University, Saudi Arabia [Grant No. 3272].

Institutional Review Board Statement: Not applicable.

Informed Consent Statement: Not applicable.

Data Availability Statement: The data presented in this study may be obtained on request from the corresponding author.

Acknowledgments: The research undertaken was made possible by the equal scientific involvement of all the authors concerned.

Conflicts of Interest: The authors declare no conflict of interest.

References

1. Aranburu, I.; Plaza, B.; Esteban, M. Sustainable Cultural Tourism in Urban Destinations: Does Space Matter? *Sustain. Sci. Pract. Policy* **2016**, *8*, 699. [CrossRef]
2. Stoica, G.D.; Andreiana, V.-A.; Duica, M.C.; Stefan, M.-C.; Susanu, I.O.; Coman, M.D.; Iancu, D. Perspectives for the Development of Sustainable Cultural Tourism. *Sustain. Sci. Pract. Policy* **2022**, *14*, 5678. [CrossRef]
3. To, C.M. Assessment of factors affecting spiritual cultural tourism an Giang Province, Vietnam. *GeoJ. Tour. Geosites* **2023**, *46*, 227–233. [CrossRef]
4. Wendt, J.A.; Ersöz Tüğen, A. The art of terracotta from the perspective of cultural geography: Tavas case (Denizli/Turkey). *GeoJ. Tour. Geosites* **2022**, *43*, 1005–1012. [CrossRef]
5. Lajçi, D.; Kuqi, B.; Fetahaj, A.; Osmanollaj, S. The values of cultural heritage in the Rugova region in promoting the development of tourism in Kosovo. *GeoJ. Tour. Geosites* **2022**, *41*, 502–508. [CrossRef]
6. Gamal, R.; Abdelkafy, J.; Soliman, A. The atmospheric elements of the egyptian museums and their effect on the egyptians' intentions to revisit. *GeoJ. Tour. Geosites* **2023**, *46*, 148–155. [CrossRef]
7. Krakowiak, B. Museums in cultural tourism in Poland. *Turyzm* **2013**, *23*, 23–32. [CrossRef]
8. Vassiliadis, C.; Belenioti, Z.-C. Museums & Cultural Heritage via Social Media: An Integrated Literature Review. *Tourismos* **2017**, *12*, 97–132. [CrossRef]
9. Palumbo, R. Enhancing Museums' Attractiveness through Digitization: An Investigation of Italian Medium and Large-sized Museums and Cultural Institutions. *Int. J. Tour. Res.* **2022**, *24*, 202–215. [CrossRef]
10. Herman, G.V.; Grama, V.; Ilieș, A.; Safarov, B.; Ilieș, D.C.; Josan, I.; Buzrukova, M.; Janzakov, B.; Privitera, D.; Dehoorne, O.; et al. The Relationship between Motivation and the Role of the Night of the Museums Event: Case Study in Oradea Municipality, Romania. *Sustain. Sci. Pract. Policy* **2023**, *15*, 1738. [CrossRef]

11. Broström, T.; Klenz Larsen, P. *Climate Control in Historic Buildings*; National Museum of Denmark: Copenhagen, Denmark, 2015.
12. Ortiz, R.; Párraga, M.; Navarrete, J.; Carrasco, I.; de la Vega, E.; Ortiz, M.; Herrera, P.; Jurgens, J.A.; Held, B.W.; Blanchette, R.A. Investigations of biodeterioration by fungi in historic wooden churches of Chiloé, Chile. *Microb. Ecol.* **2014**, *67*, 568–575. [CrossRef] [PubMed]
13. Estrada, A.R.; Torres, E.M.; Vázquez, M.A.A.; Piñero, J.L.H.; Lucio, M.A.G.; Martínez, S.M.S. Fungal spores in four catholic churches in the metropolitan area of Monterrey, Nuevo León State, Mexico–First study. *Ann. Agric. Environ. Med.* **2015**, *22*, 221–226. [CrossRef] [PubMed]
14. Di Carlo, E.; Chisesi, R.; Barresi, G.; Barbaro, S.; Lombardo, G.; Rotolo, V.; Sebastianelli, M.; Travagliato, G.; Palla, F. Fungi and bacteria in indoor cultural heritage environments: Microbial-related risks for artworks and human health. *Environ. Ecol. Res.* **2016**, *4*, 257–264. [CrossRef]
15. de Carvalho, H.P.; Mesquita, N.; Trovão, J.; Rodríguez, S.F.; Pinheiro, A.C.; Gomes, V.; Alcoforado, A.; Gil, F.; Portugal, A. Fungal contamination of paintings and wooden sculptures inside the storage room of a museum: Are current norms and reference values adequate? *J. Cult. Herit.* **2018**, *34*, 268–276. [CrossRef]
16. Prihatmanti, R.; Bahauddin, A. Indoor Air Quality in Adaptively Reused Heritage Buildings at a UNESCO World Heritage Site, Penang, Malaysia. *J. Constr. Dev. Ctries* **2014**, *19*, 69.
17. Bogdan, A.; Chambre, D.; Copolovici, D.M.; Bungau, T.; Bungau, C.C.; Copolovici, L. Heritage Building Preservation in the Process of Sustainable Urban Development: The Case of Brasov Medieval City, Romania. *Sustainability* **2022**, *14*, 6959. [CrossRef]
18. Ferdyn-Grygierek, J.; Kaczmarczyk, J.; Blaszczok, M.; Lubina, P.; Koper, P.; Bulińska, A. Hygrothermal risk in museum buildings located in moderate climate. *Energies* **2020**, *13*, 344. [CrossRef]
19. Leijonhufvud, G.; Broström, T. Standardizing the indoor climate in historic buildings: Opportunities, challenges and ways forward. *J. Archit. Conserv.* **2018**, *24*, 3–18. [CrossRef]
20. Ameen, A.; Mattsson, M.; Boström, H.; Lindelöw, H. Assessment of Thermal Comfort and Air Quality in Office Rooms of a Historic Building: A Case Study in Springtime in Continental Climate. *Buildings* **2023**, *13*, 156. [CrossRef]
21. Azmi, A.E.; Abd Rashid, A.; Abd Razak, A. An Assessment of Indoor Air Quality (IAQ) in Foundry Laboratory. In *IOP Conference Series: Earth and Environmental Science*; IOP Publishing: Bristol, UK, 2022; Volume 1019, p. 012046. [CrossRef]
22. Schwab, C.J.; Straus, D.C. The roles of Penicillium and Aspergillus in sick building syndrome. *Adv. Appl. Microbiol.* **2004**, *55*, 215–240.
23. Glevitzky, M.; Aleya, L.; Vica, M.L.; Dumitrel, G.-A.; Avram, M.; Tit, D.M.; Popa, M.; Popa, V.-C.; Behl, T.; Bungau, S. Assessing the microbiological contamination along with environmental factors of old books in the 1490-founded Bistrița Monastery, Romania. *Environ. Sci. Pollut. Res.* **2021**, *28*, 8743–8757. [CrossRef] [PubMed]
24. Syazwan Aizat, I.; Juliana, J.; Norhafizalina, O.; Azman, Z.A.; Kamaruzaman, J. Indoor air quality and sick building syndrome in Malaysian buildings. *Glob. J. Health Sci.* **2009**, *1*, 126–136.
25. Nur Fadilah, R.; Juliana, J. Indoor air quality (IAQ) and sick buildings syndrome (SBS) among office workers in new and old building in Universiti Putra Malaysia, Serdang. *Health Environ. J.* **2012**, *3*, 98–109.
26. Ilies, D.C.; Safarov, B.; Caciora, T.; Ilies, A.; Grama, V.; Ilies, G.; Huniadi, A.; Zharas, B.; Hodor, N.; Sandor, M.; et al. Museal Indoor Air Quality and Public Health: An Integrated Approach for Exhibits Preservation and Ensuring Human Health. *Sustainability* **2022**, *14*, 2462. [CrossRef]
27. Oneţ, A.; Ilieş, D.C.; Buhas, S.; Rahotă, D.; Ilieş, A.; Baias, S.; Marcu, F.; Herman, G.V. Microbial air contamination in indoor environment of University Sport Hall. *J. Environ. Prot. Ecol.* **2018**, *19*, 694–703.
28. Žuškin, E.; Schachter, E.N.; Mustajbegović, J.; Pucarin-cvetković, J.A.S.N.A.; Doko-Jelinić, J.A.G.O.D.A.; Mučić-Pucić, B.R.A.N.K.A. Indoor air pollution and effects on human health. *Period Biol.* **2009**, *111*, 37–40.
29. Abdel-Kareem, O. Monitoring, controlling and prevention of the fungal deterioration of textile artifacts in the museum of Jordanian heritage. *Mediterr. Archaeol. Archaeom.* **2010**, *10*, 85–96.
30. Łukaszuk, C.R.; Krajewska-Kułak, E.; Kułak, W. Effects of fungal air pollution on human health. *Prog. Health Sci.* **2011**, *1*, 156–164.
31. Khan, A.H.; Karuppayil, S.M. Fungal pollution of indoor environments and its management. *Saudi J. Biol. Sci.* **2012**, *19*, 405–426. [CrossRef]
32. Méheust, D.; Le Cann, P.; Reboux, G.; Millon, L.; Gangneux, J.P. Indoor fungal contamination: Health risks and measurement methods in hospitals, homes and workplaces. *Crit. Rev. Microbiol.* **2014**, *40*, 248–260. [CrossRef]
33. Hayleeyesus, S.F.; Manaye, A.M. Microbiological quality of indoor air in university libraries. *Asian Pac. J. Trop. Biomed.* **2014**, *4*, S312–S317. [CrossRef] [PubMed]
34. Mousavi, B.; Hedayati, M.T.; Hedayati, N.; Ilkit, M.; Syedmousavi, S. Aspergillus species in indoor environments and their possible occupational and public health hazards. *Curr. Med. Mycol.* **2016**, *2*, 36–42. [CrossRef] [PubMed]
35. Egbuta, M.A.; Mwanza, M.; Babalola, O.O. Health risks associated with exposure to filamentous fungi. *Int. J. Environ. Res. Public Health* **2017**, *14*, 719. [CrossRef] [PubMed]
36. Levin, H. Indoor climate and global climate change: Exploring connections. In Proceedings of the 11th International Conference on Indoor Air Quality and Climate, Indoor Air, Copenhagen, Denmark, 17–22 August 2008.
37. Brennan, T. *Indoor Environmental Quality and Climate Change*; US Environmental Protection Agency: Washington, DC, USA, 2010. Available online: https://www.epa.gov/sites/default/files/2014-08/documents/climate_change_brennan.pdf (accessed on 15 May 2023).

38. Institute of Medicine. *Climate Change, the Indoor Environment, and Health*; The National Academies Press: Washington, DC, USA, 2011; pp. 1–272. [CrossRef]
39. Nazaroff, W.W. Exploring the consequences of climate change for indoor air quality. *Environ. Res. Lett.* **2013**, *8*, 015022. [CrossRef]
40. Bungau, C.C.; Bungau, T.; Prada, I.F.; Prada, M.F. Green Buildings as a Necessity for Sustainable Environment Development: Dilemmas and Challenges. *Sustainability* **2022**, *14*, 13121. [CrossRef]
41. Bungau, C.C.; Bungau, C.; Toadere, M.T.; Prada-Hanga, I.F.; Bungau, T.; Popescu, D.E.; Prada, M.F. Solutions for an Ecological and Healthy Retrofitting of Buildings on the Campus of the University of Oradea, Romania, Built Starting from 1911 to 1913. *Sustainability* **2023**, *15*, 6541. [CrossRef]
42. Ghemis, M.T.; Scurt, A.A.; Bob, C.; Bungau, C. An analysis of sustainability about rehabilitation and reconversion of an old building. Modern technologies for 3rd Millennium. In Proceedings of the 17th National Technical-Scientific Conference on Modern Technologies for the 3rd Millennium, Oradea, Romania, 22–23 March 2018; pp. 273–278.
43. Leissner, J.; Kilian, R.; Kotova, L.; Jacob, D.; Mikolajewicz, U.; Broström, T.; Ashley-Smith, J.; Schellen, H.L.; Martens, M.; Van Schijndel, J.; et al. Climate for culture: Assessing the impact of climate change on the future indoor climate in historic buildings using simulations. *Herit. Sci.* **2015**, *3*, 38. [CrossRef]
44. Nakicenovic, N.; Alcamo, J.; Davis, G.; Vries, B.D.; Fenhann, J.; Gaffin, S.; Zhou, D. Special Report on Emissions Scenarios. 2020. Available online: https://www.ipcc.ch/site/assets/uploads/2018/03/sres-en.pdf (accessed on 15 May 2023).
45. Lankester, P.; Brimblecombe, P. The impact of future climate on historic interiors. *Sci. Total Environ.* **2012**, *417–418*, 248–254. [CrossRef]
46. Muñoz González, C.M.; León Rodríguez, A.L.; Suárez Medina, R.; Ruiz Jaramillo, J. Effects of future climate change on the preservation of artworks, thermal comfort and energy consumption in historic buildings. *Appl. Energy* **2020**, *276*, 115483. [CrossRef]
47. Vardoulakis, S.; Dimitroulopoulou, C.; Thornes, J.; Lai, K.M.; Taylor, J.; Myers, I.; Heaviside, C.; Mavrogianni, A.; Shrubsole, C.; Chalabi, Z.; et al. Impact of climate change on the domestic indoor environment and associated health risks in the UK. *Environ. Int.* **2015**, *85*, 299–313. [CrossRef]
48. Pioppi, B.; Pigliautile, I.; Piselli, C.; Pisello, A.L. Cultural heritage microclimate change: Human-centric approach to experimentally investigate intra-urban overheating and numerically assess foreseen future scenarios impact. *Sci. Total Environ.* **2020**, *703*, 134448. [CrossRef] [PubMed]
49. Yang, Y.; Javanroodi, K.; Nik, V.M. Climate change and energy performance of European residential building stocks—A comprehensive impact assessment using climate big data from the coordinated regional climate downscaling experiment. *Appl. Energy* **2021**, *298*, 117246. [CrossRef]
50. Mansouri, A.; Wei, W.; Alessandrini, J.M.; Mandin, C.; Blondeau, P. Impact of Climate Change on Indoor Air Quality: A Review. *Int. J. Environ. Res. Public Health* **2022**, *19*, 15616. [CrossRef]
51. Bandyopadhyay, B.; Banerjee, M. Decarbonization of cooling of buildings. *Sol. Compass* **2022**, *2*, 100025. [CrossRef]
52. Ilies, D.C.; Marcu, F.; Caciora, T.; Indrie, L.; Ilies, A.; Albu, A.; Costea, M.; Burtă, L.; Baias, S.; Ilies, M.; et al. Investigations of Museum Indoor Microclimate and Air Quality. Case Study from Romania. *Atmosphere* **2021**, *12*, 286. [CrossRef]
53. Orthel, B.D. Linking public health and heritage work. *Int. J. Herit. Stud.* **2022**, *28*, 44–58. [CrossRef]
54. Ilieș, D.C.; Buhaș, R.; Ilieș, A.; Gaceu, O.; Oneț, A.; Buhaș, S.; Rahotă, D.; Dragoș, P.; Baias, S.; Marcu, F.; et al. Indoor air quality issues. Case study: The Multipurpose Sports Hall of the University of Oradea. *Environ. Eng. Manag. J.* **2018**, *17*, 2999–3005.
55. Onet, A.; Ilies, D.C.; Ilies, A.; Herman, G.V.; Burta, L.; Marcu, F.; Buha, R.; Caciora, T.; Baias, S.; Indoor, C.O.; et al. Indoor air quality assessement and its perception, Case study historic wooden church, Romania. *Rom. Biotechnol. Lett.* **2020**, *25*, 1547–1553. [CrossRef]
56. Mašková, L.; Smolík, J.; Ďurovič, M. Characterization of indoor air quality in different archives—Possible implications for books and manuscripts. *Build. Environ.* **2017**, *120*, 77–84. [CrossRef]
57. Indrie, L.; Oana, D.; Ilieş, M.; Ilieş, D.C.; Lincu, A.; Ilieş, A.; Ștefan, B.; Herman, G.V.; Aurelia, O.; Costea, M.A.; et al. Indoor air quality of museums and conservation of textiles art works. Case study: Salacea Museum House Romania. *Ind. Text.* **2019**, *70*, 88–93.
58. Ilies, D.C.; Onet, A.; Marcu, F.; Gaceu, O.; Timar, A.; Baias, S.; Ilies, A.; Herman, G.V.; Costea, M.; Tepelea, M.; et al. Investigations on air quality in the historic wooden church in Oradea City, Romania. *Environ. Eng. Manag. J.* **2018**, *17*, 2731–2739. [CrossRef]
59. Saini, J.; Dutta, M.; Marques, G. Indoor air quality monitoring systems based on internet of things: A systematic review. *Int. J. Environ. Res. Public Health* **2020**, *17*, 4942. [CrossRef] [PubMed]
60. Marcu, F.; Hodor, N.; Indrie, L.; Dejeu, P.; Ilieș, M.; Albu, A.; Sandor, M.; Sicora, C.; Costea, M.; Ilieș, D.C.; et al. Microbiological, health and comfort aspects of indoor air quality in a Romanian historical wooden church. *Int. J. Environ. Res. Public Health* **2021**, *18*, 9908. [CrossRef] [PubMed]
61. Borrego, S.; Lavin, P.; Perdomo, I.; Gómez de Saravia, S.; Guiamet, P. Determination of indoor air quality in archives and biodeterioration of the documentary heritage. *Int. Sch. Res. Not.* **2012**, *2012*, 680598. [CrossRef] [PubMed]
62. Gaceu, O.; Ilieș, D.C.; Baias, Ș.; Georgiță, M.; Ilieș, A.; Caciora, T.; Indrie, L.; Albu, A.; Herman, G.V.; Baidog, A.; et al. Microclimatic Characteristics and Air Quality Inside the National Archives of Bihor County, Romania. *Environ. Eng. Manag. J. (EEMJ)* **2021**, *20*, 459–466.
63. Wysocka, M. Analysis of indoor air quality in a naturally ventilated church. *E3S Web Conf.* **2018**, *49*, 00134. [CrossRef]

64. Cincinelli, A.; Martellini, T.; Amore, A.; Dei, L.; Marrazza, G.; Carretti, E.; Belosi, F.; Ravegnani, F.; Leva, P. Measurement of volatile organic compounds (VOCs) in libraries and archives in Florence (Italy). *Sci. Total Environ.* **2016**, *572*, 333–339. [CrossRef]
65. Gonzalez-Martin, J.; Kraakman, N.J.R.; Perez, C.; Lebrero, R.; Munoz, R. A state–of–the-art review on indoor air pollution and strategies for indoor air pollution control. *Chemosphere* **2021**, *262*, 128376. [CrossRef]
66. Zorpas, A.A.; Skouroupatis, A. Indoor air quality evaluation of two museums in a subtropical climate conditions. *Sustain. Cities Soc.* **2016**, *20*, 52–60. [CrossRef]
67. Dzulkifli, S.M.; Abdullah, A.H.; Leman, A.M. Indoor Air Quality of Museum Building Environment in a Tropical Climate: Proposed Study. In Proceedings of the 3rd Scientific Conference on Occupational Safety and Health, Johor Bahru, Malaysia, 14–17 October 2014.
68. Ilies, A.; Caciora, T.; Marcu, F.; Berdenov, Z.; Ilies, G.; Safarov, B.; Hodor, N.; Grama, V.; Shomali, M.A.A.; Ilies, D.C.; et al. Analysis of the Interior Microclimate in Art Nouveau Heritage Buildings for the Protection of Exhibits and Human Health. *Int. J. Environ. Res. Public Health* **2022**, *19*, 16599. [CrossRef]
69. Azuma, K.; Ikeda, K.; Kagi, N.; Yanagi, U.; Osawa, H. Physicochemical risk factors for building-related symptoms in air-conditioned office buildings: Ambient particles and combined exposure to indoor air pollutants. *Sci. Total Environ.* **2018**, *616*, 1649–1655. [CrossRef] [PubMed]
70. Baldan, M.; Manente, S.; Izzo, F.C. The role of bio-pollutants in the indoor air quality of old museum buildings: Artworks biodeterioration as preview of human diseases. *Environ. Monit. Assess.* **2021**, *193*, 787. [CrossRef] [PubMed]
71. Sunyer, J.; Esnaola, M.; Alvarez-Pedrerol, M.; Forns, J.; Rivas, I.; López-Vicente, M.; Suades-González, E.; Foraster, M.; Garcia-Esteban, R.; Basagaña, X.; et al. Association between traffic-related air pollution in schools and cognitive development in primary school children: A prospective cohort study. *PLoS Med.* **2015**, *12*, e1001792. [CrossRef] [PubMed]
72. Tétreault, P.; Mansour, A.; Vachon-Presseau, E.; Schnitzer, T.J.; Apkarian, A.V.; Baliki, M.N. Brain connectivity predicts placebo response across chronic pain clinical trials. *PLoS Biol.* **2016**, *14*, e1002570. [CrossRef] [PubMed]
73. Carrion-Matta, A.; Kang, C.M.; Gaffin, J.M.; Hauptman, M.; Phipatanakul, W.; Koutrakis, P.; Gold, D.R. Classroom indoor PM2. 5 sources and exposures in inner-city schools. *Environ. Int.* **2019**, *131*, 104968. [CrossRef] [PubMed]
74. Ali, M.U.; Lin, S.; Yousaf, B.; Abbas, Q.; Munir, M.A.M.; Rashid, A.; Zheng, C.; Kuang, X.; Wong, M.H. Pollution characteristics, mechanism of toxicity and health effects of the ultrafine particles in the indoor environment: Current status and future perspectives. *Crit. Rev. Environ. Sci. Technol.* **2022**, *52*, 436–473. [CrossRef]
75. Wei, Y.; Jang-Jaccard, J.; Xu, W.; Sabrina, F.; Camtepe, S.; Boulic, M. Lstm-autoencoder based anomaly detection for indoor air quality time series data. *IEEE Sens. J.* **2023**, *23*, 3787–3800. [CrossRef]
76. Arar, M.; Jung, C. Analyzing the Perception of Indoor Air Quality (IAQ) from a Survey of New Townhouse Residents in Dubai. *Sustainability* **2022**, *14*, 15042. [CrossRef]
77. Aziz, N.; Adman, M.A.; Suhaimi, N.S.; Misbari, S.; Alias, A.R.; Abd Aziz, A.; Lee, L.F.; Khan, M.H. Indoor Air Quality (IAQ) and Related Risk Factors for Sick Building Syndrome (SBS) at the Office and Home: A Systematic Review. In *IOP Conference Series: Earth and Environmental Science*; IOP Publishing: Bristol, UK, 2023; Volume 1140, p. 012007. [CrossRef]
78. Pei, J.; Qu, M.; Sun, L.; Wang, X.; Yin, Y. The relationship between indoor air quality (IAQ) and perceived air quality (PAQ)—A review and case analysis of Chinese residential environment. *Energy Built Environ.* 2022, *in press*. [CrossRef]
79. Ilies, D.C.; Caciora, T.; Ilies, A.; Berdenov, Z.; Hossain, M.A.; Grama, V.; Dahal, R.K.; Zdrinca, M.; Hassan, T.H.; Herman, G.V.; et al. Microbial Air Quality in the Built Environment—Case Study of Darvas-La Roche Heritage Museum House, Oradea, Romania. *Buildings* **2023**, *13*, 620. [CrossRef]

sustainability

MDPI

Article

Administrative Aspects Regarding the Valorisation of Geothermal Waters for Balneological Purposes in Bihor County, Romania

Anca-Paula Ciurba (Pastor) [1,2], Ionel Haidu [1] and Dorina Ianc [3,*]

[1] Laboratory LOTERR-EA 7304, Department of Geography, University of Lorraine, 57045 Metz, France; anca-paula.ciurba@univ-lorraine.fr (A.-P.C.); ionel.haidu@univ-lorraine.fr (I.H.)
[2] Department of Geography, University of Oradea, 410087 Oradea, Romania
[3] Human Performance Research Center, Department of Physical Education, Sport and Physiotherapy, University of Oradea, 410087 Oradea, Romania
* Correspondence: dianc@uoradea.ro

Abstract: This study aims to analyse the development of the balneological phenomenon in Bihor County generated by its administration, consequent of which, a classification of rural settlements and new tourist resorts has been made. The objective is to identify the current trend in the exploitation of the balneological potential of the localities in Bihor County. A comparative analysis has also been carried out between the influence of the political regimes on the exploitation of geothermal water sources at the level of rural settlements and new tourist resorts during the last five decades. The comparative analysis has revealed that the development of the balneological phenomenon has been carried out sequentially. This sequence has been influenced by the political environment and the administrative factor from the period before and after the Revolution. In the period before the Revolution, all balneological sources were managed by the public administrative factor; they were maintained and kept functional, until the Revolution of 1989, against the background of a stable political environment. In the post-Revolution period, when the decision and involvement was sometimes undertaken on a small scale (private administration), and where the interest in development was focused exclusively on the valorisation of the balneological resources, it can be said that capitalist policy favourably influenced the evolution of the balneological phenomenon at the level of rural settlements and new tourist resorts. We have concluded that the capitalist period has been favourable because, when the thermal baths were found to be unanimously managed by the public administration, they were functional in greater numbers at the rural level as compared to the post-revolutionary period. We have not analysed any "policy" documents in order to make this statement.

Keywords: geothermal water; balneology; wellness and SPA; development; administration; rural settlements; Revolution of 1989; Bihor County; Romania

Citation: Ciurba, A.-P.; Haidu, I.; Ianc, D. Administrative Aspects Regarding the Valorisation of Geothermal Waters for Balneological Purposes in Bihor County, Romania. *Sustainability* **2023**, *15*, 10320. https://doi.org/10.3390/su151310320

Academic Editors: Alina Badulescu, Constantin Bungău, Delia Mirela Tit, Jin Luo, Cosmin Mihai Vesa and Dorina Camelia Ilieș

Received: 29 April 2023
Revised: 26 June 2023
Accepted: 27 June 2023
Published: 29 June 2023

1. Introduction

Bihor County is very rich in geothermal and mineral waters, and it can rely on their quality and quantity. The balneological phenomenon is based on the important aspects that contribute to the sustainability of the exploitation of geothermal resources, influenced by their administrative apparatus. There are many factors that influence the development of the balneological resource, such as economic factors (prices, income of the population, tourist offer), social factors (free time), educational factors (level of education, desire for knowledge), administrative factors (facilities and services), etc. In our study, we considered the administrative factor because it is directly involved in the sustainability of the balneological resource through the diversity and quality of the offered services, the cost of the services, the appropriate promotion, and the implementation of financing projects. A high-performing administration adapted to the changes of the present, but especially of the future, is necessary and essential for the development of balneological resources.

Bihor County has had a long history of exploiting geothermal water. Several previous studies have reported on geothermal resources and their recognition, health, well-being, and sustainability [1–3]. The development in rural localities, for the exploitation of geothermal water, began in the 1960s–1970s, as a result of prospecting carried out to identify hydrocarbon resources. In Bihor County, there are 60 localities with geothermal resources. This comes from the geothermal deposits with the perimeter at the level of the localities with balneological resources [4]. Among the localities with geothermal resources, excluding the established resorts (Băile 1 Mai, Băile Felix, Stîna de Vale, and Băile Tinca), several studies have been carried out, and we have found only 12 localities (Mădăras, Tămășeu, Săcueni, Sarcău, Chișlaz, Livada de Bihor, Sînicolau de Munte, Alesd, Beiuș, Marghita, Sântandrei, and Oradea) which capitalize on this resource for several purposes: leisure, balneological treatment, in the skincare industry, as a heating agent, and fish farming. At the scale of rural settlements, it is known that the sustainable development and well-being of the inhabitants is closely linked to economic productivity and tourist attractiveness [5,6]. The tourist attractiveness of the villages in Bihor County has been greatly favoured by its geothermal water resources and balneological potential [7,8]. One of the important aspects that contribute to the development of the balneological phenomenon in Bihor County is its administrative factor. The practical importance is demonstrated by its need of economic development, industrial production, policy application, and overall increase [9,10].

In Romania, the exploitation of geothermal water for heating purposes is supported by the Thermal Energy Law no. 325/2006 [11], which establishes the general framework for heating systems and aims to encourage the use of sustainable energy sources, through which special mention is made of the geothermal energy. According to this law, all heating systems must be public possession. However, the procedure allows for a concession to a functional private company or a public–private association [12]. These types of geothermal energy have an appreciable economic and also environmental price because of the exploitation of the Earth's natural warmth, which is an economically and ecologically attractive substitute source of energy that can satisfy the growing demands of the 21st century [13]. When it comes to using geothermal energy, the University of Oradea is the only university campus in the world that uses geothermal water as a heating agent, for sustainable purposes. This organization, along with other nine universities, is a member of the European Universities for Sustainability Alliance (EU GREEN) [1]. The benefits of thermal waters in the treatment of skin conditions such as psoriasis, atopic dermatitis, and seborrhoea [14] influenced by their richness in minerals and trace elements with established dermatological indications can justify the use of geothermal water in the skincare industry [14].

The entire evolution of capitalization, from the communist to the capitalist period, was connected to administrative changes and political goals focused on a fast capitalist exploitation of thermal baths. Initially, there were many failures as no investments were made in the development or the functional maintenance of the thermal baths, which led to a state of dereliction. Then, gradually, some entrepreneurs found a way to capitalize on the potential of thermal baths through permissive and encouraging legislation, although the state intervened timidly to support the balneological progress.

The objective is to identify the current trend in the exploitation of the balneological potential of the villages, towns, and cities from Bihor County.

Literature Review

Tourism contributes to the positive experience of both tourists and local residents [15] but it also impacts the economy, contributing to the economic development of destinations [16–18]. The sustainability, challenges, opportunities, and advantages of wellness and spa tourism are supported by the quality of the environment and resources [19]. The principles of sustainable tourism provide evidence of how integrating public health principles into destination management, destination policy, local politics, activities, and destination capital can deliver positive outcomes [20].

Tourist destinations that have the ability to become or remain attractive for wellness tourism need safety, good prices, nature-based activities, cultural attractions, a temperate climate, traditional therapies, healthy local cuisine, authenticity, sustainability, low pollution levels, and health tourism facilities, services, and experiences [19].

Studies on the role that health and wellness tourism plays in the sustainable development of the territory, by facilitating and contributing to the quality of life of communities, have demonstrated its success when it combines the different interests of the locals and of the visitors and it observes the natural, cultural, and economic resources of the territories [21–23].

Nature tourism occurs in areas where certain natural resources and climate can clearly influence well-being levels. Several studies report on the importance of balneology and the use of thermal waters for health and the role of health tourism in improving the quality of life, as well as on the facilities in health tourism and the inherent challenges in managing them in a profitable and sustainable way [24–26].

Geothermal tourism is an intensive practice worldwide, and opportunities related to geothermal bathing centres (SPA) are emerging for municipalities and administrations that decide to introduce them [27]. Water is the symbol of health tourism through its two major components—spa tourism, and wellness tourism—by supporting leisure or treatment tourism [28].

The concept of wellness is understood as the sum of all experiences located within the destination that promote health and well-being, including the holistic enrichment of tourists' physical resources, such as mineral waters, enhanced by other services such as hotels, restaurants, entertainment, and cultural activities [29].

There are many articles that analyse the cost-effectiveness of spa therapies with regard to various pathologies that also analyse patient reimbursement services after a spa therapy, and conclude a positive impact on patients' health and their own perception of their quality of life [30–34].

As the use of geothermal water in heating, Romanov and Leiss (2022), report in their study the recent developments in the construction sector, in the field of geothermal technology for heating. This indicates trends for more sustainable and environmentally friendly geothermal water supply systems for heating [35]. More studies have discussed the alternative use of geothermal water; in addition to balneology, it is used as a dermatological treatment in the cure of various conditions, through its anti-inflammatory effects and acceleration of the healing process [36–39]. Another form of valorisation that has not been researched enough is the application of geothermal energy in different areas of the agricultural sector, such as irrigation, heating of greenhouses and soil, drying of agricultural products, and cultivation of algae [40,41]. Regarding the use of geothermal water for fish farming as a source of cheap and clean energy in sustainable development, this would allow fish to grow at optimal temperatures throughout the year, which would result in the best environment conditions for rapid growth, as shown by several studies [42,43].

2. Materials and Methods

Bihor County is located in the northwest part of Romania. The advantage of the county's location in North-West Romania is that it is a border county from an economic spectrum, which offers multiple possibilities for development and collaboration. The morphological support is arranged in three relief steps (plain, hill, and mountain). Due to its geographical position, Bihor County belongs to the moderate temperate-continental climate, with an average annual air temperature that decreases with altitude: from 10.5 °C in the plain area, 8–10 °C in the hills, and 7–2 °C in the mountains [44]. As an administrative-territorial division, Bihor County has 101 territorial administrative units, grouped as follows: 10 cities, 18 suburban localities, and, in the rural area, 91 communes (with 430 villages, including commune residences). In our study, we included the localities with the balneological resources.

For this study, we collected data by terrain as well as bibliographic documentations. Field documentation included approaching and reviewing each location and administrative entity, by visiting the location, inventory of the restoration infrastructure, accommodation, treatment, and leisure infrastructure. We conducted an interview with the administrative representatives of the balneological source (mayors, private administrators, locals). We filled in the "Thermal water location sheet" at each location visited (Appendix A). A total of 10 interviews were conducted at the 10 less popular locations in Bihor County. The questions represent the items in the Location Sheet. Thus, we analysed the real situation of the balneological source. The interview consisted of questions about the history of the thermal baths, up to the Revolution of 1989, their evolution, and the form of administration until now, as well as the analysis of the current situation of valorisation of the balneological resource. Bibliographic documentation was carried out by collecting data from a literature review in Scopus, Web of Science, PubMed, and ScienceDirect involving health tourism documents, renewable energy, balneotherapy, wellness and spa, and administration. Based on the practical approach from the field and the bibliographic analysis, we claim that the evolution of the balneological phenomenon is influenced by its administrative factor, analysing the development of accommodation, restoration, leisure and treatment infrastructures. For this purpose, we present the hypothesis that the development of the balneological phenomenon at the level of rural settlements and new tourist resorts in Bihor County is conditioned by the involvement of the administrative factor in the prioritization of this sector, but also by the form of its administration—public or private. This hypothesis is verified by analysing the development of the balneological phenomenon generated by its administration, starting from the moment of the initial arrangement until the Revolution of 1989, and from the Revolution to the present. As mentioned by Cornea (2007) [45] in his study on the influence of the existing political regime on the administration of the respective state, the major differences in the way of organization and functioning of the administrative system were dictated by two categories of political regimes—non-democratic and democratic. As a result, we analysed the development of the balneological phenomenon generated by its administration, with the aim of highlighting the impact of the change in ownership due to the transition from socialism to capitalism. We developed and used the methodology for classifying rural localities and new tourist resorts, depending on the current status of exposure to the balneological resource, which is as follows: the classification of new tourist resorts of national and local interest with a specific balneological classification of localities that valorised balneological resource; the classification of localities with a balneological resource that utilized geothermal water before the Revolution of 1989; classification of localities with an unused balneological resource; the classification of rural settlements with a balneological factor as well as the new tourist resorts with a balneological factor in Bihor County, according to the form of administration, in the pre- and post-revolution period, and up to now.

3. Analysis of the Development of the Balneological Phenomenon According to the Administration Type

3.1. The Evolution of the Types of Administration in the Case of Balneological Resources of Local Interest

In Romania, until the Revolution of 1989, the form of administration of balneological resources was exclusively public. The impact of the change in ownership through the transition from socialism to capitalism, for some balneological locations, meant the transfer of the old administrative units of the balneological sources from state property to private administration (Figure 1). They were taken over as a management location (by concession), and their form of administration did not lead to any investment—only to exploitation. The management did not increase the capital; they only increased wages and invented expenses, which led to bankruptcy and ruin. The transition period in Romania can be divided into two phases: the first phase was dominated by rapid, predominantly destructive privatization that lasted from 1990 until 2000; the second phase covers the period 2001–2004 and was characterized by the privatization of large companies that occupied strategic positions in the Romanian economy. The return of property confiscated under communism completes the privatization process. During this process, fraud caused serious damage to the economy and to society alike. Middlemen exploited bribes to make tremendous profits, many of which were costly [46]. This was most likely one of the causes of the unsustainability of rural thermal baths. The administration of the balneological resources in rural localities and new tourists resorts with balneological factors in Bihor Country, from the Revolution of 1989 to the present day (2023), has been predominantly managed by private administrative factors, accounting for 61% of the total. This percentage has positively influenced the performance of balneological resources in the mentioned localities.

The form of administration of the spas from the Revolution of 1989 to the present is represented by 61% private administrative factors and 39% public administrative factors.

At first, the local community looked after the balneological water springs in a very simple manner. Afterwards, the Agricultural Production Cooperatives took them in administration, and managed them until the end of the socialist period, in 1989. After the 1989 Revolution, the country's return to capitalism generated an unclear background for the entire national tourist area, with questionable privatizations within this sphere of interest. These were accompanied by a lack of clear government policies regarding the Romanian balneological heritage, which meant that the spa towns and villages were left close to ruin [47]. Along with privatization, enterprise advanced and diverse internally (in 2010, Government Decision No. 120/2010 was issued regarding the list of investment programs and projects in tourism, sources of financing for technical documentation, and works of execution of investment programs and objectives in tourism, as well as the approval of the eligibility criteria (www.turism.gov.ro (accessed on 16 January 2023)) and external (European funding) financing source were initiated. A few of those localities thrived, certainly, where the administrative factor was involved and concerned with their evolution and improvement by adding modern accommodation structures (three-star hotels, three-star guest houses, wooden cabins, wooden houses, cand ampsites): Sânicolau de Munte, Sarcău, Mădăras, and Livada de Bihor. Other sites have been permanently maintained to function even today: in Chișlaz, Săcueni, and Tămășeu. However, with the land transfer process, a few of the thermal baths entered the authority of individuals (Rabagani commune), who solicited the local authorities to destroy the thermal baths, or they were sold to firms. Others remained in the property of the local town halls, but failed to remain functional due to high operating costs, having fallen in disrepair: Mihai Bravu, Valea lui Mihai, Cadea, and Ciocaia [48]. At the Bihor County level, in the period 2020–2022, in addition to the 4 established balneological resorts (Băile 1 Mai, Băile Felix, Stîna de Vale, and Băile Tinca), 13 localities (Oradea, Beiuș, Ștei, Marghita, Salonta, Mădăras, Borș, Vadu Crișului, Săcuieni, Bratca, Șuncuiuș, Pietroasa- Budureasa, and Padiș), following the feasibility study, met the criteria for certification as local level tourist resorts according to the annexes to Government Decision no. 852/2008 [49]. Of these, eight resorts (Oradea, Beiuș, Ștei, Marghita, Salonta, Mădăras, Borș, and Săcueni) have balneological resources. The status of a tourist resort facilitates their access to European funding axes; thus, we can foresee a positive development in their future.

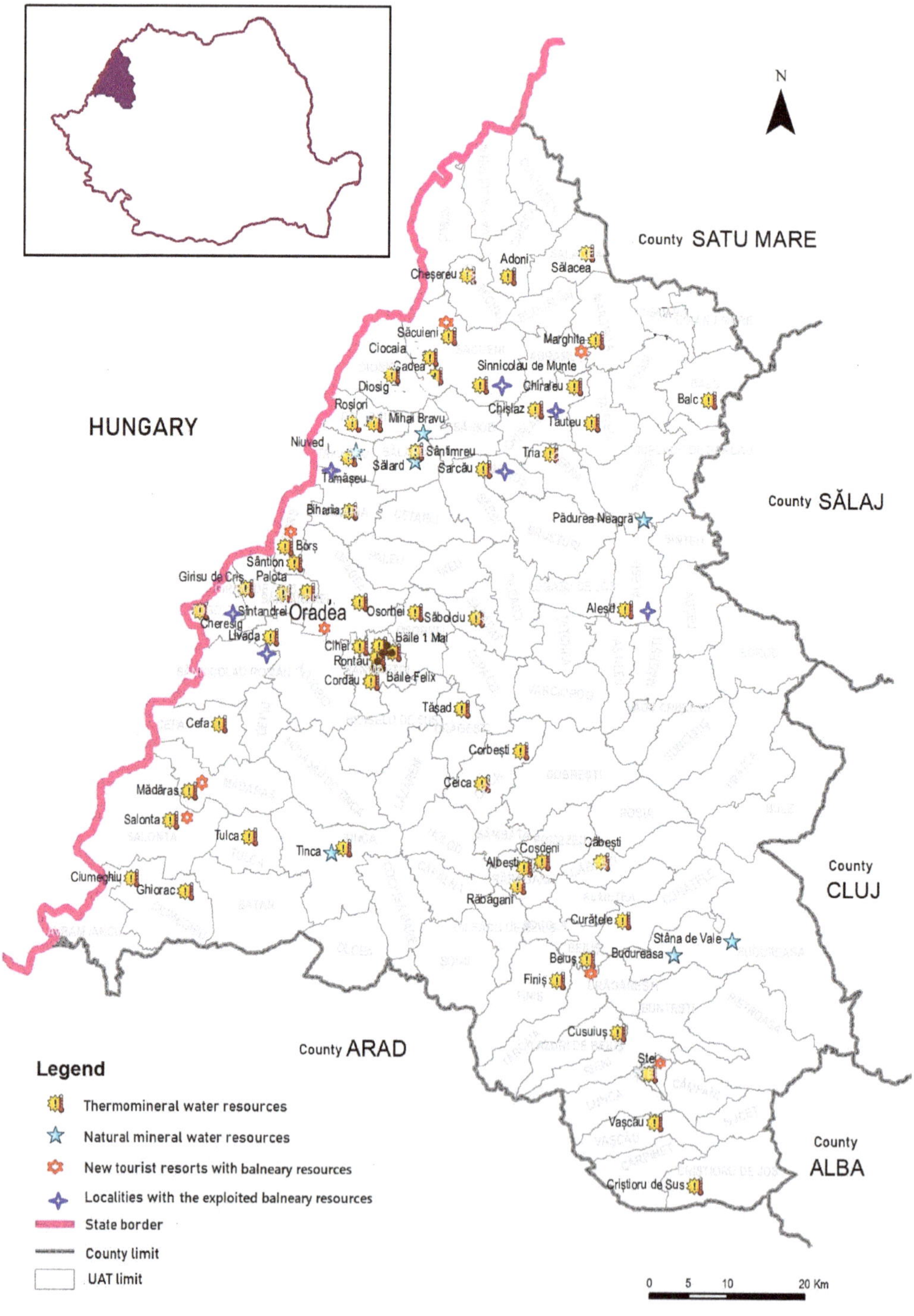

Figure 1. The current status of the valorisation classes of geothermal sources for balneological purposes. Sources: based on own survey, 2023.

3.2. Ways of Alternative Exploitation of Geothermal Water at the Rural Level and in the New Tourist Resorts

At the rural level, from the development of balneological resources until the Revolution of 1989, in addition to individual consumption and rudimentary and recreational balneological purpose (Mădăras, Sarcău, Tămășeu, Săcuieni, Sînicolau de Munte, and Chișlaz), since the 1980s, geothermal water has been used for heating (Oradea, Livada de Bihor). From 1985 until 1989, it was used in the fish industry, serving the largest station for rearing fish in thermal water in Europe (Livada de Bihor). Today, in Bihor County, geothermal water is used for fish farming in Sântandrei.

In the period following the Revolution of 1989 and until today, new boreholes have been drilled, both in the rural localities (Sântandrei) and in urban ones (Oradea, Beiuș, Ștei). After the Revolution of 1989, the number of localities that use geothermal water for heating increased (Beiuș, Livada de Bihor, Oradea, and Săcuieni). This proved to be a profitable and renewable source, which has contributed to the extension in the internal economics due to the low price of GCAL, which has also led to an increased thermal comfort for the population. As the balneological resources were under the supervision of a stable administration, new workplaces were created, which participated in the growth of the local economy by taxes paid, and as a result, contributing to the improvement of the quality of life of the villagers, through the accumulation of capital [50]. The geothermal water has also been used in agriculture, serving the local tomato greenhouses (Livada de Bihor, Roșiori). At Livada de Bihor, the greenhouses operated from 2006 to 2014, and in 2022, a tomato production restarted. Another niche for the exploitation of geothermal water, starting in 2021, is in the skincare industry (Beiuș, www.beyuscosmetics.com (accessed on 12 January 2023)). The exploitation of geothermal water in the skincare industry from Beiuș, through product and process innovation, has contributed to economic growth by creating jobs and increasing turnover through the sale of products. In addition, after the Revolution, the "wellness and spa" sector developed with geothermal water at the Elite Hotel and at the Nymphaea Aquapark in Oradea, along the Beiuș Lagoon, but also in the rural villages of Sarcău and Sânicolau de Munte. The town of Ștei has as a preliminary plan for the exploitation of geothermal water, the heating of educational institutions as well as the expansion, rehabilitation, and modernization of the balneology section of the Ștei City Hospital (SIDDU Ștei) [51].

3.3. Risk of Unsustainability of the Exploitation of Balneological Resources in Rural Localities

Lack of investment in the care of thermal baths, in order to keep them functional, has resulted in their derelict state or in their disappearance by demolition. That situation began with denationalization after the 1989 Revolution, when the thermal baths came under public or private management, and irrespective of financial support or lack of interest in their value, they were allowed to disappear or to fall into oblivion: Cadea, Mihai Bravu, Valea lui Mihai, Rabagani, Ciocaia, and Balc. This is due to the administration's lack of prioritization of local public projects, which implies that a recreation base in a rural locality does not have priority over sewage infrastructure, water supply, or the paving of village roads. The threats that endanger the sustainability of geothermal water exploitation in rural localities could be administrative risks; risks of unpopularity or risks of depopulation of villages. Administrative risks could be a cause of the unsustainability of the exploitation of geothermal waters by changing the administrative factor, especially towards the private one, which pursues a profit too fast by increasing the fees without increasing the quality of services. The risks of unpopularity could be a cause of unsustainability due to not modernizing the infrastructure, and so they have finally become unattractive for tourists, who are becoming more and more informed. The risk of depopulation of villages, determined by the migration of the young population towards localities with educational and entertainment potential, which determines perspectives and opportunities for development, could determine the unsustainability of capitalizing on the rural balneological potential.

4. Results

Following the analysis of the development of the balneological phenomenon generated by its administrative factor, we can classify the localities according to their current status of exposure to the balneological potential (Figure 1).

4.1. Localities with Simple Pools That Used Geothermal Water before the Revolution of 1989

In all these localities (Mihai Bravu, Răbăgani, Valea lui Mihai, Cadea, Balc, and Ciocaia), the balneological resource was identified in the period 1960–1970, during which, geological prospecting works were carried out in order to identify oil or deposits of natural gas. The works highlighted the presence of valuable geothermal sources, which were exploited for balneological purposes by the local residents of that time. Through their own initiative and effort, they constructed swimming pools. Even if these were simple pools with thermal water, their connection to the hydrographic network or the sampling of cooling water required works that referred to adductions, discharges, or delimitations of weir pools [52]. The administration of the balneological source in these rural locations was assigned to the Agricultural Production Cooperatives, and they managed them until the Revolution of 1989 [48]. After this period, against the backdrop of an unstable political environment, the process of privatization began, the thermal baths were managed by various management teams, and due to the fact that there was a low interest, low experience or insufficient financial aid, they remained in disrepair.

4.2. Localities with Previously Valorised Balneological Resources

In all localities with previously valorised balneological resources (Aleșd, Livada de Bihor, Sânicolau de Munte, Sarcău, Chișlaz, Tămășeu, and Sântandrei), the balneological resource was divided after development, from the 1970s, for the administration of the Agricultural Production Cooperatives or the municipalities (Aleșd), until the Revolution of 1989. At that time, these thermal pools were among the few, even the only, forms of recreation, especially in rural areas. Since the Revolution of 1989, these resources have been managed by different forms of public or private administration.

Aleșd is a city that capitalizes on geothermal water through a borehole that was drilled between 1979 and 1981 and which feeds two entities for therapeutic and recreational purposes: the thermal baths and the day centre for the elderly. The administration of the balneological resources belongs to the public sector. It has as a forecast plan for the valorisation of the hydro-therapeutic resource for the purpose of renewable resource for energy efficiency and the rehabilitation and modernization of the Aleșd Thermal Baths [53].

Livada de Bihor is a rural locality that has used geothermal water as a heating agent since 1980. Since 1985 it has served as the largest station for fish rearing in thermal water in Europe. In 1982, the thermal bath was built, which operated until 2007. Then, starting from the same year, the geothermal water served the local tomato greenhouses, until 2014. The restart of the thermal baths took place in the year 2020, when it was leased to a local private company and improved with a three-star accommodation structure and restaurant structure. Now, the administration of the balneological site is both public and private. As a preliminary plan for developing the balneological potential, 1500 sq m of land has been purchased in order to expand the thermal baths, by an indoor pool, an outdoor pool, and a boarding house (thermal baths administration).

Sânicolau de Munte is one of the few rural localities, which, with the privatization after the Revolution of 1989, developed the local balneological potential by improving the old thermal baths, by building an accommodation infrastructure, from caravans, wooden houses, to a three-star hotel, holding a capacity of 64 people. It also has a restoration infrastructure for 100 people and a modest balneological area. In addition to these local thermal baths, the Thermal Balneological- Beauty Medical Centre KRE was established in the summer of 2022. The administration of the balneological locations is managed by private factors.

Sarcău is a rural locality in Bihor, which has had thermal baths since the 1970s. Four years after the Revolution of 1989, it became the Adorianis Complex, with a complex treatment base, equipped with equipment, through an accelerated development process specialized and provides medical assistance and treatment, accommodation structure consisting of two modern three-star guesthouses and a villa; conference room; balneological structure; and a restoration structure. The administration of the balneological resource at the level of Sarcău locality has been carried out by a private factor, since 2003.

Chișlaz is a village with balneological resources that uses geothermal water at the local thermal baths. After the Revolution in 1989, the thermal baths came under the administration of a private company, until 2010. Since that period until present day, it has been managed by a public entity

Tămășeu is a village that has had mineral water since 1887, highlighted with the opening of the Oradea-Valea lui Mihai railway. The hydrothermal resource has been exploited since the 1960s. This was bottled for sale under the name Lithium, then Tămășeu. Today this well flows freely. At the level of the locality, the geothermal water is utilized at the thermal baths, which has been operating since its arrangement by the locals, in the 1970s. The administration of the balneological source is carried out by a private factor. As a preliminary project to capitalize on the balneological resources, 10 hectares of land have been purchased and the project has a treatment base with accommodation and a water park.

Sântandrei is a village that has been using geothermal water for fish farming since 1993, which is when the drilling was carried out. The administration of the balneological source is carried out by a private factor. As a preliminary plan for the development [54], a project is proposed for the establishment of a treatment base and leisure thermal baths.

4.3. New Tourist Resorts, Towns, or Villages of National and Local Interest with Balneological Specificity

Bihor County has a total of 13 new tourist and balneoclimatic resorts (Oradea, Beiuș, Ștei, Marghita, Săcueni, Mădăras, Salonta, Borș, Vadu Crișului, Bratca, Șuncuiuș, Pietroasa-Budureasa, and Padiș). Eight of these tourist resorts have resources with a balneological factor (Oradea, Beiuș, Ștei, Marghita, Săcueni, Mădăras, Salonta, and Borș). These localities were certified as tourist resorts during the period 2020–2022 (Table 1). The resorts are new due to their certification as tourist resorts in the period 2020–2022.

Table 1. The new tourist resorts with balneological potential of Bihor County.

Locality	Year of Certification as a Tourist Resort	Interest Level	Resort Type
Oradea	2022	National	Balneological and balneoclimateric
	2020	National	Tourist
Marghita	2022	Local	Tourist
Săcueni	2021	Local	Tourist
Salonta	2021	Local	Tourist
Borș	2021	Local	Tourist
Beiuș	2020	Local	Tourist
Ștei	2020	Local	Tourist
Mădăras	2020	Local	Tourist

Source: based on own survey.

Oradea is the seat of Bihor County, which acquired its status as a tourist resort of national interest [55] in 2020 (H.G. no. 377/2020) [56], through the central historical area and along Crișul Repede river. It was certified as a balneological and balneoclimatic resort in 2022 (H.G. no. 898/2022) [57]. In the area of the municipality, geothermal water is used for heating, leisure (Nymphaea Thermal Water Park, Ioșia Thermal Baths, thermal pools (hotels) and recently, for balneological purposes (Elite Hotel). The administration of the balneological facilities is both public and private. As a forecast plan for capitalizing on the balneological source, the development of the Ioșia Thermal Garden [58] can be found in the municipality's project portfolio.

Marghita is the municipality that has been certified as a tourist resort of local interest since 2022 (H.G. no. 910/2022) [59]. The geothermal water is used for therapeutic purposes (at the "Dr. Pop Mircea" Municipal Hospital, balneology section) and for leisure at the Hotel—Complex Thermal Baths. The administration of the balneological is both public and private. According to the Integrated Urban Development Strategy [60], it has projects for the realization of a balneological tourism complex and rehabilitation of a new baths with a pool covered with geothermal water for the winter period and an outdoor thermal garden.

Beiuș is the municipality that was certified as a tourist resort of local interest in 2020 (H.G. no. 1073/2020) [61]. The main use of geothermal water is in heating, designating it as the only city in the country in which the heating agent is made exclusively on the basis of geothermal water—a fact that has given it the name of the Green City (www.primariabeius.ro (accessed on 12 January 2023)), secondary for leisure (baths with thermal water) and in the cosmetics industry (www.beyuscosmetics.com/ro/ (accessed on 12 January 2023)). The administration of the balneological is public and private. As a preliminary development plan, it has a project that foresees the valorisation of the hydro-therapeutic resource by building a complex water park based on the local geothermal water resource and the local tourist potential [62].

Ștei is an urban locality that obtained the status of tourist resort of local interest in 2020 (H.G. no. 887/2020) [63]. The city of Ștei has balneological resources that are currently unused. Geothermal water drilling was carried out after the 1989 Revolution. The management of balneological resources belongs to the public administration. As the expected exploitation of geothermal water [51], it has projects in order to improve energy efficiency, expand, rehabilitate, and modernize the Ștei balneology section.

Săcueni is a locality certified as a tourist resort of local interest in 2021 (H.G. no. 343/2021) [64]. The geothermal water is used for heating (in institutions) and leisure (the thermal baths in the locality). The administration of the balneological is both public and private. As a prospective exploitation it has a project that provides for the exploitation of geothermal water by building a balneological/healing centre [65].

Mădăras is a rural locality that was certified as a tourist resort of local interest in 2020 (H.G. 1073/2020) [61]. The geothermal resource is used for recreation (Mădăras Thermal Baths). The administration of the balneological source is carried out by a private factor.

Salonta and Borș, municipality of Salonta and the village of Borș, have been certified as tourist resorts of local interest: Borș in 2020 (H.G. no. 887/2020) [63] and the Municipality of Salonta in 2021 (H.G. no. 343/2021) [64]. These localities have thermal waters that are currently not being exploited. As a preliminary plan for the exploitation of geothermal water, it provides projects for the establishment of Thermal Complex Borș, an extension of the Hotel Iris to include a treatment base, conference room, and balneological centre [66] and in Salonta for the development of a thermal balneological-baths complex [67].

4.4. Localities with Untapped Balneological Resources

At the level of Bihor County, there are 44 localities that have untapped geothermal resources. A part of this forms the reserve balneological fund, and in many localities there are boreholes: 37 thermal water boreholes (Adoni, Albești, Biharia, Borș, Căbești, Cefa, Ceica, Cheresig, Cheșereu, Chiraleu, Cihei, Ciumeghiu, Corbești, Cordău, Coșdeni, Criștioru de Sus, Curățele, Cusuiuș, Diosig, Finiș, Ghiorac, Girișu de Criș, Oșorhei, Palota, Rontău, Roșiori, Salonta, Săcădat, Săbolciu, Sîntimreu, Sîntion, Ștei, Tășad, Tăuteu, Tria, Tulca, Vașcău); 3 mineral water boreholes (Tămășeu, Sălard, Sîntimreu), and 3 mineral water springs (Tinca, Budureasa, Voivozi (holiday village Pădurea Neagră)).

4.5. Forms of Administration and Capitalization of the Balneological Potential in the Pre- and Post-Revolution Period

After the arrangement of the balneological sources in the thermal pools in the pre-Revolution period, in some villages (Sarcău, Livada de Bihor, Săcueni, Sânicolau de Munte, Mădăras, Chislaz, Tămășeu, Mihai Bravu, Cadea, Rabagani, Balc, and Ciocaia), these were assigned to a local public administrative forum (Agricultural Production Cooperatives). At the level of urban localities (Oradea, Marghita, and Alesd), the balneological sources were also managed by the public administrative factor—but by the town halls. In the post-Revolution period, both the balneological resources at the level of the rural localities and the urban ones came to be managed by public and private administrative factors (Table 2).

Table 2. Forms of administration and capitalization of the balneological potential in the pre- and post-revolution period.

The Locality with a Balneological Potential	The Form of Administration during the Period 1970–1989	The Form of the Balneological Valorisation until the Revolution of 1989	The Form of Administration from the 1989 Revolution to the Present (2023)	The Form of Valorisation from the Revolution of 1989 to the Present (2023)
Mădăras	Public	Leisure	Private	Leisure
Tămășeu	Public	Leisure	Private	Leisure
Săcueni	Public	Leisure	Private	Leisure
Sarcău	Public	Leisure	Private	Therapeutic/leisure/Spa
Chișlaz	Public	Leisure	Public	Leisure
Balc	Public	Leisure	Private	Heating
Livada de Bihor	Public	Leisure/agriculture/Heating/fish farming	Public/private	Leisure/heating
Mihai Bravu	Public	Leisure	Private	Heating
Ciocaia	Public	Leisure	Public	Heating
Cadea	Public	Leisure	Public	Heating
Sînicolau de Munte	Public	Leisure	Private	Leisure/spa
Răbăgani	Public	Leisure	Private	Demolished
Aleșd	Public	Leisure	Public	Leisure
Beiuș	The drilling was carried out after the Revolution of 1989 (2003)	Leisure	Private/public	Heating/leisure/Skincare Industry/SPA
Marghita	Public	Leisure	Public/private	Leisure/therapeutic
Valea lui Mihai	Public	Leisure	Public	Heating
Sântandrei	The drilling was carried out after the Revolution of 1989 (1993)		Private	Fish farming
Oradea	Public	Leisure/heating	Private/public	Leisure/therapeutic/Spa/heating

Source: based on own survey.

5. Discussion

The characteristic of the balneological phenomenon in Bihor County is based on important aspects that contribute to the sustainability of the exploitation of the geothermal resource, influenced by their administrative apparatus. These aspects are related to the characteristics, starting with the geothermal resource, which is easily exploited and is present in many localities of the county, the need for tourism, the need for treatment and the need for leisure.

Based on the above classifications, we made a comparative analysis of the exploitation of the geothermal resource in the rural localities and the new tourist resorts of Bihor County, from the development to the Revolution of 1989 and from the Revolution of 1989 to the present (Figure 2). The comparative analysis parallelizes the variables formed by the forms and the number of forms of valorisation of the spa source (thermal heating, agriculture, leisure, treatment, spa, fish farming, mineral water spring, thermal water drilling, mineral water drilling) and the public or private administrative factor involved in its exploitation during the period of the 1970s until the Revolution of 1989, and from the Revolution to the present.

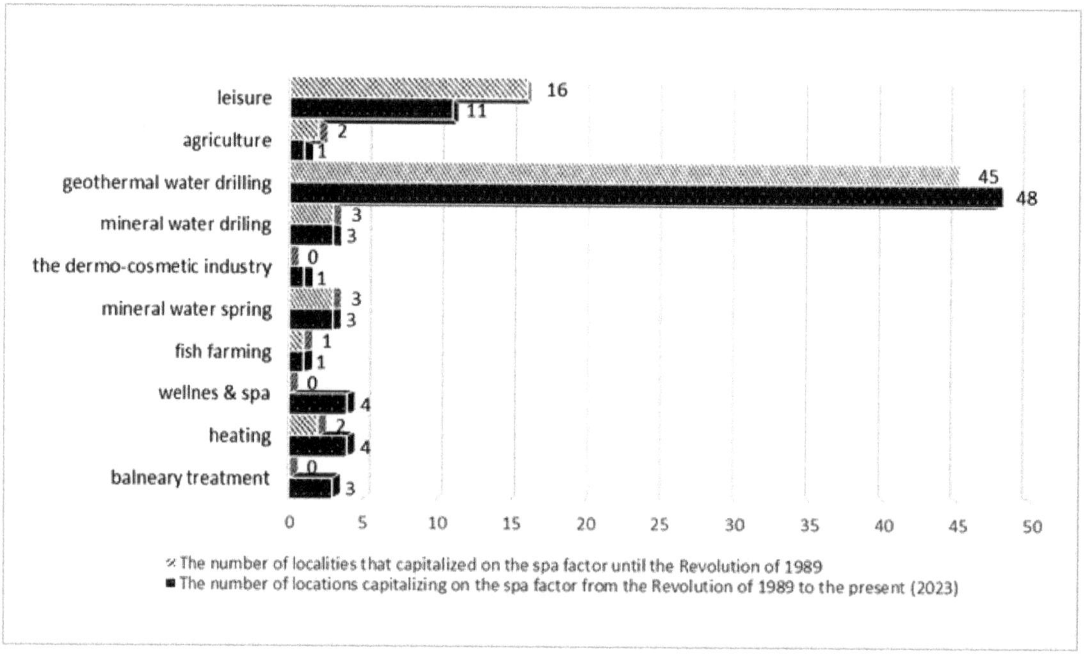

Figure 2. Comparative analysis of the exploitation of the geothermal resource in the rural localities and in the new tourist resorts of Bihor County, from the development to the Revolution of 1989, and from the Revolution of 1989 until the present day. Source: based on own survey.

Based on this, we have identified a development of the balneological phenomenon during the democratic period, as the number of localities exploiting geothermal water for the purpose of balneological treatment, wellness, and spa and thermo-cosmetic uses has increased. New forms of exploitation of geothermal water, which took shape after the Revolution, are in the balneological sector, wellness and spa and in the skincare industry. However, while some balneological resources developed, or were established, starting with new drilling, other balneological resources remained in ruins. From the Revolution to the present (2023), with the exception of Oradea, two new thermal baths have been built, in Beiuș and Sânicolau de Munte for leisure. In addition, new thermal water boreholes have been drilled (Ștei, Beiuș, and Sîntandrei).

Rural balneological localities support balneological tourism of local and, less often, regional importance, relying on a series of exploited tourist resources [5]. After 1990, the political, economic and legislative changes generated the necessary premises for the relaunch of the rural touristic phenomenon. According to Matei's statement, in Bihor County, rural localities such as Sarcău, Sânicolau de Munte, Mădăras, Livada de Bihor, and Chișlaz, support local tourism. For example, in Sînicolau de Munte and Livada de Bihor, there are over 3000 tourists per year, and the taxes that come from tourist services contribute to the local budget. In Sarcău, where the accommodation base has a much larger accommodation capacity (two guesthouses and a three-star villa) (see Tables 3–5) there seems to be a much higher contribution to the development of local tourism. Firdaus and

Hardjosoekarto (2021) [68] highlights the fact that the local administration is capable to fulfil all the attributes in the evolution of tourism (coordination, planning, regulation, entrepreneurship, stimulation and promoting, the role of tourism and the protecting interests). Thus, in Bihor County, the local administration proves to be able to fulfil its functions in the development of tourism, as demonstrated by the achievements completed by it: heating with geothermal water in Beiuș, Livada de Bihor, and Săcuieni. There are also future projects for the development of new balneological resources, some of which are the development of the Ioșia Thermal Garden, in Oradea; the construction of a balneological tourism complex and rehabilitation of a new baths with a covered pool with geothermal water for the winter period and an outdoor thermal garden, in Marghita; water park with geothermal water in Beiuș; energy efficiency, expansion, rehabilitation and modernization of the Ștei balneology section; building a balneological/healing centre in Săcueni.

Table 3. The accommodation infrastructure of the thermal baths.

Locality with a Balneological Resource	Camping Number of Places	Wooden Houses Number of Beds	Equipped Rooms Number of Beds	Tourist Guesthouse 3 *** Number of Beds	Villa Number of Beds	Hotel 2 ** Number of Beds	Hotel 3 *** Number of Beds	Total Beds
Beiuș							24	24
Chișlaz		43						43
Livada de Bihor	12	8						20
Sînicolau de Munte	8	64					12	84
Mădăras	4		22					26
Marghita	60	100				100		260
Sarcău				48	28			76
Tămășeu	none							
Aleșd	none							
Săcuieni	15	32						47
Total	99	247	22	48	28	100	36	580

** and *** indicate the number of stars. Source: based on own survey.

Table 4. The restoration infrastructure of the thermal baths.

Locality with Balneological Source	Restaurant Number of Places	Terrace Number of Places	Snack-Bar Number of Places	Total
Aleșd	200	50		250
Beiuș	200	300		500
Marghita	200	50		250
Săcueni	100	100		200
Chișlaz		20	30	50
Sarcău	60	50		110
Tămășeu	100	50		150
Sînicolau de Munte	100	140		240
Livada de Bihor	30	170		200
Mădăras		20	20	40
Total	990	950	50	1990

Source: based on own survey.

At the level of the localities with the balneological resource, excluding Oradea, Băile 1 Mai, Băile Felix, Băile Tinca, and Stîna de Vale) we have centralized 10 localities with balneological resources exploited for balneological purposes. In 8 of these locations, we found an existing accommodation infrastructure with a total number of 580 beds. The restaurants infrastructure is present in each of the locations where the balneological resource is exploited, with a total of 1990 places. SPA services can be found in 4 locations, and the treatment site owns 3 of the 10 locations studied.

Table 5. The leisure and treatment infrastructure of the thermal baths.

Locality with Balneological Source	Leisure Base				Treatment Base
	Thermal Pools	Playground	Sports Field	SPA	Treatment Procedures
Aleșd	x	x			Balneotherapy Physiotherapy Electrotherapy
Beiuș	x	x		x	
Marghita	x	x	x	x	Balneotherapy Physiotherapy Electrotherapy
Săcueni	x		x		
Chișlaz	x	x	x		
Sarcău	x		x	x	Balneotherapy Physiotherapy Electrotherapy Lymphatic drainage Magnetodiaflux
Tămășeu	x				
Sînicolau de Munte	x	x	x	x	
Livada de Bihor	x	x	x		
Mădăras	x				

x—indicates the existence in the endowment. Source: based on own survey.

According to Cohen's study (2001) [69], the purpose of the private administration is to be more effective than public administrations, due to the influence of the profit. From the political and historical backgrounds where the privatization comes from, three streams of privatization theory have appeared. The first stream maintains that private administration is superior to public administration. This fact also emerges from the comparative analysis of our study regarding the evolution of the balneological factor influenced by the type of administration. At the level of some of the rural localities, where the balneological resource is managed by private administration, the improvements are consistent, in the form and quality of the services, with accommodation structures classified with three stars, treatment, and balneological sites and even the establishment of new open-air swimming pools and baths (Sânicolau de Munte and Beiuș).

Our article demonstrates the fact that the current trend of exploiting the balneological potential of the villages in Bihor County is in the direction of tourism, leisure, balneological treatment, wellness, and spa.

We do not detail the very old rural balneological resorts (Băile Felix, Băile 1 Mai, Băile Tinca, and Stîna de Vale), as extensive studies have already been conducted on these resorts, and they are widely known [70–73].

As strengths of this study, we have highlighted balneological localities that are less known as well as the forms of administration of spa resources that have not yet been researched.

A limitation of this study is the susceptibility of the administrative factor to provide information and the delivery of incomplete answers from the administrative factor.

The topics discussed in our study are also found in several international scientific papers. Thus, the use of geothermal water in heating is also reported by Romanov and Leiss (2022) in their study [35]. In his article, Donati (2022) mentions the management of swimming pools, a theme that is also found in our study [52]. On the topic of the importance of balneology and the use of thermal waters for health, as well as on the role of health tourism and management, there are several studies [25,26,74]. The valorization of geothermal energy in different areas of agriculture is also discussed in studies such as Kępińska, 2021, and Skrzypczak et al., 2021 [40]. The new forms of valorisation of geothermal water outlined after the Revolution, wellness and spa, and the skincare sector are also discussed in other research [36–39].

These studies addressed health tourism facilities and their profitable and sustainable management, or developments in geothermal technology. In contrast to these, we addressed the importance of involving the administrative factor in the exploitation of geothermal water.

The results so far have shown that it would be worthwhile to implement a quantitative procedure for evaluating the administrative aspects capable of influencing the way geothermal waters are used for balneological purposes. Nevertheless, the quantitative modelling of social and economic processes, which to the greatest extent are influenced by numerous political factors, is always very difficult. Sandu (2011) [75] developed a local human development index for Romanian villages, which could be a variable cause, but besides that, the degree of dependence on the exploitation of geothermal waters should be quantified depending on the nature of the involvement of the administrative factor. Such an approach could be the objective of further studies.

6. Conclusions

The development of the balneological phenomenon at the level of rural localities and the new tourist resorts in Bihor County, proved to be conditioned by the involvement of the administrative factor in the prioritization of this sector. It was concluded that at the level of rural localities, where the sewage infrastructure, the drinking water network, or the asphalting of roads is almost non-existent in some places, the development of a leisure base would be among the few, or even the last priorities of the local public administrative unit. In many cases, barely functional pools under the auspices of the former Agricultural Production Cooperatives, after the Revolution of 1989 and by changing the form of administration to a private one, benefited from important financing, which led to the creation of three-star balneological complexes: Sarcau, Sânicolau de Munte, Beius, etc. The weight of the forms of private administration of the balneological sources is higher, a fact that denotes the modern forms of exploitation of geothermal water, existing in this type of administration.

The comparative analysis revealed that the development of the balneological phenomenon, from the development until now, was carried out sequentially. This sequence was influenced by the political environment and the administrative factor from the period before and after the Revolution. In the period before the Revolution, all balneological sources were managed by the public administrative factor. They were maintained and kept functional, until the Revolution of 1989, as there was a stable political environment. In the post-Revolution period, when sometimes the decision and involvement was undertaken on a small scale (private administration), and where the interest in development was focused exclusively on the valorisation of the balneological factor, it can be said that capitalist policy favourably influenced the evolution of the balneological phenomenon at the level of rural settlements and the new tourist resorts in Bihor County. A high-performing administration adapted to the changes of the present, but especially of the future, is necessary and essential for the development of the balneological phenomenon.

In the development of the spa resource before the 1989 Revolution, capitalization was based on state investments, which were directed to cities. Nowadays, there are local entrepreneurs, and the free market is more involved in capitalizing. Valorisation is favourable because there are private investors, who use their own financial sources, thus being more involved in the development of the spa site that they manage.

Author Contributions: Conceptualization and designed the experiment, A.-P.C. and I.H.; methodology, A.-P.C. and I.H.; validation and data curation, I.H. and D.I.; resources, A.-P.C.; writing—original draft preparation, A.-P.C., I.H. and D.I.; writing—review and editing, supervision, I.H. and D.I. All authors have read and agreed to the published version of the manuscript.

Funding: This research received no external funding.

Institutional Review Board Statement: Not applicable.

Informed Consent Statement: Not applicable.

Data Availability Statement: The data sets supporting the conclusions made are included in this article.

Conflicts of Interest: The authors declare no conflict of interest.

Appendix A

Location with thermal waters sheet
1. Name/public/private property.
2. Access roads (roads DC, DJ, DN; railway, airport, etc.).
3. Public transport.
4. Current water, sewerage, electricity.
5. Thermal water source:

- Spring water temperature;
- Drilling: how many drillings, the depth of the drilling, the year the drilling was executed, the temperature of the water, if the well supplies only the strand or something else;
- Test report (thermal water analysis report);
- Some photos of the spa complex, etc.

6. The existence of studies and documents that testify the presence and value of natural cure factors (mineral waters, mud, therapeutic lakes, salt mines, bioclimate, etc., from a qualitative and quantitative point of view).
7. Constitution of the ecological, hydrogeological, and sanitary protection perimeters of the natural healing factors, in accordance with the legislation in force.
8. First aid point and means of transport for medical emergencies.
9. Pharmaceutical point.
10. Spa medical assistance, as applicable, accredited in accordance with the legal rules in force.
11. Basis for the valuation of natural therapeutic resources.
12. Arrangements and equipment for revival services in built spaces, rooms for maintenance (fitness and others).
13. Arrangements and equipment for the practice of sports/leisure (water sports, tennis, football, etc.).
14. Baths: establishment of the baths, total area, how many pools, depth of pools, water temperature, other facilities at the beach, other information.
15. Treatment base (what it consists of).
16. Tourist reception facilities.
17. Hotel, motel, guesthouse, villa, camping, number of places to stay, number of stars, flowers, daisies.
18. Dining space (terrace, restaurant, number of seats).
19. Arrangements and equipment for outdoor relaxation and walking (pedestrian roads, promenade places).
20. Playgrounds for children.
21. Signage with orientation and information indicators, printed, electronic or web site, etc.
22. Landscape park.
23. The organization of tourist, cultural, sports events (repeatable on a calendar basis if applicable).
24. Tourist information and promotion centre (with permanent staff to exclusively serve the centre if necessary).
25. Basis for the valuation of natural therapeutic resources.
26. Arrangements and equipment for revival services in built spaces, rooms for maintenance (fitness and others).
27. Arrangements and equipment for the practice of sports/leisure (water sports, tennis, football, etc.).

Source: based on our own survey.

References

1. Bungau, C.C.; Bungau, C.; Toadere, M.T.; Prada-Hanga, I.F.; Bungau, T.; Popescu, D.E.; Prada, M.F. Solutions for an Ecological and Healthy Retrofitting of Buildings on the Campus of the University of Oradea, Romania, Built Starting from 1911 to 1913. *Sustainability* **2023**, *15*, 6541. [CrossRef]
2. Mocan, C.F. Sustainable Use of Geothermal Waters in Bihor County from Leisure Baths To Electricity Production. *Ann. Fac. Econ.* **2016**, *1*, 374.
3. Ilies, D.; Ilieș, A.; Herman, G.; Baias, S.; Morar, C. *Geoturist Map of the Băile Felix–Băile 1 Mai-Betfia Area (Bihor County, Romania)*; University of Oradea: Oradea, Romania, 2011; Volume IV, p. 219.
4. Pricăjan, A. *Substanţele Minerale Terapeutice Din România. Therapeutic Mineral Substances from Romania*; Ştiinţifică şi Enciclopedică: București, Romania, 1985.
5. Matei, D. Rural Tourism. In *Theory and Reality*; Publisher Terra Nostra: Iaşi, Romania, 2005.
6. Dulău, M.M. *Managementul Cererii Şi Ofertei Turistice*; Translate: Management of Tourism Supply and Demand; Babes-Bolyai University: Cluj-Napoca, Romania, 2014.
7. Marușca, A. Potenţialul Turistic Balnear Din Carpaţii Occidentali. Câmpia Şi Dealurile Banato-Crişene. Translate: The Spa Tourism Potential of the Western Carpathians. The Banato-Crisene Plain and Hills. Ph.D. Thesis, University of Oradea, Oradea, Romania, 2008.
8. Marian, E.C. Evoluţia Funcţiei Balneoclimaterice În Câmpia Și Dealurile de Vest. Translate: The Evolution of the Balneoclimatic Function in the Western Plain and Hills. Ph.D. Thesis, University of Bucureşti, București, Romania, 2012.
9. Cingolani, L. *The State of State Capacity: A Review of Concepts*; Evidence and Measures-Merit Working Papers; United Nations University—Maastricht Economic and Social Research Institute on Innovation and Technology (MERIT): Maastricht, The Netherlands, 2013.
10. Shamsul, H.; Ramesh, M.; Puppim de Oliveira, J.; de Ávila Gomide, A. Introduction Éditoriale. Renforcer La Capacité Administrative Pour Le Développement: Limites et Perspectives. *Rev. Int. Sci. Adm.* **2021**, *87*, 215–223. [CrossRef]
11. *Thermal Energy Law No. 325 2006*; Republished Official Monitor No. 217 of 16 March 2023; The Romanian Parliament: Bucharest, Romania, 2006.
12. Roșca, M.; Antal, C.; Bendea, C. Geothermal Energy in Romania: Country Update 2005–2009. In Proceedings of the World Geothermal Congress, Bali, Indonesia, 25–30 April 2010.
13. Vicidomini, M.; D'Agostino, D. Geothermal Source Exploitation for Energy Saving and Environmental Energy Production. *Energies* **2022**, *15*, 6420. [CrossRef]
14. Figueiredo, A.C.; Rodrigues, M.; Mourelle, M.L.; Araujo, A.R.T.S. Thermal Spring Waters as an Active Ingredient in Cosmetic Formulations. *Cosmetics* **2023**, *10*, 27. [CrossRef]
15. Smith, M.K.; Diekmann, A. Tourism and Wellbeing. *Ann. Tour. Res.* **2017**, *66*, 1–13. [CrossRef]
16. Pyke, S.; Hartwell, H.; Blake, A.; Hemingway, A. Exploring Well-Being as a Tourism Product Resource. *Tour. Manag.* **2016**, *55*, 94–105. [CrossRef]
17. Eusébio, C.; Carneiro, M.J.; Kastenholz, E.; Alvelos, H. The Economic Impact of Health Tourism Programmes. *Quant. Methods Tour. Econ.* **2013**, 153–173. [CrossRef]
18. Zhu, Y. The Economic Effects of Medical Tourism Industry on Kwangwon Province. *J. Tour. Manag. Res.* **2011**, *15*.
19. Andreu, M.G.N.; Font-Barnet, A.; Roca, M.E. Wellness Tourism—New Challenges and Opportunities for Tourism in Salou. *Sustainability* **2021**, *13*, 8246. [CrossRef]
20. Hartwell, H.; Hemingway, A.; Fyall, A.; Filimonau, V.; Wall, S. Tourism Engaging with the Public Health Agenda: Can We Promote 'Wellville' as a Destination of Choice? *Public Health* **2012**, *126*, 1072–1074. [CrossRef]
21. Quintela, J.; Costa, C.; Correia, A. The Role of Health and Wellness Tourism in Sustainable Territorial Development. *Tour. Hosp. Int. J.* **2017**, *9*, 113–121.
22. Wray, M.; Weiler, B. Wellness Tourism: The Factors and Processes That Drive Sustainable Regional Destinations. *Wellness Tour.* **2013**, 78–98. [CrossRef]
23. Pforr, C.; Pechlaner, H.; Locher, C.; Jochmann, J. Health Regions: Building Tourism Destinations through Networked Regional Core Competencies. In *Wellness Tourism*; Routledge: Milton Park, UK, 2013; pp. 99–111.
24. Smith, M.K.; Puczkó, L. *The Routledge Handbook of Health Tourism*; Taylor & Francis: Milton Park, UK, 2016; ISBN 1-317-43750-0.
25. Frost, W.; Laing, J. History of Spa Tourism: Spirituality, Rejuvenation and Socialisation. In *The Routledge Handbook of Health Tourism*; Routledge: Milton Park, UK, 2016; pp. 37–47. ISBN 1-315-69377-1.
26. McCarthy, M. Loss of Potential Reward Is Best Motivator in Workers' Wellness Programme, Study Finds. *BMJ* **2016**, *352*, i932. [CrossRef]
27. Kurek, K.A.; Heijman, W.; van Ophem, J.; Gędek, S.; Strojny, J. Geothermal Spas as a Local Development Factor, the Case of Poland. *Geothermics* **2020**, *85*, 101777. [CrossRef]
28. Nistoreanu, P.; Aluculesei, A.C. Can Spa Tourism Enhance Water Resources and Turn Them into a National Brand? A Theoretical Review about the Romanian Case. *Information* **2021**, *12*, 270. [CrossRef]
29. Prideaux, B.; Berbigier, D.; Thompson, M. Wellness Tourism and Destination Competitiveness. In *Wellness Tourism: A destination Perspective*; Routledge: London, UK, 2014; pp. 45–60.

30. Allard, P.; Deligne, J.; Van Bockstael, V.; Duquesnoy, B. Is Spa Therapy Cost-Effective in Rheumatic Disorders? *Rev. Rhum. (Engl. Ed.)* **1998**, *65*, 173–180.
31. Fioravanti, A.; Valenti, M.; Altobelli, E.; Di Orio, F.; Nappi, G.; Crisanti, A.; Cantarini, L.; Marcolongo, R. Clinical Efficacy and Cost-Effectiveness Evidence of Spa Therapy in Osteoarthritis. The Results of "Naiade" Italian Project. *Panminerva Med.* **2003**, *45*, 211–217.
32. Van Tubergen, A.; Boonen, A.; Landewé, R.; Rutten-Van Mölken, M.; Van Der Heijde, D.; Hidding, A.; Van Der Linden, S. Cost Effectiveness of Combined Spa–Exercise Therapy in Ankylosing Spondylitis: A Randomized Controlled Trial. *Arthritis Care Res.* **2002**, *47*, 459–467. [CrossRef]
33. Karie, S.; Gandjbakhch, F.; Janus, N.; Launay-Vacher, V.; Rozenberg, S.; Mai Ba, C.; Bourgeois, P.; Deray, G. Kidney Disease in RA Patients: Prevalence and Implication on RA-Related Drugs Management: The MATRIX Study. *Rheumatology* **2008**, *47*, 350–354. [CrossRef]
34. Ciani, O.; Pascarelli, N.A.; Giannitti, C.; Galeazzi, M.; Meregaglia, M.; Fattore, G.; Fioravanti, A. Mud-Bath Therapy in Addition to Usual Care in Bilateral Knee Osteoarthritis: An Economic Evaluation Alongside a Randomized Controlled Trial. *Arthritis Care Res.* **2017**, *69*, 966–972. [CrossRef] [PubMed]
35. Romanov, D.; Leiss, B. Geothermal Energy at Different Depths for District Heating and Cooling of Existing and Future Building Stock. *Renew. Sustain. Energy Rev.* **2022**, *167*, 112727. [CrossRef]
36. Liang, J.; Kang, D.; Wang, Y.; Yu, Y.; Fan, J.; Takashi, E. Carbonate Ion-Enriched Hot Spring Water Promotes Skin Wound Healing in Nude Rats. *PLoS ONE* **2015**, *10*, e0117106. [CrossRef] [PubMed]
37. Huang, A.; Seité, S.; Adar, T. The Use of Balneotherapy in Dermatology. *Clin. Dermatol.* **2018**, *36*, 363–368. [CrossRef] [PubMed]
38. Inaka, K.; Kimura, T. Comfortable and Dermatological Effects of Hot Spring Bathing Provide Demonstrative Insight into Improvement in the Rough Skin of Capybaras. *Sci. Rep.* **2021**, *11*, 23675. [CrossRef]
39. Inaka, K.; Kimura, T. Hot Spring Bathing Accelerates Wound Healing and Enhances Heat Retention Effect in Guinea Pigs. *J. Vet. Med. Sci.* **2022**, *84*, 1653–1664. [CrossRef]
40. Kępińska, B. Wykorzystanie energii geotermalnej w Polsce w latach 2019–2021. *Przegląd Geol.* **2021**, *69*, 559.
41. Skrzypczak, R.; Kępińska, B.; Pająk, L.; Bujakowski, W. Prospects for the Application of Geothermal Resources in Agriculture in Poland Taking Account of the Natural Functions of the Countryside. *Geotherm. Energy* **2021**, *9*, 23. [CrossRef]
42. GharibiAsl, S.; Abbaspour-Gilandeh, Y. Evaluation of the Direct Use of Geothermal Energy on Heat Factors Required for Cold-Water Fish Pisciculture. *Aquaculture* **2019**, *512*, 734291. [CrossRef]
43. Joao, S.A.; Salie, K.; Jackson, T.K.; Guwa, O.; Wilson, L. The Utilisation of Geothermal Spring Water for Tilapia Aquaculture to Promote Food Security and Skills Development at the Brandvlei Correctional Services Centre. Water Research Commission. 2022. Available online: https://www.wrc.org.za/wp-content/uploads/mdocs/TT%20892%20final%20web.pdf (accessed on 10 February 2023).
44. Axinte, A.; Baias, S.; Banto, N.; Biriș, M.; Blaga, L.; Bocoi, L.; Brădău, C.; Buhaș, R.; Caciora, T.; Dumbravă, R.; et al. *Atlasul Orizontului Geografic Local al Județului Bihor. Translate: Atlas of the Local Geographical Horizon of Bihor County*; Agentia de Management al Destinatiei Bihor: Oradea, Romania, 2020; ISBN 978-973-0-32759-5.
45. Cornea, S. Considerations Regarding the Relations of the Public Administration with the Political Environment. 2007. Available online: https://www.researchgate.net/publication/327107435 (accessed on 19 April 2023).
46. Zamfir, C. Tranziția La o Românie Modernă de Tip Occidental the Transition to a Modern and Western-like Romania. *Sociol. Românească* **2018**, *16*, 101–154.
47. Borș, S. Sîngiorz Băi-Între Imaginar Și Dispariție. Translate: Sîngiorz Băi-Between Imaginary and Disappearance. Ph.D. Thesis, Technical University, Cluj-Napoca, Romania, 2020.
48. Ciurba, A.P.; Haidu, I.; Gaceu, O.; Aur, C. Therapeutic Valorisation of Thermal Waters from Rural Locations of Bihor County. In Proceedings of the Air and Water—Components of the Environment, Conference Proceedings, University Babes-Bolyai, Cluj-Napoca, Romania, 17–19 March 2023; pp. 200–212. [CrossRef]
49. Hotarâre de Guvern (Government Decision) No. 852 2008. The Government of Romania, Official Monitor No. 613 of 20 August 2008. Available online: http://www.mdlpl.ro/_documente/turism/legislatie/hg_852_2008.pdf (accessed on 3 December 2022).
50. Ciurba, A.P.; Haidu, I.; Gaceu, O.; Biriș, M.; Meșter, C.; Ianc, D. The Thermal Water in Bihor County and Its Benefits for Treating Arthritis. A Case Study: Băile 1 Mai Resort. In Proceedings of the Air and Water—Components of the Environment, Conference Proceedings, University Babes-Bolyai, Cluj-Napoca, Romania, 18–20 March 2022; pp. 225–240. [CrossRef]
51. Strategie Integrată de Dezvoltare Urbană Orașul Ștei 2021–2030 Conform H. Nr. 136 Din 17 June 2022. Translate: Integrated Urban Development Strategy for the City of Ștei 2021–2030 According to H. No. 136 of 17 June 2022. City Hall of Stei. Available online: https://www.primariastei.ro/ProiecteEU/anunt%20consultare%20SIDU%20.pdf (accessed on 3 December 2022).
52. Donati, F. A Direct Method to Delimit Weir Pools. *Geogr. Tech.* **2022**, *17*, 47–53. [CrossRef]
53. Strategia Integrată de Dezvoltare Urbană a Orașului Aleșd 2021–2030, Conform Anexă La HCL Nr. 108. Translate: The Integrated Urban Development Strategy of the City of Aleșd 2021–2030, According to the Annex to HCL No. 108. 2021. City Hall of Aleșd. Available online: https://www.alesd-bihor.ro/strategia-orasului/HCL%20108%20p%20I.pdf (accessed on 3 December 2022).
54. Strategia de Dezvoltare Durabilă a Comunei Sântandrei 2021–2027. Translate: The Sustainable Development Strategy of the Commune of Sântandrei 2021–2027. City Hall of Sîntandrei. Available online: https://www.primariasantandrei.ro/documente/0ecf7c5e882971a9b39ef85640464f07.pdf (accessed on 3 December 2022).

55. Tătar, F.C.; Dincă, I.; Linc, R.; Stupariu, M.; Bucur, L.; Stașac, M.; Nistor, S. Oradea Metropolitan Area as a Space of Interspecific Relations Triggered by Physical and Potential Tourist Actvities. *Sustainability* **2023**, *15*, 3136. [CrossRef]
56. Hotărâre de Guvern (Government Decision) Nr. 377 2020. The Government of Romania, Official Monitor No Nr. 384, 13 May 2020. Available online: https://legislatie.just.ro/Public/DetaliiDocumentAfis/225539 (accessed on 16 December 2022).
57. Hotărâre de Guvern (Government Decision) Nr. 898 2022. Available online: https://legislatie.just.ro/Public/DetaliiDocument/257580 (accessed on 16 December 2022).
58. Strategia Integrată de Dezvoltare Urbană a Municipiului Oradea Și a Zonei Metropolitane Oradea Pentru Perioada 2021–2027. Translate: The Integrated Urban Development Strategy of the Municipality of Oradea and the Oradea Metropolitan Area for the Period 2021–2027. Oradea City Hall. Available online: https://www.oradea.ro/fisiere/userfiles/Europene/Strategia%20integrata.pdf (accessed on 16 December 2022).
59. Hotărâre de Guvern Nr. 910 Din 13 Iulie 2022. The Government of Romania, Official Monitor No 718. 18 July 2022. Available online: https://legislatie.just.ro/Public/DetaliiDocumentAfis/257598 (accessed on 16 December 2022).
60. Strategia Integrată Dedezvoltare Urbană a Orașului Marghita 2021–2027. Translate: The Integrated Urban Development Strategy of the City of Marghita 2021–2027. City Hall of Marghita. Available online: https://marghita.ro/uncategorized/strategia-integrata-de-dezvoltare-urbana-a-municipiului-marghita-2021-2027/ (accessed on 16 December 2022).
61. Hotărâre de Guvern (Government Decision) Nr. 1073 2020. The Government of Romania, Official Monitor No 1288. 24 December 2020. Available online: https://legislatie.just.ro/Public/DetaliiDocumentAfis/235613 (accessed on 16 December 2022).
62. Strategia Integrată de Dezvoltare Urbană a Orașului Beiuș 2021–2030, Conform HCL. Nr. 8969, Din 9 August 2022. Translate: The Integrated Urban Development Strategy of the City of Beiuș 2021–2030, According to HCL. No. 8969, from 9 August 2022. City Hall of Beiuș. Available online: https://www.municipiulbeius.ro/wp-content/uploads/2022/08/ANEXA-SIDU-BEIUS-2021-2030.pdf (accessed on 16 December 2022).
63. Hotărâre de Guvern (Government Decision) Nr. 887 2020. The Government of Romania, Official Monitor No 719. 18 July 2022. Available online: https://legislatie.just.ro/Public/DetaliiDocument/257607 (accessed on 16 December 2022).
64. Hotărâre de Guvern (Government Decision) Nr. 343 2021. The Government of Romania, Official Monitor No 291. 23 March 2021. Available online: https://legislatie.just.ro/Public/DetaliiDocumentAfis/239877 (accessed on 16 December 2022).
65. Strategia Integrată de Dezvoltare Urbană a Orașului Săcueni 2020–2030. Translate: The Integrated Urban Development Strategy of the City of Săcueni 2020–2030. City Hall of Săcuieni. Available online: https://www.sacueni.ro/wp-content/uploads/2021/12/PDF_SIDU_StrategieIntergartaDezvoltareUrbaniSacueni_FINAL.pdf (accessed on 16 December 2022).
66. Strategia de Dezvoltare Durabilă a Comunei Borș 2021–2027. Translate: The Sustainable Development Strategy of the Commune of Borș 2021–2027. City Hall of Borș. Available online: https://primariabors.ro/wp-content/uploads/2020/01/Strategie-Dezvoltare-Locala-Comuna-Bors-2016-2020.pdf (accessed on 16 December 2022).
67. Strategia de Dezvoltare a Municipiului Salonta 2021–2027 Conform HCLMS Nr. 269 29 November 2022. Translate: The Development Strategy of Salonta Municipality 2021–2027 According to HCLMS No. 269 2022. City Hall of Salonta. Available online: https://docplayer.ro/226528528-Strategia-de-dezvoltare-a-municipiului-salonta.html (accessed on 16 December 2022).
68. Firdaus, F.; Hardjosoekarto, S. The Role of Local Government on Rural Tourism Development: Case Study of Desa Wisata Pujonkidul, Indonesia. *Int. J. Sustain. Dev. Plan.* **2021**, *16*, 1299–1307. [CrossRef]
69. Cohen, S. A Strategic Framework for Devolving Responsibility and Functions from Government to the Private Sector. *Public Adm. Rev.* **2001**, *61*, 432–440. Available online: https://doi-org.bases-doc.univ-lorraine.fr/10.1111/0033-3352.00047 (accessed on 16 December 2022). [CrossRef]
70. Ilieş, D.C.; Josan, I. The Tourist Complex Spa Băile Felix-Băile 1 Mai–Personality, Distinctiveness by Protection versus Depersonalization and Nonspecific by Globalization. Conceptual and Practical Aspects Regarding the Role of the Natural and Anthropic Setting in Tourist Fitting. *GeoJ. Tour. Geosites* **2009**, *2*, 179–185.
71. Porowski, A. Selected Papers from the 2007 Yearly Meeting of the Commission on Mineral and Thermal Waters (CMTW), Stana the Vale, Romania, 7-11 October 2007. *Environ. Geol.* **2009**, *58*, 1627–1822. [CrossRef]
72. Moțica, A.M. Spa Tourism Economy Băile Felix Balneoclimateric Resort-Short Presentation. *Manager* **2018**, 18–29. Available online: https://manager.faa.ro/archive/manager_27/971_m_27_18_29.pdf (accessed on 16 December 2022).
73. Herman, G.V.; Peptenatu, D.; Grama, V.; Pintilii, R.-D. Tourism and Local Development. Study Case: Băile Felix-Băile 1 Mai Tourism System, Bihor County, Romania. *Ann. Univ. Oradea Geogr. Ser.* **2018**, *28*, 131–137.
74. Smith, M.K.; Puczkó, L. Balneology and Health Tourism. In *The Routledge Handbook of Health Tourism;* Routledge: Milton Park, UK, 2016; pp. 299–310. ISBN 1-315-69377-1.
75. Sandu, D. Local Human Development Index for Romanian Villages (LHDIv): SPSS Data File 2017. Available online: https://www.researchgate.net/publication/314079025_Local_human_development_index_for_Romanian_villages_LHDIv?ev=prf_pub (accessed on 2 June 2023).

sustainability

MDPI

Article

Relationship between Population and Ethno-Cultural Heritage—Case Study: Crișana, Romania

Luminița Anca Deac [1], Grigore Vasile Herman [2,*], Maria Gozner [2], Gheorghe Codruț Bulz [1] and Emilia Boc [2]

1 Department of Physical Education, Sport and Physical Therapy, Faculty of Geography, Tourism and Sport, University of Oradea, 410087 Oradea, Romania; anca_deac@yahoo.com (L.A.D.); bulz.codrut@uoradea.ro (G.C.B.)
2 Department of Geography, Tourism and Territorial Planning, Faculty of Geography, Tourism and Sport, University of Oradea, 410087 Oradea, Romania; mgozner@uoradea.ro (M.G.); emiliaboc04@yahoo.com (E.B.)
* Correspondence: gherman@uoradea.ro; Tel.: +40-751236226

Abstract: Crișana Region is one of the most representative and wide historical regions of Romania, which encompasses several "lands" and ethnographic areas, each of them being defined by a series of features, among which those of demographic nature (ethnicity, religion) and ethno-cultural features stand out. In this context, the aim of the current study is to identify, assess and emphasize the relationship between the demographic features and those related to the ethno-cultural heritage. The accomplishment of this work required the use of the multi-criteria analysis method, successfully applied in various activity areas, a method which is characterized by a high level of complexity. The obtained results emphasized the spatial distribution on 'territorial administrative unit' (TAU) level of the aggregated synthetic values and of the relationship types which were determined between population and ethno-cultural heritage in Crișana Region, Romania.

Keywords: demography; ethno-cultural heritage; multi-criteria analysis; spatial distribution; relationship

Citation: Deac, L.A.; Herman, G.V.; Gozner, M.; Bulz, G.C.; Boc, E. Relationship between Population and Ethno-Cultural Heritage—Case Study: Crișana, Romania. *Sustainability* **2023**, *15*, 9055. https://doi.org/10.3390/su15119055

Academic Editor: Wen-Hsien Tsai

Received: 17 May 2023
Revised: 29 May 2023
Accepted: 1 June 2023
Published: 3 June 2023

1. Introduction

Ethno-cultural heritage is the creative expression of the people in a given area [1]. Among the socio-demographic features which have a significant influence on contouring and promoting the ethno-cultural heritage, the ethnic and religious structures stand out [2–4]. Demographic data (referring to ethnicity and religion) are used in demographic anthropology [5–7] to answer certain questions of evolution and cultural type. The correlation between ethnography and demography is not recent [8], but the researchers concerned with demography were criticized for not using to the fullest the information provided by ethnography [9]. Furthermore, it was noted that researchers use ethnography only as a background for their work, and that the ethnographic articles are only cited as references or are used to corroborate the researchers' conclusions [9]. Ethnography is generally associated with social and cultural anthropology [10], and the anthropological observations provide a better understanding of the contemporary world, a fact which is emphasized by the demographic changes caused by culture, economy, and politics [11,12].

The study was carried out in the Crișana Region, a region situated in the North-Western part of Romania, at the border with Hungary in the West and neighboring other important Romanian regions: Maramureș in the North, Transylvania in the East, and Banat in the South. From a geographical point of view, the region features diverse forms of relief, from plains to hills and mountains, and a dense hydrographic network. The Eastern part of the region is bordered by mountains: Meseș and Plopiș Mountains, Pădurea Craiului Mountains, and, in the South-East, it continues with Bihor-Vlădeasa Mountains and Codru Moma Mountains. The Southern part is bordered by Zarandului Mountains. The hill formations and depressions are, from North to South, as follows: Crasnei Hills, Silvania

Depression, Barcăului Hills, Oradiei Hills, Crișul Repede Depression, Tășadului Hills, Crișul Negru Depression, Codrului Piedmont, and Zarand Depression. Among the major plains in Crișana, we mention, also from the North to the South: Ierului Plain, Barcăului Plain, Miersigului Plain, Crișului Negru Plain, and Crișului Alb Plain. From the dense hydrographic network of the region, we mention the Crișul Repede, Crișul Negru, and Crișul Alb Rivers, and Er and Barcău Rivers, together with their tributaries. There are also many lakes, most of them being made by man (e.g., Cefa, Inand).

The human settlements are varied from hamlets and small villages, spread in higher mountain areas, to larger villages gathered in lower hill or plain areas, to towns and cities, the city of Oradea being the largest one and with the highest number of inhabitants in the region. Throughout time, the region has gone through various significant historical moments and administrative and territorial changes, which have all influenced population migration and, implicitly, the ethnic and religious structures. At present, the administrative organization of Romania consists of counties that are formed of TAUs (territorial administrative units), but this organization dates back only to the first half of the 20th century. Before that, there were shires that, previously, had replaced the districts [13]. Representatives for this study are also the "lands" and ethnographic areas. The communist regime had a major impact on population dynamics since one of its politics was industrialization and forced urbanization, resulting in massive migration from villages to towns. After the communist regime fell in 1989, the population migrated from one region to another and also abroad.

The socio-political and historical circumstances have always been of utmost importance for ethnography [14]. Crișana Region overlaps, from spatial point of view, the area of three Lands (Silvania Land, Beiuș Land, and Zărand Land) and three ethnographic areas (Crișurilor Plain, Crișul Repede Valley, and Ier and Barcău Valley), each of them being characterized by populations of various ethnicities and religions and ethno-cultural heritage elements, correlated with the physical-geographic support of the region and with the ethnic and religious characteristics of the population [13].

The population of the Crișana Region dates to the Neolithic era [15], the proof of its presence in the territory becoming richer over time. The demographic data used in this study are those resulting from the census made by the National Institute of Statistics of Romania in 2011. Another census was completed in 2021. However, when this study was accomplished, those data were not available yet. According to the 2011 census, the total population of Crișana is 940,061 inhabitants [16]. In 2002, the year of the previous census before 2011, the total population, on the region level, was 1,034,539 inhabitants [16], the decrease of 94,478 people from 2002 until 2011 not being significant, we assume that the difference between the results of the census from 2011 and those from 2021 is not significant for this study. From 2002 until 2011, there were also no major changes recorded regarding ethnicity and the religious structure, respectively. From an ethnical and religious point of view, the population is eclectic; there are various ethnicities (Romanian, Hungarian, Rroma, German, Slovak, and Ukrainian) and religions (Orthodox, Roman-Catholic, Greek-Catholic, Calvinist, Baptist, Pentecostal, and the Seventh-day Adventist) encountered here. Those mentioned here do not encompass the entire range of ethnicities and religions; however, they are the most representative from the point of view of number and continuity in the territory. A general characteristic is the fact that, on a regional level, the Romanian ethnicity and Orthodox religion represent the highest proportions [16].

Among the representative elements of the region and the population in Crișana, we mention the ethno-cultural heritage. Ethno-cultural heritage represents all the material and spiritual cultural values overlapping the traditions and habits of a population [17]. More precisely, ethno-culture refers to the heritage elements resulting from the traditional activities of a population, specific to everyday living, in close connection with people's creativity, translated into daily realities expressed through various types of culture [18]. This phenomenon has been going on for a very long time, and it can, thus, be stated that the ethno-cultural heritage is in close connection with the ethnicities and religions of the region and that it has an essential role in perpetuating the traditions and crafts specific to Crișana Region, the purpose of anthropological demography being to understand demographic phenomena in the socio-cultural context that they exist in [5].

The ethno-cultural heritage elements studied here are representative of the population's occupations and crafts [19,20] and they are wooden churches, the works of craftsmen from various domains (wood processing, pottery, traditional clothing and folk costumes, crafting traditional musical instruments, etc.), ethnographic museums and collections, fairs with ethno-cultural specifics and traditional music and dance festivals.

A general tendency or characteristic of contemporary societies is globalization [21]. This phenomenon affects culture as well [22], fading away the particularities specific to a certain "land" or ethnographic area. Ethno-culture has a dynamic character [23] and is influenced by several social, economic, and political factors. Though some of these factors contribute to its globalization, we consider that certain ethno-religious characteristics contribute to highlighting some specific ethno-cultural elements.

Constantin (2014) accomplished a comparative study regarding the distribution on Romanian, respectively Bulgarian, territories of certain ethnic groups, studying, at the same time, their demographic dynamics with the purpose of identifying the ethnic features which define these groups [24]. The author emphasizes the importance of ethno-cultural and ethno-religious elements reassertion in the ethnic revitalization process. In this context, the present study has the purpose of establishing the relationship between the population and ethno-cultural heritage within the Crișana Region, Romania.

The working hypothesis from which the study started off relates to the fact that between the ethnic and religious structures, on the one hand, and the ethno-cultural heritage, on the other, there are close inter-conditioning relationships of qualitative and quantitative types (the present study addresses the quantitative ones). Hence, a more complex and diverse ethnic and religious structure will lead to a larger diversity of ethno-cultural heritage and vice versa. Considering the complexity of this research, the chosen working method is a multi-criteria comparative analysis, a method which has been successfully applied in numerous studies and research regarding the assessment and preservation of cultural heritage [25,26], the identification of cultural landscapes and values [27,28], assessment of a population's socio-economic development [29,30], the quality of environmental factors [31,32], etc.

2. Materials and Methods

In order to establish the relationship types between population and ethno-cultural heritage, the multi-criteria method [33,34] was used, taking into study, in the light of more or less strong connections, the ethnic criterion (with eight variables: total population and population of the following ethnicities: Romanian, Hungarian, Rroma, German, Slovak, Ukrainian and other ethnicities); the religious criterion (with nine variables: total population, population of the following religions: Orthodox, Roman Catholic, Greek Catholic, Calvinist, Baptist, Pentecostal, the Seventh-day Adventist, and other religions) and the ethno-cultural heritage criterion (with five variables: ethnographic museums and collections, folk festivals, fairs with ethno-cultural specific, traditional craftsmen, and wooden churches). Based on these variables, value standardization is accomplished to obtain an aggregate value for each criterion [35].

The method is also known as Min-Max Normalization Method or Value Mapping Method [33], and it is applicable by using the following techniques: Min-Max normalization using the two values (minimum and maximum); N score normalization uses the difference between X value and the arithmetic average of all variable values; and reported to the standard deviation and decimated normalization [33]. The newly obtained values, resulting from normalization, are comprised in the range [0,1].

The Min-Max normalization method has several stages [34] (Figure 1):

1. Preparation and setting variables that are specific and representative of the analyzed domains: ethnic structure, religious structure, and ethno-cultural heritage elements

(Table 1). Regarding the ethnic, respectively the religious structures, the following was considered: total population at TAU level and the ethnicities, respectively, the religions, which are representative of the entire studied territory (Crișana), and the data used are those from the 2011 population census made in Romania. The data regarding the ethno-cultural elements are not comprehensive. However, the most representative ones were selected as follows: ethnographic museums and collections, folk festivals, ethno-cultural fairs, traditional craftsmen, and wooden churches.

Table 1. Variables selected for the studied criteria.

Criterion	Variables	Measurement Unit	Type of Data
C_1—Ethnic structure	X_1—Total population	No./TAU	Quantitative
	X_2—Romanian	No./TAU	Quantitative
	X_3—Hungarian	No./TAU	Quantitative
	X_4—Rroma	No./TAU	Quantitative
	X_5—German	No./TAU	Quantitative
	X_6—Slovak	No./TAU	Quantitative
	X_7—Ukrainian	No./ TAU	Quantitative
	X_8—Other ethnicities	No./TAU	Quantitative
C_2—Religious structure	Y_1—Total population	No./TAU	Quantitative
	Y_2—Orthodox	No./TAU	Quantitative
	Y_3—Roman Catholic	No./TAU	Quantitative
	Y_4—Greek Catholic	No./TAU	Quantitative
	Y_5—Calvinist	No./TAU	Quantitative
	Y_6—Baptist	No./TAU	Quantitative
	Y_7—Pentecostal	No./TAU	Quantitative
	Y_8—The Seventh-day Adventist	No./TAU	Quantitative
	Y_9—Other religions	No./TAU	Quantitative
C_3—Ethno-cultural heritage elements	Z_1—Ethnographic museums and collections	No./TAU	Quantitative
	Z_2—Folk festivals	No./TAU	Quantitative
	Z_3—Ethno-cultural fairs	No./TAU	Quantitative
	Z_4—Traditional craftsmen	No./TAU	Quantitative
	Z_5—Wooden churches	No./TAU	Quantitative

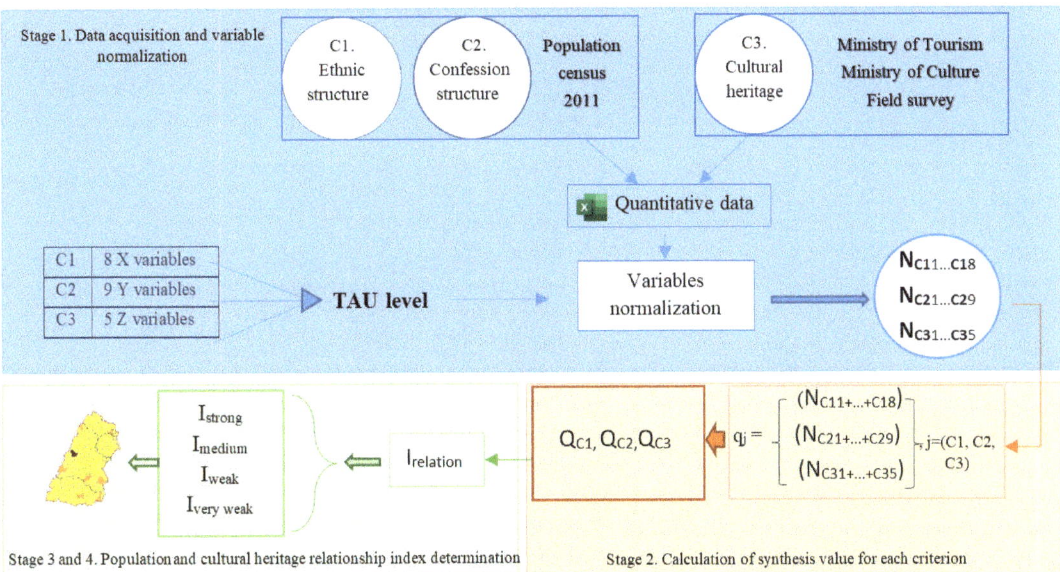

Figure 1. Methodology workflow chart.

The variable values are shown under the form of an observation matrix [35]:

$$X = [x_{ij}] = \begin{bmatrix} x_{11} & x_{12} & \cdots & x_{1n} \\ x_{21} & x_{22} & \dots & x_{2n} \\ \vdots & \vdots & \vdots & \vdots \\ x_{r1} & x_{r2} & \cdots & x_{rn} \end{bmatrix} \tag{1}$$

where: x_{ij} represents the variable value for object O_i.

The same matrix is applied for the variable values of the other criteria as well.

The normalization formula of stimulant variables is [30]:

$$N_{ij} = (X_{iJ} - \min X_{iJ})/(\max X_{iJ} - \min X_{iJ}) \quad X_j \in S, \quad N_{ij} = [0, \ldots, 1] \tag{2}$$

where X_{ij} is the value of variable j for criterion i; N_{ij} is the normalized value of variable j for criterion i; min X_{ij} is the minimum value of value X of the variable i; and max X_{ij} is the maximum value X of variable j for criterion i.

Even though the variables can be stimulant, de-stimulant, or neutral, in this study, the stimulant variables (S) are of interest because they have higher values and indicate a stronger connection between criteria.

2. The second stage implies the calculation of sum value (aggregate value), after normalizing the 22 variable indicators. First, it is obtained the sum value for the first two criteria—the ethnic and the religious structures—with their 17 variable indicators, then, the obtained values are aggregated with the values of the 5 variable indicators of the ethno-cultural heritage elements. The newly obtained values are aggregated in a unique value q_j:

$$q_j = \sum_{j=1}^{n} N_{ij} \qquad (i = 1, \ldots r) \tag{3}$$

The criterion assessment through the variable value is achieved with the synthesis value Qi:

$$Qi = \frac{1}{n}\sum_{j=1}^{n} q_i \qquad (i = 1, \ldots r), Qi \in [0, \ldots, 1] \tag{4}$$

3. The calculation of relation index value between population and ethno-cultural heritage is a normalized value, provided by the following equation:

$$I = \frac{COEFpop - COEFpat}{COEFpop + COEFpat}, \text{ where } -1 \leq I \leq 1, \tag{5}$$

I = Relation index between population and ethno-cultural heritage
COEFpop = population coefficient
COEFpat = ethno-cultural heritage coefficient

4. The fourth stage consists of determining the constant value k to categorize the objects (the 207 TAUs) in value groups according to the relationship type existing between the three criteria (domains): very weak, weak, average, and strong [35]:

$$R(Qi) = \max Qi - \min Qi \qquad k = \frac{R(Qi)}{4} \tag{6}$$

Group 1: Qi ϵ (max{xij}-k, max{xij}]—strong relationship (Qi ϵ (0.5, . . . , 1])
Group 2: Qi ϵ (max{xij}-2k, max{xij}-k]—average relationship (Qi ϵ (0, . . . , 0.5])
Group 3: Qi ϵ (max{xij}-3k, max{xij}-2k]—weak relationship (Qi ϵ (−0.5, . . . , 0])
Group 4: Qi ϵ [min{xij}, max{xij}-3k]—very weak relationship (Qi ϵ ([−1, . . . , −0.5])

The first and fourth groups define an indirect relationship based on the presence of some extreme values of the indexes compared to those calculated and a significant difference between the values of population indexes and those of ethno-cultural heritage.

The first group represents the positive inverse relationships, where the population index values are considerably higher than those of the ethno-cultural heritage, unlike the fourth group, where the situation is reversed. The second and third groups can be approached together (Figure 2) because the values of both indexes are moderate, and the difference between values are small. Within this value range (−0.5, 0.5) there are strong relationships between the two indexes, resulting from the quantitatively moderate presence in the territory of both ethno-religious features of the population and ethno-cultural heritage ones.

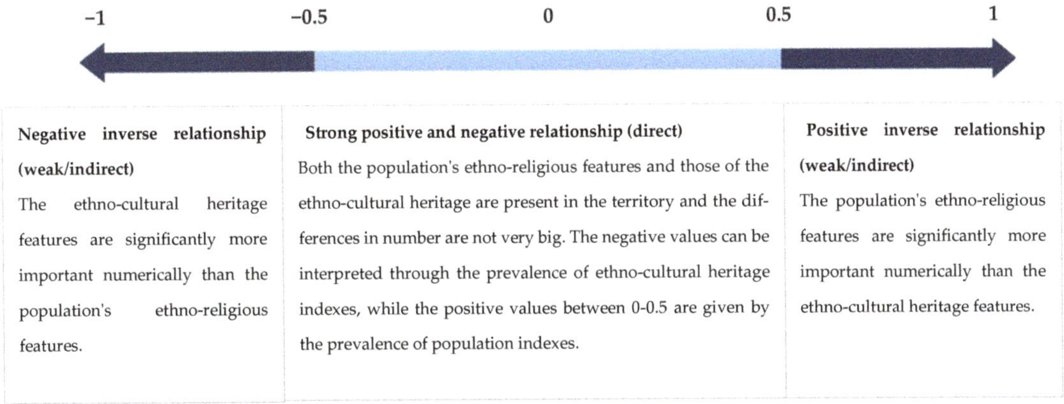

Figure 2. I index values range.

The data processing was achieved with the help of ArcGis 10.6 and Excel software. The data spatial analysis was accomplished on a polygon level, 207 territorial administrative units, in order to establish the spatial relationships between population and ethno-cultural heritage.

3. Results and Discussion

3.1. Synthetic Values of Population Characteristics (Ethnicity and Religion)

The spatial distribution analysis of population features' synthetic values (ethnicity and religion) emphasized the existence of two categories of administrative-territorial units: the first category is characterized by very low values (99.51% of the TAUs), while the other one is characterized by very high values (only 0.48% of the TAUs) (Figure 3).

It is worth mentioning the fact that values over 0.75 are found only in one TAU, Oradea City (index of 0.870124), which is the largest city in the region, and, more importantly, it has the most numerous population in the region: 183,123 inhabitants at the 2011 census. All ethnicities and religions are represented in Oradea, and, furthermore, the indexes obtained for ethnicity and religion are, each of them, over 0.7.

All the other TAUs in the region (be they towns or communes) are characterized by very low synthetic values for population (between 0.001752 and 0.25). These are TAUs with a lower number of inhabitants compared to Oradea, and not all ethnicities and religions are represented in a significant percentage. For example, Ignești commune, Arad County, has a population of only 669 inhabitants and features a synthetic value of ethno-religious index of 0.01. The tendency in this group is for one of the ethnicities, respectively, one of the religions, to be in absolute or relative majority on TAU level. The situation in which none of the ethnicities and/or religions predominates in a TAU, yet the synthetic values obtained are very low, is determined by a very low number of inhabitants recorded in the respective TAU. The average and weak ethno-religious synthetic values are not present in any TAU of the region.

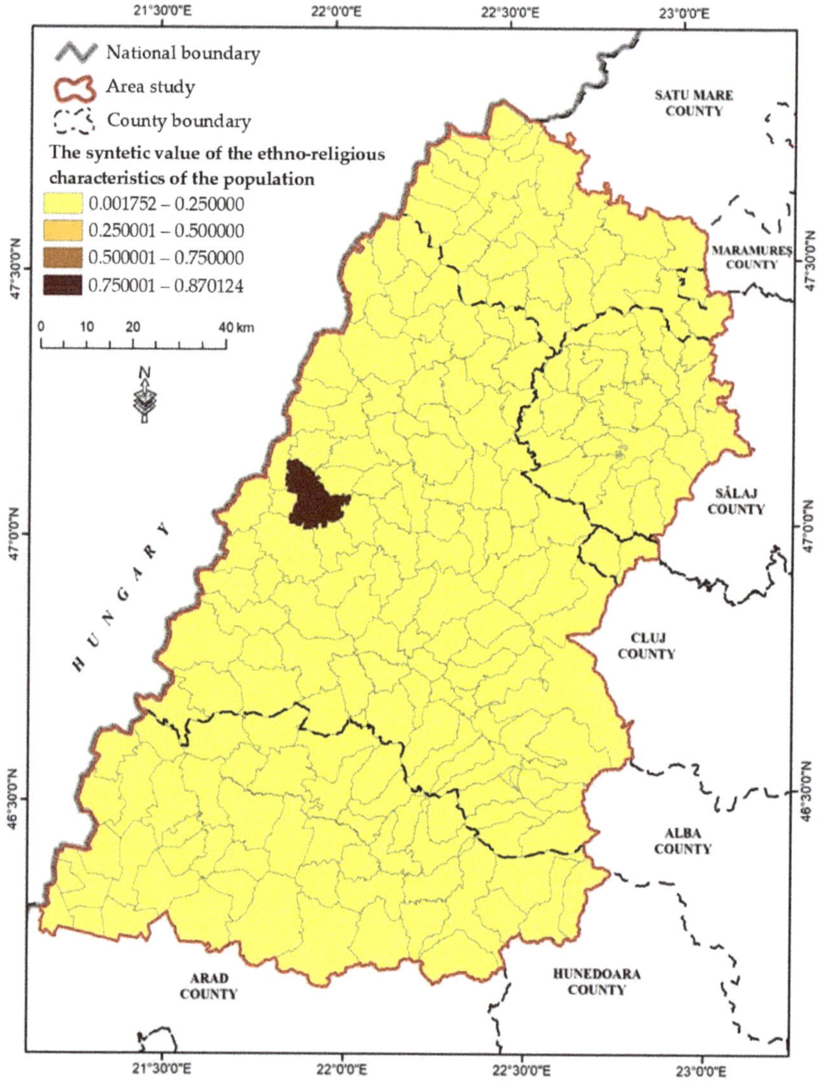

Figure 3. Distribution of the synthetic value of the ethno-religious characteristics of the population.

3.2. Synthetic Values of Ethno-Cultural Heritage Characteristics

The synthetic values of ethno-cultural heritage elements were calculated based on their incidence in number on the TAU level. The methodology also establishes four value groups: very high values, high values, small, and very small values; however, the group of high values (between 0.5 and 0.75) is not representative of any TAU (Figure 4).

Considering the fact that variables of the ethno-cultural heritage criterion are five in number, even though the TAUs feature certain ethno-cultural characteristics, many of them do not feature any of the studied variables, hence the high number of TAUs with very low synthetic values of the ethno-cultural heritage elements.

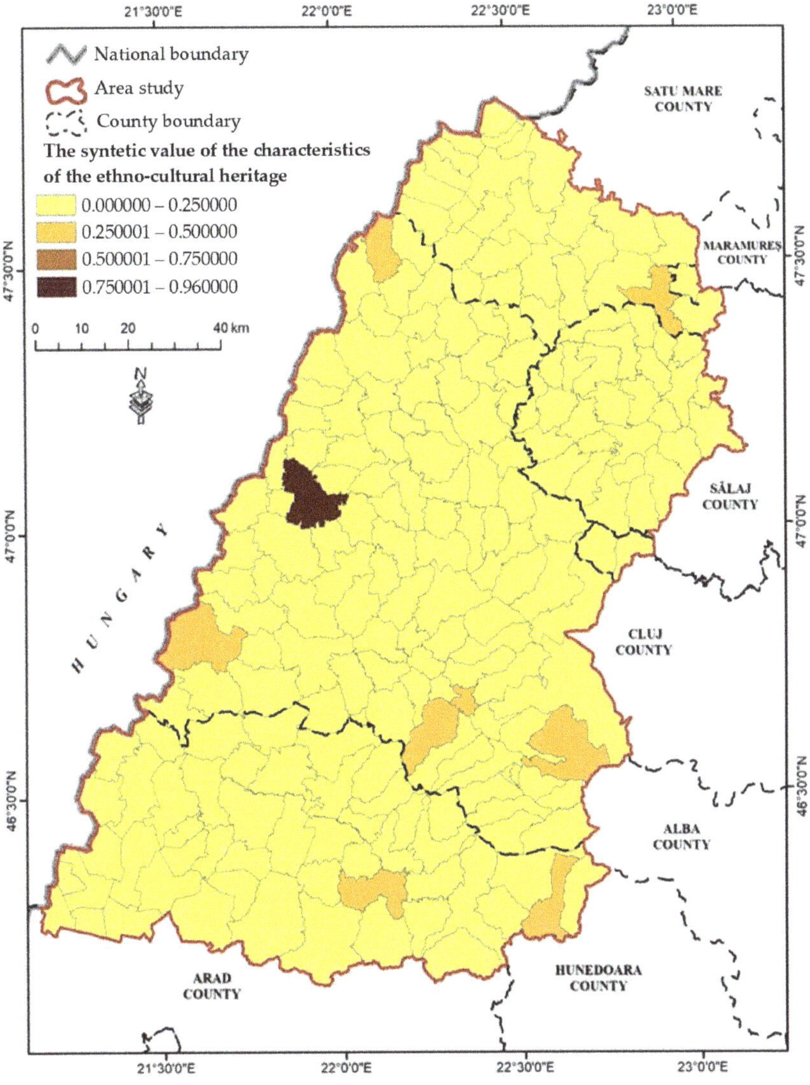

Figure 4. Distribution of the synthetic value of the characteristics of the ethno-cultural heritage.

Very high values are encountered only in Oradea City. This phenomenon is explained by the quantitative assessment of ethno-cultural heritage elements, which are 30 in Oradea (the highest number from the entire region). There are 4 wooden churches here, a museum that exhibits an important ethnographic collection, 8 folk festivals, and 2 ethno-cultural fairs are organized in Oradea every year.

The city is also represented by 14 traditional craftsmen. The high number of elements leads to a very high synthetic value (0.96). Low values of the ethno-cultural heritage elements (between 0.26 and 0.5) are noticed only on the level of 8 TAUs: Valea lui Mihai, Bogdand, Salonta, Finiș, Beiuș, Pietroasa, Buteni, and Hălmagiu, three of them are towns, and the other five are communes. Very low values are obtained for all the other TAUs from the region. This, however, does not mean that these TAUs lack completely such ethno-cultural heritage elements (except some of them), but that the number of such elements is very low, the result being very low synthetic values (between 0.0 and 0.25).

3.3. *Type of Relationships between Population Characteristics and Ethno-Cultural Heritage Elements*

The types of relationships between the demographic characteristics of the population and the ethno-cultural heritage elements are determined according to the indexes obtained for the studied criteria. These indexes are encompassed in the [−1; 1] range of values, and within this range, they are further divided into four categories, each TAU being included in one of the four categories (Figure 5).

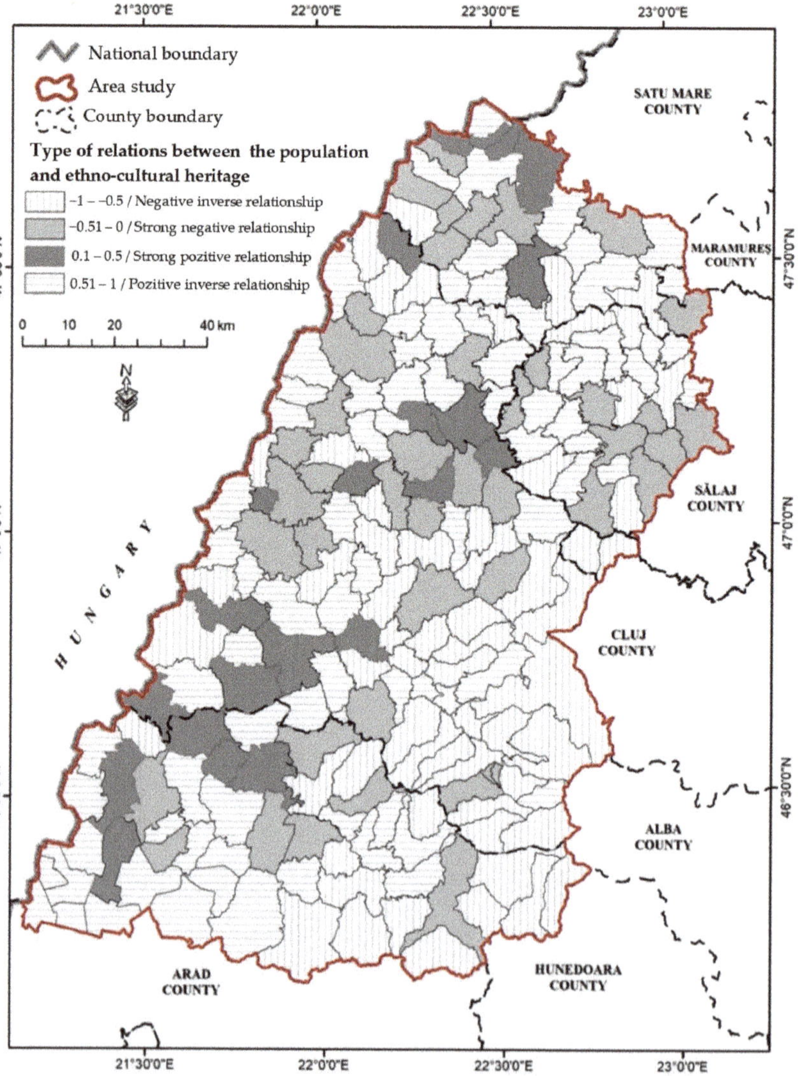

Figure 5. Type of relations between the population and ethno-cultural heritage in Crișana, Romania.

(1) The group of negative inverse relationships is characterized by index values comprised in the [−1; −0.5] range (Figure 5). These values indicate the fact that the

ethno-cultural heritage elements predominate numerically by comparison with the ethno-religious features. For example, the town Valea lui Mihai has a total population of 9688 inhabitants, out of which 7851 are of Hungarian ethnicity, while the rest of 1836 are, in various proportions, of other ethnicities. From a religious point of view, most inhabitants are of Calvinist religion (4701). However, the other ethnicities taken into consideration are represented. All these things reflect ethno-religious homogeneity to the TAU, with a synthetic value of 0.065372. The ethno-cultural heritage elements are 12 in this town, with an index value of 0.278333, significantly higher than the ethno-religious one. The calculated relationship index is −0.6196, thus placing the town Valea lui Mihai in the group of negative inverse relationships. Most TAUs belong to the group of this type of relationship.

(2) The group of strong negative relationships includes heterogeneous TAUs from an ethno-religious point of view, with all ethnicities and religions being represented, without significant differences in number. The ethno-cultural heritage elements are also present and also predominate slightly by comparison with the ethno-religious one. Within this category, there are TAUs with a relationship index value between −0.51 and 0. We are analyzing, as an example, the Șuncuiuș TAU, which is characterized from ethno-religious point of view by high homogeneity: from the total population of 3200 inhabitants, 2852 (89%) are of Romanian ethnicity, 262 are Rroma and 64 are of Hungarian ethnicity. The other ethnicities taken into the study are not represented here. From a religious point of view, diversity is slightly higher, meaning that all religions taken into study are represented, but the Orthodox one is in absolute majority with 2949 people. The aggregate value for the ethno-religious features is 0.017982. There are three ethno-cultural heritage elements, with an index value of 0.04, lower than the ethno-confessional one. The relationship index value for the two types of features is −0.37973, indicating the strong negative relationship, the ethno-cultural heritage features prevailing.

(3) The third group is that of strong positive relationships, with relationship index values comprised in the 0.1 and 0.5 range. In this category are included the TAUs characterized by the prevalence of the population features. For example, the Mișca TAU has a total population of 3588 inhabitants and, except for the Ukrainian one, all ethnicities are represented, but none of them has a majority, be it absolute or relative. The religions are also represented in variable proportions, except for the Greek-Catholic one; hence it can be stated that the TAU is characterized by ethno-religious heterogeneity, with an index value of 0.056737. The ethno-cultural heritage elements are 4, and their aggregate value is 0.05. The relationship index value is 0.063122.

(4) The positive inverse relationships group is the group that includes TAUs with relationship index values comprised between 0.51 and 1 and in which the ethno-religious features are significantly prevalent to the ethno-cultural heritage ones. The Second TAU, for example, has a total population of 2543 inhabitants of various ethnicities, without any of them being in the majority, either absolute or relative. The predominant religion is the Orthodox one, in an absolute majority (2066 inhabitants), and the other religions are also represented, however, very poorly. The aggregate index value for the two features is 0.042288. No studied ethno-cultural heritage element can be found in this TAU, the value for this category being 0, while the relationship index value for the ethno-religious and ethno-cultural features is 1.

4. Conclusions

The hypothesis from which this study started off was that there is a relationship between the ethno-religious and ethno-cultural heritage elements in Crișana Region on the TAU level, considering that the entire territory is characterized by ethno-religious and ethno-cultural diversity. With the help of applied quantitative analysis, it has been proven that there are relationships between the studied elements, and they are determined by the prevalence of one of the considered criteria or the other.

The relationship between ethnography and demography has been a preoccupation of researchers since the second half of the 20th century [9], and even though demography is a branch that uses quantitative analysis methods, the connection between the two is studied mostly by using qualitative research methods [36]. In this study, a quantitative analysis method was used to be able to quantify this relationship.

The ethno-religious diversity is higher in all countries, and this characteristic is also accompanied by globalization; interestingly, the result of globalization is the fading away of the characteristics specific to an ethno-religious group. The importance of establishing the existing relationship types between the ethno-confessional and ethno-cultural criteria consists of emphasizing their mutual influences, which can lead to keeping and promoting the ethno-religious and ethno-cultural features. The spatial distribution of relationship types (Figure 5) between ethno-religious and ethno-cultural features reveals a heterogenic dispersion. It cannot be stated that one "land" or one ethnographic area is characterized only by a certain type of relationship. All four types of relationships are present in each of the "lands" and ethnographic areas belonging to Crișana Region.

Author Contributions: Conceptualization, L.A.D. and G.V.H.; methodology, G.V.H.; software, E.B.; validation, E.B., M.G. and L.A.D.; formal analysis, L.A.D.; investigation, G.C.B.; resources, E.B.; data curation, M.G.; writing—original draft preparation, G.V.H.; writing—review and editing, M.G.; visualization, G.V.H.; supervision, L.A.D.; project administration, G.C.B.; funding acquisition, L.A.D. All authors have read and agreed to the published version of the manuscript.

Funding: This research received no external funding.

Institutional Review Board Statement: Not applicable.

Informed Consent Statement: Not applicable.

Data Availability Statement: The data presented in this study may be obtained on request from the corresponding author.

Acknowledgments: The research undertaken was made possible by the equal scientific involvement of all the authors concerned.

Conflicts of Interest: The authors declare no conflict of interest.

References

1. Herman, G.V.; Caciora, T.; Ilies, D.C.; Ilies, A.; Deac, A.; Sturza, A.; Sonko, S.M.; Șuba, N.S.; Nistor, S. 3D Modeling of the Cultural Heritage: Between Opportunity and Necessity. *J. Appl. Eng. Sci.* **2020**, *10*, 27–30. [CrossRef]
2. Ilieș, M.; Herman, G.; Hodor, N.; Baias, Ş; Ilieș, A. The Dynamics, Structure and Spatial Distribution of the Religious Choices of the Romanian Ethnic Community in the Post-Socialist Period (1992–2011). *J. Study Relig. Ideol.* **2020**, *19*, 163–185.
3. Brunn, S.D.; Matlovičová, K.; Mušinka, A.; Matlovič, R. Policy implications of the vagaries in population estimates on the accuracy of sociographical mapping of contemporary Slovak Roma communities. *Geojournal* **2018**, *83*, 853–869. [CrossRef]
4. Lechowski, Ł. The socio-economic transformations of municipalities in Lodz metropolitan area in the context of the con-struction of motorways and expressways. *Folia Geogr.* **2021**, *63*, 40–63.
5. Dewitte, S.N. Demographic anthropology. *Am. J. Phys. Anthr.* **2018**, *165*, 893–903. [CrossRef] [PubMed]
6. Kóti, T. Efficiency of Active Labour Market Policy in Hungary: Detransitive Settlement Structure of Supported Public Employment. *Folia Geogr.* **2019**, *61*, 45–70.
7. Pachura, P.; Nitkiewicz, T.; Matlovičová, K.; Matlovič, R. Identification of intellectual capital performance using data envelopment analysis. In *Knowledge Spillovers in Regional Innovation Systems: A Case Study of CEE Regions*; Springer: Berlin/Heidelberg, Germany, 2018; pp. 115–130.
8. Lorimer, F. *Culture and Human Fertility Paris*; UNESCO: Paris, France, 1954; pp. 101–102.
9. Coast, E. An evaluation of demographers' use of ethnographies. *Popul. Stud.* **2003**, *57*, 337–346. [CrossRef] [PubMed]
10. Atkinson, P.A.; Coffey, S.; Delamont, J.; Lofland, L.H. *Handbook of Ethnography*; Sage: London, UK, 2001; pp. 62–71.
11. Crawford, D. Ethnography and Demography: Moroccan Households and Cultural Change. *Hespéris-Tamuda* **2020**, *LV*, 469–491.
12. Stupariu, M.I.; Tătar, C.F.; Stașac, M.; Linc, R.; Bucur, L.; Dincă, I.; Nistor, S.; Filimon, C.A.; Filimon, L.A. A tourist flow study of the rural metropolitan area of Oradea compared to Bihor County (Romania). *Folia Geogr.* **2022**, *64*, 21–45.
13. Ilieș, A.; Josan, I. Țările-regiuni geografice specific. In *Crișana—Maramureș Atlas Geografic al Patrimoniului Touristic*; Editura Universității din Oradea: Oradea, Romania, 2014; pp. 18–19.
14. Hart, D. *The Aith Waryaghar of the Moroccan Rif: An Ethnography and History*; University of Arizona Press: Tucson, AZ, USA, 1976; pp. 83–92.
15. Josan, I. *Țara Silvaniei, Studiu de Geografie Regional*; Editura Universității din Oradea: Oradea, Romania, 2009; pp. 44–58.

16. National Institute of Statistics. Available online: https://www.recensamantromania.ro/rpl-2011/rezultate-2011/ (accessed on 6 February 2023).
17. Ruxăndoiu, P. *Folclorul Literar în Contextul Culturii Populare Românești*; Editura Grai și Suflet—Cultura Națională: București, Romania, 2001; pp. 48–50.
18. Bungau, C.C.; Bungau, T.; Prada, I.F.; Prada, M.F. Green Buildings as a Necessity for Sustainable Environment Development: Dilemmas and Challenges. *Sustainability* **2022**, *14*, 13121. [CrossRef]
19. Deac, L.-A.; Gozner, M.; Sambou, A. Ethnographic museums in the rural areas of crișana region, Romania—Keepers of local heritage, tradition and lifestyle. *Geoj. Tour. Geosites* **2019**, *27*, 1251–1260. [CrossRef]
20. Bogdan, A.; Chambre, D.; Copolovici, D.M.; Bungau, T.; Bungau, C.C.; Copolovici, L. Heritage Building Preservation in the Process of Sustainable Urban Development: The Case of Brasov Medieval City, Romania. *Sustainability* **2022**, *14*, 6959. [CrossRef]
21. Herman, G.V. The man between regionalization and globalization. *Rev. Română Geogr. Politi* **2022**, *24*, 1–9. [CrossRef]
22. Tomlinson, J. *Globalization and Culture*; Polity Press: Cambridge, UK, 1999; pp. 189–204.
23. Smith, K.M. *Issues in Cultural Tourism Studies*, 2nd ed.; Routledge: London, UK, 2009; pp. 5–14.
24. Constantin, M. The Ethno-Cultural Belongingness of Aromanians, Vlachs, Catholics, and Lipovans/Old Believers in Romania and Bulgaria (1990–2012). *Rev. Română Sociol.* **2014**, *25*, 255–285.
25. Kukuła, K.; Bogocz, D. The Zero Unitarization Method and its Application in Agricultural Ranking Studies. *Econ. Reg. Stud.* **2014**, *7*, 5–13.
26. Della Spina, L.; Giorno, C. Cultural Landscapes: A Multi-Stakeholder Methodological Approach to Support Widespread and Shared Tourism Development Strategies. *Sustainability* **2021**, *13*, 7175. [CrossRef]
27. Szmygin, B. *Heritage Value Assessment Systems—The Problems and the Current State of Research*; Polish National Committee of the International Council on Monuments and Sites ICOMOS: Warszawa, Poland; Lublin University of Technology: Lublin, Poland, 2015; pp. 18–24.
28. Król, K. Assessment of the Cultural Heritage Potential in Poland. *Sustainability* **2021**, *13*, 6637. [CrossRef]
29. Mościbrodzka, M. The Use of Methods of Multidimensional Comparative Analysis in Evaluation of the Standard of Living of Poland's Population in Comparison with Other Countries of the European Union. *Oeconomia Copernic.* **2014**, *5*, 29–47. [CrossRef]
30. Rymuza, K.; Bombik, A. Multidimensional Analysis of Social and Economic Development of Some Counties in Mazovia Voivodeship. *Acta Sci. Polonorum. Oeconomia* **2014**, *13*, 81–93.
31. Radzka, E.; Rymuza, K.; Jankowska, J. The assessment of drinking water quality using zero unitarization method. *Arch. Environ. Prot.* **2015**, *41*, 91–95. [CrossRef]
32. Zhang, Y.; He, S.; Gu, Z.; Wei, N.; Yu, C.W.; Li, X.; Zhang, R.; Sun, X.; Zhou, D. Measurement, normalisation and mapping of urban-scale wind environment in Xi'an, China. *Indoor Built Environ.* **2019**, *28*, 1171–1180. [CrossRef]
33. Patro, S.G.K.; Sahu, K.K. Normalization: A preprocessing stage. *Iarjset* **2015**, *2*, 20–22. [CrossRef]
34. Kiselakova, D.; Stec, M.; Grzebyk, M.; Sofrankova, B.A. Multidimensional Evaluation of the Sustainable Development of European Union Countries—An Empirical Study. *J. Compet.* **2020**, *12*, 56–73. [CrossRef]
35. Boc, E.; Filimon, A.L.; Mancia, M.-S.; Mancia, C.A.; Josan, I.; Herman, M.L.; Filimon, A.C.; Herman, G.V. Tourism and Cultural Heritage in Beiuș Land, Romania. *Heritage* **2022**, *5*, 1734–1751. [CrossRef]
36. Matlovičová, K.; Tirpáková, E.; Mocák, P. City brand image: Semiotic perspective a case study of Prague. *Folia Geogr.* **2019**, *61*, 120–142.

sustainability

MDPI

Article

Socio-Professional Implications of Sports Events: A Perspective from Dual-Career Students

Raluca Buhaş [1,*], Alexandru Ilieş [2], Sorana Săveanu [1], Paul Szabo-Alexi [2], Mariana Szabo-Alexi [2] and Sorin Buhaş [2]

[1] Faculty of Social Sciences, University of Oradea, 410087 Oradea, Romania; soranasav@gmail.com

[2] Faculty of Geography, Tourism and Sport, University of Oradea, 410087 Oradea, Romania; ilies@uoradea.ro (A.I.); szaboalexipaul@gmail.com (P.S.-A.); mariszabo@yahoo.com (M.S.-A.); sorin.buhas@gmail.com (S.B.)

* Correspondence: ralubuhas@gmail.com; Tel.: +40-773797515

Abstract: Background: The organization of a sports event can be analyzed from the perspective of its impact on the participants, organizers and community. University championships have a sustainable character, with several tournaments being planned over time, receiving increased interest from all stakeholders. Methods: This research aimed to describe aspects regarding the organization and planning of sports events, the academic and professional profiles of student-athletes and the implications of their participation in championships. We used two research methods: one based on geography, and another based on a sociological survey of 139 dual-career students participating in the national university championship of men's football and women's volleyball organized by the University of Oradea (Romania). Results: The research results show that academic studies provide professional competencies, while participation in sports events contributes to students' personal and professional development. The results highlight that dual-career students are more prone to develop their sports careers; therefore, participation in these types of events represents a milestone in their careers. Conclusions: Our study can be considered when organizing these types of sports events and can provide directions for the development of sustainable strategies and programs to support students with dual careers.

Keywords: sports events; dual-career students; academic and professional profile; university championship

Citation: Buhaş, R.; Ilieş, A.; Săveanu, S.; Szabo-Alexi, P.; Szabo-Alexi, M.; Buhaş, S. Socio-Professional Implications of Sports Events: A Perspective from Dual-Career Students. *Sustainability* **2023**, *15*, 7813. https://doi.org/10.3390/su15107813

Academic Editor: Kyle Maurice Woosnam

Received: 15 March 2023
Revised: 4 May 2023
Accepted: 8 May 2023
Published: 10 May 2023

1. Introduction

Sports events are important, taking into account the impact they can have on the community, participants, organizers and local population. The magnitude of these events is attributable to their continuous sequence, their history, the number of tournaments, their reputation, the local or international level of participation, the length of the development period, the number of participants and their characteristics, the temporal sequence, etc.

The relevance of studying sports events is attributable to the importance of the event itself based on its reputation, sustainability and the impact it has on the participants, organizers and community. On the other hand, the participants are equally important, focusing on their profiles and the impacts that their defining characteristics have on the planning and implementation of these types of events. All these aspects can be seen as a starting point for the proper organization and planning of such events in the future. These initiatives and studies ensure the sustainable character of sports championships.

This paper analyzes the national university championship of men's football and women's volleyball organized by the University of Oradea (Romania) from the perspective of its reputation and sustainability and the impact it has on the participants and their socio-professional status development, taking into account their particularities as dual-career students.

Our approach was based on the following research questions: What are the characteristics of student-athletes participating in university championships? What is the role of competitors in the planning, organization and development of sports events? Can a small-scale sports event be analyzed from the perspective of its mutual benefits generated by the event–competitor relationship?

The first research objective focused on investigating various aspects related to the planning and organization of a sports event. Events can be analyzed from the perspective of the impact that the competitors, participants, organizers and the community have, and in this sense, an important aspect is the reputation of the event and its continuity, resulting in its sustainable character. We approached these aspects by using a geographical framework.

We also outline the implications that the event has on the competitors in terms of their dual-career development. In this sense, the profiles of student-athletes participating in the championship are important. Participating in the event represents an opportunity to strengthen performance, to develop a professional career, to expand social networks, and to increase territorial mobility and visibility as athletes. Taking these into consideration, two other research objectives were pursued through a sociological survey: a description of the academic and sports careers of student-athletes participating in the national university championship and the analysis of the implications of participation in sports events.

1.1. Sports Events: Organization and Participants

Effectively organizing a sports event requires paying attention to all details, such as the selection of participants while taking into account their territorial distribution, planning the schedule of competitions, establishing the locations, ensuring human resources and logistics, budgeting, sponsorship, marketing, etc. [1]. In national university championships, the participants are usually dual-career students. Thus, their needs (their academic programs and the frequency of sports training) must be considered when planning the event. Both the territorial distribution of the participants, as well as their level of training/sports performance, give weight to the event itself, contributing to the increase in interest for possible future tournaments. A method that can be used in the organization and planning of sports competitions is cartography, which results in spatial planning [2]. A successful sports event for participants, organizers and the community can be an example of good practice in planning future tournaments. The magnitude of the event and its history generate increased interest from the participants and the community.

Most studies that look at the impact of sports events tend to focus on large-scale events, particularly looking at their economic effects [3,4]. Such events have a high level of reputation and continuity, and the effects are reflected both directly, on aspects related to infrastructure development and spatial planning, and indirectly, from the perspective of ensuring the international visibility of the host city, the competitiveness and attractiveness of tourist destinations, the creation of new jobs, the development of local entrepreneurship, the increase in local attachment and the promotion of cultural values. Most of the time, studies highlight the positive effects of these kinds of mega-events, in contrast to the negative ones, which particularly relate to the costs generated by the organization of such a large event, the construction of the infrastructure and urban crowding [4]. It is difficult to measure the long-term effects of sports events, and this requires a complex approach that evaluates all the structures involved. Gratton and Preuss [4] propose tracking the planned, unplanned, positive, negative, tangible and intangible effects of the structures created post-event. These structures resulting from the organization of sports events have a sustainable character due to their significant impact on the community, on tourism, on the economy and industry, and on how future events will be organized. However, there are few studies that focus on smaller sports events, such as university championships [5–7]. The profiles of both spectators and athletes participating in these sports events are rather different from those who participate in large-scale events. Additionally, the event itself has a smaller reach if we take into account the network built around it and the stakeholders involved. If smaller sports events aim to develop and increase their reputation and visibility, have a certain continuity due to the organization of several tournaments and thus generate longer-term effects at the regional level, one of the priorities of its strategy should relate to the growth and consolidation of the network of stakeholders. The complexity of this network has implications for social, personal and economic relationships [8]. Therefore, small-scale events contribute to the strategic development of the region [7]. Anyhow, studies show that small sports events organized in cities at the local level tend to have an important impact on the community, economy and local tourism [5,9]. This category also includes university championships. According to Getz [10], sports events fall into the category of those planned for the purpose of competition, leisure and socialization. Sports activities include, on the one hand, recreational ones (participation of individuals in various sports activities during traveling) and, on the other hand, sports competitions (sports events) entailing spectators and amateur or professional athletes.

Athletes are directly involved in the sports competition, so they are part of the network built around the event. The size of the event itself is significant in clarifying the role of the athletes. In the case of small events, namely, those in which it is possible for the athletes to outnumber the spectators, where the media interest is limited or where there are fewer economic effects [11], the focus is especially on sports competitors. At the same time, another important aspect is that sports events can be a source of social capital and community networks [12]. Thus, student-athletes can identify new contacts and directions for personal and professional development.

1.2. Dual-Career Students: Opportunities and Challenges

A dual career encompasses two important sectors of students' lives: a sports career combined with the pursuit of an academic or work path [13,14]. Therefore, student-athletes represent a distinct social group with particular needs that must be addressed. Several studies have focused on investigating this topic [15–19]. In the case of this social category, it is necessary to develop sustainable government policies that ensure the proper combination of the sports system, the educational system and the labor market to reduce all of the risks to talented student-athletes, especially those arising when they have to make the decision to give up a certain career path.

An optimal dual career is a challenge for students, and it is frequently associated with difficulties in managing both sports and academic paths in parallel, together with different aspects of private life [20,21]. A dual career also encompasses a challenging transition process (junior-to-senior, post-athletic career, and a higher level in sports or education) [17]. The transitions that take place in a sports career are often determined by, generate or overlap with transitions in other areas of life (academic transitions, psycho-social changes or professional changes). The sports career is shaped by interactions between all levels of development and transitions from one level to another [14]. Therefore, for some transitions, a certain coping strategy can be established, whereas, for others, it is more difficult. Student-athletes need to acquire and develop competencies to successfully prepare for and pursue a dual career in sports and education [22]. On the other hand, a dual career combining sports and studies has proven to be an adequate solution to balance sports with other spheres of students' lives [20] and provides benefits for them in terms of social and psychological as well as academic development.

A dual career is always associated with certain benefits; however, it also has some serious costs. The idea is to ensure a balance between them, because it is precisely this balance that has a decisive impact on their motivation, namely, the decision to give up a dual career. Research stresses that internal/personal resources and characteristics, external support (from family, peers and coaches) and coping strategies are some of the most important key factors for an optimal dual-career balance [17,23]. Time, energy and self-management play a major role in the sports–education balance [17,20]. At the same time, the social pressure that these students feel puts them in the situation of needing to find coping strategies. Government policies, as well as different organizations and stakeholders, must support student-athletes in the implementation of coping strategies and ensure their welfare [14].

Moreover, the educational system must take into account the particular situation of dual-career students and properly respond to their real needs. In this sense, the EU Guidelines on Dual Careers of Athletes was developed as a basis to introduce sustainable dual-career programs and states that both non-governmental and governmental stakeholders in sports and education (educational system, sports federations and sports organizations) have a major role in ensuring student-athletes' welfare [14].

2. Materials and Methods

The development of the methodological framework used in this research involved an interdisciplinary approach, focusing on two components: the analysis of sports events from the geographical perspective using cartography and spatial planning [24–28] and from the sociological perspective using a survey [29–31]; the aim was to describe the academic and professional profiles of students with dual careers who participate in sports championships, highlighting their opinions regarding the impact that the event has on them.

The geographical component of the study is relevant for the territorial impact of the sports event, highlighting several quantitative aspects (such as the number of universities and sports clubs involved) and also qualitative ones (such as the performance and quality of the participants). The sociological component is relevant for describing the profiles of students with dual careers participating in sports events and for outlining their opinions regarding these types of events. There are several studies that focus on the profiles of student-athletes participating in university championships [32,33] and the influence of participation in this type of event [34,35].

The cartographic representation, namely, the way in which the map is elaborated, with its content including the number and type of quantitative and qualitative elements represented chronologically and territorially, reflects the distribution, amplitude and spatial diversity of university sports events in general and of women's volleyball and men's football in particular for the 2022 tournament. Thus, the map, as a complex and visually suggestive cartographic product, represents basic support in the development of action plans for the present and especially for future planning strategies for similar events.

The sociological study adopted a quantitative approach with an exploratory-descriptive character using a questionnaire-based survey. The research tool was applied to the competitors in the National University Championship of Romania of Football and Volleyball (held in May 2022). The questionnaire was applied during the championship. The questionnaire was structured in accordance with the research objectives, aiming to obtain responses regarding the educational path, sports careers of the students and opinions regarding the sports events they participate in. The questionnaires were self-administered, anonymous and did not involve the collection of personal data. The questionnaires were voluntarily completed, and the collected data were analyzed only for scientific purposes through statistical processing, thus respecting all the ethical and deontological norms of sociology.

The survey aimed to be an exhaustive one, targeting all participants in this event. The final sample comprised 139 subjects who participated in the national university championships (volleyball or football). The subjects were students who chose a dual career: an academic university career and an athletic career. A total of 87 of them were football players, and 52 of them played volleyball. The football players were all male students, and those who played volleyball were female. Taking into account the field of study in which students who participated in the championship were enrolled, 101 (72.7%) of them were in a sports profile study program, and 38 (27.3%) were in other types of study programs. All participants in the study were students enrolled in a bachelor's (81.5%) or master's degree program (18.5%). The average age of the subjects was 21.5 years, and the majority resided in urban areas (74.8%).

The study combined interdisciplinary elements that could generate strategies for similar cases [21,36,37]. This paper analyzes student-athletes' participation in sports events and their perceptions of these types of events [38,39], highlighting aspects regarding the academic profiles of students [40].

3. Results

3.1. Organization of Sports Event

National university championships have brought the sports movement back to the forefront, both at the national and European levels. The reputation of these competitions at the national level is attributable to the unique annual character and the history of their tournaments. Regularly held during the communist period, with sporadic attempts to resume in the post-communism era, this type of sports event, focused on men's football, was successfully restarted only in 2015 at the initiative of the Timișoara Polytechnic University and under the supervision of the School and University Sports Federation, completing six consecutive tournaments to date (without the pandemic years 2020 and 2021) (Figure 1).

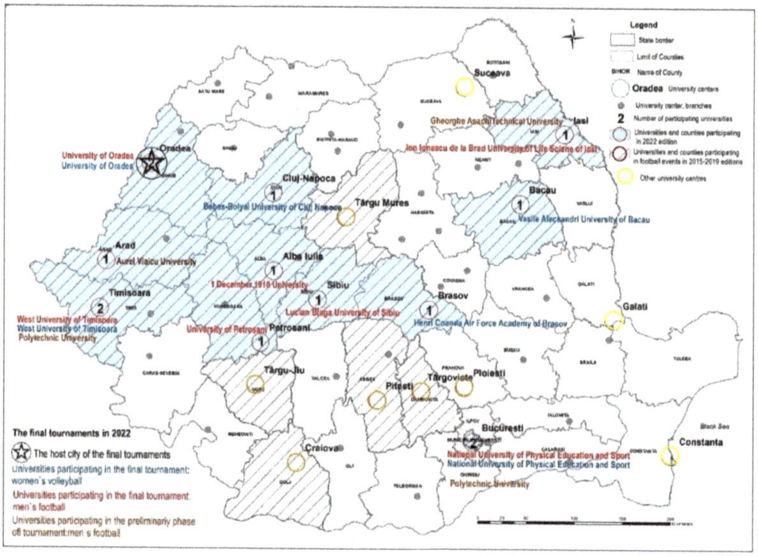

Figure 1. Romania. Universities and counties participating in the final tournaments of the national university championships of men's football and women's volleyball—Oradea, May 2022 and 2015–2019 football event tournaments.

The number of selected participating universities and university centers differed from one tournament to another (Table 1).

Table 1. Number of participating universities/tournaments.

Year	2015	2016	2017	2018	2019	2022
No. of universities	20	18	8	8	6	12

We mention that each university team used a group of 25 sports students on average, of which more than 95% were among those with a dual-career perspective, carrying out regular sports activities at professional and amateur clubs in Romania. During all six tournaments of the football event (2015–2019 and 2022), 72 participating university teams implicitly increased the importance of the national event through the contribution of approximately 2000 students. A total of 95% of students with dual careers are officially registered and represent, at the same time, over 500 professional and amateur sports clubs in Romania.

The championships were specific to the academic year 2022–2023, after two years of a "pandemic" break. For two weeks, the University of Oradea (220,000 inhabitants; approx. 30,000 students) became the host of the two major events in the lives of Romanian students: the final tournaments of the National University Championship of Romania of Football and Volleyball. Approximately 300 students and team staff members participated as main actors, representing 12 universities and 9 university centers from 8 counties and Bucharest (Figure 1), representing both the female teams (6 volleyball teams) and the male teams (6 football teams). The universities participating in the events were the University of Oradea, *Babes-Bolyai* University from Cluj-Napoca, the West University of Timisoara, the University of Petrosani, the *Vasile Alecsandri* University of Bacau, the *Henri Coanda* Air Force Academy of Brașov, the National University of Physical Education and Sports of Bucharest, the *Lucian Blaga* University of Sibiu, and the *Ion Ionescu de la Brad* University of Life Sciences of Iași. The first post-pandemic tournament, analyzed in this study, brought into competition dominant universities from Transylvania, three from Moldova and two from Bucharest (Figure 1).

The importance of the event, from the history of its six tournaments, has grown every year through the participation of an increasing number of students who are active as professional players at clubs from three Romanian professional football leagues (Steaua and Dinamo Bucharest; Politehnica and Ripensia from Timișoara; Cluj-Napoca University, UTA Arad, FC Hermanstadt Sibiu; Politehnica Iași, University of Craiova, Academica Clinceni, CSC Șelimbăr, etc). The situation is even more obvious in the case of the women's volleyball competition.

In the case of the 2022 tournament of the football event, the 12 selected universities involved over 300 student-athletes, with an average of 25 players on each team. A total of 95% of them were dual-career students from 122 professional and amateur sports clubs. At the same time, in the women's volleyball competition, all six participating universities had their teams entirely formed of female students engaged in dual activities, also officially registered in clubs from the national leagues of Romania.

Students who were competitors at the national university championships (volleyball or football) stated their opinions about this type of event. Among respondents, 61.6% of them stated that while they are away with the sports team, they participate in other sports events in the area (Figure 2). This aspect shows students' interest in their athletic career. Additionally, their participation in a sports competition outside their area of residence gives students the opportunity to be involved in other types of activities: 66.3% of them visit parks and natural attractions in the area, 60.5% of students visit places for fun and food, and 37.6% participate in socio-cultural events (festivals, concerts and fairs).

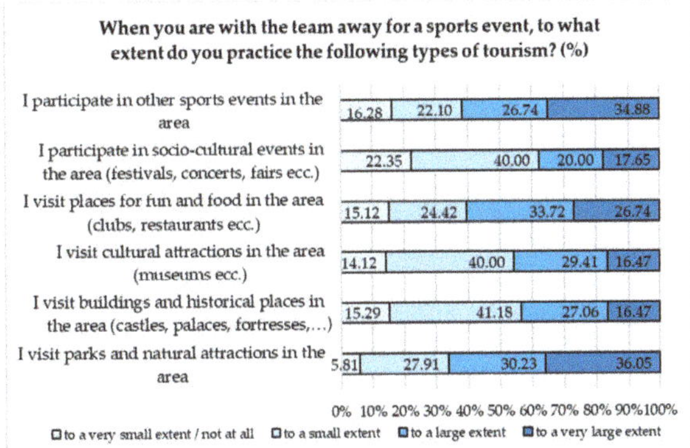

Figure 2. Student-athletes as tourists.

The majority of students (94.5%) declared that this event contributes to their personal development, 82.4% of students believed that it helped them in their current educational activities, and 75.6% of them thought that this sports event provided support in their athletic career (Figure 3).

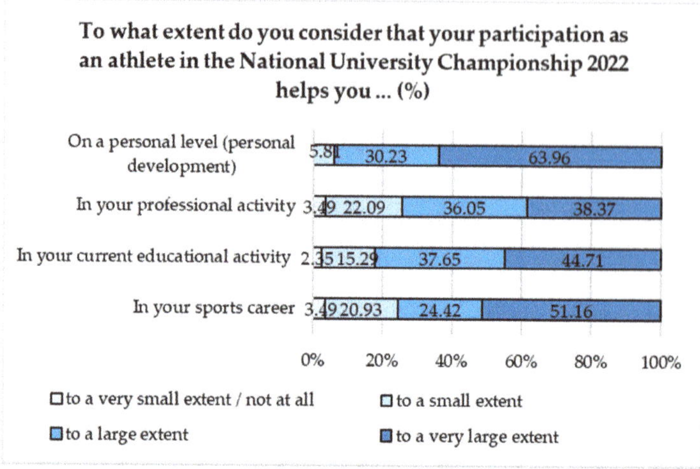

Figure 3. Students' opinions regarding their participation in the national university championship in 2022.

3.2. *Academic Profiles of Student-Athletes Participating in University Championships*

Our data show consistency regarding the academic paths of athletes. A total of 39.9% of them graduated from a sports program in high school. According to the value distribution, the other two study areas mentioned are real science (26.1%) and humanistic (24.6%), while other programs registered values less than 3% (military, technical and services, pedagogy and the environment).

Both in terms of deciding to pursue university and choosing a study program (Figure 4), the data show that students were mostly oriented toward the development of their professional careers. Therefore, they aimed to obtain specific skills and to increase their chances of finding a job, particularly a well-paid and prestigious one. Lower values were registered in terms of the connection with their previous sports activity and the influence that other people (coaches, family and friends) have on their decision regarding their field of study. However, the answers to the question regarding the impact that their athletic career had on their decision in choosing a university study program indicate that most of them were strongly influenced by it. A total of 74.3% of students declared that their athletic career had influenced their academic path to a very great or great extent.

Subjects considered their academic studies and athletic career to be strongly connected. Therefore, students thought that they would use about 67% of the knowledge and skills acquired in the study program further in their athletic careers. Additionally, the percentage of the knowledge and competencies acquired in the study program that they believed they would use in their further professional career is around 81.6%.

The results indicate that students are more oriented toward an athletic career. Being asked what they would do if they had to choose between their academic or sports activities, 57.6% of students stated that they would opt for an athletic career at the expense of their educational activities. Therefore, sports are an important part of their lives.

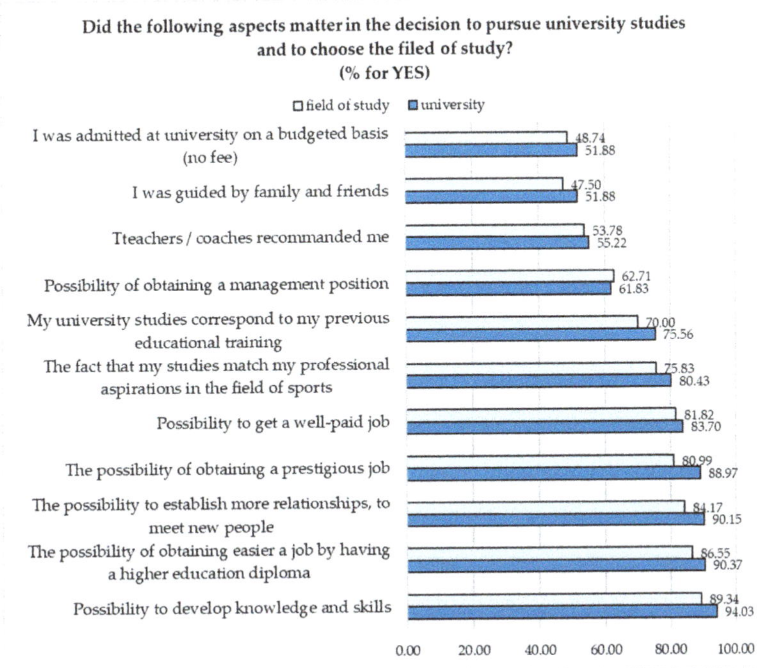

Figure 4. Aspects regarding the decision to pursue university studies.

3.3. The athletic Careers of Students

The average age from which students practiced football or volleyball was 8.96 years, with a standard deviation of 3.1 years. The minimum age was 4 years, and the maximum age was 19 years. A total of 52% of students practiced sports at the performance level, evolving to the I, II or III League, while 48% of them were in League IV.

We need to stress that 74.8% of students received a financial benefit from the sports activities they practiced, as they received a monthly income. It is possible that this situation influenced the decision to continue their athletic career in the future, after completing their studies. The majority of students (72.1%) declared that they wanted to continue their athletic career to a very large or large extent.

As mentioned above, there was a connection between university studies and an athletic career (Figure 5). Most of the students wanted to find a job after graduation, but they wanted their professional activities to continue in sports. There were quite a number of them who looked at their current sports activity as a real job and considered that they would use their sports skills in their future professional activities. At the same time, even after finishing their athletic careers, they aimed to find a job in a related field.

Regarding their current professional activities, 44.8% of students were employed, and 68.2% of them worked in the sports field. It should be mentioned that 56.7% of students had a job in a field that was not related to the studies they were enrolled in.

The majority of students (91.9%) declared that the following is totally true or rather true: the sport that they practice at a performance level gives them the opportunity to attend other sports events in addition to those in which they already participate. Sports activities can be seen as a possible means of students' personal and professional development, as 95.3% of them believed that they had the opportunity to visit new places, and 94.2% of participants believed that the sport that they practiced gave them the opportunity to meet new people.

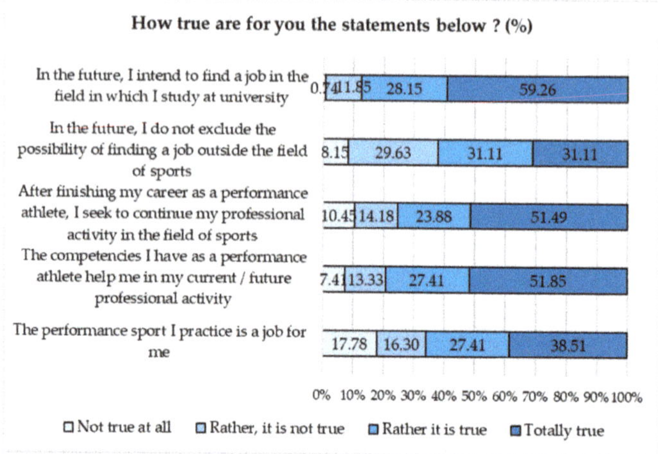

Figure 5. The relationship between university studies and athletic career.

3.4. Research Limitations

A number of research limitations are associated with the analyzed event. In particular, they are related to the fact that it is a small-scale event that took place at the level of universities in Romania. The study was limited to the football and volleyball championship that was held in Oradea in 2022. A more consistent perspective on the results could be obtained by adding the data related to all previous tournaments of the university championships, and future research could adopt a comparative approach by including all participants from the next tournaments. The profiles of dual-career students could be more detailed by a longitudinal approach in this regard. The quantitative approach could be complemented by a qualitative perspective by using in-depth interviews with dual-career students for a more comprehensive image of their status.

The data collected in this study do not offer a comparative perspective regarding the participation of dual-career students in different types of events. Comparing the data with aspects observed in the case of participation in large sports events could bring more clarification regarding how students relate to these events, their motivation to participate and the contribution that these events have to students' career development. The exploratory-descriptive character of the study gives us the possibility to describe the main characteristics of dual-career students; however, it does not provide details about how different characteristics shape their academic and athletic careers, or how participation in sports events is influenced by these characteristics.

We must mention that another research limitation is imposed by the way in which the data were collected, namely, through self-administered questionnaires. Self-reported data could generate biases, such as social desirability, and could affect the research results, which we can control only to a small extent. Even in the case of a small-scale event, a series of stakeholders are involved. A future study approach in which all stakeholders from all levels are taken into consideration could contribute to providing a more comprehensive picture of the link between sports events and competitors.

4. Discussion

The importance of national university championships is determined, on the one hand, by the characteristics of the competitors. Their reputation attributable to their status as professional players is reflected in the event itself. At the same time, we emphasize the dual careers of the participants, as students enrolled in one of the participating or organizing universities of the championship. These aspects increase the importance of the event and stimulate the interest of the public. As this interest rises, the sustainability of the event is ensured, and its repetitiveness generates effects on other related fields, such as the economy or tourism. At the community level, the impact of small-scale events is smaller compared to that of large-scale events. However, these events (such as university championships) generate economic and tourism effects that cannot be neglected, as the organization and development of the event involve specific costs and resource consumption [5]. Equally, from the opposite point of view, the event itself offers benefits to the participants. Participation in the event represents an opportunity, offering the possibility of developing sports and professional careers. In the case of dual-career students, the effects also extend to their educational path [21].

The interdisciplinary analysis of sports events highlights this link between the event and competitors. By using geographically specific methods (cartography and spatial planning), quantitative aspects are presented regarding the spatial organization of events, while the sociological approach outlines the profiles and opinions of the athletes.

From the geographical perspective, this study indicates the reputation of this type of university sports event, attributable to its history, the number of participating universities, the degree of coverage of the phenomenon at the national level (based on the map), the number of participants and professional clubs they represent, the organization of preliminary phases at the level of university centers and especially the final tournaments by considering the importance and cultural characteristics of the host city chosen for each tournament.

The research results stress the continuity of students' athletic careers and highlight the strong relationship between students' choice of university study programs and their sports careers. They consider that the competencies they acquire during the university study program they follow will be useful in their future professional activities. Most of them want to continue their athletic path; however, at the same time, the majority of them (90%) did not think of giving up their studies. Following both a higher education path and a sports career can be complex and challenging for students [19,21]. Universities must be aware of the difficulties faced by this category of students and must define their status more clearly and transparently [14,41]. First of all, a relevant aspect that universities must take into account is the fact that these students still want to remain students, but at the same time, they want to invest in their sports careers. Universities must offer support to students in finding this balance. There are several measures that can be taken to ensure this balance: academic monitoring, counseling, tutoring sessions, etc. Universities must pay more attention to these students in order to be able to identify which activities on campus provide the most benefits and to encourage better student–faculty and coach–faculty communication [42].

Because the studies highlight the fact that dual-career students report difficulties in adapting to the academic environment and to the sports environment and face different conflicting roles [43] and need to deal with the lack of support [41,44], our study outlines the importance of stimulating participation in sports events in a way that contributes to shaping their status and creating a better image in the sports community. At such events, students have the opportunity to share experiences with other students who have similar positions. Sports and school institutions are responsible for providing the necessary support to these students. The experience that these students gain in universities is a beneficial one, as is the experience gained in sports events and competitions. Both make significant contributions to their personal and professional development [45]. Academic institutions and policies that address dual-career status must first ensure that these students benefit from positive academic experiences that increase their chances of obtaining a higher education degree in parallel with maintaining their sports careers [41].

The sport they practice gives students the chance to develop their athletic careers due to the chance they have to participate in other sports competitions, to meet new people and to establish new relationships and networks. Students consider that their sport contributes to their personal development and also gives them other opportunities, such as their involvement in different activities outside of competitions.

This study encountered several limitations, most of them being generated by the small number of participants included in the survey, as well as the approach involving a single small-scale event, making the comparative approach difficult.

5. Conclusions

From a descriptive-exploratory perspective, this research approaches the implications of planning and participating in a small-scale sports event. On the one hand, the sports event itself is analyzed, including the way it is organized and its spatial planning, taking into account the territorial distribution of the participants and its continuity, and on the other hand, it places an emphasis on sport-athletes who participate in the competition. In this sense, this paper offers a new perspective on the social actors involved in sports events. A network with a complex structure is built around the event, including different categories of stakeholders. The importance of the event is also reflected by the type and number of participants—spectators and competitors. In the case of small events, the role played by athletes is especially highlighted. Taking into account their behavior, their number, their geographical representation and the activities in which they are involved outside of the sports competition itself, they make certain contributions to several sectors, such as community development, the economy or tourism.

This study focuses on a special category of competitors participating in a small-scale sports event, namely, dual-career students. Taking into account their characteristics and academic and professional profiles, this paper captures the role they have in the reputation of the university championship. This paper aimed to analyze the national university championship and focused on outlining the profiles of the student-athletes, emphasizing the implications that their specific characteristics have, namely, their ability to have dual careers, for the way in which such an event is planned, the network formed around the event and the way it is carried out. Dual-career students see the championship as a generator of certain benefits. This study reveals that the students who participate in these events pursue their personal and professional development. Thus, they manage to combine the two components of their career—sports and studies.

Although the university championship is a small-scale sports event, it is still important from the perspective of the social, economic and tourism impacts it has on the local community. The importance of the national university championship, even if it is a small-scale event, is attributable to the profiles of the competitors and their number, the level of sports activity and the sports clubs involved (football students from the professional leagues in Romania present a major interest for this type of event), the duration of the competition, the level of representation of the universities and the competition at the national level.

The analysis of sports events such as university championships can contribute to the development of strategies for the development of similar events that can be perceived as generating sources of sustainable sports activities.

Author Contributions: Conceptualization, R.B., A.I., S.S. and S.B.; methodology, R.B., A.I., S.S. and S.B.; software, R.B., A.I., S.S. and S.B.; validation, R.B., A.I., S.S., P.S.-A., M.S.-A. and S.B.; formal analysis, R.B., A.I. and S.S.; investigation, R.B., A.I., S.S., P.S.-A., M.S.-A. and S.B.; resources, A.I., P.S.-A., M.S.-A. and S.B.; data curation, R.B., A.I., S.S. and S.B.; writing—original draft preparation, R.B., A.I. and S.S.; writing—review and editing, R.B., A.I., S.S., P.S.-A., M.S.-A. and S.B.; visualization, R.B., A.I., S.S. and M.S.-A.; funding acquisition, P.S.-A. and M.S.-A. All authors have read and agreed to the published version of the manuscript.

Funding: This research received no external funding.

Institutional Review Board Statement: Not applicable.

Informed Consent Statement: Not applicable.

Data Availability Statement: The data presented in this study may be obtained on request from the corresponding author.

Acknowledgments: The research undertaken was made possible by the equal scientific involvement of all the authors concerned.

Conflicts of Interest: The authors declare no conflict of interest.

References

1. Greenwell, T.C.; Danzey-Bussell, L.A.; Shonk, D.J. *Managing Sport Events*; Human Kinetics: Champaing, IL, USA, 2020.
2. Barovic, G.; Vujacic, D.; Spalevic, V. Cartography in Sports and Sports in Cartography. *Sport Mont* **2021**, *19*, 115–118. [CrossRef]
3. Porter, P.K.; Fletcher, D. The economic impact of the Olympic Games: Ex ante predictions and ex post reality. *J. Sport Manag.* **2008**, *22*, 470–486. [CrossRef]
4. Gratton, C.; Preuss, H. Maximizing Olympic impacts by building up legacies. *Int. J. Hist. Sport* **2008**, *25*, 1922–1938. [CrossRef]
5. Turco, D.M.; Papadimitrou, D.; Berber, S. Athletes as tourists: Consumer Behaviour of Participants at the 2007 and 2009 World Universiade Games. *Phys. Cult. Sport. Stud. Res.* **2011**, *51*, 72–79. [CrossRef]
6. Ritchie, B. Small-scale sport event tourism: The changing dynamics of the New Zealand masters game. In *Niche Tourism: Contemporary Issues, Trends and Cases*; Novelli, M., Ed.; Elsevier: London, UK, 2005; pp. 157–170.
7. Bazzanella, F.; Peters, M.; Schnitzer, M. The perceptions of stakeholders in small-scale sporting events. *J. Conv. Event Tour.* **2019**, *20*, 261–286. [CrossRef]
8. Sallent, O.; Palau, R.; Guia, J. Exploring the Legacy of Sport Events on Sport Tourism Networks. *Eur. Sport Manag. Q.* **2011**, *11*, 397–421. [CrossRef]
9. Scott, A.K.S.; Turco, D.M. VFRs as a Segment of the Sport Event Tourist Market. *J. Sport Tour.* **2007**, *12*, 41–52. [CrossRef]
10. Getz, D. Event tourism: Definition, evolution, and research. *Tour. Manag.* **2008**, *29*, 403–428. [CrossRef]
11. Wilson, R. The economic impact of local sport events: Significant, limited or otherwise? A case study of four swimming events. *Manag. Leis.* **2006**, *11*, 57–70. [CrossRef]
12. Misener, L.; Mason, D.S. Creating community networks: Can sporting events offer meaningful sources of social capital? *Manag. Leis.* **2006**, *11*, 39–56. [CrossRef]
13. Ryba, T.V.; Stambulova, N.; Ronkainen, N.; Bundgaard, J.; Selanne, H. Dual career pathways of transnational athletes. *Psychol. Sport Exerc.* **2015**, *21*, 125–134. [CrossRef]
14. European Commission. EU Guidelines on Dual Careers of Athletes: Recommended Policy Actions in Support of Dual Careers in High-Performance Sport. Brussels: Sport Unit, European Commission. Education, Culture and Sport. 2012. Available online: https://ec.europa.eu/assets/eac/sport/library/documents/dual-career-guidelines-final_en.pdf (accessed on 17 June 2022).
15. Gaston-Gayles, J.L. The factor structure and reliability of the student athletes' motivation toward sports and academics questionnaire (SAMSAQ). *J. Coll. Stud. Dev.* **2005**, *46*, 317–327. [CrossRef]
16. Aquilina, D.; Henry, I. Elite athletes and university education in Europe: A review of policy and practice in higher education in the European Union Member States. *Int. J. Sport Pol.* **2010**, *2*, 25–47. [CrossRef]
17. Stambulova, N.B.; Engstrom, C.; Franck, A.; Linn'er, L.; Lindahl, K. Searching for an optimal balance: Dual career experiences of Swedish adolescent athletes. *Psychol. Sport Exerc.* **2015**, *21*, 4–14. [CrossRef]
18. Stambulova, N.B.; Wylleman, P. Psychology of athletes' dual careers: A state-of-the-art critical review of the European discourse. *Psychol. Sport Exerc.* **2019**, *42*, 74–88. [CrossRef]
19. Vidal-Vilaplana, A.; Valantine, I.; Staskeviciute-Butiene, I.; Gonzalez-Serrano, M.H.; Capranica, L.; Calabuig, F. Combining sport and academic career: Exploring the current state of student-athletes' dual career research field. *J. Hosp. Leis. Sport Tour. Educ.* **2022**, *31*, 100399. [CrossRef]
20. Lopez de Subijana, C.; Barriopedro, M.; Conde, E. Supporting dual career in Spain: Elite athletes' barriers to study. *Psychol. Sport Exerc.* **2015**, *21*, 57–64. [CrossRef]
21. Condello, G.; Capranica, L.; Doupona, M.; Varga, K.; Burk, V. Dual-career trough the elite university student-athletes'lenses, The international FISU-EAS survey. *PLoS ONE* **2019**, *14*, e0223278. [CrossRef]
22. De Brandt, K.; Wylleman, P.; Torregrossa, M.; Defruyt, S.; Van Rossem, N. Student-athletes' perceptions of four dual career competencies. *Rev. De Psicol. Del Deporte* **2017**, *26*, 28–33. Available online: https://www.redalyc.org/pdf/2351/235152047006.pdf (accessed on 6 February 2023).
23. Baron-Thiene, A.; Alfermann, D. Personal characteristics as predictors for dual career dropout versus continuation—A prospective study of adolescent athletes from German elite sport schools. *Psychol. Sport Exerc.* **2015**, *21*, 42–49. [CrossRef]
24. Bale, J. *Sport, Space and the City*; Blackburn Press: Caldwell, NJ, USA, 2001.
25. Bale, J. *Sports Geography*; Routledge: London, UK, 2003.
26. Ilieș, A.; Dehoorne, O.; Wendt, A.J.; Kozma, G. For Geography and Sport, Sport Geography or Geography of Sport. *GeoSport Soc.* **2014**, *1*, 7–18.
27. Ilieș, M.; Ilieș, D.C.; Ilieș, A.; Josan, I.; Ilieș, G. The Gateway of Maramures Land: Geostrategical implications in space and time. *Ann.-Anal. Istrske Mediter. Stud.-Ser. Hist. Et Sociol.* **2010**, *20*, 469–480.
28. Haughton, G.; Allmendinger, P.; Counsell, D.; Vigar, G. *The New Spatial Planning. Territorial Management with Soft Spaces and Fuzzy Boundaries*; Routledge: London, UK; Taylor & Francis Group: New York, NY, USA,, 2010.
29. Babbie, E. *The Basics of Social Research*, 5th ed.; Wadsworth CENGAGE Learning Publishing: Belmont, CA, USA, 2001.

30. Flick, U. *Introducing Research Methodology*; SAGE Publications: London, UK, 2011.
31. Vlăsceanu, L. *Introducere în Metodologia Cercetării Sociologice (Introduction in Methodology of Sociological Research)*; Polirom Publishing House: Iași, Romania, 2013.
32. Ilieș, A.; Deac, A.L.; Wendt, A.J.; Bulz, G.C. Romanian university sports-cultural landscape defined by the sportive space determined by national competitions (in 2015) in team sports. *Geosport Soc.* **2015**, *3*, 61–87.
33. Oros, S.; Hantiu, I. Sports career versus educational career. *GeoSport Soc.* **2018**, *8*, 30–39.
34. Gayles, G.J.; Hu, S. The Influence of Student Engagement and Sport Participation on College Outcomes Among Division I Student Athletes. *J. High. Educ.* **2009**, *80*, 315–333. [CrossRef]
35. Ilieș, A.; Caciora, T. Mapping the Scottish university football competitions. A dual performance model: Organized sports and professional training. *Geosport Soc.* **2020**, *12*, 72–90. [CrossRef]
36. Böheim, R.; Lackner, M. Returns to education in professional football. *Econ. Lett.* **2012**, *114*, 326–328. [CrossRef]
37. Caput-Jogunica, R.; C´urković, S.; Gordana, B. Comparative analysis: Support for student-athletes and the guidelines for the university in southeast Europe. *Sport Sci.* **2012**, *5*, 21–26.
38. Guidotti, F.; Cortis, C.; Capranica, L. Dual career of European student-athletes: A systematic literature review. *Kinesiol. Slov.* **2015**, *21*, 5–20.
39. Ilieș, A.; Buhaș, S.D. Dual academic system education and sport-a premise for the existence a regular university football championship in Romania. *Geosport Soc.* **2020**, *13*, 149–164. [CrossRef]
40. Christensen, M.K.; Sørensen, J.K. Sport or school? Dreams and dilemmas for talented young Danish football players. *Eur. Phys. Educ. Rev.* **2009**, *15*, 115–133. [CrossRef]
41. Capranica, L.; Doupona, M.; Abelkalns, I.; Bisenieks, U.; Sanchez-Pato, A.; Canovas-Alvarez, F.J.; Figueiredo, A.J.; Garcia-Roca, A.F.; Leiva-Arcas, A.; Meroño, L.; et al. Understanding dual career views of European university athletes: The more than gold project focus groups. *PLoS ONE* **2022**, *17*, e0264175. [CrossRef] [PubMed]
42. Simiyu, W.N. Individual and Institutional Challenges Facing Student Athletes on U.S. College Campuses. Health and Kinesiology Faculty Publications and Presentations. University of Texas at Tyler Scholar Works at UT Tyler, Paper 1 (2010). Available online: http://hdl.handle.net/10950/457 (accessed on 20 April 2023).
43. Steele, A.R.; van Rens, F.E.C.A.; Ashley, R.A. Systematic Literature Review on the Academic and Athletic Identities of Student-Athletes. *J. Intercoll. Sport* **2020**, *13*, 69–92. [CrossRef]
44. Costa, F.R.d.; Miranda, I.S.d.; Figueiredo, A.J. Sport and education: How to develop a proper dual career. *Cult. Cienc. Y Deporte* **2021**, *16*, 49–58.
45. Vogel, B.L.; Kress, J.; Jeske, D.R. Student-Athletes vs. Athlete-Students: The academic success, campus involvement, and future goals of Division I student athletes who were university bound compared to those who would not have attended a university had they not been an athlete. *Sport J.* **2019**, *24*. Available online: https://thesportjournal.org/article/student-athletes-vs-athlete-students-the-academic-success-campus-involvement-and-future-goals-of-division-i-student-athletes-who-were-university-bound-compared-to-those-who-would-not-have-attended/ (accessed on 21 April 2023).

MDPI
St. Alban-Anlage 66
4052 Basel
Switzerland
www.mdpi.com

Sustainability Editorial Office
E-mail: sustainability@mdpi.com
www.mdpi.com/journal/sustainability

www.ingramcontent.com/pod-product-compliance
Lightning Source LLC
Chambersburg PA
CBHW061453220926
51850CB00004B/81